Principles of Geology

Principles of Geology

Editor
Richard Renneboog, PhD

SALEM PRESS
A Division of EBSCO Information Services, Inc.
Ipswich, Massachusetts

GREY HOUSE PUBLISHING

Cover photo: Lone hiker at Vermillion Cliffs Wilderness, Arizona. Image by wanderluster. (Via iStock)

Principles of Geology, published by Grey House Publishing, Inc., Amenia, NY, under exclusive license from EBSCO Information Services, Inc.

For information contact Grey House Publishing/Salem Press, 4919 Route 22, PO Box 56, Amenia, NY 12501.

∞ The paper used in these volumes conforms to the American National Standard for Permanence of Paper for Printed Library Materials, Z39.48 1992 (R2009).

Publisher's Cataloging-In-Publication Data
(Prepared by The Donohue Group, Inc.)

Names: Renneboog, Richard, editor.
Title: Principles of geology / editor, Richard Renneboog, PhD.
Description: Ipswich, Massachusetts : Salem Press, a division of EBSCO Information Services, Inc. ; Amenia, NY : Grey House Publishing, [2021] | Series: [Principles of science] | Includes bibliographical references and index.
Identifiers: ISBN 9781642650426
Subjects: LCSH: Geology. | LCGFT: Reference works.
Classification: LCC QE26.3 .P75 2021 | DDC 550—dc23

First Printing
Printed in the United States of America

Contents

Publisher's Note . . . ix
Introduction . . . xi
Contributors . . . xv

Andesitic Rocks . . . 1
Aquifers . . . 4
Asteroids . . . 7
Athabasca Oil Sands . . . 10
Basaltic Rocks . . . 15
Batholiths . . . 19
Beaches and Coastal Processes . . . 24
Biogenic Sedimentary Rocks . . . 28
Bolides . . . 33
Boltwood Uses Radioactivity to Determine Ages of Rocks . . . 35
Carbonate Compensation Depths . . . 39
Carbonate Minerals . . . 43
Carbonatites . . . 46
Clays . . . 50
Clays and Clay Minerals . . . 53
Contact Metamorphism . . . 57
Coriolis Effect . . . 61
Crystals . . . 64
The "Deep-Impact" Mission: Geology of a Comet . . . 69
Deep-Sea Sedimentation . . . 72
Deglaciation . . . 76
Deltas . . . 78
Desert Landforms . . . 82
Desert Pavement . . . 85
Desertification . . . 90
Deserts . . . 93
Diagenesis . . . 97
Diamond . . . 101
Diatomite . . . 104
Doell and Dalrymple Discover the Magnetic Reversals of Earth's Poles . . . 107
Drainage Basins . . . 110
Earthquake Prediction . . . 115
Earthquakes . . . 120
Earth's Crust . . . 123
Earth's Internal Structure . . . 126
Earth's Moon . . . 129
Economic Geology . . . 130
Engineering Seismology . . . 137
Evaporites . . . 142
External Processes . . . 145
Extraterrestrial Resources . . . 147
Floods . . . 153
Fluvial and Karst Processes . . . 157
The Gaia Hypothesis . . . 163
Gem Minerals . . . 165
Geochemical Cycles . . . 170
Geologic Timescale . . . 172
Geology . . . 175
German *Meteor* Expedition Discovers the Mid-Atlantic Ridge . . . 178
Glaciations . . . 181
Glaciers . . . 183
Gondwanaland and Laurasia . . . 187
Granitic Rocks . . . 190
Gutenberg Discovers Earth's Mantle-Outer Core Boundary . . . 195
Heezen and Ewing Discover the Midoceanic Ridge . . . 199
Hess, Harry Hammond . . . 202
Hess Identifies the Cause of Continental Drift . . . 204
Hutton Proposes the Geological Theory of Uniformitarianism . . . 208
Igneous Processes, Rocks, and Mineral Deposits . . . 213
Internal Geologic Processes . . . 217
International System of Units . . . 220
Kimberlites . . . 227
Lehmann Discovers Earth's Inner Core . . . 233
Limestone . . . 235
Lithosphere . . . 236
Lomonosov Issues the First Catalog of Minerals . . . 238
Lunar Exploration . . . 241
Magma Crystallization . . . 247
Marble . . . 249
Mars Exploration Rovers . . . 251
Mean Sea Level . . . 255

Metamictization 258
Metamorphic Processes, Rocks, and Mineral Deposits. 260
Mineralization and Hydrothermal Solutions . . . 264
Mineralogy 269
Mohs Hardness Scale 275
Mountain Building 276
Nesosilicates and Sorosilicates 281
New Horizons Pluto-Kuiper Belt Mission. 285
Nonsilicates. 288
Nucleosynthesis 292
Ocean Acidification. 297
Ocean Margins 299
Oceans' Origin 301
Oil Shale and Tar Sands 305
Oldham and Mohorovičić Discover Earth's Interior Structure 309
Ophiolites 311
Orbicular Rocks 313
Paleontology 319
Physical Properties of Rocks 325
Placer Deposits 330
Planet Formation 333
Planetary Exploration. 339
Planetary Interiors 345
Plate Tectonics 351
Plutonic Rocks and Mineral Deposits. 354
Pumice 356
Pyroclastic Flows 358
Quartz and Silica 363
Reefs 367
Regional Metamorphism 371
Richter, Charles Francis 376
Richter Develops a Scale for Measuring Earthquake Strength. 379
Riverbed Forms. 382
Salt Domes 387
Sand and Gravel 388
Sandstone 391
Seafloor Spreading 393
Secondary Enrichment of Mineral Deposits. . . . 394
Sedimentary Processes, Rocks, and Mineral Deposits 396
Shale 400
Silicates 402
Siliciclastic Rocks 403
Slate 409
Soils. 410
Steno Presents His Theories of Fossils and Dynamic Geology 416
Streams and Rivers 418
Structure and Physical Properties of Minerals . . 424
Subduction and Orogeny 430
Sub-Seafloor Metamorphism 435
Teilhard de Chardin, Pierre 439
Turbidity Currents and Submarine Fans. 444
Ultrahigh-Pressure Metamorphism 449
Ultrapotassic Rocks. 452
Volcanoes 457
Weathering 461
Wegener, Alfred. 464
Wegener Proposes the Theory of Continental Drift 468
Xenoliths. 473
Zeolites 479

Bibliography 483
Glossary. 510
Organizations 525
Subject Index 527

PUBLISHER'S NOTE

Grey House Publishing is pleased to add *Principles of Geology* to its Salem Press collection of Principles of Science titles. Covering more than twenty topics, including *Energy*, *Marine Science*, *Astronomy*, *Computer Science*, *Physical Science*, *Biology*, *Botany*, *Mathematics*, and *Information Technology*, this series introduces students and researchers to the fundamentals of these scientific topics using easy-to-understand language for a solid background and a deeper understanding and appreciation. Major categories in *Principles of Geology* include principles, historical discoveries, the Big Bang event, continental drift, geochemistry, resource recovery, paleontology, and extraterrestrial geology.

The 130 entries contain the following helpful sections:

- **Abstract** provides a brief, concrete summary of the topic and how the entry is organized;
- **Key Concepts** introduce the terminology used in the entry;
- **Images and charts** illustrate key concepts such as the geologic time scale, continental drift, the Earth's mantle and crust, and magnetic pole reversal;
- **Further Reading** lists additional sources related to the study of geology.

This work begins with a comprehensive Editor's Introduction to the topic of geology written by volume editor Richard Renneboog. This discussion explains geology's unique role in the sciences and discusses the field's transition in the early nineteenth century from a hobby to the scientific study of the physical world. Stretching back to the Big Bang event and forward to the study of life on Mars, major milestones in geology are touched on, along with how geology both drew from and contributed to other scientific fields like chemistry, biology, physics, and thermodynamics, to name a few. Entries, arranged in A to Z order, range from Andeistic Rocks to Zeolites and everything in between.

Following the entries, *Principles of Geology* ends with helpful appendices: Bibliography; Glossary; and Subject Index.

Salem Press and Grey House Publishing extend their appreciation to all involved in the development and production of this work. A list of contributors to this work follow the Editor's Introduction.

Principles of Geology, as well as all Salem Press reference books, is available in print and as an e-book. Please visit www.salempress.com for more information.

Introduction

In 1823, Sir Walter Scott published a book titled Saint Ronan's Well, in which the inimitable character Meg Dodds, an outspoken, sarcastic woman, described people from the city visiting the country estate of her neighbor, a lady of the upper class:

> *Some rin up hill and down dale knapping the chucky stanes to pieces wi' hammers like sae many road—makers run daft—they say it is to see how the warld was made!*[1]

Obviously geologists. The early nineteenth century was an important stage in the history of geology, as it marked the transition from geology as an amateur hobbyist pastime to the serious scientific study of the physical world.

It is easy for those unacquainted with the field of geology to see geologists in the same way as Meg Dodds saw them: a bunch of people running about collecting odd bits of rock or looking to strike it rich with a big find of gold or gemstones. In a more recent popular portrayal of geology, from the television series "The Big Bang Theory," character Dr. Sheldon Cooper famously stood up in the middle of a paintball fight against a team from the Department of Geology and loudly proclaimed "Geology is not a real science!," upon which he was instantly plastered with paintballs fired from the guns of the geologists.[2] Ironically, Cooper was right; geology is not a real science. It is a scientific practice that is an amalgamation of principles from many different "real" sciences.

Geologists are a rather recent invention in the long history of the universe, which by all reckonings extends some 14.5 billion years into the past. The first "geologist" might be defined as the first hominid that ever picked up a rock and wondered what it was made of, what could be done with it, how it got there, and where it came from. As a scientific endeavor, the practice of geology is much younger still, by a few million years. In the present day, the practice of geology is built upon the application of the scientific principles and technologies of a number of different fields.

The progression of geology from its humble beginnings to its present-day sophistication began in earnest in 1793 with a canal digger named William Smith, who discovered what has been called Earth's fourth dimension, the order of the unseen content below the surface of the ground. Few people, if any, thought of this underworld as anything other than a jumbled mass of earth and rocks. It was obvious that a hole, such as a well, dug at a random location would reveal a succession of soil types and a random assortment of rocks and stones, but such holes were only local samplings of what lies in the ground below, seldom extending below the loose soils to bedrock. What was not obvious from such isolated samples is the continuity of those underground layers of rock. The perception was that what came out of a particular hole was unique to that hole and had no relationship to any other hole.

Canals disproved this perception. Canals are not isolated vertical holes but holes that extend horizontally for some distance through various and different terrains. The layers revealed in a canal excavation might seem as unique as those in a vertical hole, but Smith realized that he could identify the same layers in the same order in widely separated locations in a canal excavation, implying a continuity to the underground world that could be identified. From this realization, which Smith spent the next 22 years verifying, he generated "the map that changed the world"—showing the distribution of underground layers, soil types, rock structures, and all the other components of the underground. The first of its kind, Smith's map eventually encompassed all of Britain and set a precedent that advanced the profession of geologists everywhere.[3]

Imagine how valuable it would be to suddenly have a map that shows where to find the resources you need and how to find other resources...an actual treasure map. Implicit in such a map is a deeper understanding of the past written in the underground world. As scientific exploration progressed into the nineteenth century, how certain underground materials originated in the oceans would provide a logical explanation for their formation when those same materials were found inland, away from any ocean. It would imply that those inland locations were at some point in the distant past beneath the waters of an ancient sea. The questions that arise from such a simple deduction lead the geologist deeper into the quest to understand how the world was made.

The beginning of geology, as we know it today, begins with the origin of the universe in the "Big Bang" event—when all that energy produced matter. Arguably, those laws that came into existence at the same instant as the matter of the universe apply to this day throughout the universe, which has important implications for questions regarding carbon-based life elsewhere in the universe and regarding geology.

The likelihood that all matter created in the Big Bang was simple hydrogen is bolstered by the fact that 99 percent of the known matter in the universe is hydrogen, concentrated in stars and thinly dispersed throughout interstellar space. Over the millions of years since the Big Bang, the universal property called "gravity" has drawn, and continues to draw, massive quantities of hydrogen atoms together, eventually to form stars. But a mass of hydrogen atoms isn't automatically a star. A star must evolve from that mass. At some point in that evolutionary process, the force of gravity within the mass of hydrogen atoms forces them to combine—to undergo fusion—the beginning of the process of nucleosynthesis.

In nucleosynthesis, atoms of elements heavier than hydrogen atoms are formed from the fusion of hydrogen atoms. Each new element that is formed is naturally more massive than the ones from which it was formed. First helium was created, then lithium, and so on, until all of the elements including iron are created over the multi-billion-year lifetime of a star. Eventually, a star of the proper mass may become unstable, causing it to explode. Astrophysicists believe that all of the elements heavier than iron are formed in the high pressure exerted within the star as it explodes. The explosion spews all or most of those heavier atoms into space in all directions where they continue on their course through space until they encounter other atoms, or masses of atoms, and form objects. The high kinetic energies of the atoms involved in such collisions drive them into new combinations, and after a few billion years, rocks, asteroids, planets, and even nebulae—and new stars—are born, a process that continues to this day. Because light propagates through space at a finite speed, structures seen today in far outer space may actually have ceased to exist thousands, perhaps millions, of years ago. Even Earth's own star, the Sun (Sol), is known to exist with certainty up until about eight minutes before observation, the length of time it takes for the light from the Sun to reach observers on Earth.

By this process, all of the rock and water and gases that make up the geological world, and by extension the galaxy, have come to be. The practice of geology is informed by essentially all of the fundamental fields of science. Chemistry, biology, physics, thermodynamics, fluid dynamics, and so on. All of these have played a part in how the world was made in the four-and-a-half-billion-year history of this planet.

Chemistry and Geology

Chemistry is the science of how atoms interact to create the billions of different molecules that can exist. The chemistry that is unique to geology is called geochemistry, and the "stars" of geochemistry are the elements carbon, oxygen and silicon, with a supporting cast of all the other elements. There is something special about these three elements, however. Carbon and oxygen, along with nitrogen, are called Period 2 elements due to their position in the periodic table of the elements. Silicon is a Period 3 element in the same group as carbon. Silicon and carbon have the unique ability to form extensive chains and networks both with themselves and with oxygen, due to their characteristic electron arrangements.

The number of different compounds that can be formed by chains and three-dimensional networks of carbon atoms is infinite. Many thousands exist naturally in the form of petroleum, making them important geologically as recoverable resources. Given that petroleum was produced through processes of heat and pressure acting on carbon-based organisms—the huge amounts of bacteria and algae that existed on Earth for hundreds of millions of years before any other creatures existed here—the amount of petroleum existing in the world today is a limited resource that will one day become unrecoverable. It is, however, a resource that is still in high demand and the role of geologists in locating and identifying those remaining resources is increasingly important, as is geological expertise in developing methods of producing alternative products to naturally occurring petroleum from renewable resources such as algae.

Carbon-based compounds of interest to the energy industry begin with the simplest hydrocarbons, methane (CH_4) and ethane (C_2H_6), the components of natural gas, by-products of the petroleum formation process and generally found as geological deposits accompanying petroleum deposits. Methane, and perhaps ethane, form stable hydrates under the proper, low-temperature conditions. Large quantities are thought to exist on deep ocean floors, entrapped in the permafrost soils of the northern hemisphere. Certain geological features—large holes—that have occasionally suddenly appeared in Siberia are thought to be the result of the sudden explosive release of methane from the permafrost. The energy industry encourages geologists and engineers to find a way to capture permafrost methane rather than have it escape into the atmosphere.

Complex carbon compounds are formed by concatenation (or linking together) of carbon atoms into chains, and they are the basis of a structure as geometrically perfect as diamond. Every single perfect natural diamond crystal is a single molecule of diamond, no matter its size. Diamond is the hardest known natural substance.

Silicon is also able to concatenate in a manner similar to carbon, and silicon forms extensive molecular structures when oxygen atoms are interposed between the silicon atoms. The simplest silicon-oxygen material is silica, SiO_2, which is the foundation unit of silicate rocks and minerals, one of the two rock types that make up Earth's crust (the other being the carbonate rocks and minerals.) Silicate rocks and minerals come in many different forms, from the most mundane rock of mountains to the finest of gemstones, with quartz being the most common of all. Almost all sand consists of quartz particles.

When carbon interacts with oxygen atoms, carbonates are produced. Carbonates begin as carbonic acid, which readily dissociates to produce bicarbonate (HCO_3^-) and carbonate (CO_3^{2-}) ions. When metal ions are present, inorganic salts—solid deposits—are formed. The most common carbonate salt is calcium carbonate. Over the billions of years of Earth's geologic history, calcium carbonate eventually turned into limestone through geochemical and geological processes.

The accretion of calcium carbonate was greatly accelerated as many marine creatures secrete calcium carbonate to form protective body parts—a process that has evolved over time and adds discarded calcium carbonate shells to the ocean. These shells drifting down the water column for hundreds of millions of years, as they still do today, continue to build up the fossiliferous layers of chalk and other limestone forms. The carbonic acid formed naturally in rain water dissociates to produce a mildly acidic organic (carbon-based) acid solution that can, over time, break down carbonate rock and dissolve away the remaining oxides. Other gases emitted by volcanic activity, such as hydrogen chloride, produce acidic mineral acid solutions that dissolve limestone more effectively. What this produces in the natural environment is the many karst landscapes found around the world. Where other metal atoms have accumulated with the carbonate, significant ore bodies also form, all fodder for the geologist.

Biology and Geology

Geology is also informed by the science of biology. When organisms, which excrete calcium carbonate to form the tests and shells that protect and support their bodies, die, they become part of the mineral composition of the environment. Other organisms also incorporate silica in their structures, and they too drift down to the floors of lakes and oceans. In freshwater bodies, the smaller particles become part of the silt load carried by the water as it flows through. Only in the deepest freshwater lakes, where there is little to no effective current flow, do the remnants build up as bottom ooze which, in time, may become layers of siliceous rock.

In the salt water oceans, these carbonaceous and siliceous particles drift slowly downward to the bottom, which can take many years. At a certain depth in the saline environment, the calcium carbonate begins to slowly dissolve as it drifts downward through this "carbonate compensation depth." By the time the debris reaches the ocean floor it consists primarily of the remaining silicate debris rather than the carbonate debris that was dominant at the surface.

Where the carbonate compensation depth is in effect, the seafloor acquires a coating of siliceous ooze that is transported over time by the seafloor spreading to the nearest subduction zone where the siliceous ooze and the basaltic seafloor crust carrying the ooze are eventually returned to the mantle and recycled into new rock, a process that can take millennia. Where the carbonate compensation depth is not in effect, the seafloor acquires a coating of both carbonaceous and siliceous ooze (primarily carbonate) that also is eventually recycled by tectonic activity and becomes rock.

The geologist examining terrestrial or even extraterrestrial rock structures may discern important clues to the origins of those rocks and their likely history from their carbonate and silicate mineral composition, which allows a broader understanding of the geologic history of entire regions. Consider, for example, the implications of large beds of carbonate rock such as fossiliferous limestone in a present-day desert far from any ocean water, or perhaps on the surface of Mars.

Creatures other than marine organisms are also affected by the formation of sedimentary rock. Everything from the smallest of bacteria to the greatest of dinosaurs, as well as plants, have left their marks in rock and entered the annals of Earth's geologic past. Paleontology is the science of finding, identifying, and understanding ancient creatures. Where they are found, the type of rock structures they are found in, even the position of their remains provide a wealth of information about past geologic conditions, as well as clues to the geologic future.

One of the most significant contributions fossils have made to geology was the discovery of fossils of the same species found in identical rock formations on opposite sides of the present-day Atlantic Ocean. The reason for this wasn't known until the mid-twentieth century, when seafloor spread was discovered and explained how the continents of Africa and South America were long ago joined as parts of a single great continent. While recognition of the complimentary shapes of continental outlines has been around since the time of Sir Francis Bacon, it was not until Alfred Lothar Wegener proposed the theory of continental drift as a workable, though now known to be erroneous, explanation. Wegener based his theory in part on those identical fossils and geological environments found on opposite sides of the Atlantic. Wegener had proposed that the continental masses themselves drifted on the dense basalt crust of Earth, like boats on water, due to their different densities. In the theory of tectonic plates that evolved from Wegener's theory, it is large plates of the crust itself that move and carry the continents along with them as they move. Accordingly, the crustal plates that carry North and South America are being driven apart from the crustal plates carrying Africa and Europe.

Some twenty years or so after Wegener's untimely death on the Greenland icecaps, researchers identified the Mid-Atlantic Ridge and observed the separation of the seafloor that is occurring there. Harry Hess identified this process as seafloor spreading, the mechanism for continental movement.

During seafloor spreading, the oceanic crust and continental crust are also involved during this tectonic activity. At mid-ocean ridges, tectonic plates are being separated, allowing basaltic magma from the mantle to ooze up into the space between the separating plate. This creates new oceanic crust as the preceding crustal material moves away from the ridge. But Earth is not becoming larger by this expansion of crustal material. When the front of the oceanic crust plate encounters either a denser plate or a continental plate, it is forced to subduct, or go beneath it, in order to keep moving as the seafloor spreads. Subduction drives it back down into the mantle to be melted. The immense pressure exerted by the opposing plates in subduction causes buckling of the other plate, and major cracks and faults that engender both earthquakes and volcanoes. Material ejected through fissures and volcanoes forms igneous rock and mineral deposits that often have important economic value. Diamonds, for example, are recovered from kimberlite pipes that were once the channel of a volcano that had exploded violently, allowing kimberlite magma to rise up to fill the empty channel.

The collision of continental crust plates has a somewhat different effect than seafloor spreading. Rather than one plate subducting to the other plate, the rock of both plates begins to fold and buckle, creating great mountain ranges such as the Alps of Europe and the Himalayas of India and China. This orogeny is not accompanied by volcanic activity generally, and seismic activity (earthquakes) caused by the process tends to be less powerful than that along subduction zones.

Geochemistry and Geology

Paleontologists study the geochemical process of how living bone becomes mineralized into a stone fossil, another source of information about the geologic past. This process is believed to occur by replacing calcium in bone with minerals in water that seep through the sediments and rock in which the bones are found. The identity of the minerals found in the fossils can tell a great deal about the terrain and even the climate conditions of the region that was inhabited by the bones' original owners. Interment of a creature's remains may be extremely rapid, as was the case of an oviraptor and a velociraptor found near the Red Cliffs in the Gobi Desert. It seems that the two were interred in the midst of battle, likely by the collapse of a sand dune, and their fossilized remains are locked in battle to the present day. Alternatively, interment of remains may be slow, giving scavengers ample time to dismember a carcass and scatter the bones over a wide area.

However, the vast majority of ancient creatures do not become fossils. Fossilization requires the special circumstance of anaerobic (lacking oxygen) conditions following interment, as oxygen quickly decomposes animal remains. Once anaerobically interred, however, the fossilization process by mineral replacement can begin. Whether trapped under a sand dune or buried under the fine silt carried by a flood, there is still no guarantee that a fossil will be formed, or that it will survive intact. Seismic activity, floods, and other natural phenomena can and do expose fossil material every day, allowing weather and time to destroy the evidence. While geologists, paleontologists and others have recovered many thousands of fossils, only a small percentage have been sufficiently complete to allow identification and reconstruction of the creatures that produced them. The search goes on, however, and new fossils are found every day.

The fields that have contributed to, and benefitted from, the practice of geology include fundamental chemistry, astrophysics, geochemistry, volcanology, paleontology, oceanography, ecology, climatology and biology. Geology is a modern practice not only concerned with Earth's past, but very much concerned with the present dynamics of planet Earth. All of the fields that have contributed to modern geology in the past continue to play a role in the present day. Stars and planets continue to be formed in space, and the planets of the solar system are coming to be known in real time. There are present-day efforts to establish a human presence on the Moon and undertake first-hand examination of Mars. New physical environments unfamiliar to humans can only become known through geological study and methods. Geologists of all kinds will be important explorers of these new environments. Yet on Earth, the role of geologists is even more important precisely due to its dynamic nature. There is something new for geologists to find or learn every day.

—Richard Renneboog, PhD

Notes

1. Scott, Sir Walter. *Saint Ronan's Well.* 1823. Oxford UP, 1912, p. 25.
2. "The Big Bang Theory," season 5, episode 1, 22 Sept. 2011.
3. Winchester, Simon. *The Map That Changed The World: William Smith and the Birth of Modern Geology.* Perennial, 2001.

Contributors

Richard Adler
University of Michigan-Dearborn

Arthur L. Alt
College of Great Falls

Michael P. Auerbach
Marblehead, MA

Victor R. Baker
University of Arizona

Anita Baker-Blocker
Applied Meteorological Services

Melissa A. Barton
Westminster, Colorado

Raymond D. Benge, Jr.
Tarrant County College

Alvin K. Benson
Brigham Young University

John L. Berkley
State University of New York, Fredonia

David M. Best
Northern Arizona University

Tyler Biscontini
Independent Scholar

Howard Bromberg
University of Michigan

Alan Brown
Livingston University

Michael A. Buratovich
Spring Arbor University

Scott F. Burns
Louisiana Tech University

Robert E. Carver
University of Georgia

Dennis Chamberland
Independent Scholar

H. G. Churnet
University of Tennessee at Chattanooga

James R. Craig
Virginia Polytechnic Institute and State University

Anna M. Cruse
Oklahoma State University

Robert L. Cullers
Kansas State University

Larry E. Davis
Washington State University

Ronald W. Davis
Western Michigan University

Albert B. Dickas
University of Wisconsin

Dean A. Dunn
Independent Scholar

Steven I. Dutch
University of Wisconsin, Green Bay

Sherry L. Eaton
San Diego City College

David Elliott
Northern Arizona University

David G. Fisher
Lycoming College

Robert G. Font
Independent Scholar

Annabelle M. Foos
University of Akron

Charles I. Frye
Northwest Missouri State University

Roberto Garza
Independent Scholar

Noreen A. Grice
Boston Museum of Science

Pamela J. W. Gore
DeKalb Community College

Edward C. Hansen
Hope College

Angela Harmon
Independent Scholar

Clayton D. Harris
Middle Tennessee State University

Paul A. Heckert
Western Carolina University

David F. Hess
Western Illinois University

Stephen C. Hildreth Jr.
Pickens, South Carolina

J. D. Ho
Independent Scholar

René A. De Hon
University of Louisiana at Monroe

Robert M. Hordon
Rutgers University

Earl G. Hoover
Independent Scholar

Paul F. Hudson
University of Texas at Austin

Micah L. Issitt
Independent Scholar

Pamela Jansma
Jet Propulsion Laboratory

Richard C. Jones
Texas Women's University

Karen N. Kähler
Independent Scholar

David Kasserman
Independent Scholar

Kyle L. Kayler
Richard H. Gorr and Associates

Diann S. Kiesel
University of Wisconsin Center

David T. King Jr.
Independent Scholar

Ludwik Kowalski
Montclair State University

Ralph L. Langenheim Jr.
University of Illinois at Urbana-Champaign

Gary G. Lash
State University of New York College, Fredonia

Denyse Lemaire
Rowan University

Donald W. Lovejoy
Palm Beach Atlantic College

Gary R. Lowell
Southeast Missouri State University

Michael Mattei
Independent Scholar

Rubén A. Mazariegos-Alfaro
University of Texas/Pan American

Richard Means
Independent Scholar

Randall L. Milstein
Oregon State University

Otto H. Muller
Alfred University

Mysore Narayanan
Miami University

Richard L. Orndorff
University of Nevada, Las Vegas

Michael R. Owen
St. Lawrence University

Robert J. Paradowski
Rochester Institute of Technology

Robert T. Pennock
Independent Scholar

Rex D. Pieper
New Mexico State University

C. Nicholas Raphael
Eastern Michigan University

Donald F. Reaser
University of Texas at Arlington

Robert V. Reeves
Independent Scholar

Jeffrey C. Reid
Independent Scholar

Raymond U. Roberts
Oklahoma Department of Environmental Quality

Lisa A. Roble
Independent Scholar

Charles W. Rogers
Southwestern Oklahoma State University

Neil E. Salisbury
University of Oklahoma

Rose Secrest
Independent Scholar

John F. Shroder Jr.
University of Nebraska at Omaha

Paul P. Sipiera
Harper College

Billy R. Smith Jr.
Anne Arundel Community College

John Brelsford Southard
Massachusetts Institute of Technology

Robert J. Stern
University of Texas at Dallas

Dion C. Stewart
Adams State College

Anthony N. Stranges
Independent Scholar

Charles R. Sullivan
Independent Scholar

E. Summers
Independent Scholar

Frederick M. Surowiec
Society of Professional Journalists

Randa Tantawi
Independent Scholar

Stephen M. Testa
California State University, Fullerton

Keith J. Tinkler
Brock University

Roger Dale Trexler
Southern Illinois University

Ronald D. Tyler
Independent Scholar

Kristen L. Zacharias
Independent Scholar

A

ANDESITIC ROCKS

Fields of Study: Geology; Volcanology; Geoinformatics

ABSTRACT
Andesite is an intermediate extrusive igneous rock. Active volcanoes on the earth erupt andesite more than any other rock type. Andesites are primarily associated with subduction zones along convergent tectonic plate boundaries.

KEY CONCEPTS

groundmass: the fine-grained material between phenocrysts of a porphyritic igneous rock

intermediate rock: an igneous rock that is transitional between a basic and a silicic rock, having a silica content between 54 and 64 percent

phenocryst: a large conspicuous crystal in a porphyritic rock

plutonic rock: igneous rock formed at a great depth within the earth

porphyry: an igneous rock in which phenocrysts are set in a finer-grained groundmass

stratovolcano: a volcano composed of alternating layers of lava flow and ash; also called a "composite volcano"

subduction zone: a convergent plate boundary

viscosity: a substance's ability to flow; the lower the viscosity, the greater the ability to flow

INTEREST TO GEOLOGISTS

Andesite takes its name from lavas in the Andes mountains of South America. To most geologists, andesites are light gray porphyritic volcanic rocks containing phenocrysts of plagioclase, very little quartz, and no sanidine or feldspathoid. Despite their lackluster appearance, andesites are of great interest to geologists for several reasons. First, active volcanoes on Earth erupt andesite more than any other rock type; andesite is the main rock type at 61 percent of the world's active volcanoes. Second, andesites have a distinctive tectonic setting. They are primarily associated with convergent plate boundaries and occur elsewhere only in limited amounts. Of the active volcanoes that occur within 500 kilometers of a subduction zone, 78 percent include andesite; only three active volcanoes not near a convergent plate boundary erupt it. Third,

Andesite pillar. Photo by Doronenko, CC BY 3.0, via Wikimedia Commons.

andesites have bulk compositions similar to estimates of the composition of continental crust. This similarity, in association with the tectonic setting of andesites, suggests that they may play an important role in the development of terrestrial crust. Fourth, the development and movement of andesitic magma seem to be closely related to the formation of many ore deposits, including such economically important ores as molybdenum and porphyry copper deposits.

VISCOSITY

Rocks are classified chemically according to how much silica they contain; rocks rich in silica (more than 64 percent), such as rhyolite, are called "silicic." They consist mostly of quartz and feldspars, with minor amounts of mica and amphibole. Rocks low in silica (less than 54 percent), with no free quartz but high in feldspar, pyroxene, olivine, and oxides, are called "basic." Basic rocks, free of quartz, tend to be dark, while silicic rocks are lighter and contain only isolated flecks of dark minerals. Basalts are examples of basic volcanic rocks. Andesites, having a silica content of about 60 percent, are volcanic rocks termed "intermediate." Andesite's plutonic equivalent is diorite. There is no cut-and-dried difference between basalt and andesite, or between andesite and rhyolite. Instead, there is a broad transitional group of rocks that carry names such as basaltic andesite or andesitic rhyolite.

Nevertheless, some generalizations can be applied to the lavas and magmas that form these rocks. One generalization has to do with viscosity. Andesite lavas are more viscous than basalt lavas and less viscous than rhyolite lavas. This difference is primarily an effect of the lava's composition, and to a certain extent it is a result of the high portion of phenocrysts present in the more viscous lavas.

Different minerals crystallize at different temperatures. As a basalt magma cools, a sequence of minerals appears. The first mineral to crystallize is usually olivine, which continues to crystallize as the magma cools until a temperature is reached at which a second mineral, pyroxene, begins to crystallize. As the temperature continues to drop, these two continue to crystallize. Cooling continues until a temperature is reached when a third mineral, feldspar, crystallizes. This chain of cooling and mineral crystallization is known as Bowen's reaction principle. Often, olivine and pyroxene crystallize out early in the process, so they may be present in the final rock as large crystals up to a centimeter across. These crystals are called "phenocrysts." The size of these phenocrysts is in direct contrast to the fine-grained crystals of the groundmass. Igneous rocks with phenocrysts in a fine-grained groundmass are known as porphyries. Most volcanic rocks contain some phenocrysts. The groundmass crystals form when the magma cools upon reaching the surface. If the magma has a low viscosity, reaches the surface, spreads out as lava, and cools quickly, individual crystals do not have enough time to grow. The overall rock remains fine-grained. The phenocrysts crystallized out much earlier, while the magma was still underground. There, they had plenty of time to grow and were then carried to the surface with the magma during eruption.

ANDESITE FLOWS

Basalts, as a result of their low viscosity, tend to produce thin lava flows that readily spread over large areas. They rarely exceed 30 meters in thickness. Andesite flows, by contrast, are massive and may be as much as 55 meters thick. The largest single andesite flow described, which is in northern Chile, has an approximate volume of 24 cubic kilometers. Because of their low viscosity, basalt flows can advance at considerable rates; speeds up to 8 kilometers per hour have been measured. Andesite flows often move only a matter of tens of meters over several hours. As a result of their higher viscosity, andesite flows show none of the surface features of more liquid lavas. Flow features such as wave forms, swirls, or the ropy textures often associated with basalt flows never occur in andesite flows. Andesite flows tend to be blocky with large, angular, smooth-sided chunks of solidified lava. The flow tends to behave as a plastic rather than a liquid. An outer, chilled surface develops on the slow-moving flow, with the interior still molten. Plasticity within the flow increases toward the still-molten inner portion. As the flow slowly shifts, moves, and cools, the hard, brittle outer layer breaks into the large angular blocks characteristic of andesite lava flows. The flow slowly moves, with the blocks colliding and overriding one another to form piles of angular andesite blocks.

The viscous nature of andesite lavas is responsible for many classic volcanic features. Most notable is the symmetrical cone shape of the stratovolcanoes of the circum-Pacific region. Short, viscous andesitic flows pouring down the flanks of these volcanoes are alternately covered by pyroclastic material and work to build the steep central cone characteristic of composite stratovolcanoes. When an andesite's silica content rises to a point that it approaches rhyolite composition, it is termed a "dacite." Dacite is often so viscous that it cannot flow and blocks the vent of the volcano. This dacite plug is called a "lava dome." Such a dome can be seen in pictures of Mount St. Helens. If the plugged volcano becomes active again, the lava dome does not allow for a release of accumulating pressure and explosive gases. Pressure builds until the volcano finally erupts with great force and violence.

ANDESITE LINE

The differences between basalts and andesites reflect their differences in composition, which is a function of the environment in which their source lavas occur. Basalts are typically formed at midocean ridges and build up oceanic crust. The generation of andesite magma is characteristic of destructive plate margins. Here, oceanic plates are being subducted below continental plates. Destructive plate boundaries tend to produce a greater variety of lavas than spreading zones (ocean ridges). The close association of andesite with convergent plate boundaries is the significance of the "Andesite Line" often drawn around the Pacific Ocean basin. This fairly well-defined line separates two major petrographic regions. Inside this line and inside the main ocean basin, no andesites occur. All active volcanoes inside the line erupt basaltic magma, and all volcanic rocks associated with dormant volcanoes within this region are basaltic. Outside the line, andesite is common. The Andesite Line parallels the western and northern boundary of the Pacific plate and the eastern boundary of the Juan de Fuca, Cocos, and Nazca plates. The Andesite Line parallels the major island arc systems, the subducting edges of the tectonic plates listed above, and a chain of prominent and infamous stratovolcanoes known as the Ring of Fire. These stratovolcanoes are exemplified by the following: Mount Rainier and other volcanoes of the western United States, El Chichón of Mexico, San Pedro of Chile, Mounts Egmont and Taupo of New Zealand, Krakatau and Tambora of the Indonesian Arc, Fujiyama of Japan, Bezymianny of Kamchatka, and the Valley of Ten Thousand Smokes in Alaska.

BENIOFF ZONE

The Benioff zone is a plane of seismic activity dipping at an angle of about 45 degrees below a continent, marking the path of a subducting oceanic plate at the convergent plate margin. At the Benioff zone, the subducting plate heats up and gives off water, which lowers the melting point of the overlying mantle and causes it to melt. As magma is formed at the Benioff zone and begins its slow rise toward the surface, it passes through and comes in contact with regions of the mantle and continental crust. During its rise, the magma also comes in contact with circulating meteoric water (water originating on the surface), and new fluids derived from magma are formed. Because andesitic magma rises through a variety of host rocks, the variety of minerals that enrich the new fluids is increased. The rising magma also stresses the surrounding host rocks, causing them to fracture. These fractures and similar open spaces are filled with mineral-rich water and fluids derived from magma. As the fluids cool, mineral ores precipitate from the solutions and fill the open spaces. These filled spaces often become mineral-rich dikes and sills.

FIELD STUDY

The study and interpretation of andesitic rocks are accomplished in three basic ways: fieldwork, in which researchers travel to locations to assess and interpret a specific region of Earth that is known or suspected of being andesitic terrain; petrological studies, which use all available methods of study to ascertain the history, origin, conditions, alterations, and weathering of the rocks collected during fieldwork; and mineralogical studies, which through intensive laboratory investigations identify the specific mineral characteristic of a sample.

If a scientist suspects that a region of Earth is andesitic terrain, the first step is to read the information already available about the region. If travel to that region for fieldwork is considered useful, the researcher will most likely collect a large number of samples. The location of each sample is carefully recorded on a map and additional data describing the geologic setting of the sample are recorded. These data include thickness, areal extent, weathering, and strike and dip of the location. Photographs are also commonly used to record the surrounding environment of the sample location. Samples are usually given preliminary study at the researcher's field camp.

PETROLOGICAL AND MINERALOGICAL STUDIES

When the fieldwork is over, the researcher returns with the samples to a laboratory setting, where petrological and mineralogical investigations begin. A petrological study of an andesitic sample will include both petrographic and petrogenetic analyses. The petro- graphic analyses will describe the sample and attempt to place it within the standard systematic classification of igneous rocks. Identification of the sample is accomplished by means of examining a thinly sliced and polished portion of the sample under a petrographic microscope—a process known as thin-section analysis. The information obtained by microscopic examination gives a breakdown of the type and amount of mineral composition within the sample. More advanced analysis techniques, such as X-ray fluorescence and electron microprobe, are often used to obtain very precise chemical compositions of samples. Knowing the conditions under which these minerals form allows the researcher to make a petrogenetic assessment.

Petrogenesis deals with the origin and formation of rocks. If a mineral is known to form only at certain depths, temperatures, or pressures, its presence within a sample allows the researcher to draw some specific conclusions concerning the rock's formation and history. Mineralogical studies of field-gathered samples aid in the petrogenetic portion of the analysis. Mineralogy involves the study of how a mineral forms—its physical properties, chemical composition, and occurrence. Minerals can exist in a stable form only within a narrow range of pressure and temperature. Experimental confirmation of this range enables the researcher to make a correlation between the occurrence of a mineral in a rock and the conditions under which the rock was formed. In this sense, mineralogical and petrological analyses complement each other and work to formulate a concise history of a given sample.

When data gathered from the field and laboratory are combined, the geological history of a given field area can begin to be interpreted. When areas of similar igneous rock types, ages, and mineral compositions are plotted on a map, they form a petrographic province. A petrographic province indicates an area of similar rocks that formed during the same period of igneous activity. On a global scale, the Andesite Line marks the boundary between two great provinces: the basaltic oceanic crust and the andesitic continental crust. Both crustal forms have distinctly different geological histories and mineral compositions.

—Randall L. Milstein

Further Reading

Aiello, Gemma. *Volcanoes: Geological and Geophysical Setting, Theoretical Aspects and Numerical Modeling, Applications to Industry and Their Impact on Human Health.* IntechOpen, 2018.

Faure, Gunter. *Origin of Igneous Rocks: The Isotopic Evidence.* Springer-Verlag, 2010.

Gill, Robin. *Igneous Rocks and Processes: A Practical Guide.* Wiley-Blackwell, 2011.

Gomez-Tuena, A., S. M. Straub, and G. F. Zellmer, editors. *Orogenic Andesites and Crustal Growth.* Geological Society, 2014.

Harker, Alfred. *The Natural History of Igneous Rocks.* Cambridge UP, 2011.

Klein, Cornelis, and Cornelius S. Hurlbut, Jr. *Manual of Mineralogy.* 23rd ed., John Wiley & Sons, 2008.

Ross, Pierre-Simon, et al. "Basaltic to Andesitic Volcaniclastic Rocks in the Blake River Group, Abitibi Greenstone Belt: 2. Origin, Geochemistry, and Geochronology." *Canadian Journal of Earth Sciences,* vol. 48, 2011, pp. 757-77.

AQUIFERS

Fields of Study: Geology; Hydrogeology; Stratigraphy

ABSTRACT

An aquifer is a body of earth material that can store and transmit significant amounts of water. The earth material can be in either consolidated or unconsolidated form as long as it has sufficient permeability for the movement of water. In terms of groundwater occurrence, all the rocks found on and below Earth's surface are associated with either aquifers or confining beds.

KEY CONCEPTS

anisotropic: a structure in which a physical property has different values depending on direction

Darcy's Law: the rate of flow of water in an aquifer is dependent on the hydraulic conductivity of the aquifer matrix, its perpendicular cross-sectional area, and the hydraulic gradient of the aquifer

hydraulic conductivity: the ability of a porous matrix to permit the passage of water

isotropic: a structure in which a physical property has the same value regardless of direction

water table: the depth of the uppermost level of an aquifer relative to the ground surface

BACKGROUND

An aquifer is a typically a rock or soil layer that is permeable enough to yield water in usable amounts to a well or a spring. In geologic usage, the term "rock" includes unconsolidated sediments such as sand, silt, and clay, as well as what is commonly considered to be rock. A confining bed is a rock unit that has such low hydraulic conductivity (poor permeability) that it restricts or severely impedes the flow of groundwater into or out of nearby aquifers.

UNCONFINED AND ARTESIAN AQUIFERS

There are two major types of groundwater occurrence in aquifers. The first type pertains to those aquifers that are only partially filled with water. In those cases, the upper surface (or water table) of the saturated zone can rise or decline in response to variations in precipitation, evaporation, and pumping from wells. The water in these aquifers is classified as unconfined, and such aquifers are called "unconfined," or "water- table," aquifers.

The second type occurs when water completely fills an aquifer that is overlain by a confining bed. In this case, the water in such an aquifer is classified as confined and the aquifers are called "confined," or "artesian," aquifers. In some fractured rock formations, such as those that occur in the west-central portions of New Jersey and eastern Pennsylvania, local geologic conditions result in semiconfined aquifers that, as one might expect, have hydrogeologic characteristics of both unconfined and confined aquifers.

Wells that are drilled in unconfined, water-table aquifers are simply called "water-table wells." The water level in these unconfined wells indicates the depth of the water table below the ground surface, which is the top of the saturated zone. Wells that are drilled into confined aquifers are called "artesian wells." The water level in an artesian well is generally located at a height above the top of the confined aquifer but not necessarily above the land surface. Flowing artesian wells occur when the water level in an artesian well stands above the land surface. The water level in tightly cased wells in unconfined or artesian aquifers is called the potentiometric surface of the aquifer.

AQUIFER PERMEABILITY

Water flows (very slowly) in aquifers from recharge areas in interstream areas at higher elevations to discharge areas along streams and adjacent floodplains at lower elevations. Thus, aquifers function as "pipelines" filled with various types of earth material. Darcy's law governing groundwater flow was developed by Henry Darcy, a French engineer, in 1856. In brief, Darcy's law states that the amount of water moving through an aquifer per unit of time is dependent on the hydraulic conductivity (or permeability) of the aquifer, the cross-sectional area that is at a right angle to the direction of flow, and the hydraulic gradient. The hydraulic conductivity depends upon the size and interconnectedness of the pores and fractures in an aquifer. It ranges through an astonishing twelve orders of magnitude. There are few other physical parameters that have such a wide range of values. For example, the hydraulic conductivity ranges from an extremely low 10-7 and 10-8 meters per day in unfractured igneous rock such as diabase and basalt to as much as 103 and 104 meters per day in cavernous limestone and coarse gravel. Typical low-permeability earth materials include unfractured shale, clay, and glacial till. High-permeability earth materials include lava flows and coarse sand.

In addition to this wide range of values, hydraulic conductivity varies widely in place and in

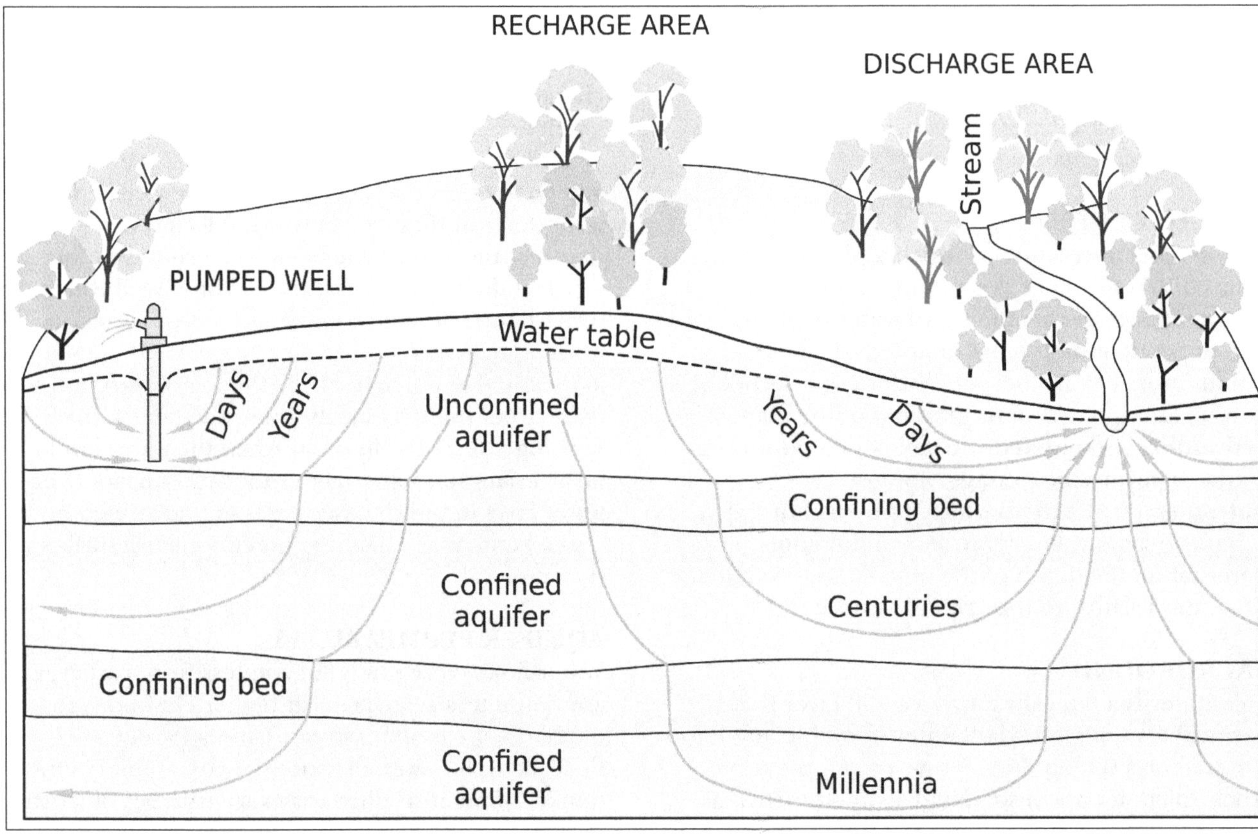

Schematic of an aquifer showing confined zones, groundwater travel times, a spring and a well. Image via Wikimedia Commons. [Public domain.]

directionality within the same aquifer. Aquifers are isotropic if the hydraulic conductivity is about the same in all directions, and anisotropic if it is different in different directions. As a result of all of these factors, groundwater yield is extremely variable both within the same aquifer and from one aquifer to another when they are composed of different rocks.

AQUIFER TESTS

In order to determine the groundwater yield and contaminant transport characteristics of an aquifer, it is necessary to obtain sufficient geologic and hydrologic information. One of the most important hydrologic investigations in such a study is the analysis of the change over time of the water levels in an aquifer as a consequence of well pumpage. This type of study is called an "aquifer test" and usually involves pumping a well at a constant rate for several hours to several days while changes in the water levels of one or more observation wells located at different distances from the pumped well are measured. The test provides invaluable information about the ability of an aquifer to yield sufficient water under the stress of constant pumping.

—Robert M. Hordon

Further Reading

Appelo, C. A. J., and D. Postma. *Geochemistry, Ground Water, and Pollution.* 2nd ed., Balkema, 2005.

Bailey, Heather. *Aquifers: Properties, Roles and Research.* NOVA Science Publishers, 2017.

Fetter, C. W. *Applied Hydrogeology.* 4th ed., Prentice Hall, 2001.

Javaid, Muhammad Salik, and Shaukat Ali Khan, editors. *Aquifers: Matrix and Fluids.* IntechOpen, 2018.

Maliva, Robert G. *Aquifer Characterization Techniques.* Springer, 2016.

Sharp, John. *Fractured Rock Hydrogeology.* CRC Press, 2014.

Stevanovic, Zoran, editor. *Karst Aquifers—Characterization and Engineering.* Springer, 2015.

Todd, David Keith, and Larry W. Mays. *Groundwater Hydrology*. 3rd ed., Wiley, 2005.
Younger, Paul L. *Groundwater in the Environment: An Introduction*. Blackwell, 2007.
Zektser, Igor S., and Lorne G. Everett, editors. *Ground Water Resources of the World and Their Use*. UNESCO, 2004. Reprint. National Ground Water Association Press, 2006.

ASTEROIDS

Fields of Study: Astrogeology; Astronomy; Sub-planet Astronomy

ABSTRACT
Asteroids are small, airless, cratered astronomical objects that are made of mostly rock. Most are irregularly shaped, but some have regular spherical shapes. Most asteroids can be found orbiting the sun between the planets Mars and Jupiter. Ceres is the largest known asteroid, at approximately 950 kilometers (590 miles, or 3,115,000 feet) wide, and 1991 BA is among the smallest known, at 6 meters (20 feet) wide. Asteroids are a potential source of high-value elements that are not abundant on Earth and their geological characteristics are indicative of their formation.

KEY CONCEPTS

asteroid belt: an area between the planets Mars and Jupiter where the majority of asteroids can be found

dwarf planet: a relatively large astronomical object that orbits the sun and has enough mass to be roughly spherical in shape but has not cleared its orbit of other astronomical objects

meteor: a "shooting star," generally a small particle of orbiting material that enters Earth's atmosphere and incinerates before reaching the surface

meteorite: a meteor that has passed through the atmosphere and struck the surface of the planet

near-Earth object (NEO): an asteroid whose orbit brings it close enough to Earth's orbit that an orbital perturbation could result in a collision

small solar system body: an astronomical object other than a planet, dwarf planet, or satellite that orbits the sun

WHAT ARE ASTEROIDS?

Asteroids are celestial bodies located in the solar system. While they share some characteristics with planets, they are much smaller in size. To put their size into perspective, the total mass of all known asteroids combined is less than the mass of Earth's moon.

Asteroids are sometimes referred to as planetoids or minor planets, which are defined as astronomical objects other than planets or comets that orbit the sun. However, new discoveries in the twenty-first century began to blur the differences between planets, comets, asteroids, tiny meteoroids, and distant trans-Neptunian objects (TNOs). In 2006, the International Astronomical Union (IAU) created the umbrella term small solar system bodies (SSSBs). The designation includes all celestial objects except for planets, satellites, and dwarf planets. The term "asteroid" is still widely used, however, especially for objects in the inner solar system. All inner solar system asteroids are classified as SSSBs except for Ceres, which is a dwarf planet.

Asteroids were formed about 4.6 billon years ago during the formation of the solar system. Basically, asteroids are the leftover pieces that did not form into planets. When the planet Jupiter formed, it prevented other planets from forming between itself and the planet Mars. Small objects between these planets collided with one another until they broke apart into asteroids.

The area between Mars and Jupiter, where the majority of asteroids can be found, is called the asteroid belt. Billions of asteroids can be found in the asteroid belt. Of these, more than 200 are over 100 kilometers (62 miles) wide, and more than 750,000 are over 1 kilometer (0.62 mile) wide. Other celestial objects, such as comets and dwarf planets, also can be found in the asteroid belt.

Some asteroids lie beyond the asteroid belt. Trojan asteroids are a group of asteroids that share the orbital path of Jupiter. Three other types of asteroids—Aten, Amor, and Apollo asteroids—orbit close enough to Earth to pose a potential threat.

Asteroids are irregularly shaped objects composed mainly of rock. Some are made of solid rock, while others are merely balls of dust and rubble held together by gravity. They are often categorized based on their composition. C-type (carbon-rich) asteroids are the most common and have remained largely unchanged since their formation. S-type (stony or siliceous) asteroids have melted and reformed. M-type (metal-rich) asteroids are a reddish color due to their nickel-iron content. Most asteroids fall into these three categories, although other types exist. All types are typically airless, dust covered, and pitted or cratered on the surface. Scientists have also found asteroids with characteristics similar to those of other celestial bodies, such as comets and planets. Rings, tails, moons, and water vapor have all been found on asteroids.

Asteroids rotate and orbit around the sun. Most follow an elliptical path, although some asteroids spin erratically due to their irregular shapes. Some asteroids are binary systems, in which two asteroids of similar size orbit a common center of mass. A few triple asteroid systems exist.

Sometimes asteroids get caught up in a planet's gravitational pull and become moons. Astronomers believe that Mars's moons, Phobos and Deimos, were once asteroids before they were pulled into Mars's orbit. Many of the distant moons of Jupiter, Saturn, Uranus, and Neptune are thought to be captured asteroids as well.

Asteroids are very cold, which makes them inhospitable to life. Their average surface temperature is -73° Celsius (-99° Fahrenheit). Some have conditions that have not changed since they formed billions of years ago, making them valuable for learning about the solar system's history.

DISCOVERY OF ASTEROIDS

Astronomers in the eighteenth century used the Titius-Bode law, or simply Bode's law, to predict the

253 Mathilde, a C-type asteroid measuring about 50 km (30 mi) across, covered in craters half that size. Photograph taken in 1997 by the NEAR Shoemaker probe. Photo via NASA/Wikimedia Commons. [Public domain.]

spacing of planets in the solar system. German astronomer Johann Titius (1729-96) first proposed the theory in 1766, and Johann Elert Bode (1747-1826) formulated it as a mathematical expression in 1778. The law relates the distances of the planets from the sun to a progression of numbers. In 1781 Sir William Herschel (1738-1822) discovered the seventh planet, Uranus, which seemed to confirm Bode's law. However, the law also predicted that a planet should exist between Mars and Jupiter. Near the end of the eighteenth century, a group of astronomers in Germany set out to find the missing planet. But the Italian priest and astronomer Giuseppe Piazzi (1746-1826) beat the group and found what was thought to be the missing planet on January 1, 1801.

Scientists named the new object Ceres, for the Roman goddess of agriculture. However, they quickly determined that Ceres (later officially dubbed 1 Ceres), which was only about 950 kilometers (590 miles) wide, should not be considered a planet. As astronomers continued to search for the predicted planet, they found more objects even smaller than Ceres. The asteroids 2 Pallas, 3 Juno, 4 Vesta, and 5 Astraea were observed and named. It was established that no large body lay between Jupiter and Mars. Scientists determined that Jupiter's gravitational pull prevented the loose material in the area from forming a planet. This loose material was called the asteroid belt, and scientists began to study the new class of objects.

Ceres was the first asteroid discovered and is the largest known. It accounts for about one-quarter of the total mass of all asteroids in the asteroid belt, yet it is fourteen times less massive than Pluto. In 2006 Ceres was classified as a dwarf planet. It is the only asteroid in the inner solar system to be so designated. Pluto and 136199 Eris are other examples of dwarf planets.

EXPLORING ASTEROIDS

Since asteroids were discovered in the nineteenth century, scientists have continued to study them and search for new ones. New technology in the twentieth century enabled closer study and images of asteroids. In 1971, the National Aeronautics and Space Administration (NASA) spacecraft *Mariner 9* took the first detailed images of Phobos and Deimos, which are believed to be captured asteroids.

The first close-up images of nonsatellite asteroids were taken by the NASA spacecraft *Galileo*. In 1991, *Galileo* photographed 951 Gaspra, which is located on the inner edge of the asteroid belt. The spacecraft took images of a second asteroid, 243 Ida, in 1993. These images revealed the presence of a moon orbiting Ida. The moon, which was named (243) Ida I Dactyl, was the first confirmed asteroid moon.

In the beginning of the twenty-first century, NASA's *NEAR* spacecraft studied 433 Eros, a near-Earth asteroid, for nearly a year. *NEAR* landed on Eros in 2001, becoming the first spacecraft to successfully land on an asteroid. In 2005 the Japanese spacecraft *Hayabusa* landed on the asteroid 25143 Itokawa, took samples, and then relaunched. The spacecraft returned to Earth in 2010 with the first samples taken from an asteroid.

In 2007 NASA sent the space probe *Dawn* to explore Vesta. It reached the asteroid in 2011 and spent a year in orbit, collecting information. It then left for Ceres, entering the dwarf planet's orbit in March 2015. *Dawn* was the first spacecraft to successfully visit either of these asteroids.

In addition to space probe missions, astronomers on Earth continue to study asteroids. In 2013 they spotted an asteroid with six tails located in the asteroid belt. Dubbed P/2013 P5, the asteroid is thought to have gained the six tails as result of intense spinning. The following year scientists discovered 10199 Chariklo, an asteroid with a set of rings. Considered small at 250 kilometers (155 miles) wide, Chariklo is located between Neptune and Uranus and is also thought to have an undiscovered moon. Researchers believe more asteroids such as these exist, and they continue to search the skies for them.

EARTH IMPACTS

Even though asteroids are not very large in comparison to Earth, they have the potential to be dangerous due to the amount of kinetic energy that is released on impact. Many asteroids have hit Earth and other planets since their formation, and they will continue to do so in the future. Scientists study the orbits and physical characteristics of asteroids to determine if they pose a danger to Earth.

According to NASA, asteroids more than a quarter-mile wide pose a risk of disrupting life on Earth. These large asteroids have the potential to collide with Earth once every one thousand centuries. Smaller asteroids can still cause tsunamis or destroy large areas. Such an asteroid is estimated to strike Earth every one thousand to ten thousand years.

Asteroid impacts are believed to be responsible for a number of major extinctions throughout Earth's history, including that of the dinosaurs. Less powerful impact events have been recorded by humans. On June 30, 1908, an asteroid exploded in the atmosphere near the Podkamennaya Tunguska River in Siberia, Russia. It caused a shock wave that destroyed a large area of the Siberian forest and registered effects throughout Europe and Asia. On February 15, 2013, another asteroid crashed into the atmosphere over Chelyabinsk, Russia. The explosion caused a shock wave that injured more than one thousand people. The asteroid was estimated to be twenty meters (nearly sixty-six feet) wide when it entered the atmosphere.

Scientists spend much time tracking asteroids and have cataloged numerous asteroids as "potentially hazardous" to the Earth. These near-Earth objects (NEOs) could crash into the planet if their orbits shift. In 2001 a United Nations team was formed to explore joint actions to respond to an asteroid threat. As of 2015, NASA was developing technology to redirect asteroids. In 2013, NASA's Asteroid Initiative began developing plans for the Asteroid Redirect Mission, a robotic mission to visit an NEO, collect materials from its surface, and redirect it.

—Angela Harmon

Further Reading

"Asteroids: The Discovery of Asteroids." *Space for Europe*. European Space Agency, n.d. Accessed 11 May 2015.

"Bode's Law." *Astro 2201: Our Home in the Universe*. Cornell University, n.d. Accessed 11 May 2015.

Burbine, Thomas H. *Asteroids: Astronomical and Geological Bodies*. Cambridge UP, 2017.

"Ceres: Overview." *Solar System Exploration*. NASA, 6 Jan. 2015. Accessed 6 May 2015.

Cessna, Abby. "Planetoid." *Universe Today*. Fraser Cain, 10 Aug. 2009. Accessed 6 May 2015.

Choi, Charles Q. "Asteroids: Fun Facts and Information about Asteroids." *Space.com*. Purch, 21 Nov. 2014. Accessed 6 May 2015.

Gannon, Megan. "Bizarre Asteroid with Six Tails Spotted by Hubble Telescope." *Space.com*. Purch, 7 Nov. 2013. Accessed 6 May 2015.

Nugent, Carrie. *Asteroid Hunters*. TED Books, 2017.

Phillips, Tony, editor. "The Tunguska Impact: 100 Years Later." *NASA Science*. NASA, 30 June 2008. Accessed 6 May 2015.

Redd, Nola Taylor. "Asteroid Found with Rings! First-of-Its-Kind Discovery Stuns Astronomers." *Space.com*. Purch, 26 Mar. 2014. Accessed 6 May 2015.

"Small Worlds: The Neighborhood." *Discovery Program*. NASA, n.d. Accessed 6 May 2015.

"What Is NASA's Asteroid Redirect Mission?" *Asteroid Redirect Mission*. NASA, 30 Apr. 2015. Accessed 11 May 2015.

ATHABASCA OIL SANDS

Fields of Study: Geology; Petroleum Geology

ABSTRACT

The Athabasca oil sands are located in the Athabascan basin of northeastern Alberta, Canada, near the Saskatchewan border. The Athabasca River runs through the region—hence its name. With two smaller oil sands deposits located elsewhere in the province, Alberta has a total of about 142,200 square kilometers of oil sands. Alberta's oil sands have the third largest oil reserves in the world, following Venezuela and Saudi Arabia.

KEY CONCEPTS

bitumen: the most viscous grade of crude petroleum, having a relatively low content of lighter, volatile components; typically used for road paving when mixed with fine aggregate materials

energy-intensive: requiring the input of large amounts of energy to produce the desired output products

greenhouse effect: the elevation of average atmospheric temperature due to the ability of certain gas molecules to absorb infrared (heat) energy from surface emissions and return a significant portion of that energy to the atmosphere, thus preventing it from radiating back out into space

greenhouse gas: any atmospheric gas, typically carbon dioxide, methane, water and others, that tends to increase the greenhouse effect in the atmosphere

oil sands: also called "tar sands," a combination of bitumen and sand, clay and other soil types, ranging in consistency from semisolid to solid such as might be obtained by mixing used motor oil with soil and allowing the mixture to set

WHERE FOUND

The Athabasca oil sands are located in the Athabascan basin of northeastern Alberta, Canada, clustering around the town of Fort McMurray, near the Saskatchewan border. The Athabasca River runs through the region—hence its name. With two smaller oil sands deposits located elsewhere in the province, Alberta has a total of about 142,200 square kilometers (54,900 square miles) of oil sands. Alberta's oil sands have the third largest oil reserves in the world, following Venezuela and Saudi Arabia.

PRIMARY USES

With industrial processes, bitumen can be extracted from the oil sands and upgraded into light crude oil. Although the First Nations used the tarlike substance to waterproof and patch their boats, in modern times oil sands have one commercial use: the production of crude oil. To produce crude oil, an energy-intensive process is required to extract the bitumen from the thick, sludgy, sandy substance. Bitumen is a heavy, viscous oil that can be industrially upgraded into synthetic crude oil, which in turn is refined into gasoline and diesel fuels.

TECHNICAL DEFINITION

The Athabasca oil sands consist of a tarlike mixture of about 80 to 85 percent sand and rich mineral clays, 10 to 12 percent bitumen, and 4 to 6 percent water. The valuable resource in this mixture is the bitumen. Bitumen is a heavy, black, asphalt-like substance that has been pressurized underground for millions of years but not long enough to be concentrated into coal or light sweet crude oil. Nonetheless, it is a form of crude oil that can be processed for commercial use. Technically, bitumen is a mix of petroleum hydrocarbons with a density greater than 960 kilograms per cubic meter.

DESCRIPTION, DISTRIBUTION, AND FORMS

Oil sands are a viscous mix of hydrocarbons and can be found throughout the world. The nature of the material may be approximated by mixing old, used motor oil with a quantity of sand, powdered clay and other small particulate matter, and allowing the mixture to set. While the Athabasca region has the only surface quantities, there are also deposits buried in the Peace River and Cold Lake regions of Alberta. (Although the United States has some oil sand deposits, it also has massive quantities of oil shale, rocklike formations containing crude oil that can also be refined at high industrial cost.) The oil sands contain a mix of sands, clays, water, and bitumen, but the bitumen, a form of heavy crude oil, is what

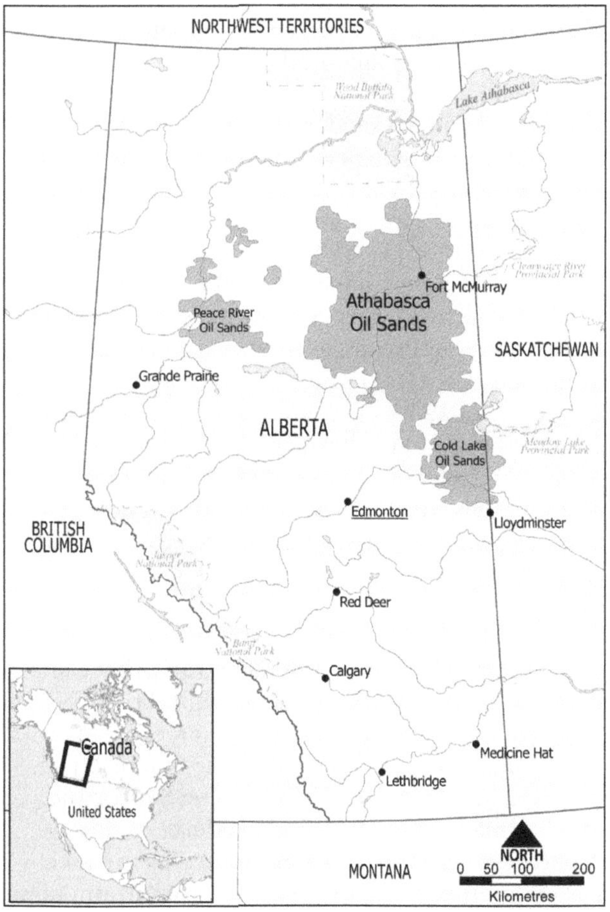

This map shows the extent of the oil sands in Alberta, Canada. The three oil sand deposits are known as the Athabasca Oil Sands, the Cold Lake Oil Sands, and the Peace River Oil Sands. Image via Wikimedia Commons. [Public domain.]

gives oil sands their distinctive properties. Many ancient cultures made use of bitumen for its sticky, adhesive qualities. It was used as a sealant for boats, a building mortar, and an ingredient for mummification. In modern times, bitumen is valuable as a crude oil that can be refined into commercial petroleum. Semisolid at normal temperatures, the bitumen must be heated or diluted with hydrocarbons to make it flow through supply pipelines. The bitumen is extracted through a steam separation process. Hot water is injected into the mined oil sands, causing the bitumen to float to the surface, where it can be recovered. The bitumen is then upgraded into synthetic oil and petroleum products.

HISTORY

The First Nations knew about the oil sands deposits from ancient times and used the tarlike material to bind their canoes. Early Canadian explorers such as Peter Pond and Alexander Mackenzie wrote of the fluid bitumen pooling near the Athabasca River. In 1882, geologist Robert Bell surveyed the basin and oil fields. However, oil production did not become possible until Karl Clark first developed a process for separating the bitumen from the sands. Clark's process, developed between 1922 and 1929, relied on hot water and steam to turn the sand into a soupy substance from which the bitumen could be extracted. In 1930, the Canadian government leased a large portion of the Athabasca basin for development to petroleum engineer Max Ball and his Abasand Oils company. Abasand's separation process was primitive, however, and even by the 1940s Abasand processed less than 20,000 metric tons of sand per year.

The technological challenges of extraction remained daunting, but in the 1960s the Great Canadian Oil Sands (GCOS) company built the first large-scale oil sands production plant, capable of producing about 24,000 barrels of synthetic crude oil per day. GCOS eventually became Suncor Inc., which remains the leading oil producer in the Athabasca region. In 1978, prompted by the oil embargoes of the 1970s, the Syncrude consortium of oil companies built a second major oil sands plant, followed by Imperial Oil's Cold Lake plant in 1987, Alsands Project Group's facility in 1988, and Shell's Albian Sands mine in 2003. In the twenty-first century, oil production increased, with new technologies reducing the cost of bitumen extraction. Dozens of oil companies and industrial consortia opened plants and increased supply. The proven feasibility of oil sands production led the U.S. Energy Information Administration to estimate that Canada's oil reserves totaled approximately 180 billion barrels.

OBTAINING OIL SANDS

The bulk of the oil sands are located near the surface and can be obtained through massive open-pit mining operations. Because of the heavy mineral clays that make up most of the sand, the operations use some of the largest shovels and trucks in the world to dig up and move the sands. Deposits that are located deeper—below 75 meters—are recovered by *in situ* methods, which include cyclic steam stimulation and steam-assisted gravity drainage. About 75 percent of the valuable resource, the bitumen, is obtained during the recovery. After the sand is processed and the bitumen removed, Canadian environmental laws require the processed sand to be returned to the pit and the site restored to its original condition. Much of the waste—"tailings"—is deposited in tailings ponds. These toxic basins need to be cleaned up and reclaimed.

USES OF OIL SANDS

The Athabasca oil sands represent one of the great oil reservoirs in the world, estimated at 174 to 180 billion barrels. Canada has approximately 10 percent of the world's proved oil reserves, of which 97 percent is oil sands. There are currently thousands of agreements in Alberta between Canadian governments and oil companies for production of oil. More than 1 million barrels of oil were produced each day in 2005, which increased to 2.6 million barrels per day by 2016, and is expected to increase to 4.5 million barrels by 2040. Because of the thickness of the oil sands, it takes a tremendous amount of energy to produce oil flow. In the winter, temperatures in the Athabasca region fall to as low as -40° Celsius and the extraction machinery can easily freeze up and break down. Massive amounts of surface material are moved, sifted, and heated. Thus, two major issues must be addressed for the world to exploit this resource: first, the high cost of extraction and, second, the profound impact on the environment.

The profitability of the oil sands depends directly on the price of oil. While light crude oil flows easily from conventional oil wells, oil sand producers must expend fixed costs for mining and extracting the bitumen and converting it to liquid crude. Thus the oil sands may be profitable if the price of oil is more than forty-four dollars per barrel, for example, but noncompetitive below that. With new technology developed and implemented in the first decade of the twenty-first century, production costs were lowered to about thirty-three to thirty-seven dollars per barrel, spurring an oil sands boom. As to the environment, the massive processing of oil sands is bound to leave a scarring of the earth, despite Canadian restoration legislation. Oil sands production releases carbon dioxide, which is believed to contribute to the greenhouse effect. It is estimated that oil sands production represents nearly 10 percent of Canada's total greenhouse gas emissions, although emissions per barrel of oil produced in the oil sands in 2014 were 31 percent lower than 1990 levels, due to improved environmental regulations and technologies. The recovery of bitumen itself requires the consumption of tremendous amounts of resources, mostly in the form of natural gas and water.

In 2003, with the rise of oil prices, the oil sands operation became consistently profitable for the first time, and the three major mines—Suncor, Syncrude Consortium, and Shell Canada—began producing about 1.2 million barrels of synthetic crude oil per day. In 2010, oil sands production exceeded conventional production in Canada for the first time. By 2015, oil sands production had grown to more than 2.4 million barrels per day, making Canada one of the world's leading oil-producing countries in the world. Cumulative investment in oil sands production in the first decade of the twenty-first century was estimated to be about $70 billion, resulting in about 275,000 jobs. Between 2010 and 2015, government revenues from the oil and gas industry in Canada averaged $20.3 billion each year. Oil sands production represented 61 percent of Canada's total oil production in 2015, with an estimated total capital investment of $271 billion, including $22.5 billion in 2015 alone.

Since 2015, the Athabasca oil sands industry has faced significant difficulties. Annual investment in oil sands has dropped about 45 percent from its 2014 peak of $30 billion. These challenges flow from a sharp decline in oil prices, the growth of U.S. energy oil production, particularly natural gas (99 percent of Athabasca oil production is sold to the United States), and lack of export pipelines, especially with the delayed construction of the Keystone XL pipeline, which would send oil directly to U.S. refineries. Overproduction of Canadian oil and the water and air pollution caused by oil sands extraction have led to an uncertain regulatory and legal regime for the large oil producers operating in Athabasca.

—Howard Bromberg

Further Reading

Banerjee, Dwijen. *Oil Sands, Heavy Oil & Bitumen: From Recovery to Refinery*. Pennwell, 2012.

Dawson, Chester. "Canadian Oil-Sands Producers Struggle." *The Wall Street Journal*, 19 Aug. 2015, www.wsj.com/articles/oil-sands-producers-struggle-1440017716. Accessed 15 Jan. 2019.

Energy Fact Book, 2016-2017. Natural Resources Canada, June 2016, www.nrcan.gc.ca/sites/www.nrcan.gc.ca/files/energy/pdf/EnergyFactBook_2016_17_En.pdf. Accessed 15 Jan. 2019.

George, Rick. *Sun Rise: Suncor, the Oil Sands and the Future of Energy*. HarperCollins, 2012.

Gonzalez, George. *American Empire and the Canadian Oil Sands*. Palgrave MacMillan, 2016.

Hicks, Brian, and Chris Nelder. *Profit from the Peak: The End of Oil and the Greatest Investment Event of the Century*. John Wiley & Sons, 2008.

McCarthy, Shawn, and Jeff Lewis. "Canadian Oil Sands Are Crucial If Producers Cut Costs, Emissions: IEA." *The Globe and Mail*, 25 Feb. 2016, www.theglobeandmail.com/report-on-business/industry-news/energy-and-resources/iea-cautiously-optimistic-about-future-of-canadian-oil-sands/article28920041. Accessed 15 Jan. 2019.

Nikiforuk, Andrew. *The Tar Sands: Dirty Oil and the Future of a Continent*. Greystone, 2009.

Sweeney, Alastair Black. *Bonanza: Canada's Oil Sands and the Race to Secure North America's Energy Future*. John Wiley and Sons, 2010.

Tertzakian, Peter A. *Thousand Barrels a Second: The Coming Oil Break Point and the Challenges Facing an Energy Dependent World*. McGraw-Hill, 2006.

Turner, Chris. *The Patch: The People, Pipelines, and Politics of the Oil Sands*. Simon & Schuster, 2017.

B

Basaltic Rocks

Fields of Study: Geology; Seismology; Geophysics

ABSTRACT

"Basalt" is the term applied to dark, iron-rich volcanic rocks that occur everywhere on the ocean floors, as oceanic islands, and in certain areas on continents. It is the parent material from which nearly any other igneous rock can be generated by various natural processes.

KEY CONCEPTS

augite: an essential mineral in most basalts, a member of the pyroxene group of silicates

crust: the upper layer of the earth and the other "rocky" planets; it is composed mostly of relatively low-density silicate rocks

lithospheric plates: giant slabs composed of crust and upper mantle; they move about laterally to produce volcanism, mountain building, and earthquakes

magma: molten silicate liquid, including any crystals, rock inclusions, or gases trapped in that liquid

mantle: a layer beginning at about 5 to 50 kilometers below the crust and extending to the earth's metallic core

oceanic ridges: a system of mostly underwater rift mountains that bisect all the ocean basins; basalt is extruded along their central axes

olivine: a silicate mineral found in mantle and some basalts, particularly the alkaline varieties

peridotite: the most common rock type in the upper mantle, where basalt magma is produced

plagioclase: one of the principal silicate minerals in basalt, a member of the feldspar group

subduction zones: areas marginal to continents where lithospheric plates collide

BASALT COMPOSITION

Basalt is a dark, commonly black, volcanic rock. It is sometimes called "trap" or "traprock," from the Swedish term *trapp*, which means "steplike." On cooling, basalt tends to form hexagonal columns, which in turn form steplike structures after erosion. Excellent examples can be seen at Devils Postpile, in the Sierra Nevada in central California, and the Giant's Causeway, in County Antrim, Ireland. The term "trap," however, is used mostly by miners and nonspecialists; scientists prefer the word "basalt" for the fine-grained, volcanic rock that forms by the solidification of lava flows. Basaltic magma that crystallizes more slowly below Earth's surface, thus making larger mineral grains, is called "gabbro."

The importance of basalt to the evolution of planets like Earth cannot be overemphasized. Basalt is considered to be a "primary silicate liquid," in that the first liquids to form by the melting of the original minerals that made up all the so-called rocky planets (those composed mostly of silicates) were basaltic in composition. In turn, basalt contains all the necessary ingredients to make all the other rocks that may eventually form in a planet's crust. Furthermore, many meteorites (which are believed to represent fragments of planetoids or asteroids) are basaltic or contain basalt fragments, and the surfaces of the Moon and the planets Mercury, Venus, and Mars are known to be covered to various degrees by basaltic lava flows.

Like most rocks in Earth's crust, basalt is composed of silicate minerals, substances whose principal structural component is silica. Compared with other silicate rocks, basalts contain large amounts of iron and magnesium and small amounts of silicon. This characteristic is reflected in the minerals in basalts, which are mostly pyroxene minerals (dark-colored, calcium-iron-magnesium silicates such as augite) and certain feldspar minerals called "plagioclase" (light-colored, calcium-sodium-aluminum silicates). Pyroxenes and plagioclase are essential minerals in basalts, but some types of basalt also contain the mineral olivine (a green, iron-magnesium silicate). The high abundance of dark green or black pyroxene and, in some cases, olivine gives basalts their characteristic dark color. This color can be mainly attributed to the high iron content of pyroxene and olivine.

OCEANIC BASALT DISTRIBUTION

Basalt is the most common type of igneous rock (i.e., rock formed by the crystallization of magma) on or near the surface of Earth. It is the principal rock in the ocean floors and is common, though less abundant, on the continents as well. Drilling into the seafloor by specially designed oceanographic research ships reveals that basalt invariably lies just below a thin cover of fine, sedimentary mud. Ocean floor basalt flows out of midocean ridges, a system of underwater mountain ranges that spans the globe. These ridges commonly trend roughly down the middle of ocean basins, and they represent places where Earth's lithospheric plates are being literally split apart. In this process, basalt magma is generated below what is referred to as rift mountains; it

Columnar basalt flows in Yellowstone National Park, USA. Photo by Brocken Inaglory, CC BY-SA 3.0, via Wikimedia Commons.

flows onto the cold ocean floor and solidifies. Although oceanic ridges are normally hidden from view under the oceans, a segment of the Mid-Atlantic Ridge emerges above the waves as the island of Iceland.

Oceanic islands are also composed of basalt. The islands of Hawaii, Fiji, Mariana, Tonga, and Samoa, among others, are large volcanoes or groups of volcanoes that rise above water from the ocean floor. Unlike the basalts that cover the ocean floors, however, these volcanoes do not occur at oceanic ridges but instead rise directly from the seafloor. Basalt is also a fairly common rock type on island arcs, volcanic islands that occur near continental margins. These curvilinear island chains arise from melting along subduction zones, areas where Earth's lithospheric plates are colliding. This process generally involves material from the ocean basins diving under the less dense continental masses; andesite (a light-colored rock) volcanoes are the main result, but some basalt erupts there as well. The Japanese, Philippine, and Aleutian island chains are examples of island arcs, as are the island countries of New Zealand and Indonesia.

CONTINENTAL BASALT DISTRIBUTION

Basaltic lava flows are not nearly as common on continents as in oceanic areas. Andesites, rhyolites, and related igneous rocks are far more abundant than basalt in continental settings. In North America, basaltic lava flows and volcanoes are most exposed to view in the western United States and Canadian provinces, western Mexico, Central America, and western South America. The greatest accumulations of basaltic lava flows in the United States are in the Columbia Plateau of Washington, Oregon, and Idaho. In this large area, a series of basaltic lava flows have built up hundreds of feet of nearly flat-lying basaltic flows over a few million years. These "fissure flows" result from lava pouring out of long cracks in the crust, or fissures. They are similar in many respects to the basalt flows produced at oceanic ridges, because no actual volcanic cones are produced—only layer after layer of black basalt. The Snake River Plain in southern Idaho has a similar origin, and other extensive basalt plateaus occur in the Deccan area of southwestern India, the Karroo area of South Africa, and Paraná State in Brazil.

Not all basalt is erupted as lava flows. If the lava is particularly rich in volatiles such as water and carbon dioxide, it will be explosively ejected from the volcano as glowing fountains of incandescent particles that rain down on the surrounding area. Conical volcanoes composed almost exclusively of basalt ejecta particles are called "cinder cones." Good examples of cinder cones are Sunset Crater in northern Arizona, and the numerous cinder cones in Hawaii and Iceland.

BASALT CLASSIFICATION

Although basaltic rocks may all look alike to the nonspecialist, there are actually many different kinds of basalt. They are arranged by scientists into a generally accepted classification scheme based on chemistry and, to some extent, mineralogy. To begin with, basalt can be distinguished from the other major silicate igneous rocks by its relatively low silica content (about 50 percent). Within the basalt clan itself, however, other means of classification are used. Basalts are divided into two major groups: the alkaline basalts and the subalkaline basalts. Alkaline basalts contain large amounts of the alkali metal ions potassium and sodium but relatively small amounts of silica. In contrast, subalkaline basalts contain less potassium and sodium and more silica. As might be expected, this chemical difference translates into differences in the mineral content of the basalt types as well. For example, all alkaline basalts contain one or more minerals called "feldspathoids" in addition to plagioclase feldspar. They also commonly contain significant olivine. Subalkaline basalts, in contrast, do not contain feldspathoids, although some contain olivine, and they may be capable of crystallizing very tiny amounts of the mineral quartz. The presence of this very silica-rich mineral reflects the relatively high silica content of subalkaline basalt magmas versus alkaline basalt magmas. Within these two major groups are many subtypes, too numerous to discuss here.

STUDY TECHNIQUES

Like other igneous rocks, basalts are analyzed and studied by way of many techniques. Individual studies may include extensive field mapping, in which the distribution of various types of basalt is plotted on maps. The history of magma generation and its relation to tectonic history (earth movements) can be reconstructed for a particular area by correlating. Geologic maps of basalt types with absolute ages are determined through radiometric dating techniques. Good examples of such studies are those conducted in recent years on the Hawaiian Islands. These studies indicate that the alkaline basalts on any given island are generally older than the subalkaline basalts, showing that magma production has moved upward, to lower-pressure areas, in the mantle with time. This finding supports the idea that oceanic island basalts such as those in Hawaii are generated within so-called mantle plumes—roughly balloon-shaped, slowly rising masses of mantle material made buoyant by localized "hot spots."

Samples of basalt are also analyzed in the laboratory. The age of crystallization of basalt is obtained by means of radiometric dating techniques that involve the use of mass spectrometers to determine the abundances of critical isotopes, such as potassium-40 (40K) and argon-40 (40Ar), or rubidium-87 (87Rb) and strontium-86 (86Sr) and -87 (87Sr). To obtain information on how basalt magma is generated and how it subsequently changes in composition before extrusion as a lava flow, scientists place finely powdered samples in metallic, graphite, or ceramic capsules and subject them to heating and cooling under various conditions of pressure. Such procedures are known as experimental petrology. Such studies prove that nearly any other igneous magma composition can be derived from basalt magma by the process of crystal fractionation. Widely believed to be the major factor influencing chemical variation among igneous rocks, this process results in ever-changing liquid compositions, as the various silicate minerals crystallize and are thus removed from the liquid over time. Basalt's parental role gives it enormous importance in the discipline of igneous petrology.

TRACE ELEMENT ANALYSIS

Another useful avenue of research is the trace element analysis of basalts. Trace elements occur in such low abundances in rocks that their concentrations must usually be expressed in terms of parts per million or even parts per billion. Among the most useful substances for tracing the history of basalt are the rare-earth elements. The elements chromium, vanadium, nickel, phosphorus, strontium, zirconium, scandium, and hafnium are also used. There are many methods for measuring these elements, but the most common, and most accurate, is neutron activation. This method involves irradiating samples in a small nuclear reactor and then electronically counting the gamma-ray pulses produced by the samples. Since different elements tend to emit gamma rays at characteristic energies, these specific energies can be measured and the intensity of gamma pulses translated into elemental concentrations.

Once determined for a particular basalt sample, trace element abundances are sensitive indicators of events that have transpired during the evolution of the basalt. There are two reasons for this sensitivity. First, trace elements are present in such low concentrations (as compared with major elements—iron, aluminum, calcium, silicon, and the like) that any small change in abundance caused by changes in the environment of basalt production will be readily noticed. Second, different minerals, including those crystallizing in the magma and those in the source peridotite, incorporate a given trace element into their structures or reject it to the surrounding liquid to widely varying degrees. Therefore, trace element concentrations can be used to show the minerals that were involved in producing certain observed chemical signatures in basalts and those that were likely not involved.

For example, it is well known that the mineral garnet readily accepts the rare-earth element lutetium into its structure but tends to reject most lanthanum to any adjacent liquid, even though these elements are very similar chemically. Basalts with very little lutetium but much lanthanum were therefore probably derived by the melting of garnet-bearing mantle rocks. Since garnet-bearing mantle rocks can exist only at great depths, basalts with such trace element patterns must have originated by melting at these depths in the mantle. In fact, that is one of the most important lines of evidence to support the origin of alkaline basalts at high-pressure regions in the mantle.

ECONOMIC APPLICATIONS

Basaltic islands, particularly in the Pacific basin, are some of the most popular tourist stops in the world.

More importantly, however, basalt magma contains low concentrations of valuable metals that, when concentrated by various natural processes, provide the source for many important ores. Copper, nickel, lead, zinc, gold, silver, and other metals have been recovered from ore bodies centered in basaltic terrains. Some of the richest mines of metallic ores in the world are located in Canada, where ores are found associated with extremely old basaltic rocks, called "greenstones," from long-vanished oceans. The richest of these mines is Kidd Creek in northern Ontario. These ore-bearing basalts were first extruded more than 2 billion years ago, during what geologists call Precambrian times (the period from 4.6 billion to about 600 million years ago). Other notable ore deposits include the native, or metallic, copper in late Precambrian basalts that were mined for many years in the Keweenaw Peninsula of northern Michigan. The island of Crete in the Mediterranean Sea has copper mines that were mined thousands of years ago during the "copper" and "bronze" ages of human history. The basalt enclosing these ores is believed to have erupted from an ancient midocean ridge formerly located between Africa and Europe.

Basalt can also be used as a building stone or raw material for sculptures, but its high iron content makes it susceptible to rust stains. It is also ground up to make road gravel, especially in the western United States, and it is used as decorative stone in yards and gardens.

—*John L. Berkley*

Further Reading

Aiello, Gemma. *Volcanoes: Geological and Geophysical Setting, Theoretical Aspects and Numerical Modeling, Applications to Industry and Their Impact on Human Health.* IntechOpen, 2018.

Faure, Gunter. *Origin of Igneous Rocks: The Isotopic Evidence.* Springer-Verlag, 2010.

Philpotts, Anthony, and Jay Aque. *Principles of Igneous and Metamorphic Petrology.* 2nd ed., Cambridge UP, 2009.

Reidel, Stephen P., Victor E. Camp, and Martin E. Ross, editors. *The Columbia River Flood Basalt Province.* Geological Society of America, 2013.

Ross, Pierre-Simon, et al. "Basaltic to Andesitic Volcaniclastic Rocks in the Blake River Group, Abitibi Greenstone Belt: 2. Origin, Geochemistry, and Geochronology." *Canadian Journal of Earth Sciences*, vol. 48, 2011, pp. 757-77.

Tarbuck, Edward J., Frederick K. Lutgens, and Dennis Tasa. *Earth: An Introduction to Physical Geology.* 13th ed., Pearson Education, 2019.

West, Janet P. *Basalt: Types, Petrology and Uses.* NOVA Science Publishers, 2012.

BATHOLITHS

Fields of Study: Geology; Geochemistry; Volcanology; Stratigraphy

ABSTRACT

Batholiths are gigantic bodies of granitic rock located in mobile belts surrounding the ancient cores of the continents. The growth of continental crust during the past 2.5 billion years is intimately related to the origin and emplacement of major volumes of granitic magma that solidify as batholiths.

KEY CONCEPTS

crystallization: the solidification of molten rock as a result of heat loss; slow heat loss results in the growth of crystals, but rapid heat loss can cause glass to form

granitic/granitoid: descriptive terms for plutonic rock types having quartz and feldspar as major mineral phases

I-type granitoid: granitic rock formed from magma generated by partial melting of igneous rocks in the upper mantle or lowermost crust

magma: molten rock material that crystallizes to form igneous rocks

migmatite: a rock exhibiting both metamorphic and plutonic textural traits

mobile belt: a linear belt of igneous and deformed metamorphic rocks produced by plate collision at a continental margin; relatively young mobile belts form major mountain ranges; synonymous with orogenic belt

partial melting: a process undergone by rocks as their temperature rises and metamorphism occurs; magmas are derived by the partial melting of preexisting rock; also known as ultrametamorphism or anatexis

pluton: a generic term for an igneous body that solidifies well below Earth's surface; plutonic rocks are coarse-grained because they cool slowly

S-type granitoid: granitic rock formed from magma generated by partial melting of sedimentary rocks within the crust

BATHOLITH FORMATION

Batholiths are large composite masses of granitoid rock formed by numerous individual bodies of magma that have risen from deep source areas in molten form and solidified near enough to the surface to be exposed by erosion. The resulting rocks are relatively coarse-grained in texture and markedly heterogeneous in chemical and mineralogical composition. A well-studied example is the coastal batholith of Peru, which forms an almost continuous outcrop 1,100 kilometers long and 50 kilometers wide along the western flank of the Andes. This enormous body has steep walls and a flat roof. It is composed of more than one thousand individual plutons emplaced along a narrow belt parallel to the present coastline during a volcanic-plutonic event that extended over a period of 70 million years. Many such batholiths are known in the mountainous areas of the world, but few are as large or as magnificently exposed to view as that in Peru. Geological glossaries often define a batholith as a "coarse-textured igneous mass with an exposed surface area in excess of 100 square kilometers." This description has the virtue of simplicity, but it is misleading; the description encompasses granitic plutons, and even nongranitic plutons, which form under conditions quite removed from those associated with the world's major batholiths.

In most instances, there is evidence to indicate that the individual plutons of a batholith were emplaced as hot, viscous melts containing suspended crystals. This molten material is called "magma." Cooling and crystallization occur during the ascent of magma toward the surface and gradually transform it to solid rock, which prevents further upward movement. The depth at which total solidification occurs varies and is strongly dependent upon the initial temperature and water content of the magma. Extreme levels of ascent, within 3 to 5 kilometers of the surface, are possible only for very hot magmas with very low initial water contents. Most granitic plutons complete their crystallization at depths in the range of 8 to 20 kilometers. The characteristic coarse textures observed in most granitic plutons are the result of slow cooling, which, in turn, implies that the rate of magma ascent is also slow. These traits distinguish plutonic rocks from their volcanic counterparts. As would be expected, the formation of a batholith is a complex and lengthy event that is the sum of the processes responsible for the emplacement of each member pluton. Each member pluton has an individual history involving the generation of magma in the source region, ascent and partial crystallization, the physical displacement of overlying solid rock, chemical interaction with the solid rocks encountered during ascent, and the terminal crystallization phase. Consequently, each member pluton of a batholith can be expected to exhibit a unique combination of textural, mineralogical, and chemical variations.

MOBILE BELTS

It has long been recognized that major batholiths are confined to narrow zones elongated parallel to present or former continental margins. In such zones, granitic melts intrude either thick sequences of chemically related volcanic rocks or highly deformed and metamorphosed sedimentary rocks. These granite-dominated zones are called "mobile belts." The ancient cores of continents are all more than 2.5 billion years old. They are surrounded by mobile belts that become successively younger away from the core. The most recent mobile belts form major mountain chains along continental margins. The resulting age pattern clearly shows that continents grow larger with time by the marginal accretion of mobile belts. In the late 1960s, the emergence of plate tectonic theory provided a basis for understanding how mobile belts form and are accreted to preexisting continent margins. The impetus provided by this theory sparked intensive study of the world's mobile belts. These studies amply show that logical time-space relationships exist between plate collisions, deformation styles, and rock types that occur in mobile belts.

Two distinct types of mobile belts are now recognized, and each is dominated by granitic batholiths. These batholiths, however, are very different in

Half Dome, a granite monolith in Yosemite National Park and part of the Sierra Nevada Batholith. Photo via Wikimedia Commons. [Public domain.]

terms of granitic rock types, modes of pluton emplacement, rock associations, metamorphic effects, and the metallic ores they host. The batholiths of the two mobile belt types are called "I-type" and "S-type" batholiths. I-type batholiths are those that seem to have derived mostly from igneous sources—that is, by melting of the mantle—and S-type batholiths seem to have derived from remelting of metamorphic rocks that were originally sedimentary.

Mobile belts along the eastern margin of the Pacific Ocean contain I-type batholiths exclusively. Their size and collective volume is staggering. The Peruvian batholith is an example already mentioned. Others of this type include the Sierra Nevada batholith, the Idaho batholith, and the tremendous Coast Range batholith, which extends from northern Washington to the Alaska-Yukon border. In contrast, the western margin of the Pacific Ocean is dominated by mobile belts with S-type batholiths, although some I-types are also present. The batholiths of Western Europe are also mainly of the S-type. In southeastern Australia, where the two types of batholiths were first recognized, S-type and I-type granitoids form a paired belt parallel to the coastline. Although their geographical distribution is uneven, both I-type and S-type batholiths occur worldwide.

I-TYPE BATHOLITHS

The most distinctive trait of I-type batholiths is the broad range of granitic rock types they contain. In these batholiths, the rock types gabbro-diorite, quartz diorite-granodiorite, and granite occur in the approximate proportions of 15:50:35. This means that quartz diorite (also called "tonalite") and granodiorite jointly compose 50 percent of I-type batholiths, and true granite is a subordinate component in them. This wide compositional spectrum not only characterizes an entire I-type batholith but also is typical of the individual member plutons. Usually the major plutons are concentrically zoned with small central cores of true granite enveloped by extensive zones of granodiorite, which grade outward into margins of quartz diorite. Small plutons in the compositional range of gabbro-diorite are common but subordinate to the zoned granitic bodies.

Most member plutons of I-type batholiths have domal or cylindrical shapes and very steep contacts with the surrounding rock. Others may have a steeply tilted sheetlike form, but regardless of shape most I-type plutons cut through the preexisting rock layers at a steep angle. The emplacement of these plutons appears to be controlled by near-vertical fractures that may extend downward to the base of the crust. In younger I-type batholiths such as that in Peru, the granitoids have intruded into a roof of chemically related volcanic rocks that show the same compositional spectrum as the granitic plutons. This volcanic pile, dominated by andesite, may be 3 to 5 kilometers thick at the time of pluton emplacement. Gradually, this volcanic roof is stripped away by erosion, so that volcanic rocks are generally absent in older, deeply eroded batholiths. The grade of regional metamorphism in the rocks enclosing I-type batholiths is relatively low, and there is little evidence of large-scale horizontal compression or crustal shortening. Structural displacements and the movements of rising plutons are dominantly vertical and typically occur over a time span of 50 to 100 million years.

S-Type Batholiths

S-type batholiths contrast with I-types in almost every respect. To begin with, the ratio of gabbro-diorite to quartz diorite-granodiorite to granite is 2:18:80 in S-type batholiths. These plutonic complexes are very much dominated by true granite, and gabbro-diorite plutons are rare or absent. In many cases, S-type granites are the distinctive "two mica granites," which contain both biotite and muscovite, and are frequently associated with major tin and tungsten ore deposits. The batholiths of northern Portugal are typical examples of this association. S-type batholiths, as well as their member plutons, lack the concentric zoning that characterizes I-type plutons. Compositional homogeneity is their trademark. S-type plutons are intruded into thick sequences of regionally metamorphosed sedimentary rocks. The metamorphic grade ranges from moderate to very high, and, frequently, the granites are located within the zone of highest metamorphic grade. In such cases, migmatites are often present. The enclosing metamorphosed rocks are intensely folded in response to marked crustal compression, and volcanic rocks are conspicuously absent. S-type batholiths are smaller in volume and form over a shorter period of time (usually less than 20 million years) than their I-type counterparts.

Differing Origins of Batholith Types

The many contrasting traits of I-type and S-type batholiths are an indication that the conditions of magma generation and emplacement are very different in the mobile belts in which they are found. In the case of I-type batholiths, it appears that magmas are generated at relatively great depths and above the subduction zones that form at destructive plate boundaries. The magmas are derived by partial melting of upper-mantle basic igneous rocks and, perhaps, lower crustal igneous rocks. The melts rise along the steep fractures produced by crustal tension over the subduction zone. The igneous ancestry of these melts is the reason for calling the resulting plutons "I-type." The hottest and driest of these I-type magmas will reach the surface to produce extensive fields of volcanic rocks and large calderas. In some cases, like the Peruvian batholith, the rise of magma was "passive" in the sense that room was provided for the rising plutons by gravitational subsidence of the overlying roof rock. This is the process of cauldron subsidence. In the case of the Sierra Nevada batholith, however, it appears that I-type magmas were emplaced by "forceful injection." In this process, rising magma makes room for itself by forcing aside the surrounding solid rock.

Conversely, the evidence suggests that S-type magmas originate by partial melting of metamorphosed sedimentary rocks. This sedimentary parentage of the magmas is the reason for designating them as "S-type." Melting is made possible by dehydration of water-bearing minerals under conditions of intense metamorphism. The frequent presence of migmatites (the complex intermingling of igneous and metamorphic rocks) is evidence for this transition from metamorphic conditions to magmatic conditions. The essential requirements for relatively high-level crustal melting are high temperatures, intense horizontal compression to produce deep sedimentary basins, and a thick pile of sediments to fill these basins. Such conditions are best met in back-arc basins, which form at subduction margins but considerably inland from the volcanic-plutonic environment of I-type batholiths. This may explain why some I-type and S-type batholiths occur in paired belts parallel to a continental margin, as in southeastern Australia.

The collision environment that arises when continent meets continent in the terminal stage of subduction may also provide suitable conditions for S-type magma production. The magmas that result will be relatively cool and wet and will not be able to rise far above their zone of melting. During this limited ascent, the S-type magmas tend to assume the shape of a light bulb, with a neck tapering down to the zone of melting. Because of their limited capacity for vertical movement, S-type plutons require no special mechanisms to provide additional space for them.

FIELD STUDY

The study of a batholith begins with the study of its individual member plutons. This always involves fieldwork, laboratory analysis of rock samples returned from the field, and comparison of the resulting data with those obtained from other batholiths. Because of their great size, batholiths present special problems for field study. The most informative studies are those in mountainous areas, such as the Peruvian Andes, where erosion has exposed the batholith roof contact and cut steep canyons between and through individual plutons. High topographic relief is desirable if the geologist is to learn anything about the variation in shape and composition within the plutons.

Study of even a small portion of a major batholith requires several well-trained geologists working intensively during short field seasons over a period of several years. Geologists traverse on foot across and around the individual plutons as topography permits, and they record the textural, mineralogical, and structural features observed on maps or aerial photographs of the area. These maps eventually reveal the overall shape of plutons and the patterns of concentric zoning within them. Special maps are prepared to show the distribution of fractures and flow structures within individual plutons. These features indicate how fluid the magma was at the time of emplacement. Contacts between the plutons and older enclosing rocks are closely examined for deformation effects and evidence of thermal and chemical interaction with the magma. Fragments of older rock engulfed by magma are often preserved in a recrystallized state, and these are scrutinized carefully, since they provide clues as to whether the emplacement process was passive or forceful.

As the end of the field season draws near, the mapped plutons are sampled. Large, fresh samples must be collected from each recognized zone of each pluton for subsequent laboratory study. The number of samples collected from a single pluton depends upon its size and homogeneity but is frequently in the range of one hundred to five hundred samples. A smaller number of samples is collected from the host rocks at varying distances from the plutonic contact in order to study the thermal effects produced by the pluton. The field description, identifying number, and exact location of each sample site must be meticulously recorded. If, at the end of the field season, several plutons have been studied and sampled, there may be several thousand rock samples to label, pack securely, and ship to the laboratory, where they will receive further study.

LABORATORY STUDY

At the laboratory, the samples are usually cut in half and labeled in a permanent fashion. One half of each sample is stored for future reference, and the other is prepared for the laboratory procedures. Paper-thin slices of each sample are glued to glass slides for examination under a petrographic microscope. The microscopist identifies the mineral phases present in each slide and determines the abundance of each. The texture of each rock, as revealed under the microscope, is carefully described and interpreted in terms of crystallization sequence and deformation history. When the microscopic study is complete, certain samples, perhaps fifty to one hundred, are chosen for chemical analysis. Most will be analyzed because they are judged to be representative of major zones of a pluton; a few may be analyzed because they exhibit unusual minerals or some peculiar trait not explained by the microscopic study. If the age of a pluton is not known, a few samples (one to ten) will be shipped to a laboratory that specializes in age determinations by radioisotope methods.

DETERMINING ORIGIN AND EMPLACEMENT

Finally, on the basis of the field observations, microscopic examinations, and chemical data, the investigators will assemble rival hypotheses or scenarios for the origin and emplacement of the plutons that have been studied. Any scenario that conflicts seriously with known facts is discarded. Those remaining are compared with well-known laboratory melting-crystallization experiments on synthetic and natural rock systems. The size and shape of the plutons, as determined by the field mapping, can be compared with those of "model plutons" derived through sophisticated, but idealized, centrifuge experiments in laboratory settings. The investigators will compare their data in detail with data reported in the geological literature by workers in other parts of the world. They will also compare their results with earlier studies of the same plutons, or studies in the same region, if they exist. As additional plutons are studied in detail, more constraints on the mode of origin and emplacement of the batholith are obtained.

ROLE IN CRUSTAL GROWTH

Mobile belts, dominated by immense granitic batholiths, have been systematically accreted to the ancient continental cores for the last 2.5 billion years of Earth history. Modern plate tectonic theory has provided the basis for understanding the periodic nature of mobile belt accretion and the ways in which crustal and mantle materials are recycled. It is evident that the emplacement of batholiths is at present—and has been for at least 2.5 billion years—the major cause for progressive crustal growth. It is also clear that the rate at which batholiths formed during this lengthy period has far exceeded the rate of continental reduction by erosion. The generation of large volumes of granitic magma and its subsequent rise to form batholiths a few kilometers below the crustal surface must be viewed as fundamental to crustal growth. Batholiths play a major role in the formation of mountain systems and are the most important element in the complex rock and metallic ore associations of mobile belts. The very existence of continents is, in fact, a result of the long-standing process of batholith emplacement.

—Gary R. Lowell

Further Reading

Bennison, George M., Paul A. Olver, and A. Keith Moseley. *An Introduction to Geological Structures and Maps.* 8th ed., Hodder Education, 2011.

Gill, Robin. *Igneous Rocks and Processes: A Practical Guide.* Wiley-Blackwell, 2011.

Grotzinger, John, and Thomas H. Jordan. *Understanding Earth.* 6th ed., W. H. Freeman, 2010.

Hildebrand, Robert S., and Joseph B. Whalen, *The Tectonic Setting and Origin of Cretaceous Batholiths within the North American Cordillera: The Case for Slab Failure Magmatism and Its Significance for Crustal Growth.* American Geological Society, 2017.

Hill, Mary. *Geology of the Sierra Nevada.* Rev. ed., U of California P, 2006.

Jerram, Dougal, and Nick Petford. *The Field Description of Igneous Rocks.* 2nd ed., Wiley-Blackwell, 2011.

Klein, Cornelis, and Cornelius S. Hurlbut, Jr. *Manual of Mineralogy.* 23rd ed., John Wiley & Sons, 2008.

Twidale, C. R., and J. R. Vidal Romani. *Landforms and Geology of Granite Terrains.* A.A. Balkema Publishers, 2005.

Winter, John D. *Principles of Igneous and Metamorphic Petrology.* 2nd ed., Pearson Education, 2010.

Beaches and Coastal Processes

Fields of Study: Geology; Oceanography; Marine Engineering; Marine Ecology

ABSTRACT

The shoreline is the meeting place for the interaction of land, water, and atmosphere, and rapid changes are the rule rather than the exception. Ancient beaches that have been subsumed into the geological structure of Earth over time are features that provide significant information for geologists working to determine the detailed geological characteristics of a region.

KEY CONCEPTS

beach: an accumulation of loose material, such as sand or gravel, that is deposited by waves and currents at the border of a body or stream of water

longshore current: a slow-moving current between a beach and the breakers, moving parallel to the beach; the current direction is determined by the wave refraction pattern

longshore drift: the movement of sediment parallel to the beach by a longshore current

oscillatory wave: a wind-generated wave in which each water particle describes a circular motion; such waves develop far from shore, where the water is deep

tsunami: a low, rapidly moving wave created by a disturbance on the ocean floor, such as a submarine landslide or earthquake

wave base: the depth to which water particles of an oscillatory wave have an orbital motion; generally the wave base is equal to one half the distance between successive waves

wave refraction: the process by which the vertical angle of a wave moving into shallow water is changed or bent

BEACH COMPOSITION

Beaches are a dynamic feature of Earth's surface and have existed since the first accumulations of liquid water on the planet. The processes that create, erode, and modify beaches are many, and have been functioning unchanged since the beginning. Marine processes, such as waves, wave refraction, currents, and tides, work concurrently. This suggests that a beach is a sensitive landform, and indeed it is. Generally a beach is a sedimentary deposit made by waves and related processes. Beaches are often regarded as sandy deposits created by wave action; however, beaches may be composed of broken fragments of lava, seashells, coral reef fragments, gravel, or even large stones. A beach is composed of whatever sediment is available. Beaches have remarkably resilient characteristics. They are landforms made of loose sediment and are constantly exposed to wave and current action. On occasion, the coastal processes may be very intensive, such as during a hurricane or tropical storm. Yet, in spite of the intensity of wave processes, these rather thin and narrow landforms, although perhaps displaced, restore themselves within a matter of days. Coastal scientists are inclined to believe that the occurrence and maintenance of beaches are related to their flexibility and rapid readjustment to the varying intensity of persistent processes.

Sediment deposited on beaches is often derived from the continents. Rivers are one source of beach sediment, and the coast is another. As rivers erode the land, the sediment they carry ultimately finds its way to the shore. Because the sediment may be transported several tens or hundreds of kilometers, it is refined and broken down even further to finer sized particles. Once the river reaches the sea, the sediment is distributed by longshore currents along the shoreline as a beach. Beaches also occur where the shoreline is composed of cliffs, such as along the Pacific coast of North America. Here, waves erode the sea cliffs, and the sediment is deposited locally as a beach. Under these conditions, the beach deposit is most often gravelly because the sediment is transported only a short distance and has not had an opportunity to break up into finer sized sediment such as sand. Some island beaches are made up entirely of shell fragments.

WAVES

The most obvious mechanical energy source working on beaches is waves. Waves approaching a beach are generally created by winds at sea. As wind velocities increase, a wave form develops and radiates out from the storm. An oscillatory motion of the water occurs as the wave form moves across the water surface. It is important to note that the water movement within a wave is not the same as the movement of a wave form. In a wave created in deep water that is approaching a beach, the water particles move in a circular orbit, and little forward movement of the water occurs. The water at the surface moves from the top of the orbit (the wave crest) to the base of the orbit (the wave trough) and then back up. Thus the water particles move in an oscillatory wave; this motion continues down into the water. Although the size of the orbits in the water column decreases, motion occurs to a depth referred to as the wave base. At this point, the depth of the wave base is less than the depth of the water. As wave crests and troughs move into shallow water, the water depth decreases to a point where it is equal to the wave base. From this point, the orbital motion is confined because of the shallowness of the water and takes on an elliptical path. The ellipse becomes more confined as the waves enter ever shallower water and eventually becomes a horizontal line. At this point, there is a net forward movement of water in the form of a breaker.

As a wave enters shallow water, many adjustments occur, such as a change in the orbital path of water particles already described and a change in the velocity of the wave form. Since the wave base is "feeling the bottom" in shallower water, friction occurs, slowing the wave down. As seen from an airplane, waves entering shallow water do so at an angle, not parallel to the shoreline. Therefore, one part of the wave enters shallow water and slows down relative to the rest of the wave. Thus, a part of the wave crest is feeling bottom sooner than the rest of the wave, causing the breaking action of the wave to appear to peel itself along the beachfront. Because the wave crests and troughs have different velocities, the wave refracts or bends. In so doing, the wave crests and

Beach formed on a wild, untamed rocky coastline. Photo by Elizaveta Kovaleva, CC BY 3.0, via Wikimedia Commons.

troughs try to parallel the shallow bottom topography that they have encountered.

The wave refraction is seldom completed, and the breaking wave surges obliquely up the slope of the beach and then returns perpendicular to the shoreline. The result is a current that basically moves water in one direction parallel to the beach in a zigzag pattern (a longshore, or littoral, current). It is a slow moving current that is located between the breaking wave out in the water and the beach. Because longshore current movement operates in shallow water and along the beach, it is capable of transporting sediment along the shoreline. "Longshore drift" refers to the movement of sediment along beaches. In a sense, the longshore current is like a river moving sand and other material parallel to the shore. Beaches are always in a state of flux. Although they appear to be somewhat permanent, they are constantly being moved in the direction of the longshore current. Along any shoreline, several thousand cubic meters of sediment are constantly in motion as longshore drift. Along the beaches of the eastern United States, about 200,000 cubic meters of sediment is transported annually within a longshore drift system.

A different type of ocean wave is one that is generated on the ocean floor rather than by the wind. Such waves are properly termed "tsunamis." Although they popularly have been coined "tidal waves," they are completely unrelated to tides or the movement of planets. Some type of submarine displacement, such as the creation of a volcano, a landslide on the seafloor, or an earthquake beneath the ocean bottom, displaces a volume of water, which triggers waves. Tsunami waves are low, subdued forms traveling thousands of kilometers over the ocean surface at extremely high velocities, exceeding 800 kilometers per hour under the proper conditions. As a tsunami approaches shallow water or a confined bay, its height increases. There is no method for direct measurement of the heights of tsunami waves. However, in 1946, a lighthouse at Scotch Cap, Alaska, located on a headland 31 meters above the Pacific Ocean, was destroyed by waves caused by a landslide-generated tsunami.

The changing character of a beach is very dynamic because the properties of waves are variable. During storm conditions or when more powerful waves strike a beach during the winter season, for example, the beach commonly is eroded, has a steeper slope, and is composed of a residue of coarser sediment, such as gravel, that is more difficult to remove. During fair weather or summer conditions, however, beaches are redeposited and built up.

STUDY OF BEACHES AND COASTAL PROCESSES

The study of beaches and coastal processes is not particularly easy because of constant wave motion and changes in the beach shape. Scientists have, however, developed field methods as well as laboratory techniques to study these phenomena. To study wave motion and related current action, several techniques have been devised that include tracers, current meters, aerial photography, and satellite-based monitoring. Two types of tracers are commonly used to determine the direction and velocity of sand movement: radioactive isotopes and fluorescent coatings to produce luminophors. (Luminophors are sediment particles coated with selected organic or inorganic substances that glow under certain light conditions.) In the former type, a radioactive isotope such as gold, chromium, or iridium is placed on the surface of the grains of sediment. Alternatively, grains of glass may be coated with a radioactive element. The radioactive sediment can then be detected relatively easily and quickly using just a Geiger counter. Both techniques can trace the direction and abundance of sediment along the seashore.

Bright colored dyes or current meters can also be used to document the direction of longshore currents. Surveying instruments are used to determine the high and low topography of beaches and adjacent sand bars. By measuring beach topography before and after a storm, for example, scientists can record immediate changes. In this way, they can document the volume and dimensions of beach erosion or deposition that has taken place during a storm. Measurement of wave height and other wave characteristics can be done with varying degrees of sophistication. Holding a graduated pole in the water and visually observing wave crest and troughs is both the simplest and the cheapest method. To achieve more refined measurements, however, scientists place pressure transducers or ultrasonic devices on the shallow seafloor to record pressure differences or fluctuations of the sea surface. These more precise instruments also record other wave data for later study and analysis.

All these methods are detailed field techniques. By comparing aerial photographs, detailed maps, satellite pictures, and in some cases government studies in a coastal sector, changes over long periods of time may be detected. Finally, because wave motion cannot be controlled on a shoreline, wave tank studies are used to derive wave theories. Normally, an elongated glass lined tank containing water from 0.3 to 1 meter in depth is used. A mechanical device called a "wave machine" displaces water at one end of the wave tank to create waves and sediment is introduced at the other end. The heights and other characteristics of the waves can be varied, as can the type of sediment, to form the beach. In this controlled way, the various beach and wave relationships can be studied.

SIGNIFICANCE

More than 70 percent of the population of the United States lives in a shoreline setting, and thus an understanding of how waves interact with beaches is important. Shoreline property is prized for residential use and hence highly valuable. On many beaches, the great investments made in hotels and condominiums suggest that beaches are the most sought after environment on Earth, and shoreline frontage is sold by the foot or meter, not by the acre or hectare. Currently, however, such demand and investment are threatened with rising sea levels and continued coastal erosion.

Beaches represent the buffer zone that mitigates wave erosion. Waves generated in the open sea slow down in shallow water, and the beach deposit absorbs the impact of waves. Beaches are therefore constantly changing and are one of the most ephemeral environments on Earth. Thus a sound knowledge of longshore currents and beach development is necessary prior to nearshore or marine construction. Seawalls and groins, for example, interfere with waves and long-shore currents and may cause considerable erosion in selected areas, while inducing abnormal sedimentation in others. Seawalls are often constructed perpendicular to a shoreline to slow longshore currents so that a beach can be deposited. Downcurrent, beyond the area of beach deposition, erosion will take place. Similarly, rivers, which are a major source of the sediment that creates beaches, are sometimes dammed, thus depriving beaches of sediment and resulting in their erosion. Unless planners, developers, and builders understand these processes, major damage can result: Failure to understand coastal processes has, on occasion, caused a property owner to sue a neighbor who caused beach erosion to take place.

—C. Nicholas Raphael

Further Reading

Blumberg, Alan F., and Michael S. Bruno. *The Urban Ocean: The Interaction of Cities with Water.* Cambridge UP, 2018.

Ciavola, Paolo, and Giovanni Coco. *Coastal Storms: Processes and Impacts.* Wiley-Blackwell, 2017.

Davidson-Arnott, Robin, Bernard Bauer, and Chris Hauser. *Introduction to Coastal Processes and Geomorphology.* 2nd ed., Cambridge UP, 2019.

Davis, Richard Jr., and Duncan Fitzgerald. *Beaches and Coasts.* John Wiley & Sons, 2009.

Haslett, Simon K. *Coastal Systems.* 2nd ed., Routledge, 2009.

Hobbs, Carl H. *The Beach Book: Science of the Shore.* Columbia UP, 2012.

Jackson, Derek, and Andrew D. Short, editors. *Sandy Beach Morphodynamics.* Elsevier, 2020.

Komar, Paul D., and J. Robert Moore, editors. *Handbook of Coastal Processes and Erosion.* CRC Press, 2018.

Masselink, G., and M. G. Hughes. *Introduction to Coastal Processes and Geomorphology.* Hodder Arnold Publication, 2003.

Micallef, Anton, and Allan Williams. *Beach Management: Principles and Practice.* Earthscan, 2009.

Schwartz, Maurice L. *Encyclopedia of Coastal Science.* Springer, 2005.

BIOGENIC SEDIMENTARY ROCKS

Fields of Study: Geology; Sedimentology; Stratigraphy; Geochemistry

ABSTRACT

Biogenic sedimentary rocks represent the accumulation of skeletal material, produced by the biochemical action of organisms, to form limestone, chert, and phosphorites, or the

accumulation of plant material to form coal. Biogenic sedimentary rocks are economically important as sources of energy, fertilizers, and certain chemicals, and they contain evidence of former life on Earth.

KEY CONCEPTS

carbonate rocks: the general term for rocks containing calcite, aragonite, or dolomite

chert: multicolored, fine-grained sedimentary rock composed of silica and formed in the ocean

coal: dark brown to black sedimentary rock formed from the accumulation of plant material in swampy environments

diagenesis: the chemical, physical, and biological changes undergone by a sediment after its initial deposition

limestone: light gray to black sedimentary rock composed of calcite and formed primarily in the ocean

lithification: the process whereby loose material is transformed into solid rock by compaction or cementation

opal: a form of silica containing a varying proportion of water within the crystal structure

phosphorite: sedimentary rock composed principally of phosphate minerals

polymorph: minerals having the same chemical composition but a different crystal structure

LITHIFICATION AND COALIFICATION

Numerous organisms, particularly those living in the oceans, produce a shell or some type of skeletal material composed of the minerals calcite or aragonite (polymorphs of calcium carbonate), silica (silicon dioxide), or phosphate minerals, principally apatite. After the organism dies and the soft tissue decays, the skeletal material accumulates as sediment. This sediment then undergoes lithification, a process involving compaction and chemical cementation. The time necessary for lithification is highly variable and ranges from a few decades to millions of years. Biogenic sedimentary rocks composed of calcite or aragonite minerals are commonly referred to as limestones; those composed of silica are cherts, and those composed of phosphate minerals are known as phosphorites or sedimentary phosphate deposits. Limestones and cherts may form as the result of direct chemical precipitation of calcium carbonate or silica from a saturated solution, but the primary mode of formation is biochemical. Limestones are the most abundant of the biogenic sedimentary rocks and, economically, the most important. Iron formations, the principal source of iron ore, were also precipitated biogenically early in Earth's history.

Coal also is a biogenic sedimentary rock. Most coal forms from the accumulation of woody plant material in anoxic (oxygen-lacking) environments, such as stagnant lakes, swamps, and bogs. The formation of coal is referred to as coalification and represents microbiological, physical, and chemical processes whereby the percentage of carbon is increased and the volatile content decreased. Limestones, cherts, phosphorites, and coal are often closely associated and may occur as alternating beds in a sequence of sedimentary rocks.

PHOSPHORITES

Because of their biological affinities, biogenic sedimentary rocks contain important evidence of former life on Earth, a record of organic evolution (changes in organisms through geologic time), and clues to the past environments in which the rocks were formed. Limestones may serve as reservoir rocks for oil and gas, which form from the chemical breakdown of microscopic organisms that accumulate with the sediment at the bottom of the ocean and in some lakes. Phosphates, which are contained in phosphorites, are one of the chief constituents of fertilizers and are widely used in the chemical industry. Important evidence of past swamp communities, including both plants and animals, has been preserved in coal beds.

Phosphorus is one of the essential elements of life, and when chemically linked with other elements to form phosphate minerals (especially apatite) it becomes one of the major constituents of all vertebrate skeletons and some invertebrate hard parts. Phosphate is a primary nutrient in marine waters and, therefore, controls organic productivity. Phosphorites often occur as nodules, which are highly variable in size and shape; they may be several centimeters in diameter and up to a few meters in length. These nodules are usually rich in vertebrate skeletal debris, especially that of fish, and fecal material. Bedded phosphorites are commonly interbedded with limestones. Thin beds of phosphorite are rich in bones, fish scales, and fecal material. Modern phosphorites are forming in areas where cold, nutrient-rich waters rise from the ocean depths toward the surface, such as off the west coasts of Africa and North and South America.

CHERTS

Cherts, which are occasionally referred to as flint or novaculite, are not nearly as abundant as limestones; however, chert has two properties that have made it one of the most important rocks to humans. Chert has the same chemical composition as quartz but is cryptocrystalline; that is, its crystals are submicroscopic in size. Like quartz, chert is very hard and it tends to break along smooth, curved surfaces. With a little practice, one can produce a sharp edge by chipping a piece of chert with another hard object. Although early humans used numerous rock types for tools and weapons, chert was the most important for the production of arrowheads, knives, and scrapers. Consequently, chert might be considered to have contributed as much to the rise of civilization as did the development of the steam engine. In addition to chert's utility, Precambrian cherts (1.8 to 3.4 billion years old) are interbedded with important iron deposits known as banded iron formations. These cherts contain evidence of some of the earliest life to evolve on Earth.

Chert, which may be any color depending on the presence of impurities, occurs principally as spherical to oblong nodules 2 to 25 centimeters in diameter in limestones and dolostones, or as thin beds 2 to 25 centimeters deep. Nodular cherts are commonly parallel to bedding or stratification. Bedded cherts may extend laterally for great distances and are commonly interbedded with limestones and dolostones. Most geologists believe that nodular chert forms from the replacement of limestone and that bedded chert forms either by the complete replacement of carbonate-rich beds or by the diagenetic alteration of siliceous ooze.

The source of the silica may be the chemical precipitation of silicon dioxide, volcanic ash, or skeletal material. Biogenic silica is opaline, which means that it may contain up to 10 percent water. The principal biogenic sources of silica for chert are sponges, diatoms, and radiolarians. Skeletal materials from these organisms are common constituents of oceanic sediments. As this siliceous skeletal material accumulates, the opaline silica is diagenetically converted to crystalline opal and reprecipitated as bladed crystals. Continued diagenesis results in the formation of quartz chert, a mosaic of microscopic quartz crystals. The final diagenetic change often obliterates the shape or structure of the original skeletal material.

Sponges, which are abundant in most marine environments, contain as part of their supportive structure microscopic rods of opaline silica called "spicules." When the sponge dies, these spicules accumulate and become part of the sediment. During burial, the opaline silica undergoes diagenesis and precipitation of crystalline opal occurs within pores in the sediment. As crystalline opal is converted into quartz chert, chemical replacement of the surrounding sediment (usually calcite) occurs, resulting in the formation of chert nodules.

Radiolarians and diatoms are microscopic organisms with disk-shaped, elongate, or spherical tests (shells) with spines and surface ornamentation composed of opaline silica. Radiolarians occur as part of the marine zooplankton, and diatoms are part of the marine and nonmarine phytoplankton. Radiolarian and diatom oozes accumulate on the deep ocean floor. In time, these beds of silica-rich ooze are diagenetically converted into thin, bedded cherts. Diatoms, in contrast, may occur in great numbers in lakes and accumulate to form diatomaceous earths, or diatomites. Diatomaceous earths have a wide variety of uses, such as in filtering agents, absorbents, and abrasives.

LIMESTONES

Limestones are the most abundant of the chemical sedimentary rocks. A minor amount of limestone forms from the inorganic precipitation of calcite from seawater or the deposition of calcite in caves and around hot springs. The majority of limestones form as the result of biological and biochemical processes that produce aragonite or calcite. Later, this material becomes part of the carbonate sediment. Once deposited, the carbonate sediment is often modified by the chemical and physical processes of diagenesis.

Numerous animal phyla, such as mollusks, brachiopods, echinoderms, bryozoans, coelenterates, and certain protozoans, produce aragonite or calcite as part of their skeletal structure. The skeletal remains of these organisms are important constituents of carbonate sediments and, eventually, of most limestones. Some marine, bottom-dwelling algae secrete aragonite or calcite. These calcareous algae represent a significant contribution to carbonate sediment. Calcareous algae called "coccoliths" and calcareous foraminifera are found in great numbers in the plankton of the open ocean. Accumulation of this calcareous material at the bottom of the ocean contributes greatly to the formation of chalk, a type of biogenic limestone. Modern marine organisms principally secrete aragonite; consequently, modern calcareous sediments are initially composed predominantly of aragonite. Aragonite, however, is unstable at low pressures and quickly undergoes diagenesis to calcite.

It is not uncommon to find that ancient limestones have been partially or completely transformed into dolostones; this process is known as dolomitization. Dolomitization occurs when calcium carbonate minerals are diagenetically converted into dolomite. Diagenesis may take place soon after the calcite or aragonite has been deposited or a long time after the deposition. The diagenesis is the result of the magnesium-bearing waters (seawater or percolating meteoric water) moving through the carbonate sediment or limestone. One important aspect of dolomitization is that it often leads to the formation of pores, cavities, and fissures, which enable the rock to serve as a reservoir for oil, gas, and water. Recent research suggests that much dolomitization is caused by the circulation of deep, warm fluids long after burial.

Under certain conditions, limestones are relatively easy to dissolve. In areas such as Florida or Kentucky, where limestones are abundant and there is adequate rainfall, cave systems may develop. Regions with extensive cave systems and related features are referred to as karst.

STUDY OF ROCK SAMPLES

Initially, the study of biogenic sedimentary rocks involves field investigations. Where outcrops of these rocks occur, geologists plot and map their distribution and thicknesses; note changes in the rocks' character, both vertically and horizontally; observe their association with other types of rocks; and collect samples of both the rock and any fossils that may be present. When drilling for oil, gas, or water, engineers and geologists study the same phenomena.

Chalk. Photo by James St. John, CC BY 2.0, via Wikimedia Commons.

Samples, which are brought back to the laboratory, are analyzed with a vast array of techniques, including X-ray analysis, electron microprobe analysis, scanning electron microscopy, cathodoluminescence, and thin-section examination with a polarized microscope. Samples containing large fossils are observed with a magnifying glass or dissecting microscope. Some samples, particularly limestones, are dissolved in acid to recover both megafossils and microfossils. When samples have been ground thin enough to allow light to pass through the rock, microfossils can be observed with the help of microscopes. Scanning electron microscopy is particularly useful for observing extremely small fossils, especially radiolarians, diatoms, and coccoliths.

AREAS OF RESEARCH

In general, the chemistry of biogenic sedimentary rocks is simple; however, diagenesis can alter the structure, texture, and mineralogy of a sediment during its deposition, lithification, and burial. The analysis of diagenetic changes, particularly in limestones, is one of the most important avenues of investigation. Limestones are greatly influenced by diagenetic modifications and undergo changes in sediment size, porosity (the spaces between sediment grains), and mineralogy. Early diagenetic changes include the conversion of unstable aragonite—which is produced by most calcium-carbonate-producing organisms—into calcite and changes in the calcite's magnesium content. (Since the atoms of calcium and magnesium have similar properties, they can often substitute for each other in the crystal structure.) Dolomitization may also occur. The replacement of carbonate minerals by silica, so that chert is formed, is another important aspect of diagenesis. Stages of diagenesis can best be observed by using cathodoluminescence, which bombards a thin slice of rock with an electron beam. This process causes minerals within the rock to luminesce (emit light energy for a short interval after the energy source has been removed). The luminescences indicate various stages of diagenesis and are particularly important in studying the details of cements and crystal growth. X-ray analysis and electron microprobe techniques allow scientists to detect subtle differences in chemistries and the presences of trace elements or rare-earth elements.

Another major area of research is the classification of biogenic sedimentary rocks. The classification of limestones is of primary importance. Limestones have been classified using several criteria, but in general, classifications have focused on chemical and mineralogical composition; fabric features, such as fossils and cements; and special physical parameters, such as porosity. When examining chert, the geoscientist must ask: What was the source of the silica (volcanic ash, skeletal grains, chemical precipitate)? What was the time and rate of conversion of siliceous ooze to chert? What was the environment of deposition?

ECONOMIC AND EVOLUTIONARY SIGNIFICANCE

Biogenic sedimentary rocks are among the most important of all sedimentary rocks. Because of their biological affinities, they have preserved an important record of past life on Earth. In addition, some of these rocks were used for toolmaking by early humans; others are important resources for construction materials, fertilizers, and chemicals. Oil and natural gas, which also have biological affinities, are inseparably linked to biogenic sedimentary rocks.

Although not widespread, biogenic phosphorites are economically important, particularly to the fertilizer and chemical industries. Biogenic chert was of major importance in the past. Early humans used chert, or flint, to make tools and weapons. Biogenic limestone is the most common and widespread biogenic sedimentary rock. Economically, limestones are very important. Hydrocarbons (oil and gas) are commonly recovered from porous limestones and dolostones, and in some areas, limestones are reservoirs for groundwater supplies. Limestones often serve as host rocks for important mineral deposits, such as lead. Limestones are quarried as building stone; crushed to form construction materials, such as gravel; or processed into lime and cement. Biogenic limestones are the most important sedimentary rock containing fossils and, therefore, are the most important record of the evolution of life.

—Larry E. Davis

Further Reading

Blatt, Harvey, Robert J. Tracy, and Brent Owens. *Petrology: Igneous, Sedimentary, and Metamorphic.* 3rd ed., W. H. Freeman, 2005.

Boggs, Sam, Jr. *Petrology of Sedimentary Rocks.* Cambridge UP, 2009.

Chernicoff, Stanley. *Geology: An Introduction to Physical Geology.* 4th ed., Prentice Hall, 2006.

James, Noel P., and Yvonne Bone. *Biogenic Rocks in a Cold Cenozoic Ocean: Neritic Southern Australia.* Springer International Publishing, 2021.

———, and Brian Jones. *Origin of Carbonate Sedimentary Rocks.* Wiley, 2016.

Nichols, Gary. *Sedimentology and Stratigraphy.* 2nd ed., Wiley-Blackwell, 2009.

Prothero, Donald R., and Fred Schwab, editors. *Sedimentary Geology.* W.H. Freeman, 2013.

Ramkumar, Muhammad, editor. *Chemostratigraphy: Concepts, Techniques and Applications.* Elsevier, 2015.

Riding, Robert E., and Stanley M. Awramik. *Microbial Sediments.* Springer-Verlag, 2010.

Tucker, Maurice E. *Sedimentary Rocks in the Field.* 4th ed., John Wiley & Sons, 2011.

Bolides

Fields of Study: Astrometry; Astrophysics; Astronomy; Astrogeology

ABSTRACT
A bolide is a meteor originating in outer space that explodes as it falls to Earth. Astronomers usually use the term to describe especially brightly burning objects or fireballs. Bolides are sometimes dangerous, but they usually land away from people.

KEY CONCEPTS

asteroid: a minor planet in the inner solar system that orbits the Sun

comet: an icy, small solar system body that heats up and releases gases when it passes near the Sun in its orbit

fireball: a meteor that is unusually bright

meteor: a small body of rock or metal that enters Earth's atmosphere and emits light as friction with Earth's atmosphere causes it to heat up to incandescence

meteoroid: a small body of rock or metal that travels through space

TYPES OF BOLIDES

Bolides start as one of three types of objects found in space: a comet, an asteroid, or a meteoroid. Comets are balls of ice and dust, asteroids are minor bodies that are too small to be classified as actual planets, and meteoroids are small chunks of rock or metal. When these objects enter Earth's atmosphere, they heat up. If the object becomes hot enough to emit light, it is called a "meteor." If the becomes brighter than the planet Venus in the night sky, it is called a "fireball." If that fireball gets so hot that it explodes before reaching Earth's surface, the object is called a "bolide." Especially bright bolides may be known as superbolides.

WHY BOLIDES EXPLODE

On Earth, the fluid medium of air slows down any moving object passing through it. An object's terminal velocity is the fastest that object can free-fall before the resistance of the atmosphere prevents it from accelerating further. However, there is no atmosphere in outer space. Because of this, many objects in outer space move much faster than objects on Earth.

Meteors that enter Earth's atmosphere are usually traveling much faster than their terminal velocity. Because of this, the atmosphere in front of the meteor is compressed, creating pressure and friction and causing the object to slow down. As the meteor slows, the air around it quickly heats up. This causes the meteor to burst into flames. When the heat and pressure become stronger than the meteor can bear, the meteor violently breaks apart.

NOT ALL BOLIDES EXPLODE IN THE ATMOSPHERE

The term "bolide" is also used in reference to meteorites that impact the surface of Earth before they can explode in the atmosphere. Several examples are known of impact craters formed in the past where large meteorites or bolides have struck the surface of the planet. The tell-tale structure of the impact zone can often be identified in geological maps by a characteristic circular pattern that may show a 'double ring' pattern about the point of impact. Well-known among these are the Chesapeake Bay impact crater, the Sudbury impact crater in Northern Ontario, Canada, and Lake Manikouagan, also in Canada. Perhaps the most well-known of all is the Chicxulub Crater in the Caribbean Sea at Belize, generally accepted as "the big one" that ended the reign of the dinosaurs. Such impacts greatly alter the geology of the immediate region through the incorporation of the metal content of the meteor, primarily iron and nickel. The Sudbury Crater, for example, has been the site of productive nickel mining operations for many years, but the vast surrounding extent of the Canadian Shield geological structure is not known to possess quantities of nickel ore that can be mined profitably. The extent of impacts can also be identified in mineralogical mappings of an area, as the mapping will reveal a pattern of enhanced concentrations of certain metals and other elements as well as impactites, mineral forms that are only produced under the force of impact.

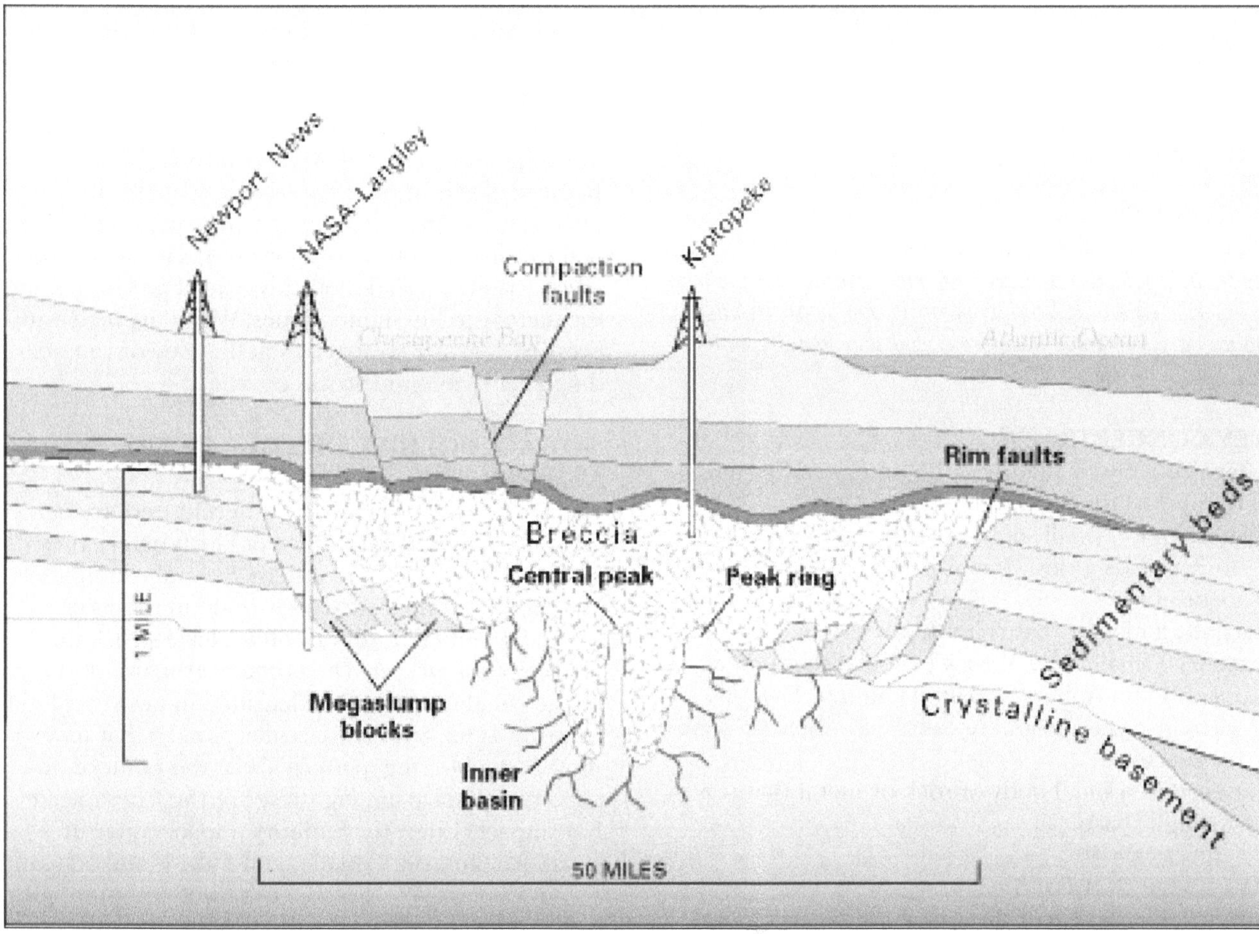

Side view of the Chesapeake Bay impact crater. The crater was formed by a bolide that impacted the eastern shore of North America about 35.5 ± 0.3 million years ago. Image via Wikimedia Commons. [Public domain.]

BOLIDES AND HUMANS

When bolides explode near humans, the results can be catastrophic. On February 15, 2013, a bolide exploded near Chelyabinsk, Russia. According to the National Aeronautics and Space Administration (NASA), the bolide exploded with almost five hundred kilotons of force. For comparison, the atomic bomb dropped at Hiroshima exerted only sixteen kilotons of force. The explosion injured hundreds of people and damaged more than 297 buildings.

More than a century earlier, an even larger bolide exploded over Russia's Tunguska region. Scientists estimate that blast was one thousand times as powerful as the Hiroshima bomb. The Tunguska blast flattened trees for miles around the blast site and caused so much light and reflective dust to be thrown into the atmosphere that people as far away as western Europe reported glowing night skies.

—Tyler Biscontini and Richard M. Renneboog

Further Reading

Allain, Rhett. "Why Does a Meteor Explode in the Air?" *Wired.* Condé Nast, 18 Feb. 2013. Accessed 11 Mar. 2015.

"Bolide." *Cosmos: The SAO Encyclopedia of Astronomy.* Swinburne U of Technology, n.d. Accessed 27 Mar. 2015.

Bryner, Jeanna. "Meteor Explosion in Russia Hurts Hundreds of People: Reports." *LiveScience.* Purch, 15 Feb. 2013. Accessed 11 Mar. 2015.

Osinski, G. R., and E. Pierazzo. *Impact Cratering: Processes and Products.* Blackwell Publishing, 2013.

Poag, C. Wylie, Christian Koeberl, and Wolf Uwe Reimold. *The Chesapeake Bay Crater: Geology and Geophysics of a Late Eocene Submarine Impact Structure.* Springer, 2004.

Popova, Olga, and Ivan Nemchinov. "Bolides in the Earth Atmosphere": *Catastrophic Events Caused by Cosmic Objects*, edited by Adushkin and Nemchinov, pp. 131-62, Springer, 2008.
Reynolds, Mike D. *Falling Stars: A Guide to Meteors & Meteorites.* Stackpole, 2010.
Richardson, James, and James Bedient, comps. "Fire-ball FAQs." *American Meteor Society* American Meteor Society, n.d. Accessed 11 Mar. 2015.

BOLTWOOD USES RADIOACTIVITY TO DETERMINE AGES OF ROCKS

Fields of Study: Geology; Geochemistry; Radiochemistry

ABSTRACT
Bertram Borden Boltwood pioneered the radiometric dating of rocks, giving impetus to the use of nuclear methods in geology and establishing a new chronology of the earth.

KEY CONCEPTS
geochronology: the measurement of the ages of rocks
lead-to-uranium ratio: the amount of lead relative to the amount of uranium contained in a uranium-bearing rock
nonradiogenic: atoms that are not produced as a result of a radioactive decay process
radioactive decay: the conversion of an atom of a particular element into atoms of other elements by the spontaneous loss of one or more protons and neutrons from its nucleus, accompanied by the release of energy
radiogenic: atoms that are produced through the process of radioactive decay
radiometric dating: determination of the ages of materials base on the measurement of the amount of a particular radioactive isotope they contain

BOLTWOOD AND THE THEORY OF RADIOACTIVITY
Bertram Borden Boltwood, educated at Yale University and in Germany, began his career as an instructor of analytic chemistry at Yale University. He remained at Yale until 1890, when he left to establish a private laboratory (Pratt and Boltwood, Consulting Mining Engineers and Chemists) in New Haven, Connecticut. He was interested in radium and was fascinated by the theory of radioactive disintegrations proposed in 1903 by two McGill University scientists, Ernest Rutherford and Frederick Soddy. According to that theory, radioactivity is always accompanied by the production of new chemical elements on an atom-by-atom basis. In 1904, Boltwood impressed Rutherford by demonstrating that all uranium minerals contain the same number of radium atoms per gram of uranium. This confirmation of the theory of radioactivity marked the beginning of a close collaboration between the two scientists. Actually, Boltwood's involvement with research in radioactivity can be traced to 1899, when he was supervising a student project on the extraction of radium from uranium ores by Marie Curie's method.

The significance of Boltwood's work is best appreciated in the light of a chronological controversy raging at that time between geologists and physicists. It had been accepted generally that Earth, at some time in its history, had been a liquid ball, and that its solid crust was formed when the temperature was reduced by cooling. The geological age of Earth was defined, accordingly, as the period of time necessary to allow it to cool down from the melting point to the present temperature. Several estimates of that time were made in the second half of the nineteenth century by the famous physicist Baron Kelvin, Sir William Thomson (known as Lord Kelvin), who claimed that the age of the planet did not exceed 100 million years. In 1897, Lord Kelvin specified that the actual figure was probably close to 20 million and certainly not as large as 40 million. His calculations were mathematically correct, but they did not take into account radioactivity, which had been discovered the year before.

HELIUM
Geologists believed that 40 million years simply was not enough time to create continents, erode mountains, and supply oceans with minerals and salts. Studies of sequences of layers (stratigraphy) and of fossils (paleontology) led them to believe that Earth was older than 100 million years, but they were not able to prove it. Boltwood and Rutherford proved that geologists were right. This came as a byproduct

of their research on the nature of radioactivity. The gaseous element helium played an important role in that line of research. Rutherford knew that helium is always present in natural deposits of uranium, and this led him to believe that in radioactive minerals, alpha particles somehow are turned into ordinary atoms of helium. Accordingly, each rock of a radioactive mineral is a generator of helium. The accumulation of the gas proceeds more or less uniformly, so that the age of a rock can be determined from the amount of trapped helium. In that sense, radioactive rocks serve as natural clocks.

In line with this model, and knowing how much helium is produced from each gram of uranium per billion years, Rutherford and his collaborators were able to see that naturally radioactive rocks are often older than 100 million years. The accumulated amounts of helium in the rocks they studied were simply too large to allow other interpretations. The ages of some of the samples exceeded 500 million years. Moreover, Rutherford was aware that the ages they assigned would have to be revised upward in order to account for helium that was escaping from the rocks.

LEAD

Impressed by these results, and trying to eliminate the uncertainties associated with leakage of helium, Boltwood decided to work on another method of dating the rocks. This resulted from his earlier attempts to demonstrate that, as in the case of radium, all uranium minerals contain the same number of atoms of lead per gram of uranium. Chemical data, however, did not confirm this expectation; the mea-

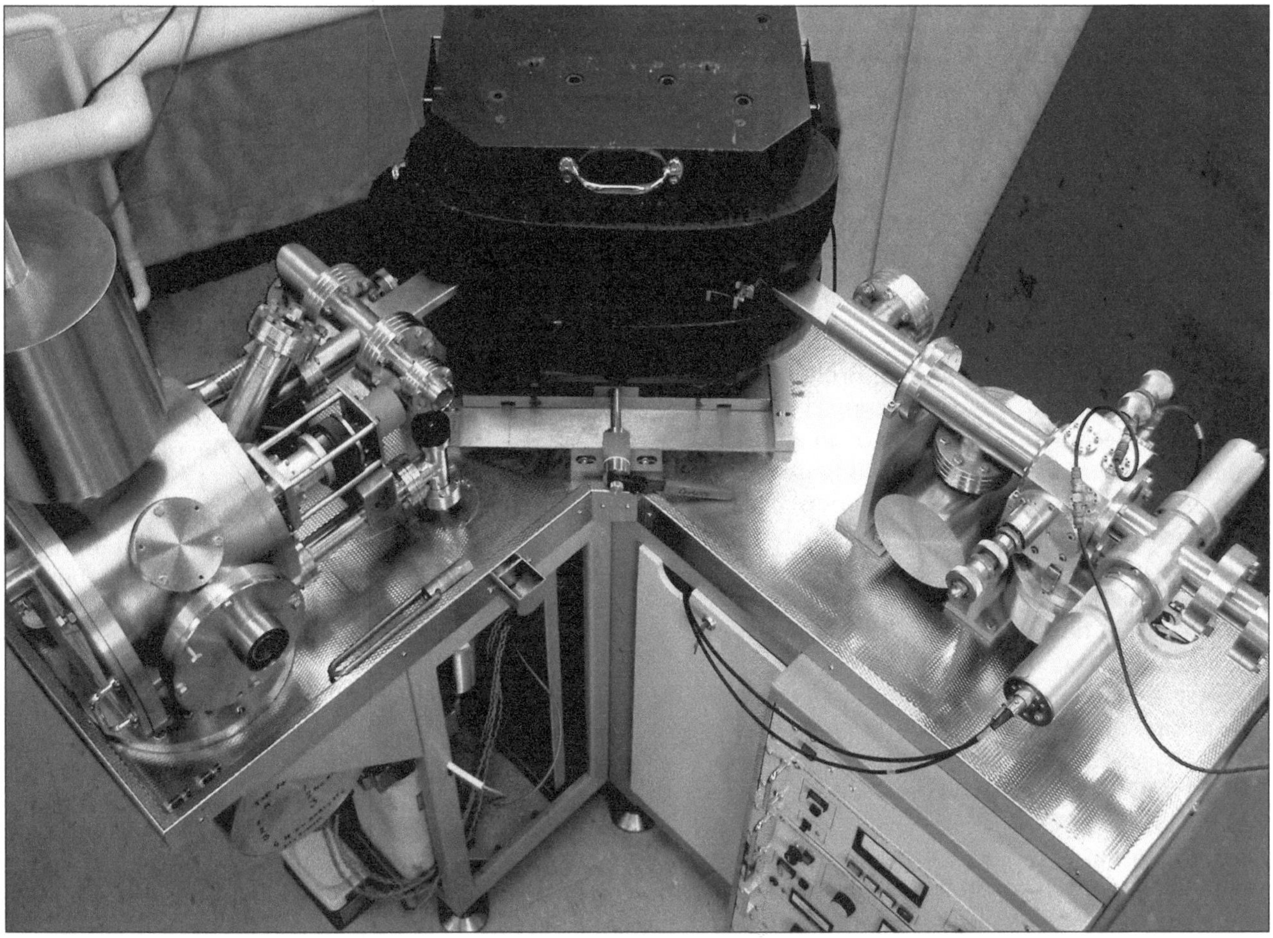

Example of a thermal ionization mass spectrometer used in radiometric dating. Photo via Wikimedia Commons. [Public domain.]

sured lead-to-uranium ratios were found to be different in minerals from different locations on Earth. Yet this still needed to be interpreted. According to Rutherford and Soddy's theory, a spontaneous transformation of uranium into a final product proceeds through a set of steps in which alpha particles and electrons are emitted, one after another. Boltwood realized that lead must be the final product and that its accumulation could be used to date minerals. By focusing on lead rather than on helium, he hoped to reduce the uncertainties associated with the leakage. He reasoned that lead is less likely than helium to escape from rocks because, once trapped, it becomes part of a solid structure.

Motivated by these ideas, Boltwood proceeded with the development of the uranium-lead method of dating. To accomplish this, he had to determine the rate at which lead is produced from uranium. That rate is about 0.15 grams of lead per gram of uranium, per billion years. Suppose, for example, that a chemical analysis of a rock yields 0.03 grams of lead for each gram of uranium. The age of the rock can be found by setting $0.03 = 0.15 \times t$, where t is the age of the rock, in billions of years. The solution of this simple equation is $t = 0.2$, or 200 million years. A slightly smaller t, 192 million, would be deduced if the exponential nature of radioactive decay were taken under consideration.

Boltwood was aware that even the most carefully performed experiments can lead to highly unrealistic conclusions. In his case, two sources of interpretational errors were possible, one having to do with the preexisting (nonradiogenic) lead and another having to do with the fact that chemicals can migrate through rock boundaries. Suppose, Boltwood argued, that 50 percent of the lead found in a sample was already there at the time of solidification, but a researcher was not aware of that fact. In such a case, the derived age of the sample would be exaggerated by a factor of two. In other words, preexisting lead would make rocks appear older than they actually were. On the other hand, the opposite would be true if one-half of the lead generated from uranium were able to escape through the rock boundaries. Recognizing the dangers of interpretational errors, Boltwood was very careful in selecting samples. He analyzed only deep portions of samples to minimize uncertainties associated with the possibility of lead leakage. Boltwood started his investigations in 1905, and before the end of the year he had analyzed twenty-six samples. One of them was identified as 570 million years old. In a formal publication that appeared in 1907, he described forty-six minerals collected in different locations; their reported ages were between 410 and 2,200 million years old.

Similar results had been reported earlier by Rutherford, from his laboratory in Montreal, and by Robert John Strutt (also known as Fourth Baron Rayleigh), from the Imperial College in London. It is true that similarities were mostly qualitative, rather than quantitative; however, all data showed that the absolute time scale was undeniably longer than anticipated, even by some geologists. In other words, the large margins of error that characterized early results did not interfere with the basic overall conclusion that many rocks were at least ten times older than Lord Kelvin had calculated. It was significant that helium and lead methods, based on different chemical procedures and applied to a large number of very different minerals, yielded similar results at three institutions.

SIGNIFICANCE

The main result of Boltwood's pioneering work and that of his successors was the realization that geological time must be expressed in hundreds and thousands of millions of years rather than in tens of millions, as advocated by Lord Kelvin. This was particularly significant for the acceptance of the theory of evolution and, in general, for better understanding of many long-term processes on Earth. Geochronology (the study of geological time) has been used, for example, in investigations of reversals of the terrestrial magnetic field. Such reversals occurred many times during Earth's geological history. Scientists discovered and studied them by dating pieces of lava that were naturally magnetized during solidification. The most recent reversal took place approximately 700,000 years ago.

It is clear, in retrospect, that the discovery of radioactivity affected geochronology in two ways: by providing tools for radiometric dating and by invalidating Lord Kelvin's thermodynamic calculations. These calculations were based on the assumption that the geothermal energy lost by Earth is not replenished. The existence of radioactive heating, discovered in 1903 in France, contradicted that assumption and prepared scientists for the acceptance of Boltwood's findings. Lord Kelvin died in the same year in which these findings were published, but he knew about Rutherford's findings as early as spring, 1904. He was very interested in radioactive heating but never came forward with any public retraction of his earlier pronouncements.

In its modern form, the uranium-lead method of dating is very different from what it was at the beginning of the twentieth century, primarily because radioactive transformations are better understood and because instruments have been developed that were not available in Boltwood's day. The most important tool of modern geochronologists is the mass spectrometer. Its widespread use after World War II is responsible for the development of new methods of dating. The availability of several methods is important when redundancy can be used to minimize, and sometimes eliminate, the kinds of interpretational errors recognized by Rutherford, Strutt, and Boltwood.

By the beginning of the twenty-first century, a large number of samples from different geological strata, all over the world, plus lunar samples and meteorites, had been dated by modern methods. The data showed that the oldest known terrestrial rocks solidified 3.7 billion years ago; this is 0.9 billion years less than the ages of lunar samples. The ages of most meteorites, between 4.5 and 4.7 billion years, are essentially the same as the ages of lunar samples. This is consistent with the overall picture, also backed by astronomical data, that the Solar system was formed 4.7 billion years ago, and that Earth remained liquid in the first 0.9 billion years of its existence. Boltwood would be proud to be counted among the contributors to that knowledge.

—Ludwik Kowalski

Further Reading

Allégre, Claude J. *Isotope Geology*. Cambridge UP, 2008.

Dickin, Alan P. *Radiogenic Isotope Geology*. 2nd ed., Cambridge UP, 2005.

Gorst, Martin. *Measuring Eternity: The Search for the Beginning of Time*. Broadway Books, 2001.

Holland, H. D., and K. K. Turekian, editors. *Radioactive Geochronometry from the Treatise on Geochemistry*. Academic Press, 2011.

O'Hara, Kieran D. *A Brief History of Geology*. Cambridge UP, 2018.

Reiners, Peter W., Richard W. Carlson, Paul R. Renne, et al. *Geochronology and Thermochronology*. John Wiley & Sons, 2017.

C

Carbonate Compensation Depths

Fields of Study: Oceanography; Marine Geochemistry; Marine Geology

ABSTRACT

Carbonate solubility increases with increasing pressure and decreasing temperature. Carbonate compensation depths are the depths in ocean basins that separate calcium carbonate-containing seafloor sediments from carbonate-free sediments. The carbonate compensation depth (CCD) may be different in different ocean basins, and it may rise or fall at different times in the same ocean basin as a result of the balance between surface production of carbonate and deep-water carbonate dissolution.

KEY CONCEPTS

deposition: the process by which loose sediment grains fall out of seawater to accumulate as layers of sediment on the seafloor

paleodepth: an estimate of the water depth at which ancient seafloor sediments were originally deposited

plankton: microscopic marine plants and animals that live in the surface waters of the oceans; these floating organisms precipitate the particles that sink to form biogenic marine sediments

precipitation: the formation of solid mineral crystals from their chemical components that are dissolved in water

productivity: the rate at which plankton reproduce in surface waters, which in turn controls the rate of precipitation of calcareous or siliceous shells or tests by these organisms

red clays: fine-grained, carbonate-free sediments that accumulate at depths below the CCD in all ocean basins; their red color is caused by the presence of oxidized fine-grained iron particles

test: an internal skeleton or shell precipitated by a one-celled planktonic plant or animal

MARINE SEDIMENTS

Marine sediments are composed of a variety of materials of biological or mineral origin. Individual sediment grains in oceanic deposits may be either clastic or biogenic particles. Clastic sediments are materials derived from the weathering, erosion, and transportation of exposed continental rocks, and these grains are classified by particle diameter into gravels, sands, silts, and clays. Clastic sediment particles become less common with increasing distance from the continental landmasses and are nearly absent from deep-water sediments on the abyssal plains. Biogenic sediment particles are composed of skeletons and tests precipitated by planktonic plants and animals living in the shallowest waters of the ocean. Biogenic particles composed of calcium carbonate or of opaline silica make up the majority of oceanic sediments, which are deposited at great distances from land. Deposition of biogenic sediments is controlled by two factors: the biological productivity of surface waters and the dissolution of biogenic particles by corrosive bottom waters. Increased levels of atmospheric carbon dioxide lead to an increase in the corrosiveness of bottom waters.

Biogenic sediment particles are produced in shallow, well-lighted surface waters as a result of the biochemical activity of microscopic plants and animals, which precipitate solid shells and tests formed from minerals dissolved in seawater. Biological productivity is a measure of the number of organisms present and their rate of reproduction. The productivity of planktonic organisms is directly related to chemical and physical conditions in the surface waters. High productivity waters have abundant supplies of oxygen, with dissolved nutrients and chemicals needed for the formation of shells and tests, and enough light for photosynthesis by plants. Generally, high productivity is found in warm water areas with abundant dissolved oxygen and nutrients.

When planktonic plants and animals die, their shells and tests fall through the water column to be deposited on the bottom of the ocean. This "planktonic rain" causes deposition of seafloor sediments as the biogenic particles produced in the surface waters sink and accumulate on the seafloor, and it moves chemicals from surface waters to deep waters. The higher the productivity values in surface waters, the greater the supply of biogenic sediment particles to the seafloor. As biogenic particles sink through the water column to the seafloor, they may be dissolved by corrosive seawater. Surface waters are saturated with dissolved calcium carbonate, so most dissolution takes place in deep waters, which are undersaturated with carbonate.

Calcareous sediments are produced by the accumulation of biogenic particles of calcium carbonate that survive the fall through the water and are deposited and buried on the seafloor. Accumulation rates of biogenic sediments are controlled by the balance between surface productivity and deep-water dissolution: The higher the biological productivity at the surface, the greater the number of shells and tests that will sink to the ocean bottom. In certain high productivity areas associated with upwelling of cold, nutrient loaded water masses to the surface, seafloor sediment accumulation rates may be as high as 3 to 5 centimeters per thousand years. In areas with higher dissolution rates, fewer calcareous particles will survive to be deposited on the ocean floor. In these low productivity areas, all the carbonate produced at the surface may be dissolved, and sediment accumulation rates may be as low as 1 millimeter per million years.

LEVEL OF CARBONATE COMPENSATION DEPTH

The carbonate compensation depth (CCD) marks the boundary between carbonate rich sediments (calcareous oozes and chalks) and carbonate-free sediments (red clays) in the oceans. It is the result of deep-water dissolution rates exceeding the rate of supply of calcium carbonate to the deep sea by surface productivity. The depth of the CCD marks the zone in which the supply of carbonate sinking in the planktonic rain from surface waters is exactly balanced by the rate of removal of carbonate dissolution in deep waters. Calcareous sediments will be deposited on the seafloor in water depths shallower than the CCD, because in these areas, the rate of supply of calcium carbonate is higher than the dissolution rate. In these areas, individual particles of calcium carbonate will survive the trip through the water column and be deposited as biogenic sediments on the seafloor.

Below the compensation depth, dissolution exceeds the rate of supply, so all carbonate particles supplied from the surface become dissolved, and seafloor sediments are carbonate-free. Surface sediments in water depths below the CCD tend to be red clays, or combinations of fine-grained materials derived from continental sources and carried to the deep sea by wind, mixed with micrometeorites and other particles from extraterrestrial sources. Red clays generally lack fossils, as a result of complete dissolution of carbonate and opaline silica, so only a few dissolution resistant fossils, such as phosphatic fish teeth and whale ear bones, are found in these sediments. The reddish-brown color of deep-sea clay deposits is a result of the presence of iron particles, which have reacted with oxygen in seawater to form the rust brown color.

Much information on the past history of the compensation depth in different ocean basins has been provided by ocean floor drilling programs. The level of the CCD has changed dramatically throughout geologic history, with fluctuations of up to 2,000 meters being recorded in deep-sea sediments. Changes in the CCD are believed to be caused by changes in either the rate of supply of carbonate to the oceans or the rate of carbonate dissolution in the deep sea, which may be caused by changes in the shape of ocean basins, by changes in the location of carbonate deposition within different ocean basins, or by changes in the concentration of carbon dioxide in the atmosphere. During the last 100,000 years, carbonate sediments formed at greater depths during glacial intervals than during interglacials, indicating a less corrosive deep-water environment resulting from lower concentrations of atmospheric carbon dioxide.

CALCIUM CARBONATE DEPOSITION

Calcium carbonate is delivered to the oceans by rivers carrying particles eroded from continental rocks. Ocean sediments are the primary geochemical reservoir for calcium carbonate, so most of the calcium carbonate on Earth remains dissolved in ocean water or in the form of calcareous sediments on the seafloor. The amount of carbonate deposition on the seafloor depends on the input of calcium carbonate derived from continental weathering and delivered to the oceans by rivers. Because oceanic plankton can precipitate solid calcium carbonate at a much faster rate than the rate of input of dissolved carbonate from rivers, most of the calcium carbonate that is deposited in the oceans must dissolve in order to maintain the chemical balance between dissolved carbonate and solid carbonate in oceanic sediments.

Any changes in the locations of carbonate deposition may cause corresponding changes in the level of the CCD as the compensation depths in different oceans change as a result of bathymetric fractionation or basin-basin fractionation. In bathymetric fractionation, a balance is established between the rates of carbonate deposition in shallow-water and deep-water sedimentary basins. Greater deposition of calcium carbonate in shallow waters atop the continental shelves will cause a shallowing of the CCD as more deep-water carbonate deposits are dissolved so as to balance the shallow-water deposition. Similarly, the level of the compensation depth may vary by basin-basin fractionation of carbonate, which establishes a balance between the compensation depths in different ocean basins. For example, greater deposition of calcium carbonate in the Pacific Ocean will cause the Pacific CCD to move to a greater depth, while at the same time the Atlantic compensation depth must become shallower, because greater deposition of carbonate in Pacific sediments will leave less dissolved carbonate for precipitation in the Atlantic Ocean.

Even within an ocean basin, the level of the CCD may vary, depending on the balance between carbonate productivity and dissolution in a local area. For example, in the equatorial Pacific Ocean, the compensation depth is 500 to 800 meters deeper than in areas immediately to the north and south of this high productivity area as a result of the greater supply of carbonate in the planktonic rain below these high productivity surface waters. Also, the compensation depth tends to shoal near the edges of ocean basins, because higher biological productivity in shallow water near the continents causes rapid sinking of large amounts of organic carbon produced by planktonic plants and animals. Breakdown of this organic carbon by seafloor bacteria produces increased amounts of dissolved carbon dioxide, which reacts with water molecules to form carbonic acid; carbonic acid is corrosive to solid calcium carbonate. Greater carbonic acid concentrations lead to increased carbonate dissolution in bottom waters and cause upward migration of the CCD into shallower waters.

STUDY OF CARBONATE COMPENSATION DEPTHS

Carbonate compensation depths may be studied by obtaining a series of deep-sea sediment samples from different depths within an ocean basin to determine the relationship between sediment type and water depth. In 1891, a study was published describing the global patterns of seafloor sediment type in each ocean basin, based on sediment samples obtained on the HMS *Challenger* oceanographic expedition. It was discovered that virtually no calcium carbonate was present below depths of 4,500 meters as a result of dissolution of carbonate.

The CCD in the modern ocean was first described in detail in 1935 based on sediment core transects taken across the South Atlantic Ocean by the 1925-1927 German "Meteor" expedition. Similar studies of sediment cores from different depths in the Pacific Ocean revealed that calcareous oozes are common seafloor sediments in water depths to 4,400 meters, with noncalcareous red clays being found in surface sediments deeper than 4,400 meters. Cores with calcareous oozes underlying red clays were obtained, however, in depths well below the present CCD, indicating that this geochemical boundary has migrated vertically throughout geologic history.

One innovative experiment to measure the rate of calcium carbonate dissolution with increasing water depth involved the placement of a stationary mooring for a period of months in deep water in the Pacific Ocean. Calcite spheres and calcareous microfossils were hung in permeable nylon bags at different water depths on the mooring; the nylon bags allowed seawater to come in contact with the calcium carbonate and thus permitted carbonate dissolution to occur. By measuring the weight loss of spheres suspended for a few weeks to months on the mooring, the rate of dissolution was determined for different water depths. In this experiment, little carbonate dissolution was observed at water depths shallower than 3,700 meters, while a rapid transition from minimal dissolution to extreme dissolution was seen at depths below 3,700 meters. Rapid loss of carbonate by dissolution occurred in carbonate spheres suspended between the lysocline (the depth at which carbonate dissolution first begins to occur) and the compensation depth. Below 4,500 meters, the carbonate compensation depth, all carbonate was removed within a matter of weeks, demonstrating the ability of bottom waters to dissolve calcium carbonate.

Information on compensation depths may also be provided from microfossils preserved in seafloor sediments. Calcite dissolution can be measured by enrichment of dissolution resistant forms of planktonic organisms, by benthic planktonic foraminiferal ratios (foraminifera are one-celled animals that secrete a calcium carbonate internal test), by fragmentation indices (the percentage of broken planktonic tests compared to whole tests), or by the coarse fraction ratios of seafloor sediments. Different microfossils will have differing susceptibilities to dissolution, depending on the thickness of the walls of the microfossil tests. Thin-walled plankton will be more susceptible to dissolution, while thicker-walled tests will resist dissolution. Thin-walled planktonic tests in seafloor sediments deposited in water depths between the lysocline and the CCD will be removed by dissolution more rapidly than thicker-walled tests, leading to greater enrichment of dissolution resistant fossils with greater carbonate dissolution. Relative dissolution rates may be measured by the ratio between dissolution susceptible and dissolution resistant planktonic shells in seafloor sediments.

Similarly, the deeper water bottom dwelling benthic foraminifera tend to have thicker walls than the tests of planktonic foraminifera, which live floating in the shallow surface waters. Seafloor sediments may contain both benthic and planktonic varieties of foraminifera. Dissolution of calcium carbonate will preferentially remove the thinner-walled planktonic foraminifera, thus leading to enrichment in the proportion of benthic forms remaining in the sediment. The relative amount of carbonate dissolution in sediments may be determined by measuring the proportion between benthic and planktonic foraminifera in sediment samples.

Finally, fragmentation of planktonic tests may provide an indication of dissolution of carbonate from marine sediments. The percentage of broken tests to whole tests will increase with greater dissolution of carbonate. Also, coarse fraction percentages of sediments provide information on carbonate dissolution, because unbroken foraminiferal tests are sand-sized, with particle diameters greater than 63 microns in size (1 millimeter is equal to 1,000 microns). As foraminifera are dissolved, they tend to break into smaller fragments, so dissolution tends to break down sand-sized particles into smaller silt-sized fragments. By measuring the proportion between sand-sized and silt-sized particles (the coarse fraction percentage) in calcareous sediments, it is possible to obtain an indicator of the relative amount of dissolution that has affected those sediment deposits. The more dissolution that has occurred, the smaller will be the percentage of coarse (sand-sized) sediment particles.

DEPTH ESTIMATES

All these methods provide similar depth estimates for the CCD at approximately 4,500 meters below the sea surface, about halfway between the crests of the midocean ridges and the abyssal plains. Individual compensation depths may vary in the different ocean basins of the world, however, as a result of basin-basin fractionation of carbonate. For example, in the Pacific Ocean, the CCD is typically between 4,200 and 4,500 meters, while in the Atlantic and Indian Oceans the CCD is deeper, being found near a depth of 5,000 meters. Even within an ocean basin, the level of the CCD may vary, depending on the balance between carbonate productivity and dissolution in a local area. In the centers of the North and South Pacific Oceans, the average compensation depth is between 4,200 and 4,500 meters, while near the equator it is found near 5,000 meters because of the higher biological productivity of equatorial surface waters.

Analyses of deep-sea cores drilled by the *Glomar Challenger* have revealed that the level of the carbonate compensation depth has changed by up to 2,000 meters in the South Atlantic, Indian, and Pacific

Ocean basins. For example, one of the results of Deep Sea Drilling Project Leg 2 was the discovery of significant vertical excursions in the compensation depth of the Atlantic Ocean. In seafloor boreholes drilled in the North Atlantic, 2 to 5 million year old calcareous ooze sediments were found atop older (5to 23 million year old) red clays, which in turn were deposited atop carbonate deposits older than 23 million years. In order for these sediments to have accumulated in that order, large vertical changes must have occurred in the compensation depth, starting when the seafloor was shallower than the CCD more than 23 million years ago. Red clays were deposited between 23 and 5 million years ago, when the CCD became shallower. After these red clays accumulated, deepening of the compensation depth allowed the deposition of younger calcareous sediments atop the carbonate-free red clays.

In order to study the past history of the compensation depth within an ocean basin, it is necessary to obtain a series of cores that were deposited at different paleodepths at the same time in the past. (The paleodepth is the depth at which ancient seafloor sediments were deposited.) Paleodepth estimates for seafloor sediments are calculated by studying the cooling history of seafloor basement rocks. After new seafloor is produced by volcanic activity at the mid ocean ridge system, these rocks cool and contract, thus sinking to greater water depths as they move away from the ridge system. The older the seafloor, the greater its water depth will be, so sediments deposited atop volcanic seafloor will accumulate in progressively deeper water. Once a series of sediment cores deposited at the same time in the past has been obtained, it is possible for the oceanographer to determine the paleodepth of the compensation depth by finding the paleodepth below which no calcium carbonate is present in ancient sediment deposits.

—Dean A. Dunn

Further Reading

Ahr, Wayne M. *Geology of Carbonate Reservoirs.* John Wiley & Sons, 2008.

Burdige, David J. *Geochemistry of Marine Sediments.* Princeton UP, 2020.

Emerson, Steven, and John Hedges. *Chemical Oceanography and Marine Carbon Cycle.* Cambridge UP, 2008.

Garrison, Tom S. *Oceanography: An Invitation to Marine Science.* Brooks/Cole, Cengage Learning, 2010.

Middelberg, Jack J. *Marine Carbon Biogeochemistry: A Primer for Earth System Scientists.* SpringerOpen, 2019.

Rigo, M., et al. "A Rise in the Carbonate Compensation Depth of Western Tethys in the Carnian (Late Triassic): Deep Water Evidence for the Carnian Pluvial Event." *Palaeogeography, Palaeoclimatology, Palaeoecology,* vol. 246, 2007, pp. 188-205.

Segar, Douglas. *An Introduction to Ocean Sciences.* 2nd ed., W. W. Norton, 2007.

Sverdrup, Keith A., Alyn C. Duxbury, and Alison B. Duxbury. *An Introduction to the World's Oceans.* 8th ed., McGraw Hill, 2004.

Swart, Peter K., Gregor Eberli, and Judith A. McKenzie, editors. *Perspectives in Carbonate Geology.* Wiley Blackwell, 2009.

Talley, Lynne D., George L. Pickard, William J. Emery, and James H. Swift. *Descriptive Physical Oceanography: An Introduction.* 6th ed., Elsevier, 2011.

Williams, Richard G., and Michael J. Follows. *Ocean Dynamics and the Carbon Cycle: Principles and Mechanisms.* Cambridge UP, 2011.

Carbonate Minerals

Fields of Study: Geology; Mineralogy

ABSTRACT

Calcite, a common carbonate mineral, dominates the metamorphic rock marble and the sedimentary rock limestone. Calcite has been used in building (cement, structural and ornamental stone), as a flux in smelting various types of metal ores, in agriculture, in the chemical industry for the manufacture of various products, for polishing, and as a filler in paint and rubber. Other carbonate minerals are used as ores of various metals, in manufacturing, as ornamental stone, or in jewelry.

KEY CONCEPTS

calcrete, caliche: commonly called "hardpan," sand and other particulates that have become cemented together by the accretion of calcium carbonate in the interstices between the particles

metamorphic: rocks whose original nature has been altered by the action of heat and pressure

polymorphs: different crystal forms having the same chemical composition

rhombohedral cleavage: cleavage pattern of a mineral crystal structure that occurs according to the form of a rhombohedron

travertine: calcium carbonate deposits formed by deposition from mineral-laden water in hot springs and caves

tufa: calcium carbonate deposits formed by deposition from mineral-laden water from hot springs

WHERE FOUND

Calcite, a common carbonate mineral, dominates the metamorphic rock marble and the sedimentary rock limestone. It also occurs in cave and hot spring deposits, some dry lake deposits (including oolitic sands of Great Salt Lake, Utah), and modern marine sediment in some tropical areas such as the Great Bahama Bank, Florida, Mexico, the Persian Gulf, and Australia. Shells of many marine invertebrates are made of calcium carbonate (including corals, molluscs such as bivalves and snails, echinoderms such as sand dollars and sea urchins, and planktonic organisms whose microscopic shells accumulate to form chalk). In arid climates calcium carbonate accumulates in soil to form calcrete or caliche (hardpan). Carbonates other than calcium carbonate occur in sedimentary deposits and in association with ore veins.

PRIMARY USES

Calcite has been used in building (cement, structural and ornamental stone), as a flux in smelting various types of metal ores, in agriculture, in the chemical industry for the manufacture of various products, for polishing, and as a filler in paint and rubber. Other carbonate minerals are used as ores of various metals, in manufacturing, as ornamental stone, or in jewelry.

TECHNICAL DEFINITION

Carbonate minerals contain the carbonate anion, $(CO_3)^{-2}$, in their chemical formula. There are approximately sixty carbonate minerals, but many are rare. Among the more common carbonates are calcite and aragonite (both $CaCO_3$), dolomite ($CaMg(CO_3)_2$), magnesite ($MgCO_3$), and siderite ($FeCO_3$). Carbonate minerals effervesce in hydrochloric acid, but with some carbonates, the acid must be hot or the mineral must be powdered to obtain the reaction. Most carbonates are soft, and rhombohedral cleavage is common.

DESCRIPTION, DISTRIBUTION, AND FORMS

Carbonate minerals may be divided into three groups, each of which has a similar crystal structure: the calcite group, the dolomite group, and the aragonite group. Some carbonate minerals are "polymorphs" of one another, with identical chemical formulas but different crystal structures. An example is $CaCO_3$, which exists in nature as three different crystal structures: calcite (hexagonal system), aragonite (orthorhombic system), and vaterite (hexagonal, also called "μ-calcite").

The calcite group belongs to the hexagonal crystal system, hexagonal-scalenohedral class. This group includes calcite ($CaCO_3$), magnesite ($MgCO_3$), siderite ($FeCO_3$), rhodochrosite ($MnCO_3$), and smithsonite ($ZnCO_3$). The dolomite group belongs to the hexagonal crystal system, rhombohedral class. This group includes dolomite ($CaMg(CO_3)_2$) and ankerite ($CaFe(CO_3)_2$). The aragonite group belongs to the orthorhombic crystal system, rhombic-dipyramidal class. This group includes aragonite ($CaCO_3$), witherite ($BaCO_3$), strontianite ($SrCO_3$), and cerussite ($PbCO_3$).

The basic copper carbonates, malachite ($Cu_2CO_3(OH)_2$) and azurite ($Cu_3(CO_3)_2(OH)_2$), belong to the monoclinic crystal system, prismatic class. Other monoclinic carbonates are trona [$Na_3H(CO_3)_2 \bullet 2(H_2O)$], hydromagnesite [$(Mg_5(CO_3)_4(OH)_2 \bullet 4(H_2O))$], and artinite [$Mg_2(CO_3)(OH)_2 \bullet 3H_2O$].

The most abundant carbonate mineral is calcite ($CaCO_3$), which comprises limestone, chalk, travertine, tufa (sedimentary rocks), and marble (metamorphic rock). Most limestone forms in warm, shallow seas, far from sources of land-derived sediment. Chalk is made of the shells of microscopic floating organisms which once lived in the sea. Spring deposits are travertine or tufa, and cave deposits (stalactites and stalagmites) are travertine. These deposits form from the evaporation of groundwater carrying dissolved calcium carbonate. Marble is limestone which has been changed by heat and pressure. Mala-

Rhodochrosite, Sweet Home Mine, Alma, Colorado. Photo by Rob Lavinsky, iRocks.com, CC-BY-SA-3.0, via Wikimedia Commons.

chite and azurite are associated with the oxidized portions of copper deposits and with copper veins through limestone deposits.

Sodium carbonate minerals are present in association with dry salt lake deposits in some parts of the world. These include trona, natron [$(Na_2CO_3)\bullet 10H_2O$)], thermonatrite [$(NaCO_3\bullet H_2O)$], nahcolite ($NaHCO_3$), gaylussite [$(Na_2(CO_3)_2\bullet 5H_2O)$], pirssonite [$(Na_2(CO_3)_2\bullet 2H_2O)$], and shortite [$(Na_2Ca_2(CO_3)_3$].

OBTAINING CARBONATE MINERALS

The most important use of calcite is in the production of cements and lime. When limestone is heated to about 900° Celsius, it loses CO_2 and is converted to quicklime or lime (CaO). Mixed with sand, quicklime forms mortar. When mixed with water, it hardens or "sets," swelling and releasing heat. The most widely produced cement is Portland cement (used in concrete), which is generally made from limestone and silica- and alumina-bearing material such as clay or shale. The raw materials are ground together, and the mixture is heated in a kiln until it fuses into a "clinker," which is then crushed to a powder.

USES OF CARBONATE MINERALS

Lime (CaO) is also used in agriculture to neutralize acid in soils, in the manufacture of paper, glass, and whitewash, and in tanning leather. It is used in refining sugar, as a water softener, and as a flux for smelting various types of ores. Fine-grained limestone has been used in lithography (printing). Blocks of cut limestone and marble are used as building stone and ornamental stone and may be polished. Crushed limestone is used as aggregate in concrete and as road-building material. Dolomite has uses similar to those of calcite.

Several carbonates are metal ores: dolomite and magnesite (ores of magnesium), rhodochrosite (manganese), siderite (iron), smithsonite (zinc), strontianite (strontium), witherite (barium), cerrusite (lead), malachite and azurite (copper), and trona (sodium).

Magnesite is used in the manufacture of refractory materials capable of withstanding high temperatures, for special types of cements, and in the paper, rubber, and pharmaceutical industries. Strontianite is also used in the manufacture of fireworks, producing a purplish-red flame. Malachite (green) and azurite (blue) are used as pigments. Sodium carbonate and sodium bicarbonate are important in the manufacture of washing soda (or sal soda) and are used as cleaning agents and water softeners. They are used in the manufacture of glass, ceramics, paper, soap, and sodium-containing compounds (such as sodium hydroxide) as well as in petroleum refining. Sodium bicarbonate, also known as baking soda, is an important part of baking powder, is a source of carbon dioxide in fire extinguishers and is used medicinally to neutralize excess stomach acid. Several carbonates are used as ornamental stone and in jewelry, including malachite, azurite, aragonite (alabaster), rhodochrosite, and smithsonite.

—Pamela J.W. Gore

Further Reading

Alexander, Pramod O. *A Handbook of Minerals, Crystals, Rocks and Ores.* New India Publishing, 2009.

Klein, Cornelis, and Barbara Dutrow. *The Twenty-third Edition of the Manual of Mineral Science.* J. Wiley, 2008.

Schumann, Walter. *Minerals of the World.* 2nd ed., Sterling, 2008.

Tegethoff, F. Wolfgang, editor. *Calcium Carbonate from the Cretaceous Period into the 21st Century.* Springer, 2001.

CARBONATITES

Fields of Study: Geology; Volcanism; Geochemistry

ABSTRACT

Carbonatites are composed of carbonate minerals that appear to have formed from carbonate liquids. They typically contain many minerals of unusual composition that are seldom found elsewhere. Carbonatites have been mined for a variety of elements, including the rare-earth elements niobium and thorium. The rare-earth elements are used as phosphors in cathode ray tubes and in high-quality magnets; niobium is used in electronics and in steel to make it resist high temperatures.

KEY CONCEPTS

calcite: a mineral composed of calcium carbonate

dike: a tabular igneous rock formed by the injection of molten rock material through another solid rock

dolomite: a mineral composed of calcium magnesium carbonate

igneous rocks: rocks formed from liquid or molten rock material

ijolite: a dark-colored silicate rock containing the minerals nepheline (sodium aluminum silicate) and pyroxene (calcium, magnesium, and iron silicate)

isotopes: different atoms of the same element that have different numbers of neutrons (neutral particles) in their nuclei

limestone: a sedimentary rock composed mostly of calcium carbonate formed by organisms or by calcite precipitation in warm, shallow seas

mineral: a naturally occurring element or compound with a more or less definite chemical composition

rock: a naturally occurring consolidated material that usually consists of two or more minerals; sometimes, as in carbonatites, rocks may consist mainly of one mineral

silicate mineral: a mineral composed of silicon, oxygen, and other metals, such as iron, magnesium, potassium, and sodium

COMPOSITION OF CARBONATITES

Carbonatites are unusual igneous rocks because they are not primarily composed of silicate minerals, as are most other rocks formed from molten rock material. Instead, carbonatites are composed mostly of carbonate minerals and of minor amounts of other minerals that are rare in other rocks. The carbonate minerals composing carbonatites are usually calcite (calcium carbonate) or dolomite (calcium magnesium carbonate). The other minerals found in carbonatites often contain large concentrations of elements that rarely become concentrated enough to form these minerals in other rocks.

Carbonatites are often associated with rare silicate rocks injected at about the same time as the carbonatites. These silicate rocks are unusual, as they normally lack feldspars (calcium, sodium, and potassium aluminum silicate minerals), which are abundant in most igneous rocks formed within Earth. Instead of feldspar, many of these silicate rocks contain varied amounts of the minerals nepheline (sodium aluminum silicate) and clinopyroxene (calcium, magnesium, iron silicate) and minor amounts of minerals not commonly found in other igneous rocks. The silicate rock consisting of more or less equal amounts of nepheline and clinopyroxene is called "ijolite."

OCCURRENCE OF CARBONATITES

Carbonatites often occur as small bodies of varying size and shape that cut across the surrounding rocks. Often, they occur as dikes that may be only a few feet to tens of feet wide, but they may be greater in length. An example of carbonatites occurring as

Carbonatite from Jacupiranga, Brazil. This rock is a mixture of calcite, magnetite and olivine. Photo by Eurico Zimbres Zimbres, CC BY-SA 3.0, via Wikimedia Commons.

dikes is found at Gem Park near Westcliffe, Colorado. Carbonatites sometimes occur as somewhat equidimensional bodies that are larger than those forming dikes. The Sulfide Queen carbonatite at Mountain Pass, California, for example, is about 800 meters long and 230 meters wide. Often, carbonatites contain foreign rock fragments. The cross-cutting relations and foreign rocks found within the carbonatites suggest that they form by the injection or intrusion of molten carbonate into the surrounding solid rock. The foreign rock fragments within the carbonatite could be solid rocks ripped off the walls by the moving molten carbonate material as it was injected through solid rock.

The occurrence of molten carbonate at the active volcano at Oldoinyo Lengai in northern Tanzania, Africa, confirms that such material exists. The abundance of recently formed volcanic material composed of carbonatite at Oldoinyo Lengai and other locations may indicate that some carbonatite can form near the surface. The carbonatite liquids at the surface may flow as lava out of the volcano or be ejected explosively into the air, similar to that of other volcanoes formed mostly from silicate liquids. Carbonatites of volcanic origin are especially abundant in Africa along a portion of the continent called the" East African rift zone." In this location, Africa is slowly being split apart by forces driven by the convection of magma within Earth. Some carbonatites that intruded below the surface as dikes may also at one time have fed volcanoes at the surface. Erosion of the extinct volcano and associated silicate rocks may have exposed dikes composed of solidified carbonatite.

COMPLEXES

The carbonatites, ijolites, and other associated igneous rocks that formed at the same place and were injected at about the same time are called "complexes." These complexes have small areas of exposure at the surface; most are exposed over a surface area of only about 1 to 35 square kilometers. The Magnet Cove complex in Arkansas, for example, has about a 16-square-kilometer exposure. The Gem Park complex in Colorado is only about 6 square kilometers in area. The associated silicate rocks usually compose most of the area of the exposed complex; only a small portion is carbonatite. The overall shape of these complexes is often circular, elliptical, or oval, but departures from these shapes are common. Often, the different rock types within a complex are built of concentric zones much like the layers of an onion.

Examples of these complexes include Seabrook Lake, Canada, where the complex is roughly circular and is about 0.8 kilometer across. Its "tail," however, extends from the main circular body to about 1.2 kilometers to the south. The central core of carbonatite, composed mostly of calcite, is about 0.3 kilometer across. Smaller carbonatite dikes can be found within other rocks. The largest carbonatite body is surrounded by a dark rock containing angular blocks composed mostly of carbonate, clinopyroxene, or biotite (a dark, shiny potassium, iron, magnesium, and aluminum silicate). This latter body is surrounded by a mixture of ijolite and pyroxenite (a rock containing only pyroxene). The ijolite and pyroxenite also compose much of the tail to the south.

Another complex is located at Gem Park in Colorado. Gem Park is oval in shape, roughly 2 by 3.3 kilometers. There is no central carbonatite there. Instead, scores of small, dolomite-rich carbonatites have intruded as dikes across the silicate rocks of the complex. The main silicate rocks of the complex are pyroxenite and a feldspar-clinopyroxene rock called "gabbro." The pyroxenite and gabbro form concentric rings with one another. The large amount of gabbro makes this complex unusual, as gabbros are seldom present with carbonatite in the same complex.

The complexes at Gem Park and Seabrook Lake are rather simple, as they contain very few rock types. A wide variety of minerals can be found in some complexes, resulting in a large number of rock types. The Magnet Cove complex in Arkansas, for example, has twenty-eight major rock types listed on the geologic map. (The reading accompanying the map extends the total rock types to an even greater number.) A large proportion of the rock names in geology are generated by the wide variation of minerals found in these complexes that compose merely a tiny portion of the earth's surface.

HYPOTHESES OF CARBONATITE FORMATION

Up until the 1950s, geologists believed that carbonatites were limestones that melted and were intruded as molten rock material, or that circulating waters formed carbonatites by replacing silicate minerals with carbonate minerals. Some geologists thought that carbonatites could have formed from limestones that were remobilized by a solid, plastic flow—much like the flow of toothpaste squeezed out of a tube.

A major problem with the suggestion that carbonate material could melt, however, was the apparently high melting point of pure calcite or dolomite. Few geologists could believe that the temperature within Earth was high enough to melt calcite or dolomite. Another problem concerning the belief that limestones were the source of carbonatite was that in some areas, no limestones could be found anywhere near the occurrences of carbonatites. Also, the concept of the intrusion of limestones by the plastic flow of a solid was difficult to reconcile with many observations of carbonatites, including the occurrence of foreign igneous rock fragments composed of silicate minerals within them. The absence of fossils in the "limestone" also was noted as unusual, as most limestones have abundant fossils. The lack of fossils could be explained, however, by the melting hypothesis: the fossil evidence would have been destroyed during the melting.

Experiments in furnaces at temperatures and pressures similar to those expected deep within Earth have done much to support the melting hypothesis for the igneous formation of carbonatites. Several experiments in the late 1950s showed that carbonate minerals could melt at reasonably low temperatures (about 600 degrees Celsius) if abundant carbon dioxide and water vapor coexisted with the carbonate minerals. This dispelled the notion that molten carbonate could not exist within the earth. Also, the discovery in 1960 that the Oldoinyo Lengai volcano was extruding carbonate lavas confirmed that carbonate liquids could exist within Earth. Similar experiments at high temperature and pressure relative to the composition of carbon dioxide or water vapor suggested that their composition could not produce carbonatites by the replacement of silicate minerals with carbonate minerals.

These experiments, combined with field observations (including the way carbonatites cross-cut surrounding rocks and the presence of foreign rock fragments), confirmed that most carbonatites formed by the intrusion of carbonate liquids. Even so, the experiments fell short of dispelling the notion that carbonate liquids could have been derived from melted limestone. The melted limestone hypothesis met objections, because isotopic and element concentrations in carbonatites were much different from those observed in limestone. For example, the elements lanthanum and niobium are hundreds of times more concentrated in carbonatites than they are in limestone. There is no known process that can produce this magnitude of element enrichment, either by melting or leaching the carbonate liquid from the solid rock through which it moved. Such observations have caused the limestone origin of carbonatites to be rejected by most geologists.

CONTINUING STUDY OF CARBONATITE FORMATION

Scientists continue to try to understand how carbonatites form. Experiments in furnaces suggest that some carbonate liquids may separate from some silicate liquids similar to those occurring with carbonatites. This process would be like the separation of oil and water as immiscible liquids. Other experiments in furnaces suggest that rocks more than 80 kilometers deep within Earth may melt in small amounts and produce the carbonate liquids and associated silicate liquids similar in composition to those observed in the natural rocks. Although these experiments fail to prove that carbonate liquids form in these ways, several other lines of evidence have convinced geologists that either of these possibilities could produce carbonatites in nature. For example, some possible source rocks that could melt and produce carbonatites or associated silicate rocks are sometimes carried up with lava from deep within Earth. The strontium-87 (^{87}Sr) to strontium-86 (^{86}Sr) ratios and the element contents of these possible source rocks have been measured and are similar to those expected for rocks that would melt and produce carbonate and silicate liquids. Strontium-87 forms from the decay of rubidium-87 (^{87}Rb), and the ratio of strontium-87 to strontium-86 is a measurement of the age of the possible source rocks, and can help to identify the rocks that gave rise to the carbonatite magma.

Geologists want to understand how carbonatites form partly because of their economic importance. With this knowledge, they can design better strategies to find carbonatites or the associated silicate rocks that are as yet undiscovered. For example, carbonatites are very small targets to find on the surface of Earth, but if the silicate rocks associated with carbonatites contain abundant magnetic minerals, an aerial survey of the area can detect magnetic fields due to the rocks by using a magnetometer device. Once geologists find areas with magnetic anomalies, they can collect soil or stream samples in the area to see if any unusual minerals associated with carbonatites or the associated silicate rocks are present. They can also drill the area to see if any carbonatites are below the surface.

ECONOMIC VALUE OF CARBONATITES

The unusually high concentrations of some elements in certain minerals make carbonatites potential ores for these elements. Carbonatites have high concentrations of niobium, thorium, and the rare-earth elements of lower atomic number (such as lanthanum and cerium). Iron, titanium, copper, and manganese also have been mined from carbonatites.

Niobium has been economically extracted from the mineral pyrochlore at Fen, Norway, and at Kaiserstuhl, Germany. Niobium is used as an alloy in steel to render it resistant to high temperature; this steel is then used in gas turbines, rockets, and atomic power plants. Rare-earth elements have been mined from the carbonatite at Mountain Pass, California. The reserves of rare-earths are enormous at Mountain Pass (averaging about 7 percent) compared to other carbonatites. This carbonatite hosted the largest rare-earth mine in the United States. The mine closed for a few years, but rising demand for rare-earth elements prompted plans to reopen in 2011. Other carbonatites enriched with rare-earths occur in Malawi in Africa. The rare-earth elements are concentrated in many minerals, including perovskite, monazite, xenotime, and a variety of rare-earth carbonate minerals. The rare-earths are used as color phosphors and as components in high-quality magnets. Thorium, often enriched along with the rare-earths in many carbonatites, tends to concentrate in the same minerals and deposits as do the rare-earth elements. Thorium is radioactive and has been used as a source of atomic energy. It has also been used for the manufacture of mantles for incandescent gas lights. The Oka Complex west of Montreal, Canada, has also been mined for rare-earth elements. Rising demand for rare-earths for electronics, plus dependency on China as the leading refiner of rare-earths, has spurred additional exploration for new resources.

—Robert L. Cullers

Further Reading

Bell, Keith, and Jörg Keller, editors. *Carbonatite Volcanism: Oldoinyo Lengai and the Petrogenesis of Natrocarbonatites.* Springer Science + Business Media, 2012.

Faure, Gunter. *Origin of Igneous Rocks: The Isotopic.* Evidence Springer, 2010.

Grotzinger, John, and Thomas H. Jordan. *Understanding Earth.* 6th ed., W. H. Freeman, 2010.

Jerram, Dougal, and Nick Petford. *The Field Description of Igneous Rocks.* 2nd ed., Wiley-Blackwell, 2011.

Kresten, Peter, and Valentin R. *Troll: The Alnö Carbonatite Complex, Central Sweden.* Springer, 2018.

Winter, John D. *An Introduction to Igneous and Metamorphic Petrology.* Prentice Hall, 2001.

Woolley, Alan R. *Alkaline Rocks and Carbonatities of the World, Parts 1-4.* Geological Society, 2019.

CLAYS

Fields of Study: Geology; Soil Science; Geochemistry; Sedimentology

ABSTRACT

The term "clay" may be used to describe a group of fine-grained minerals, a type of rock, or a range of particle size, generally less than four micrometers. As a rock term, clay is generally understood to mean an earthy, fine-grained material formed largely of crystalline minerals known as the clay minerals.

KEY CONCEPTS

fuller's earth: the type of clay used in "fulling" wool by removing lanolin and other contaminants

lenticular: lens-shaped

octahedral sheet: the two-dimensional layer consisting of octahedral arrangements of silica structural units

slurries: mixtures in which fine solid particles are mixed with an amount of liquid to allow the material to flow

tetrahedral sheet: the two-dimensional layer consisting of tetrahedral arrangements of silica structural units

BACKGROUND

Clays can be found throughout the world, but economically valuable deposits are limited in extent and distribution. Major kaolin deposits in the United States are found in Georgia and South Carolina. The world's major bentonite deposits are found in Wyoming and Montana, and large fuller's earth deposits can be found in Georgia and Florida. Ball and refractory clays are abundant in Kentucky and Tennessee.

Clays are used in a number of applications requiring the incorporation of fine-grained materials that contribute to a product's physical or chemical properties. Uses include fillers in paint, paper, and plastics, additives to drilling muds, the manufacture of ceramics and brick, carriers for pesticides and insecticides, the manufacture of catalysts, and cosmetic and pharmaceutical uses.

Clays are considered "industrial minerals," a group of minerals composed of geological materials having commercial value and of a nonmetallic, nonfuel character. They may be marketed in a natural, as-mined state or as processed materials. Clays can vary widely in composition and physical characteristics. Certain similarities exist among a number of clays, however, and they can be categorized in broad terms as kaolin, bentonite or fuller's earth, ball clay, and refractory clay based on similarities in either composition or functional performance. Clays that do not fall into any of the major categories are generally referred to as common clay or shale.

MINERALOGY AND CHEMISTRY

Clays are hydrous (water-containing) aluminum silicates containing ions of alkalies or alkaline earth elements. Magnesium or iron may substitute wholly or partially for aluminum in the clay mineral structure. Clay minerals are composed of alternating layers of two different atomic structures. The first is an aluminum-bearing octahedral sheet structure, and the second is a layer of silica tetrahedrons. The aluminum and silicon atoms are chemically bonded to oxygen in these layers, which are held to one another by weaker electrostatic bonds. Interlayer sites in many clays contain water molecules or cations such as calcium, sodium, potassium, magnesium, lithium, or hydrogen. The presence or absence of interlayer molecules affects both the physical and chemical properties of the clay.

KAOLIN

Kaolin is a clay consisting predominantly of pure kaolinite or related clay minerals. Most major deposits of kaolin are referred to as either primary (residual) or secondary. Primary deposits are formed in place as the weathering products of granite or other feldspar-rich rocks. Other minerals associated with deposits of this type include quartz, micas, amphiboles, tourmaline, and unweathered feldspars. Primary deposits are irregular in shape, grading downward into unaltered parent (source) rock.

Secondary deposits of kaolin are sedimentary accumulations of kaolinitic material that has been transported from its source area. Deposits of this type may contain up to 95 percent kaolinite; in contrast, primary deposits may contain as little as 10 percent kaolinite. Associated minerals may include quartz, micas, other clay minerals, and a variety of high-density "heavy minerals." Secondary deposits are generally lenticular or tabular in shape, with thicknesses up to sixty meters and areal dimensions of up to about two kilometers.

Kaolin is also found as a product of hydrothermally altered rocks. Deposits of this nature are of limited size and extent. They occur as irregularly elongated pods or pipelike bodies along faults, joints, and other conduits along which hot solutions have flowed. Kaolin is generally soft and plastic, although harder silica-bearing varieties also exist. Crystals of kaolinite are hexagonal, composed of individual platelets stacked in an accordion-like manner. There is little ionic substitution in the crystal lattice.

Kaolin has numerous industrial uses and is perhaps best known for its use in the manufacture of china and porcelain. Its chemical inertness, high brightness, white color (either naturally or resulting from processing and beneficiation), and crystal shape make it useful in other applications as well.

Kaolin is used as a filler or coating in the manufacture of paper, as a filler in paint, plastics, and pharmaceuticals, and in the manufacture of rubber, tile, brick, ink, adhesives, detergents, cosmetics, pencils, pastes, and other consumer products.

BALL CLAY AND REFRACTORY CLAY

Ball clays are composed of up to 70 percent kaolin. They generally occur in secondary sedimentary deposits characterized by the presence of organic matter along with varying amounts of other clays, quartz, feldspar, calcite, and heavy minerals. Sedimentary deposits of ball clay represent accumulations of clay materials that were derived from a number of sources and that were deposited in nonmarine environments. Most deposits are lenticular, with areal dimensions of up to 850 meters and thicknesses of up to 10 or more meters.

Ball clays are plastic or semiplastic and are used to provide strength and malleability to ceramic bodies prior to firing. They fuse during firing, also acting as a "cement" to bind together the refractory, nonshrinking component of a ceramic body. Ball clay is used to manufacture tableware, stoneware, tiles, plumbing fixtures, and bricks. It is also used as a sealant in landfills.

Refractory clays are generally kaolin, containing only small quantities of mica or iron-bearing minerals that might combine with other materials during firing to form low-melting-point glasses. Refractory clays have a high heat resistance. Other properties that affect overall quality include shrinkage, warping, cracking, and abrasion. Refractory clays can be soft and plastic or hard like flint. They generally occur as sedimentary deposits that are lenticular or tabular in shape. They are mined for use in the manufacture of firebrick, insulating brick, and other heat-resistant clay products.

BENTONITE AND FULLER'S EARTH

"Bentonite" is generally understood to mean a clay consisting of minerals from the montmorillonite group, regardless of origin or occurrence. The most important commercial montmorillonites are the sodium and calcium varieties. Sodium montmorillonite (Wyoming or western bentonite) has high swelling capabilities when added to water. Calcium montmorillonite (southern bentonite) has a lower swelling capacity, and it generally crumbles when added to water. Other montmorillonites include those rich in lithium (hectorite), magnesium (saponite), or iron (nontronite).

Bentonite can be both physically and chemically reactive. It shrinks or swells as it releases or absorbs interstitial water or organic molecules, and it has important cation exchange and chemical sorption properties. Bentonite's physical and chemical properties account for its usefulness in modifying fluid viscosity or plasticity; it also has a variety of other uses. Bentonites can be modified through chemical treatment to enhance selected physical or chemical properties.

Wyoming bentonites are suitable for use in an as-mined condition. They are used as an additive in drilling mud to increase viscosity and aid in the removal of drillhole cuttings. The clay also helps maintain cuttings in suspension and creates an impervious coating on the wall of the drillhole to prevent fluid loss during drilling. Southern bentonites can be modified to have properties similar to Wyoming bentonites, but their use is generally restricted to other applications such as binding iron ore during pelletizing and the manufacture of catalysts and no-carbon-required (NCR) copy papers. Bentonites are also used to refine, decolor, and purify oils and beverages; to manufacture fire retardants; and as hydraulic barriers.

"Fuller's earth" refers to clays (generally bentonitic) suitable for bleaching and absorbent or other special uses. The term was first used to describe materials used for cleansing or fulling wool (removing lanolin and dirt), but it is now used more broadly to include decolorizers or purifiers in filtering applications. Fuller's earth products include cat litter, pesticide and insecticide carriers, soil conditioners, light-weight aggregate, and pharmaceuticals.

MINING AND PROCESSING

Most clay deposits are mined from open pits, although some are mined by underground methods. Open-pit mining generally involves the stripping of overburden, excavation of the clay, and transport of mined material to the processing plant. Some operations may require blasting.

The simplest operations involve excavation, transport to the plant, drying, and shipment to the customer. More complex operations may require that the mined material first be put into a slurry form for removal of grit or sand, with transport to the plant by pipeline. Clay slurries can be chemically or physically treated to remove contaminants that contribute to discoloration or poor chemical or physical performance. They can then be filtered and dried prior to packaging and shipment to the customer. Some clays are put back into slurry form prior to shipment, depending upon a customer's needs.

—Kyle L. Kayler

Further Reading

Bergaya, Faïza, Benny, K. G. Theng, and Gerhard Lagaly, editors. *Handbook of Clay Science.* Elsevier, 2006.

Cavallaro, Giuseppe, Rawil Fakhrullin, and Pooriah Pasbakhsh, editors. *Clay Nanoparticles: Properties and Applications.* Elsevier, 2020.

Chatterjee, Kaulir Kisor. "Clay." *Uses of Industrial Minerals, Rocks, and Freshwater.* NOVA Science Publishers, 2009.

Humphrey, Justin P., and Daniel E. Boyd, editors. *Clay: Types, Properties and Uses.* NOVA Science Publishers, 2011.

Kogel, Jessica Elzea, et al., editors. "Clays." In *Industrial Minerals and Rocks: Commodities, Markets, and Uses.* 7th ed., Society for Mining, Metallurgy, and Exploration, 2006.

Meunier, Alain. *Clays.* Trans. Nathalie Fradin. Springer, 2010.

Morari do Nascimento, Gustavo, editor. *Clays, Clay Minerals and Ceramic Materials Based on Clay Minerals.* IntechOpen, 2016.

Murray, Haydn H. *Applied Clay Mineralogy: Occurrences, Processing, and Application of Kaolins, Bentonites, Palygorskite-Sepiolite, and Common Clays.* Elsevier, 2007.

Wesley, Liam R. *Clays and Clay Minerals: Geological Origin, Mechanical Properties and Industrial Applications.* NOVA Science Publishers, 2014.

Clays and Clay Minerals

Fields of Study: Geology; Geochemistry; Sedimentology

ABSTRACT

Clays are fine-grained materials with unique properties, such as plastic behavior when wet. They form by the weathering of silicate rocks at the earth's surface, by diagenetic reactions, and by hydrothermal alteration. An understanding of clays is important to solving problems in petroleum geology, engineering, and environmental science.

KEY CONCEPTS

authigenic minerals: minerals that formed in place, usually by diagenetic processes

cation exchange capacity: the ability of a clay to adsorb and exchange cations, or positively charged ions, within its environment

chemical weathering: a change in the chemical and mineralogical composition of rocks by means of reaction with water at the earth's surface

detrital minerals: minerals that have been eroded, transported, and deposited as sediments

diagenesis: the conversion of unconsolidated sediment into consolidated rock after burial by the processes of compaction, cementation, recrystallization, and replacement

hydrolysis: a chemical weathering process that produces clays by the reaction of carbonic acid with aluminosilicate minerals

phyllosilicate: a mineral with silica tetrahedra arranged in a sheet structure

shale: a sedimentary rock with a high of clays

DEFINITION AND PROPERTIES

The definition of clays varies depending on the scientific discipline or application. An engineer's definition, which is based on particle size, differs from a mineralogist's definition, which is based on crystal structure. In the broadest sense, clays are materials that have a very fine grain size (less than 0.002 millimeter) and behave plastically when wet. A more specific definition of a clay mineral is a hydrous aluminum phyllosilicate, or, more simply stated, a mineral that contains water, aluminum, and silicon, and has a layered structure. The term "clays" will be used for the broad definition, and "clay mineral" for the specific definition. Rock flour, or material that was ground to a fine powder by glaciers, would fit the definition of clays; however, it may contain minerals such as quartz and feldspars that do not fit the definition of a clay mineral. Mica is a hydrous aluminum phyllosilicate, but it often occurs as large crystals, so it does not fit the definition of clays. Certain minerals such as zeolites and hydroxides (goethite and gibbsite) have a very fine grain size and physical properties similar to clay minerals, so they are often included with the clay minerals. Unique properties of clays, including their plastic behavior when wet and ability to adsorb water and ions in solution, can be attributed to their small crystal size, large surface area, and unique crystal structure.

CLAY MINERAL STRUCTURE

There are two basic elements of the clay mineral structure: a tetrahedral sheet and an octahedral sheet. A silicon atom surrounded by four oxygen atoms forms the basic building block of all silicate minerals, the four-sided silica tetrahedron. In phyllosilicate minerals, the silica tetrahedra are linked together by sharing the three oxygen atoms at the corners of the tetrahedra, forming a continuous sheet. The tetrahedral sheet has a negative charge and a general chemical formula of $Si_4O_{10}^{4-}$.

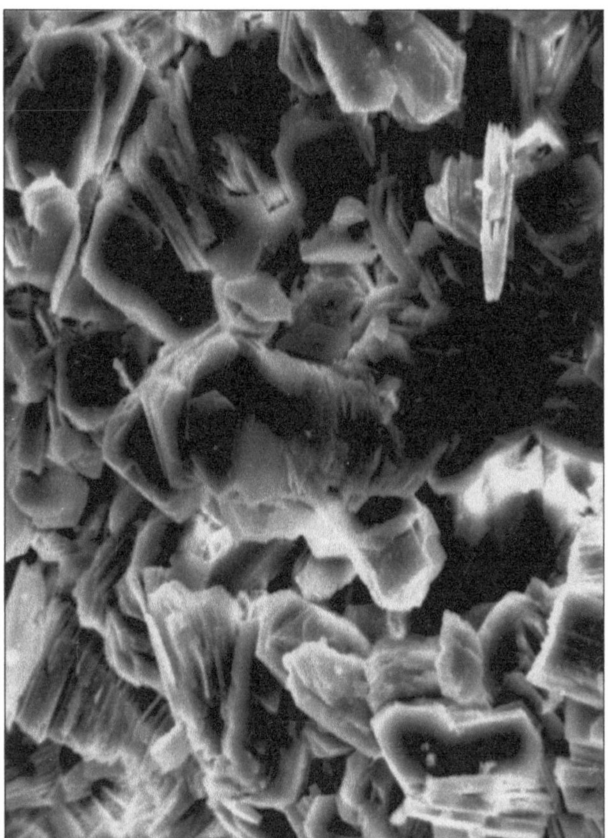

Hexagonal sheets of a clay mineral (kaolinite). Photo via Wikimedia Commons. [Public domain.]

The octahedral sheet consists of a sandwich of positively charged atoms (cations), like magnesium, iron, or aluminum, between sheets of oxygen and hydroxyl anions, or negatively charged ions (OH^-). Each cation is surrounded by six anions that lie at the corners of an octahedron, a shape made of eight triangles. The octahedra are linked together to form a sheet by sharing the anions on the edges of the octahedra.

There are two types of octahedral sheets: trioctahedral and dioctahedral. The prefixes refer to the fraction of octahedra occupied by cations: 2/3 for dioctahedral and 3/3 for trioctahedral. The trioctahedral sheet is composed of divalent cations such as Mg^{2+} and Fe^{2+}. For every six hydroxyl anions, the trioctahedral sheet contains three cations, resulting in a sheet where all the available octahedral sites contain a cation. The dioctahedral sheet contains trivalent cations such as Al^{3+} and Fe^{3+}. The dioctahedral sheet has only two cations per six hydroxyl anions, resulting in only two-thirds of the available octahedral sites being filled with cations. The general chemical formulas for the trioctahedral and dioctahedral sheets are Mg_3OH_6 and Al_2OH_6, respectively.

Layers in clay minerals are made of combinations of tetrahedral and octahedral sheets. The unshared oxygen of the silica tetrahedra take the place of some of the hydroxyl anions in the octahedral sheet, resulting in neutralization of the negative charges on the tetrahedral sheet. There are two types of layers: a 1:1 or T-O layer made up of one tetrahedral sheet and one octahedral sheet, and a 2:1 or T-O-T layer, which contains one octahedral sheet sandwiched between two tetrahedral sheets. Not all minerals with these layered structures are considered clay minerals.

The prototype clays have neutral layer charges, with the layers held together by weak van der Waals bonds where positive portions of one sheet are attracted to negative portions of another sheet. Kaolinite and serpentine are prototype 1:1 minerals. Kaolinite is dioctahedral and serpentine is trioctahedral. Kaolinite, a pure white clay, is the major constituent of fine porcelain. Serpentine is not generally considered a clay mineral. It can occur as chrysotile asbestos, which was widely used as insulation material before it was recognized as a health hazard. Talc and pyrophyllite are the trioctahedral and dioctahedral 2:1 prototype minerals. The waxy or slippery feel of talc (also not a clay mineral) is the result of cleavage along the weak van der Waals bonds between layers.

Because of the various types of ionic substitutions in the tetrahedral and octahedral sheets, the layers may develop a negative charge, which needs to be balanced by interlayer materials. A lower-valence cation may substitute for a higher-valence cation in the tetrahedral or octahedral sheets. Layers can also develop a negative charge if some of the sites in a trioctahedral sheet are left vacant. In the mica group of minerals, which are also not considered clays, the charge on the 2:1 layer is -1 per repeat unit and is balanced by a positively charged potassium ion that occurs between the layers. Muscovite is a dioctahedral 2:1 phyllosilicate, and biotite is trioctahedral. The perfect cleavage of micas is in the direction parallel to the tetrahedral sheets and allows them to be peeled into paper-thin sheets. Micas usually occur as larger crystals and, therefore, are not considered true clay minerals; however, illite, a common clay mineral, has a structure and chemical formula similar to muscovite, a mica. The chlorite clay minerals have an extra octahedral sheet between the 2:1 layers to balance the excess negative charges. Two groups of clay minerals known as vermiculite and smectite have negatively charged layers that are balanced by hydrated cations between the layers; a hydrated cation is a positively charged ion, such as sodium, that is surrounded by water. In smectites, this water is held very loosely between the 2:1 layers and can be easily lost or gained depending on the humidity of the environment, causing these clays to shrink or swell.

FORMATION AND DEPOSITION

Clay minerals occur in soils, sediments, sedimentary rocks, and some metamorphic rocks. Sedimentary rocks cover approximately 80 percent of Earth's surface, and shales are the most common type of sedimentary rock. Because shales are composed predominantly of clays, their abundance makes clays one of the most important constituents of Earth's surface.

Most clays form through the breakdown and weathering of minerals rich in aluminum and silicon at Earth's surface. Physical weathering is the breaking and fragmentation of rocks with no change in mineralogical or chemical composition. This process can form clay-sized particles, but it does not form clay minerals. Physical weathering, however, increases the surface area of minerals, which then favors chemical weathering. A change in the chemical and mineralogical composition of rocks by reaction with water at Earth's surface is called "chemical weathering." One chemical weathering reaction that results in the formation of clays is called "hydrolysis." Water and carbonic acid, a weak acid, react with aluminum silicate minerals, resulting in the production of a clay mineral plus ions in solution. (The weak acid is called "carbonic acid"; it forms when carbon dioxide, a common gas in the atmosphere, is dissolved in rainwater.) The type and amount of clay that forms by this reaction depend on the nature of the rock being weathered (the parent material) and the intensity of weathering.

Clays produced by weathering are eventually eroded, transported, and deposited as sediments. Most clays are transported in suspension. The brown, muddy waters of rivers are a reflection of the clays being carried in suspension. Clays are deposited and accumulate in quiet water environments, where the energy is low enough to allow the clays to settle out of suspension. Several processes enhance the deposition of clays. When fresh river water mixes with salty ocean water, the negative charges on clay surfaces are neutralized, causing them to flocculate (clump together). Clay floccules—or aggregates of clay-sized particles that behave like larger silt- or sand-sized grains—rapidly settle out of suspension. Biodeposition is a process whereby organisms ingest clays with their food; the resultant fecal pellets settle to the bottom as sand-sized particles.

TYPES OF CLAY MINERALS

There are two types of clays in sedimentary rocks: detrital and authigenic. Minerals that are transported and deposited as sediments are called "detrital." Authigenic minerals form within the rocks during diagenesis, a process whereby sediments buried within Earth's crust undergo compaction and cementation into sedimentary rocks. Water expelled from the sediments during this process may react with other minerals in the sediment to form clay minerals. Kaolinite, chlorite, illite, and smectite formed by diagenesis have been observed in sandstones. Diagenesis may also result in one clay mineral being converted into another clay mineral. One reaction that commonly occurs is the alteration of smectite to form illite as a result of an increase in temperature. This reaction is important to petroleum geologists because intermediate mixed-layered illite/smectite clays form at different temperatures. The clay mineralogy of shale can be used to determine the maximum burial temperature of a rock and, in turn, help to predict the chemistry of any petroleum encountered.

Clay minerals may also form in metamorphic rocks as the result of hydrothermal alteration. Hydrothermal fluids are hot, chemically active fluids that accompany igneous intrusions. In addition to forming clay minerals as they pass through a host rock, hydrothermal fluids are responsible for producing important ore deposits such as copper ores. Economic geologists may use the distribution of hydrothermal clay minerals to locate valuable ore deposits.

IDENTIFICATION AND ANALYSIS

Clay minerals are difficult to identify and analyze because of their small crystal size. Because they can be observed neither with the naked eye nor by standard petrographic microscopes, they require the use of sophisticated equipment for their identification.

Analysis is further complicated by the fact that it is difficult to obtain pure samples of the clay minerals because they often occur as mixtures with other minerals. Before a clay mineral can be analyzed, it must be isolated from the sample by means of special physical and chemical techniques.

The tool a clay mineralogist uses most often is an X-ray diffractometer (XRD). This instrument focuses X-rays onto the sample. The crystal structure of the minerals acts as a diffraction grating, and the instrument records X-rays that are diffracted from the mineral. Mineralogists can determine the spacing between planes of atoms in the crystal structure by measuring the position of "reflections" produced by X-rays diffracted by the mineral. Clay mineralogists prepare specially oriented samples that enhance the reflections between the layers of clay minerals called "basal reflections." The basal reflections are used in determining the type of clay mineral present.

Other instruments that are used to investigate clay minerals include the scanning electron microscope (SEM) and the transmission electron microscope (TEM). The very high resolution of these microscopes allows the scientist to observe clay minerals at a very great magnification. The scientist is thus able to observe the outward crystal form of clay minerals and their texture or orientation with respect to other grains in the sample. The SEM is very helpful in distinguishing between detrital and authigenic clays in sedimentary rocks—that is, clays transported from elsewhere versus clay minerals that formed in place.

A property that is helpful in identifying clays and understanding their behavior is the cation exchange capacity (CEC). Because of their small size and unique crystal structure, clay surfaces are negatively charged and have the ability to adsorb, or attract, positive ions on their surfaces and within clays between the layers. These cations are easily exchanged with solutions. If clay that is saturated with cations is placed in a solution saturated with sodium ions, it will exchange its cations for the sodium ions. The ability of clay to adsorb and exchange cations is called the "cation exchange capacity," and depends on the type of clay mineral. Kaolinite, chlorite, and illite have relatively low CECs; smectite has a relatively high CEC. This property is especially important to soil scientists, because it controls the availability of nutrients necessary for plant growth.

INDUSTRIAL APPLICATIONS

Clays have a variety of applications in industry. They are a readily available natural resource and are relatively inexpensive. Approximately 50 million tonnes of clay materials worth more than $1 billion are used industrially each year. The petroleum industry uses kaolinite as a cracking catalyst in the refinement of petroleum, for triggering chemical reactions.

Sedimentary rocks that produce oil and gas by the heating of organic matter after it is buried are called "source rocks." The source rock for most oil and gas is shale. By determining the type of diagenetic clay minerals present, the petroleum geologist can determine if the source rock has been heated to a temperature that is high enough to produce oil or gas.

Rocks that contain oil or gas that can be easily extracted are called "reservoir rocks." The best reservoir rocks are sandstones with a high porosity and permeability. Some sandstones contain clay minerals that occur between the sand grains as a cement or within the pores. Clay minerals have the potential to reduce the porosity and permeability of a reservoir. It is important to know the type and amount of clay minerals in a reservoir in order to evaluate its quality.

A knowledge of clays is also important to engineering because the concentration of clays in a soil determines it stability. Soils that contain high percentages of smectites could cause damage to the foundations of buildings because smectites swell when saturated with water and subsequently shrink when dried out.

Clays may be used as liners for sanitary landfills because the small grain size of clays allows them to be packed together very closely, forming an impermeable layer. The liner prevents toxic leachate (which forms when rainwater reacts with solid waste) from moving out of the landfill and contaminating surface water and groundwater supplies. Chemical engineers have developed what are called "designer clays" by altering the properties of naturally occurring clays. These clays act as catalysts in the breakdown of toxic substances to form less toxic products. Designer clays are helpful in the destruction and disposal of toxic wastes such as dioxin, and in the cleaning of existing toxic waste sites.

Clays are the basic raw material of the ceramics industry. When clays are mixed with water, they become plastic and are easily molded. A hard ceramic material is produced by firing the molded clay. In addition to the familiar pottery and dinnerware, fired clays are used in the production of brick, tiles, sewer pipes, sanitary ware pottery, kiln furniture, cement, and lightweight aggregates. Kaolinite is used as a coating on fine paper, in paints, and as a filler in plastics and rubber. Swelling clays such as smectite are used as binders in animal feed and iron ore pellets (taconite), drilling muds, industrial absorbents, and pet litter.

—Annabelle M. Foos

Further Reading

Bergaya, F., B. K. G. Theng, and G. Lagaly, editors. *Handbook of Clay Science*. Elsevier, 2006.

Gates, W. P., J. T. Kloprogge, J. Madejová, and F. Bergaya, editors. *Infrared and Raman Spectroscopies of Clay Minerals*. Elsevier, 2017.

Klein, Cornelis, and Cornelius S. Hurlbut, Jr. *Manual of Mineralogy*. 23rd ed., John Wiley & Sons, 2008.

Maurizio, Galimberti, editor. *Rubber-Clay Nanocomposites*. John Wiley & Sons, 2011.

Meunier, Alain. *Clays*. Translated by Nathalie Fradin. Springer, 2010.

Morari do Nascimento, Gustavo. *Clays, Clay Minerals, and Ceramic Materials Based on Clay Minerals*. IntechOpen, 2016.

Theng, Benny K. G. *Clay Mineral Catalysis of Organic Reactions*. CRC Press, 2019.

Wesley, Liam R. *Clays and Clay Minerals: Geological Origin, Mechanical Properties and Industrial Applications*. NOVA Science Publishers, 2014.

CONTACT METAMORPHISM

Fields of Study: Geology; Geophysics; Geochemistry

ABSTRACT

Contact metamorphism is caused by the temperature rise in rocks adjacent to magmatic intrusions of local extent that penetrate relatively shallow, cold regions of the earth's crust. Many economically important metallic mineral deposits occur in contact metamorphic zones.

KEY CONCEPTS

aureole: a ring-shaped zone of metamorphic rock surrounding a magmatic intrusion

contact metamorphic facies: zones of contact metamorphic effects, each of which is characterized by a small number of indicator minerals

facies: a part of a rock or group of rocks that differs from the whole formation in one or more properties, such as composition, age, or fossil content

hornfels: the hard, splintery rocks formed by contact metamorphism of sediments and other rocks

lithology: the general physical type of rocks or rock formations

CONTACT METAMORPHIC FACIES

Contact metamorphic rocks are moderately widespread; they occur at or near Earth's surface, where magmas of all kinds intrude into lower-temperature rocks. Minerals in contact metamorphic rocks are similar to those in regional metamorphic rocks of comparable metamorphic grade. Contact metamorphic effects are divided into facies (zones), each of which is characterized by a small number of concentric indicator mineral rings surrounding an intrusive rock. Nonsymmetrical zones imply special conditions, such as less (or more) chemically or thermally reactive rocks or a nonvertical intrusive body.

Development of contact metamorphic facies reflects both the history of pressure and temperature changes and the bulk-rock chemistry. Thus, by stating that a rock belongs to a particular facies, scientists convey much about the rock's history. This information is vital to exploration for metallic and industrial minerals that commonly occur in contact metamorphic aureoles and for general understanding of regional geology.

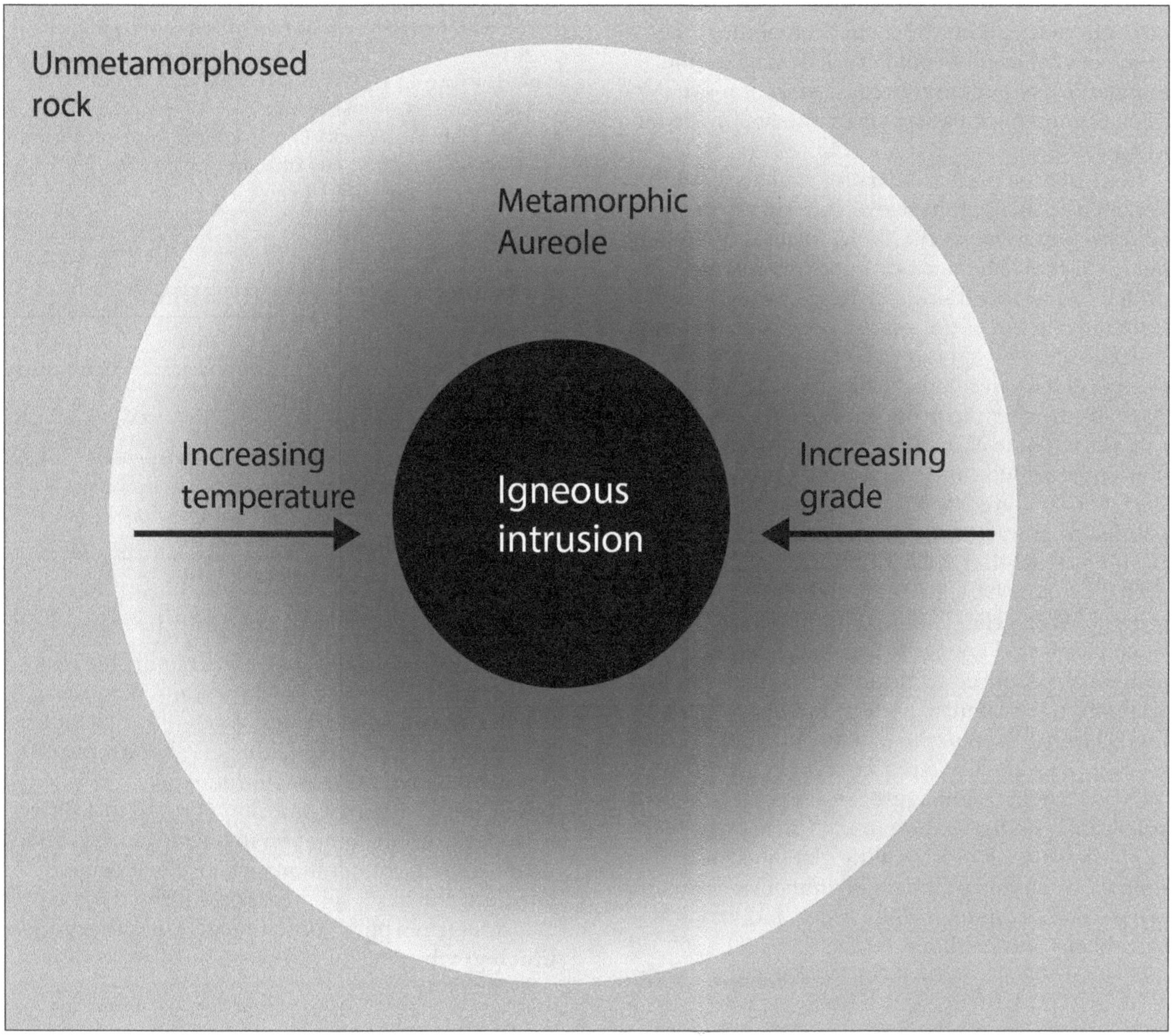

Contact metamorphism occurs often in proximity to an igneous intrusion. Image by Jillcurie, CC BY-SA 3.0, via Wikimedia Commons.

Contact metamorphic mineral facies have counterparts with regional metamorphic zones. In addition to the bulk chemical composition of a rock, temperature and pressure are two variable factors. These two factors can be independent. For example, one can find low-pressure, high-temperature facies or, alternatively, high-pressure, low-temperature facies. With contact metamorphism, these facies occur very close to the intrusive rock. With regional metamorphism, however, the effect is widespread and may not be related to an intrusive rock. It can be hard to tell if metamorphism around deep intrusive rocks is contact or regional metamorphism, or a combination of the two.

AUREOLES

Contact metamorphic rocks are recognized by their location adjacent to igneous bodies and by evidence indicating a genetic temporal relationship. Contact metamorphic rocks are commonly massive. Granitic rocks are the most common intrusive material. The most frequent depth of the solidification of a granitic magma is 3 kilometers, corresponding to a load pressure of 800 to 2,100 bars (a bar represents approximate atmospheric pressure, or pressure under 10 meters of water). There are intrusions that solidify at a greater or shallower depth; a depth of 1 kilometer corresponds to a load pressure of 250 bars. Consequently, the load pressures effective during contact metamorphism range from 200 to 2,000 bars in most cases. In contrast, load pressures prevailing during regional metamorphism are generally greater.

When a magma intrudes into colder rock structures, the adjacent rocks are heated. If the heat content of the intruded magma is high and the volume of the magma is not too small, there will be a temperature rise in the bordering rock that lasts long enough to cause mineral reactions to occur. The rocks adjacent to small intrusions of dikes and sills are not metamorphosed (only baked), whereas larger plutonic rocks give rise to a distinct contact aureole of metamorphic rocks. Several zones of increasing temperature are recognized in contact aureoles.

The contact metamorphic zones surround the intrusion in generally concentric rings that approximate the shape of the intrusion. Those zones that correspond to the highest-temperature minerals are closest to the intrusion; the zone corresponding to the lowest temperature is located farthest from the intrusion. The lowest-temperature contact metamorphic zone gradually grades into unmetamorphosed country rock.

MINERAL DEVELOPMENT

Contact metamorphic rocks are characteristically massive because of lack of deformation; most are fine-grained except for a special variant called a "skarn." Skarns may contain metallic mineralization in sufficient concentration that they can be worked for a profit. Such mineral concentrations are called "ores."

Contact metamorphic rocks lack the schistosity or platy texture characteristic of most metamorphic rocks. The very fine-grained, splintery varieties are called hornfels. The large metamorphic gradient, decreasing from the hot intrusive contact to the unaltered country rock, gives rise to zones of metamorphic rocks differing markedly in mineral constituents. The intensity and mineral assemblage in contact metamorphic zones depend on several factors: the chemical composition of the intrusive rock; its temperature and volatile content; the composition and permeability of the host units; the structural or spatial relationship of the reactive units to intense contact effects and solutions conduits, or traps; and the pressure or depth of burial.

Argillaceous (clay rich) limestone is usually susceptible to contact metamorphism, as its diverse rock chemistry allows mineral development over a broad set of physical and chemical conditions. A typical contact metamorphic mineral assemblage formed from rocks originally of this composition is a plagioclase-garnet-epidote rock. This kind of rock frequently hosts important ore deposits. Shale usually converts to hornfels, which is characterized by the minerals cordierite, biotite, and chlorite and is essentially nonreactive and nonpermeable. This lithology usually does not host ore deposits. Porous limestone is often exceptionally susceptible to solution and to replacement by ore. Rocks formed from this material contain magnetite, garnet, and pyroxene.

Pure, massive limestone has only thin, contact metamorphic zones developed in it. Pure limestone merely recrystallizes, forming coarse-grained marble in the highest-temperature contact metamorphic zones adjacent to an intrusion. Dolomite is usually poorly mineralized and nonreactive. Siliceous dolomite, however, may act as a good host, with assemblages characterized by tremolite, diopside, serpentine, and talc. Mafic rocks, such as andesites, diabases, and diorites, can be reactive hosts capable of producing some types of ore deposits. Secondary biotite is the key alteration mineral in mafic rocks and may be associated with ore. The biotite zone may be very broad.

DIAGNOSTIC MINERALS

Contact metamorphic facies are identified by the mineral assemblages that are developed in the metamorphosed rocks. From lowest to highest temperature, the contact metamorphic facies are named albite-epidote, hornblende hornfels, pyroxene hornfels, and sanidinite.

Albite and chlorite are the contact metamorphic minerals restricted to the albite-epidote hornfels facies. Calcite, epidote, and talc also occur in this facies. Andalusite may occur in the highest-temperature part of the albite-epidote hornfels.

The amphiboles anthophyllite and cummingtonite are restricted to the hornblende hornfels facies. Muscovite is present in this facies as well as in the albite-epidote hornfels facies, and sometimes in the lowermost pyroxene hornfels facies. Grossular andradite garnet and idocrase, sometimes with biotite and almadine garnet, are present here as well as in the pyroxene hornfels facies. Sillimanite may occur in the highest-temperature part of the facies at higher pressure; staurolite may occur in high-pressure, iron-rich rocks. Calcite is present, but not with tremolite-actinolite, epidote, or plagioclase. Under certain chemical compositions of the original rocks, other minerals that may be present include anthophyllite, cummingtonite, phlogopite, biotite, diaspore, and scapolite.

Orthoclase with andalusite or sillimanite is restricted to the pyroxene hornfels facies. Sillimanite is present here and also in the upper hornblende hornfels facies. Hyperstine and glass present in the pyroxene hornfels may also be present in the sanidinite facies. Muscovite is present only in the lowest-temperature part of the facies. For silica-deficient rocks in the pyroxene hornfels facies, dolomite, magnesite, and talc may occur only in the lowest-temperature part of the facies or at pressures of high carbon dioxide. The pyroxene hornfels facies is characterized by the breakdown of the last water-bearing minerals, muscovite and amphibole.

Diagnostic minerals of the sanidinite facies are sanidine, mullite, tridymite, and pigeonite. In silica-deficient rocks at this facies, wollastonite, grossularite, and plagioclase are present. Other minerals that may occur under special conditions are perovskite, spinel, diopside, and pseudobrookite. This is the highest-temperature facies of contact metamorphism.

STUDY AND EXPLORATION

Contact metamorphic zones are studied by standard geologic techniques, including the preparation of geologic maps through field study. Aerial photographs and satellite images are frequently used to identify contact metamorphic zones through detecting rock alteration in the contact metamorphic zone. Satellite sensors measure, analyze, and interpret electromagnetic energy reflected from Earth's surface for subsequent computer analysis. Geophysical techniques (gravity and electrical methods) are used to locate mineralized zones containing relatively heavy metallic ore minerals. The gravity contrast of these zones can be measured with surface instruments and then mapped. Some of these same minerals, particularly the ore minerals, transmit electric current in an anomalous manner. These anomalies plotted on base maps may be an additional clue to the presence of ore deposits in contact metamorphic zones.

Subsequent laboratory work involves the determination of mineral relationships by studying thin slices of rock through which light passes under a microscope (petrography). Frequently, the chemistry of entire rock samples is determined to help scientists understand the presence or absence of minerals as a guide to mineral exploration and to the composition of the original rock. Because of significant advances in laboratory instruments, detailed mineral chemistry analyses are routinely performed that determine chemical makeup of microvolumes of minerals. These determinations permit earth scientists to understand the conditions under which the contact metamorphic zones formed and to interpret the history of the contact metamorphic zone.

This information provides critical insight into the potential for metallic ore deposits within the contact metamorphic zone. Such deposits are explored by means of surveys that detect and record variations in geochemistry, gravity, and plant types and abundances. Anomalously high element concentrations in rocks, soils, and plant tissues may indicate mineral deposits that are not exposed. Many ore deposits have an anomalous gravity signature because of heavy associated silicate minerals and metallic minerals. Maps of localized variations may lead to subsurface exploration, which is conducted by drilling techniques. Continuous cylindrical samples are taken from depths of up to several thousand feet and the study of such subsurface samples leads to the evaluation of possible minable concentrations of certain metals and other elements used by modern civilizations.

ECONOMIC AND GEOLOGIC VALUE

Contact metamorphic zones frequently contain metallic mineral deposits without which modern civilization could not exist. Many significant mineral deposits worldwide occur in these zones. The metals extracted from deposits of this type include tin, tungsten, copper, molybdenum, uranium, gold, silver, and, in some cases, refractory industrial metals. Specialized surveys assess the potential of metallic ore deposits before expensive subsurface sampling by drilling.

Recognition and understanding of the contact metamorphic environment can lead to a significantly improved understanding of regional geology and the regional geologic history. In some areas, contact metamorphic zones are associated with igneous intrusives of only one geologic age. Specialized laboratory studies provide supporting data to determine the geologic history and potential for economic mineralization.

—*Jeffrey C. Reid*

Further Reading

Best, Myron G. *Igneous and Metamorphic Petrology*. 2nd ed., Blackwell Science Ltd., 2003.

Bucher, Kurt, and Martin Frey. *Petrogenesis of Metamorphic Rocks*. 7th ed., Springer, 2002.

Fettes, Douglas, and Jacqueline Desmons, editors. *Metamorphic Rocks: A Classification and Glossary of Terms*. Cambridge UP, 2007.

Philpotts, Anthony, and Jay Aque. *Principles of Igneous and Metamorphic Petrology*. 2nd ed., Cambridge UP, 2009.

Vernon, Ron H., and Geoffrey Clarke. *Principles of Metamorphic Petrology*. Cambridge UP, 2008.

Winter, John D. *Principles of Igneous and Metamorphic Petrology*. 2nd ed., Pearson Education, 2010.

Coriolis Effect

Fields of Study: Physical Geography; Geology; Fluid Dynamics; Atmospheric Science; Oceanography

ABSTRACT

The Coriolis effect is an apparent angular acceleration of a moving object as seen in a rotating system. The acceleration is not a true change in velocity, but an illusion caused by the rotation of the system beneath the moving object.

KEY CONCEPTS

anticyclonic: the clockwise rotation of an air mass in the Northern hemisphere, counterclockwise in the Southern hemisphere

cyclonic: the counterclockwise rotation of an air mass in the Northern hemisphere, clockwise in the Southern hemisphere

inertia: the tendency of physical objects to remain at rest or with unchanging speed and direction unless acted upon by an external force

inertial frame: the frame of reference by which speed and distance measured

relative motion: proportional speed and distance of two or more objects moving within the sane inertial frame of reference

RELATIVE MOTION

An unconscious tendency to regard Earth as a fixed frame of reference generates an unrecognized expectation on the part of many people that objects free of forces will move with unchanging direction and speed. That is, they will be unaccelerated. This expectation is consistent with the First Law of Motion, proposed by Sir Isaac Newton (1642-1727), which states that a body in motion will remain in

A low-pressure system swirls off the southwestern coast of Iceland. Because this low-pressure system occurred in the Northern Hemisphere, the winds spun in toward the center in a counter-clockwise direction, a phenomenon known as the Coriolis force. Photo by NASA/GSFC, MODIS Rapid Response Team, Jacques Descloitres, via Wikimedia Commons. [Public domain.]

motion, in a constant direction, with constant speed, unless acted upon by an external force.

Newton's First Law of Motion, however, applies only in frames of reference that are themselves not accelerating. These so-called inertial frames must be moving in a constant direction with constant speed. This condition is not met by rotating frames, such as Earth's surface, where every point in the frame (though traveling at a constant speed) is constantly changing direction as it completes a circular path around the axis of rotation.

The farther an object is from the axis of rotation, the faster it will travel on its circular path. An object on the equator, for example, completes a path of 40,000 kilometers in one day, while an object at 60° north latitude travels only half that distance in the same time. Thus, the object at 60° north latitude travels at half the speed of the object on the equator.

Wind and water that leave the equator headed due north carry their equatorial speed with them. As they move north, they pass over territory that is traveling eastward more slowly than they are. As a result, the wind and water move eastward relative to the ground or the seafloor. Conversely, wind and water starting at northern latitudes and moving due south will cross ground that is moving eastward faster than they are; they will move westward relative to the ground or seafloor. This deflection from the original direction of motion is the Coriolis effect, named for Gustave-Gaspard Coriolis (1792-1843).

SIGNIFICANCE FOR GEOLOGY

It is important to remember that the Coriolis effect applies to all systems in motion on a rotating surface. The most classic demonstration of the effect is two people tossing a ball to each other while standing at opposite points on the edge of a rotating platform such as might be found in a playground. While the two people are stationary relative to each other, so that there is a direct straight line between them that the ball would normally follow, they are in motion relative to the fluid medium of the air above the platform. What is observed when the ball is tossed from one to the other as the platform rotates is that as the ball travels along that straight line trajectory, but the platform beneath it changes the position of the person to whom the ball has been tossed. Relative to the platform surface, the path of the ball as it crosses the platform is not straight, but curved. The same hold true for a ball rolling across the surface of the rotating platform, as it also does for a stream of water or other fluid flowing across the surface of the platform.

The surface of Earth rotates beneath the fluid layer of the atmosphere, giving rise to the most apparent Coriolis effect of the atmosphere. As air moves in the form of wind, it moves massive quantities of dust and other fine particulates that can accrete as aeolian sediment deposits or produce aeolian erosion patterns. The traces of these effects can be seen in geologic strata, and in sedimentary rock formations, and their interpretation can provide clues to the history and mechanics of formation in geologic structures.

The movement of the heavier fluid medium of water is also subject to the Coriolis effect, especially if the water is in the form of ice-age glacial ice. The advance and retreat of glacial ice caps is consistent with the flow of a liquid across a rotating surface in that the Coriolis effect plays a significant role in the direction of the advance or retreat of the ice cap. Many lakes in North America, particularly the Great Lakes and the other major lakes of Canada, were formed as the ice caps of the last Ice Age ground their way back toward the north polar region, gouging out great depressions as they did so. On a flat map, these lakes occupy a generally curved path from south to north. But it is surprising to see that these lakes follow a straight line path across the curved surface of Earth. Similarly, the gouge marks left in the underlying rock as advancing glaciers pushed other rocks across it tell the direction of motion of the glaciers as they advanced. While those marks appear as straight lines on a small local scale, extending the lines across greater distances by matching their patterns shows a general curvature that evidences the influence of the Coriolis effect due to the rotation of the planet's surface as the glaciers moved across them.

As glaciers move, they also transport massive quantities of rock particles of all sizes, dropping the load periodically as the motion stops and starts under seasonal variation. This is the source of large deposits of gravel that is valued today as a construction resource. Over thousands of years, these deposits become buried under an overburden of sand and other material, and understanding the movement of their glacial sources aids geologists in locating gravel and other glacially sourced deposits.

—Billy R. Smith Jr. and Richard M. Renneboog

Further Reading

Cossu, Remo, and Matthew G. Wells. "The Evolution of Submarine Channels under the Influence of Coriolis Forces: Experimental Observations of Flow Structures." *Terra Nova*, vol. 25, no. 1, 2013, pp. 65-71. *Academic Search Complete*. Accessed 19 Mar. 2015.

Denny, Mark. *How the Ocean Works: An Introduction to Oceanography*. Princeton UP, 2012.

Dunkel, Ged. *Understanding the Jet Stream: Clash of the Titans*. Authorhouse, 2010.

Goh, Gahyun, and Y. Noh. "Influence of Coriolis Force on the Formation of a Seasonal Thermocline." *Ocean Dynamics*, vol. 63, nos. 9-10, 2013, pp. 1083-92. *Energy & Power Source*. Accessed 19 Mar. 2015.

Pidwirny, Michael. *Understanding Physical Geography.* Our Planet Earth Publishing, 2017.

Crystals

Fields of Study: Crystallography; Geology; Geochemistry; Geophysics

ABSTRACT

Crystals are composed of regularly repeating three-dimensional patterns of atoms or ions; a crystal is therefore a highly ordered structure. Crystals have a number of electronic and scientific applications, including uses in optics and in radio transmitters (piezoelectric quartz crystals). Well-formed crystals are also prized by collectors, and crystals of gem minerals are cut into jewelry.

KEY CONCEPTS

cleavage: the characteristic splitting of a crystal in accord with the crystal's internal structure

cube: a geometric form having six equal square faces

melt: the molten form of a material within which solid crystals of the material can form, essentially the same as crystal formation from a solution but in a melt the molten liquid takes the place of the solvent (typically water)

octahedron: a geometric form having eight equal triangular faces

prism: a geometric form having three rectangular faces between two triangular faces

pyramid: a geometric form having four equal triangular faces on a square base

tetrahedron: a geometric form having four equal triangular faces

twinning: the intergrowth of two crystals to form a single crystal structure, essentially by growing from each other at the same time

BACKGROUND

Crystals are solids that naturally display smooth planar exterior surfaces called "faces," which form during the growth of the solid. These faces collectively produce a regular geometric form that mimics the orderly internal atomic arrangement of the elements present in the solid. Some scientists use the term "crystal" to refer to any solid having an ordered internal atomic structure regardless of whether the solid displays faces. However, most scientists use the word "crystalline" for such solids when no faces are present. Many solids display a cleavage, a flat planar surface formed when the solid is broken; cleavage fragments are sometimes mistaken for crystals.

Crystals are described and classified according to the symmetrical relationship existing between the faces. The fundamental way of describing a crystal is to list the "forms" that it displays. Scientists recognize a total of forty-eight different forms, many designated by common geometric terms such as cube, octahedron, tetrahedron, pyramid, and prism. Most crystals display multiple forms. For example, quartz crystals display one prism and at least two sets of pyramids. Considering every possible symmetrical arrangement of faces, every crystal can be placed into one of thirty-two groupings called "crystal classes." These classes are further grouped into six crystal systems based on similar symmetry characteristics. The names of the six systems, from most to least symmetrical, are isometric, hexagonal, tetragonal, orthorhombic, monoclinic, and triclinic.

WHERE CRYSTALS ARE FORMED

Large crystals can develop when the faces growing in a melt, solution, or gas are unimpeded by other surrounding solids. This situation commonly occurs where open cracks and cavities exist in rock and the liquid or vapor from which the crystal is growing has free access to the open space. The largest crystals are found in igneous pegmatites. The Etta pegmatite in the Black Hills of South Dakota contained a 12-meter crystal weighing more than 18 tonnes. The largest known crystal was a single feldspar from a pegmatite in Karelia, Russia, that weighed several thousand tonnes. Crystals are also found along fault planes, in hot springs areas, around vents for volcanic gases, and in cavities within igneous and sedimentary rocks where underground water is circulating. Another mechanism for the growth of crystals occurs during the process of metamorphism. Preexisting rocks that are subjected to elevated temperatures and pressures within Earth can recrystallize while still solid. During this metamorphism some of the new minerals that form have a strong surface energy and will develop faces even while in contact with other growing minerals.

The growth conditions discussed above are so common within Earth that crystals can be found in almost every state in the United States and every country in the world. It is impossible to specify all the important occurrences of large, well-formed crystals. Some of the more notable classic localities in the United States include quartz in Hot Springs, Arkansas, and Herkimer County, New York; galena in the tristate district of Missouri, Kansas, and Oklahoma; zinc-bearing minerals in Franklin, New Jersey; garnets at Gore Mountain, New York; iron oxides in the upper peninsula of Michigan; and fluorite and celestite at clay Center, Ohio.

USES OF CRYSTALS

Particularly well-formed crystals are highly prized by collectors and museums. Most crystals, however, are more valuable for their chemistry or as crystalline solids. Many crystals are crushed during the processing of ore minerals. It was a common practice for miners to save the larger, better-formed crystals from the crushing mill because they were worth more as specimens for collectors than they were worth as ore material. Most crystals of gem minerals are cut and faceted to make jewelry. A large diamond crystal, for example, is worth more as a well-faceted gemstone than as a crystal specimen.

There are a growing number of technological uses of "crystalline solids" where the systematic internal arrangement of atoms can produce a variety of desirable physical phenomena useful in the fields of electronics and optics. As an example, very pure untwinned quartz is called "optical grade crystal" even though it lacks faces. Quartz crystal is cut, ground, and made into lenses and prisms for optical instruments and is also used in radio oscillators, tim-

A range of crystals. Photo via iStock.com/Boogich [Used under license.]

ing devices, and pressure gauges in the electronics industry.

CRYSTAL DEFECTS AND GROWTH RATES

Crystal defects occur naturally as crystals are formed; they are also sometimes introduced artificially, as they have useful electrical, mechanical, and optical qualities. A growing crystal typically requires the proper placement of trillions of atoms per hour. About one atom in every one hundred thousand is misplaced to form a defect. These defects can be point disorders, or they can geometrically combine to form line, plane, or three-dimensional disorders. The Schottky defect is a point disorder in which an atom is missing from the spot it should occupy, leaving a hole in the pattern. The Schottky defect results when a second layer of atoms is quickly deposited before all the positions can be filled in the first layer. The Frenkel defect occurs when an atom is out of its proper position and can be found nearby, inappropriately stuck between other atoms. The impurity defect is yet another point disorder, occurring when an atom of a foreign element (an impurity) either substitutes for the normal atom or is stuffed between the proper atoms of the structure.

Coloration can be caused by various point defects. When an electron is captured by the hole of a Frenkel defect it will absorb energy from passing light and become what is known as a "color center." An abundance of Frenkel color centers in fluorite will give the crystal a purple color. An impurity defect can be accompanied by a shift in electrons, also causing a color center. Smoky quartz is caused by color centers resulting from impurity defects. The electron shifts are either induced by low levels of radiation in the Earth over geological time or by artificial exposure to an intense X-ray or gamma-ray beam for a few minutes. A significant number of the smoky quartz crystals on the market began as natural colorless quartz that has been irradiated.

Line disorders are linear defects and are commonly called "dislocations" because they create an offset within the crystal. The most common is an edge dislocation resulting when an entire plane of atoms is pinched out and terminated as adjacent planes on either side begin to bond directly together. When crystals are stressed they will often deform by slipping along linear disorders.

Crystals can also become deformed or malformed because of variations in the growth rates of different faces or different parts of the crystal. When the chemistry of the growing solution begins to lack the atoms needed by the crystal, then the faces can stop growing while the edges where faces meet will continue to grow. In extreme instances the resulting malformed crystal has a skeletal look, showing a network of edges without any faces, yet all the symmetrical forms are still evident, allowing proper classification of the crystal.

TWINNING

During formation, a solid may produce a symmetrical intergrowth of two or more crystals. When the intergrowth is crystallographically controlled, the resulting composite is called a "twinned crystal." The individual crystals within the twinned aggregate are related to one another by a different symmetrical element—one that is not seen in any of the individual parts. This often results in a symmetrical, exotically shaped aggregate that does not appear to belong to any single crystal class. Crystals displaying exceptional twins can be more valuable for their twinning than as mineral specimens.

HISTORY OF CRYSTALS

Crystals have a history that reaches back into the realm of legends and myths. An important early work that combined legend with the first sound science was the thirty-seven-volume *Historia Naturalis*, written by Pliny the Elder in the first century of the Common Era. Pliny described many real as well as nonexistent crystals, which he stated were formed by such exotic processes as "the light of the moon" or "the purge from the sea." Nicolaus Steno (1638-1686) established the first law of crystallography in 1669, known as the law of constancy of interfacial angles. The law holds that for all crystals of a given mineral the angles measured between similar faces are always exactly the same. This law allows for the positive identification of deformed or malformed crystals simply by measuring the angles between existing faces.

In 1781, René-Just Haüy (1743-1822) was the first to recognize that a crystal is composed of a large number of smaller particles arranged in a regular geometric order such that it fills space without gaps. This was a remarkable advance, considering that it preceded a workable concept of the atom in chemistry by more than twenty years. In 1830, based on graphical and mathematical considerations, Johann Hessel (1796-1872) predicted the existence of thirty-two classes of symmetry corresponding to modern crystal classes. In the 1920s, two crystallographers, C. H. Hermann and Charles-Victor Mauguin (1878-1958), developed the notation that is used to designate the symmetrical arrangement of faces found on any crystal.

—Dion C. Stewart

Further Reading

De Graef, Marc, and Michael E. McHenry. *Structure of Materials: An Introduction to Crystallography, Diffraction, and Symmetry*. Cambridge UP, 2007.

Glazer, A. M. *Crystallography: A Very Short Introduction*. Oxford UP, 2016.

Hoffmann, Frank. *Introduction to Crystallography*. Springer, 2020.

Klein, Cornelis, and Barbara Dutrow. *The Twenty-third Edition of the Manual of Mineral Science*. Wiley, 2008.

Read, Peter G. *Gemmology*. 3rd ed., Elsevier, 2005.

Szwacki, Nevill Gonzalez, and Teresa Szwacka. *Basic Elements of Crystallography*. 2nd ed., CRC Press, 2016.

Tilley, Richard J. D. *Crystals and Crystal Structures*. John Wiley, 2006.

Wenk, Hans-Rudolf, and Andrei Bulakh. *Minerals: Their Constitution and Origin*. Cambridge UP, 2004.

D

The "Deep-Impact" Mission: Geology of a Comet

Fields of Study: Astrogeology; Astronomy; Sub-planet Astronomy

ABSTRACT

Deep Impact was the first space mission designed to study the interior of a primitive celestial body directly. This was accomplished by slamming a heavy impactor into Comet Tempel 1, creating a fresh crater. Never before has any space mission tried to make an impact crater of this size in any object.

KEY CONCEPTS

ejecta: the material explosively ejected from a crater

impactor: essentially a massive body designed for the sole purpose of delivering a load of kinetic energy to a selected target

perihelion: the point in an elliptical orbit at which the satellite is at its closest approach to the Sun

SUMMARY OF THE MISSION

In July, 1999, the National Aeronautics and Space Administration (NASA) approved funding under its Discovery Program for a mission to slam an object into a comet, and to study the ejecta from that impact. This proposed mission became the Deep Impact project. By November, 1999, serious work had commenced in design of the project. Except for minor variations, the plans were finalized by May, 2001. These plans called for development of a flyby spacecraft to study the impact ejecta, and for a smart impactor that would guide itself to the target. The comet selected as the target was Tempel 1, discovered in 1867 by Ernst Tempel (1821-1889). The actual project began on January 12, 2005 and lasted to August 8, 2013.

COMET TEMPEL 1

Comet Tempel 1 orbits the Sun every 5.5 years. It approached perihelion, the closest point to the Sun in its orbit, in early summer, 2005. A target date for comet rendezvous was selected as July 4, 2005. Earth's orbit placed it in position to launch a mission to arrive at Tempel 1 on this date during a period of only a few weeks in January, 2004, and again for several weeks a year later. Periods during which a launch is possible are known as launch windows.

Initially, plans called for the mission to be launched during the first launch window, in January of 2004. However, the spacecraft was not ready for launch at that time, and so launch was postponed

Comet Tempel 1 67 seconds after it collided with Deep Impact, taken by high-resolution camera on flyby spacecraft. The image reveals ridges, scalloped edges and possibly impact craters formed long ago. Photo by NASA/JPL-Caltech/UMD, via Wikimedia Commons. [Public domain.]

until December 30, 2004, the first day of the second launch window, initially reserved as a backup in case the launch could not occur in January, 2004. The launch window was only about three weeks long, and if the spacecraft was not launched within this time, then no rendezvous mission would have been possible. In November, 2004, launch was further postponed until January 8, 2005, at 19:40 Coordinated Universal Time (UTC). This date was well into the last launch window, and thus it gave very little leeway if problems developed with weather or with the launch vehicle.

Deep Impact provided tremendous celestial fireworks on the Fourth of July, striking very close to its target at a relative speed of 10.2 kilometers per second (22,800 miles per hour). The time of impact (5:52 UTC) was selected in order for the impact to happen above the horizon simultaneously for two of NASA's Deep Space Network radio telescopes, allowing both receivers to record real-time data transmitted from the spacecraft. The impact excavated a crater in the comet nucleus, allowing the flyby spacecraft to study the material from inside the comet. The flyby spacecraft began imaging one minute before impact, and it was 10,000 kilometers (6,200 miles) away at the moment of impact. The flyby spacecraft continued collecting data as it closed with the comet nucleus, eventually passing within 500 kilometers (310 miles) of it.

Selected images were sent back to Earth in near-real time; however, data were collected far more quickly than the communication system could send the information to Earth. Thus, recorded data began to be sent back to Earth starting about fifty minutes after impact, with most of the data returned within a day with provisions to extend data playback for nearly a month if necessary. Shortly after this highly successful encounter, Deep Impact program officials sought funds to continue the mission, and redirect the flyby spacecraft toward another comet encounter. Preliminary permission was given, but budget consideration held the potential for cessation of the extended mission.

SPACECRAFT AND IMPACTOR

The flyby spacecraft was somewhat irregular in shape, measuring about 1.7 meters wide, 2.3 meters high, and 3.3 meters long and having a mass of 650 kilograms. Power was provided by a 7.5-square-meter solar panel located on one side of the spacecraft bus. The two main science instruments were located on the other side of the main bus. The spacecraft bus housed the propulsion, communication, and computer systems. Mounted to the top of the bus, a 1-meter-diameter dish provided primary communication with Earth. A separate antenna communicated with the impactor.

The principal science instruments on the flyby spacecraft were the High Resolution Instrument (HRI) and the Medium Resolution Instrument (MRI). Both instruments were Cassegrain-type telescopes, and CCD (charge-coupled device) cameras with filter wheels to permit observations in different colors of light. The HRI was one of the largest instruments dedicated to solar system astronomy ever put aboard a spacecraft. With its telescope having a diameter of 30 centimeters, the HRI had five times the resolution of the MRI and was capable of resolving features smaller than 2 meters when nearest the comet nucleus. The HRI also was able to make measurements in infrared light. The MRI, with its 12-centimeter-diameter telescope, also had a filter wheel to allow observations in different spectral ranges, and it acted as a backup to the HRI. Additionally, at close approach its larger field of view provided images of the entire nucleus, rather than merely a portion like the HRI, which saw only a small portion of the nucleus at a time. Additionally, in the last ten days before the encounter, the MRI acted as a navigation camera for the spacecraft.

The impactor was a 350-kilogram cylindrical device about 1 meter in diameter and 1 meter in length. The impactor was attached to the flyby spacecraft bus until twenty-four hours before impact, and it received all electrical power from the flyby spacecraft's solar panel. After separation, it operated entirely on battery power. The impactor was destroyed upon impact with the comet nucleus. It was made of mainly copper, because copper provides the least interference with spectral measurements of the impact ejecta. The impactor contained a small propulsion system to make corrections to its trajectory, in order to impact at the proper point on the comet nucleus, as well as a communication system to send data back to the flyby spacecraft. The main instrument on board the impactor was the Impactor

Target Sensor (ITS), which was virtually a duplicate of the MRI, except without the MRI's filter wheel. Thus, in addition to guiding the impactor to the correct target zone, the ITS was capable of taking useful science images on its way to impact with the comet nucleus. Upon impact with the comet, the impactor delivered an energy of 19 gigajoules (equivalent to about 4.8 tons of dynamite) and excavated a sizable crater on the surface of the comet nucleus.

CONTRIBUTIONS

Comets form in the outer solar system, and are made mostly of ice with some dust and rock mixed in. They may be virtually unchanged since the formation of the solar system. For most of the time since their formation, comets have remained either in a zone beyond the planet Pluto called the "Kuiper Belt" or in a cloud around the solar system called the "Oort Cloud." Occasionally the orbits of comets are altered, sending the icy bodies into the inner solar system, where heat from the Sun begins to vaporize the ices that make up a comet's nucleus, shedding dust debris, and forming a cloud around the nucleus and generating the tail of the comet. This cloud prevents astronomers from observing the nucleus of a comet with great detail. The ITS optics aboard Deep Impact were expected to be destroyed prior to impact by collisions with dust particles from the comet, but they still provided the most detailed images of a comet nucleus to date.

By blasting into the comet, Deep Impact permitted the inner layers of a comet to be studied for the first time. The outer portions of a comet in the inner solar system have been altered by the heat from the Sun, but the inner parts may remain as they were when the comet formed. Also, in striking the comet, Deep Impact helped astronomers determine the consistency of the comet nucleus. It is not known if the nucleus is rock-solid, like an iceberg, or is soft and fluffy, like a snowflake. The size of the crater answered some of these questions. Preliminary analysis of images of the comet and data collected by examination of the plume created during crater excavation indicated a surface with much more dust and dirty water ice than had been expected. Scientists would continue to study Deep Impact data for years to come.

CONTEXT

Knowing more about comets permits astronomers to know more about the formation of the solar system. Comet studies also help geologists and other scientists understand Earth, because some of the water and organic materials needed for life are believed to have been brought to Earth by comets. However, understanding comets better has a more practical application as well, because comets can impact planets. If a comet were on a collision course with Earth, knowing the structure and consistency of the nucleus of the comet might permit a strategy to be developed to deflect or destroy the comet, given sufficient time.

Deep Impact and its sister comet mission, Stardust, were both funded under NASA's Discovery Program. Discovery missions are relatively inexpensive space missions with a narrow focus. An earlier Discovery mission, Deep Space 1 (1998), passed by a comet, but that mission's primary goal was development of a new ion drive for space probes. Other than the Comet Nucleus Tour mission (CONTOUR) of 2002, which lost contact with its spacecraft, Stardust and Deep Impact were the first U.S. missions designed primarily to explore comets. They form part of an international effort to study comets, including Rosetta, launched in 2004 by the European Space Agency (ESA) toward Comet Churyumov-Gerasimenko.

Deep Impact also marked a first in space exploration, in that it attempted to alter the body being studied, in this case by blasting a crater into it. All other space missions have been more passive in their approach, with only minimal impact on the body being studied.

—Raymond D. Benge, Jr.

Further Reading

Brandt, John C., and Robert D. Chapman. *Introduction to Comets.* Cambridge UP, 2004.

Crovisier, Jacques, and Thérèse Encrenaz. *Comet Science.* Trans. Stephen Lyle. Cambridge UP, 2000.

Fletcher, Charles. *Physical Geology: The Science of Earth.* Wiley, 2017.

Davies, John. *Beyond Pluto: Exploring the Outer Limits of the Solar System.* Cambridge UP, 2001.

Russell, Christopher T., editor. *Deep Impact Mission: Looking Beneath the Surface of a Cometary Nucleus.* Springer, 2005.

Deep-Sea Sedimentation

Fields of Study: Oceanography; Marine Geology; Geology; Stratigraphy

ABSTRACT

Deep-sea sedimentation occurs by the settling of particles to the ocean floor and by the transport of material from shallow to deep water. Knowledge of the distribution of sediments and the processes of sedimentation will help to evaluate the potential of the world's oceans for the mining of natural resources and the storage of waste.

KEY CONCEPTS

calcareous ooze: sediment in which more than 30 percent of the particles are the remains of plants and animals composed of calcium carbonate

carbonate compensation depth (CCD): the depth in the oceans at which the rate of supply of calcium carbonate equals the rate of dissolution of calcium carbonate

clay: a sediment particle less than 0.1 millimeter in diameter

clay minerals: a group of hydrous silicate minerals characterized by a structure of layers of thin sheets that are held together loosely

pelagic: "of the deep sea"; refers to sediments that are fine-grained and are deposited very slowly at great distances from continents

siliceous ooze: fine-grained sediment in which more than 30 percent of the particles are organic remains of plants and animals composed of silica

turbidity current: a mass of water and sediment that flows downhill along the bottom of a body of water because it is denser than the surrounding water; common on continental slopes

upwelling: an ocean phenomenon in which warm surface waters are pushed away from the coast and are replaced by cold waters that carry more nutrients up from depth

INVESTIGATION OF DEEP-SEA SEDIMENTS

Sedimentation in the deep sea differs substantially from that in any other environment. Deep-sea sedimentation is pelagic. It typically occurs at great distances from continents and at depths in excess of 1,000 meters, primarily by the settling of particles through the overlying water. Every 100,000 liters of seawater contains about 1 gram of extremely small particles. Their settling is extremely slow, and sediments accumulate at rates on the order of a few millimeters per thousand years. Transport of material by turbidity currents along the bottom of the ocean from shallow to deep regions is also a mechanism by which deep-sea sediments are deposited, but this is far less important than the settling of particles from the surface. In contrast, current transport dominates sedimentation in all other environments, where sedimentation rates typically exceed several centimeters per year.

The oceans cover 70 percent of Earth, and, as a result, deep-sea sedimentation may be the most common sedimentary process. Early investigations of the seas were limited, however, to coastal processes and oceanic circulation because of the lack of techniques available to study the deep ocean bottom. Knowledge of deep-sea sedimentation was profoundly increased by the voyage of HMS *Challenger* from 1872 to 1876, which made the first systematic study of the ocean floor. Prior to the *Challenger* expedition, practical investigations of the oceans concentrated on winds, tides, and water depths in harbors to provide information for commercial navigation and for the laying of submarine telegraph cables. The *Challenger* crew collected data on the temperature and depth of the oceans, marine plants and animals, and sediments on the ocean bottom. Deep-sea sediments were described by Canadian oceanographer John Murray (1841-1914), naturalist to the expedition. He noted that sediments in the deepest parts of the ocean are fine-grained and are composed of particles that settled from the surface. Comparison of rocks exposed on land with those dredged from the ocean led to the assertion that deep-sea sediments could not be found on land. This conclusion caused great debate, and the interpretation of some rocks as deep-sea deposits by a few scientists persisted despite opposition by most of the scientific community.

Investigation of deep-sea sedimentation flourished again during and immediately after World War II. Despite the voluminous amount of data available by 1950, the prevailing doctrine in the first half of the twentieth century was similar to that in the previous century: The continents and oceans were generally believed to be permanent features of Earth's surface that had developed to their present form near the beginning of geologic time. Thus, the ocean basins were billions of years old, ocean bottom sediments were tens of kilometers thick, having accumulated from early in the geological history of Earth to the present, and deep-sea sediments were not exposed on land, a point that was still severely contested. Then, during the 1950s two discoveries shocked the oceanographic community. First, fossils taken from seamounts (submarine mountains) in the Pacific Ocean at depths of 2 kilometers were only 100 million years old, suggesting that subsidence of the ocean floor was recent. Second, deep-sea sediments were found to be less than 200 meters thick. The advent of the theory of plate tectonics in the 1960s profoundly changed ideas about the ocean basins by providing both an explanation for the two observations of the previous decade and a mechanism for the emplacement of deep-sea sediments onto continents. In addition, the motion and cooling of the oceanic plates predicted by plate tectonics explain the global distribution of deep-sea deposits.

PELAGIC SEDIMENTS

Pelagic sediments are characterized by their fine-grain size. Few particles are larger than 0.025 millimeter, and most are smaller than 0.001 millimeter. Particles include terrigenous debris (derived from the continents), such as clay minerals, quartz, and feldspar; volcanogenic grains (derived from volcanoes), such as volcanic glass, pumice, and ash; biogenic material (derived from living organisms), such as fecal pellets (waste produced by organisms inhabiting the surface waters) and skeletons of planktonic plants and animals; and cosmogenic matter (derived from space or the cosmos), such as "space dust" and pieces of meteorites. Pelagic sediments that contain more than 30 percent biogenic material are called "oozes." If the biogenic debris is composed of calcium carbonate, the sediment is a calcareous ooze. If the biogenic debris is composed of silica, the sediment is a siliceous ooze. Most calcareous oozes consist of foraminiferans, which are single-celled animals that secrete calcium carbonate and live in the surface waters. In contrast, siliceous oozes contain diatoms (green, unicellular algae) in cold water and radiolarians (silica secreting single-celled animals) in warm water. Pelagic, or red, clays are sediments that contain less than 30 percent biogenic debris.

The two major controls on pelagic sedimentation are the calcium compensation depth (CCD) and the fertility of the surface waters. The CCD reflects the interplay between the release of carbon dioxide during the decay of surface organisms and the dissolution of the calcium carbonate skeletons as they descend through the ocean. Above the CCD, calcareous oozes are dominant, whereas below the CCD, siliceous oozes and pelagic red clays are common. For example, in one region of the ocean, sediments decreased from 90 percent calcium carbonate at a depth of 1,000 meters to 20 percent calcium carbonate at a depth of 5,000 meters. Oozes accumulate below areas of high fertility where upwelling brings nutrient rich bottom water to the surface. Because upwelling creates increased productivity of both calcareous and siliceous organisms, siliceous oozes form only where the water depth exceeds the CCD. Calcareous oozes, the most abundant biogenic sediments in the ocean, cover only 15 percent of the Pacific Ocean bottom but blanket 60 percent of the Atlantic Ocean floor. This difference reflects the combination of a shallower CCD and greater water depth in the Pacific Ocean than in the Atlantic Ocean. Siliceous oozes are in areas of elevated CCDs such as the equatorial regions, the subarctic and subantarctic zones, and the continental margins of northwestern South America and eastern Africa.

Sedimentation of pelagic clays is restricted to the central portions of the oceans where the fertility is low. This reflects the difference in settling rates between nonbiogenic particles and fecal pellets. Because of its small size, a particle may take many years to sink through the ocean. The concentration of many particles into fecal pellets by organisms feeding at the surface, therefore, is thought to be an important mechanism by which debris settles to the bottom. The greater size of the pellets allows them to fall much faster than unconsolidated particles. It is only in areas of low productivity, therefore, where biogenic material does not overwhelm the sediment. Wind is the primary means by which nonbiogenic particles reach the sea. Dust is carried in the atmosphere to great distances from continents and constitutes as much as 10 percent of pelagic sediment.

OTHER PROCESSES

Additional processes in deep-sea sedimentation include the rafting of glacial debris by ice, the transport of terrigenous and shallow water material by turbidity currents along the ocean floor, and the precipitation of nodules on the ocean bottom. Pieces of ice calve off glaciers in the polar regions and float toward warmer waters where they eventually melt, releasing glacial sediment into the ocean. Although this process is most common in the south polar region today, glacial sediments greater than ten thousand years old elsewhere on the ocean floor suggest the process extended over a much greater area in the past. Turbidity currents generated along continental slopes may traverse hundreds of kilometers of ocean basin where ridges and depressions do not occur. Far away from continents, the currents commonly contain only silt but may deposit the grains over great distances. These currents also may locally erode the seafloor, creating breaks in the continuity of the sediment pile by removing the most recent sediments and exposing older sediments at the abraded surface. Chemical reactions of seawater with the surface of the sediment cause the precipitation of ferromanganese (composed of iron and manganese) nodules in areas where sediments accumulate slowly. The nodules form by nucleating around a manganese source in the sediment and by accreting minerals from the seawater. Such nodules may cover more than 40 percent of the seafloor.

Sedimentation in the deep sea is a complex interaction between water depth, fertility of the surface waters, and current transport. To illustrate how the depth dependence of sedimentation affects the global distribution of sediments, one must understand the relationship between plate tectonics and the topography of the ocean floor. At a midocean ridge, which rises above the surrounding ocean basin, molten rock is extruded to form new seafloor. Calcareous oozes accumulate on the new seafloor because of its depth above the CCD. As more seafloor forms, the older seafloor moves away from the midocean ridge toward the deep ocean basin (i.e., "seafloor spreading"), sinking to lower depths as it does so. The old seafloor gradually descends below the CCD, where siliceous oozes accumulate above the calcareous oozes. Eventually the motion of the seafloor carries it below a region of low surface productivity such that pelagic clays are deposited above the siliceous oozes. This vertical sequence has been documented in several localities throughout the world's oceans.

STUDY OF DEEP-SEA SEDIMENTATION

Techniques used to study sedimentation in the deep ocean are diverse and include piston coring, seismic reflection profiling, the use of deep-diving submarines, and experiments performed both at sea and in the laboratory. Piston coring is the most important technique because it enables direct sampling of the ocean floor. It uses the same kind of rotary drilling equipment that is standard in petroleum exploration to drill into the top few hundred meters of the seafloor. The cores are 7 centimeters in diameter and provide samples that are large enough to preserve features that yield insights into the processes by which the sediment accumulated. For example, current produced features in some samples were the first direct evidence of currents in the deep seas. In addition, piston coring confirmed the vertical sequence of sediments predicted by plate tectonics.

To map the ocean floor using a 7 centimeter drill core is time consuming and inefficient. Seismic reflection profiling has the advantage of providing information about the thickness and distribution of layers in the sediments on the seafloor relatively quickly. The two major disadvantages are that it cannot determine the composition of the sediment and that it identifies only layers that are tens of meters thick. In seismic reflection profiling, a low frequency sound source is towed over the bottom. The sound waves penetrate and are reflected by the various layers within the seafloor to several kilometers below the boundary between the sediment and the water. The returning signals are displayed on strip charts that reveal the water depth and the thickness and pattern of layers in the sediments. One of the most significant early contributions of seismic reflection profiling was the evidence that sediments on the ocean floor were too thin to have been accumulating since early in Earth's history, an observation explained by plate tectonics.

The use of deep-diving submarines has allowed scientists to observe the ocean floor directly. Previously the ocean floor was only photographed. Deep-diving submarines allowed scientists to confirm or reject hypotheses based on other indirect techniques. The ranges and endurance times of the submersible vessels, however, are limited. In addition, they frequently cannot be used in areas where current velocities are greater than 1 meter per second.

Simple experiments either at sea or in the laboratory also can be illuminating. For example, the existence of the CCD was verified by lowering spheres of calcium carbonate to various water depths in the Pacific Ocean. When the spheres were examined, it was discovered that the spheres in the deepest water had dissolved the most. Scientists in the laboratory conduct experiments on settling rates by dropping particles of various sizes in large beakers and measuring the time necessary for the particles to reach the bottom. Thus the amount of time represented by the thickness of sediment on the ocean floor was roughly estimated at a few hundred million years.

One of the most important developments in deep-sea sedimentation was the creation in 1965 of the Deep Sea Drilling Project (DSDP), now known as the Ocean Drilling Project (ODP), to drill and investigate the ocean floor. The project began in earnest in 1968 with the maiden voyage of the American ship the *Glomar Challenger*, which was outfitted with state of the art navigational, positioning, and drilling equipment. International teams of scientists boarded the ship in ports all over the world and remained at sea for two months. Since its inception, ODP has undertaken more than sixty-five voyages and contributed significantly to scientists' understanding of the ocean.

SIGNIFICANCE

The oceans cover nearly three-fourths of the surface of Earth. To study deep-sea sedimentation, therefore, is to understand one of the most globally prevalent processes. Understanding deep-sea sedimentation is important for several other reasons, however. First, sediments on the ocean floor yield insight into bottom currents and water depths, which affect the navigation of submarines and the installation of underwater communications cables. Second, the sediments preserve the climatic record of the past few hundred thousands of years. For example, glacial deposits approximately ten thousand years old are found over a much greater area of the bottom of the ocean than recent glacial deposits, suggesting that transport of glacial debris by ice was very common ten thousand years ago. This implies that the temperature of Earth's environment was lower in the past, allowing the formation of significant quantities of ice. Scientists can use this information to gauge present day climate patterns and predict future fluctuations in the global climate. Third, the seas receive a significant amount of the garbage produced by modern society. Knowledge of sedimentation rates and currents in the deep sea helps to establish the length of time necessary to bury the waste, the direction the waste will travel prior to its burial, and the effect of dumping waste on the overall health of the ocean.

Fourth, and most important, understanding deep-sea sedimentation and the global distribution of deep-sea sediments provides a framework within which one can estimate the economic potential of the deep sea and minimize the damage from ocean exploitation. As the reserve of natural resources on the continents dwindles because of expanding demand from high technology industries, exploding population growth, and increasing consumer appetite, attention focuses on the oceans as a possible source of raw materials. Important resources that may be found locally in the deep sea include oil and gas. The decay and accumulation of plants and animals are essential to the generation of hydrocarbons and are processes that occur over a large area of the seafloor. Emphasis also is placed on mining the deep sea for important materials such as manganese, gold, cadmium, copper, and nickel. Similar to most new industries, large scale exploitation of the natural resources of the deep sea is limited at present by the current state of technology. The incentive to mine the oceans, however, will grow as cheaper and more sophisticated technologies are created.

—*Pamela Jansma*

Further Reading

Blatt, Harvey, Robert Tracy, and Brent Owens. *Petrology: Igneous, Sedimentary, and Metamorphic*. 3rd ed., W. H. Freeman, 2005.

Coe, Angela L., editor. *The Sedimentary Record of Sea-Level Change.* Cambridge UP, 2003.
Grotzinger, John, and Thomas H. Jordan. *Understanding Earth.* 6th ed., W. H. Freeman, 2010.
Gurvich, Evgeny G. *Metalliferous Sediments of the World Ocean.* Springer Verlag, 2010.
Huneke, Heiko, and Thierry Mulder. *Deep Sea Sediments.* Elsevier, 2011.
Joseph, P. *Deep Water Sedimentation in the Alpine Basin of SE France.* Special Publication 221. Geological Society of London, 2004.
McKinney, Frank. *The Northern Adriatic Ecosystem: Deep Time in a Shallow Sea.* Columbia UP, 2007.
Seibold, Eugen, and Wolfgang Berger. *The Sea Floor: An Introduction to Marine Geology.* Springer, 2017.
Stein, Ruediger. *Arctic Ocean Sediments: Processes, Proxies and Paleoenvironment.* Elsevier, 2008.
Trujillo, Alan P., and Harold V. Thurman. *Essentials of Oceanography.* 10th ed., Prentice Hall, 2010.

DEGLACIATION

Fields of Study: Cryology; Glaciology; Geology; Geomorphology

ABSTRACT

Deglaciation is the uncovering or exposure of a land surface that was previously covered by glacial ice. It results from ice melting or subliming. Deglaciation accompanies the end of a glacial stage. As deglaciation occurs, several processes are effected, such as meltwater stream flow, development of meltwater lakes, addition of water to the world's oceans, exposure of the land, and rebound of the land. In addition, faunal and floral changes accompany deglaciation in response to changes in landscape and ecology.

KEY CONCEPTS

albedo: the percentage of incoming light that is reflected from the surface of a planet or other body

disequilibrium condition: a condition in which a relatively small change can render a system unstable, causing it to collapse to a condition that is more stable

interglacial stage: the warm period between two periods of glaciation

positive feedback loop: a condition in which the output from a system affects the system in such a way that the particular output is increased

rebounding: raising of the land's elevation as a result of the removal of the weight of the overlying glacial ice

sedimentary material: particulate matter that may be carried by a moving fluid and settles out from the fluid when the flow slows or stops

GLACIATION AND DEGLACIATION

Deglaciation is the uncovering or exposure of a land surface that was previously covered by glacial ice. It results from ice melting or subliming (transforming

Deglaciation is the transition from full glacial conditions during ice ages to warm interglacials, characterized by global warming and rising sea levels. Photo via iStock.com/RistoArnaudov [Used under license.]

from solid directly into vapor). Deglaciation, therefore, accompanies the end of a glacial stage. As deglaciation occurs, several processes take effect. Among these processes are meltwater stream flow, development of meltwater lakes, addition of water to the world's oceans (raising the global sea level), exposure of the land, and rebound of the land (lifting of the land's elevation as a result of the removal of the weight of the overlying ice). In addition, faunal and floral changes accompany deglaciation in response to changes in landscape and ecology.

GLACIERS AND THEIR RETREAT

Glaciers have a seemingly infinite capacity to entrain and transport sedimentary material, from tiny clay particles to giant rock boulders. Glaciers move these materials within and upon the ice, but when they melt all this sediment is deposited in the area where the ice melts. For this reason, areas that have experienced deglaciation are typically covered by glacially transported sediment. Alternatively, glaciers may sweep an area clean of loose material, creating deglaciated areas of bare bedrock. Where deglaciation has formed modern shorelines, those shorelines tend either to be laden with glacial sediment or to present bare bedrock to the waves. In some places, rebound has lifted the land along the modern shore, forming sea cliffs.

Deglaciation accompanies the transition from a glacial stage to an interglacial or warm stage. There have been several such transitions over the past two million years. In addition, deglaciation—to a lesser extent—accompanies minor warming events that occur during glacial stages. Prior to the current epoch of glacial and interglacial stages, Earth experienced several periods during which glaciers episodically covered large parts of its surface. There have been at least four such glacial periods during the past one billion years.

DEGLACIATION AND CLIMATE CHANGE

Deglaciation accompanies climate change and can be a cause of climate change. The geological record indicates that, when deglaciation commences, there is typically a climatic turn toward global warming. In other words, after climatic warming initiates deglaciation, the deglaciation itself can create a positive feedback loop engendering further warming. Glaciers are highly reflective of sunlight, contributing to Earth's albedo (the percentage of sunlight reflected back into space from the planet's surface). As glaciers melt, white ice and snow are replaced by the rock or earthen surface of the planet. As the planet warms, more glaciers melt, the albedo decreases with darker surface elements. Earth's albedo decreases, and more solar radiation is absorbed and retained even more, and the process continues. Melting glaciers also contribute water to lakes and oceans, which help retain atmospheric heat. Rising sea levels due to glacial meltwater contribute to global climate change as well.

Using the modern deglaciation as an example, loss of glacial ice cover on the land has had and continues to have profound consequences for the global climate. For example, release of water locked up in glaciers has affected the amount of water in the oceans as well as on land, in rivers and streams, and in the form of groundwater. The geological record has revealed that mean sea level has in the past been as much as two hundred meters lower than it is at present, as well as significantly higher, due to the amount of water contained in ice or released from it. This has affected coastal and interior ecosystems, which are dependent upon water for life. Changing patterns in the distribution of water cause both climates and ecosystems to change.

Deglaciation, as compared to glaciation, can be a relatively rapid process, once the melting triggers feedback mechanisms that increase its pace. The rapid nature of this change has led to disequilibrium conditions on land, such as unstable slopes, high gradients in rivers and streams, and unstable lakes that drain catastrophically. In the biotic realm, rapid deglaciation has led to mass death and mass extinction among plant and animal groups, as well as mass migration of animal populations.

Deglaciation leaves behind profound physical effects of ice movement, such as landscapes altered by the erosive forces of massive ice sheets, depositional landforms created by sediment released from melting ice, and lakes created by the meltwater—including waters from marooned blocks of ice that melt long after the main glacial mass is gone. The resulting altered landscapes generally have low levels of vegetation (at least initially), as well as areas of low elevation where water can accumulate. This type of landscape has a higher capacity to retain radiated heat from the Sun and therefore also contributes to atmospheric warming.

—*David T. King Jr.*

Further Reading

Alley, R. B., and P. U. Clark. "Deglaciation of the Northern Hemisphere." *Annual Reviews,* vol. 28, 2000, pp. 149-82.

Anderson, Robert S., and Suzanne P. Anderson. *Geomorphology: The Mechanics and Chemistry of Landscapes.* Cambridge UP, 2010.

Bender, Michael L. *Paleoclimate.* Princeton UP, 2013.

Drzyzga, Scott A. *Mapping and Modeling Glacial Lake Algonquin in Northern Michigan and Western Ontario with Models of Uncertainty.* Michigan State U, 2007.

Ehlers, J., and P. L. Gibbard, editors. *Quaternary Glaciations: Extent and Chronology.* 3 vols. Elsevier, 2004.

Shakun, Jeremy D., Peter U. Clark, Feng He, et al. "Global Warming Preceded by Increasing Carbon Dioxide Concentrations during the Last Deglaciation." *Nature,* vol. 484, no. 7392, 2012, pp. 49-54.

Stanley, Steven. *Earth Systems History.* 3rd ed., W. H. Freeman, 2009.

Wood, Gillen D'Arcy. *Land of Wondrous Cold: The Race to Discover Antarctica and Unlock the Secrets of Its Ice.* Princeton UP, 2020.

Deltas

Fields of Study: Geology; Geoarcheology; Marine Biology

ABSTRACT

Deltas are dynamic sedimentary environments that undergo rapid changes over very short periods of time. Found in lakes and shallow ocean waters, they are rich in organic material and provide food and shelter for fish and wildlife. Their traces left in the geological record provide significant information for geologists about the paleogeography of an area and clues to the possible presence of mineral resources.

KEY CONCEPTS

crevasse: a break in the bank of a distributary channel causing a partial diversion of flow and sediment into an interdistributary bay

delta: a deposit of sediment, often triangular, formed at a river mouth where the wave action is low and the river's current slows suddenly

distributary channel: a river that is divided into several smaller channels, thus distributing its flow and sediment load

geoarchaeology: the technique of using ancient human habitation sites to determine the age of landforms and when changes occurred

interdistributary bay: a shallow, triangular bay between two distributary channels; over time, the bay becomes filled with sediment and colonized with marsh plants or trees

natural levee: a low ridge deposited on the flanks of a river during a flood stage

prodelta: a sedimentary layer composed of silt and clay deposited under water; it is the foundation on which a delta is deposited

sediment: fragmented rock material such as gravel, sand, silt, or clay that is deposited by a river to form a delta

wave energy: the capacity of a wave to erode and deposit; as wave energy increases, erosion increases

FORMATION AND SHAPE

Deltas contain many valuable resources. Government agencies such as the U.S. Fish and Wildlife Service study the surface properties of deltas because of the enormous wetlands and abundant wildlife that occupy these landforms. Geologists study deltas because they are favored places for the accumulation of oil and gas resources. This low topographical feature serves society in many ways, and that is why it has been the object of intense study.

Deltas are deposits of sediments, such as sand or silt, that are carried by rivers and deposited at the shoreline of a lake, estuary, or sea. As the river meets the water body, its velocity is greatly decreased, which reduces its ability to support the sediment load that it has been carrying. As a result, the river sediment becomes deposited. If the accumulated sediment is not removed by waves or currents, a delta will accumulate and continue to extend itself into the lake or ocean.

The term "delta" is used to describe this depositional landform because it is often triangular. It is believed that the Greek historian Herodotus

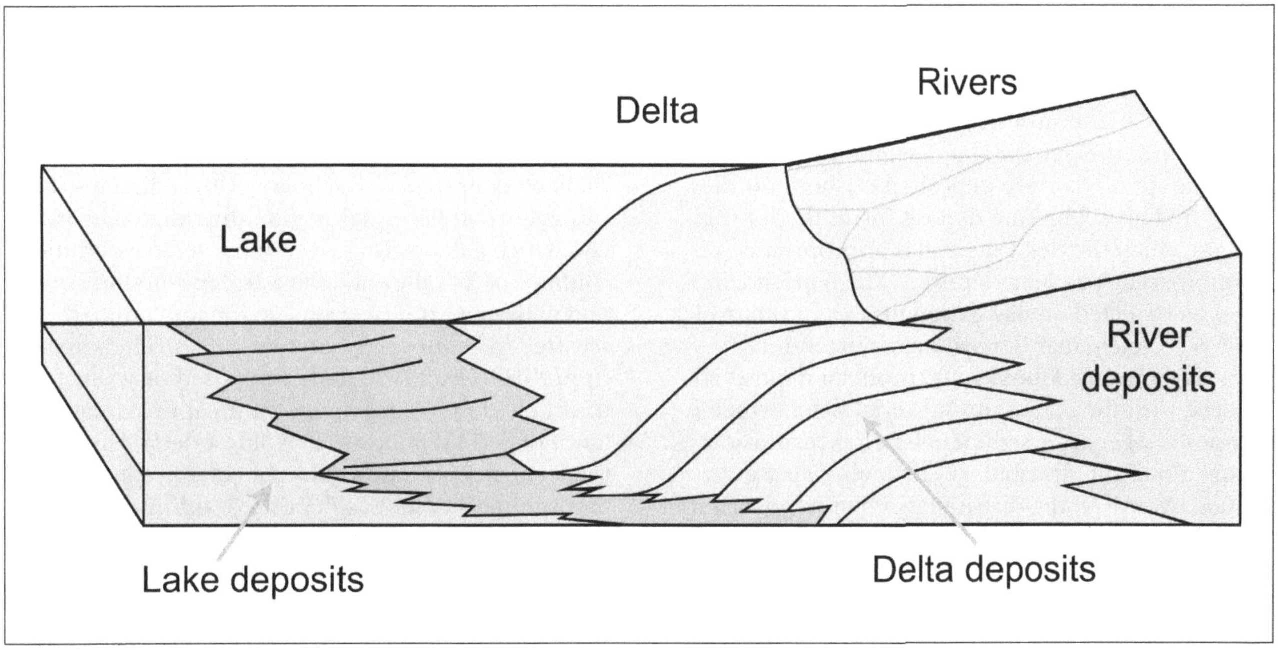

A delta forms where a river meets a lake. Image by AntanO, CC BY 4.0, via Wikimedia Commons.

coined the term with the shape of the capital Greek letter delta in mind. Herodotus visited Egypt, where the Nile Delta is located, and he correctly defined the shape of that delta. Not all deltas are triangular, however.

The Ganges Delta, the Colorado River Delta, the Mississippi River Delta, and many other deltas have different shapes. The shape postulated by Herodotus is, in fact, somewhat unusual, but it is applicable to the Nile River Delta. The Nile Delta has a smooth but curved shoreline—and so is designated an "arcuate" delta, whereas the Mississippi Delta has been extended into the Gulf of Mexico and has spreading channels resembling the digits of a bird's foot. Current and wave action at the shoreline can cause sediment to be distributed to the left and right of a river channel, forming smooth beaches on either side. Such a delta acquires a shape more like a cone, the point of which projects toward the sea, and is called a "cuspate" delta. The Tiber River, which empties into the Mediterranean Sea, is a classic example of this type of delta. Rivers such as the Seine, in France, may deposit sediment in elongated estuaries, forming shoals and tidal flats.

Earth scientists have noted that the shapes of deltas are associated with several conditions, such as the character of river flow, the magnitude of wave energy and tides, and the geologic setting. The bird foot delta of the Mississippi River has extended itself well into the Gulf of Mexico because the river carries and deposits a high volume of sediment on a shallow continental shelf. The wave and tidal forces are low, and the delta deposit is not redistributed along the shoreline or swept away. Conversely, a cuspate delta, such as that of the Tiber, is a product of strong waves moving over a steep continental shelf. Persistent high wave energy redistributes the sediment often, forming beaches and sand dunes along the delta shoreline. The Nile Delta is an arcuate delta characterized by moderately high wave energy and a modest tide range. Occasional high wave conditions deposit beaches and sand dunes along the arc shaped delta front at the Mediterranean Sea. Tides also play a direct role in the creation of deltas. Deltas in estuaries are formed because of a high tidal range coupled with low wave conditions. The Seine estuary, with its distinctive mud flats exposed at low tide, provides a good example.

LANDFORMS

Although deltas have different shapes that reflect differences in the intensity of river, wave, current, and tidal processes, certain landforms may be identified as characteristic of delta formation. Submarine features are deposited below sea level, and subaerial features form at or just above sea level. As a river empties into the sea, the finest sediments, usually very fine silt or clay, are deposited offshore on the seafloor. This submarine deposit forms the foundation on which the delta sits and is appropriately referred to as a "prodelta deposit." The deposit can often be detected on navigation maps as a relatively shallow, semicircular deposit under the water.

As deposition continues, the prodelta deposit is covered with the extending subaerial delta, which is composed of coarser sediments. Deltaic extension occurs along the distributary channels. During higher river flow, the distributary channels overflow, depositing natural levees along their sides. The digitate distributary pattern of the Mississippi Delta illustrates this process well. As the distributaries extend to deeper water, the shallow areas between the distributaries are better developed. These areas, known as interdistributary bays, are shallow landforms colonized by aquatic plant life. Over time, deposition occurs in the interdistributary bays through breaches in the natural levees. As the river mouth distributaries enter a flood stage, the lower regions of a natural levee are broken, and fine suspended sediments are introduced into the interdistributary bay area. Such overbank splays, or crevasse splays, are primarily responsible for the infilling of a delta. The crevassing is usually a very rapid but short lived process, occurring during a high river stage and operating over a ten to fifteen year period. With the passage of time, the open water interdistributary bays become silt filled and colonized. Eventually, however, the marshy bays may subside due to compaction of the sediment, creating water areas once again.

GEOLOGIC HISTORY

Although the geologic history of large deltas such as the Mississippi is complex, the succession and behavior of shifting deltas have been determined in some detail. Over the past twenty thousand years, the large continental glaciers that occupied much of the upper midwestern United States began to melt. As Earth's climate continued to warm, the meltwater was returned to the oceans, and the sea level rose some 100 meters, inundating valleys that had previously been cut by streams. The oceans reached their present approximate level about five thousand years ago. The Mississippi and similar valleys were flooded and became elongated bays. Over time, the shallow water bays were choked with sediments that formed broad floodplains extending down the valleys. Once a depositing river extended beyond the confines of its valley, a delta was deposited in deeper water. Because the river was no longer confined, it was free to shift over greater distances. The Mississippi River Delta is actually composed of seven distinct delta lobes extending over an approximate distance of 315 kilometers. The oldest delta, Saleì Cypremort, was deposited some 4,600 years ago; the most modern delta was deposited within the past 550 years. Older Mississippi Delta lobes, such as the Teche Delta, have subsided since they were deposited, giving an opportunity for a more recent delta (in this case, the Lafourche) to be deposited on top and more seaward of the older feature. The different delta lobes making up the enormous deltaic plain have resulted from a shifting of the Mississippi River well upstream in its valley. This process may be visualized as a hand movement occurring because of a shoulder movement.

Because of significant changes in the shoreline environments, many deltas in marine coastal zones are eroding. The Mississippi River is at the edge of the continental shelf and cannot build out into deep water. Also, subsidence and rising sea levels are causing the delta to erode. The Nile Delta is eroding as well. With the construction of the High Aswan Dam upstream, there has been a significant decrease in the sediment supplied to the Nile Delta. This lack of sediment, along with a slight rise in sea level, has led to erosion. Some earth scientists have suggested that the wave action in coastal Egypt is cutting back the Nile Delta at a rate of 15 to 30 meters per year in some areas.

STUDY OF DELTAS

Deltas are difficult to study because many of the features are very flat, marshy, or under water. Because deltas change rapidly over time, however, maps are an important tool with which to determine changes. Navigation maps and maps that illustrate the topography of coastal areas around the world have been made for generations. By comparing the size and location of a delta on old maps and new maps, changes can be analyzed. Also, aerial photographs and pictures taken from satellites aid in identifying the erosion and deposition of delta landforms.

Often, older delta lobes were settled by ancient peoples. Through the science of geoarchaeology, it can be determined when changes occurred. As deltas such as the Mississippi shift from side to side over time, the human population follows the deltas from place to place. By examining the location of archaeological sites over the past fifteen hundred years, scientists have determined the minimum age of the several deltas forming the Mississippi deltaic plain. The Indigenous pottery found there reveals that the delta framework was deposited very recently. Cultural remains indicate other environmental changes—in salinity, subsidence, and delta deterioration.

By boring holes into the soft sediments of a delta, geologists can decipher its subsurface aspects. Because oil and natural gas are often associated with deltas, oil companies have bored holes in many delta landforms. Information derived from this method of study reveals the composition of the thick delta sediments and the rate of delta accumulation. In fact, boreholes in some deltas have indicated that older deltas once existed and are now buried beneath younger deltas.

Finally, deltas can be created in the laboratory. In nature, deltas are very large and complex. To make the study of deltas less difficult, scientists use tanks filled with water and sediments. Experiments can be performed that, for example, control the amount of sediment used to build deltas. Relationships between sediments and current velocities may be studied to gather information on such properties as the rate of delta growth. By controlling the phenomena that cause deltas to form, geologists can gain an overview of the behavior and processes of delta development.

SIGNIFICANCE OF DELTAS

Deltas, with their marshes and bogs, are not generally aesthetically pleasing; however, depositional landforms have been useful to prehistoric and historic populations in many ways. Deltas, along with estuaries, are among the most biologically productive areas on Earth. Most deltas are colonized with wetland swamps or marshes, which are breeding areas for wildlife. In many countries, for example, duck hunting is a popular sport associated with wetlands. In marine deltas where there is tidal influence, freshwater and saltwater mix. The river brings oxygen and nutritive substances into the delta, and the result is an enormous production of sea life. High biological productivity attracted humans to this land feature. Deltas have often been centers of civilization; the deltas of the Nile, in Egypt, and the Tigris-Euphrates, at the head of the Persian Gulf, have supported important societies. Soils in delta regions are nourished through seasonal flooding, and water tables are high, guaranteeing adequate water with which to irrigate crops, even in the dry season. Food and crop production from tropical deltas is significant because most tropical soils are not very productive. Deltas such as those of the Mekong and the Ganges are outstanding examples.

Deltas are transition zones between the land and the sea, and between river and marine processes. Their rivers are also links between ocean and continent. Cities such as New Orleans, Venice, Amsterdam, and Rotterdam owe their prosperity to their delta geography. Such cities, known as "entrepôts", thrive on marine traffic entering a country or on overland traffic exiting the country.

Because deltas are areas of vast accumulations of sediments, they generate building material for future mountains. The young mountains of the world, such as the Alps and the Himalayas, parallel coastal areas and are composed of sedimentary rock. Marine fossils frequently found in such rock reveal that they not only are composed of sediment but also were once deposited under water, later to be thrust upward to great heights.

—*C. Nicholas Raphael*

Further Reading

Bianchi, Thomas S. *Deltas and Humans: A Long Relationship Now Threatened by Global Change.* Oxford UP, 2016.

Levinton, Jeffery S., and John R. Waldman. *The Hudson River Estuary.* Cambridge UP, 2006.

Middleton, Gerard V., editor. *Encyclopedia of Sediments and Sedimentary Rocks.* Springer, 2003.

National Research Council. *Drawing Louisiana's New Map: Addressing Land Loss in Coastal Louisiana.* National Academies Press, 2006.

Nett, Mary T., Martin A. Locke, and Dean A. Pennington. *Water Quality Assessments in the Mississippi Delta: Regional Solutions, National Scope*. American Chemical Society, 2004.
Newson, Malcolm. *Land, Water and Development: Sustainable and Adaptive Management of Rivers*. 3rd ed., Routledge, 2008.
Nicholls, Robert J., W. Neil Adger, Craig W. Hutton, and Susan E. Hanson, editors. *Deltas in the Anthropocene*. Palgrave MacMillan, 2020.
Schmidt, Paul E. *River Deltas: Types, Structures and Ecology*. Nova Science Publishers, 2011.
Wohl, Ellen. *A World of Rivers*. U of Chicago P, 2011.

Desert Landforms

Fields of Study: Geology; Geomorphology

ABSTRACT
Deserts are often striking in color, form, or both. The underlying lack of water in deserts produces unique features not found in humid regions. Arid lands include deserts and surrounding steppes, semiarid regions that act as transition zones between arid and humid lands. The north and south polar regions typically receive no more precipitation than deserts in hot climates, and there are more than 5 million square kilometers of polar deserts on Earth. The same geomorphic factors function in both hot and cold deserts.

KEY CONCEPTS
aeolian erosion: erosion caused by abrasion due to sediment particles being transported by wind
hoodoo: an erosion structure consisting of a pillar-like base capped by a flat cap of harder material
saltation: a process in which solid particles being transported within a moving fluid (water, air, etc.) bump into other particles and are lifted over them as a result, eventually producing an accumulation structure with a gradual incline on the upstream side and a steep dropoff on the downstream side
ventifacts: landscape forms produced by wind erosion
yardangs: bedrock outcrops that have been eroded to a streamlined shape by wind-borne particles

HOT DRY LANDSCAPES
Deserts are often striking in color, form, or both. The underlying lack of water in deserts produces unique features not found in humid regions. Arid lands cover approximately 30 percent of Earth's land surface, an area of about 40 million square kilometers (15.4 million square miles). Arid lands include deserts and surrounding steppes, semiarid regions that act as transition zones between arid and humid lands.

Many of the world's largest and driest deserts are found between 20 and 40 degrees north and south latitude. These include the Mojave and Sonoran Deserts of the United States, the Sahara in northern Africa, and the Great Sandy Desert in Australia. In these deserts, the subtropical high prevents cloud formation and precipitation while increasing rates of surface evaporation.

Some arid lands, like China's Gobi Desert, form because they are far from oceans that are the dominant source for atmospheric water vapor and precipitation. Others, like California's Death Valley, are arid because mountain ranges block moisture from coming from the sea. The combination of mountain barriers and very low elevations makes Death Valley the hottest, driest desert in North America.

SAND DUNES
Many people envision deserts as vast expanses of blowing sand. Although wind plays a more important role in deserts than it does elsewhere, only about 25 percent of arid lands are covered by sand. Broad regions that are covered entirely in sand (such as portions of northwestern Africa, Arabia, and Australia) are referred to as sand seas. Why is wind more effective here than elsewhere?

The lack of soil water and vegetation, both of which act to bind grains together, allows enhanced aeolian (wind) erosion. Very small particles are picked up and suspended within the moving air mass, while sand grains bounce along the surface. Removal of material often leaves behind depressions called "blowouts" or "deflation hollows." Moving grains abrade cobbles and boulders at the surface, creating uniquely sculpted and smoothed rocks known as ventifacts. Bedrock outcrops can be streamlined as they are blasted by wind-borne grains to form features called "yardangs." As these rocks are ground away, they contribute additional sediment to the wind.

Desert sand dunes are not stationary features—instead, they represent accumulations of moving sand. Wind blows sand along the desert floor. Where it collects, it forms dunes. Typically, dunes have relatively shallow windward faces and steeper slip faces. Sand grains bounce up the windward face in a process called "saltation," and then eventually cascade down the slip face, the movement of individual grains driving movement of the entire dune in a downwind direction.

Four major dune types are found within arid regions. Barchan dunes are crescent-shaped features, with arms that point downwind. They may occur as isolated structures or within fields. They form where winds blow in a single direction and where the supply of sand is limited. With a larger supply of sand, barchan dunes can join with one another to form a transverse dune field. There, ridges are perpendicular to the predominant wind direction. With quartering winds (i.e., winds that vary in direction throughout a range of about 45 degrees) dune ridges form that are parallel to the average wind direction. These so-called longitudinal dunes have no clearly defined windward and slip faces. Where winds blow sand from all directions, star dunes form. Sand collects in the middle of the feature to form a peaked center with arms that spiral outward.

BADLANDS, MESAS, AND BUTTES

As scarce as it may be, water is still the dominant force in shaping desert landscapes. Annual precipitation may be low, but the amount of precipitation

Rock pedestal. Photo via iStock.com/AlexPapp. [Used under license.]

in a single storm may be a large fraction of the yearly total. An arid landscape that is underlain by poorly cemented rock or sediment, such as that found in western South Dakota, may form badlands as a result of the erosive ability of stormwater runoff. Overall aridity prevents vegetation from establishing the interconnected root system that holds soil particles together in more humid regions.

Cloudbursts cause rapid erosion that forms numerous gullies, deeply incised washes, and hoodoos, which are created when rock or sediment that is more resistant protects underlying material from erosion. Over time, protected sections stand as prominent spires, typically with a prominent "cap" of harder material, while surrounding material is removed. Landscapes like those found in Badlands National Park in South Dakota are devoid of vegetation and erode rapidly during storms.

Arid regions that are underlain by flat-lying rock units can form mesas and buttes. Water follows fractures and other lines of weakness, forming ever-widening canyons. Over time, these grow into broad valleys. In northern Arizona's Monument Valley, remnants of original bedrock stand as isolated, flat-topped structures. Broad mesas are marked by their flat tops (made of a resistant rock like sandstone or basalt) and steep sides. Buttes are much narrower, with a small resistant cap, but are often as tall and steep as neighboring mesas.

DESERT PAVEMENT AND DESERT VARNISH

Much of the desert floor is covered by desert pavement, an accumulation of gravel and cobbles that forms a surface fabric that can interconnect tightly. Fine material has been removed by wind and water, leaving behind larger fragments that inhibit further erosion. In many areas, desert pavements have been stable for long periods of time, as evidenced by their surface patina of desert varnish. Desert varnish is a thin outer coating of wind-deposited clay mixed with iron and manganese oxides. Varying in color from light brown to black, these coatings are thought to adhere to rocks by the action of single-celled microorganisms. Under a microscope, desert varnish can be seen to be made up of very fine layers. A thick, dark patina means that a rock has been exposed for a long time.

PLAYAS

Where neither dunes nor rocky pavements cover the desert floor, one may find an accumulation of saline minerals. A playa is a flat surface that is often blindingly white in color. Playas are usually found in the centers of desert valleys and contain material that mineralized during the evaporation of a lake. Dry lake beds are a common feature of the Great Basin in the western United States. During glacial stages, the last of which occurred about 20,000 years ago, lakes grew in what are now arid, closed valleys. As the climate warmed, these lakes shrank, and many dried completely. As a lake evaporates, minerals that were held in solution crystallize, forming salts, including halite (table salt) and borax. These salt deposits frequently are mined for useful household and industrial chemicals.

California's Death Valley is the driest desert in the United States, with an average rainfall of only 38 millimeters (1.5 inches) per year at the town of Furnace Creek. It is also consistently one of the hottest places on Earth, with a record high of 57°C (134°F). In the distant past, however, Death Valley held lakes that formed in response to global cooling. Over 120,000 years ago, Death Valley hosted a 90-meter-deep (295 feet) body of water called Lake Manley. Evidence of this lake remains in evaporite deposits that make up the playa in the valley's center, in wave-cut shorelines, and in beach bars.

COLD DESERTS

As deserts are defined by the amount of precipitation they receive annually, it is easy to overlook the fact that the north and south polar regions typically receive no more precipitation than deserts in hot climates. There are, in fact, more than 5 million square kilometers of polar deserts on Earth. Their mean temperature during the warmest months is less than 10°C, and their annual precipitation is less than 250 millimeters. It is interesting to note also that, unlike hot climate deserts that experience consistent, blazing sunlight throughout the year, polar deserts are devoid of sunlight for a period of nine to ten months each year.

While the apparent differences between polar and hot deserts are obvious, they are in many ways the same. Wind is the most important factor that affects polar landscapes. Snowflakes and ice granules have the same scouring effect as grains of sand because they are practically as hard as sand grains at the coldest temperatures. Wind, sometimes blowing with hurricane force with some regularity, transports snow and ice granules across the polar landscapes, producing polished rock surfaces. As sand does in hot deserts, snowflakes and ice granules in polar deserts can also produce dune-like accumulations by essentially the same mechanism. Similarly, they may also pack together to form hard snow surfaces. However, whereas sand grains in hot deserts do not "weld" to each other, the energy of high-velocity impacts of snow and ice granules with each other has the capability of partially melting them, allowing them to combine as they almost instantly refreeze. This is a factor in the formation of glaciers and dense ice caps.

The climatic conditions in a polar desert do not permit of much melting, although this is observed to be changing as the global climate slowly warms and record amounts of meltwater runoff occur each year. Typically, less of the solid precipitation melts, evaporates or sublimes than is received annually. Because of this, the precipitation that does fall in a polar desert, though very little, builds up over time and the accumulating weight compresses the underlying snow into ice, resulting in the formation of an ice cap that may become several kilometers thick after sufficient time has passed. Such ice caps, once they begin to form, prevent any further erosion of the bedrock underlying them. In addition, as the weight of ice increases, so too does the pressure it exerts on the underlying ice. The molecules of water in that ice are forced to adopt different crystal structure configurations as a result, with different phases being called ice-1, ice-2, and so on. The ice-9 made famous in a Kurt Vonnegut novel does indeed exist, but unlike its literary counterpart real ice-9 just melts when it contacts water rather than freezing the whole world solid.

—Richard L. Orndorff and Richard M. Renneboog

Further Reading

Aleshire, Peter. *Deserts.* Chelsea House, 2008.

Maghsoudi, Mehran. *Desert Landforms and Landscapes of Iran.* Springer, 2020.

Parsons, Anthony J., and Athol D. Abrahams, editors. *Geomorphology of Desert Environments.* 2nd ed., Springer, 2009.

Seppälä, Matti. *Wind as a Geomorphic Agent in Cold Climates.* Cambridge UP, 2004.

Desert Pavement

Fields of Study: Geology; Geomorphology; Climatology; Ecology

ABSTRACT

Desert pavements are concentrations of stones on the land surface of arid areas, produced by wind and running water erosion and upward movement of stones through the soil. Stone pavements may signal serious soil erosion that must be addressed before the land can be irrigated for agriculture.

KEY CONCEPTS

creep: the slow, gradual downslope movement of soil materials under gravitational stress

deflation: the sorting out, lifting, and removal of loose, dry, silt- and clay-sized soil particles by turbulent eddy action of the wind

dirt cracking: a process in which clays accumulate in rock cracks, take on water, and expand to rupture the rock

expansion-contraction cycles: processes of wetting/drying, heating/cooling, or freezing/thawing, which affect soil particles differently according to their size

salt weathering: the granular disintegration or fragmentation of rock material affected by saline solutions or by salt crystal growth

thermal fracture: the formation of a fracture or crack in a rock as a result of temperature changes

ventifact: any stone or pebble that is shaped, worn, faceted, cut, or polished by the abrasive action of windblown sand, generally under desert conditions

OCCURRENCE OF DESERT PAVEMENT

Desert pavements are extensive stony surfaces in arid areas that occur not only on slopes but also on a range of lowland surfaces, including water eroded and deposited areas. The stones tend to be closely packed on flat or moderately inclined plane surfaces. Stone pavements are rendered prominent by

Desert pavement showing desert varnish on pebbles, Gibber plains of central Australia. Photo by Mark Marathon, CC BY-SA 4.0, via Wikimedia Commons.

their lack of vegetation but, in any case, they form most readily by processes typical of arid regions where plant roots are not normally established to bind the soil to any extent. The pavements are most striking in areas of low relief, where it is the largely flat, stony surface itself, rather than the contours of the land, that impresses the observer.

Desert stone pavements range from rocky- or boulder-strewn surfaces to smooth plains of fine gravel. Terminology for different types commonly derives from the Arabic language, as one might expect, given the prevailing aridity across North Africa and the Middle East. Thus, the term *hamada* (Arabic for "unfruitful") describes a boulder-strewn terrain, and *reg* (meaning "becoming smaller") indicates a finer pavement of small stones. The term *serir* is a synonym for "reg" in the central Sahara.

Most hamada pavements are residual, consisting of stones derived from the bedrock beneath, or constituted of boulders transported only short distances. Most regs consist of transported stones. These distinctions of size and transport distance become rather blurred over time, as the larger rock fragments are progressively weathered to finer sizes and transported farther by wind and water. Nevertheless, a residual origin of many hamada pavements is clearly evident in a number of places by the angularity and lack of sorting of the rock fragments and by the similarity to the bedrock beneath. The water deposited nature of a reg pavement can be indicated by sorted and rounded gravels of mixed composition and distant origin and also by the pavement's occurrence close to dry stream channels. Residual regs, which are usually closely associated with hamadas, commonly consist of angular flakes of local bedrock or of a less weatherable residue. Reg pavements can be composed of two elements in varying proportions: a compact mosaic of small stones embedded in soil and a more uneven component of larger fragments lying loose on that surface or protruding through it.

WIND AND WATER EROSION

Desert pavements are polygenetic, which is to say there is more than one process by which they can be produced. The most traditional explanation, however, is that the stones are concentrated by means of wind erosion or deflation of the fine particles. Deflation may eventually settle the pebbles into such stable positions that they fit together almost like the pieces of a mosaic or the blocks of a cobblestone street.

The effectiveness of deflation on desert pavements is less, however, than has been commonly believed. Under natural conditions, soils that have been tested by scientists can resist wind erosion because they are silty and cohesive and tend to form crusts as a result of repeated wetting and drying. A larger grain size of the soil also makes it more resistant to wind erosion. Soil cohesiveness is an important factor. Even where particles are spherical, the magnitude of the cohesive forces between particles less than 0.1 millimeter in diameter is greater than the weight of the particle. Small particles, however, are usually more irregular in shape, or even plate like, a factor that increases their cohesion. Since cohesion is so important between small particles, silts and clays should be quite resistant to wind erosion. That is observed to be true, but only where the surfaces are smooth, the clays and silts are uniformly fine, and no other material is blown onto them. Otherwise, large grains are dislodged from fine soils as aggregates of particles or knocked loose by the impact of other large sand grains, a process known as "saltation." In any case, the amount of lowering and stone concentration by wind action is limited. Deflation diminishes markedly as the protective stone cover increases, and deflation becomes virtually ineffective when stones cover 50 percent of the surface. Undisturbed stone pavements are among the most windstable of desert surfaces.

Rain and water flow appear to be more effective than wind in eroding fine-textured soils on sloping desert pavements. In test plots cleared of stones in one experiment on 5-degree slopes, water wash accounted for most of the 5- to 50-centimeter surface lowering that occurred in a period of five years. During this time, the stone pavement was renewed with stones from below this differential erosion.

UPWARD DISPLACEMENT

Where subsoils are so clay rich that they do not erode easily by either wind or water and the subsoils are also largely stone free, the formation of stone pavements by deflation is more difficult to understand. In these situations, the mechanism of formation seems to have been forces of expansion and contraction within the soil that cause upward displacement and concentration of stones on the surface. Soils that exhibit this phenomenon contain expansive clays in alkaline chemical conditions and are subject to swelling and heaving upon wetting and to shrinkage and cracking upon drying. Periods of heating followed by cooling and of freezing followed by thawing also contribute to the cycles of expansion and contraction, which cause stones to move upward and concentrate on the surface.

The exact mechanism for upward stone movement in deserts is not precisely known, although by analogy to known stone movement in areas of intense freeze and thaw cycles, some details are understood. It is possible, for example, that the stones shift upward as the underlying soils swell in wet periods and that, as the soil shrinks in dry periods, fine particles fall into cracks beneath the stones and prevent the return of the rock fragments. Stones may also induce differential swelling in the soil by speeding the downward advance of a wetting front around and over them. The stones are then likely to be displaced upward, away from the dry zone beneath them, and held tightly by the wet and sticky soil above. Soil may squeeze into the space left and thus prevent a return movement of the stones. These processes also resemble those postulated for certain types of sorted, patterned ground, where stones become arranged in polygons or striped zones. In these cases, it is clear that no process other than upward movement along certain zones could produce such patterns.

Stone pavements of the upward movement type occur on and in the topsoils of weakly salty soils on the stony tablelands of arid Australia, where the subsoils are almost stone free. The pavement stones are silica rich, originally precipitated within the soils by mobilization and concentration of silica in unusual chemical reactions. For those stone pavements to have originated as a residue from erosion would have required a stripping of more than 1.5 meters of erosion resistant, clay rich soil. Such a process is unlikely, so an upward movement is far more probable, given the swelling and cracking potential of salty clays. Similar pavements have been noted in deserts in Nevada and California as well as in the Atacama Desert of Peru.

An alternative form of displacement is possible where the soil material has been transported, particularly by the wind. For example, in South Australia, a stone pavement occurs on a layer of clay rich in gypsum (calcium sulfate), which is thought to have been blown into place by the wind. The stone pavement on the surface resembles a buried pavement. It is possible that the original pavement first trapped a small amount of windborne dust among the stones. Some of the rock fragments were then displaced upward, little by little, through wetting and swelling of the aeolian (wind deposited) clays during the course of their accumulation, and thus the stones were never deeply buried.

Concentration of stone pavement through winnowing by wind or wash can be relatively rapid, but the contribution by movement through the soil may take longer. Once formed, a pavement is relatively stable. The closely spaced stones act as a drag on the surface wind, restrict the entrainment of finer intervening materials, and so limit deflation. On moderate slopes, runoff water is spread over the surface by the stone mantle and thus does not tend to cut a deep gully. Selective erosion is countered further as the stone concentration increases. Whether residual or transported, pavements naturally consist of materials resistant to weathering and therefore serve as "armor" for deserts.

SURFACE WEATHERING

Another manner in which stone pavements can be generated is by relative concentration through surface weathering. Rock fragments in the relatively moist subsurface environment of a desert soil are more susceptible to weathering than those on the arid surface, particularly where the soil is impregnated with salt or gypsum. Consequently, a stone pavement may survive above a soil that itself has few stones because they disintegrate at depth over time. The phenomenon is pronounced in granitic gravels. Some terrace sediments, for example, are known to have larger-sized stone pavements above horizons of small granite fragments formed by the chemical weathering breakdown of boulders beneath the surface. These subsurface zones of fine particles are deepest and most free of subsurface stones on the highest and oldest terraces, where they may be more than 50 centimeters thick. In such cases, it is thought that stone pavements first formed from the abundant rock fragments in the area, but as subsurface weathering progressed all the buried stones were destroyed, while those on the surface remained relatively unweathered.

In spite of their status as a protective armor on desert surfaces, stone pavements are subject to further evolution over thousands of years as the stones weather and their secondary products are redistributed. This evolution is generally toward an increasingly even surface of particles of small grain size and characterized by greater compaction. Also common is progressive darkening by surface weathering and the formation of rock or desert varnishes of manganese and iron stains, precipitated on the stone surfaces by microbial and weathering action.

Weathering of pavement stones occurs not only beneath the stones, where they are in contact with the protected, mildly corrosive soil layers, but also on their exposed surfaces. Pavement stones can be wetted frequently by dew, which also contributes salts to both the surface and the subsurface in the weathering process. Because pavements are generally unchanneled and provide little runoff, the features are particularly subject to episodic or seasonal cycles of shallow wetting by rainfall and evaporative drying, through which weathering is activated. Bare and generally dark colored pavements, among the most strongly heated surfaces of the desert, are areas of considerable evaporation, as (mostly saline) moisture is drawn upward through narrow cracks. As a result, thin salt crusts are widespread, the soil itself is impregnated with chlorides and sulfates of calcium and sodium, and salt weathering is significant. In this process, the growth of various salt crystals in the pores of the stones generates forces sufficient to disrupt the rocks.

Pavement stones also trap windborne dust in fissures and cracks, and dirt cracking can result from expansion of the dust particles when they are wetted. Lichens and algae also exploit the shaded and relatively moist environments under the stones, adding an organic element of chemical decomposition to the weathering process.

In the breakdown of pavement stones, there tends to be a further selective concentration of resistant fine grained material. For example, siliceous flint or chert pebbles may accumulate on the surface as relatively soft limestones are weathered. Such stones can eventually break down by incorporation of water into their microcrystalline atomic structures, by spalling (chipping), by blocky fracturing (crazing), or by complete cleavage and radial splitting. Coarser grained stones undergo granular disintegration and pitting. Fracturing of pavement stones also has been attributed to differential expansion and contraction caused by solar heating, but the idea is controversial. Such thermal fracturing may be only a partial cause, as much of the broken stone has been superficially altered chemically, and dirt cracking expansion may be a more important factor. In some cases, stones below the surface, where solar heating is not a possibility, can also be seen to have been pried apart in this fashion.

Wind abrasion is a form of natural sandblasting. Its effectiveness is related to wind velocity, the hardness of the sand and dust carried by the wind, and the hardness of the rock fragments being eroded. It is probably most effective in certain polar areas where cold, dense air can carry large particles at very high velocities and where, at winter temperatures, even ice has the hardness of some minerals. The "dry valleys" of Antarctica, for example, have extensive stone pavements that have been affected in this way.

BOULDER AND PEBBLE FORMS

Boulders and pebbles in stone pavements may be fluted, scalloped, and faceted by wind abrasion. Flute and scallop forms vary in length from a few millimeters to several meters, but large flutes are not common in hard rocks. It is thought that they are produced by turbulent helical (helix shaped) flows carrying dust and sand and that the scallops grow downwind. Where they are cut on large, immobile boulders, they are clear indicators of the strongest (dominant) wind directions. For this reason, they are most commonly reported from places where wind directions have remained largely unchanged for thousands to hundreds of thousands of years.

Some of the best-known wind eroded stones in pavements are "ventifacts" (a general term for wind-faceted pebbles and boulders). A rock face abraded by wind may be pitted if there is a range of hardness in the minerals of the face, or it may be smooth where the rock is fine grained or composed of only one mineral. Thus, rocks such as coarse grained granite have pitted surfaces, but fine grained quartzite generally produces only smooth, polished surfaces. Ventifacts have a great range of surface shapes, with plane and curved faces and two or more facets. The German term *dreikanter* is used for ventifacts with three facets, and *einkanter* is used for two-faceted stones. Multiple facets indicate either that there was more than one wind direction or that the stones have been turned over through time. As the sizes of a stone pavement's fragments are progressively reduced, a matrix of increasingly fine particles is supplied to the pavement. The proportion of such secondary material reveals the maturity of development of a desert stone pavement, although in practice it may be difficult without sophisticated analysis to distinguish between a new pavement in the process of being formed and an old one being degraded.

The fine particles in a degraded pavement are redistributed by wash and rain, a process that contributes to the smoothing and compaction of the pavement. Any exposed soil is puddled and sealed by heavy rainfall or runoff water so that a saturated layer flows into hollows between the stones. Bare soil interspersed through a pavement is generally crusted above a bubbly or vesicular horizon, 1 to 3 centimeters thick, that also extends around and beneath the pavement stones. The bubbles result from the escape of entrapped air. Equally important in compaction is the gravitational settling of the stones during the expansion and contraction of the surface upon heating and cooling and (especially in saline and expansive clay soils) upon wetting and drying. Pavements of this type are generally soft and puffy after rain, but, upon drying, the stones become ever more firmly embedded in a tight mosaic.

The slow downhill creep of water-saturated surface materials can also assist in either the smoothing or the roughening of a sloping stone pavement, particularly where dispersal of the matrix is accentuated by salinity. Microrelief on a stone pavement can be reduced by the differential flow of the fine sediments, and the pavement stones can thereby become more evenly distributed and further embedded. In some cases the stone pavements can even move into a series of rough steps, aligned along the contours of a slope.

—John F. Shroder Jr.

Further Reading

Laity, Julie J. *Deserts and Desert Environments.* John Wiley & Sons, 2008.

Maghsoudi, Mehran. *Desert Landforms and Landscapes of Iran.* Springer, 2020.

Mainguet, Monique. *Aridity: Drought and Human Development.* Springer Verlag, 2010.

Middleton, Nick. *Deserts: A Very Short Introduction.* Oxford UP, 2009.

Nicholson, Sharon E. *Dryland Climatology.* Cambridge UP, 2011.

Parsons, A. J., and Athol D. Abrahams. *Geomorphology of Desert Environments.* Springer Science+Business Media b.v., 2009.

Schaetzl, Randall J., and Sharon Anderson. *Soils: Genesis and Geomorphology.* Cambridge UP, 2005.

Summerfield, Michael A. *Global Geomorphology: An Introduction to the Study of Landforms.* Routledge, 2013.

Thomas, David S. G. *Arid Zone Geomorphology: Process, Form and Change in Drylands.* 3rd ed., Wiley Blackwell, 2011.

Desertification

Fields of Study: Geology; Ecology; Agriculture

ABSTRACT

Desertification is the degradation of semiarid lands and includes the reduction in the biological diversity of the ecosystem, the reduction in soil nutrients, and soil erosion.

KEY CONCEPTS

anthropogenic: caused by or as a result of human activity

drought: a prolonged period without rain, generally with sunshine and hot temperatures, in an area that normally receives sufficient annual rainfall to maintain crops and support animal herds

monoculture plantation: a plantation in which only crops of a single species

BACKGROUND

Desertification occurs over a period of years, with numerous factors interacting to stress the environment. Natural desertification in the distant past leaves its trace in the geologic record and in some cases the progression and expanse of ancient desertification can be traced over a large area as strata are defined and analyzed. Anthropogenic desertification results when social, political, and economic forces cause agricultural exploitation beyond the carrying capacity of marginal arid lands. Airborne and waterborne erosion strip vulnerable topsoil, rendering the land less productive over time. People living in areas undergoing desertification have poor harvests and are increasingly unable to feed themselves.

Although desertification has resulted from natural causes over eons, it has also occurred in relatively short periods of time, hurried along by anthropogenic influences, notably as a result of improper land-use management—usually a combination of deforestation and overgrazing on semiarid grasslands. Nonnative cash crops and monoculture plantations quickly deplete soils, impoverishing the ecosystem and accelerating desertification. In developed countries, boom and bust agricultural practices in arid lands cause overexpansion during favorable climatic conditions, followed by drought. People living on marginal lands in less developed countries depend on subsistence agricultural production. Drought causes these people severe hardships, as their crops fail and animal herds die. Sustainable development of semi-arid lands in less developed countries would benefit many impoverished people worldwide.

DESERTIFICATION IN THE UNITED STATES

In 1806, Zebulon Pike characterized the southern plains as a sandy wasteland, an area later called the "Great American Desert" on maps. Following the Homestead Act of 1862, 65 hectares of shortgrass could be claimed by settlers living and working on the land, and some people thought that to be sufficient incentive to migrate there. Amish and Mennonite settlers found the 65 hectares sufficient for their type of agriculture, but the concept of large commercial ranches and farms, coupled with the introduction of machinery, prompted the passage of the Enlarged Homestead Act of 1909, resulting in a land rush.

By the time World War I began, wheat was the favored crop in the Great Plains. Previously uncultivated land throughout Kansas, Colorado, New Mexico, Oklahoma, and Texas was plowed and planted with wheat by residents and absentee farmer-speculators between the end of World War I and 1930. When a lengthy drought began in 1931, precipitation was spotty, and wheat crops began failing. Fields were abandoned, and the airborne soil erosion which characterized the Dust Bowl began.

Within the United States, elimination of homesteading, massive purchase of marginal land, large public works projects, and agricultural subsidies helped reclaim marginal desert lands throughout the Great Plains and Southwest. The "New Deal" of President Franklin Delano Roosevelt began land reclamation using the Soil Conservation Service and the Civilian Conservation Corps throughout the Dust Bowl region, sowing sorghum and planting shelterbelts of trees in an effort to limit soil erosion. The introduction of contour plowing limited some of the worst disk-plowing problems for agriculture.

THE SAHEL: SOUTHERN ENCROACHMENT OF THE SAHARAN DESERT

The African region known as the Sahel, which roughly follows the 15° north parallel, is an example of dry woodland and dry wooded grassland undergoing desertification. Its northern border with the Sahara Desert receives 150 millimeters of annual rainfall; its southern border receives 600 millimeters of rain yearly.

Between 1931 and 1960 the Sahel experienced greater than normal levels of precipitation, coupled with a doubling in population. During this time, a large part of the Sahel was populated by nomadic peoples herding small livestock flocks and practicing subsistence agriculture. In 1970, a period of decades-long drought began. This drought signaled a period of overgrazing by livestock and deforestation to feed cooking fires, two key factors in accelerating desertification. Starting in 1968, the Sahel began to move southward from its 1931-1960 boundaries. Warmer ocean waters changed rainfall patterns, resulting in less precipitation in the northern Sahel in the late twentieth century. More than 60 percent of foreign aid in the region was expended on road construction, which consumed valuable water resources. Development of large-scale dams in the Senegal River Valley during the 1980s caused population displacement because much of the irrigated land was taken over by large plantations growing cash crops of peanuts, rice, and cotton. Deforestation to expand plantations was widespread in Senegal.

Annual rainfall began to increase in the mid-1990s. Satellite data show that vegetative growth in the Sahel increased beyond what would be expected from observed rainfall. In Niger, where felling of nitrogen-fixing gao acacia trees has stopped, an estimated 49,000 square kilometers of savanna showed increased vegetation by 2008.

In 2008, the population of the Sahel was estimated at fifty million. Emigration via the port of Dakar to Europe, or overland to North Africa, continues to reduce the population, as does continuing warfare in the Darfur region of Sudan, which claimed an estimated 200,000 lives between 2003 and 2008. Deforestation has been rapid around Darfur refugee camps, as trees have been cut to fuel refugee cooking fires and brick kilns for houses for United Nations aid workers.

MITIGATION OF DESERTIFICATION

Identification of the problems caused by desertification by national governments and international agencies led to concerted efforts to halt and reverse the process during the late twentieth and early twenty-first centuries. Developed nations, including

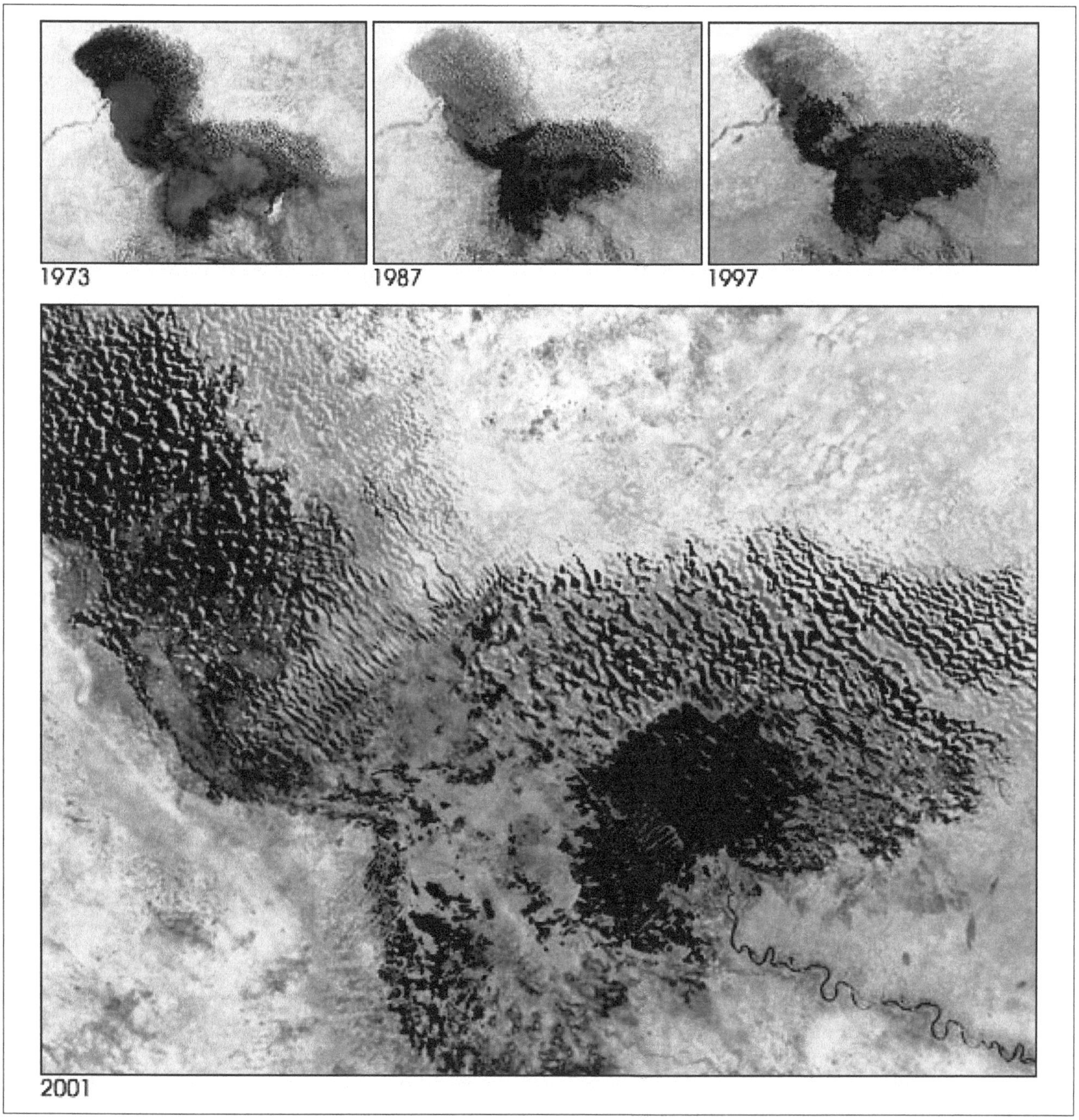

Lake Chad in a 2001 satellite image. The lake lost more than 90% of its surface area between 1987 and 2005. Photo via NASA/Wikimedia Commons.

the United States and Australia, have relied heavily on technology to combat desertification. Ranchers have formed effective organizations to lobby the central governments for aid. Drilling wells, building dams, installing irrigation canals, using chemical fertilizers on depleted soil, using crop-dusting airplanes to limit pest damage, and implementing large-scale reforestation and revegetation programs are some of the methods used to combat desertification.

In Australia, ranchers of large areas import feed in drought years and move herds of cattle and sheep out of desert environments to more productive grazing land. Government-sponsored selective breeding programs produced special cattle breeds—the Belmont Red and the Australian Milking Zebu—better suited to production in arid lands.

Less developed countries rarely have the resources to purchase technology. International aid efforts focused on building dams have been found to result in dislocation of hundreds of thousands of people while worsening public health problems like schistosomiasis and malaria. Because of these problems, beginning in the 1990s, international aid organizations increasingly emphasized the need for introducing low technology to combat desertification and permit sustainable development. Important initial steps included involvement of the local populace to establish sustainable land-use policies and then to teach agricultural techniques to reclaim land. Natural revegetation, the self-seeding by plants remaining in the area, is encouraged, as is the use of direct seeding of crops and planting of trees to limit erosion. Where overgrazing by livestock is removing most of the sparse natural vegetative covering, reducing and limiting herd size is important. Concentration of land ownership accelerates growth of unsustainably large herds and encourages overgrazing. Farmers at the subsistence level own few animals, using them mainly for milk production.

Minimizing the use of wood for cooking fuel through the introduction of more efficient cooking stoves and solar cookers has lessened desertification in the Sahel. Other successful low-technology methods include digging shallow basins for planting in eastern Mali and stone line construction to retain water and limit erosion in Burkina Faso. In 1984, after stone lines were completed in Yatenga Province in Burkina Faso, crop yields doubled.

In less developed countries where a few landowners control large, semiarid plantations utilized for nonsustainable cash crops, such as peanuts and cotton, land redistribution to smaller, privately owned farms growing drought-resistant crops for local consumption is ideal. However, the World Bank underwent a paradigm shift in the 1980s that favored commercial farmers over small farmers. This accelerated desertification in many arid regions of less developed countries, especially in Africa. Large farmers raise unsustainable large herds and force poor people onto marginal land.

Mitigation of desertification becomes difficult when severe periodic droughts occur. Expansion of agriculture above subsistence levels is inevitable during good times when crop yields increase. When drought struck the U.S. Great Plains from 1954 to 1957 and again from 1974 to 1977, airborne erosion identical to the 1930s Dust Bowl occurred, with dramatic economic consequences. Deep-well irrigation was introduced, lessening the apparent risks of large-scale mono-crop agriculture in the region. Technology has become essential to maintaining agriculture in the Great Plains.

—Anita Baker-Blocker

Further Reading

Hazell, Peter, and Stanley Wood. "Drivers of Change in Global Agriculture" *Philosophical Transactions of the Royal Society of London,* vol. 363, no. 1491, Feb. 12, 2008, pp. 495-515.

Liebesman, Lawrence R., and Rafe Petersen. *Endangered Species Deskbook.* Environmental Law Institute, 2003.

Worster, Donald. *Dust Bowl: The Southern Plains in the 1930s.* 25th anniversary ed., Oxford UP, 2004.

DESERTS

Fields of Study: Geology; Ecology; Climatology

ABSTRACT

Deserts are regions defined by the scarcity of a crucial resource, water. They are unique ecosystems with their own types of plant and animal life; a number of desert plants have been used by humans for thousands of years. Mining for minerals and petroleum also occurs in desert regions around the world.

KEY CONCEPTS

alluvial fan: a fan-shaped deposit of silt, sand, rocks and other debris that is carried downslope by heavy runoff or a flash flood

desertification: the slow expansion of a desert area tending to increase the area it encompasses rather than just a shifting of its location

interdune area: the low area between dunes in a desert

rain-shadow effect: a mountain range parallel to a coastline causes incoming warm, moist air on the prevailing winds from the ocean to rise and cool, resulting in heavy rainfall on the ocean-facing side of the mountain range and dry, cool air on the leeward side of the mountain range; the arid leeward region is described as the mountain range's rain shadow

BACKGROUND

Although deserts are characterized by general aridity, there is no universal definition of a desert. *Webster's* dictionary defines a desert as an "arid land with usually sparse vegetation; especially such land having a very warm climate and receiving less than 25 centimeters of sporadic rainfall annually." Deserts are generally thought of as hot and dry, but heat is not necessarily a requisite of most definitions of a desert. Low precipitation does not, alone, characterize a desert. Arctic tundra landscapes often receive scant precipitation, yet tundra soils are often saturated from low evaporation and restrictive permafrost below a narrow active thawed layer during the summer. The common perception of a desert is a trackless expanse of sand such as the Sahara or the Arabian Desert. In reality, however, deserts are often fairly well vegetated. The characteristic that all deserts have in common is aridity. Aridity refers to a general dryness, not to short drought periods.

Deserts are located on all continents; Antarctica is considered a desert. They occur on a variety of substrates but often are characterized by historical erosion patterns leaving alluvial fans on the foothills of small mountains or hills and isolated islands of more resistant material. These remnants are called "buttes" in the western United States. Sandy soils are

Image of a desert. Photo via iStock.com/Mlenny [Used under license.]

common, but wind-blown soils and ash deposits from volcanoes are also prevalent in many deserts. Desert soils often lack structural aggregation and are subject to erosion. They often have surface crusts from raindrop impact and deposits from sediments that are left as water infiltrates into the soil.

CHARACTERISTICS OF DESERTS

In the United States, cool deserts (sagebrush grass and salt-desert shrub types) occur in old lake beds. Ancient Lake Bonneville in northern Utah and Lake Lahontan in Nevada once occupied much larger areas and dominated the landscape. These deserts are also influenced by the Sierra Nevada and Cascade Mountain ranges, which cast a rain-shadow effect on their eastern valleys. Intermittent drainages (wadis or arroyos) often cut across desert landscapes; they contain running water only during or immediately after a rainfall event. These drainages support unique vegetation and serve as important habitats for birds and other animals.

Another major feature of deserts is that nutrients are often limited. Soil nitrogen and organic matter are especially low in these ecosystems. Free-living and symbiotic nitrogen-fixing bacteria are scarce in deserts. Research indicates that lichens and algae, forming crusts on desert soils, may be the major source of soil nitrogen for plant growth.

DESERT ADAPTATIONS

Rainfall events occur infrequently in deserts. Both plants and animals must have adaptations to take advantage of these episodic periods of available water to survive in these harsh environments. Thus one sees flushes of desert flowers during spring and summer months, especially in years when rainfall is abundant. These same flowers may not be seen again for several years. Desert plants exhibit several adaptations that allow them to exist successfully under these stressful conditions. Some plants, such as mesquite, have deep root systems that allow access to deep sources of water. They do not have to rely on rainfall during the growing season. Some plants have dense, shallow root systems that allow them to tap soil water in the soil surface from light showers. Some plants have both types of root systems. Cactus plants and other succulents have the ability to store water in their tissues for use during periods of low rainfall. Some plants, such as the creosote bush, shed many leaves to reduce transpirational requirements. Others have few stomata on their leaf surfaces, thereby reducing transpirational stress. These stomata tend to close and restrict transpiration. Desert soils are often high in salt content from surface evaporation of rainfall that does not penetrate far into the soil, and plants growing on saline soils have special adaptations for coping with these conditions.

Animals also have many adaptations to desert conditions. Some desert animals can live in dormant stages during unfavorable periods. Many exhibit no definite breeding season and can breed whenever conditions are favorable. Some, including birds, can conserve water by reducing loss through concentrated urine. Some desert animals obtain most of their water through the food they eat. Many are nocturnal, thereby avoiding the high temperatures of the day. Intermediate-sized and small mammals often burrow to escape the heat and find more favorable conditions.

CONSERVATION AND PREVENTION ISSUES

Desert plants and animals are frequently under threat from a variety of natural and human-related activities. Since water resources are so scarce, any external factor that changes the water cycle or the availability of water at critical times may threaten organisms. In some cases the existence of desert plants and animals is not obvious, and adequate information on their status is lacking.

Many desert landscapes were formed by erosion, but changes may be very slow and difficult to ascertain. For example, there has been much concern over the increase in desert areas in the world. The word “desertification” has been used to describe the degradation of arid and semiarid areas into desert like environments. For some time nearly all observers believed that the Sahara Desert was expanding, but this has become a subject of debate. Detailed remote sensing analyses have suggested that the Sahara Desert may expand and contract, perhaps in response to climatic cycles. In many areas in Africa, such as the Sahel, livestock populations are increasing to support the expanding human population. During the dry season, woody plants supply needed high-quality forage for these animals. These same woody plant resources are also heavily exploited for fuel wood despite restrictions placed on their harvest. Depletion of these valuable woody plant resources could have serious consequences for these delicate ecosystems. Other examples of resource problems in desert areas include overhunting of native ungulates for animal products such as tusks.

NATURAL RESOURCES

Deserts of the world hold most of the reserves of oil, natural gas, and coal, and they therefore serve as a source of natural wealth for many countries in the Middle East. These resources remained sufficient to supply energy needs into the twenty-first century, but with escalating costs, alternative energy sources will become more important. Deserts also provide other minerals, such as silver, lead, diamonds (in southern Africa), and copper, but many deserts such as the Sahara are not important for mineral resources.

Desert plants are also used as food by many native people. In the southwestern United States, American Indians used cactus fruits as a staple food. In the Sahel in northern Africa, the following types of foods were available in substantial quantities: rhizomes and fleshy stems, the seeds of forbs, grass seeds, fruits, edible gums, and mannas. Fibers from desert plants are also used for basket weaving in many countries, and plant pigments are used as natural dyes.

VEGETATIONAL DYNAMICS

In desert regions of the United States and elsewhere, shrubby plants are increasing at the expense of grass land. Reasons for this change are not clear, but several factors are probably important. Some workers have suggested that climatic changes have favored a shift from grasses to shrubs. Increased carbon dioxide concentration of the atmosphere from burning fossil fuels is one hypothesis. Others insist that the introduction of domestic livestock beginning in 1850 disrupted the ecological balance in favor of shrubs. Lack of fire to restrict development of shrubs in grassland has also been advanced as a possible cause. Several studies in southwestern deserts in the United States have shown that native mammals and rabbits can exert considerable influence on vegetation and can reduce grass cover and abundance. It is possible that all these factors, and probably others not considered, have together been responsible for changes in desert vegetation.

Increases in shrubs, such as mesquite in the United States, eventually alter nutrient distribution patterns. In relatively uniform grasslands, soil nutrients are distributed fairly evenly. As shrubs such as mesquite increase at the expense of grasses, however, the shrubs are able to take up nutrients from a much larger volume of soil. These nutrients then tend to become concentrated around the individual mesquite trees and to form islands of nutrient concentration. Interdune areas often suffer soil loss through wind erosion, and the soil accumulates around individual mesquite plants. If the soil is deep enough, mesquite dunes eventually form. Interdune areas with little surface soil are deficient in nutrients, especially nitrogen, and lack water-holding capacity. These changes also affect animal life. Some animals, such as banner-tailed kangaroo rats and pronghorn antelope, utilize grassland habitat and are favored by grassland. Others, such as those that feed on mesquite (insects and arthropods), are favored by the mesquite dunelands.

Mining for coal and other minerals often disturbs desert ecosystems. Mines are limited in size and in the area affected, but they can leave conspicuous scars on the landscape, especially with deep open-pit mines. The area disturbed by the mines can be re stored, but restoration is difficult and expensive.

—Rex D. Pieper

Further Reading

Dimmitt, Mark A., Patricia Wentworth Comus, and Linda M. Brewer, editors. *A Natural History of the Sonoran Desert.* 2nd ed., Arizona-Sonora Desert Museum Press/U of California P, 2015.

Laity, Julie. *Deserts and Desert Environments.* Wiley-Blackwell, 2008.

Maghsoudi, Mehran. *Desert Landforms and Landscapes of Iran.* Springer, 2020.

Nevill, Simon. *Australia's Great Western Deserts: The Land-The People-The Explorers.* Woodslane Pty. Limited, 2021.

Rabinowitz, Dan. *The Power of Deserts: Climate Change, the Middle East and the Promise of a Post-Oil Era.* Stanford UP, 2020.

DIAGENESIS

Fields of Study: Geology; Geochemistry; Thermodynamics

ABSTRACT

Diagenesis refers to the physical, chemical, and biological changes that sediment undergoes after it is deposited. These processes change loose sediment into sedimentary rock and occur in the upper several hundred meters of Earth's crust.

KEY CONCEPTS

bar: a unit of pressure equal to 100 kilopascals and very nearly equal to 1 standard atmosphere

carbonate: a mineral with the carbonate anion CO_3^{2-} in its chemical formula, such as calcite ($CaCO_3$)

lithification: the hardening of sediment into a rock through recrystallization, compaction, cementation, or other processes

pore fluids: fluids, such as water (usually carrying dissolved minerals, gases, and hydrocarbons), in pore spaces in a rock

porosity: the amount of space between the sedimentary grains in a rock or sediment

sediment: loose grains of solid, particulate matter resulting from the weathering and breakdown of rocks, chemical precipitation, or secretion by organisms

sedimentary rock: a rock resulting from the consolidation of loose sediment that has accumulated in flat-lying layers on the earth's surface

DISTINGUISHING DIAGENESIS FROM METAMORPHISM

Diagenesis refers to the physical, chemical, and biological processes that occur in sediment after deposition as it is buried and transformed into sedimentary rock. These processes alter the texture, porosity, fabric, structure, and mineralogy of the sediment. Through diagenesis, sand is changed into sandstone, mud is changed into shale, and carbonate sediments are changed into limestone and dolomite. The processes and degree of alteration depend in part on the initial sediment composition and the depth of burial.

As sediment is buried to increasing depths, the temperatures and pressures increase, and diagenesis grades into metamorphism. Although the exact limits separating diagenesis and metamorphism are not strictly defined, diagenesis can be considered to occur under temperatures ranging from those at Earth's surface up to nearly 300°C and under pressures ranging from atmospheric pressure to at least 1 kilobar (1,000 bars). These conditions occur at depths approximating 10 kilometers. Some classifications restrict the zone of diagenesis to about 0 to 1 kilometers and define zones of catagenesis (several kilometers deep, with temperatures of 50°C to 150°C and pressures of 300 to 1,000 or 1,500 bars), metagenesis (up to about 10 kilometers deep), and finally metamorphism. Temperature is an important control on many diagenetic processes, because it influences chemical reactions such as the dissolution and precipitation of minerals, recrystallization, and authigenesis.

COMPACTION

The primary physical diagenetic process is compaction. Compaction presses sedimentary grains closer together under the load of overlying sediment, causing pore space to be decreased or eliminated and squeezing out pore fluids. In sandstones, compaction occurs by the rotation and slippage of sand grains, the breakage of brittle grains, and the bending and mashing of ductile (soft, easily deformed) grains. Brittle grains include thin shells, skeletal fragments, and feldspar grains. Ductile grains include clay or shale chips and fecal pellets. Compaction also causes some mineral grains to interpenetrate, producing irregular stylolitic contacts where the grains have intricately complex boundaries. In general, sands compact much less than muds. That is true since the average sandstone has a high percentage of hard grains, such as quartz; muds typically have a high initial water content, and water is squeezed out during compaction. The compaction of sands, however, is influenced by the nature of the sand grains present; sands with a large percentage of ductile grains are more susceptible to compaction.

CEMENTATION, AUTHIGENESIS, AND REPLACEMENT

Chemical diagenetic processes include cementation, the growth of new minerals (authigenesis), replacement, neomorphism (recrystallization and inver-

Pyritized Lloydolithus lloydi fossil from lower Ordovician. Photo via Wikimedia Commons. [Public domain.]

sion), and dissolution. Cementation is the precipitation of minerals from pore fluids. These minerals cement the grains of sediment together, forming rock. The most common cements are quartz (silica), calcite, and hematite, but such cements as clays, aragonite, dolomite, siderite, limonite, pyrite, feldspar, gypsum, anhydrite, barite, and zeolite minerals occur as well. The type of cement is controlled by the composition of the pore fluids.

Authigenesis refers to the growth of new minerals in the sediment and the transformation of one mineral into another. Some of the most common authigenic minerals in sandstones are calcite, quartz, and clay cements. Other than cements, authigenic minerals include glauconite, micas, and clay minerals. Glauconite is a green mineral that forms on the seafloor when sedimentation rates are low. Authigenic micas and clays typically form in the subsurface at higher temperatures and pressures. Often, one clay mineral is transformed into another as a result of dehydration (water loss) or chemical alteration by migrating fluids. Clays may also be formed from the alteration of feldspars or volcanic ash and rock fragments. Authigenesis also includes the alteration of iron-bearing minerals (such as biotite, amphibole, or pyroxene) to pyrite, under reducing conditions, or to iron oxide (limonite, goethite, or hematite), under oxidizing conditions.

Replacement is the molecule-by-molecule or volume-for-volume substitution of one mineral for another. Replacement generally involves the simultaneous dissolution of an original mineral and precipitation of a new mineral in its place. Fossils that were originally calcium carbonate may be replaced by different minerals, such as quartz, pyrite, or hematite. Many minerals are known as replacement minerals, including calcite, chert, dolomite, hematite, limonite, siderite, anhydrite, and glauconite. Factors controlling replacement include ph, temperature, pressure, and the chemistry of the pore fluids.

NEOMORPHISM AND DISSOLUTION

Neomorphism is a term meaning "new form"; it refers to minerals changing in size, shape, or crystal structure during diagenesis. The chemical composition of the minerals, however, remains the same. Neomorphism includes the processes of recrystallization and inversion. Recrystallization alters the size or shape of mineral grains without changing their chemical composition or crystal structure. Recrystallization can occur in any type of sedimentary rock, but it is most common among the carbonates. Limestones are commonly recrystallized during diagenesis, producing a coarsely crystalline rock in which original sedimentary textures and structures may be fully or partially obliterated. The reason that minerals recrystallize is not well understood, but it may be related to energy stored in strained crystals or to a force arising from the surface tension of curved crystal boundaries.

Inversion is a process in which one mineral is changed into another with the same chemical composition but a different crystal structure. The two minerals involved are called "polymorphs", meaning "multiple forms." Aragonite and calcite are polymorphs. Both have the same chemical composition, but each has a different crystal structure: aragonite is orthorhombic and calcite is rhombohedral. Aragonite, with time, will become calcite by inversion. Inversion may occur along a migrating film of liquid, causing the simultaneous dissolution of one mineral and precipitation of its polymorph, or by solid-state transformation (switching of the positions of ions in the crystal lattice).

Dissolution refers to the dissolving and total removal of a mineral, leaving an open cavity or pore space in the rock. This pore space may persist, or it may become filled by another mineral at a later time. Some of the more soluble minerals are the carbonates and the evaporites, such as halite and gypsum.

Large-scale dissolution of limestone leads to the formation of caves and caverns. Pressure solution is the dissolution of minerals under the pressure of overlying sediment. Stylolites, a common result of pressure solution, commonly occur in carbonate rocks. Stylolites are thin, dark, irregular seams with a zigzag pattern that separate mutually interpenetrating rocks. The dark material along the seam is a concentration of insoluble material such as clay, carbon, or iron oxides. Pressure solution can result in a 35 to 40 percent reduction in the thickness of carbonate rocks. The carbonate removed by pressure solution is frequently a source of carbonate cements.

Finally, minerals in solution may be deposited on existing grains. This phenomenon is very common with quartz grains. Silica in solution builds on the existing crystalline structure of the grain, causing new quartz to form. Often the sand grain develops actual crystal faces as a result. Such new mineral growths are called "overgrowths."

BIOLOGICAL DIAGENETIC PROCESSES

Biological diagenetic processes occur soon after sediment is deposited and consist of the activities of organisms in and on the sediment. Bacteria are particularly important to the chemical diagenetic processes. Bacteria living in the sediment control many chemical reactions involving mineral precipitation or dissolution; they are involved in the breakdown or decomposition of organic matter (one of the steps in the formation of oil and gas), and can cause the pH of the pore fluids to increase or decrease, depending on the kinds of microorganisms, organic matter, decomposition products, and availability of oxygen. For example, in aerobic environments (those where oxygen is present), decay of organic matter generally causes decreasing pH (increasing acidity), which may lead to the dissolution of carbonate minerals such as calcite. Under anaerobic conditions, organic decay generally raises the pH and may lead to the precipitation of calcite cement. The formation of pyrite is also influenced by the activity of bacteria. Sulfate-reducing bacteria in anoxic environments change sulfate ion into sulfide ion. If iron is present, it reacts with the sulfide to form iron sulfides, such as pyrite.

Bioturbation is the disturbance of the sediment by burrowing (excavation into soft sediment), boring (drilling into hard sediment), the ingestion of sediment and production of fecal pellets, root penetration, and other activities of organisms. Bioturbation generally occurs shortly after deposition and causes mixing of sediment that was originally deposited as separate layers, destruction of primary sedimentary structures and fabrics, and breakdown or clumping of grains. In some cases, chemical alteration of the sediment accompanies bioturbation. For example, light-colored halos may form around burrows or roots, particularly in red or brown sediments, because of the reduction of iron.

Diagenesis may decrease or increase the porosity and permeability of the sediment. Porosity is decreased by compaction, the precipitation of cements in pore spaces, and bioturbation. Porosity is increased by dissolution. Zones of increased porosity are particularly favorable for oil and gas accumulations.

STUDY TECHNIQUES

Diagenesis is primarily studied using sedimentary petrography, which is the microscopic examination of thin sections of sedimentary rocks. Thin sections are slices of rock, typically 30 micrometers thick, bonded to glass slides, which are examined with a petrographic microscope. In this way, minerals can be identified based on their optical properties, and textural relationships can be studied, such as the size, shape, and arrangement of grains; the geometry of cements and pore spaces; the character of contacts between grains; the presence of dissolution features; and mashing or fracturing of grains. Thin sections may be enhanced, to allow easier identification of minerals, with various staining and acid etching techniques. In addition, acetate peels may be prepared from etched and stained rock surfaces for examination with the microscope.

There are a number of other techniques that can be used in conjunction with petrography to obtain more specific diagenetic data. Cathodoluminescence microscopy can provide information about the spatial distribution of trace elements in rocks. Luminescence is the emission of light from a material that has been activated or excited by some form of energy. Cathodoluminescence works by activating various parts of a polished thin section with a beam of electrons. The electron beam excites certain ions, producing luminescence. This technique can reveal small-scale textures and inhomogeneities of particles and cements through differences in their luminescence, which are related to differing concentrations of trace element ions.

Scanning electron microscopy can magnify images 70,000 times or more, permitting detailed study of extremely small particles that cannot be adequately examined using a petrographic microscope. Scanning electron microscopy involves reflections of an electron beam from a rock or mineral surface. Fine details of cements and grains may be readily observed and photographed.

X-ray diffraction is used to determine the mineralogy of sedimentary rocks, particularly fine-grained rocks such as shales. The technique is based on reflections of X-rays from planes in the crystal structure of minerals. Each mineral has a characteristic crystal structure and produces a distinctive X-ray diffraction pattern consisting of peaks of different position and intensity, which are plotted on chart paper by the X-ray diffractometer.

Fluid inclusions are extremely small droplets of fluid encased within crystals or mineral grains. The fluids are a small sample of the original pore fluids from which the mineral was precipitated. By examining fluid inclusions using heating and freezing devices attached to a microscope, the geologist can determine the composition of the original pore fluids and the temperature at which the mineral was precipitated. This technique reveals that many minerals and cements were precipitated from hot, saline pore fluids.

Stable isotopes of oxygen and carbon are commonly used to determine the chemistry of the pore fluids and the temperatures under which precipitation of cements or authigenic minerals occurred. Studies of the stable isotopes of microfossils have also provided information on past climatic changes. Isotopes are different forms of elements that vary in the number of neutrons present in the nucleus; hence, the various isotopes of an element have different atomic weights. By comparing the ratios of oxygen-16 and oxygen-18 or carbon-12 and carbon-13 in minerals such as calcite, it is possible to determine whether the minerals precipitated from freshwater or marine water, or to determine the temperature of the fluid from which the mineral precipitated.

—Pamela J. W. Gore

Further Reading

Boggs, Sam, Jr. *Petrology of Sedimentary Rocks.* Cambridge UP, 2009.

———. *Principles of Sedimentology and Stratigraphy.* 5th ed., Prentice Hall, 2011.

James, Noel P., and Brian Jones. *Origin of Carbonate Sedimentary Rocks.* Wiley, 2016.

Leeder, Mike R. *Sedimentology and Sedimentary Basins: from Turbulence to Tectonics.* Wiley-Blackwell, 2011.

Mackenzie, F. T., editor. *Sediments, Diagenesis, and Sedimentary Rocks.* Elsevier, 2005.

Middleton, Gerard V., editor. *Encyclopedia of Sediments and Sedimentary Rocks.* Springer, 2003.

Selley, Richard C., and Stephen A. Sonnenberg. *Elements of Petroleum Geology.* 3rd ed., Academic Press, 2015.

Diamond

Fields of Study: Geology; Geomorphology; Geochemistry; Gemology; Crystallography

ABSTRACT

Diamond is one of the world's most important minerals and gemstones; it is the element carbon (C) crystallized in the isometric system. Each atom of carbon in a diamond is covalently bonded to four other carbon atoms in a tetrahedral arrangement, effectively making every single diamond crystal, no matter how large, a single molecule.

KEY CONCEPTS

covalently bonded: of atoms that are connected to each other by sharing of two electrons without charge separation into positive and negative ions

diamond pipes: the chimneys of explosive volcanic eruptions that have allowed diamond-bearing kimberlite to rise from deep in the mantle and fill the available space within the chimney

faceting: a specialist skill in which a rough diamond is meticulously positioned against a rotating lap in order to create flat surfaces in a particular design

lap: a device consisting of a hard and perfectly flat disc that can be impregnated with abrasive grit, typically diamond dust, used to produce a polished surface on a gemstone

Mohs hardness scale: a relative scale based on the hardness of ten naturally occurring minerals, with talc being the least hard and diamond being the hardest

refractive index: a measure of the extent to which a transparent material can alter the direction of, or "bend," rays of light

A BRIEF HISTORY OF DIAMONDS

Diamond is the hardest known substance, natural or artificial, and is number 10 on the Mohs hardness scale. The atoms in a diamond are each covalently bonded to four other carbon atoms in a tetrahedral arrangement, which gives diamond its unique hardness. It also has the highest thermal conductivity of any known substance.

Historical records of diamonds date back to 3000 BCE. In recent centuries, Golconda diamonds of India dominated diamond production until the early eighteenth century. In 1725, Brazilian diamond mines gained prominence. South Africa's "great diamond rush" began in 1867, and in 1890, the De Beers company consolidated dozens of mining

A diamond. Photo via iStock.com/enot-poloskun [Used under license.]

communities in Africa. Diamond derives its name from the Greek word *adamas*, which means "unconquerable."

Almost all of the world's diamond production comes from Africa, most notably South Africa. Other diamond-producing countries include Angola, Borneo, Canada, Ghana, Guyana, Namibia, Sierra Leone, Tanzania, Venezuela, Congo, Brazil, and Russia (Siberia). In the United States diamonds have been found in Arizona, Arkansas, Montana, and Nevada.

TECHNICAL DEFINITION OF DIAMOND

The atomic number of carbon is 6, and its atomic weight is 12.011. It belongs to Group IVA of the periodic table of elements. Gem diamonds have a density of 3.52, although "black diamonds" have a density of about 3.15. Diamond slowly burns to carbon dioxide at a very low temperature (900° Celsius).

Diamond's high refractive index (2.417) and strong dispersion property (0.058) guarantee its supremacy as a gemstone. However, only about one-fifth of all the diamonds mined qualify as gems. Most of the remaining uncut diamonds are used by industry. Tunnel boring and oil-well drilling equipment uses diamond-studded rotary bits. Carbide grinding wheels, abrasion-resistant cutting tools, and glass-etching and glass-cutting equipment use industrial-quality diamonds. Some dentists and surgeons use diamond-headed scalpels to cut delicate bones and tissue. Diamond coatings are used in integrated circuits, prosthetic devices, and biosensors. Diamond is the most important industrial abrasive, and industry uses about 80 percent (by weight) of all diamonds produced. However, this represents only about 30 percent by value.

CREATION AND PROPERTIES OF DIAMOND

About 30 kilometers inside Earth, exceedingly high pressures and temperatures (more than 1,400° Celsius) promote crystallization of magnesium-rich rock melts, resulting in the formation of diamonds. Samples of deep mantle material contain diamonds as a natural component. The reaction of groundwater with hot, magnesium-rich, deep mantle material aided by carbon dioxide leads to the formation of a rock called "kimberlite." Kimberlite is an igneous rock that is ultrabasic and contains very little silica. Kimberlite is the world's principal source of diamonds. Explosive eruptions create craters filled with deep mantle rock formations and permit diamond-containing rock to surface through cracks. These are known as diamond pipes (sometimes incorrectly called "volcanic necks").

In addition to being the hardest known substance, diamond is an excellent conductor of heat. Because diamonds possess the highest thermal conductivity of any known substance, industrial-quality diamonds are used in abrasion-resistant cutting tools. Almost all diamonds are nonconductors of electricity. However, some diamonds permit the passage of electric current when they are altered by bombardment with radiation. Diamond crystals form as octahedrons, dodecahedrons, and cubes. A well-cut gem can reflect almost all the light that it receives. This quality is called "luster." In addition, it can disperse or separate the colors of the spectrum while reflecting the incident light. This quality is called "fire."

OCCURRENCE OF DIAMONDS

Mining experts have discovered hundreds of diamond-containing dikes and pipes in Transvaal, Kimberley District, and Free State (formerly Orange Free State), South Africa; Yakutsk, Siberia; Shinyanga, Tanzania; Mbuji-Mayi, Democratic Republic of the Congo; Yengema, Sierra Leone; Murfreesboro, Arkansas; the Yukon and Northwest Territories of Canada, and several other locations. However, it is not economically feasible to carry out "mine-at-depth" procedures in most of these pipe mines.

The erosion of diamond pipes over millions of years has resulted in secondary deposits called "alluvial" or "placer" deposits. These deposits contribute significantly to the world's total diamond production. Most alluvial diamonds are recovered from stream gravel, but beach gravel is also a good source. Diamond-containing beach gravel extends to the depths of the ocean floor, although there is no economical means of recovering diamonds from ocean depths. Diamonds are also found in glacial tills. Minute quantities of microscopic diamonds have been found in meteorites as well.

SYNTHETIC DIAMONDS

On February, 15, 1955, the General Electric Company announced its success in creating a synthetic diamond. Since then synthetic diamonds have become widely used in grinding wheels and a number of other applications. They are normally single crystals, usually octahedral in shape. Since they have several cutting edges, they are preferred over natural diamonds for industrial purposes. To make them, graphite (another form of crystalline carbon) is subjected to very high temperatures and pressures. Extreme pressures as high as 296,076 atmospheres (about 30 billion pascals) and temperatures as high as 3,037° Celsius (t boils at 100° Celsius) have been used, depending upon the actual process. Two common procedures are shock conversion and static conversion. Synthetic diamonds are also manufactured by the static crystallization of certain alloys and molten metals.

Synthetic diamonds, normally black in color, are produced in grain sizes that are about one-hundredths of a centimeter in diameter. It is possible to "grow" larger, gem-quality synthetic diamonds, but the process is too costly to be feasible. Synthetic diamonds are chemically and crystallographically identical to the naturally occurring diamond gemstone. "Imitation diamonds," on the other hand, are completely different from either synthetic or genuine diamonds. Imitation diamonds do not possess either the hardness or the crystallographic structure of the genuine diamond; they are chemically different. They are made of glass or other material and are simply intended to imitate the appearance of a diamond.

CUTTING DIAMOND

After mining and recovery, gem-quality diamonds are separated from industrial-quality ones. A rough, uncut diamond looks like a dull piece of glass. Precise cutting, artful grinding, and skillful polishing of the diamonds yield outstanding gems, and some have attained historic fame. Diamond cutting began in India and was later perfected in Italy. Only a diamond can cut a diamond: Diamond crystals are cut, cleaved, shaped, and polished by "diamond dust on a lap." World-famous diamond cutting establishments are concentrated in Antwerp, Belgium, and in Amsterdam, the Netherlands. India and Israel have also emerged as world leaders in diamond cutting. The most popular cut is the "brilliant cut," which has a round shape with fifty-eight facets. Gem-quality diamonds are classified according to their weight, clarity, color, and absence of flaws. The weight of a diamond is measured in carats; a carat equals 0.2 gram, or about 0.00704 ounce. Transparent, colorless, and light blue diamonds are extremely rare and are considered to be highly valuable gems. There are red, pink, blue, and green diamonds, the colors usually being due to various atoms becoming trapped within the crystal lattice of diamond as it is formed, although "holes" in the crystal lattice may also have an effect on the color of the stone. Diamonds with a yellow tint are more common. As the tint becomes increasingly yellowish, the value decreases. Industrial-quality diamonds are gray, brown, or black and are almost opaque. They are gems of poor quality.

FAMOUS DIAMONDS

The largest diamond ever found, the Cullinan, was found in 1905 in the Premier Mine, Transvaal, South Africa, and weighed 3,106 carats. This stone was cut and polished into several gems, two of them world famous: The 530-carat Star of Africa and the 309-carat Star of Africa II are among the British crown jewels, housed in the Tower of London. These are the world's largest cut diamonds. The cutting of the Cullinan also resulted in another seven large gems and ninety smaller ones.

The 109-carat Koh-i-Noor ("mountain of light"), set in the British crown itself, is the oldest diamond gemstone known to historians; its history has been traced back to 1304. This diamond had its origin in India, and it originally weighed 186 carats before Queen Victoria had it recut in 1852. Many believe that the largest blue diamond, the 44.5-carat Hope diamond, presently in the Smithsonian Institution, adorned the eye of an Indian god. Other world-famous diamonds India has contributed include the Regent or Pitt (140 carats, presently in the Louvre, France); the Orlov (200 carats, presently in Russia); the Florentine (137 carats, location unknown); and the Great Mogul (280 carats, location also unknown).

—Mysore Narayanan

Further Reading

Balfour, Ian. *Famous Diamonds.* 4th ed., Christie's, 2000.

Chatterjee, Kaulir Kisor. "Diamond." *Uses of Industrial Minerals, Rocks, and Freshwater.* Nova Science, 2009.
Dundek, Marijan. *Diamonds.* Noble Gems International, 2009.
Green, Timothy. *The World of Diamonds.* Morrow, 1981.
Pierson, Hugh O. *Handbook of Carbon, Graphite, Diamond and Fullerenes.* William Andrew, 2012.
Prelas, Mark A., Galina Popovici, and Louis K. Bigelow, editors. *Handbook of Industrial Diamonds and Diamond Films.* Routledge, 2018.
Tappert, Ralf, and Michelle C. Tappert. *Diamonds in Nature: A Guide to Rough Diamonds.* Springer, 2011.
Hart, Matthew. *Diamond: A Journey to the Heart of an Obsession.* Walker, 2001.
Nazaré, M. H., and A. J. Neves, editors. *Properties, Growth, and Applications of Diamond.* IEE, 2001.
O'Donoghue, Michael. *Gems.* 6th ed., Robert Hale, 2008.
Zoellner, Tom. *The Heartless Stone: A Journey Through the World of Diamonds, Deceit, and Desire.* St. Martin's Press, 2006.

DIATOMITE

Fields of Study: Geology; Geomorphology; Geochemistry; Materials Science

ABSTRACT

Diatomite, or diatomaceous earth, is found in deposits near present-day or ancient bodies of water, because it is composed of the silica shells of water-dwelling diatoms. Diatomite deposits are found throughout the world; major producers include the United States, China, Denmark, and Japan. The United States is the main producer of diatomite, accounting for at least 50 percent of the world's diatom exports every year.

KEY CONCEPTS

calcinating: a process of heating a material in air
diatom: a microscopic single-cell alga found in all parts of the world; the large deposits of fossil diatoms are composed of a type of sedimentary rock called "diatomite"
frustule: the overlapping valves of a diatom, typically containing silica
lacustrine: relating to freshwater lakes

PRIMARY USES

Uses for diatomite fall into four main categories: for filtering, for insulating and building, as a filler material, and as a mild abrasive. Diatomite is commonly used to filter a wide variety of substances, ranging from oils to drinking water. As an abrasive, known by woodworkers as "rottenstone," it is used in toothpastes and metal polishes. Many products, ranging from ceramics to paints, use diatomite as a filler to add volume.

TECHNICAL DEFINITION

Chemically, diatomite consists primarily of silica with trace amounts of magnesium, sodium, iron, and other elements, although exact proportions vary. Purified diatomite is essentially silica (SiO_2), with an average molecular mass of 60.8. Diatomite has a melting point of 1,710°C and a density of 2.3 grams per cubic centimeter. Heating it to high temperatures forms crystalline silica.

Diatomite is usually white (if pure), buff, gray, and rarely black. *In situ*, it is generally found as a soft sedimentary rock or as powder. Raw diatomite is typically processed by a series of crushing, drying, size reduction, and calcining procedures to produce different grades of diatomite for different specialized applications.

DESCRIPTION, DISTRIBUTION, AND FORMS

Diatomite is a soft, chalklike, fine-grained sedimentary rock composed primarily of the fossilized silica shells of microscopic algae called "diatoms." It is finely porous, is low in density, and has low thermal conductivity. Diatom frustules are composed of two symmetrical silica valves, which can be elaborately ornamented with tiny holes and protrusions. These tiny holes are what make diatoms an ideal material for filtration. The word "diatom" comes from Greek *diatomos*, meaning "cut in half," because of the two valves.

Diatoms live in a wide range of moist environments, although most abundantly in marine (oceanic) and lacustrine (freshwater) environments. Three main types of diatomite deposits are recognized in the United States: marine rocks near continental margins, lacustrine rocks formed in ancient lakes or marshes, and sedimentary rocks in modern lakes, marshes, and bogs. Another commonly used term for diatomite, diatomaceous earth, more properly refers to unconsolidated or less lithified forms of diatomite. One of the most important marine diatomite deposits is near Lompoc, California, reported to be the world's largest producing district by volume. Economically important lacustrine deposits in the United States are found in Nevada, Oregon, Washington, and eastern California. In 2007, the United States produced 33 percent of the world's diatomite. Other leading producers were China (20 percent), Denmark (11 percent; all moler diatomite, containing 30 weight percent clay), Japan (6 percent), and France (4 percent).

HISTORY

Some of the earliest references to diatomite are to the ancient Greeks' probable use of it to form lightweight bricks for building; they also used diatomite as an abrasive. In 535 CE, the Roman Emperor Justinian I used diatomite bricks in building the church of St. Sofia in Constantinople (now Istanbul).

Diatomite use became industrially important to Western Europe after 1867, when Alfred Nobel invented dynamite. Pulverized diatomite was commonly used to absorb and stabilize the nitroglycerine used to make dynamite. By the late 1800s, the United States had become the primary producer of diatomite. By 1900, diatomite's uses had expanded to include many of its present-day uses, including beer filtration and building materials.

During the 1920s, techniques for calcining (thermally treating) and grading diatomite enabled a wider variety of uses for this resource. By World War II, the U.S. Army and Navy made wide use of diatomite to purify drinking water, to remove oil from boiler and engine water, and to create low-light-reflectance paints for ships.

OBTAINING DIATOMITE

Because of its abundance and usual occurrence near the surface in the United States, most diatomite produced is obtained from open-pit mines. The diato-

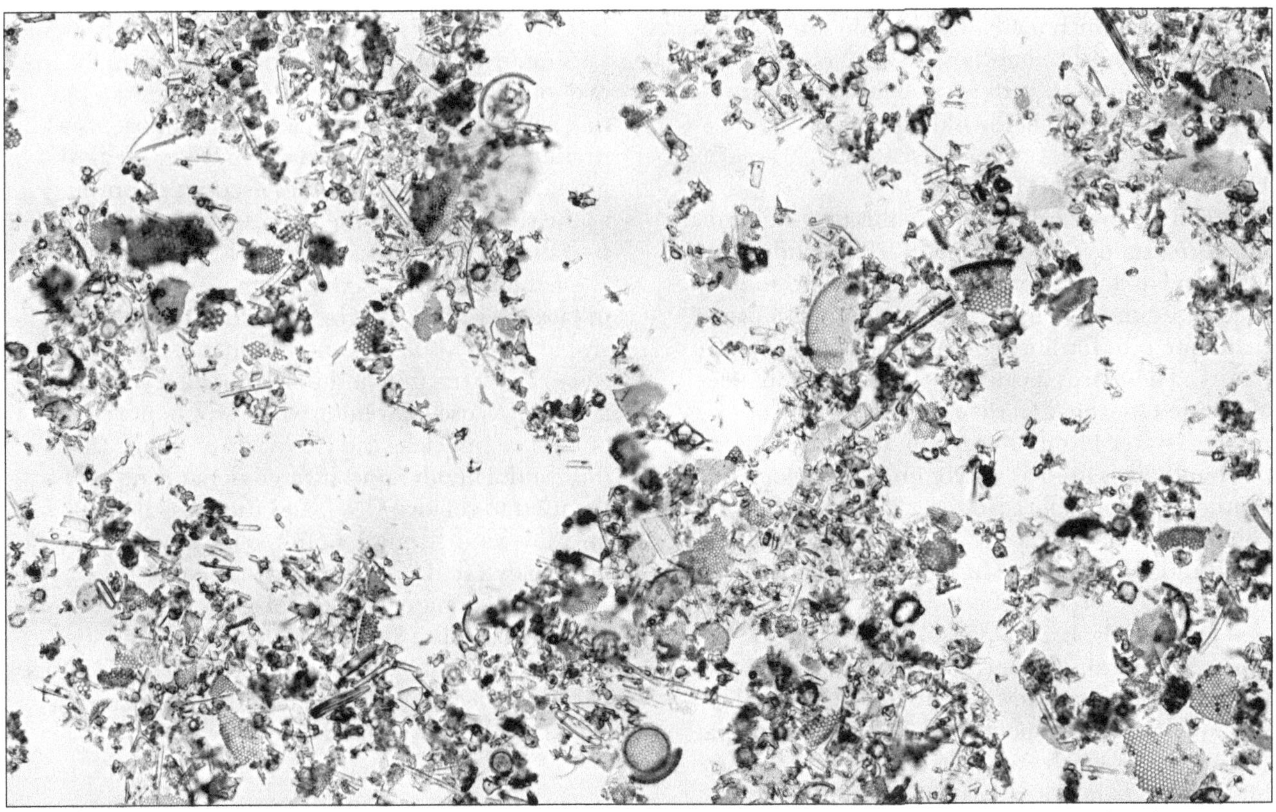

Diatomaceous earth as viewed under bright field illumination on a light microscope. Photo by Zephyris, CC BY-SA 3.0, via Wikimedia Commons.

mite is excavated by machine after the overburden is removed. Outside the United States—particularly in China, Chile, and France—underground diatomite mining is fairly common. These mines are usually pit-and-pillar mines excavated by machine, although some small mines are excavated using hand tools. In Iceland, diatomaceous mud is dredged from Lake Myvatn. Diatomite is often dried in the open air near the mine before processing.

Diatomite processing is often carried out near the mine from which it is extracted. Raw diatomite may contain up to 65 percent water and is expensive to transport. Primary crushing of ore is usually done with spiked rolls and hammer mills, reducing the ore to 1.25-centimeter pieces while limiting damage to the diatom structure.

Passage through heated air, milling fans, and air cyclones further dries the diatomite and begins to classify for size as well as remove impurities of different density. Processing aims to separate individual diatom valves without destroying their structure, which is key to filtration uses.

Calcining, which increases filtration rate, specific gravity, and particle hardness, as well as oxidizing iron, is usually done with rotary kilns. Calcining is particularly important for filter grades.

USES OF DIATOMITE

Diatomite is primarily used as a filtration medium but also is used for insulation, as a filler and absorbent, and as a mild abrasive, in addition to some specialized medical uses. The most common use of diatomite is in filtration, because of its finely porous nature. These uses include water purification, beer and wine filtering, and the removal of oils from water. As a water filtration element, diatomaceous earth usually is used as a layer on a filter element or septum (a permeable cover over interior collection channels), called "precoat" filtration. Diatomite water filtration systems are lightweight, cheap, and simple and can remove bacteria and protozoans as well as cysts, algae, and asbestos. This usage of diatomite first became important during World War II, when the U.S. Army needed a water filter suitable for mobile military operations. The first municipal diatomaceous-earth water filtration system was set up in 1948, and more than two hundred operate presently in the United States. Diatomite is also used to filter nonpotable water, such as that which is used in swimming pools.

Diatomite began to be used after Prohibition to filter beer and wine in the United States, replacing wood pulp in filters. It is also used to filter liquid sweeteners, oils and fats, petroleum and other chemicals, and pharmaceuticals. Another major use of diatomite is in building, where it is used for lightweight blocks and bricks and for thermal insulation (high clay-content Danish moler in particular). Diatomite is also a frequent cement additive; diatomite for cement requires less processing.

As a filler, diatomite has many uses. In addition to providing bulk, diatomite can reduce reflectivity in paints, reduce caking in granular mixtures, and provide a variety of effects in plastics, including preventing film sticking. Diatomite is absorbent and often used for cleaning industrial spills and in cat litter.

As an insecticide, diatomite is less toxic than chemical pesticides, as it works by absorbing lipids from insects' exoskeletons, causing dehydration. However, it harms beneficial insects as well as pests. Diatomite also is used as a growing medium for hydroponics and an additive in various types of potting soil, because it retains water and nutrients while draining quickly, similar to vermiculite. Medical-grade diatomite is sometimes used for deworming, as the sharp edges of the frustules are thought to kill parasites, but the efficacy of this is questionable.

Diatomite is also used in cosmetics—for example, in facial masks to absorb oil—and as a minor abrasive in jewelry polishes and toothpastes. Some processes for extracting and purifying deoxyribonucleic acid (DNA) use diatomite, which will remove DNA but not ribonucleic acid (RNA) or proteins. Diatomite and a highly concentrated denaturing agent are used to remove DNA, and then a slightly alkaline, low ionic strength buffer (such as water) can be used to extract DNA from the diatomite.

While diatomite can be replaced by other materials—such as silica sand, perlite, talc, ground lime, ground mica, and clay—for most of its applications, its abundance, availability, and low cost make it a popular and heavily used resource.

—Melissa A. Barton

Further Reading
U.S. Geological Survey. *Minerals Yearbook*. Author, 2008.

DOELL AND DALRYMPLE DISCOVER THE MAGNETIC REVERSALS OF EARTH'S POLES

Fields of Study: Geology; Geomagnetism; Stratigraphy

ABSTRACT
Richard Rayman Doell and Brent Dalrymple's discovery of the magnetic reversals of Earth's poles represented a breakthrough in understanding the history of the planet and a major missing piece of the theory of continental drift.

KEY CONCEPTS

Curie temperature: the temperature at which metal atoms are no longer able to maintain magnetic susceptibility; above this temperature the magnetic moment of the atoms is lost, below it the magnetic moment of the atoms becomes fixed

midoceanic ridges: a line of volcanic peaks roughly bisecting the ocean floor where convection currents in the magma of Earth's mantle tend to force separation of the tectonic plates as basaltic magma is forced upward into the rift

paleomagnetism: generally, the embedded magnetism in ancient rocks and its relation to the geologic time scale

remanent magnetization: the embedded magnetism in specific ancient rock samples with regard to its directionality

stratigraphic horizon: a particular stratum within Earth's crust corresponding to a corresponding period in the geologic time scale

SUMMARY OF THE DISCOVERY

In the early twentieth century, the idea was proposed that the polarity of the earth's magnetic field experienced changes. This was based on studies by geophysicists of remanent magnetization of volcanic rocks and baked earth. In 1906, Bernard Brunhes, a French physicist, found volcanic rocks magnetized in the opposite direction with reference to Earth's present magnetic field. He concluded that the magnetic field had reversed. His conclusion was accepted, but scientists were not interested in this type of research at that time.

Computer simulation of the Earth's field in a period of normal polarity between reversals. The lines represent magnetic field lines. Image via Wikimedia Commons. [Public domain.]

Laboratory studies showed that volcanic rocks and baked earth, when heated to their Curie temperature (the temperature above which a substance is no longer magnetic) and allowed to cool, acquired a weak and stable remanent magnetization (thermoremanent magnetism develops in igneous rocks when they cool below the Curie temperature), which was parallel to Earth's present magnetic field. Scientists also found bricks and baked earth in archaeological sites that had their magnetism reversed because of fires made by humans. Lava flows also change the magnetism of rocks (referred to as baked earth) that they have covered.

In 1925, R. Chevallier studied lava flows from Mount Etna, which contained remanent magnetizations that correlated with the magnetic field of Earth at the time of their eruption. In 1926, Paul L. Mercanton suggested that if magnetized rocks showed a reversed magnetic field for Earth, then those reversals should be recorded worldwide in other rocks. Motonori Matuyama, in 1929, was the first paleomagnetist to determine the times when the magnetic field had reversed itself by determining the ages of rocks. He found that the youngest rocks with reversed polarity were of early Quaternary period. After this early work on geomagnetic reversals, little was done in this area from 1928 to 1948. This resulted from the fact that science did not have an adequate theory to explain Earth's magnetic field, much less magnetic reversals. The study of magnetic reversals resumed around 1950. This resulted from the theory of how the earth's magnetic field was produced. The mechanism proposed was that of a magnetohydrodynamic dynamo.

The theory developed from the work of Walter M. Elsasser and Sir Edward Bullard. The dynamo idea suggested that Earth's outer core consisted of molten iron and nickel and was similar to the electrical conductors of a dynamo. It was hypothesized that the rotation of the planet produced convection currents in the outer core and that the resulting electric currents created an eternal magnetic field in Earth. The magnetic lines of force that would develop would be similar to those formed as if there was a bar magnet inside Earth. This, however, did not explain how Earth's magnetic field could reverse itself.

Scientists have suggested that a period of strong solar-flare activity could cause the magnetic field to die and that after this activity was over the magnetic field would come back, but in a reversed manner. The question has arisen as to whether the magnetism in a rock can go through self-reversal. Laboratory experiments show that this occurs in less than 1 percent of thousands of rock samples tested from around the world. Therefore, if reversal of the magnetic field is the correct explanation for geomagnetic reversals, transitions from reversed to normal polarization, or the opposite, must occur at the same stratigraphic horizon worldwide.

Many geologists and geophysicists have done work in paleomagnetism, but the work of Richard Rayman Doell and Brent Dalrymple provided a complete picture of paleomagnetic reversals of the past 5 million years. Allan V. Cox, Doell, and Dalrymple spent five years (1959-1964) working on continental volcanic rocks from Alaska, Hawaii, Idaho, California, and New Mexico, to determine if the last magnetic reversal from reversed polarity (named the Matuyama reversed epoch) to the present normal polarity (named the Brunhes normal epoch) occurred contemporaneously around the world approximately 1 million years ago. Paleomagnetists had up to this time not found magnetic reversals in the upper Pleistocene.

Geologically, 1 million years ago marks the transition from the Pleistocene epoch (the last Ice Age) to the Holocene (or recent) epoch of time. (Some geologists refer to the Holocene as the upper Pleistocene.) Reversals first appeared in the recent geologic record in the middle and lower Pleistocene (1 million to 2 million years ago). By 1964, Cox, Doell, and Dalrymple had found no evidence for contemporaneous reversals to conform with the change from the Matuyama reversed epoch to the Brunhes normal epoch. This problem eluded them for another year.

In 1965, while Cox was doing research abroad, Doell and Dalrymple continued their study of volcanic rocks from the Valles Caldera in New Mexico, which is part of the Jemez Mountains located 56 kilometers northwest of Santa Fe. On November 4, 1965, at a meeting of the Geological Society of America in Kansas City, Missouri, Dalrymple presented a paper, written by himself, Doell, and Cox, in which preliminary results from their work in New Mexico were reported. The paper was entitled "Recent Developments in the Geomagnetic Polarity Epoch Time-Scale," and it contained the following statement: "There may be another event at about 0.9 million years, although it is yet not confirmed. Because of the short duration of these events, the chances of finding another one are rather small; thus subsequent work may turn up additional events that are as yet unrecognized."

On May 20, 1966, *Science* published a two-page article by Doell and Dalrymple, "Geomagnetic Polarity Epochs: A New Polarity Event and the Age of the Brunhes-Matuyama Boundary." Doell and Dalrymple determined the radiometric age using the potassium-argon method and paleomagnetism of nineteen Pleistocene volcanic rock units collected during the summer of 1964 in the Valles Caldera. Six samples had ages that ranged from 0.7 to 1.0 million years. These dates coincided with the age of the Bishop Tuff of California, which had been dated at between 0.87 and 1.2 million years.

The new dates were valuable in defining the Matuyama-Brunhes boundary, which was established at 0.7 million years. Their study of the rock samples from New Mexico showed that a magnetic field reversal took place between 0.8 and 0.9 million years ago. This event occurred toward the end of the Matuyama reversed polarity epoch. These rocks showed a normal polarity, and the event has been named the Jaramillo normal event after a creek in the area where the rocks had been collected. The rocks dated between 0.7 and 0.8 million years ago showed a reversed polarity. From 0.7 million years ago to the present, the rocks from New Mexico indicated a normal polarity. The type of brief reversals, as the Jaramillo normal event, occurred earlier in the Matuyama reversed epoch and is known as the Olduvai normal event, which lasted 0.32 million years. During the Gauss normal epoch, which precedes the Matuyama reversed epoch, a brief reversal event called the "Mammoth reversed event" occurred, which lasted 0.12 million years.

SIGNIFICANCE OF THE DISCOVERY

In 1963, Fred J. Vine, Drummond H. Matthews, and Lawrence W. Morley independently proposed the hypothesis that the rock of the ocean floor showed evidence of magnetic field reversals. These magnetic stripes mirrored each other on opposite sides of midoceanic ridges and were of the same thickness. In February, 1966, the new geomagnetic polarity-reversal time scale that contained the Jaramillo event developed by Doell and Dalrymple was correlated with the magnetic-anomaly profiles across midoceanic ridges. The Vine-Matthews-Morley hypothesis (better known as the Vine-Matthews hypothesis) was confirmed and the theory of seafloor spreading and continental drift became inevitable.

About the same time, from a third source of independently derived data, polarity reversals were demonstrated in deep-sea sediment cores. In February, 1966, while Vine visited Neil Opdyke at the Lamont-Doherty Geological Observatory, Opdyke mentioned to Vine that he and his colleagues had discovered a new magnetic anomaly at about 0.9 million years in deep-sea sediment cores from the South Pacific's East Pacific Rise. These scientists had found the same magnetic reversal as Doell and Dalrymple had found on land. Other oceanic cores from various parts of the world also showed the reversal. This confirmed further that the Jaramillo event was a worldwide event.

At one time the question was asked if the poles had wandered (known as polar wandering) throughout geologic time or if the continents had drifted. Paleomagnetism has proved that the magnetic field, despite its reversals, shows it having had a common North Pole. Studies of rocks of various geologic ages from North America and Europe have been correlated and show that the paleomagnetism preserved in the rocks points toward a common North Pole. When Europe and North America are rotated back together to a predrift position, the apparent polar wandering curves of both continents come together and point toward a common North Pole.

Paleomagnetism was the keystone piece of the continental drift puzzle. It also added credence to the plate tectonics theory, formally called the" continental drift theory." Other scientific data that had been collected and studied also helped confirm the idea of seafloor spreading. A revolution in the earth sciences had been triggered by the work of Doell and Dalrymple.

—Roberto Garza

Further Reading

Continents Adrift: Readings from "Scientific American." W. H. Freeman, 1972. Introductions by J. Tuzo Wilson.

Cox, Allan, editor. *Plate Tectonics and Geomagnetic Reversals.* W. H. Freeman, 1973.

Cox, Allan, G., Brent Dalrymple, and Richard R. Doell. "Reversals of the Earth's Magnetic Field." *Scientific American*, vol. 216, Feb. 1967, pp. 44-54.

Glen, William. *The Road to Jaramillo: Critical Years of the Revolution in Earth Science.* Stanford UP, 1982.

Gradstein, Felix M., James G. Ogg, Mark D. Schmitz, and Gabi M. Ogg. *Geologic Time Scale 2020.* Vol. 1 Elsevier, 2020.

McElhinny, Michael W., and Phillip L. McFadden. *Paleomagnetism: Continents and Oceans.* Academic Press, 2000.

Menard, H. W. *The Ocean of Truth: A Personal History of Global Tectonics.* Princeton UP, 1986.

Merrill, Ronald T. *Our Magnetic Earth: The Science of Geomagnetism.* U of Chicago P, 2010.

Mitchell, Alanna. *The Spinning Magnet: The Force That Created the Modern World-and Could Destroy It.* OneWorld Publications, 2018.

Vrielynck, Bruno, and Philippe Bouysse. *The Changing Face of the Earth: The Break-Up of Pangaea and Continental Drift over the Past 250 Million Years in Ten Steps.* UNESCO, 2003.

DRAINAGE BASINS

Fields of Study: Geology; Geography; Geomorphism; Hydrodynamics

ABSTRACT

A drainage basin collects water from a large area and delivers it to a channel or lake. Drainage basins reflect the operation of physical laws affecting water flow over the ground surface and through rocks. These basins concentrate groundwater flow from a broad area into rivers, which, in turn, carry away both water and sediment.

KEY CONCEPTS

basin order: an approximate measure of the size of a stream basin, based on a numbering scheme applied to river channels as they join together in their progress downstream

channel: a horizontal depression in the ground surface caused and enlarged by the concentrated flow of water

erosion: the displacement of sediment from one location to another on the ground surface

groundwater: water that sinks below the ground surface to join the water table, slowly flowing toward river channels

hydrological: relating to the systematic flow of water in accordance with physical laws

limestone: a rock composed primarily of the mineral calcite or dolomite, which can be dissolved by water that is acidic

CHARACTERISTICS OF DRAINAGE BASINS

A drainage basin is an area that collects water, which accumulates on the surface from rain or snow. Its slopes deliver the water to either a channel or a lake. Normally, the channel that collects the water leads to the ocean. In this case, the drainage basin is defined as the entire area upstream whose slopes deliver water to that channel or to other channels' tributary to it. Thus, strictly speaking, drainage basins are defined as natural units only when streams enter bodies of water such as lakes or the ocean or when two streams join.

Less often there is no exit to the ocean: this type of basin is called a "basin of inland" or "interior drainage." Notable examples are the basins containing the Great Salt Lake in Utah, the Dead Sea in Israel and Jordan, and the Caspian Sea in Asia. The basin and range province in the Rockies (an area of about 1 million square kilometers extending from southern Idaho and Oregon through most of Nevada, western Utah, eastern California, western and southern Arizona, southwestern New Mexico, and northern Mexico) has at least 141 basins of inland drainage. The center of these basins is usually marked by a "playa," a level area of fine-grained sediments, often rich in salts left behind as inflowing waters evaporate. At certain times in the geological past, when the annual rainfall was heavier, some of these basins completely filled with water to the point of overflowing, at which point the drainage system may have connected to another interior basin or connected to a river system that drained to the sea. The basin now containing the Great Salt Lake (known to geologists as Lake Bonneville) overflowed at Red Rock Pass about 15,000 years ago. The overflowing waters discharged into the Snake River system, and thus to the Columbia River and the Pacific Ocean. Drainage basins may change in character over relatively short periods of geological time. There is some evidence that the entire Mediterranean Sea was a basin of inland drainage for a period of time about 3 to 5 million years ago: substantial salt deposits are found on its bed, and traces of meandering rivers have been seen in certain geological sections.

GROUNDWATER AND EROSION

Although the term "drainage basin" is normally thought of as applying to the surface, an important component of the basin as a hydrological unit is the rock beneath the surface. Much of the water that arrives on the surface sinks into the soil and the underlying rock, where it is stored as soil water in the "unsaturated zone" and as groundwater. Soil water

Digital terrain map of the Latorita River's drainage basin in Romania. Image via Wikimedia Commons. [Public domain.]

either sinks farther to become groundwater, or flows through soil and back out onto the surface downslope when and where the soil is saturated. Groundwater moves very slowly through the rock (millimeters to centimeters per day), but it eventually seeps into stream channels and so provides the base flow of water in rivers long after rain has finished falling.

This characteristic of groundwater can lead to a circumstance that alters the definition of a drainage basin when the rocks are primarily composed of limestone or any rocks susceptible to solution. Because limestone is soluble in acidic water (natural rain is slightly acid and contact with various airborne pollutants increases the acidity), over thousands of years, percolating groundwater dissolves substantial volumes of rock and causes a system of underground channels to develop, which may eventually become enlarged into caverns. Yugoslavia is famous for its underground cave systems; the U.S. states of Kentucky, Florida, and New Mexico are well known for their limestone terrain and underground drainage systems.

Essentially every karst landscape on earth is home to a system of caverns that has been formed through chemical erosion over long periods of time. In these cases, the route of the water underground may bear little or no relation to the pattern of channels and slopes seen on the surface, some or all of which may have become completely inactive. The determination of the drainage basin is then very difficult as various types of tracers (colored dyes or other readily detectable chemicals) have to be placed in the water in order to identify the points of egress of the water so that an interpretation of the underground channels may be made. The pattern of water flow to any given site will also depend on the location of the stormwaters causing the flow. The very slow solution of limestone to form underground river systems and caves emphasizes the fact that water moving through the basin removes solid rock. As the rivers dissolve their way downward, they leave some caves "high and dry" above the general level of the underground water (the water table), which points out the fact that rivers work down through the rock with the passage of time.

This aspect of drainage basins is harder to observe in areas of less soluble rock, even though water flowing out of the drainage basin carries sediment (small particles of soil and rock) and has been doing so for long periods of geologic time. Thus, in the long run, the surface of earth is gradually lowered and, even in the short run, enormous amounts of sediment may be removed from a basin every year. The Mississippi River system removes a total of about 296 million metric tons per year, or 91 tons per square kilometer, though this is small compared to the Ganges, which takes out 1,450 million metric tons per year, or 1,520 tons per square kilometer. Both of these are dwarfed by the average annual sediment load carried by the Yellow River (Huang He), in China, of 1.6 billion tons. If there were no corresponding uplift of the drainage basins or other interference, this removal would lead to the leveling of entire basins within 10 to 50 million years, depending on the lowering rate and the mean altitude of the basin.

The process of erosion proceeding at different rates in adjacent basins may cause the drainage divide (the line separating different flow directions for surface waters) to migrate toward the basin with the lower erosion rate. This is most common in geologically "new" terrain when stream systems are not deeply incised into the rocks. Drainage diversions may be simulated, as in the Snowy Mountain diversion in Australia, where waters are diverted across a divide by major engineering works in order to provide irrigation water for the Murray Darling River basin and also for hydroelectric power generation.

FLOODS

When winter snowmelts or severe storms bring heavy rain to large areas, the water that falls to the surface flows into channels and generates floods in the rivers. Floods are not abnormal; rather, they are an expectable occurrence in drainage basins. It is easy to understand that when basin relief is high and slopes are steep, as in the Rockies or the Appalachians, floods tend to generate higher flood peaks than when slopes are gentle. A basin that is round in shape tends to concentrate floodwater quickly because the streams tend to converge in the middle, whereas a long, narrow stream has the effect of attenuating the flow peak, even when the total amount of water falling on the basin may be the same. Similarly, a forest tends to attenuate flood peaks and to promote higher river flows between flood peaks than does open farmland. With the latter, there is a tendency for water to flow rapidly off the surface into channels, whereas in a forest much water is intercepted by the leaves of trees and the impact of rain on the surface is weaker, in part because leaf cover protects the soil. Because the soil is not so well protected, sediment loss from the surface into streams is greater from farmland than from forests and is higher again from land disturbed by major building projects.

CLASSIFICATION AND MEASUREMENT OF DRAINAGE BASINS

Various methods have been devised to classify basins according to size. The most common method depends on a numbering system applied to the streams that drain them. All the "fingertip" streams are labeled with 1. When two of these tributaries meet the channel, it is termed a "second-order channel" and is labeled with 2. Subsequently, the order of a stream increases by one only when two streams of equal order join. Otherwise, if two streams of unequal order join, the order given to the downstream segment is that of the larger of the two orders. The order of the drainage basin is then the order of the stream in the basin. In this type of numbering system (called "Horton/Strahler ordering"), the Mississippi drainage basin is an eleventh- or twelfth-order basin. The exact number depends on the detail (map scale) with which the fingertip streams are defined. The larger orders are rare because of the requirement for another river of roughly similar magnitude to join in to make the next higher-order basin.

Basin order may be used as a relatively natural basis for the collection of other data about the basin. The simplest measure is of the area in square kilometers. In addition, the basin relief, or the height difference between the lowest and the highest points, and the mean relief, or the average height of the basin above the outlet, may be recorded. The most precise method of recording basin relief is by computing the hypsometric (height) curve for the basin, which requires an accurate topographic map. When constructed, it shows, for any altitude, the proportion of the basin area above that particular altitude and, for comparative purposes, it may be produced in a dimensionless form by dividing both the height and the area measures by their maximum values or by the difference between the maximum and minimum heights if zero is not the minimum height.

Basin shape and basin dimensions (length and width) may also be recorded, although the notion of basin shape suffers from the problem that no completely unambiguous numerical measure exists that can be used to define the shape of an area in the plane (that is, on a map), and the problem is especially intractable if there are indentations in the edge of the basin. All measures are dependent to a considerable degree on the accuracy of source maps. In mountainous terrain, such maps may often be much less than perfect, if they exist at all. Even with automated drafting aids and digitizers (which automatically record positions on maps and save them as a data file), the measurement of basin properties is a tedious and time consuming process. Unless there are pressing reasons for a new analysis, it is common to rely on data tabulations made by hydrological or environmental agencies whenever possible.

Measurements are made of drainage basin properties because they are often used in statistical analyses together with the known flow of the gauged rivers in order to predict flow characteristics for rivers that have not been metered. The direct measurement of stream flow, while straightforward in principle, is time consuming, especially in the early stages, and a flow record is not very useful for predictive purposes until it has recorded at least twenty years of flow (preferably much more). Because of the high capital and maintenance costs involved in collecting river records, there has been an understandable emphasis on records for large rivers; the economic benefits from prediction (and eventual control) of the flow are more obvious, and measured flow records can sometimes be supplemented by anecdotal evidence of historic large floods, those flows that are often of most interest in land use planning (for example, zoning of land for residential use). It has been acknowledged that the hydrological behavior of low-order basins is less well understood, and more information has been collected on them, especially for urban areas where the routing of the large quantities of water that run off from impermeable surfaces in the city (roofs and roadways) has been recognized as a serious planning problem, especially in regard to groundwater recharge and the flow systems that connect with urban sewage systems.

SIGNIFICANCE OF DRAINAGE BASINS

The control of water outflow from drainage basins is necessary in some regions in order to promote irrigation, to supply domestic and industrial water, to generate power, and to implement flood control. The Hoover Dam on the Colorado River was originally conceived as a control dam, but hydroelectrical generators were also included in order to help defray its costs by selling power. There are nineteen major dams in the Colorado basin. Aside from the legal technicalities of water ownership and redistribution, difficulties arise from the fact that to control substantial amounts of water, large areas of the basin have to be regulated. In addition, there are economies of scale in large projects, particularly in the construction of large dams and reservoirs. A single control dam strategically placed may regulate flow downstream for hundreds of kilometers, whereas it would require hundreds of small dams on first- and second-order streams to achieve the same effect.

Large control dams do generate problems. The reservoirs trap sediment coming from upstream, which will eventually fill them, at which point they will become useless. Small reservoirs may fill within a few years. An original estimate for the Hoover Dam suggested that it would take four hundred years to fill Lake Mead; after only fourteen years, surveys revealed that the water capacity had been reduced by 5% and that sediment on the lake bottom had reached a maximum of 82 meters where the upstream river entered the still waters of the lake.

Downstream of a dam, the reduced sediment content and the regulated water flow often seriously affect riparian environments. There may be a variety of channel responses, often unpredictable, to the interference in the river regime caused by the dam. The stream may cut into its bed, it may change the dimensions of its channel, or it may even aggrade its bed. In the case of the Hoover Dam, the water downstream, deprived of its sediment by the dam, had an increased ability to remove fine sediment from the river bed but left coarser rocks behind because the flood peaks that would normally have removed them were now controlled and much reduced. The result is an "armoring" of the stream bed with coarse rocks, an effect that extends 100 kilometers downstream in the case of the Colorado River below the Hoover Dam. In the Colorado system as a whole, the net effect of controlling flood peaks has been for rapids to stabilize and to increase in size as sediment becomes trapped in them. A corollary of the "winnowing" of fine material has been the disappearance of river beaches and an increased propensity to pollution as sediment becomes much less mobile and more concentrated in space.

—Keith J. Tinkler

Further Reading

Christopherson, Robert W., and Mary-Louise Byrne. *Geosystems: An Introduction to Physical Geography.* Pearson/Prentice Hall, 2006.

Dodds, Walter K., and Matt R. Whiles. *Freshwater Ecology: Concepts and Environmental Applications of Limnology.* 2nd ed., Academic Press, 2010.

Foresman, Timothy, and Alan H. Strahler. *Visualizing Physical Geography.* John Wiley & Sons, 2012.

Newson, Malcolm. *Land, Water and Development: Sustainable and Adaptive Management of Rivers.* 3rd ed., Routledge, 2008.

Rhoads, Bruce L. *River Dynamics: Geomorphology to Support Management.* Cambridge UP, 2020.

Rodriguez Iturbe, Ignacio, and Andrea Rinaldo. *Fractal River Basins: Chance and Self-Organization.* Cambridge UP, 2001.

Strahler, Alan H. *Modern Physical Geography.* 4th ed., Wiley India, 2008.

E

Earthquake Prediction

Fields of Study: Geology; Oceanography; Geophysics; Seismology; Structural Engineering; Materials Science; Soil Science

ABSTRACT

Earthquakes are among the most potentially devastating catastrophes human beings can face, and ways of predicting their occurrence in advance are urgently needed. Short-range predictions are needed so that people can evacuate dangerous locations or shutdown critical services, such as nuclear power plants and transportation systems. Long-range predictions are needed to help communities upgrade their building codes and identify at-risk areas where construction or other land uses should not be permitted.

KEY CONCEPTS

elastic rebound: sudden release of strain built up by fault creep over a long period of time

fault: crack in the Earth's crust along which differential movement has occurred

foreshock: preliminary vibration before a major earthquake

locked fault: fault along which anticipated movement is overdue

moment magnitude: scale for measuring earthquake size based on the amount of energy released

primary wave: first group of vibrations to arrive from an earthquake

Richter scale: scale for measuring earthquake size based on the amount of shaking

secondary wave: second group of vibrations to arrive from an earthquake

DEFINITION AND BASIC PRINCIPLES

Earthquake prediction is the act of determining in advance that an earthquake is likely to take place. The prediction must address four factors: the time period during which the earthquake is expected to occur, the area that the earthquake is expected to affect, the type of shaking that will occur, and the likelihood that the earthquake will actually take place. Short-range predictions estimate that an earthquake will occur within the next few hours, or at most, the next few days. These are based on premonitions of an impending earthquake, such as changes in groundwater levels in wells, an increased number of foreshocks, or unusual animal behavior. Long-range predictions, which estimate that an earthquake may take place within a certain number of years, are generally based on the past history of the area, augmented by Global Positioning System (GPS) data or the digging of trenches in critical areas. Short-range warnings are valuable if the earthquake occurs, but if the anticipated earthquake does not materialize, they will have disrupted people's lives unnecessarily. Long-range predictions are useful primarily because they often result in reviews of building codes and reconsideration of the suitability of some land areas for construction and development.

BACKGROUND AND HISTORY

The scientific understanding of earthquakes and the development of methods for predicting them date only to the time of the great San Francisco earthquake in 1906. Before that time, the relationship between faulting and earthquakes was not clearly understood. H. F. Reid, a professor at The Johns Hopkins University, was appointed to California's state-funded commission to study the great earthquake. Based on his examination of the terrain surrounding the city, he was able to show that points on opposite sides of the San Andreas fault had moved differentially as much as 6 meters before breakage. Reid coined the term "elastic rebound" to describe what then happened. Strain had built up in the rocks by fault creep, as the two sides of the fault drifted in opposite directions. When breakage finally

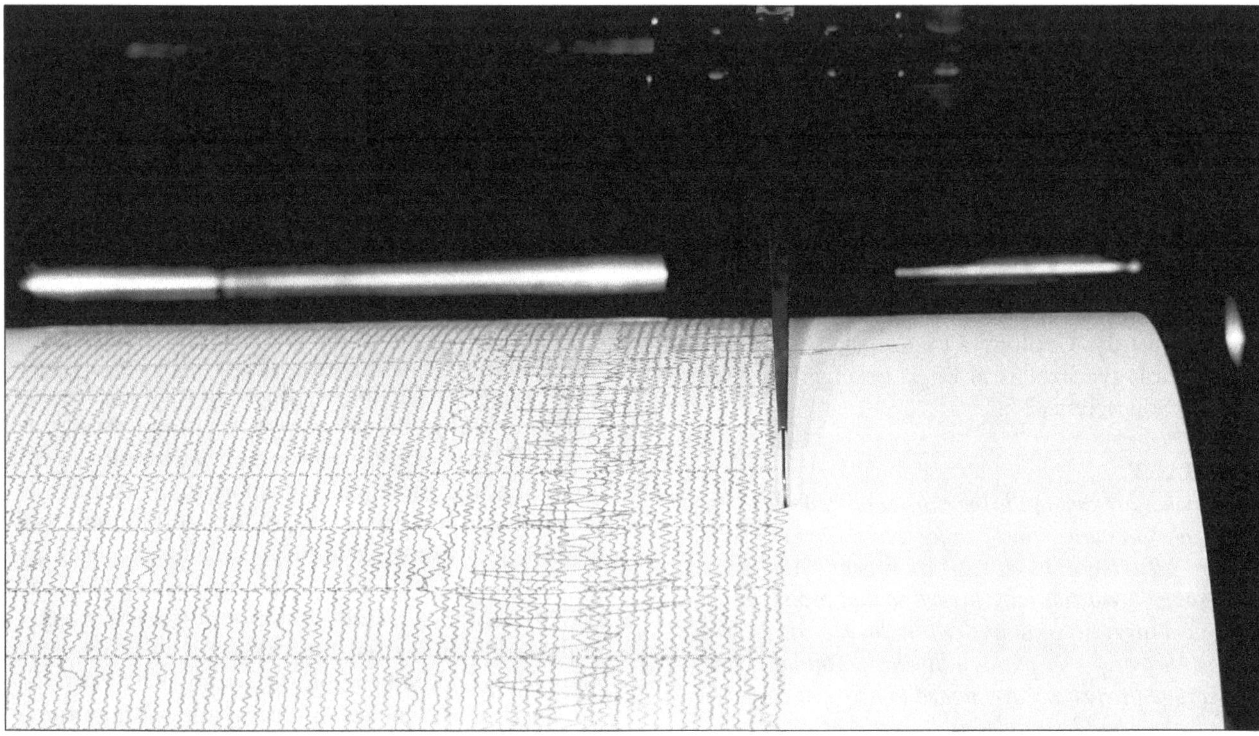

Seismograph. Photo via iStock.com/gpflman. [Used under license.]

occurred, the rocks returned to their unstrained positions just as a rubber band returns to its original shape after the band snaps, and this snapping back of the rocks was what had caused the earthquake. Reid led the way to an understanding of how the breakage of faults causes earthquakes, thereby enabling scientists to turn their attention to ways of predicting them in advance.

HOW IT WORKS

Short-range predictions. A variety of methods has been used to predict earthquakes on a short-term basis, and several of them are related to changes in underground water. Levels in wells may change, turbidity may appear, or water temperatures may rise. Chemical changes have also been noted. Kobe, Japan, experienced a devastating earthquake in 1995. Subsequent to the earthquake, an analysis of the water produced at the city's bottling plant revealed that the chemistry of the water had been steadily changing for two weeks before the earthquake. The water at the bottling plant is being studied daily for warnings of the next earthquake.

Mexico City uses a different method of predicting earthquakes. Fault lines are located along the Pacific coast, 300 kilometers away from Mexico City. A network of seismographs is used to determine that an earthquake has taken place, and that information is transmitted by satellite radio to the city, warning residents 50 to 80 seconds before the damaging waves arrive. Fault lines lie near and under the Los Angeles and Tokyo metropolitan areas, so there is less time to prepare for an earthquake, but seismographs around the city can detect the less destructive primary waves 10 to 15 seconds before the destructive secondary waves arrive. This gives computerized control systems enough time to shut off the power to specified industries and close down the mass transit systems.

Changes in the behavior of animals before an earthquake has been noted in some instances, most notably before the 1906 San Francisco earthquake. The earthquake occurred around five o'clock in the morning, and afterward residents reported that dogs at the pound had howled incessantly from midnight on and that horses had stamped and neighed in their livery stables during that same period. Presumably the animals heard high-pitched sounds, not audible to humans, caused by dilation of the rock preliminary to the final breaking.

Dilatancy, the opening of small cracks in the rocks of a fault zone before breakage, may cause other signals that an earthquake is about to occur. These signals include the tremors known as foreshocks, which often precede major earthquakes, as well as changes in the bedrock's magnetism, electrical resistance, or ability to transmit seismic waves. Radon gas may also be released into wells, and in some instances, changes in the land surface will also be noted. These changes can be monitored using sensitive instruments known as tiltmeters. Because earthquakes often accompany volcanic eruptions, any volcano overlooking a major city—such as Vesuvius beside Naples, Italy, or Mount Rainier next to Seattle, Washington—is monitored daily by seismographs, watching for the slightest change in vibrations that might indicate that an eruption is imminent. Other warning signs from a volcano include steam clouds coming from the crater, ash falling in the surrounding area, tilting of the volcano's flanks as magma swells into the crater, and changes in water features around the mountain, such as springs and wells.

Long-range predictions. The historical record of an area's previous earthquakes is useful in making long-range predictions of future shocks and is helpful in the preparation of seismic risk maps. A seismic risk map of the United States would show the West Coast, Alaska, and Hawaii as the highest risk zones. A repeated pattern of earthquakes is also helpful in making predictions. Parkfield, California, has experienced a magnitude 6 or 8 earthquake about every twenty-two years since 1857. Parkfield experienced an earthquake on September 28, 2004, so the town might expect its next earthquake in 2026. Not all segments of a fault break at the same time, and the inactive segments are known as seismic gaps. People living in areas where the seismic risk is high need to remember that an earthquake is possible at any time.

Careful fieldwork by geologists is another way to predict future shocks. The study of a fault trace across the landscape may reveal offset features that provide clues as to how often and when earthquake movements have taken place. Trenches may be dug for an examination of the soil layers, and a pilot hole is being drilled at Parkfield for a study of the San Andreas fault at depth.

Seismologists have begun using GPS measurements to analyze the differential creep on two sides of a fault. Using such measurements, in April, 2008, scientists predicted that the Enriquillo fault, near Haiti's capital city of Port-au-Prince, was capable of a magnitude 7.2 earthquake. Just twenty months later, in January, 2010, a devastating 7.0 earthquake did occur.

APPLICATIONS AND PRODUCTS

Construction in earthquake-prone areas requires special precautions, and attention must be given to the type of structure planned. Some structures, such as nuclear power plants and large dams, should never be built in areas with active faults. Engineers and developers must determine the type of earthquakes affecting the area, how their vibrations might affect the intended structure, and how long the shaking could last. Knowledge of the area's geology is critical because soil and bedrock characteristics can influence how an earthquake will affect structures. For earthquakes with magnitudes up to 5, damage should be slight; for earthquakes of magnitudes 5 to 7, damage should be easily repaired; and for magnitude 7 or higher earthquakes, the structure must not collapse although significant repairs, or even demolition, may be required.

Nature of the substratum. Structures built on solid bedrock invariably suffer less damage in an earthquake than structures built on filled land or water-saturated sediments. Filled land or water-saturated sediments lose their cohesiveness because their loose structure amplifies the vibrations from the earthquake. They can begin behaving like a liquid, in a process known as liquefaction, causing severe damage to the structures on the surface. Liquefaction occurred during a 7.5 magnitude earthquake in Niigata, Japan, on June 16, 1964, and during a magnitude 9.2 earthquake in Anchorage, Alaska, on March 18, 1964. In Niigata, part of which was built on sandy soil, hundreds of buildings, including multistory concrete-reinforced apartment buildings, tilted and fell over. In Anchorage, clay beds beneath the Turnagain Heights subdivision liquefied, and more than 200 acres covered with homes slid toward the sea. The damage to homes, streets, sidewalks, water and sewer lines, and electrical systems was massive.

Proper building materials. Adobe and mud-walled structures are the weakest buildings and invariably collapse during severe earthquakes. Stucco, unreinforced brick, and concrete-block structures also suffer severe damage; their walls prove inflexible to shaking once the mortar that binds them is broken. Surprisingly, wooden structures survive earthquakes quite well. Wood is lighter weight than metal or concrete, and if the frame of the building is properly secured to the ground, the nails and screws in the wood provide enough flexibility to resist shaking. Wood is preferable for smaller private homes in earthquake-prone areas, although concrete can be used for larger structures, provided it is heavily reinforced with steel bars.

Construction of larger buildings. The most important consideration in earthquake-proof construction is that the foundation, floors, walls, and roof of a building be tied together to withstand both horizontal and vertical stresses. Open lower floors are hazardous and should be reinforced with added pillars or diagonal bracing. Tall buildings and skyscrapers need to sway to prevent cracking, and the use of diagonal steel beams will permit this. Such buildings may even be placed on a layered steel or rubber structure that acts as a shock absorber to decrease such swaying. A good example of this is the pyramid-shaped Transamerica building in San Francisco. In addition to its tapering shape, which reduces the amount of mass in the higher floors, the skyscraper sits on a two-story-high cushion of massive diagonal trusses, designed to help the building absorb earthquake vibrations.

Retrofitting older structures. Many older structures are being reinforced to provide protection during earthquakes using a process known as retrofitting. The object is to make the existing structure more resistant to stresses, ground motion, and possible soil failure. High-strength steel, fiber-reinforced polymers, or fiber-reinforced concrete can be used to build new structural walls, and columns can be jacketed with steel. Stronger columns can be added and diagonal bracing placed on all floors. Foundation work might include jacking up the building to place springs or other flexible materials between it and the ground.

Quake preparedness. People in earthquake-prone zones need to be ready to live for long periods without electricity and other services following a severe shock. Battery-powered radios, flashlights, and first-aid kits are essential, and homeowners need to know the location of the fuse box and shut-off valves for water and gas. Heavy appliances must be securely anchored to the floor, and shelves, bookcases, and cupboards firmly attached to walls. Heavy objects should not be placed on top shelves. Families need to have a plan for reuniting in the event that members become separated during an earthquake. During the earthquake itself, people who are outside should remain outside, away from possible falling objects, and people who are inside should seek shelter under a heavy piece of furniture. Those driving cars should come to a stop, away from overhead objects.

IMPACT ON INDUSTRY

Most research related to earthquake prediction is centered in the United States, Russia, Japan, and China because these countries have regions of significant infrastructure and dense population, and have experienced serious earthquakes in the past. Research includes laboratory and field studies of rock behavior before, during, and after earthquakes, as well as the monitoring of activity along major faults. Both government agencies and universities conduct this research.

Government agencies. The United States Geological Survey (USGS) plays a major role in earthquake studies and earthquake prediction for the United States. In addition to providing general information regarding earthquakes, it makes available data on the latest earthquakes in the United States and around the world. For selected areas, such as California and Nevada where the potential for earthquakes is high, the USGS provides an index map, updated constantly, showing where the latest earthquake activity has been observed. The agency also makes short-range predictions of future earthquake activity, such as the probability of strong shaking in critical areas within the next twenty-four hours, and estimates of the probability of aftershocks in areas where earthquakes have recently occurred. The

USGS also makes long-range predictions of future earthquake activity. It forecasts a 62 percent probability of a magnitude 6.7 or greater earthquake, capable of causing widespread damage, striking the San Francisco Bay Area before 2032.

The Alaska Earthquake Information Center (AEIC) provides services similar to those provided by the USGS, which serves primarily the lower forty-eight states. The center is important for Alaska residents because the state is just as prone to earthquakes as California, having had the world's second largest earthquake in 1964. The AEIC receives daily reports of earthquakes from a network of more than four hundred seismograph stations and makes these available on its website. The center makes predictions of future earthquakes, although shocks of magnitude 8 and lower are so common that no effort is made to predict them. For earthquakes of magnitude 8 and greater, predictions are made when possible, giving the expected location, magnitude, and time.

A third government agency involved in earthquake monitoring and prediction is the National Oceanographic and Atmospheric Administration (NOAA), which operates tsunami warning centers in Hawaii and Palmer, Alaska. Because of the great loss of life during the Indian Ocean tsunami of 2004, the Hawaii warning center has expanded its coverage to the Indian Ocean and the Caribbean until centers for those two areas can be created. The Alaska center serves the coastal areas of Canada and the United States, but not Hawaii.

Academic research. The California Institute of Technology (Caltech) is a leading institution involved in earthquake research and prediction on the West Coast of the United States. Charles Richter and Beno Gutenberg developed the Richter scale for measuring the magnitude of earthquakes at Caltech in the 1930s. Geophysicists in the division of geological and planetary sciences supervise Caltech's seismological laboratory and also help maintain the Southern California Seismic Network and the Southern California Earthquake Data Center.

On the East Coast of the United States, Columbia University's Lamont-Doherty Earth Observatory also has a long tradition of earthquake research. Besides performing observational studies in the field, scientists in the seismology, geology, and tectonophysics division operate the Lamont-Doherty Cooperative Seismograph Network and also contribute to research for the Center for Hazards and Risk at Columbia's Earth Institute.

SOCIAL CONTEXT AND FUTURE PROSPECTS

The importance of predicting earthquakes in advance of their occurrence cannot be overstressed. Many of the world's best known cities lie near, or directly on top of, one or more major through-going faults, and the destruction resulting from movement on such faults can be catastrophic. San Francisco and Oakland, California, for example, lie in a narrow sliver of crust, bordered on one side by the San Andreas fault and on the other by the equally dangerous Hayward fault. Destructive earthquakes are likely to take place in Mexico City, Los Angeles, and Tokyo, which have had major earthquakes in the past and need adequate warning systems. Cities that have not had severe shocks in the recent past but are thought to be at high risk are Seattle and Istanbul. Seattle has not had a major earthquake in several hundred years, but geologists doing fieldwork around the city have found evidence that a massive earthquake took place there in the 1700s. Seattle residents need to prepare for a repeat of such an earthquake. Istanbul has not had a major earthquake in many years, but geologists point out that it is at the west end of a fault similar to the San Andreas. Fourteen earthquakes with magnitudes of 6 or higher have taken place along this fault since 1936, with each earthquake moving closer to the city. Despite all the advances that have been made in long-range earthquake prediction, scientists can still give people only a vague idea of when a large earthquake will hit their area.

—Donald W. Lovejoy

Further Reading

Bolt, Bruce A. *Earthquakes.* 5th ed., W. H. Freeman, 2006.

Chester, Ray. *Furnace of Creation Cradle of Destruction: A Journey to the Birthplace of Earthquakes, Volcanoes, and Tsunamis.* AMACOM, 2008.

Hough, Susan Elizabeth. *Predicting the Unpredictable: The Tumultuous Science of Earthquake Prediction.* Princeton UP, 2010.

———. *Richter Scale: Measure of an Earthquake, Measure of a Man.* Princeton UP, 2007.

Pulinets, Sergey, and Dimitar Ouzounov. *The Possibility of Earthquake Forecasting: Learning from Nature.* Institute of Physics Publishing, 2019.
Taranath, Bungale S. *Tall Building Design. Steel, Concrete and Composite Systems.* CRC Press, 2017.
Winchester, Simon. *A Crack in the Edge of the World: America and the Great California Earthquake of 1906.* HarperCollins, 2006.

EARTHQUAKES

Fields of Study: Geology; Seismology; Geophysics; Volcanology

ABSTRACT
An earthquake is a shaking or trembling of the ground caused by energy coming from volcanic eruptions or from rocks breaking during the geologic process known as "faulting." This process takes place as a result of differential movements of segments of the Earth's crust known as plates. When the frictional forces holding blocks of rock together are overcome, the movement of the blocks produces cracks, which can infill with minerals. Earthquakes result when a buildup of stress in brittle rock within Earth's crust is released suddenly.

KEY CONCEPTS

elastic rebound: the movement that returns a strained material in a high-energy state to its normal low-energy state, such as the release of an elastic band

faulting: the formation of cracks and sheer planes through a body of solid rock

focus: the location within Earth's crust at which the energy released in a seismic event originates

isostatic rebound: the upward movement of a region of continental mass and the underlying crustal rock when the weight of glacial ice above it has been removed

plates: segments of Earth's crust that have formed in response to forces exerted by movement of magma in Earth's mantle, somewhat resembling the fragments of a cracked eggshell

strain energy: energy built up within a plate due to an inability to move in response to forces exerted by movement of underlying mantle material

DEFINING EARTHQUAKES

An earthquake is a shaking or trembling of the ground caused by energy coming from volcanic eruptions or from rocks breaking during the geologic process known as "faulting." This process takes place as a result of differential movements of segments of Earth's crust known as plates. The plate tectonics theory has identified seven major crustal plates, as well as several minor plates, which are thought to be colliding with one another, separating from one another, or sliding past one another like books on a tabletop. The sudden movements of these plates, as they separate, collide, or slide past one another, are believed to cause most of the planet's earthquakes.

Earthquakes occur following the rapid release of strain energy stored in rocks. Rock beneath Earth's surface is continually subjected to forces in all directions by the movement of underlying magma and the rotation of the planet. When the forces exceed the limits the rocks can sustain, they respond by either folding or breaking. If the forces are relatively rapid and the rocks are brittle, then the rocks actually break. The result is a shaking of the ground. This shaking is most prominent on Earth's surface.

OVERVIEW OF EARTHQUAKES

An earthquake first originates at a point called the "focus," which is beneath the Earth's surface. This fracture, which begins at a point, grows from a microscopic crack into a large fault that can extend for many kilometers. However, as mentioned, this fracture will be propagated only through brittle material. In other words, faults will not extend indefinitely into the Earth's subsurface. Nor will they extend indefinitely through brittle material, because as the energy released at the focus is 'consumed' by the formation of the extended fracture, there will be a point at some distance from the focus where there is insufficient energy remaining to fracture rock far removed from the initial source of a fracture. Focal depths of earthquakes occur over a range of depths, extending from just below the Earth's surface to a depth of approximately 700 kilometers. Below this great depth rocks are no longer brittle and thus cannot break.

In addition to the more obvious effects of seismic activity on the surface, earthquakes cause a considerable amount of subsurface activity. Seismic energy

An apartment building in ruins after an earthquake, Turkey 2011. Photo via iStock.com/adlaphotography. [Used under license.]

passing through brittle rock produces faults and cracks of varying sizes throughout the rock. These fissures serve as conduits for fluids, which can move through the rock much more readily than they could before the rock was broken. If the fluids contain dissolved minerals, these will be deposited in concentrated amounts. Such is the case when molten rock rises below the surface and is injected into cracks. Concentrated deposits of gold, silver, and other valuable metals are commonly found filling cracks that were produced by earthquakes that occurred in the recent geologic past.

Large-scale faulting can move massive blocks of rock closer to the Earth's surface. If these blocks are later exposed by erosion of the overlying material, new minerals are exposed. Layers containing coal, limestone, and gravels become available for mining.

Volcanoes can cause earthquakes too, but these earthquakes are generally minor and occur only during periods of heightened volcanic activity. Some additional causes of earthquakes include landslides that drop large quantities of earth suddenly on underlying surfaces, and human activities, such as blasting during surface or underground mining operations. The shock waves from underground nuclear tests can generate vibrations similar to earthquakes, as well as trigger actual earthquakes.

EARTHQUAKES AND CLIMATE CHANGE

Concerns have been raised that climate change may lead to earthquakes in Alaska, or elsewhere, and these concerns may not be entirely unwarranted. Many glaciers and ice sheets are located in regions of past or continuing geologic activity, where crustal movements are still taking place. Removal of ice cover during periods of global warming might be sufficient to activate ancient faults in these areas and set off earthquakes of considerable magnitude. The rate at which the ice melts may not need to be rapid to cause such activity. Rocks can build up strain over a period of years without breaking at all, until the tension is suddenly released in a devastating shock. Scientists describe this process as "elastic rebound." While the rocks on either side of a fault slowly slide past each other, the fault itself may remain "locked," with no movement taking place. Meanwhile, the rock masses on both sides of the fault begin to stretch, in a fashion similar to the stretching of an elastic band. When the strain in the rocks becomes too great, the fault will break, and the rocks on either side will snap back to their original dimensions.

There is clear evidence that the removal of ice from Earth's surface following the ice ages is causing significant upward movement of the crust in several places. In parts of Scotland, Scandinavia, and Canada, the amount of rise has been dramatic. Interior Scandinavia has risen as much as 250 meters since the ice disappeared, and elevated beaches around Hudson Bay in northern Canada indicate the land there has risen more than 330 meters during the same period. Farther south, the slowly rising Niagara Escarpment also evidences the rebounding and rising landscape. Scandinavia is still rebounding at a rate of one meter per century, and some coastal cities have been uplifted so rapidly that docks constructed several hundred years ago are now inland, while harbors have grown so shallow that they can no longer accommodate the ships that once docked there.

Scientists refer to this uplifting process as "isostatic rebound." When the weight of a continental ice sheet is added to Earth's crust, the crust responds by subsiding. When the ice is removed, the crust begins to rise again, just as a cargo ship that sank lower in the water when it was loaded floats

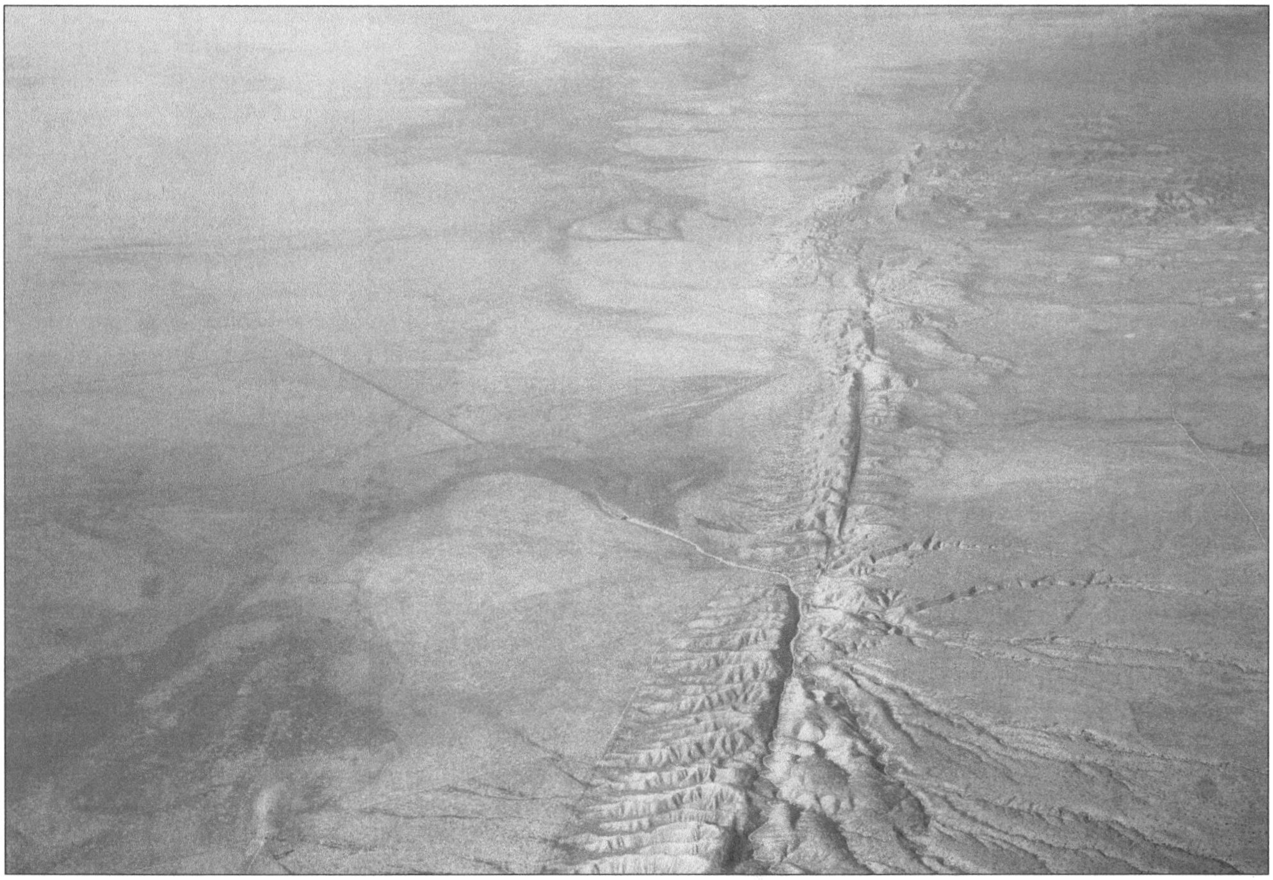

Aerial photo of the San Andreas Fault in the Carrizo Plain, northwest of Los Angeles. Photo by Ikluft, CC BY-SA 4.0, via Wikimedia Commons.

higher after its cargo is removed. The weight of the three-kilometer-thick ice sheets that once covered North America and Europe apparently caused down-warping of the crust by hundreds of meters. Now that the ice has been removed, the crust is rebounding to its original height.

Whether faults that were once active can be reactivated sufficiently by this rebounding process to cause earthquakes is difficult to determine, but such rebounding may well explain a series of mysterious earthquakes that plagued upper New York State during the late 1900s, as well as earthquakes that rattled Boston at the same time. Both areas were once buried beneath the North American ice sheet, and Boston's jolts have continued into the twenty-first century, with low-energy earthquakes being experienced almost continually in the corresponding region of Canada and the United States. By contrast, the continent of Antarctica still has its ice cover, and Antarctica's continental shelves, which would normally be found 120 meters below sea level, are now found 330 meters down. Presumably, the shelves were formed in water depths of 120 meters when Antarctica was ice free, and they will rise again to normal depths if the weight of the ice now covering the continent is someday removed.

—David M. Best and Donald W. Lovejoy

Further Reading

Anderson, David E., Andrew S. Gaudie, and Adrian G. Parker. *Global Environments Through the Quaternary: Exploring Environmental Change.* Oxford UP, 2007.

Hough, Susan Elizabeth. *Earthshaking Science: What We Know and Don't Know About Earthquakes.* Princeton UP, 2002.

Matson, W. R. *Earthquakes: The Sound of Multi-Modal Waves.* Morgan & Claypool, 2016.

McGuire, Bill. *Waking the Giant: How a Changing Climate Triggers Earthquakes, Tsunamis, and Volcanoes.* Oxford UP, 2012.

Stein, Seth, and Michael Wyesession. *An Introduction to Seismology, Earthquakes, and Earth Structure.* Wiley, 2013.

Stimpson, Ian. "Japan's Tohoku Earthquake and Tsunami." *Geology Today*, vol. 27, no. 3, 2011, pp. 96-98.

Sykes, Lynn R. *Plate Tectonics and Great Earthquakes: 50 Years of Earth-shaking Events.* Columbia UP, 2019.

Zeilinga de Boer, Jelle, and Donald Theodore Sanders. *Earthquakes in Human History: The Far-Reaching Effects of Seismic Disruptions.* Princeton UP, 2005.

Earth's Crust

Fields of Study: Geology; Geochemistry' Geomorphology; Geophysics; Seismology

ABSTRACT

Earth's crust is the outer hard layer of the planet. The crust overlies Earth's mantle and is separated from it by the Mohorovičić discontinuity, or Moho. There are two great classes of crust on Earth, oceanic and continental, which differ in thickness, composition, density, age, mode of formation, and significance for mineral resources.

KEY CONCEPTS

andesite: the general type of rock typical of the continents, of a lesser density than the underlying basaltic crust

basalt: the general type of rock typical of oceanic seafloor, produced by upwelling magma from Earth's mantle

convergent plate boundary: the line of demarcation between two tectonic plates where they are being forced into contact with each other

divergent plate boundary: the line of demarcation between two tectonic plates that are being driven apart from each other

Mohorovičić discontinuity: the interfacial region between the solid silicate rock material of the crust and the fluid basaltic magma of the mantle

subduction zone: the convergent plate boundary between seafloor and continental mass, where the denser basaltic seafloor is forced to go beneath or subduct to the continental mass

BACKGROUND

Earth consists of a nested set of spheres of different composition and of decreasing density with distance from the center of the planet. The crust is the outermost and lowest-density hard shell, significantly less dense (2.7 to 3.0 grams per cubic centimeter) than the underlying mantle (3.3 grams per cubic centimeter). Earth's two distinct types of crust—continental and oceanic—differ in five fundamental aspects: thickness, density, composition, age, and mode of formation.

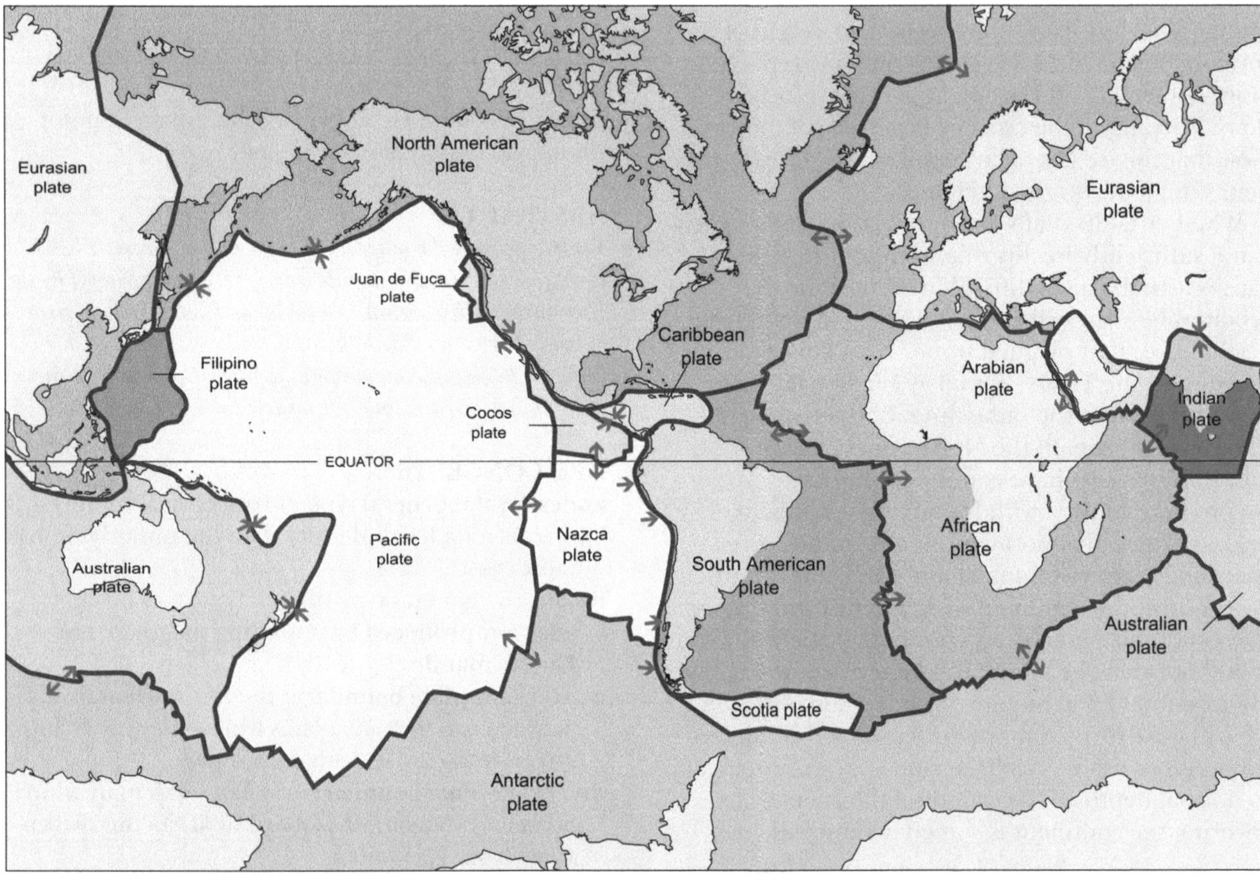

Plates in the crust of the earth. Image by Scott Nash, via USGS/Wikimedia Commons. [Public domain.]

CONTINENTAL AND OCEANIC CRUST

Continental crust is generally found beneath the exposed parts of Earth's surface known as continents. In addition, continental crust is submerged and makes up the continental shelves and submerged continental platforms. Correspondingly, a larger proportion of Earth's surface is composed of continental crust (40 percent) than is exposed above sea level as continents (25 percent). Oceanic crust makes up the floor of the oceans; in rare cases it rises above sea level, such as in Iceland and Ethiopia. Our store of nonrenewable natural resources is produced and kept in the crust. Hydrothermal systems associated with oceanic crust formation at mid-ocean ridges produce metal deposits. Nearly all economic ore deposits are extracted from the continental crust. Basins in the continental crust and along the continental margins are the principal sites for the formation and storage of oil and gas deposits.

Typical continental crust is about 40 kilometers thick, has a density of about 2.7 grams per cubic centimeter, and has a bulk composition similar to the volcanic rock andesite; it is about 60 percent silicon dioxide (SiO_2). Continental crust as old as 4 billion years has been found, and 2.5-billion-year-old continental crust is common. Earth is about 4.5 billion years old, and continental crust from Earth's first 500 million years has not been preserved. This contrasts with the situation for Earth's moon, where the lunar highlands preserve crust that formed shortly after the moon itself. Oceanic crust is about 6 kilometers thick, has a density of about 3.0 grams per cubic centimeter, and has a bulk composition similar to the volcanic rock basalt (about 50 percent SiO_2). Although ophiolites may be much older, the oldest *in situ* oceanic crust is about 170 million years old.

The large difference in age between oceanic and continental crust reflects the greater density of the former, which allows it to slide back into the mantle along subduction zones. In contrast, buoyant continental crust is difficult to subduct. The formation of oceanic and continental crusts is fundamentally different as well: Oceanic crust forms by seafloor spreading at midocean ridges, whereas continental crust forms at island arcs lying above subduction zones (such as Japan or the Mariana Islands in the western Pacific). Although the area of oceanic crust is much larger than that of continental crust (60 versus 40 percent of Earth's surface), the volume of continental crust is much larger than that of oceanic crust (80 versus 20 percent).

METAL AND HYDROCARBON DEPOSITS

The two types of crust play different roles in the formation of nonrenewable natural resources such as metallic ores and hydrocarbons. Metallic ores are predominantly produced at divergent or convergent plate boundaries—that is, where oceanic crust is either produced or destroyed. Vast deposits of manganese and cobalt exist on the deep seafloor in the form of manganese nodules. Hydrocarbon deposits form principally in basins on continental crust or beneath continental margins, at the boundary between oceanic and continental crust. The configuration of continents may also be important for controlling oil and gas deposits, because it can cause the formation of restricted basins where oxygen-poor waters allow organic matter to be preserved and buried. The relatively thin sedimentary sequences typically deposited on oceanic crust are not conducive to formation and preservation of hydrocarbon deposits.

The distribution of mineral and hydrocarbon resources is strongly controlled by the age of the crust and the sedimentary basins that these harbor. In spite of the fact that the oceanic crust is the principal factory for generating ore deposits, a minuscule proportion of these are presently exploited, largely for economic reasons. Because of its age and mode of formation, the continental crust acts as a warehouse for ore deposits produced over Earth's history, especially those deposits produced at convergent plate boundaries. Particularly rich ores are preserved in crust produced in the first 2 billion years of Earth history, and those nations which have large tracts of such ancient crust (among them are Australia, Canada, Russia, and South Africa) are blessed with especially rich metal deposits.

RESOURCE FRONTIERS

A wide range of mineral and hydrocarbon resources are sought on all continents except Antarctica. This search benefits increasingly from abundant technological resources, including satellite remote sensing, geophysical surveys, geochemical studies, and traditional field mapping, and from the tremendous increase in computing power available to process large and complex data sets. These nonrenewable resources are likely to be depleted in the future, leading to a rise in prices that will reward exploitation of "frontier" deposits. Resource frontiers pertaining to Earth's crust include mining and drilling for oil deeper below the continental surface, drilling for oil in deeper water offshore, the mining of deep-sea resources, and exploiting geothermal and hydro thermal resources for energy, including the tremendous heat energy stored in the deep continental crust and vented from hydrothermal sites along the midocean ridges.

—Robert J. Stern

Further Reading

Anderson, Robert S., and Suzanne P. Anderson. *Geomorphology: The Mechanics and Chemistry of Landscapes*. Cambridge UP, 2010.

Brown, Michael, and Tracy Rushmer, editors. *Evolution and Differentiation of the Continental Crust*. Cambridge UP, 2006.

Condie, Kent C. *Earth as an Evolving Planetary System*. 2nd ed., Academic Press, 2011.

Davis, Earl E., and Harry Elderfield, editors. *Hydrogeology of the Ocean Lithosphere*. Cambridge UP, 2004.

Fowler, C. M. R. *The Solid Earth: An Introduction to Global Geophysics*. 2nd ed., Cambridge UP, 2005.

Grotzinger, John, and Thomas H. Jordan. *Understanding Earth*. 6th ed., W. H. Freeman, 2010.

Mathez, Edmond A., and James D. Webster. *The Earth Machine: The Science of a Dynamic Planet*. Columbia UP, 2004.

Molnar, Peter. *Plate Tectonics: A Very Short Introduction*. Oxford UP, 2015.

Rogers, John J. W., and M. Santosh. *Continents and Supercontinents*. Oxford UP, 2004.

Van der Pluijm, Ben A., and Stephen Marshak. *Earth Structure: An Introduction to Structural Geology and Tectonics.* W.W. Norton, 2010.
White, William M. *Geochemistry.* Wiley, 2020.

Earth's Internal Structure

Fields of Study: Geology; Seismology; Geophysics; Geodynamics

ABSTRACT
Heat released by colliding particles and the heat energy released by the decay of radioactive elements maintain Earth's molten interior. This enabled the process of differentiation, which allowed the heavier elements to sink toward Earth's center while the lighter elements moved upward, as a result of the contrast in density of the earth's forming elements. Earth's layers can be classified either by their composition (the traditional method) or by their mechanical behavior (strength).

KEY CONCEPTS

aluminosilicates: silicate rocks in which the major metal present is aluminum

differentiation: the trend of elements to separate according to density under the force of gravity within the molten interior of Earth

ferro-magnesium silicates: silicate rocks in which the major metals present are iron and magnesium

Mohorovičić discontinuity: the interfacial region between the solid silicate rock material of the crust and the fluid basaltic magma of the mantle

P-wave: a longitudinal compression wave originating at the focus of an earthquake or other seismic event

S-wave: a transverse sheer wave originating at the focus of an earth-shaking or other seismic event

EARLY FORMATION OF EARTH

Earth is one of the nine known planets in the Solar system that formed from a giant cloud of cosmic dust called a "nebula." This event is thought to have happened between 4.44 billion years ago (based on the age of the oldest-known Moon rock) and 4.56 billion years ago (the age of meteorite bombardment). During and after Earth's early formation, heat released by colliding particles combined with the heat energy released by the decay of radioactive elements maintained Earth's molten interior. This enabled the process of differentiation, which allowed the heavier elements, mainly iron and nickel, to sink toward Earth's center while the lighter elements moved upward, as a result of the contrast in density of the earth's forming elements. This process is similar to the differentiation of components that occurs in the present-day industrial method called "zone refining."

This process of differentiation was probably the most important event of Earth's early history. It changed the planet from a homogeneous mixture with neither continents nor oceans to a planet with three layers: a dense core beginning at 2,900 kilometers (1,800 miles) deep and ending at Earth's center, 6,400 kilometers (3,977 miles) below the surface; a mantle beginning between 5 and 70 kilometers (3 and 44 miles) deep and ending at Earth's core; and a crust going from Earth's surface to about 5 to 10 kilometers (3 to 6 miles) deep for oceanic crust and 35 to 70 kilometers (22 to 44 miles) deep for continental crust.

LAYERED STRUCTURE OF EARTH

Earth's layers can be classified either by their composition (the traditional method) or by their mechanical behavior (strength). Compositional classification identifies several distinct concentric layers, each with its own properties. The outermost layer of Earth is the crust or skin. This is divided into continental and oceanic crusts. The continental crust varies in thickness between 35 and 40 kilometers (22 and 25 miles) under flat continental regions and up to 70 kilometers (44 miles) under high mountains. The oceanic crust is made up of igneous rocks rich in iron and magnesium, such as basalt and peridotite. The upper continental crust is composed mainly of alumino-silicates. The oldest continental crustal rock exceeds 3.8 billion years, while oceanic crustal rocks are not older than 180 million years. The oceanic crust is denser than the continental crust.

Earth's next layer is the mantle, which is made up mostly of ferro-magnesium silicates. It is about 2,900 kilometers (1,800 miles) thick and is separated into the upper and lower mantle. Most of Earth's internal heat is contained within the mantle. Large convective cells in the mantle circulate heat and may drive plate-tectonic processes.

The last layer is the core, which is separated into the liquid outer core and the solid inner core. The outer core is 2,300 kilometers (1,429 miles) thick, twice as thick as the inner core. The outer core is mainly composed of a nickel-iron alloy, while the inner core is almost entirely composed of iron. Earth's magnetic field is believed to be controlled by the liquid outer core.

In the mechanical layering classification of the earth's interior, the layers are separated based on mechanical properties or strength (resistance to flowing or deformation) in addition to composition. The uppermost layer is the lithosphere (sphere of rock), which comprises the crust and a solid portion of the upper mantle. The lithosphere is divided into many plates that move in relation to each other due to tectonic forces. The solid lithosphere floats atop a semiliquid layer known as the asthenosphere (weak sphere), which enables the lithosphere to move around.

EXPLORING EARTH'S INTERIOR

Volcanic activity provides natural samples of the outer 200 kilometers (124 miles) of Earth's interior. Meteorites—samples of the solar system that have collided with Earth—also provide clues about Earth's composition and early history. The most ambitious human effort to penetrate Earth's interior was made by the former Soviet Union, which drilled a superdeep research well, named the Kola Well, near Murmansk, Russia. This was an attempt to penetrate the crust and reach the upper mantle. The reported depth of the Kola Well is a little more than 12 kilometers (7.5 miles). Although impressive, the drilled depth represents less than 0.2 percent of the distance from the earth's surface to its center.

A great deal of knowledge about Earth's composition and structure has been obtained through computer modeling, high-pressure laboratory experi-

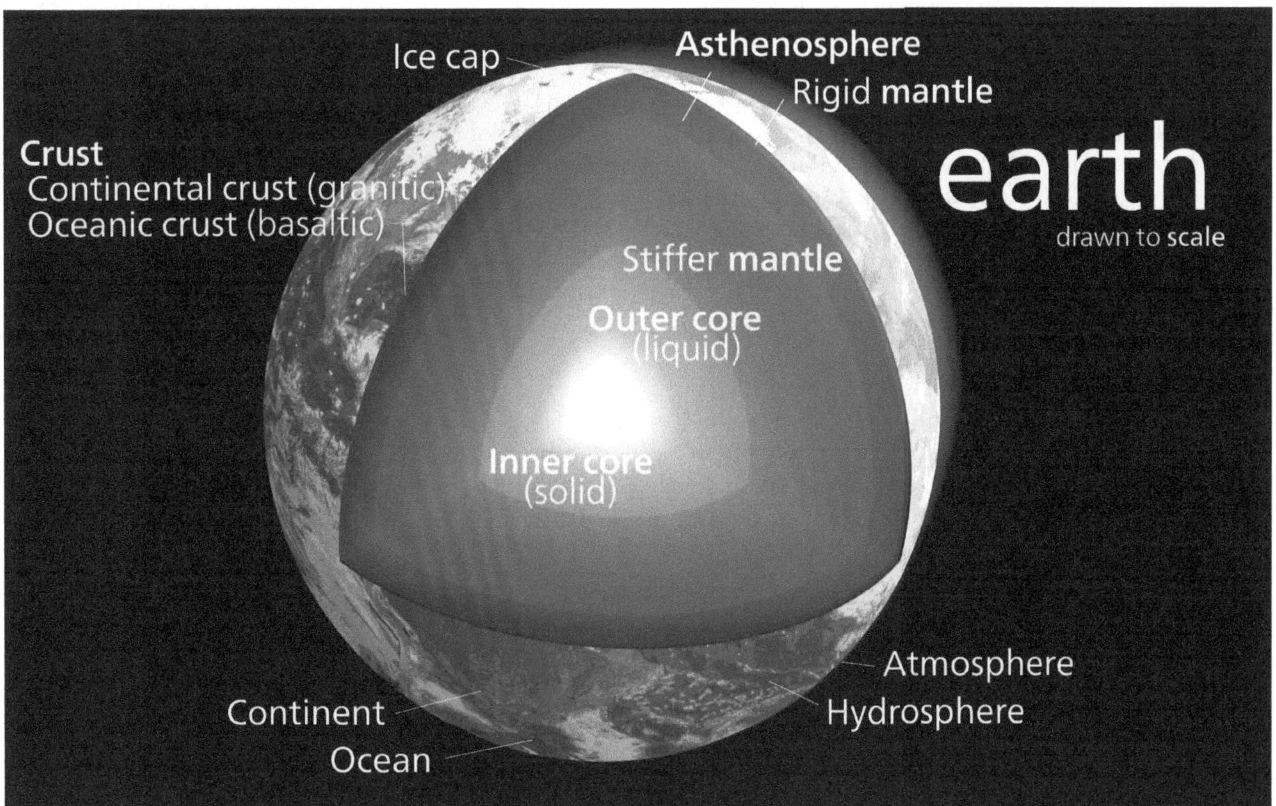

Earth's internal structure. Image by Kelvinsong, CC BY-SA 3.0, via Wikimedia Commons.

ments, and meteorites, but most of what is known about Earth's interior has been acquired by studying seismic waves generated by earthquakes and nuclear explosions. As seismic waves are transmitted, reflected, and refracted through the planet, they carry information to the surface about the materials through which they have traveled. Seismic waves are recorded at receiver stations (seismographic stations) and processed to provide a picture-like image of Earth's interior.

Seismic waves are designated as "P-waves" and "S-waves" according to their mode of transmission. P-waves are longitudinal waves that transmit through the planet as a compression or Pressure wave. S-waves are transverse Sheer waves. The corresponding analogies are the sudden compression of a spring that travels back and forth along the central axis of the spring, and the sudden lateral displacement of one end of a spring which then travels back and forth along the central axis of the spring in a sinusoidal manner. Changes in P- and S-wave velocities within Earth reveal the sequence of layers that make up Earth's interior. P-wave velocity depends on the elasticity, rigidity, and density of the material. By contrast, S-wave velocity depends only on the rigidity and density of the material. There are sharp variations in velocity at different depths, which correspond to boundaries between the different layers of Earth. P-wave velocity within crustal rocks ranges from 6 to 7 kilometers (3.6 to 4.2 miles) per second.

The boundary between the crust and the mantle is called the "Mohorovičić discontinuity" or just the Moho. At the Moho, P-wave velocity increases from 7 to 8 kilometers (4.2 to 4.8 miles) per second. Beyond the crust-mantle boundary, P-wave velocity increases gradually up to about 13.5 kilometers (8.1 miles) per second at the core-mantle boundary. At this depth, S-waves are not transmitted and P-wave velocity, decreases from 13.5 to 8 kilometers (8.1 to 4.8 miles) per second, which strongly supports the concept that the outer core is liquid, since S-waves cannot travel through liquids. As P-waves enter the inner core, their velocity again increases, to about 11.3 kilometers (6.8 miles) per second.

Earth's interior seems to be characterized by a gradual increase with depth in temperature, pressure, and density. Extensive experimental and modeling work indicates that the temperature at 100 kilometers (62 miles) is between 1,200°C and 1,400°C (2,192°F to 2,552°F). The temperature at the core-mantle boundary—about 2,900 kilometers (1,802 miles) deep—is calculated to be about 4,500°C (8,130°F). At Earth's center the temperature may exceed 6,700°C (12,092°F). Although at Earth's surface, heat energy is slowly but continuously lost as infrared radiation and as a result of outgassing, as from volcanic eruptions, its interior remains hot.

SEISMIC TOMOGRAPHY AND FUTURE EXPLORATION

Seismic tomography is one of the newest tools that earth scientists are using to develop three-dimensional velocity images of Earth's interior. In seismic tomography, several crossing seismic waves from different sources (earthquakes and nuclear explosions) are analyzed in much the same way that computerized axial tomography (CAT) scanners are used in medicine to obtain images of human organs. Seismic tomography is providing two-and three-dimensional images from the crust to the core-mantle boundary. Fast P-wave velocities have been correlated to cool material—for example, a piece of sinking lithosphere (cool rigid layer) such as in regions underneath the Andes Mountains (subduction zone); slow P-wave velocities have been correlated with hot materials—for example, rising plumes of magma in the mantle at hot spots such as the one responsible for the nearly constant volcanic activity in the Hawaiian Islands.

—Rubén A. Mazariegos-Alfaro

Further Reading

Christiansen, Eric H., and W. Kenneth Hamblin. *Dynamic Earth: An Introduction to Physical Geology.* Jones & Bartlett, 2015.

Fletcher, Charles. *Physical Geology: The Science of Earth.* 3rd ed., Wiley, 2017.

Kennett, Brian L. *Seismic Wave Propagation in Stratified Media.* Australian National UP, 2009.

Stein, Seth, and Michael Wyesession. *An Introduction to Seismology, Earthquakes, and Earth Structure.* Wiley, 2013.

Turcotte, Donald L., and Gerald Schubert. *Geodynamics.* 3rd ed., Cambridge UP, 2014.

Van der Pluijm, Ben A., and Stephen Marshak. *Earth Structure: An Introduction to Structural Geology and Tectonics.* W.W. Norton, 2010.

Earth's Moon

Fields of Study: Geology; Astrogeology; Astronomy

ABSTRACT

The Moon is the fourth-largest natural satellite in the Solar system and orbits Earth with mutual rotation such that observers on Earth only ever see the same lunar face. Between 1969 and 1972, National Aeronautics and Space Administration (NASA) astronauts set foot on the Moon in the Apollo space program. Material samples were recovered and returned to Earth for examination, providing the first authentic insights into the geology of the Moon.

KEY CONCEPTS

apogee: the point in an elliptical orbit at which the orbiting body is at its furthest distance from the main focal point of the orbit

ejecta: material thrown up from the surface as a result of the impact of a meteoroid striking the surface

maria: regions of the lunar surface that appear dark and smooth when viewed from Earth, once thought by ancient astronomers to be actual seas of water but now known to be the remnants of lava outflow resulting from large impacts

perigee: the point in an elliptical orbit at which the orbiting body is at its closest distance to the main focal point of the orbit

volcanism: the extent to which volcanic activity is exhibited on rocky planets and planetoids

EARTH'S CONSTANT COMPANION

The fourth-largest natural satellite in the Solar system, Earth's moon has a diameter of 3,475 kilometers (2,159.2 miles)—less than one-third the diameter of Earth. The Moon's mass is less than one-eightieth that of Earth.

The Moon orbits Earth in an elliptical path. When it is at perigee (when it is closest to Earth), it is 356,410 kilometers (221,473 miles) distant. When it is at apogee (farthest from Earth), it is 406,697 kilometers (252,722 miles) distant.

The Moon completes one orbit around Earth every 27.3 Earth days. Because the Moon rotates at the same rate that it orbits Earth, observers on Earth only see one side of the Moon. The changing angles between Earth, the Sun, and the Moon determine how much of the Moon's illuminated surface can be seen from Earth and cause the Moon's changing phases.

The full moon. Photo by Gregory H. Revera, CC BY 3.0, via Wikimedia Commons.

VOLCANISM

Naked-eye observations of the Moon from Earth reveal dark areas called "maria," the plural form of the Latin word *mare* for sea. The maria are the remains of ancient lava flows from inside gigantic impact craters; the last eruptions were more than 3 billion years ago. The lava consists of basalt, similar in composition to Earth's oceanic crust and many volcanoes. The maria have names such as Mare Serenitatis (15° to 40°N, 5° to 20°E), the Sea of Serenity, and Mare Tranquillitatis (0° to 20°N, 15° to 45°E), the Sea of Tranquility. Some of the smaller dark areas on the Moon also have names that are water-related: *lacus* (lake), *sinus* (bay), and *palus* (marsh).

IMPACT CRATERS

Observing the Moon with an optical aid, such as a telescope or a pair of binoculars, provides a closer view of impact craters. Impact craters of various sizes cover 83 percent of the Moon's surface. More than 33,000 craters have been counted on the Moon.

One of the easiest craters to observe from the Earth is Tycho. Located at 43.3°S longitude, 11.2 degrees west, it is about 85 kilometers (50 miles) wide. Surrounding Tycho are rays of dusty material, known as ejecta, that appear to radiate from the crater. When an object from space, such as a meteoroid, slams into the Moon's surface, it is vaporized upon impact. The dust and debris from the interior of the crater fall back onto the lunar surface in a pattern of rays. Because the ejecta is disrupted by subsequent impacts, only the youngest craters still have rays. Sometimes, pieces of the ejecta fall back and create smaller craters called "secondary craters." The ejecta rays of Tycho extend to almost 3,000 kilometers (1,865 miles) beyond the crater's edge.

OTHER LUNAR FEATURES

Near the crater called "Archimedes" is the Apennines mountain range, which has peaks nearly 60,000 meters (20,000 feet) high—altitudes comparable to South America's Andes.

The Moon also has valleys. Two of the most well known are the Alpine Valley, which is about 185 kilometers (115 miles) long; and the Rheita Valley, located about 250 kilometers (155 miles) from the Stevinus crater, which is 383 kilometers (238 miles) long, 25 kilometers (15.5 miles) wide, and 609 meters (2,000 feet) deep.

Smaller than valleys and resembling cracks in the lunar surface are features called "rilles," which are thought to be places of ancient lava flow. Many rilles can be seen near the Aristarchus crater. Rilles are often up to 5 kilometers (3 miles) wide and can stretch for more than 167 kilometers (104 miles).

A wrinkle in the lunar surface is called a "ridge." Many ridges are found around the boundaries of the maria. The Serpentine Ridge cuts through Mare Serenitatis.

EXPLORATION OF THE MOON

Robotic spacecraft were the first visitors to explore the Moon. The Russian spacecraft Luna 1 made the first flyby of the Moon in January, 1959. Eight months later, Luna 2 made the first impact on the Moon's surface. In October, 1959, Luna 3 was the first spacecraft to photograph the side of the Moon not visible from Earth. In 1994 the United States' *Clementine* spacecraft was the first probe to map the Moon's composition and topography globally.

The first humans to land on the Moon were the U.S. astronauts Neil Armstrong and Edwin "Buzz" Aldrin. On July 20, 1969, they landed in the *Eagle* lunar module, during the Apollo 11 mission. Armstrong's famous statement, "That's one small step for man, one giant leap for mankind," was heard around the world by millions of people who watched the first humans set foot on the lunar surface, at the Sea of Tranquillity. The last twentieth century human mission to reach the lunar surface, Apollo 17, landed there in December, 1972. Astronauts Gene Cernan and geologist Jack Schmitt landed in the Taurus-Littrow Valley (20°N, 31°E).

—Noreen A. Grice

Further Reading

Galimov, Erik M., and Anton M. Krivtsov. *Origin of the Moon—New Concept: Geochemistry and Dynamics.* Walter de Gruyter, 2012.

Launius, Roger D. *Apollo's Legacy: Perspectives on the Moon Landings.* Smithsonian Institution, 2019.

Levinson, Alfred A., and S. Ross Taylor. *Moon Rocks and Minerals.* Elsevier, 2015.

Ringwood, Alfred E. *Origin of the Earth and Moon.* Springer, 2012.

Schmitt, Harrison H. *Return to the Moon: Exploration, Enterprise, and Energy in the Human Settlement of Space.* Praxis, 2006.

Spudis, Paul D. *The Value of the Moon: How to Explore, Live and Prosper in Space Using the Moon's Resources.* Smithsonian Books, 2016.

Taylor, Stuart Ross. *Lunar Science: A Post-Apollo View: Scientific Results and Insights from the Lunar Samples.* Elsevier, 2016.

ECONOMIC GEOLOGY

Fields of Study: Geology; Chemistry; Physics; Structural Geology; Mineralogy; Petrology; Geochemistry; Mining Engineering; Geophysics; Stratigraphy

ABSTRACT

Economic geology is the study of the origin and distribution of mineral deposits of metals, other useful materials (such as building stone and salt), and fossil fuels (petroleum, natural gas, and coal). The economic geologist explores for these materials, predicts the mineral reserves of known deposits, and assesses the economics for the extraction of the ore.

KEY CONCEPTS

basalt: dark, fine-grained igneous rock composed mostly of plagioclase (a calcium-sodium silicate mineral) and pyroxene (a magnesium iron silicate mineral)

clay minerals: very small minerals that result from the weathering of other silicate minerals

granite: igneous rock composed of coarse crystals (greater than about 2 millimeters) of plagioclase (calcium-sodium silicate mineral), alkali feldspar (potassium-sodium silicate mineral), and quartz (all silicate with no other plus ions), and a few dark silicate minerals

hydrothermal: ore deposit in which the minerals were deposited by "water" vapor

magma: molten silicate rock below the surface with suspended minerals that have crystallized out of the molten material

mineral: naturally occurring element or compound that has a definite ordered arrangement of atoms

ores: minerals that may be economically and legally extracted from the earth

sandstone: sedimentary rock containing mostly quartz, feldspar, or rock fragments with grain sizes mostly in the range of 0.0625 to 2 millimeters

shale: very fine-grained sedimentary rock often consisting mostly of clay minerals and quartz

subduction zone: boundary where an oceanic plate (oceanic crust and part of the upper mantle) slowly slips below another oceanic or continental plate, creating earthquakes and magma

DEFINITION AND BASIC PRINCIPLES OF ECONOMIC GEOLOGY

Economic geology is the study of the origin and distribution of mineral deposits of metals, other useful materials (such as building stone and salt), and fossil fuels (petroleum, natural gas, and coal). The economic geologist explores for these materials, predicts the mineral reserves of known deposits, and assesses the economic aspects of the extraction of the ore. For instance, for an iron ore deposit, the geologist needs to assess the kinds and amount of iron minerals present, the depth of the deposit, the location of the deposit, and the ownership of the land so that estimates for the possibility of mining the iron ore profitably may be made.

Economic geologists explore for nonrenewable resources that formed so slowly, often over millions to even billions of years, that the processes of formation are much slower than the speed at which the resources can be extracted. For example, much of the oil and gas were formed by the slow, deep burial of marine, organic plankton that were alive as much as 2 billion years ago. These organisms need to be gradually buried to a certain depth long enough so that they gradually change to petroleum. If they are buried too deeply, the petroleum may break down to form natural gas (methane) or the mineral graphite (pure carbon); if they are not buried deep enough, then useful petroleum or natural gas will not form.

Nonrenewable resources include the abundant metals (such as iron and aluminum), scarce metals (gold and copper), materials used for energy (fossil fuels and uranium minerals), building materials (limestone, crushed stone, sand, and gravel), and other miscellaneous minerals (halite or natural salt). Running water, wind power, and solar power are not included among materials used for energy because they are renewable sources.

Economic geologists use their knowledge of geology to interpret where certain economic deposits might form. For instance, copper and some other associated elements may occur in what are called "copper porphyry deposits." These deposits formed as superheated water carried dissolved copper out of granite magma and upward along rock fractures in which the copper precipitated to form a variety of copper minerals. The porphyry copper deposits occur only in association with subduction zones. Therefore, the economic geologist knows to search for them only where ancient or existing subduction zones occur, such as along the west coasts of North and South America.

BACKGROUND AND HISTORY OF ECONOMIC GEOLOGY

The first use of natural resources was to obtain water, salt, and other natural materials to make tools and weapons. Larger cities were located where there was a source of water such as a river. Salt was used as a flavoring and to preserve food. A major discovery around 9000 BCE was that clay minerals could be heated to make pottery so that food and water could be stored and transported much more easily.

The first metals used were those such as gold and copper, which sometimes were found as native metals (not combined with other elements). These native elements could be shaped into useful materials by hammering or cutting. By 4000 to 3000 BCE, minerals of copper, zinc, lead, and tin were heated by burning charcoal in a very hot flame so that these elements could be separated from the ore. Much later, this process was applied to iron ore. The purification of iron had to wait until the development of blast furnaces, in which oxygen was blown through molten iron to purify it.

German physician Georgius Agricola wrote about the ways that ore minerals might form in *De Re Metallica* (1556; English translation, 1912). He divided the origins of ore deposits into those formed in streams and those formed in place. However, Latin was the language of scholars, not those actually working with ore deposits, so his writings were largely ignored.

Up to the nineteenth century, most ores were discovered accidentally because no one understood how ores and the rocks that contained them were geologically formed. Indeed, many people believed that mysterious celestial powers formed the ore minerals. During the period from the latter part of the eighteenth century to the twentieth century, the origin of rocks and ores gradually became understood so that people had a much better understanding of how to search for ores.

An open pit uranium mine in Namibia. Photo by Ikiwaner, GFDL 1.2, via Wikimedia Commons.

SELECTION OF A POTENTIAL ORE REGION

An economic geologist may first examine a geologic map, an aerial photograph, or a satellite photograph of a region, usually where other economic deposits have been found, to see if any geographical features provide clues as to where new deposits might be located. Petroleum and natural gas, for instance, usually occur in sedimentary rocks where certain geologic structures may be seen on photographs or maps. There must be shales that contain organic matter that can potentially be converted to oil or gas if buried to the right depth below the surface. Then there must be an overlying sedimentary rock, such as a permeable sandstone, sandwiched between two impermeable shales, and the sandstone must have spaces between mineral grains through which the oil may move so that it can flow to where it can be trapped. A variety of traps can be used, but one kind uses an impermeable rock to stop the oil from migrating so that it can collect. Once geologists find these circumstances, they can drill below the ground along the trap to see if any petroleum is present. Narrowing the range of possibilities to find oil is important because the cost of drilling even a single well is high.

The method for exploration differs depending on the type of deposit. For example, copper porphyry deposits often occur as veins in igneous rocks that formed along subduction zones; therefore, the exploration takes place along subduction zones. Some of the large iron ore deposits called "banded iron formations" found in the Lake Superior region formed only in shallow oceans from 1.8 billion to 2.6 billion years ago. Therefore, geologists search for new iron deposits only in other sedimentary rocks of similar ages.

GEOPHYSICAL METHODS

Remote methods of discovering an economic mineral deposit use instruments that can detect variations in magnetism, electrical conductivity, gravity, and radioactivity in rocks at or close to the surface of Earth. For example, some iron-rich minerals such as magnetite (Fe_3O_4) and ilmenite ($FeTiO_3$) will produce magnetic attractions that suggest that a region may have been hydrothermally mineralized. The variations in magnetism may be detected fairly quickly by flying a magnetometer over a region in a regular pattern. A Geiger counter may be used to quickly assess if there are any radioactive elements such as uranium or thorium present at the surface. Also instruments that detect variations in electrical conductivity can be used to determine if certain sulfide minerals might be present.

Petroleum geologists may detonate small charges along the surface so that sound waves travel through the ground for some distance. The sound waves, which can be detected at a receiver station, travel at varying speeds through different kinds of rocks and geologic structures. A computer program can then be used to construct a cross-section of the characteristics of the rocks below the surface. Sedimentary rocks were deposited in horizontal layers, but in some areas, they may be folded into arches (anticlines), warped downward (synclines), or fractured along faults that displace the sedimentary rock layers. The top portion of an anticline or the side of some fault may provide a possible trap where oil and natural gas can accumulate. Geologists can drill a well into these traps to see if oil or gas is present.

GEOCHEMICAL METHODS

Ore minerals may be weathered out of ore deposits and move into streams, where they slowly move downstream. To search for such deposits, the geologist may collect a series of sediments, often at stream junctions, and look for ore minerals in the sediments or analyze the sediments to see if any metals are present in abnormally high concentrations. This technique was used to discover a number of metal ores, such as the gold and copper porphyry ores in Papua, New Guinea.

Once a suspected ore deposit has been found in a certain region, soil samples may be collected in a systematic grid pattern over the area. The soil samples are analyzed for the elements most likely to be present in the potential ore deposit. Areas that have much higher concentrations of these elements may indicate the presence of ore directly below the surface. Geologists may drill directly into the areas with high concentrations of these elements to obtain subsurface samples to confirm the kinds of minerals and the elemental concentrations. This method, for example, found zinc deposits near Queensland, Australia.

The results of such a drilling program are used to evaluate whether mining the ore prospect might be worthwhile. The decision of whether to mine the prospect depends on many factors, such as the size and location of the deposit, the concentration and types of elements present, the finances of the company, the price obtained for the ore, the ownership of the land, and whether open-pit or underground mining can be used. For example, a gold deposit in northern Canada may not be profitable because of the costs of transporting equipment and miners to the area and of transporting the ore to a production facility. If gold prices drastically increase, then a previously uneconomical mine might be able to yield a profit. Also, large companies sometimes abandon smaller deposits and try to sell them to smaller companies that might still profitably mine the ore.

APPLICATIONS AND PRODUCTS

The mineral resources discovered by economic geologists are used by industries to produce the abundant metals, the scarce metals, fossil fuels, and other natural materials that support society. Economic geologists are not directly concerned with the use of these materials, but they must be aware of the demand for these materials so that they know the amount of money that a company may receive for them.

ABUNDANT METALS AND THEIR USES

The abundant metals are those found in the highest concentrations within Earth. Iron minerals are used to make steel, which is used to make automobiles, buildings, roads, bridges, major appliances, and construction materials such as nails, screws and other hardwares. The major iron ore minerals, hematite and magnetite (both iron oxides), are converted to steel by heating them to a high temperature with coke and limestone to form molten steel, which contains up to 2 percent carbon to produce an iron alloy that is harder than iron alone.

Aluminum is another abundant metal that is obtained from the ore bauxite, a mix of the minerals boehmite (aluminum oxyhydroxide) and gibbsite (aluminum hydroxide). The production of aluminum is very expensive because the bauxite must be heated with the compound cryolite to make a molten solution in which the aluminum is concentrated by an electric current. Aluminum is used in products such as cans, foil, windows, vehicles, and household items and mixed with other metals in alloys. Fortunately, aluminum metal is much cheaper to recycle into "new" metal than it is to produce from raw materials.

Titanium occurs in the minerals rutile (titanium oxide) and ilmenite (iron titanium oxide). Like aluminum, titanium is expensive to produce from its ore. Most titanium used is in the form of titanium oxide, which produces an intense white color in paint and paper. Titanium is also alloyed with aluminum, vanadium, and iron and used in ships, airplanes, and missiles. Magnesium oxide is often alloyed with aluminum to provide corrosion-resistant cans and materials used in vehicles. Magnesium metal is expensive to produce because like aluminum, electricity is required for its production.

Silicon is obtained by melting the mineral quartz (silicon oxide) with iron and coke in an electric furnace. Silicon is often alloyed with iron, aluminum, and copper because it improves the strength of these alloys and guards against corrosion. Silicon is also the base material for the formation of transistors on the chips used in computers and many other electronic devices, including modern automobiles.

SCARCE METALS AND THEIR USES

Scarce metals occur in Earth in much lower concentrations than the abundant metals. Some minerals, however, may locally concentrate these metals, making them potentially economic to mine. Some scarce metals may be added to steel to give it certain characteristics. For instance, chromium, molybdenum, tungsten, or vanadium may be added to make steel harder, especially at higher temperature. Stainless steel contains more than 11 percent chromium combined with nickel to help prevent the steel from rusting. Chromium and nickel may improve the strength of steel. Vanadium added to steel decreases the weight of steel, its strength, ductility, and ease of welding.

In addition, chromium is also used in chromium compounds to produce paint and ink pigments with green, yellow, and orange colors. Molybdenum is used in catalysts, pigments, and lubricants. Tungsten is often combined with carbon to produce a very hard compound, tungsten carbide, which is nearly as hard as diamond.

Copper, lead, zinc, tin, mercury, and cadmium are often grouped together simply because they are not used to alloy with iron. Copper metal conducts electricity well and can be shaped into wires, so it has been used mainly in electric lines and electric motors. Lead and mercury are used much less than in the past because of their toxicity. Lead is still used in batteries for vehicles, and mercury is used in some batteries, electrical switches, and for some chemical compounds. Zinc or tin are used in protective coatings for steel to keep it from rusting. Zinc oxide is added to paint to produce a white color and to a variety of lotions to prevent sunburn. Many so-called tin cans contain tin plated on other metals.

The precious metals are gold, silver, and the platinum metals. These metals are often used in jewelry because of their beauty. Gold and silver were also used for money until recently, and are still in industrial materials such as electronic connectors and dental fillings. Silver is also used in some electrical equipment. Platinum is used as a catalyst, for example, in catalytic converters in cars.

A variety of other metals, such as the rare earth elements, are used for industrial purposes. Rare earth elements have been used as catalysts and to color glass and ceramics and to provide some colors in old-style cathode ray tube (CRT) television screens.

CHEMICAL MINERALS AND FERTILIZERS

A number of nonmetallic minerals such as halite, baking soda, and sylvite are used in food. Halite and sylvite are salts used for flavoring. Halite is also used to soften water and to melt ice on roads. Borax is a boron compound that is used in some detergents and cosmetics.

A variety of potassium, nitrogen, and phosphorous minerals are used to fertilize crops. Commonly used nitrate compounds are ammonium nitrate, sodium nitrate and potassium nitrate. Ammonium phosphate and apatite are examples of phosphate minerals.

Gypsum (calcium sulphate) is the main mineral in wallboard. Sulfur compounds derived from petroleum are used to make many industrial compounds such as sulphuric acid.

BUILDING MATERIALS

Building materials include building stones, crushed rocks, sand, gravel, cement, plaster, bricks, and glass. Common building stones are granite, limestone, and sandstone. Granite, for example, can be polished to form an attractive surface for building facings and countertops, and headstones for graveyards.

Huge quantities of crushed rocks, sand, and gravel are used for roads, building foundations, and concrete. In the United States, more than 100 billion tons of these materials are used annually.

GEMSTONES

The most important gems are diamonds, sapphires, rubies, and emeralds. Gems must be harder than most other minerals, and they must be beautiful. Diamonds are composed of the element carbon, and rubies and sapphires are gem-quality varieties of the mineral corundum. Emeralds are gem-quality varieties of the mineral beryl.

FOSSIL FUELS

Petroleum, natural gas, and coal are the main fossil fuels. Most of the coal that is mined is burned to provide electrical power, and most of the natural gas that is recovered is burned to provide heat in homes and industry. Petroleum products such as gasoline are used in providing power for automobiles and trucks. All fossil fuels are falling into disfavor due to their role in producing anthropogenic climate change.

IMPACT ON INDUSTRY

Most of the most easily found minerals and fossil fuels have already been discovered, so economic geologists must continue to devise new ways to search for deposits. The demand for most metals, fossil fuels, and industrial minerals is likely to increase in the future.

INDUSTRY AND BUSINESS

Private companies control the search for metals, fossil fuels, and industrial minerals. The worldwide supply and demand of many of these commodities controls the prices. Many metals are easy to transport, and many are in high demand, so the prices for them continue to increase. Large groups that control the supply may dictate the prices of some commodities such as diamonds and oil. The Organization of Petroleum Exporting Countries (OPEC) is a cartel that controls the amount of oil exported from its members and hence its price. The DeBeers Group controls much of the supply of diamonds and therefore their price.

ROLE OF GOVERNMENTS IN ECONOMIC GEOLOGY

Governments do not usually explore for minerals or fuels, but they may, for instance, periodically stockpile them, thus driving up the prices. For instance, the United States has a stockpile of petroleum, to which it periodically adds. Sometimes it sells part of the petroleum to counteract rapid increases in petroleum prices if OPEC cuts back on the supply.

Governments will usually tax the income from private industry. They may also pass laws to increase the safety in mines and production facilities for a given commodity such as coal.

ROLE OF UNIVERSITIES IN ECONOMIC GEOLOGY

Geology and engineering departments in universities provide instruction in economic geology. Academic researchers in geology typically focus their research on how to better explore for mineral resources, and those in engineering concentrate on how to mine minerals and improve on extraction methods.

A person interested in exploring for metals, fossil fuels, and other industrial minerals should major in geology in college with a concentration in economic geology and exploration geology. Courses such as mineralogy, petrology, economic geology, and structural geology are required to obtain a bachelor's degree. A variety of mathematics, chemistry, and physics courses should also be taken. Further study for master's and doctoral degrees would enable the student to carry out research on a specific problem in economic geology. Those interested in solving geochemical aspects of exploration should take many supporting courses in chemistry; those interested in exploring using geophysical techniques should take supporting courses in mathematics and physics. Advanced degrees enable an individual to be given a lot of responsibility if employed by a company searching for natural resources. Those interested in developing mining operations once a potential site has been discovered can major in mining engineering with coursework on how to develop mines and supporting courses in geology, physics, and chemistry. Those interested in solving environmental problems associated with an economic site should take a variety of courses in geology, chemistry, water chemistry, and hydrology.

SOCIAL CONTEXT AND FUTURE PROSPECTS

The increased demand for most metals, oil, natural gas, coal, and other raw materials such as sand and gravel is likely to continue indefinitely. Construction materials such as sand, gravel, and limestone, which are used in large quantities, need to be sourced close to where they will be used because of the high cost of transporting them. Most metals are used in smaller quantities, and they can be shipped economically from sites around the world to where they will be used. In industrial societies such as the United States, the amount of money spent on industrial minerals is much greater than that spent on the metals.

The use of iron, manganese, aluminum, copper, zinc, gold, graphite, nickel, silver, sulphur, vanadium, and zinc have increased substantially from 1950 to the twenty-first century. The use of some metals, such as lead, increased from the 1950s to the 1980s but subsequently declined because of associated environmental and health problems.

Worldwide, the distribution of many metals and fossil fuels is very uneven, which means that many countries must import these resources. For example, the United States augments its petroleum stores by obtaining oil from countries such as Saudi Arabia and Venezuela, which are rich in petroleum. The United States must also import a great deal of aluminum, platinum, manganese, tantalum, and tungsten to meet its demand. The United States, however, has an excess of salt, lead, copper, and iron that it trades with other countries.

—Robert L. Cullers

Further Reading

Craig, James R., David J. Vaughan, and Brian Skinner. *Earth Resources and the Environment.* Prentice Hall, 2010.

Klein, Cornelis, and Barbara Dutrow. *The Twenty-third Edition of the Manual of Mineral Science.* Wiley, 2008.
Pohl, Walter L. *Economic Geology Principles and Practice.* Wiley-Blackwell, 2011.
Ridley, John. *Ore Deposit Geology.* Cambridge UP, 2013.
Robb, H. G. *Introduction to Ore Forming Processes.* Blackwell Science, 2005.
Singer, D. A., and W. D. Menzie. *Quantitative Mineral Resource Assessments: An Integrated Approach.* Oxford UP, 2010.
Wellmer, Friedrich-Wilhelm, Manfred Dalheimer, and Markus Wagner. *Economic Evaluations in Exploration.* Springer, 2008.

Engineering Seismology

Fields of Study: Engineering; Geology; Physics; Geophysics; Seismology; Geodynamics; Seismic Modeling; Stratigraphy; Structural Engineering

ABSTRACT

Engineering seismology is a scientific field focused on studying the likelihood of future earthquakes and the potential damage such seismic activity can cause to buildings and other structures. Engineering seismology utilizes computer modeling, geological surveys, existing data from historical earthquakes, and other scientific tools and concepts. Engineering seismology is particularly useful for the establishment of building codes and for land-use planning.

KEY CONCEPTS

duration: the length of time ground motion occurs during an earthquake
epicenter: the surface-level geographic point located directly above an earthquake's hypocenter
focal depth: the depth of an earthquake's hypocenter
ground motion: the ground-level shaking that occurs during an earthquake
hypocenter: the point of origin of an earthquake
Love waves: seismic waves that occur in a side-to-side motion
magnitude: an earthquake's size and relative strength
Rayleigh waves: seismic waves that occur in a circular, rolling fashion
Richter scale: logarithmic scale used to assign a numerical value to the magnitude of an earthquake
source parameters: a series of earthquake characteristics, including distance, duration, energy, and the types of waves that occur
stress drop: the amount of energy released when locked tectonic plates separate, causing an earthquake
wave propagation path: the directions in which seismic waves travel in an earthquake.

DEFINITION AND BASIC PRINCIPLES OF ENGINEERING SEISMOLOGY

Engineering seismology (also known as earthquake engineering) is a multidisciplinary field that assesses the effects of earthquakes on buildings, bridges, roads, and other structures. Seismology engineers work in the design and construction of structures that can withstand seismic activity. They also assess the damages and effects of seismic activity on existing structures. Engineering seismologists analyze such factors as quake duration, ground motion, and focal depth in assessing the severity of seismic events and how those events affect fabricated structures. They also consider source parameters, which help seismologists zero in on a seismic event's location and the speed and trajectory at which the quake's resulting waves are traveling.

Earthquake engineers also study theoretical concepts and models related to potential earthquakes and historical seismic events. Such knowledge can help engineers and architects design structures that can withstand as powerful an earthquake as the geographic region has produced (or possibly will produce). Mapping systems and programs and mathematical and computer-based models are essential to engineering seismologists' work. Such techniques are also useful for archaeologists and paleontologists, both of whom may use engineering seismology concepts to understand how Earth has evolved over millions of years and how ancient civilizations were affected by seismic events.

BACKGROUND AND HISTORY OF ENGINEERING SEISMOLOGY

Throughout human history, people have struggled to understand the nature of earthquakes and, as a result, have faced the challenges of preparing for these seismic events. Some ancient civilizations attributed earthquakes to giant snakes, turtles, and

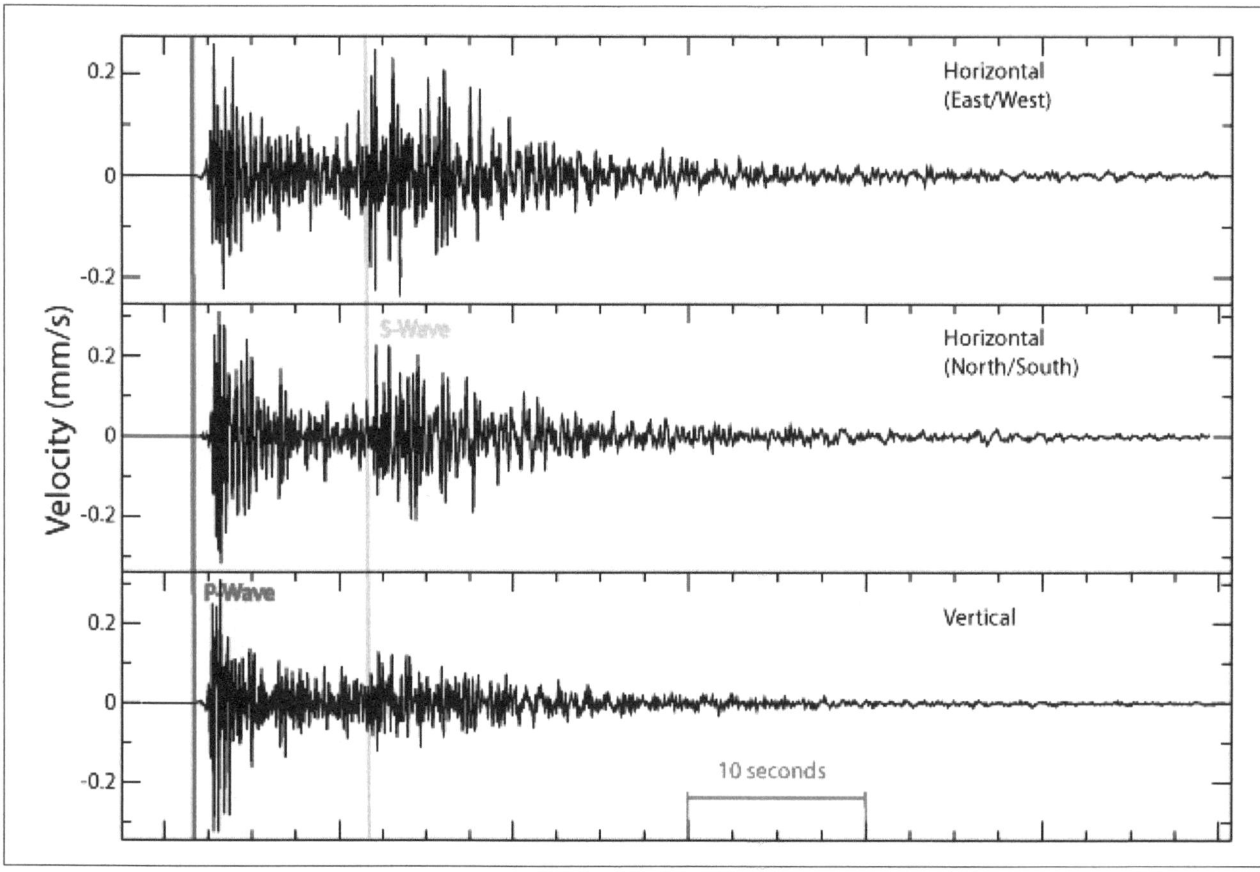

Seismograph showing S-wave and P-wave activity. Copyrighted free use, Wikimedia.

other creatures living and moving beneath Earth's surface. In the fourth century BCE, Aristotle was the first to speculate that earthquakes were not caused by supernatural forces but rather were natural events. However, little scientific study on earthquakes took place for hundreds of years, despite the occurrence of many major seismic events (including the eruption of Mount Vesuvius in Italy in 79 CE, which was preceded by a series of earthquakes).

In the mid-eighteenth century, however, the British Isles experienced a series of severe earthquakes, which created a tsunami that destroyed Lisbon, Portugal, killing tens of thousands of people. Scientists quickly developed an interest in cataloging and understanding seismic events. In the early nineteenth century, Scottish physicist and glaciologist James D. Forbes invented the inverted pendulum seismometer, which gauged not only the severity of an earthquake but also its duration.

Throughout history, seismology has seen advances that immediately followed significant seismic events. Engineering seismology, which is proactive, represents a departure from reactionary approaches to the study of earthquakes. Today, engineering seismology uses seismometers, computer modeling, and other advanced technology and couples it with historical data for a given site. The resulting information helps civil engineers and architects construct durable buildings, bridges, and other structures and assess the risks to existing structures posed by an area's seismic potential.

UNDERSTANDING ENGINEERING SEISMOLOGY

To understand engineering seismology, one must understand the phenomenon of earthquakes. Earthquakes may be defined as the sudden shaking of a portion of Earth's surface as caused by the movement of subterranean rock. Massive segments of Earth's crust, called "tectonic plates," resting on Earth's mantle layer, are driven to constant movement by convection currents in the molten magma below them. The boundaries formed by the edges of these plates are known as faults. When passing plates lock together so that their motion relative to each other is stopped, strain energy builds up gradually in the rock. The plates eventually give way and move again, causing that energy to be released suddenly and sent from the quake's point of origin (the hypocenter) outward to the surface in the form of seismic (or surface) waves. Such waves occur either in a retrograde elliptical, rolling motion (Rayleigh waves) or in a twisting, side-to-side motion perpendicular to the direction of travel of the wave (Love waves).

The field of seismology has developed only over recent centuries, largely because of major, devastating seismic events. The practice of engineering seismology has grown in demand in recent years, mainly because of the modern world's dependency on major cities, infrastructure (such as bridges, roadways, and rail systems), and energy resources (including nuclear power plants and offshore oil rigs). Earthquake engineers, therefore, have two main areas of focus: studying seismology and developing structures that can withstand the force of an earthquake.

To study seismic activity and earthquakes, engineering seismologists may use surface-based detection systems, such as seismometers, to monitor and catalog tremors. They also employ equipment—including calibrators and accelerometers—that is lowered into deep holes. Such careful monitoring practices help seismologists and engineering seismologists better understand a region's potential for seismic activity.

When earthquakes occur, engineering seismologists quickly attempt to locate the hypocenter and the epicenter (the surface point that lies directly above the hypocenter). They are able to do so by monitoring two types of waves—P- and S-waves—that move much quicker than surface waves and, therefore, act as precursors to surface waves. These engineers also work to determine the magnitude (a measurement of an earthquake's size) of the event.

Magnitude may be based on a number of key factors (or source parameters), including duration, distance to the epicenter and hypocenter, the size and speed of the surface waves, the amount of energy (known as the stress drop) that is released from the hypocenter, P- and S-waves, and the directions in which surface waves move (the wave propagation path). Analyzing an earthquake's magnitude provides an accurate profile of the quake and the conditions that caused it.

In addition to developing a profile of a region's past seismic activity, earthquake engineers use such information to ascertain the type of activity a geographic region may experience in the future. For example, scientific evidence suggests that the level of stress drop is a major contributor to the severity of seismic activity that can cause massive destruction in major urban centers. Similarly, studies show that the duration of ground motion (the "shaking" effects of an earthquake) may be more of a factor in the amount of damage to buildings and other structures than stress drop.

The field of engineering seismology is less than one century old, but in the twenty-first century, it plays an important role in urban development and disaster prevention. Earthquake seismologists work with civil engineers and architects to design buildings, roads, bridges, and tunnels that may withstand the type of seismic activity that has occurred in the past.

APPLICATIONS AND PRODUCTS

Engineering seismology applies knowledge of seismic conditions, events, and potential to the design and development of new and existing fabricated structures. Among the methods and applications employed by earthquake seismologists are the following:

Experimentation. Engineering seismologists may construct physical scale models of existing structures or proposed structures. Using data from a region's known seismic history, the engineers attempt to recreate an earthquake by placing these models on "shake tables," large mechanical platforms that simulate the motions of a wide range of earthquake types. After the "event," engineering seismologists examine the simulation's effects on the model structure, including its foundations, support beams, and walls. This approach enables the engineers and architects to directly examine the pre- and poststimulation structure and determine what sort of modifications may be warranted.

Computer modeling. One of the most effective tools utilized by engineering seismologists is computer modeling. Through the application of software, engineering seismologists can input a wide range of source parameters, ground motion velocities, wave types, and other key variables. They also can view how different structural components withstand (or fail to withstand) varying degrees of seismic activities without the expense and construction time required for a shake table simulation. Computer modeling has become increasingly useful when attempting to safeguard against earthquake damage to dams, nuclear power plants, and densely developed urban centers. Computer modeling is also used by engineering seismologists to predict the path of destruction that often occurs after an earthquake, such as that caused by fires or flooding.

Seismic design software. Earthquake engineers study seismic activity in terms of how it affects structures. To this end, engineers must attempt to predict how earthquakes will strike an area. Seismic design software is used to create a map of a region's seismic activity and how those conditions will potentially cause structural damage. The software enables government officials to establish formalized building codes for buildings, bridges, power plants, and other structures. This software is easily obtained on the Web and through the U.S. Geological Survey (USGS) and other organizations.

Mathematics. Engineering seismology is an interdisciplinary field that relies heavily on an understanding of physics and mathematics. To quantify the severity of earthquakes, to calculate the scope of seismic activity, and, in general, to create a profile of a region's seismic environment, engineering seismologists utilize a number of mathematical formulae. One of the most-recognized of these formulae is the Richter scale, which was developed in 1935 by American seismologist and physicist Charles Richter. The Richter scale assigns a logarithmic numerical value (with no theoretical limit) to the magnitude of an earthquake. The Richter scale takes into account the amplitude (the degree of change) of seismic waves and the distance between the equipment that detects the quake and the quake's epicenter.

Earthquake engineers use such mathematical data as part of their analyses when working with civil engineers on construction projects. Earthquake engineers also are increasingly called upon by government officials to use this data to assess individual structure and citywide structural deficiencies that resulted in earthquake destruction. Such forensic engineering was called into service in 2009, when Australian emergency officials intervened in Padang, Indonesia, after a magnitude 7.6 quake devastated that city. Engineers used mathematical formulae and statistical data to assess systemwide structural deficiencies in Padang rather than analyzing damage on a structure-by-structure basis. In light of the countless variables involved with studying earthquakes and their effects on fabricated structures, the use of established logarithm values, data sets, and mathematical formulae is a time-honored practice of engineering seismologists.

Sensors. Not all earthquakes cause immediate and significant damage to affected structures. According to the USGS, the Greater San Francisco area experienced more than eighty earthquakes in 2011 alone, with none of those quakes registering higher than a 2.3 on the Richter scale. However, seismic activity on a small but frequent basis can cause long-term damage to structures through structural fatigue. For example, seismic events can shift soil pressure on underground structures (such as pipes and foundations). Earthquake engineers are therefore highly reliant on sensor equipment, which enables them to gauge the effects of frequent seismic activity not only on aboveground structures but also on the ground itself.

To examine shifts in soil pressure caused by seismic activity, seismology engineers use an array of tactile pressure sensors, which were originally designed for artificial intelligence systems but were later utilized for the purposes of designing car seats and brake pad systems. The use of such equipment helps engineers study the long-term effects of seismic activity on water pipes, underground cables, and underground storage tanks.

Arguably one of the best-known types of seismic detector systems is the seismograph. The seismograph uses a pendulum-based system to detect ground motion from seismic activity. Originally, the modern seismograph was designed to detect only significant earthquakes and tremors, and it could be found only in seismically stable environments (such as a laboratory). Today, however, there are many different types of seismographs; some may be placed underground, others can be used in the field, while others are so sensitive that they can detect distant explosions or minute tremors such as the vibrations caused by heavy vehicles in motion.

IMPACT ON INDUSTRY

Engineering seismology is a relatively new combination of civil engineering and seismology, along with other fields (such as geology, emergency management, and risk management). Its uses have proven invaluable, however, as many urban centers in earthquake-prone regions, such as Tokyo and San Francisco, have benefited from the careful application of safe building practices and disaster mitigation programs that are borne of engineering seismology. Japan, the United States, and Switzerland are among the leaders in earthquake engineering, along with Australia and New Zealand. Engineering seismology also involves a range of public and private organizations.

Governments. National and regional governments play an important role in the application of the findings of engineering seismologists. The USGS, for example, offers a wide range of resources and services for studying seismic activity and the dangers it poses. Additionally, the Federal Emergency Management Agency (FEMA), the National Science Foundation (NSF), the National Institute of Standards (NIS), and the USGS combine their resources to operate the National Earthquake Hazards Reduction Program (NEHRP), which seeks to reduce property losses and human casualties caused by earthquakes through careful engineering seismology practices (including mapping seismically active areas and generating building codes).

Engineering seismology societies. Engineering seismologists share information and theories with their peers through professional associations and societies, many of which are global in nature. The International Association for Earthquake Engineering (IAEE) is one such network, holding worldwide conferences on engineering seismology every four years. This organization has branch societies in Japan, Europe, and the United States, each working locally but also contributing to the association as a whole.

Universities. Because of the contributions engineering seismology provides to the field of civil engineering, earthquake engineering continues to evolve within this educational discipline. Universities such as Stanford and the University of California, Berkeley, offer such programs. Many other universities feature coursework in environmental engineering, which includes earthquake engineering and seismology studies. Furthermore, a large number of universities house research laboratories, shake tables, and seismograph stations covering seismic activity throughout a given geographic area.

Consulting firms. There is a considerable financial benefit to constructing a building, bridge, or other structure that will survive in a seismically active environment. Oil companies and other energy corporations, mining operations, and other businesses frequently seek the advice of private engineering consultants who offer seismic monitoring services. Construction companies also look to these consultants, seeking structural analyses and other services. The person who introduced the Richter scale, Charles Richter, founded a consulting firm upon his retirement from the California Institute of Technology in 1970.

EDUCATION AND TRAINING REQUIREMENTS

Engineering seismologists must receive training in the geosciences, including seismology, geology, and physics. These fields include courses in geodynamics, plate tectonics, statistical seismology, and mineral dynamics. They must also study civil engineering, structural design, and computer science (which must include training in computer modeling, digital mapping systems, and design software, which are essential in this arena). Furthermore, engineering seismologists must demonstrate excellent mathematical skills, particularly in geometry, algebra, and calculus. Finally, earthquake engineers must be trained in the use of many of the technical systems and devices that seismologists must use to monitor earthquakes.

SOCIAL CONTEXT AND FUTURE PROSPECTS

The study of earthquakes is a practice that dates back hundreds of years. However, earthquake engineering specifically represents an evolution toward a practical application of the study of seismic activity to the design and construction of large buildings, power plants, and other structures. Engineering seismologists work closely with seismologists and civil engineers to help design and operate detection equipment and systems to help explain seismic activity. This collaboration is critical, as improved knowledge of seismic activity can save lives and property. For example, Japan has long utilized engineering seismology practices in its urban centers. The magnitude 8.9 earthquake that struck that country in March, 2011, did not devastate Tokyo because of strong building codes that, among other things, cause skyscrapers to sway with the region's seismic waves rather than stand in rigid fashion. Comparatively, the magnitude 7.0 Haiti earthquake in 2010 virtually flattened the country's capital, Port-au-Prince, and outlying areas, largely because Haiti did not have earthquake-safe building codes. Its buildings were built using insufficient structural steel reinforcement and on slopes with no reinforcing foundation or support systems. One observer in Haiti commented that Port-au-Prince would likely not have survived even a magnitude 2.0, much less the 100-thousand times more powerful 7.0 quake it did have.

The significance of the 2011 Japan disaster and of the rare 5.8 Virginia earthquake that struck the East Coast of the United States in August, 2011, continues to cast light on the need for earthquake engineering in structural design and construction. As regions with a history of major seismic activity (and those regions with the potential for such activity) continue to grow in size and population, engineering seismologists are likely to remain in high demand.

—Michael P. Auerbach

Further Reading

Chopra, Anil K. *Dynamics of Structures: Theory and Applications to Earthquake Engineering*. 3rd ed., Prentice Hall, 2007.

———. *Earthquake Engineering for Concrete Dams: Analysis, Design and Evaluation*. Wiley-Blackwell, 2020.

Gray, William G., and Genetha A. *Gray Introduction to Environmental Modeling*. Cambridge UP, 2017.

Griffith, M. C., et al. "Earthquake Reconnaissance: Forensic Engineering on an Urban Scale." *Australian Journal of Structural Engineering*, vol. 11, no. 1, 2010, p. 63.

Havskov, Jens, and Gerardo Alguacil. *Instrumentation in Earthquake Seismology*. 2nd ed., Springer, 2016.

Slak, Tomaz, and Vojko Kilar. "Development of Earthquake Resistance in Architecture from an Intuitive to an Engineering Approach." *Prostor*, vol. 19, no. 1, 2011, pp. 252-63.

Stark, Andreas. *Seismic Methods and Applications*. BrownWalker Press, 2008.

Sucuoglu, Halûk, and Sinan Akkar. *Basic Earthquake Engineering from Seismology to Analysis and Design*. Springer, 2014.

Villaverde, Roberto. *Fundamental Concepts of Earthquake Engineering*. CRC Press, 2009.

EVAPORITES

Fields of Study: Geochemistry; Geomorphology; Economic Geology

ABSTRACT

Evaporites are sedimentary deposits of salt minerals that crystallize from marine and continental brines. Common evaporite minerals include halite (sodium chloride, or table salt), gypsum (hydrated calcium sulfate), calcite (calcium carbonate), dolomite (calcium-magnesium carbonate), and various borate minerals.

KEY CONCEPTS

cementation: the binding together of sedimentary particles by the formation of crystalline material in the pores between the particles

diagenesis: the conversion of unconsolidated sediment into consolidated rock after burial by the processes of compaction, cementation, recrystallization, and replacement

evaporite: the residue remaining when mineral-laden water has evaporated

ionic compound: a neutral compound whose discrete unit consists of a combination of positive and negative ions

playa: the exposed, and generally arid, remnant of the floor of an inland lake or sea

DEFINITION OF EVAPORITES

Evaporites form in environments where evaporative water loss from a body of water exceeds, at least periodically, the rate of inflow to the body. Evaporites occur in all the major continents; the most extensive deposits are found in North and South America, Europe, and the Middle East. Notable North American localities are the Michigan Basin and the Permian Basin of Texas and New Mexico. Most rock salt (halite) is mined from evaporites, as is the gypsum used in wallboard and other construction materials. Borate minerals are used in cleaning agents and in other industrial uses. Other evaporite minerals are important sources for industrial metals such as magnesium and strontium.

OVERVIEW OF EVAPORITES

Evaporites are stratified sedimentary deposits consisting of minerals precipitated from salt brines (highly concentrated salt water). These deposits have formed on every continent and throughout geologic time, although the Silurian (438 to 408 million years ago) and Permian (286 to 248 million years ago) periods were the most prolific times of evaporite formation. In recent times evaporite deposition has been relatively rare, though human societies have used this evaporation process on a small scale for thousands of years to obtain salt from seawater. The chief factors influencing evaporite formation are aridity and a closed basin environment in which water inflow is restricted. High air temperatures generally accompany these conditions, but this is not always the case. For example, recent minor evaporites are known from the arid regions of Antarctica.

Evaporite minerals crystallize from seawater in a certain order, depending on their relative solubilities. Calcite generally crystallizes first as the amount of water is reduced by evaporation. Gypsum follows, with halite precipitating when only about one tenth of the original solution is left. More soluble minerals crystallize in the final liquid, including sylvite (potassium chloride). This process produces concentrated crystalline layers that consist of only one or two major minerals.

World evaporite bodies can be divided into marine deposits, the thickest and most extensive in origin, and continental deposits. Marine evaporites may form in marginal lagoons closed off from the sea by a sandbar or other barrier. Another important environment is the sabkha, a shallow margin of a sea or ocean in an extremely arid climate, as occurs, for example, in the Persian Gulf. Continental deposits most commonly form in temporary desert lakes called "playas.' Evaporite minerals in playas are derived from streams that leach marine brines trapped in sedimentary rocks from surrounding mountains.

Major evaporite localities in North America are the Michigan Basin (Salina deposits), the Permian Basin of Texas and New Mexico, the Midcontinent field centered in Kansas and adjacent states, and the borate deposits of Death Valley and adjacent areas. The first three of these localities are mostly known for halite and gypsum production. They are primarily marine deposits formed in shallow continental seas during Silurian times (Michigan Basin) or during the Permian period (Permian Basin and Midcontinent). The Death Valley area is famous for its borate deposits, minerals that were deposited in playas.

CHEMICAL NATURE OF EVAPORITES

Evaporite materials are, without exception, individual molecular species, as opposed to the large concatenated sheets of silicate minerals. They are ionic compounds composed of a discrete combination of positive and negative ions. Sodium chloride, for example, consists of one sodium ion (Na^+) and one chloride ion (Cl^-) in each molecular unit of sodium chloride (NaCl). Gypsum consists of a calcium ion (Ca^{2+}) and a sulphate ion (SO_4^{2-}), with two water molecules associated with, but not bonded to, each calcium sulphate ($CaSO_4$) unit. Because of the alternating ordering of positive and negative ions, evaporites are capable of forming very large crystals over time. In 2000, mining engineers and cave explorers discovered a cave 300 meters below ground near Naica, Mexico, containing crystals of pure selenite, a form of gypsum, that greatly dwarf a human being in size. They are the largest known natural crystals, as much as 18 meters long by 1 meter wide, and weighing approximately 55 tonnes. It is estimated that these crystals grew over a period of 500,000 years in the hot, mineral-rich water that once filled the cave. The hot, moist environment that presently exists in the cave is necessary for maintaining the crystals in their natural form, as loss of water from the crystal structures results in its breakdown.

EVAPORITE DEPOSIT FORMATION

The Naica gypsum crystals formed by a different mechanism than other evaporites, however. Their slow formation is the results of dynamic crystallization-solution equilibrium processes in a saturated solution within a closed system. Evaporites such as halite and borax, however, were formed in an open system by the evaporative loss of water from the solution that carried them. The process can be easily demonstrated by preparing a saturated solution of salt in water and allowing it to slowly evaporate. Observation of the solution as it evaporates will exhibit first the formation of minute crystals of sodium chloride that are in excess of the amount of salt that particular amount of water can dissolve. As time passes and evaporative water loss continues, a thin layer of fine, white particles of salt will be seen to accumulate on the bottom of the vessel holding the solution. This will continue until all of the water has evaporated. However, if one continues to add more saturated salt solution to the dry residue, two things will be observed over time. The layer of fine salt crystals will continue to grow, and the size of individual salt crystals will increase over time. This is essentially what happens in nature as evaporite deposits are formed.

Over geologic periods of time, such evaporative loss deposits build up to the point where they become subject to the effects of temperature and pressure as all sedimentary deposits will. Salt deposits in the area of Goderich and other parts of southwestern Ontario, in Canada, for example, are located at significant depths beneath solid limestone bedrock and appear as glass-like beds in many cases. Like all sedimentary deposits under conditions of temperature and pressure, the ordering of particles is changed. Salt crystals naturally form with a perfectly cubic shape, but as they precipitate out the individual crystals form a chaotic mass without structure and with significant porosity. Under conditions of

A cobble encrusted with halite evaporated from the Dead Sea, Israel. Photo via Wikimedia Commons. [Public domain.]

temperature and pressure, the porosity is effectively "squeezed out" and the individual crystals are forced into a more orderly array. With sufficient pressure, crystals may also fuse together, forming the clear-as-glass structure that is observed in underground salt deposits. Carbonates undergo a similar process under pressure called "diagenesis" and "cementation." However, in carbonates and carbonatites, the pores between the compressed particles become filled with other minerals from infiltrating water. The process in salt formations results in the intergranular pores becoming filled primarily with precipitated salt, although mineral inclusions are common.

GENERAL ASPECTS OF EVAPORITES

When a bed of evaporite material becomes exposed to surface conditions through erosion or the lack of an overlying layer of rock, it quickly breaks down due to weathering. A salt bed, for example, is readily dissolved and carried away by rain, if there is rain, or eroded and blown away by wind. Borate deposits in Death Valley were once heavily surface-mined for the borax found there, used as a laundering agent and for other industrial purposes. A work crew would St. Louis, Missouri, to go into Death Valley, in California, with a team of twenty mules pulling several large wagons, and a week's supply of water.

They would fill the carts with borax, and immediately make the return trip to St. Louis, Missouri. It took two weeks to get to Death Valley and two weeks to return to St. Louis, whereupon the team of mules would be fed and watered, and immediately sent out on another borax-gathering mission. It is a testament to the strength and durability of mules, to say the least. Trucks now do the work of the mules.

In other parts of the world, exposed evaporite beds were, and still are, the source of salt and other minerals as valuable commodities. Unexposed evaporite beds are mined for their mineral wealth. What all of these source areas have in common is that they represent places where inflowing water has no sufficient exit for outflow to an ocean, and they are typically in regions of low rainfall where temperatures are high and atmospheric pressures are low, both factors contributing to evaporative water loss. The existence of such areas millions of years ago is what produced the evaporite deposits known in the present time.

—*John L. Berkley and Richard M. Renneboog*

Further Reading

Jackson, Martin P. A., and Michael R. Hudec. *Salt Tectonics: Principles and Practice*. Cambridge UP, 2017.

Leith, C. K. *The Economic Aspect of Geology*. Good Press, 2016.

Losiak, Anna Isabela. *The Development of Evaporite Minerals During Weathering of Antarctic Meteorites*. Michigan State U Geology Department, 2009.

Nichols, Gary. *Sedimentology and Stratigraphy*. 2nd ed., Wiley-Blackwell, 2009.

Parise, M., F. Gabrovsek, G. Kaufmann, and N. Ravbar, editors. *Advances in Karst Research: Theory, Fieldwork and Applications*. The Geological Society, 2018.

Warren, John K. *Evaporites: A Geological Compendium*. 2nd ed., Springer, 2016.

External Processes

Fields of Study: Physical Geography; Geophysics; Geology

ABSTRACT

Continuous processes are at work shaping the earth's surface. These include breaking down rocks, moving the pieces, and the pieces in new locations. Weathering breaks down rocks through atmospheric agents. The process of moving weathered pieces of rock by wind, water, ice, or gravity is called "erosion". The materials that are deposited by erosion are called "sediment." Mechanical weathering occurs when a rock is broken into smaller pieces but its chemical makeup is not changed. If the rock is broken down by a change in its chemical composition, the process is called "chemical weathering."

KEY CONCEPTS

exfoliation: spalling off of the surface layer of a rock body due to weathering

mass movement: the movement of a significant amount of earthen material by any of a variety of drivers such as wind, water and gravity

oxidation: reaction of elements in rock with oxygen to produce oxide compounds such as rust

rust: a general term for the products of the oxidation of metal components in rock

slump: the collapse of a hillside, caused by loss of structural integrity due to excessive groundwater content and the force of gravity, such that the affected portion slips downhill with a rotational movement

talus: the assortment of broken rock at the base of a cliff, formed as weathering causes pieces of rock to break free from the cliff face and fall

MECHANICAL WEATHERING

Different types of mechanical weathering occur, depending on climatic conditions. In areas with moist climates and fluctuating temperatures, rocks can be broken apart by frost wedging. Water fills in cracks in rocks, then freezes during cold nights. As the ice expands and pushes out on the crack walls, the crack enlarges. During the warm days, the water thaws and flows deeper into the enlarged crack. Over time, the crack grows until the rock is broken apart. This process is active in mountains, producing a pile of rock pieces at the mountain base called "talus."

Salt weathering occurs in areas where much salt is available or there is a high evaporation rate, such as along the seashore. Salt crystals form when salty moisture enters rock cracks. Growing crystals settle in the bottom of the crack and apply pressure on the crack walls, enlarging the crack.

Thermal expansion and contraction occur in climates with fluctuating temperatures, such as deserts. All minerals expand during hot days and contract during cold nights, and some minerals expand and contract more than others. This process continues until the rock loosens up and breaks into pieces.

Mechanical exfoliation can happen to a rock body overlain by a thick rock or sediment layer. If the heavy overlying layer over a portion of the rock body is removed, pressure is relieved and the exposed rock surface will expand in response. This expanding surface will break off into sheets parallel to the surface, but the remaining rock body remains under pressure and unchanged.

When plant roots grow into cracks in rocks, they enlarge the cracks and break up the rocks.

Finally, abrasion can occur to rock fragments during transport. Either the fragments collide, breaking apart, or fragments are scraped against rocks, breaking off pieces.

CHEMICAL WEATHERING

Water and oxygen create two common causes of chemical weathering. For example, dissolution occurs when water or another solution dissolves minerals within a rock and carries them away. Hydrolysis can occur when water flows through earth materials. The hydrogen ions or the hydroxide ions of the water may react with minerals in the rocks. When this occurs, the chemical composition of the mineral is changed, and a new mineral is formed. Hydrolysis often produces clay minerals.

Some elements in minerals combine with oxygen from the atmosphere, creating a new mineral. This process is called "oxidation." Some of these oxidation minerals are commonly referred to as rust.

MASS MOVEMENT

Weathered rock pieces (sediments) are transported (eroded) by one or more of four transport processes: water (streams and oceans), wind, ice (glaciers), or gravity. Mass movement transports earth materials down slopes by the pull of gravity. Gravity, constantly working to pull surface materials down, parallel to the slope, is the most important factor affecting mass movement. There is also a force involved perpendicular to the slope that contributes to the effects of friction.

Friction, the second factor, is determined by the earth material type involved. For example, weathering may create cracks in rocks, which form planes of weakness on which the mass movement can occur. Loose sediments always tend to roll downhill.

The third factor is the slope angle. Each earth material has its own angle of repose, which is the steepest slope angle on which the materials remain stable. Beyond this slope angle, earth materials will move downslope.

Water, the fourth factor, affects the stability of the earth material in the slope. Friction is weakened by water between the mineral grains in the rock. For example, water can make clay quite slippery, causing the mass movement.

The rooting system of vegetation, the fifth factor, helps make the surficial materials of the slope stable by binding the loose materials together.

Mass movements can be classified by their speed of movement. Creep and solifluction are the two types of slow mass movement, which are measured in fractions of inches per year. Creep is the slowest mass movement process, where unconsolidated materials at the surface of a slope move slowly downslope. The materials move slightly faster at the surface than below, so evidence of creep appears in the form of slanted telephone poles. During solifluction, the warm sun of the brief summer season in cold regions thaws the upper few feet of the soil. This waterlogged soil flows downslope over the underlying permafrost.

Rapid mass movement processes occur at speeds of feet per second. Falls occur when loose rock or sediment is dislodged and drops from a steep slope, such as along sea cliffs where waves erode the cliff base. Topples occur when there is an overturning movement of the mass. A topple can turn into a fall or a slide. A slide is a mass of rock or sediment that becomes dislodged and moves along a plane of weakness, such as a fracture. A slump is a slide that separates along a concave surface. Lateral spreads occur when a fractured earthen mass spreads out at the sides.

A flow occurs when a mass of wet or dry rock fragments or sediment moves downslope as a highly viscous fluid. There are several different flow types. A debris flow is a mass of relatively dry, broken pieces of earth material that suddenly has water added.

The debris flow occurs on steeper slopes and moves at speeds of 2 to 4 kilometers (1 to 25 miles) per hour. A debris avalanche occurs when an entire area of soil and underlying weathered bedrock becomes detached from the underlying bedrock and moves quickly down the slope. This flow type is often triggered by heavy rains in areas where vegetation has been removed. An earthflow is a dry mass of clayey or silty material that moves relatively slowly down the slope. A mudflow is a mass of earth material mixed with water that moves quickly down the slope.

A quick clay can occur when partially saturated, solid, clayey sediments are subjected to an earthquake, explosion, or loud noise and become liquid instantly.

—*Sherry L. Eaton*

Further Reading

Carter, Chuck, Stephen Reynolds, Julia Johnson, Paul Morin, and Michael Kelly. *Exploring Earth Science.* McGraw Hill, 2015.

Chaparo, Juan Carlos Mora. *Mapping the Risk of Flood, Mass Movement and Local Subsidence: A New Proposal for Major Cities.* Springer, 2020.

Fu, Helin, Wei Chen, and Jiajun Fu. *Rock Mechanics and Engineering: Prediction and Control of Landslides and Geological Disasters.* Elsevier, 2021.

Hudson, Travis. *Living with Earth: An Introduction to Environmental Geology.* Routledge, 2016.

Losiak, Anna Isabela. *The Development of Evaporite Minerals During Weathering of Antarctic Meteorites.* Michigan State U Geology Department, 2009.

Reddy, D. Venkat. *Engineering Geology.* VIKAS Publishing, 2010.

Extraterrestrial Resources

Fields of Study: Geology; Astrogeology; Physics; Geochemistry; Astronautics

ABSTRACT

During the twenty-first century, new resources may well be required to build and energize civilization on other planets and in space, as well as to enhance accelerated change in a dynamic Earth-based society.

KEY CONCEPTS

ceramics: nonmetal compounds, such as silicates and clays, produced by firing the materials at high temperatures

fiber-reinforced composites: materials produced by drawing fibers of various types through a material being cast to produce a high weight-to-strength ratio

hydrolysis: the breakdown of water by energy into its constituent elements of oxygen and hydrogen

off-planet: pertaining to regions outside Earth in orbital or planetary space

permafrost: a layer of soil and water ice frozen together

photovoltaic cell: a device commonly made of layered silicon that produces electrical current in the presence of light; also, a solar cell

regolith: that layer of soil and rock fragments just above the planetary crust

superconductors: materials that pass electrical current without exhibiting any electrical resistance

CLASSIFICATION OF RESOURCES

In the largest possible sense, there will be two broad categories of resources during the twenty-first century: Earth-based resources and off-planet resources. The reason for the distinction is an economic one, based on where the materials will be used, and is driven by consideration of the gravitational field of Earth. Earth's gravitational field is very strong in comparison with extraterrestrial space, where there is little gravitational influence. It is also strong in relation to Earth's Moon and Mars, where the gravitational field is much weaker than Earth's. Hence, materials required in space will be much more economical if mined from small bodies in space (such as asteroids or comets) or the moon. The same resources, if shipped into space from Earth, would cost many times more. This example is provided as an insight into the distinctive ways of looking at future resources in the developing economy of the twenty-first century. In a more systematic approach to categorization, one could fundamentally classify tomorrow's most basic resources in the same broad classes used in recent times: agricultural products, chemicals (including petroleum and derivatives), metals, ceramics, energy, wood (and derivatives), and power.

IMPACT OF SCIENTIFIC DISCOVERY

Though the same broad resource categories will be required in the future as are required today, many of their individual identities and uses will be much different. Scientists are beginning to foresee new types of materials that will have a vital use. Such an example was afforded in 1989 when two scientists announced, erroneously, that they thought they had discovered a revolutionary power source. One major element in their design was the rare-earth metal palladium. Until their announcement, there had been few other uses for the metal; after their announcement, the price of palladium temporarily increased as a result of its presumed new importance. As new discoveries are made, new uses for existing materials, such as metals, will cause the demand for the resource to change as availability dictates.

Scientific discovery and technological development are often the key to the advancement and change of society and the resources that drive it. Examples from the past are abundant: Tungsten, presently used for the filaments in incandescent lightbulbs, had few uses other than as an alloying metal before the invention of the incandescent lightbulb; silicon (the key element of common beach sand and glass) assumed vital importance with the discovery of microelectronics. Probably no other single resource has had a more far-reaching impact on humankind or planet Earth than petroleum and its derivatives. The science of superconductivity produces materials that conduct electrical current without resistance. Late in 1986, this branch of physics took a revolutionary turn when it was discovered that synthetically produced materials could become superconductive at temperatures much higher than had been thought possible. The synthetic materials were described as "artificial rocks" and were made by mixing together various metal oxides under very specific conditions. As the ultimate goal of room-temperature superconductivity is approached, the materials used in producing the superconductors are likely to become highly sought-after and valuable resources.

Lunar sample 15415, also known as the "Genesis Rock." Photo via NASA/Wikimedia Commons. [Public domain.]

CERAMICS AND FIBER-REINFORCED COMPOSITES

The use of ceramic materials increased dramatically in the 1980s. Ceramic materials are nonmetal compounds produced by firing at high temperatures. Ceramics can be lightweight, resilient, and heat-resistant. Materials used in the production of ceramics are clays, silicates, and calcium carbonates. Uses of ceramics include refractories (heat-shielding or heat-absorbing materials, such as those used to construct spacecraft heat-shielding tiles) and electrical components. They have also been used in tests as automobile engine blocks. The resources that produce such ceramics are projected to have even broader applications in the future.

Another type of product that has become quite important and is expected to play a vital role in the future is fiber-reinforced composites. Such composites are made by casting a resin material about a designed layout of fiber materials. When the resin is cured and hardens, the fibers inside it make even very light products, such as aircraft wings and space vehicle fuselages, much stronger. The materials from which these composites are made (fiberglass, boron, tungsten, aluminum oxide, and carbon) will assume new importance with any increase in the use of fiber-reinforced composites.

ECONOMICS OF OFF-PLANET RESOURCES

Because the exploitation and possible settlement of space will probably be the most significant development of the twenty-first century, resources that drive that revolution will become highly valuable. The value of these space materials will be based on their origin. It will always be expensive to ship resources into space from Earth. A kilogram of aluminum mined and processed on the moon and shipped to Earth could one day cost a tenth of the amount of the same aluminum shipped to orbit from Earth because of launch-energy costs. Even under the best of circumstances, in a well-developed launch system, a liter of water launched from Earth into space will always cost thousands of times more than will a liter of water on the surface. One must always keep the gravitational interfaces in mind when considering the economics of space resources. Many studies have been conducted concerning the utilization of off-planet resources, their availability, and their uses. Such studies have been conducted concerning the construction of lunar and Martian bases and colonies, as well as the construction of enormous space colonies.

OFF-PLANET WATER AND POWER RESOURCES

The most valuable resources in space will be water and energy; the use of one is dependent on the other. Water and energy yield two other vital resources: oxygen and hydrogen. Water is the most basic of all resources necessary for human survival. Aside from the obvious purposes of direct consumption and hygiene, water will be necessary for cleaning, cooling, and producing food. Raw water can be broken down by a process called "hydrolysis" into elemental hydrogen and oxygen. The hydrogen may be used for energy production and the oxygen for breathing.

The mass of water is great enough that the ability to launch volumes that are sufficient to meet the needs of space resources will play a significant role in determining the use of space. Much attention has been paid to the likely sources of water off-planet, especially in deposits that can be most easily exploited. The Moon may contain deposits of water ice locked up beneath the regolith. The Clementine spacecraft bounced radar signals from the lunar surface during its 1994 exploration of the Moon. Radio telescopes on Earth monitoring those signals discovered that regions near the lunar poles were reflective in a way that was suggestive of the presence of water ice. The Lunar Prospector spacecraft carried a neutron spectrometer capable of detecting the energy of neutrons blasted from the lunar surface by the solar wind. It was announced on March 5, 1998, that neutrons from the polar regions carried the energy signature of having interacted with hydrogen. More recent remote explorations of both the Moon and Mars have revealed that water, in the form of ice, is more abundant than had been thought on both bodies. The most likely reservoir of hydrogen would be as a constituent of the water molecules in water ice. There may be as much as 300 million tonnes of ice particles mixed with the lunar soil of the permanently shadowed craters at each pole of the Moon. This is not a huge amount of water, so future lunar colonies will have to recycle water completely. Presumably, the icy particles have been delivered to the moon over billions of years as comet debris.

There is considerable evidence to suggest that Mars may contain vast amounts of water locked up beneath its regolith as permafrost (water ice mixed with soil). Moreover, some asteroids may contain water ice, and comets are generally considered to be mostly water ice. These resources may be captured and relocated to an orbit suitable for mining. Finally, pure water is produced as a waste product of fuel cells that react hydrogen and oxygen to produce power, but the hydrogen and oxygen of fuel cells will have been obtained by the hydrolysis of water in the first place.

Solar energy will be an abundant resource in space. Large reflectors can be used to concentrate heat to drive turbines and melt ores for smelting. Another source of energy is the direct conversion, by photovoltaic (solar) cells, of sunlight into electricity. Such cells may be produced off-planet by using lunar materials such as silicon, a significant constituent of lunar soil. Future space colonists may want to locate reserves of radioisotopes for nuclear power. Such reserves, if found, could be exploited with an excellent promise of safety, especially in high Earth, lunar, or solar orbit, as the isolation of the radioactive contaminants and power plant itself would be assured. Questions concerning other power sources, such as high-temperature fusion and unconventional fusion techniques, remain unresolved. Such devices, when perfected, would overcome nearly every known deterrent, allowing the fullest exploitation of space.

OFF-PLANET BUILDING MATERIAL RESOURCES

Aside from the most basic resources of survival off-planet (water, power, hydrogen, and oxygen), space colonists will require massive building material resources. Launching millions of tons of raw building materials into space is as impractical as hauling volumes of water. A Princeton University physicist, Gerard K. O'Neill, has addressed the question of producing such quantities of building materials for construction of off-planet colonies. O'Neill and his Princeton-based Space Studies Institute have designed a mass driver, a device that could catapult "buckets" of lunar soil into lunar or Earth orbit, to be processed—in automated space factories—into sheets of aluminum, magnesium, titanium, glass, and other materials for use in the construction of lunar or orbiting colonies. Lunar soil has been found to be rich enough in the necessary components to produce such materials in space. Current studies on the rocks brought back from the Moon during the Apollo program indicate that roughly 50,000 tonnes of aluminum sheets per year could be produced from 900,000 tonnes per year of moon ores catapulted into space by the mass drivers.

SURVEYS OF OFF-PLANET RESOURCES

Space probes have been utilized to recover information on the location of resources in space and their amounts. The Apollo mission astronauts returned samples from the lunar surface that were analyzed extensively and used in subsequent engineering and economic studies to reveal how the lunar material might be processed as building materials. One such study was performed by Gerald W. Driggers of the Southern Research Institute of Birmingham, Alabama. In Driggers' account, he estimated the cost of building an orbiting metals factory ($20 billion) and its power requirements (300 megawatts). He also estimated the final cost of the aluminum sheeting ($22 per kilogram) and compared it to the cost of the Earth-produced material ($2.2 per kilogram); he then compared it to the cost of Earth-based material launched into space ($88 per kilogram). Such economic studies will drive the capital outlays that will fund and justify such substantial expenditures.

The resources of Mars were surveyed at two landing sites in 1976 by the United States' Viking landers. The Viking landers were not designed to return samples directly from Mars but were sent with a device called an "X-ray fluorescence spectrometer," which would analyze the Martian soil for its inorganic materials. No probe has ever taken a direct observation of an asteroid, although many scientists are convinced that the Martian moons Phobos and Deimos are captured asteroids. Flyby studies of these bodies indicate that their densities are low enough that they could be made up of some water ice in their interiors. In 1986, several spacecraft flew past Halley's comet as it approached the sun, returning spectacular, closeup photos of the comet's body and taking measurements of its density. These flybys have supported earlier speculation that comets are largely water ices covered over by exceptionally dark, carbonaceous material.

DEVELOPMENT OF ENERGY PRODUCTION

The survivability and quality of every human's life is directly affected by the resources available to each individual. From water and power to food and building materials, humankind has been in a constant struggle to improve the availability and quality of resources while reducing the magnitude of the struggle to obtain them. The historical propensity of civilization, if not its fundamental purpose, is to ease that struggle and increase the availability of the resource base while constantly improving its superiority and basic usefulness.

The resources of the future will follow this trend, with momentous advances being made in the most elemental domain: energy. Energy production directly affects the acquisition of all other resources. Improved ways of generating power will directly influence food production for all people, but the most profound influence will affect those populations in regions fixed on the edge of continual famine. High-temperature superconductivity would change the basis of electronic devices, resulting in more efficient and cheaper electrical power production, storage, transmission, and use. The exploitation of this science would enable supercomputers that operate at unprecedented speed. Transportation on Earth and in space would be revolutionized.

—Dennis Chamberland

Further Reading

Chiras, Daniel D., and John P. Reganold. *Natural Resource Conservation: Management for a Sustainable Future.* 10th ed., Addison Wesley, 2009.

Gupta, Chiranjib, and Harvinderpal Singh. *Uranium Resource Processing: Secondary Resources.* Springer-Verlag, 2010.

Heppenheimer, T. A. *Colonies in Space: A Cand Factual Account of the Prospects for Human Colonization of Space.* Stackpole Books, 2017.

Letcher, Trevor M., editor. *Future Energy: Improved, Sustainable and Clean.* Elsevier, 2008.

Rapp, Donald. *Use of Extraterrestrial Resources for Human Space Missions to Moon or Mars.* Springer, 2013.

Schmidt, Victor. *Planet Earth and the New Geoscience.* 4th ed., Kendall/Hunt, 2003.

Sinding-Larsen, Richard, and Friedrich W. Wellmer, editors. *Non-renewable Resource Issues: Geoscientific and Societal Challenges*. Springer, 2012.

Zubrin, Robert, and Richard Wagner. *The Case for Mars: The Plan to Settle the Red Planet and Why We Must.* Free Press, 2011.

F

Floods

Fields of Study: Hydrology; Hydrodynamics; Geology; Geophysics

ABSTRACT

Floods are extreme conditions of flowing water. They generally occur because of inordinate amounts of rainfall or snowmelt, but also result from other causes such as dam failures and volcanic eruptions. Floods exert a major role in shaping river systems, and their occurrence is critical to the human use of riparian lands.

KEY CONCEPTS

discharge: the volume of water moving through a given flow cross section in a given unit of time

flash floods: rises in water level that occur unusually rapidly, generally because of especially intense rainfall

flood: a rising body of water that overtops its usual confines and inundates land not usually covered by water

hydrology: the branch of science dealing with water and its movement in the environment

jökulhlaup: a flood produced by the release of water sequestered by a glacier, most often due to the failure of some type of glacial dam or to subglacial volcanic activity

monsoon: a seasonal, reversing pattern of wind between warm ocean bodies and landmasses

recurrence interval: the average time interval in years between occurrences of a flood of a given magnitude in a measured series of floods

runoff: that part of precipitation that flows across the land and eventually gathers in surface streams

CAUSES OF FLOODS

Floods involve extremely large flows of water in rivers and streams. Some technical hydrological definitions of floods involve stages, or heights, of water above some reference level, such as the banks of a river channel. In practice, however, floods can be thought of as any extreme flow of water that exceeds usual experience, often damaging or threatening life and property.

Floods may be caused by a variety of physical factors, including dam failures and the subsidence of land. The most common kind of flood occurs when excessive precipitation and physical factors of the land combine to produce maximum runoff of water. Precipitation can yield runoff directly, or water from melting snow can produce the flow. Very intense flash floods are usually associated with heavy, short duration rainfall from thunderstorms. Such rainfall easily overwhelms the infiltration capacity of the ground, and water rapidly runs off from the local drainage area or watershed into adjacent stream channels. If enough water concentrates in the channel to exceed the carrying capacity of the channel, it will constitute a flood.

Physical factors on the land surface also determine the rate of concentration of floodwater in stream channels. For example, less permeable soils will allow water to run off faster, as will a lack of vegetation. Artificial enhancement of runoff occurs when slopes are covered by impermeable materials. This situation commonly occurs in construction, when the natural ground surface is replaced by buildings and pavement. The result of such construction is extreme enhancement of runoff. In cities, the yield of floodwater from paved surfaces may be ten times greater than the same storm's yield under natural conditions. In this way, human construction tends to exacerbate flooding problems in urban areas.

SEDIMENT AND BEDROCK

As the flow of water increases in a stream channel, it is usually associated with considerable sediment. If such water borne sediment also composes extensive deposits adjacent to the stream channel, the river is termed "alluvial." An alluvial river commonly has a floodplain of deposited sediment adjacent to the channel. The floodplain is really an intimate part of the river system, as sediment is added to it every time the river rises above its banks. As the water rises, the river's depth rapidly increases, causing an increased ability to transport sediment that is eroded from the bed and banks. If the stream is appropriately loaded with sediment, it will be deposited when the banks are overtopped and the width of the flow greatly increases. If not enough sediment is supplied by erosion, however, the increasingly energetic flood flows will be erosive, attacking the banks, widening the channel, and restoring the appropriate sediment load to the stream.

When the bed and banks of a river are composed of bedrock, these adjustments of sediment load to flow energy cannot occur. Because the bedrock can be very resistant to erosion, the energy level in the flow can rise spectacularly without being damped by sediment. The excess energy of such sediment impoverished floods goes into the development of turbulence. Turbulence at high energy levels takes on an organized structure of powerful vortices that produce immense pressure changes. These pressure effects may be sufficient to erode the bedrock boundary by a "plucking" action.

A famous geological controversy once surrounded the problem of bedrock erosion by great floods. In the 1920s, University of Chicago geologist J. Harlen Bretz proposed that immense tracts of eroded basalt bedrock in Washington State had been created by a catastrophic glacial flood. Bretz subsequently showed that the fascinating landforms of that region, known as the Channeled Scablands, were created when a great lake impounded by glacial ice burst. The lake, glacial Lake Missoula, had been

Homeowners brace against flooding. Photo via iStock.com/KarenMassier. [Used under license.]

more than 600 meters deep at its ice dam. It took Bretz nearly fifty years to convince his many critics that catastrophic flooding could explain all the bizarre features of the Channeled Scabland. Geologists now know that the physics of catastrophic flooding is completely consistent with Bretz's observations. Missoula flood flows moved at depths of 100 to 200 meters and velocities of 20 to 30 meters per second. The power (rate of energy expenditure) per unit area for such flows is thirty thousand times that for a normal river, such as the Mississippi, in flood. The reason for such immense flow power is that the Missoula flooding occurred on very steep slopes and that the potential energy of the water when the dam burst was great.

The Missoula floods occurred during the last Ice Age, more than twelve thousand years ago. There are, however, modern examples of glacial floods called "jökulhlaups." In Iceland, jökulhlaups occur where glaciers overlie active volcanoes. Volcanic heat releases water that is stored in subsurface reservoirs by melting the overlying ice. Lakes may also form adjacent to the ice masses. Because ice is less dense than water, such juxtapositions are inherently unstable. When the pressure is high enough, the water may lift the ice dam and burst out from beneath the glacier. The jökulhlaups move house-sized boulders and transport immense quantities of sediment.

DISCHARGE AND RECURRENCE INTERVAL

The volume of water released by a flood per unit of time is termed its "discharge." This quantity is the magnitude of the flow that will potentially inundate an area. The chances of experiencing floods of different magnitudes are expressed in terms of frequency. Large, catastrophic floods have a low frequency, or probability of occurrence; smaller floods occur more often. The probability of occurrence for a flood of a given magnitude can be expressed as the odds, or percent chance, of the recurrence of one or more similar or bigger floods in a certain number of years. Analyses of flood magnitude and frequency are achieved by measuring floods and statistically analyzing the data. Results are expressed in terms of the probability of a given discharge being equaled or exceeded in any one year. The reciprocal of this probability is the return period, or recurrence interval, of the flooding, expressed as a number of years.

The concept of a recurrence interval is sometimes confusing; some examples will illustrate the meaning. Suppose analysis of the flood frequency of a river flow indicates certain discharges for ten- and one-hundred-year recurrence intervals. The smaller, ten-year flood will have a probability of occurrence in one year of 0.1, or 10 percent. The larger one-hundred-year flood's probability of occurrence in one year is 0.01, or 1 percent. Note that these numbers do not preclude several such floods occurring in a given period; for example, two or more floods of the one-hundred-year magnitude could occur in the same year. Such an event, while unlikely, does have a small probability of occurring.

Flood magnitude frequency relationships vary immensely with climatic regions. In the humid temperate regions of the globe, such as the northern and eastern United States, stream flow is relatively continuous. Even rare floods are not appreciably larger than more common floods. A result of this relationship is that stream channels have developed in size to convey the relatively common, moderate-sized floods with maximum efficiency. In contrast, the more arid regions of the southwestern United States have immensely variable flood responses. Stream channels may be dry most of the time, filling with water only after rare thunderstorms. The flash floods that characterize these streams may also be highly charged with sediment. Indeed, the sediment may so dominate the flow that the phenomenon is of mud or debris flow rather than stream flow. Because extreme events dominate in these environments, the stream channels have developed in size according to these rare, great floods.

MONSOONS AND ANCIENT FLOODS

In tropical areas, some of the greatest known rainfalls are produced by tropical storms and monsoons. Monsoons are seasonal wet to dry weather patterns driven by atmospheric pressure changes over the oceans and continents. They result in alternations between dry periods, which inhibit vegetation growth, and immensely wet periods, which facilitate runoff. Tropical rivers thus have a pronounced seasonal cycle of flooding, and some floods may be immense. These rivers also show channel-sized development in accordance with rare, great flows. Some of the most populous places on Earth are situated on seasonal tropical rivers, such as the Ganges and Brahmaputra Rivers in India and Bangladesh. Immense tragedies have occurred when a particularly severe monsoon or tropical storm has produced especially great floods. In 1974, monsoon-related flooding in Bangladesh killed 2,500 people. This pales in comparison to a flood in 1970 that took 500,000 lives and another in 1991 that killed more than 100,000 people.

Studies of ancient floods (paleofloods) through geological reconstruction of past discharges show that monsoons and other flood generating systems have varied in the past. Between about ten thousand and five thousand years ago, floods were apparently much more intense in many world areas on the boundaries between the tropics and the midlatitude deserts. These intense floods may have been related to long term glacial to interglacial cycles. Because of modern increases in carbon dioxide and other greenhouse gases, it appears quite likely that tropical floods may again become more intense. Such a situation could have grave consequences for flood prone tropical countries.

STUDY OF FLOODS

Hydrologists study floods by measuring flow in streams. Measurements are taken at stream gauges, where mechanical devices are used to record the water level, or stage of the river. To transform these stage measurements into discharge values, the hydrologist must perform a rating of the stream gauge. That is accomplished by measuring velocities in the stream channel during various flow events. A velocity meter with a rotor blade calibrated to the flow rate is used for this purpose. When the average measured velocity of the stream channel is multiplied by the cross-sectional area of the channel, the result is the discharge for that flow event. When several flows at different stages are measured, the data are used to generate a rating curve for the gauge. This curve shows discharges corresponding to any stage measured at the gauge.

The discharge values obtained at a gauge are collected over many years, constituting a record that can then be used in flood frequency analysis. Several statistical procedures can be employed to plot the flood experience and to extrapolate to ideal values of the ten-year flood, the one-hundred-year flood, and so on. The discharges associated with these recurrence intervals are then used as design values for hazard assessment, dam construction, and other flood controls. "Flood control" is probably a misnomer, however, because flows larger than the design floods are always possible. Flood control really involves providing various degrees of flood protection.

Another approach to evaluating extreme flood magnitudes involves careful study of the precipitation values that generate the greatest known floods. By transposing the patterns of known extreme storms to other areas, scientists can, in theory, calculate what the runoff would be from hypothetical great storms. Such calculations involve the use of a rainfall runoff model. These models are prevalent in hydrology because they can be easily programmed. The models give idealized predictions of how water from a given storm would concentrate. The flood discharge modeled from the assumed maximum rainfall is called a "probable maximum flood."

Unfortunately, there are problems inherent to both of these traditional hydrological approaches to flood studies. Both make assumptions in calculating potential flood flows. Another procedure is to study the natural records of ancient floods, or paleofloods, that are preserved in geological deposits or in erosional features on the land. It has been found for some sections of bedrock that nonalluvial rivers act as natural recorders of extremely large flood events. These natural "flood gauges" can be interpreted only by detailed studies that combine geological analysis with hydraulic calculations of the ancient discharges.

Paleoflood hydrology generates real data on the largest floods to occur in various drainages over several millennia. The data include the floods' ages, or the periods in which they occurred, and their discharges. The information can be used directly in a flood frequency analysis, or it can be used to assess the probable validity of extrapolations from conventional data on smaller floods. Paleoflood data can also be compared with probable maximum flood estimates. In this way, the expense of overdesign and the danger of underdesign can be avoided in flood-related engineering projects.

SIGNIFICANCE OF FLOODS

People have lived with floods since the beginning of civilization. The first great civilizations on Earth developed along the fertile but flood prone valleys of rivers such as the Nile, the Tigris and the Euphrates, the Indus, and the Yangtze and the Huang He. Various means of coping with floods have been documented since the biblical accounts of Noah. Early societies merely avoided zones that tradition told them were hazardous. It has only been in the modern era that large cities have systematically developed on immense tracts of flood prone lands. Thus, the natural process of flooding has become an unnatural hazard to humans.

There are immense consequences for the human insistence on occupying those areas near rivers that infrequently receive floodwater. A single tropical storm system, Hurricane Agnes, generated more than $3 billion in flood damage to the northeastern United States in 1972. Flooding of the Red River and Mississippi River basins generated more than $12 billion worth of damage, while the city of New Orleans may never fully recover from the flooding caused by Hurricane Katrina. Floods in Bangladesh have killed millions of people. In the 1930s, a national U.S. program began to respond to such problems by constructing large dams. Despite, or perhaps because of, this expensive effort, flood damage to life and property is much greater today than it was in the 1930s.

One trend has been to manage flood hazard zones with multiple approaches that respond to the nature of the flood risk. The river is treated as a whole integrated system, rather than as individual segments, for engineering design. Management alternatives for this system are not limited solely to structural controls, such as dams and levees. Instead, options are considered for land use adjustment. Flood prone lands can be used for parks, greenbelts, and bikeways instead of industrial warehouses, stores, and housing. Even when construction must be done on floodplains, it may be possible to make provision for flood risks. Warehouses, for example, can be organized for rapid transfer of materials to second stories or to temporary, safe storage sites. Such adjustments require accurate and timely warning systems that involve measuring rainfall in headwater areas and rapidly predicting the flood consequences to downstream sites at risk.

There is a general need to educate the public that floodplains are a natural part of rivers. Living on a floodplain is really choosing to play a game of "flood-plain roulette," in that it is known with certainty that a flood will happen, but it is not known when that flood will happen. For this reason, the most accurate and reliable methods of evaluating flood magnitudes and frequencies are necessary. The choices made for land use on floodplains cannot be based merely on idealized theories of how floods behave.

—*Victor R. Baker*

Further Reading

Aldrete, Gregory S. *Floods of the Tiber in Ancient Rome.* Johns Hopkins UP, 2006.

ASCE. *Hurricane Katrina External Review Panel: The New Orleans Hurricane Protection System: What Went Wrong and Why.* American Society of Civil Engineers, 2007.

Freitag, Bob, et al. *Floodplain Management: A New Approach for a New Era.* Island Press, 2009.

O'Gorman, Emily. *Flood Country. An Environmental History of the Murray-Darling Basin.* CSIRO, 2012.

Prothero, Donald R. *Catastrophes!: Earthquakes, Tsunamis, Tornadoes, and Other Earth-Shattering Disasters.* Johns Hopkins UP, 2011.

Rossario, Frederic. *Floodscapes: Contemporary Landscape Strategies in Times of Climate Change.* NAi Uitgevers, 2021.

Schneider, Bonnie. *Extreme Weather: A Guide to Surviving Flash Floods, Tornadoes, Hurricanes, Heat Waves, Snowstorms, Tsunamis and Other Natural Disasters.* Palgrave Macmillan, 2012.

Sene, Kevin. *Flood Warning, Forecasting and Emergency Response.* Springer Science + Business Media, 2008.

Singh, Vijay P., editor. *Hydrology of Disasters.* Kluwer Academic Publishers, 2010.

Fluvial and Karst Processes

Fields of Study: Geomorphology; Physical Geography; Geochemistry

ABSTRACT

Earth's landscape has been sculptured into an almost infinite variety of forms. Earth's surface has been modified by various processes in a process of slow, continuous change for billions of years to arrive at the modern configuration

of landscapes. Each process that transforms the surface is classified as either endogenic or exogenic.

KEY CONCEPTS

drainage basin: an area in which all surface water from rainfall, snowmelt, etc., flows toward a central channel that functions as the outlet for the water on its eventual way to the ocean

endogenic: originating from or formed by forces acting within Earth

exogenic: originating from or formed by forces acting at Earth's surface

interfluve: the high-ground divide that separates one drainage basin from another

polje: a lake of irregular shape resulting from the formation of adjacent sinkholes

uvala: circular flat-bottomed depressions, often the floor of a small lake, formed when a sinkhole becomes filled with sediment

EARTH'S CHANGING LANDSCAPE

Earth's landscape has been sculptured into an almost infinite variety of forms. Earth's surface has been modified by various processes in a process of slow, continuous change for billions of years to arrive at the modern configuration of landscapes.

Each process that transforms the surface is classified as either endogenic or exogenic. Endogenic processes are driven by Earth's internal heat and energy and are responsible for major crustal deformation. Endogenic processes are considered constructional, because they build up Earth's surface and create new landforms, such as mountain systems. Conversely, exogenic processes are considered destructional because they result in the wearing away of landforms created by endogenic processes. Exogenic processes are driven by solar energy putting into motion Earth's atmosphere and water, resulting in the lowering of features originally created by endogenic processes.

The most effective exogenic processes for wearing away the landscape are those that involve the action of flowing water, commonly referred to as fluvial processes. Water flows over the surface as runoff, after it evaporates into the atmosphere and infiltrates into the soil. The water that is left over flows down under the influence of gravity and has tremendous energy for sculpting Earth's surface. Although flowing water is the most effective agent for modifying the landscape, it represents less than 0.01 percent of all the water on Earth's surface. By comparison, nearly 75 percent of Earth's surface water is stored within glaciers.

DRAINAGE BASINS

Fluvial processes can be considered from a variety of spatial scales. The largest scale is the drainage basin. A drainage basin is the area defined by topographic divides that diverts all water and material within the basin to a single outlet. Every stream of any size has its own drainage basin, and every portion of Earth's land surfaces are located within a drainage basin. Drainage basins vary tremendously in size, depending on the size of the river considered. For example, the largest drainage basin on earth is the Amazon, which drains about 5.83 million square kilometers (2.25 million square miles) of South America.

The Amazon Basin is so large that it could contain nearly the entire continent of Australia. By comparison, the Mississippi River drainage basin, the largest in North America, drains an area of about 3,200,000 square kilometers (1,235,000 square miles). Smaller rivers have much smaller basins, with many draining only an area roughly the size of a football field. While basins vary tremendously in size, they are spatially organized, with larger basins receiving the drainage from smaller basins, and eventually draining into the ocean. Because drainage basins receive water and material from the landscape within the basin, they are sensitive to environmental change that occurs within the basin. For example, during the twentieth century, the Mississippi River was influenced by many human-imposed changes that occurred either within the basin or directly within the channel, such as agriculture, dams and reservoirs, and levees.

DRAINAGE NETWORKS AND SURFACE EROSION

Drainage basins can be subdivided into drainage networks by the arrangement of their valleys and interfluves. Interfluves are the ridges of higher elevation that separate adjacent valleys. Where an interfluve represents a natural boundary between two or more basins, it is referred to as a drainage divide. Valleys contain the larger rivers and are easily

Entrance to the Puerto Princesa Underground River, Philippines. Photo by Mike Gonzalez, CC BY-SA 3.0, via Wikimedia Commons.

distinguished from interfluves by their relatively low, flat surfaces. Interfluves have relatively steep slopes and, for this reason, are eroded by runoff. The term erosion refers to the transport of material, in this case sediment that is dislodged from the surface.

Runoff starts as a broad sheet of slow-moving water that is not very erosive. As it continues to flow downslope, it speeds up and concentrates into rills, which are narrow, fast-moving lines of water. Because the runoff is concentrated within rills, the water travels faster and has more energy for erosion. Thus, rills are responsible for transporting sediment from higher points of elevation within the basin to the valleys, which are at a lower elevation. Rills can become powerful enough to scour deeply into the surface, developing into permanent channels called "gullies."

The presence of many gullies indicates significant erosion on the landscape and represents an expensive and long-lasting problem if it is not remedied after initial development. The formation of gullies is often associated with human manipulation of the earth. For example, gullies can develop after improper land management, particularly intensive agricultural and grazing practices. A change in land use from natural vegetation, such as forests or prairie, can result in a type of land cover that is not suited for preventing erosion. Such land surfaces become susceptible to the formation of gullies during heavy, prolonged rains.

At a smaller scale, fluvial processes can be considered from the perspective of the river channel. River channels are located within the valleys of basins, offering a permanent conduit for drainage. Higher in the basin, river channels and valleys are relatively narrow, but grow larger toward the mouth of the basin as they receive drainage from smaller rivers within the basin. River channels may be categorized by their planiform pattern, which refers to their overhead appearance, such as would be viewed from the window of an airplane.

The two major types of rivers are meandering and braided. Meandering rivers have a single channel that is sinuous and winding. These rivers are characterized as having orderly and symmetrical bends, causing the river to alternate directions as it flows across its valley. In contrast, braided rivers contain numerous channels divided by small islands, which results in a disorganized pattern. The islands within a braided river channel are not permanent. Instead, they erode and form over the course of a few years, or even during large flood events. Meandering channels usually have narrow and deep channels, but braided river channels are shallow and wide.

SEDIMENT AND FLOODPLAINS

Another distinction between braided and meandering river channels is the types of sediment they transport. Braided rivers transport a great amount of sediment that is deposited into midchannel islands within the river. Also, because braided rivers are frequently located higher in the drainage basin, they may have larger sediments from the erosion of adjacent slopes. In contrast, meandering river channels are located closer to the mouth of the basin and transport fine-grained sediment that is easily stored within point bars, which results in symmetrical bends within the river.

The sediments of both meandering and braided rivers are deposited within the valleys onto floodplains. Floodplains are wide, flat surfaces formed from the accumulation of alluvium, which is a term for sediment that is deposited by water. Floodplain sediments are deposited with seasonal flooding. When a river floods, it transports a large amount of sediment from the channel to the adjacent floodplain. After the water escapes the channel, it loses energy and can no longer transport the sediment. As a result, the sediment falls out of suspension and is deposited onto the floodplain. Because flooding occurs seasonally, floodplain deposits are layered and may accumulate into very thick alluvial deposits over thousands of years.

KARST PROCESSES AND LANDFORMS

A specialized type of exogenic process that is also related to the presence of water is karst. Karst processes and topography are characterized by the solution of limestone by acidic groundwater into a number of distinctive landforms. While fluvial processes lower the landscape from the surface, karst processes lower the landscape from beneath the surface. Because limestone is a very permeable sedimentary rock, it allows for a large amount of groundwater flow. The primary areas for solution of the limestone occur along bedding planes and joints. This creates a positive feedback by increasing the amount of water flowing through the rock, thereby further increasing solution of the limestone. The result is a complex maze of underground conduits and caverns, and a surface with few rivers because of the high degree of infiltration.

The surface topography of karst regions often is characterized as undulating. A closer inspection reveals numerous depressions that lack surface outlets. Where this is best developed, it is referred to as cockpit karst. It occurs in areas underlain by extensive limestone and receiving high amounts of precipitation, for example, southern Illinois and Indiana in the Midwestern United States, and in Puerto Rico and Jamaica.

Sinkholes are also common to karst regions. Sinkholes are circular depressions having steep-sided vertical walls. Sinkholes can form either from the sudden collapse of the ceiling of an underground cavern or as a result of the gradual solution and lowering of the surface. Sinkholes can fill with sediments washed in from surface runoff. This reduces infiltration and results in the development of small circular lakes, particularly common in central Florida. Over time, erosion causes the vertical walls to retreat, resulting in uvalas, which are much larger flat-floored depressions.

Where there are numerous adjacent sinkholes, the retreat and expansion of the depressions causes them to coalesce, resulting in the formation of poljes. Unlike uvalas, poljes have an irregular shape, and the floor of the basin is not flat because of differences between the coalescing sinkholes.

Caves are among the most characteristic features of karst regions, but can only be seen beneath the surface. Caves can traverse the subsurface for miles, developing into a complex network of interconnected passages. Some caves develop spectacular formations as a result of the high amount of dissolved limestone transported by the groundwater. The evaporation of water results in the accumulation of carbonate deposits, which may grow for thousands of years. Some of the most common deposits are stalactites, which grow downward from the ceiling of the cave, and stalagmites, which grow upward and occasionally connect with stalactites to form large vertical columns.

—Paul F. Hudson

Further Reading

Bierman, Paul R., and David R. Montgomery. *Key Concepts in Geomorphology*. W.H. Freeman, 2019.
Charlton, Ro. *Fundamentals of Fluvial Geomorphology*. Routledge, 2008.
Chorley, Richard J., editor. *Introduction to Fluvial Processes*. Routledge, 2019.
Peterson, James F., Dorothy Sack, and Robert E. Gabler. *Physical Geography*. Cengage, 2016.

G

The Gaia Hypothesis

Fields of Study: Climatology; Biogeography

ABSTRACT

The Gaia Hypothesis postulates that the chemical composition of Earth's atmosphere and oceans was actively molded and maintained by the planet's living organisms. It describes Earth and its associated biosphere as one large, living, self-regulating organism.

KEY CONCEPTS

coccolithophores: a type of algae that consumes carbon dioxide

feedback system: any system that produces an output result that either reduces the effect of the system (negative feedback) or increases the effect of the system (positive feedback)

greenhouse gas: any atmospheric gas that captures infrared radiation from Earth's surface and returns that energy to the atmosphere rather than to space

homeostatic: any mechanism that acts to maintain a system in its current state

self-regulating: controlled by processes that exist entirely within the system

DEFINITION OF THE GAIA HYPOTHESIS

In 1979, British atmospheric chemist James E. Lovelock and his collaborator, University of Massachusetts, Amherst, microbiologist Lynn Margulis, postulated that the chemical composition of Earth's atmosphere and oceans was actively molded and maintained by the planet's living organisms. Up to this point, scientists assumed that life adapted to the changing terrestrial environment as needed but did not mold Earth into a life-supporting planet. Lovelock's hypothesis turned this assumption around and asserted that biological organisms not only had actively transformed Earth into a life-friendly place but also continued to make it a life-friendly planet. Novelist William Golding suggested that Lovelock call his hypothesis the "Gaia hypothesis" in deference to the Greek goddess who gave birth to the gods.

Lovelock went further, describing Earth and its associated biosphere as one large, living, self-regulating organism. While many scientists ridiculed this characterization as a kind of neopagan, New Age religion, some environmental activists found in it a powerful metaphor. Speaking of Earth as alive and even sentient provided a strong emotional appeal for environmental causes. Describing environmental degradation as "wounding" Earth conveyed a strong moral impetus for ecological concerns.

SIGNIFICANCE FOR CLIMATE CHANGE

The Gaia hypothesis proposes that biologically mediated, negative feedback mechanisms contribute to environmental homeostasis and make the environment suitable for life. Perhaps the most controversial feature, the Gaia hypothesis further argues that these negative feedback mechanisms arise by means of neo-Darwinian selection.

The surface temperature of Earth provides an example of one such mechanism. The global surface temperature has remained relatively constant, even though the energy output of the Sun has increased by 25 to 30 percent since life began on the planet. Temperature regulation results from a decrease in the atmospheric concentration of the greenhouse gas (GHG) carbon dioxide (CO_2). Bacteria and plants metabolize CO_2 into biomass, bringing about CO_2 reduction. Furthermore, increased atmospheric CO_2 levels increase plant carbon sequestration and removal of CO_2 from the atmosphere. Increased carbon utilization also increases the number of marine, carbon-utilizing algae called "coccolithophores." When coccolithophores die, they release dimethyl sulfide, which nucleates cloud formation. Increased cloud cover, which was initially induced by increased atmospheric CO_2, further cools Earth. Other life-dependent homeostatic processes are also thought to regulate the composition of Earth's atmosphere and the salinity of the oceans.

Some climate scientists favor the Gaia hypothesis, while others are highly critical of it. Virtually all Earth scientists recognize that living organisms have significant effects on the physical and chemical aspects of the environment. However, biologically mediated feedback systems are not intrinsically homeostatic, since some of them can destabilize the environment and are deleterious to life. For example, increased warming increases respiration rates of soil-based organisms, which increases CO_2 release from soils. Increased soil carbon release increases the greenhouse effect and global warming. This positive feedback system destabilizes the terrestrial environment. Furthermore, genome studies of plants and animals have failed to reveal Gaian global feedback mechanisms. Likewise, natural selection affects individual organisms and is not a global process. Similarly, natural selection does not act with foresight.

Gaia, the mythical earth mother, as depicted by artist Anselm Feuerbach (1875). Image via Wikimedia Commons. [Public domain.]

With respect to global climate change, Earth, according to the Gaia hypothesis, should use a series of feedback mechanisms to regulate global temperatures. Thus, the freezing of the poles covers them with ice, which reflects the rays of the Sun and cools the planet. Likewise, increases in atmospheric CO_2 cause increased marine algal growth, which removes CO_2 and decreases greenhouse warming. However, human intervention has short-circuited these negative feedback mechanisms. Industrial pollution of the oceans has introduced increased amounts of mercury (Hg), which prevents algal growth and destroys the primary means of regulating atmospheric CO_2 levels. Increased CO_2 levels cause increased greenhouse-based heating of the planet, which in turn melts the polar ice caps and decreases the reflectivity of Earth's surface, further heating the atmosphere.

In one sense, the Gaia hypothesis is strongly optimistic, since it strongly asserts that Earth and at least some of its resident life will adapt and survive. However, this optimism does not include humanity, since the environment of Earth after such dramatic climate change will probably be highly inhospitable to human life.

—Michael A. Buratovich

Further Reading

Latour, Bruno. *Facing Gaia: Eight Lectures on the New Climatic Regime*. Polity Press, 2017.

Lovelock, James. *The Ages of Gaia: A Biography of Our Living Earth*. Norton, 1995.

———. *Gaia: A New Look at Life on Earth*. New ed., Oxford UP, 2000.

———. *The Revenge of Gaia: Earth's Climate Crisis and the Fate of Humanity*. Basic Books, 2007.
Margulis, Lynn. *Symbiotic Planet: A New Look at Evolution*. Basic Books, 1999.
Ruse, Michael. *The Gaia Hypothesis: Science on a Pagan Planet*. U of Chicago P, 2013.
Schneider, Stephen H., et al., editors. *Scientists Debate Gaia: The Next Century*. MIT Press, 2008.

GEM MINERALS

Fields of Study: Geology; Mineralogy; Gemology; Geochemistry; Geophysics

ABSTRACT

A gem mineral is any mineral species that yields a gem upon cutting and/or polishing. Gem minerals have value based on their potential to produce gems. Although gems have had various uses in the past, including special powers attributed by folklore, their principal uses have been personal adornment (as jewels), a mode of investment, and a symbol of wealth and power.

KEY CONCEPTS

gem: a cut and polished stone that possesses durability, rarity, and beauty necessary for use in jewelry and, therefore, of value

gemstone: any rock, mineral, or natural material that has the potential for use as personal adornment or ornament

inclusion: a foreign substance enclosed within a mineral; often very small mineral grains and cavities filled with liquid or gas; a cavity with liquid, a gas bubble, and a crystal is called a "three-phase inclusion"

mineral: a naturally occurring inorganic substance with a characteristic chemical composition and atomic structure, manifested in its external geometry and other physical properties

mineral species: a mineralogic division in which all the varieties in any one species have the same basic physical and chemical properties

mineral variety: a division of a mineral species based upon color, type of optical phenomenon, or other distinguishing characteristics of appearance

rough: gem mineral material of suitable quality to be used for fashioning gemstones

synthetic mineral: a human-made reproduction of the structure, composition, and properties of a particular mineral

PHYSICAL PROPERTIES OF GEM MINERALS

Gem minerals are those mineral species that have yielded the material from which specimens have been fashioned into gems. Gems and therefore gem mineral varieties must have the same properties. These properties are beauty (color, phenomenon, or clarity), durability (hardness and toughness), and value (rarity, demand, and tradition). Of the approximately five thousand mineral species known, only about ninety of these produce material in a quality and quantity suitable for gems, and this quantity of suitable material makes up generally only a very small portion of the gem mineral found. Often the quality of this gem material is such that the material is given a gem varietal name or mineral varietal name based upon color or phenomenon. A different name is given for each color variation; hence, there is a large number of gem names compared with the gem mineral names. Two or more different gem minerals may have the same gem name, as in the case of jade and the minerals nephrite (actinolite and tremolite) and jadeite. Finally, new gem varieties are found from time to time. For example, tanzanite was found in the 1960s and tsavorite during the 1970s.

BEAUTY OF GEM MINERALS

The beauty of a gem may be inherent in the gem mineral species—such as clarity, color, or phenomena (stars, eyes, and the like)—or it may be brought out as it is fashioned by cutting and polishing. Faceting is the cutting of a stone to add faces (facets) or flat surfaces, generally with a regular geometric form. Because faceting enhances the brilliance of a stone by causing reflections of both front and back facets, those mineral varieties having superior clarity became important gem varieties. The faceting of a clear gem mineral produces a color phenomenon called "dispersion." This is the ability of a particular gem mineral to break light up into a rainbow of component colors, as does a prism. This property is

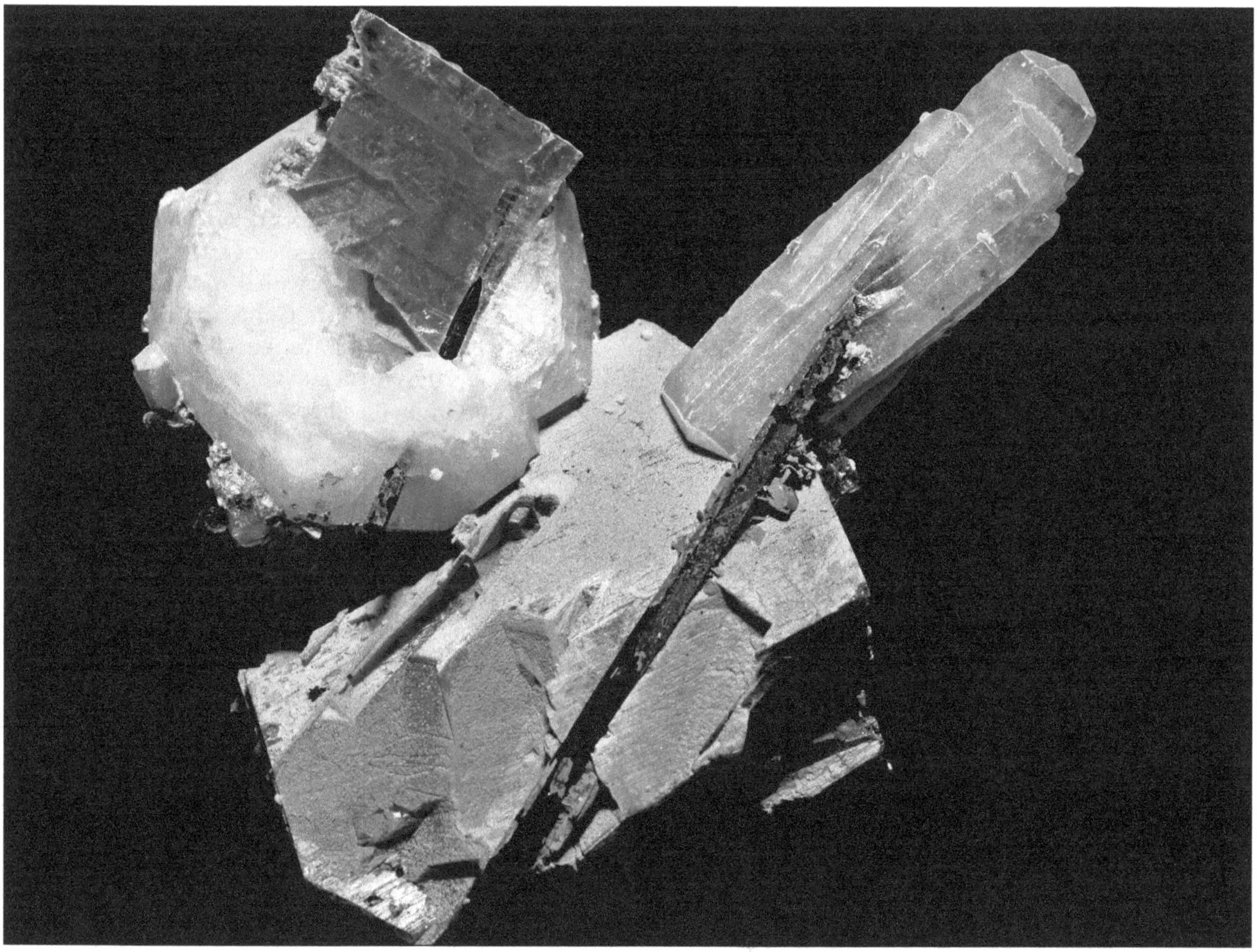

Crystals of serandite, natrolite, analcime, and aegirine from Mont Saint-Hilaire, Quebec, Canada. Photo by Géry Parent, CC BY-SA, via Wikimedia Commons.

commonly known as "fire" and is well known in diamonds.

Color is a very important property of gems. There are many gem varieties of gem minerals based solely on this property. For example, the named color varieties of beryl include aquamarine (blue-green), chrysolite (yellow-green), emerald (intense green), heliodor (brownish-green), and morganite (pink or orange). Clarity of color determines much of the beauty and value of gemstones. A "play of color" is produced by various combinations of reflection, refraction, diffraction, and interference phenomena; it is well known from such gems as opal (opalescence) and labradorite (labradorescence). Some stones are selected entirely for their color. These may be opaque or nearly so and may or may not take a high polish. They include such stones as turquoise, jade, and many varieties of quartz, such as jasper, bloodstone, agate, and carnelian.

Some stones have inclusions that produce aesthetic effects, such as stars, eyes, spangling, or even pictures. Stars and eyes are produced by minute aligned voids or needles of minerals like rutile. If they are in one direction, they produce eyes, as in chrysoberyl and tourmaline. If they are in a hexagonal or in other star-shaped patterns, they produce stars, as in star rubies, sapphires, diopside, and garnet. Spangling results from variously oriented small inclusions of reflective minerals within the gem mineral, such as mica, rutile, or tourmaline.

DURABILITY OF GEM MINERALS

Durability of a gem mineral is its resistance to scratching or breaking, a necessity for an owner of jewelry because one does not want to damage a gem and thus markedly reduce its value simply by wearing it.

Hardness is the ability of a gem mineral to resist scratching. Mineralogists and gemologists rank minerals against a hardness scale of common minerals, called the "Mohs hardness scale." From hardest to softest, the scale is 10 (diamond), 9 (corundum), 8 (topaz), 7 (quartz), 6 (orthoclase feldspar), 5 (apatite), 4 (fluorite), 3 (calcite), 2 (gypsum), and 1 (talc). The hardness of 7, the hardness of quartz, a very common mineral, is very important in the selection of gems or gem minerals. Quartz is a very common mineral, which means a stone with a hardness of less than 7 becomes easily scratched in our environment and therefore lacks durability. Most of our well-known gem minerals have a hardness greater than or equal to 7.

Resistance to breaking is referred to as tenacity. Most minerals are brittle, but native gold is malleable, ductile, and sectile.

Minerals break in two major ways called "cleavage" and "fracture." Cleavage is the breaking of minerals parallel to more weakly bonded atomic planes found within the mineral and produces flat reflecting surfaces. Fracture is the random breaking of a mineral across atomic planes. Fracture along curved surfaces is called "conchoidal" fracture. Tenacity is directly related to the bond strength between adjacent atoms. Cleavage thus reduces tenacity, as in easily cleaved topaz, while an intergrown mat of fine crystals increases tenacity of a specimen, as in the case of the jades nephrite and jadeite. Cleaving was probably the first way to facet gems and still is very important in the shaping of rough diamonds before cutting and polishing.

VALUE OF GEM MINERALS

The value of a gemstone or a particular gem is based upon the interaction of many factors that can be divided into four major groups: beauty, rarity, quality of fashioning, and demand. Beauty includes such properties as color, quality of color, and clarity (freedom of inclusions, fractures, or cloudiness); it also includes phenomena such as quality of stars, eyes, or play of color. Rarity involves not only the rarity of the variety of the gem mineral but also the rarity of that quality of stone being used. For example, size influences the value because large stones are rarer than small stones; perfection influences the value because high-quality stones are less common than flawed stones. Rarity of the gem mineral can increase value or decrease value. Very common varieties of quartz, for example, may have less value than similar qualities and sizes of corundum because corundum is less common. Some gems, however, are so rare that people have not heard of them, and there is no demand except by connoisseurs and hobbyists, which in turn decreases value. Quality of fashioning includes degrees of perfection of symmetry and customizing of facet junction angles to the properties of the gem mineral species, which maximizes brilliance, color, and dispersion, and fineness or degree of polishing. Demand includes such issues as fashion, fads, and engagement ring and birthstone traditions. These associations, in general, increase value. Good or bad luck attributed to gemstones may increase or decrease their value, respectively.

The terms "precious" and "semiprecious" used with a gemstone name are misleading in assessing the value of the stone. "Precious" has been historically applied to gems such as diamonds, rubies, sapphires, opals, and emeralds, while "semiprecious" has been applied to such gems as aquamarine, tourmaline, chrysoberyl, citrine, topaz, and amethyst. Almost any gem mineral is found in a wide variety of qualities and therefore so are gems; thus, there are high-quality stones called "semiprecious" that are often more valuable than low-quality stones called "precious." Unfortunately, the term "precious" has also been applied all too often to low-quality diamond, ruby, sapphire, opal, or emerald in order to deceive the buyer.

FORMATION OF GEM MINERALS

Gem minerals originate in the same way as all other minerals. They occur in four broad categories: igneous deposits (formed from molten rock—lava and magma), hydrothermal and pneumatolytic deposits (formed from hot water and steam, respectively), metamorphic deposits (formed by heat, pressure, and chemically active fluids with crystallization in the solid state), and sedimentary deposits (formed by processes of weathering, erosion, and deposition). Corundum is an example of a gem mineral that may be found under several conditions. A well-known igneous gem mineral formed from the cooling of molten rock is olivine (peridot), which is found in lava. Gem minerals found in plutonic rock (molten rock cooled within Earth) are diamond, various garnets, zircon, corundum, spinel, labradorite, orthoclase, and albite-oligoclase. Well-known gem minerals formed by hydrothermal and pneumatolitic processes are varieties of quartz (amethyst, citrine, crystal, smoky), varieties of beryl (aquamarine, morganite, emerald), various kinds of tourmaline, topaz, spodumene, and microcline. These minerals are commonly found in bodies or rock called "pegmatites," or "veins." Metamorphic rocks containing gem minerals form in the roots of mountain chains and/or adjacent to igneous bodies, where there are high pressures and temperatures. Well-known gem minerals formed here include corundum, kyanite, lazurite (lapis lazuli), various garnets, jade (both nephrite and jadeite), spinel, beryl, and chrysoberyl. Gem minerals found in sedimentary deposits may be divided into two major divisions: placer deposits (gravels that contain heavy minerals resistant to weathering and left by erosion and deposition of streams) and primary sedimentary minerals grown in the sediments. Placer gem minerals include diamond, corundum, spinel, tourmaline, topaz, various garnets, zircon, and chrysoberyl and may be associated with placer gold. Primary sedimentary minerals are mainly varieties of quartz (agate, chalcedony, onyx, sard, carnelian, petrified wood, heliotrope, bloodstone, jasper). Gem minerals may also result from weathering processes. A fine example is turquoise, which is produced by the weathering of copper-bearing hydrothermal veins under arid conditions.

TOOLS OF THE GEMOLOGIST

The binocular microscope is one of the most useful tools to the gemologist, whether for making identifications, grading, or preparing a stone for lapidary work. These microscopes have relatively low magnification, but the twin eyepieces allow the gemologist to view the stone in three dimensions. Other magnifying instruments used include the hand lens and loupe, which are mostly used for preliminary inspections of gems. Properties of minerals that can be determined for identification purposes with the binocular microscope include hardness—by noting polish quality or wear characteristics—and crystallography, which can be determined from traces of cleavage on the unpolished girdle of the stone, noting single or double refraction, orientation of inclusions, and other qualities. Properties of minerals that can be determined for the grading of gems include clarity, quality of cutting and polishing, symmetry, and color. Properties of minerals that can be determined for lapidary purposes include crystallographic orientation, cleavage directions, and location of color variations and inclusions. Most gems are considered flawless when no inclusions, fractures, surface blemishes, or cutting flaws are visible to the naked eye.

The refractometer is also one of the most useful tools of the gemologist for purposes of gem identification. This technique makes use of the fact that each mineral species has a distinctive and characteristic index of refraction, or indices of refraction. Different minerals may have one, two, or three indices of refraction. A refractometer such as the Rayner or Duplex refractometer can measure the indices of refraction of a gem by use of the critical angle of light for that mineral, which is dependent upon the mineral's refractive index or indices. The index or indices of refraction are measured from flat facets of cut gems or crystal faces; an average of the indices may be obtained from a curved face or from any polished surface of a fine-grained aggregate, such as jade or chalcedony. Tables of refractive indices allow gemologists to compare results of their testing with those of known gem materials.

The polariscope is a simple yet very valuable instrument. It is a light source with two Polaroid plates mounted above and below the gem. The polariscope is used to distinguish between crystalline double refractive material (minerals whose optical properties vary with direction in the mineral) and singly refractive crystalline material or amorphous materials, and to identify doubly refractive fine-grained aggregates, such as jades and chalcedony. For example, ruby is doubly refractive and will change appearance when rotated in polarized light, whereas a similarly colored garnet will not.

The spectroscope may on occasion be a useful instrument. The handheld types are difficult to use, and the table models are very expensive. They are generally not necessary for the identification of stones but are useful in confirming or determining the presence of metallic ions like iron or chromium. Light traveling through the mineral is separated into its color components in a rainbow of color by prism or diffraction. There will be dark lines on this rainbow, corresponding to light frequencies absorbed by atoms in the mineral. Genuine ruby will show absorption lines for chromium; red garnet does not.

ROLE IN ECONOMICS AND TECHNOLOGY

The study of gems or gem minerals interests the connoisseur, hobbyist, investor, and those whose business is the gem trade. Two reasons for a large public ownership of gemstones in western society are the traditional diamond engagement ring in North America and the European use of the diamond as an investment. It is particularly in this market that most of the fraudulent practices in the gem and gem mineral trade exist. Investing in gems and gem minerals has been common for several reasons, including short-term and long-term gains, and as a hedge against inflation. Gemstones have also been valued for their portability, especially in unstable political situations. As with any investment, however, there is an element of risk. The degree of risk may vary with the gem mineral; for example, finding a large new deposit drops the price. Cornering of the market by a particular group forces the price up or down for some political or economic reason. Diamond pricing is probably the most structured and stable because it has been controlled by the De Beers Central Selling Organization (CSO) in London for the last hundred years. Nevertheless, prices for a top-quality, flawless one-carat diamond (round brilliant) fluctuated from more than $7,000 in 1977 through $50,000 to $60,000 in 1980 down to about $12,000 in 1985 to more than $17,000 in 1988 in a very volatile world market. Many of the African nations where diamonds are mined are prone to civil war, and diamonds are frequently used to fund those wars. Such diamonds, called "conflict diamonds," are a major humanitarian problem because they are often mined with forced labor and battles are frequently fought for control of mining areas. Many diamonds are now engraved by laser with serial numbers that can be used to verify they did not originate in conflict zones.

Because of their perfection, gems are used by scientists for research concerned with crystals and crystallization processes. They also have many technological applications, from ruby (corundum) lasers to quartz crystals in citizen-band radios and telephone communications equipment. Because of the large demand for these minerals, synthetic substitutes are generally utilized today for their technological applications. Mineral species, of which the gem-quality varieties may make up only a very small quantity, may also have other major non-gem-related industrial uses. The mineral diamond, for example, has use as a gem not only in transparent, attractively colored, and colorless varieties but also in translucent, nearly opaque, or highly included stones, which are not pleasing to the eye yet have important industrial cutting and abrasive uses. Much high-technology grinding, cutting, and polishing is based upon diamonds, from the bits that drill oil wells and cut metals and the saws that cut stone to the fine abrasives used to polish gems and lenses. The mineral corundum, the second hardest mineral known, not only produces specimens of ruby and sapphire but is also extensively used as an abrasive. Some of the abrasive uses, such as coarse grinding with emery, have decreased because of the introduction of relatively inexpensive artificial diamonds, but other uses, such as polishing with alumina, have increased. The mineral hematite is best known as the chief ore of iron, but it is also used for black gems and has become more fashionable in recent years. The jewelry trade knows it best for yet another use: jewelers' rouge, a finely powdered hematite used to polish metals and some gems. The gem mineral quartz, with many gem varieties—amethyst, citrine, smoky, rose, crystal, agate, carnelian, bloodstone—has many nongem uses, including glassmaking and the production of abrasives (sandpaper) and sand for sandboxes, beaches, and concrete.

—Charles I. Frye

Further Reading

Campbell, Greg. *Blood Diamonds: Tracing the Path of the World's Most Precious Stones*. Basic Books, 2004.

Groat, Lee Andrew. *The Geology of Gem Deposits*. Mineralogical Association of Canada, 2007.
Hall, Cally. *Gemstones*. 2nd ed., Dorling Kindersley, 2002.
Klein, Cornelis, and Cornelius S. Hurlbut, Jr. *Manual of Mineralogy*. 23rd ed., John Wiley & Sons, 2008.
Lagomarsino, James. *A Pocket Guide to Rocks and Minerals*. Parragon Books, 2008.
Manutchehr-Danai, Mohsen. *Dictionary of Gems and Gemology*. 3rd ed., Springer-Verlag, 2005.
O'Donoghue, Michael. *Gems*. 6th ed., Robert Hale, 2008.
Rygle, Kathy J., and Stephen F. Pedersen. *The Treasure Hunter's Gem and Mineral Guides to the U.S.A.* 5th ed., GemStone Press, 2008.

GEOCHEMICAL CYCLES

Fields of Study: Geochemistry; Biogeochemistry; Environmental Studies

ABSTRACT
Geochemical cycles refer to the movement, or cycling, of elements through the biosphere and/or ecosystems. Both biotic (living) and abiotic (nonliving) components make up such systems.

KEY CONCEPTS

guano: the fecal droppings of seabirds that has accumulated over thousands of years and dried to a consistency that is amenable to mining; used in agriculture as fertilizer

gypsum: mineral forms of calcium sulfate, $CaSO_4$, generally as the dihydrate $CaSO_4 \cdot 2H_2O$

pyrite: iron sulfide, FeS, a dense mineral with a gold color; often mistaken for gold, so is known as "fool's gold"

DEFINING GEOCHEMICAL CYCLES

Geochemical cycles are generally considered to be those involving nutrient elements utilized by organisms in various ecosystems. Cycling involves both biological and chemical processes. While nearly all natural elements are cycled through both abiotic and living systems, certain elements are most commonly described in such systems. These include carbon, nitrogen, phosphorus, and a variety of lesser elements (including iron, sulfur, and trace elements such as copper and mercury).

Although the cycling of elements is often thought of as occurring in a relatively rapid fashion, many of these elements spend long periods locked in abiotic systems. For example, carbon may be found in materials that require millions of years to cycle through ocean sediment back into the atmosphere. The fate of such elements depends on many factors, including their chemical properties and their ability to erode or return to the atmosphere. Some chemical elements, such as carbon, oxygen, and nitrogen, are incorporated into organisms from the atmosphere. Other elements, such as phosphorus, potassium, sulfur, and iron, are found mainly in rocks and sediments.

CARBON AND OXYGEN CYCLES

The carbon and oxygen cycles are greatly dependent on each other. Molecular oxygen (O_2), which comprises approximately 20 percent of the atmosphere, is used by organisms through a metabolic process called "respiration." In these reactions, the oxygen reacts with reduced carbon compounds such as carbohydrates (sugars) and generates carbon dioxide (CO_2). Though carbon dioxide constitutes only a small proportion of the volume of the atmosphere (0.04 percent), it is in this form that it is used by primary producers such as plants. In the process of photosynthesis, utilizing sunlight as an energy source, plants and some microorganisms bind, or fix, the CO_2, converting the carbon again into carbohydrates, resulting in growth of the plant or replication of the microorganism. The complex carbohydrates that are generated in photosynthesis serve as the food source for consumers—organisms such as animals (including humans) that eat the plants. The carbohydrates are then broken down, regenerating carbon dioxide. In a sense, the combinations of respiration and photosynthesis represent the cycle of life. The concentration of carbon dioxide in the atmosphere is a factor in regulating the temperature of Earth. Consequently, the release of large quantities of the gas into the atmosphere through the burning of fossil fuels could potentially alter the Earth's climate.

NITROGEN CYCLE

Nitrogen gas (N_2) makes up 78 percent of the total volume of the atmosphere. However, because of the extreme stability of the bond between the two nitrogen atoms in the gas, plants, and animals are unable to use atmospheric nitrogen directly as a nutrient. Nitrogen-fixing bacteria in the soil and in the roots of leguminous plants (peas, clover) are able to convert the gaseous nitrogen into nitrites and nitrates, chemical forms that can be used by plants. Animals then obtain nitrogen by consuming the plants. The decomposition of nitrogen compounds results in the accumulation of ammonium (NH_4^+) compounds in a process called "ammonification." It is in this form that nitrogen is commonly found under conditions in which oxygen is limited. In this form, some of the nitrogen returns to the atmosphere. In the presence of oxygen, ammonium compounds are oxidized to nitrates (nitrification). Once the plant or animal has died, bacteria convert the nitrogen back into nitrogen gas, and it returns to the atmosphere.

PHOSPHORUS CYCLE

Unlike carbon and nitrogen, which are found in the atmosphere, most of the phosphorus required for biotic nutrition is found in mineral form. Phosphorus is relatively water insoluble in this form; it is only gradually dissolved in water. Available phosphorus is therefore often growth-limiting in soils (it is second only to nitrogen as the scarcest of the soil nutrients). Ocean sediments may bring the mineral to the surface through uplifting of land, as along coastal areas, or by means of marine animals. Enzymatic breakdown of organic phosphate by bacteria and the consumption of marine organisms by seabirds cycle

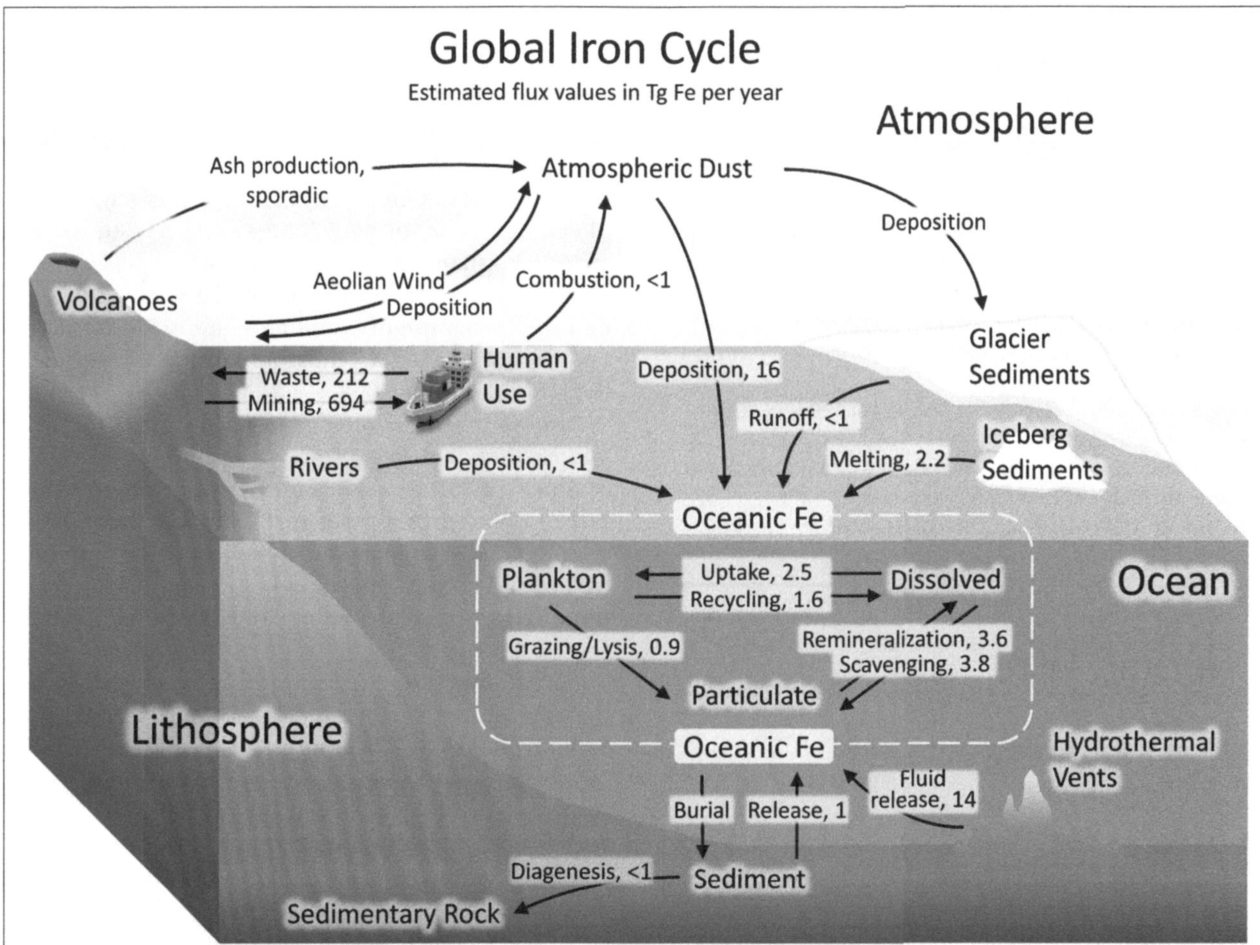

The iron cycle, one of many important geochemical cycles. Image by BeccaMGM, CC BY-SA 4.0, via Wikimedia Commons.

the phosphorus into forms available for use by plants. Deposition of guano (bird feces) along the American Pacific coast has long provided a fertilizer rich in phosphorus.

Bacteria also play significant roles in the geochemical cycling of many other elements. Iron, despite its abundance in Earth's crust, is largely insoluble in water. Consequently, it is generally found in the form of precipitates of ferric (Fe^{+3}) compounds, seen as brown deposits in water. Acids are often formed as byproducts in the formation of ferric compounds. The bacterial oxidation of pyrite (FeS_2) is a major factor in the leaching process of iron ores and in the formation of acid mine drainage. Likewise, much of the sulfur found in Earth's crust is in the form of pyrite and gypsum ($CaSO_4$). Weathering processes return much of the sulfur to water-soluble forms; in the absence of air, the bacterial reduction of sulfate (SO_4^{-2}) to forms such as hydrogen sulfide (H_2S) allows its return to the atmosphere. Since sulfide compounds are highly toxic to many organisms, bacterial reduction of sulfates is of major biogeochemical significance.

—Richard Adler

Further Reading

Knoll, Andrew H., Don E. Canfield, and Kurt O. Konhauser, editors. *Fundamentals of Geobiology*. John Wiley & Sons, 2012.

Libes, Susan. *Introduction to Marine Biogeochemistry*. 2nd ed., Elsevier/Academic, 2009.

Madigan, Michael T., et al. *Brock Biology of Microorganisms*. 12th ed., Pearson/Benjamin Cummings, 2009.

Schlesinger, William H., editor. *Biogeochemistry*. Elsevier, 2005.

Geologic Timescale

Fields of Study: Geology; Stratigraphy; Paleontology

ABSTRACT

Although certain geologic events, such as floods and earthquakes, seem to occur over short periods of time, the vast majority of observed geological features formed over a great span of time.

KEY CONCEPTS

original horizontality: the principle that any stratum was originally distributed evenly over the region of interest, such as on a lake bed, seafloor, or land area

original lateral continuity: the principle that the particular content of any stratum extended continuously to the borders of the region of interest

radiometric age-dating: determination of the age of rocks and other materials based on the relative quantities of specific isotopes related to each other by an unequivocal nuclear decay sequence, and the half-life span of that sequence

stratigraphy: the study and correlation of rock strata according to age

superposition: the principle that requires the oldest stratum to be at the base of any series of strata above it in undisturbed rock or sediment beds

SCIENCE AND GEOSCIENCE

A major difference between the geosciences (earth science) and other sciences is the great enormity of their timescale. One might compare the magnitude of geologic time for geoscientists to the vastness of space for astronomers. Every geological process, such as the movement of crustal plates (plate tectonics), the formation of mountains, and the advance and retreat of glaciers, must be considered within the context of time.

Although certain geologic events, such as floods and earthquakes, seem to occur over short periods of time, the vast majority of observed geological features formed over a great span of time. Consequently, modern geoscientists consider Earth to be exceedingly old. Using radiometric age-dating techniques, they calculate the age of Earth to be 4.6 billion years.

Early miners were probably the first to recognize the need for a scale by which rock and mineral units could be compared over large geographic areas. However, before a timescale—and even geology as a science—could develop, certain principles had to be established. This did not occur until the late eighteenth century when James Hutton (1726-1797), a Scottish naturalist, began his extensive examinations of rock relationships and natural processes at work on the earth. His work was amplified by Charles

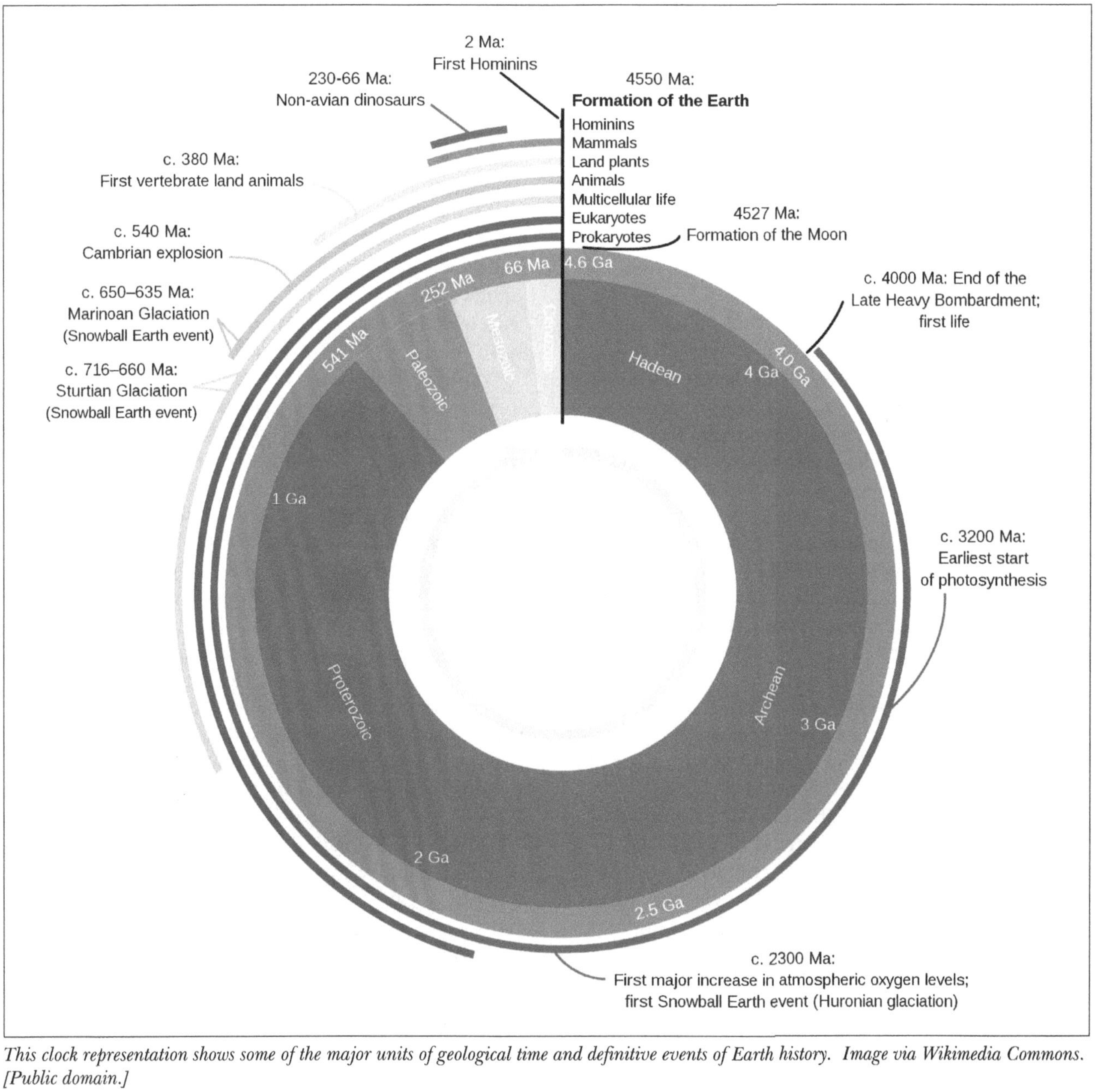

This clock representation shows some of the major units of geological time and definitive events of Earth history. Image via Wikimedia Commons. [Public domain.]

Lyell (1797-1875) in his textbook *Principles of Geology* (1830-1833). After careful observation, Hutton concluded that the natural processes and functions he observed had operated in the same basic manner in the past, and that, in general, natural laws were invariable. That idea became known as the principle of uniformitarianism.

THE BIRTH OF STRATIGRAPHY

In 1669 Nicholas Steno (1638-1686), a Danish physician working in Italy, recognized that horizontal rock layers contained a chronological record of Earth's history and formulated three important principles for interpreting that history. The principle of superposition states that in a succession of undeformed strata, the oldest stratum lies at the bottom, with successively younger ones above. The principle of original horizontality states that because sedimentary particles settle from fluids under the influence of gravity, sedimentary rock layers must be horizontal; if not, they have suffered from subsequent disturbance. The principle of original lateral continuity states that strata originally extended in all directions until they thinned to zero or terminated against the edges of the original area of deposition.

In the late eighteenth century, the English surveyor William Smith (1769-1839) recognized the wide geographic uniformity of rock layers and discovered the utility of fossils in correlating these layers. By 1815, Smith had completed a geologic map of England and was able to correlate English rock layers with layers exposed across the English Channel in France.

From the need to classify and organize rock layers into an orderly form arose a subdiscipline of modern geology—stratigraphy, the study of rock layers and their age relationships. In 1835 two British geologists, Adam Sedgwick (1785-1873) and Roderick Murchison (1792-1871), began organizing rock units into a formal stratigraphic classification. Large divisions, called "eras," were based upon well-known and characteristic fossils, and included a number of smaller subdivisions, called "periods."

The periods are often subdivided into smaller units called "epochs." Each period is defined by a representative sequence of rock strata and fossils. For instance, the Devonian period is named for exposures of rock in Devonshire in southern England, while the Jurassic period is defined by strata exposed in the Jura Mountains in northern Switzerland.

Approximately 80 percent of Earth's history is included in the Cryptozoic (meaning "obscure life") eon. Fossils from the Cryptozoic eon are rare, and the rock record is very incomplete. After the Cryptozoic eon came the Paleozoic (ancient life), Mesozoic (middle life), and Cenozoic (recent life) eras. Most of the life forms that evolved during the Paleozoic and Mesozoic eras are now extinct, whereas 90 percent of the life-forms that evolved up to the middle Cenozoic era still exist.

THE FOSSIL RECORD

The word "fossil" comes from the Latin *fossilium*, meaning "dug from beneath the surface of the ground." Fossils are defined as any physical evidence of past life. Fossils can include not only shells, bones, and teeth, but also tracks, trails, and burrows. The latter group are referred to as trace fossils. Fossils demonstrate two important truths about life on Earth: First, thousands of species of plants and animals have existed and later became extinct. Second, plants and animals have evolved through time, and the communities of life that have existed on Earth have changed.

Some organisms are slow to evolve and may exist in several geologic time periods, while others evolve quickly and are restricted to small intervals of time within a particular period. The latter, referred to as index fossils, are the most useful to geoscientists for correlating rock layers over wide geographic areas and for recognizing geologic time.

The fossil record is incomplete, because the process of preservation favors organisms with hard parts that are rapidly buried by sediments soon after death. For this reason, the vast majority of fossils are represented by marine invertebrates with exoskeletons, such as clams and snails. Under special circumstances, soft-bodied organism can be preserved, for instance the preservation of insects in amber, made famous by the feature film *Jurassic Park* (1993).

THE ROCK CYCLE

A rock is a naturally formed aggregate of one or more minerals. Three types of rocks exist in Earth's crust, each reflecting a different origin. Igneous rocks have cooled and solidified from molten material either at or beneath Earth's surface. Sedimentary rocks form when preexisting rocks are weathered and broken down into fragments that accumulate and become compacted or cemented together. Fossils are most commonly found in sedimentary rocks. Metamorphic rocks form when heat, pressure, or chemical reactions in Earth's interior change the mineral or chemical composition and structure of any type of preexisting rock.

Over the huge span of geologic time, rocks of any one of these basic types can change into either of the other types or into a different form of the same type. For this reason, older rocks become increasingly rare.

EARTH'S HISTORY COMPRESSED INTO ONE CALENDAR YEAR

One way to visualize events in Earth's history is to compress geologic events into a single calendar year. Earth's birth, 4.6 billion years ago, would occur during the first minute of January 1. The first three-quarters of Earth's history is obscure and would take place from January to mid-October. During this time, Earth gained an oxygenated atmosphere, and the earliest life-forms evolved. The first organisms with hard parts preserved in the fossil record (approximately 570 million years ago) would appear around November 15. The extinction of the dinosaurs (65 million years ago) would occur on Christmas Day. Homo sapiens would first appear at approximately 11 p.m. on December 31, and all of recorded human history would occur in the last few seconds of New Year's Eve.

—Larry E. Davis

Further Reading

Catuneanu, Octavian. *Principles of Sequence Stratigraphy.* Elsevier, 2006.

Gradstein, Felix M., James G. Ogg, Mark D. Schmitz, and Gabi M. Ogg, editors. *Geologic Time Scale 2020.* Elsevier, 2020.

Nichols, Gary. *Sedimentology and Stratigraphy.* 2nd ed., Wiley-Blackwell, 2009.

Prothero, Donald R., and Fred Schwab, editors. *Sedimentary Geology.* W.H. Freeman, 2013.

Ritter, Scott, and Morris Petersen. *Interpreting Earth History: A Manual in Historical Geology.* 8th ed., Waveland Press, 2015.

Winchester, Simon. *The Map That Changed the World: William Smith and the Birth of Modern Geology.* Perennial, 2001.

GEOLOGY

Fields of Study: Geology; Geochemistry; Geophysics; Seismology; Volcanology

ABSTRACT

The study of Earth and its geological processes is essential to the discovery, extraction, and management of natural resources, from minerals to energy resources.

KEY CONCEPTS

geothermal energy: electricity produced using steam or hot water from natural hot springs to drive generator turbines

mountaintop removal mining: as the words imply, the excavation and removal of the top portion of a low mountain that is acting as overburden to a coal bed

uniformitarianism: the principle that the geological mechanisms at work in the world today have operated in the same way throughout the past and will throughout the future

WHAT IS GEOLOGY?

Geology is the study of the planet Earth, and by extension all rocky planets and bodies: its composition, origin, and history, and the environmental, biological, chemical, and physical forces outside and within it. As a science, geology grew from the nineteenth-century study of natural features, stratigraphy, and fossils in rock outcroppings to a wide variety of scientific subspecialties covering myriad aspects of the planet. Since the early nineteenth century, geology has involved accurate mapping of Earth's topography and discovery, study, and exploitation of major mineral deposits around the world.

A guiding principle in geology has been uniformitarianism: geological processes that are observed today are the same as those that occurred in the past and those that will occur in the future. Application of this concept on a planetary scale allows scientists to prospect for minerals using remote-sensing techniques.

Catastrophic events, including meteor impacts, have been deduced from geological deposits and have been credited with causing widespread mass extinctions observed in the fossil record. Identification of the large meteor impact responsible for forming Chesapeake Bay has provided an explanation for recent earthquakes in the region and for the presence of saltwater aquifers in Virginia. Some economic geologists have postulated that the nickel and platinum deposit at Sudbury, Ontario, is a meteor impact site from billions of years ago.

The U.S. Geological Survey (USGS, part of the Department of the Interior) is the governmental agency responsible for producing official maps and reports. Most other nations have similar agencies, including the Geological Survey of Canada, Servicio Geológico Mexicano, the British Geological Survey, Geoscience Australia, the Geological Survey of Japan, and the South Africa Council for Geoscience, which all host informational websites.

Knowledge of geology is fundamental to the understanding of all inanimate resources on Earth. Geothermal energy can provide an inexpensive alternative to fossil and nuclear sources for generating electricity. Discovery of necessary mineral resources is a prerequisite to exploitation, while the mechanics of exploiting those resources also requires geological expertise. Environmental geologists are involved in mapping and investigating toxic contamination areas for possible mitigation. Search for permanent geological sites for radioactive materials is going on in countries around the world. Potential natural hazards—including unstable topography, earthquake fault lines, and volcanic activity—require geological monitoring to warn people of impending disasters.

EXPLORATION GEOLOGY

Exploration geologists focus on the discovery and exploitation of mineral and ore deposits and fossil fuels. Stone Age humans found outcroppings of flint and chert with which to make arrowheads and other tools. Eventually, humans moved on to easily worked metals such as copper, tin, silver, gold, and iron. Precious gems have been highly valued for millennia, and new sources for these ores and minerals continue to be found.

Modern industrialized society requires metals for basic construction and manufacturing. The increasing technological demand has moved geological exploration from the California gold rush era of the American West to the worldwide search for uranium for nuclear weapons to the search for rare earth ele-

Kinney Lake and Mount Whitehorn near Mount Robson, Canada. Photo by Florian Fuchs, CC BY 3.0, via Wikimedia Commons.

ments for high-tech electronics and lithium deposits for batteries. Geologists in the twenty-first century rarely engage in time-consuming initial field exploration and prospecting, relying on remote sensing from aircraft and satellites to determine where new mineral deposits might be found. Confirmation of mineral deposits and plans for exploitation require geological expertise.

The exploitation of coal, oil, and gas deposits around the world provides vital sources of energy to the billions of people on Earth. Most geologists are employed, usually by governments and private industry, in this aspect of geology.

Coal remains the most important fuel for electric power production worldwide though climate change concerns are reducing the demand for coal, with reserves of anthracite ("hard") and bituminous ("soft") coal widespread in North America, Europe, and Asia. Large-scale underground coal mining is labor intensive and expensive. Many American coal companies have opted to use cheaper methods of obtaining coal, such as strip mining and mountaintop removal mining (MTR), in which coal deposits are located at or near surface level. MTR in areas like West Virginia and Kentucky is unpopular with the general public because of the widespread environmental degradation that occurs when entire mountains are leveled and overburden (soil and noncoal rock) materials are placed in adjacent valleys. Land use after MTR may be determined by geological studies; the areas are usually left unvegetated after mining activities end. Coal slurry impoundments are used to hold huge amounts of MTR coal waste, and if the impoundment fails, aquatic wildlife in the area's streams and rivers is eradicated. In March, 2009, the U.S. Environmental Protection Agency announced that permits for MTR of coal would be carefully scrutinized.

Oil deposits occur around the world, and geological exploration teams continue to find major discoveries. Exploration and development of new oil fields are often complicated by politics, on both national and international scales. Opening up the Alaskan National Wildlife Reserve (ANWR) to oil exploration and drilling is an example of such complications. Even though President George W. Bush and Vice President Dick Cheney both strongly favored drilling, Congress was unwilling to authorize oil leases in the eight years (2001-2009) of the Bush presidency. In the 2008 U.S. presidential election, ANWR became a major campaign issue when Alaskan governor Sarah Palin, who was the Republican vice presidential candidate, strongly endorsed drilling.

Discovery and commercial exploitation of heavy bitumen oil sands, which cannot be pumped out of the ground like petroleum deposits, have become major political issues because this fossil fuel leaves a large carbon footprint. Oil sands are strip-mined or hauled from massive open-pit mines. An estimated 780,000 barrels of oil are produced per day from Canadian oil sands in Alberta, and about 60 percent of this is exported to the United States. Oil sands in Alberta are estimated to contain more than one trillion barrels of oil, 80 percent of which is not accessible through present surface mining methods. Oil sand deposits also occur in Utah, Venezuela, and Russia.

GEOLOGICAL MONITORING OF VOLCANOES AND EARTHQUAKES

The devastating Boxing Day tsunami of December 26, 2004, which engulfed Indian Ocean shorelines from Indonesia to East Africa and killed more than 225,000 people, followed an undersea event known as the Great Sumatra-Andaman earthquake. The countries most affected by the tsunami lacked geological monitoring stations. Such a seismological monitoring network could have provided many areas with several hours warning of the impending tsunami and lessened the death toll. The U.S. National Oceanic and Atmospheric Administration operates the Pacific Tsunami Warning System, which warns of potential problems for Hawaii, Alaska, and the Pacific coast of North America.

Volcano monitoring is necessary to warn people of impending eruptions. Erupting volcanoes emit clouds of ash that can be sucked into jet aircraft engines, where the ash liquefies and then deposits a solid glass coating to the rear of the jet turbine. This glass coating interferes with the jet enough to cause the aircraft to crash. Ash problems necessitate closure of airports within the reach of the erupting volcano, and aircraft must be diverted from routes that pass through the ash clouds.

ENVIRONMENTAL GEOLOGY

Environmental geologists use a variety of geological, geochemical, microbiological, and hydrological techniques to identify and mitigate hazards resulting from urban sprawl, industrialization, and mining activities. The most common environmental problems include surface water and groundwater contamination, dumping of hazardous wastes in unprotected ground, and air pollution related to improper waste handling.

A permanent geological storage site for reactor waste in the United States has been a limiting factor in public support for the nuclear power industry. (Nuclear weapons waste is stored in Carlsbad, New Mexico.) Requirements for geological storage include the absence of groundwater and total lack of seismic activity in a solid bedrock formation. Many locations have been proposed. In 1987, Yucca Flat, Nevada, was selected, but the selection met with almost immediate opposition because of unanswered geological questions. In March, 2009, President Barack Obama announced that plans to use Yucca Flat had been abandoned. Stephen Chu, the Secretary of Energy for the Obama administration, indicated that the United States might build nuclear power reactors that could utilize nuclear waste, thus dramatically lessening (but not eliminating) the amount of radioactive waste requiring permanent storage. Other countries, including Sweden, have conducted rigorous nationwide geological surveys to identify potential nuclear waste storage sites and are moving closer to final site selection. Some nuclear industry experts believe that the United States will not select a site until 2030.

COMMERCIAL POWER PRODUCTION FROM GEOTHERMAL ENERGY

Harnessing hot springs and geysers to produce electricity has been going on at Larderello, Italy, for more than a century and is well established in Iceland and the Philippines; the latter two countries produce about 20 percent of their electricity from geothermal energy. Iceland has a geothermal capacity of 1.3 terawatt-hours per year. There are twenty-seven electricity plants at The Geysers, in Northern California, producing 750 megawatts. Important geological concerns arise when harnessing geothermal sources. The major problem at generating locations like Wairakei, New Zealand, and The Geysers is local depletion of heat sources; heated zones are tapped too intensively for too long of a period to allow recharge of heat from deep within Earth. Other problems include the need for drilling deep wells and for fracturing rock around the deep wells at geothermal locations. Although the technology for drilling deep wells exists, it is a costly process.

—Anita Baker-Blocker

Further Reading

Burek, Cynthia V., and Bettie Higgs, editors. *The Role of Women in the History of Geology*. The Geological Society, 2007.

Fossen, Haakon. *Structural Geology*. 2nd ed., Cambridge UP, 2016.

Hudson, Travis. *Living with Earth: An Introduction to Environmental Geology*. Routledge, 2016.

Ritter, Scott, and Morris Petersen. *Interpreting Earth History: A Manual in Historical Geology*. 8th ed., Waveland Press, 2015.

Selley, Richard C., and Stephen A. Sonnenberg. *Elements of Petroleum Geology*. 3rd ed., Elsevier, 2015.

Timmons, J. Michael, and Karl E. Karlstrom, editors. *Grand Canyon Geology: Two Billion Years of Earth's History*. The Geological Society of America, 2012.

West, Terry R., and Abdul Shakoor. *Geology Applied to Engineering*. 2nd ed., Waveland Press, 2018.

GERMAN *METEOR* EXPEDITION DISCOVERS THE MID-ATLANTIC RIDGE

Fields of Study: Oceanography; Geooceanography; Sonography; Geology; Geophysics

ABSTRACT

Using a newly developed echo sounder, the Meteor, *a refitted German gunboat, made the first transoceanic crossing with closely spaced soundings, leading to the discovery of the Mid-Atlantic Ridge in the southern and equatorial Atlantic Ocean.*

KEY CONCEPTS

echo sounder: a device that emits a pulse of audible or ultrasonic sound waves and measures the time between emission of the pulse and reception of its echo signal to determine distance; the operating principle of the modern "fish finder"

Mid-Atlantic Ridge: a ridge or line of undersea mountains formed by the upwelling of magma into the space between two segments of the Atlantic Ocean floor that are being slowly driven apart as a divergent plate boundary

plate tectonics: the evolution of the continental drift theory, in which Earth's crust is made up of plates of crust material that slowly move relative to each other over time

theory of continental drift: the theory proposed by Alfred Wegener in 1912 to explain the apparent "fit" of South American and African continental coasts, that continents "drift" atop the denser material of Earth's basaltic crust

THE *METEOR* EXPEDITION

The initial purpose of the German *Meteor* expedition was economic. In 1872, E. Sonstadt reported that the oceans contained a gold concentration of 65 milligrams per tonne of seawater. The Treaty of Versailles (1919), which ended World War I, required that Germany repay its enormous war debt in gold, and Fritz Haber, a German chemist, proposed that gold extracted from the international seas might solve the problem.

The Treaty of Versailles prohibited the German navy from sending ships to foreign ports; however, in 1919, a member of the German admiralty, Captain Nippe, persuaded authorities to allow a German vessel to be outfitted and sent on a major peacetime oceanographic expedition. The *Meteor*, a class C gunboat, was selected for the study. Alfred Merz, an adviser to the German navy, was named the chief scientist of the *Meteor*. After Merz's death, Georg Wüst took over leadership of the oceanographic data and study, which was to take place in the Atlantic Ocean.

The *Meteor* was to be equipped with the newly developed echo sounder, a device used to find the dis-

Meteor *departing its homeport with full military honors. Photo via Wikimedia Commons. [Public domain.]*

tance and direction of objects under or partially under the water. An echo sounder works by measuring the time it takes a sonic or ultrasonic pulse emitted by the sounder to reach the object below and then return to the sounder. The time intervals and other data are then converted into numerical reference points of distance and direction. The navies of the world use echo sounding to chart the positions of foreign ships and submarines. During peacetime, echo sounders are used to locate sunken ships, find schools of fish, and map the profile of the seafloor.

The *Meteor* expedition marked the first use of echo sounders to map the profile of the deep seafloor. With the exception of the *Meteor*, no such detailed maps were available before World War II. The profiles aboard the *Meteor* were closely spaced echo soundings taken by extremely diligent workers, which only made the recordings that much more accurate and detailed. Moreover, a new anchor system had been developed that enabled the ship to anchor in deep waters.

On April 26, 1925, the *Meteor* left for Buenos Aires, Argentina, to start work on the planned sections in the South Atlantic. Actual surveys began on June 3, 1925. Merz was in charge in spite of his illness; however, by the time the vessel arrived at the fifth hydrographic station, his condition had worsened to the point that the captain ordered the ship to return to Buenos Aires so that Merz could receive medical attention. Merz then turned the leadership of the scientific part of the expedition over to the captain. In reviewing the work along the first section, the captain found that the acoustic depth sounder had revealed that much of the bottom had been incorrectly charted.

After completing the thirteenth traverse in the equatorial Atlantic Ocean, the *Meteor* sailed for Germany, arriving home on May 29, 1927. In two years, the *Meteor* had traveled more than 41,943 miles (67,500 kilometers), had collected data at 310 hydrographic stations, had anchored ten times in deep ocean, and had made approximately 70,000 soundings of ocean depths. As a result, the expedition was the first to reveal the true ruggedness of the ocean floor.

The significant discovery of the *Meteor* was that a continuous ridge (the Walvis Ridge) runs in a southwesterly direction from the vicinity of Walvis Bay, southwest Africa. (A ridge is a long, narrow elevation on the seafloor.) This in turn led to the discovery of the Mid-Atlantic Ridge. It runs along the north-south axis of the Atlantic and is basically a long, curving zone of mountains, volcanoes, and fractured plateaus. This ridge's now-familiar herringbone pattern was first suggested in 1935 by Wüst and Theodor Stocks.

SIGNIFICANCE OF THE DISCOVERY OF THE MID-ATLANTIC RIDGE

The *Meteor* expedition's discovery of the Mid-Atlantic Ridge was a significant event. Unfortunately, this finding was not recognized widely at the time as being enough to support the theory of continental drift. That theory had continued to generate controversy and debate among scientists ever since 1912, the year it was first proposed by the German geophysicist and meteorologist Alfred Wegener. Wegener's theory stated that the earth was once made up of one large ocean and one large landmass. Gradually, this supercontinent split into two parts. Additional separations and "drift" continued to occur across many millions of years, eventually resulting in the seven continents as they have come to be known.

As modern science now recognizes, these massive continental "plates" tend to drift apart at divergent boundaries. These boundaries may be seen as the midoceanic ridges such as the Mid-Atlantic Ridge. Such rending of the earth's crust brings with it great earthquakes and great flows of volcanic materials that, as they pile up in the "cracks," slowly create ridges.

In the 1950s, a new generation of equipment and instruments was introduced that led to an explosion of data supporting the theory of continental drift, or "plate tectonics," as it is now called. Continuously recording echo sounders, magnetometers, temperature probes, explosion seismometers, piston corers, dredgers, and deep-sea submersibles were used not only to discover additional evidence to support the theory of continental movement but also to add to the sum of knowledge about deep-sea activity.

The *Meteor*'s discovery was a midpoint during the development of the proposed theory of continental drift in 1912 and the concept of plate tectonics that became widely accepted in 1968. Earth scientists realize now that the positions of landmasses are not fixed. The separation of continental plates has resulted in the formation of new ocean basins, and older segments of the seafloor are being recycled continually in areas where deep-ocean trenches are found. This profound reversal of scientific opinion has been described as a scientific revolution, one in which the *Meteor* expedition played a pioneering role.

—*Earl G. Hoover*

Further Reading

Emery, William J., and Walter Zenk. *Captain F. Spiess and the German Atlantic Meteor Expedition of 1925-1927*. BrownWalker Press, 2019.
Knauss, John A., and Newell Garfield. *Introduction to Physical Oceanography*. 3rd ed., Waveland Press, 2017.
Medwin, Herman, et al. *Sounds in the Sea: From Ocean Acoustics to Acoustical Oceanography*. Cambridge UP, 2005.

GLACIATIONS

Fields of Study: Geology; Geomorphology; Cryology

ABSTRACT

Discovering glaciations and trying to understand them have occupied scientists for over a century. In the process, great progress has been made in understanding the drivers of climate change in the past and the likely drivers in the future.

KEY CONCEPTS

glaciation: the advance of a continental ice sheet
ice age: a period of time during which major continental ice sheets advanced and retreated
interglacial: the warm period between glaciations
marine isotope stage: half of a glacial cycle, as identified in the oxygen isotope data from ocean cores
Milanković (also spelled Milankovitch) cycle: period of variation in Earth's orbital parameters, including axial inclination, climatic precession, and orbital eccentricity

THE GLACIATION PUZZLE

Glaciations presented a challenge to climate science. With ice sheets today only on Greenland and Antarctica, huge continental glaciers were difficult even to imagine. As the idea took root, however, evidence for glaciations throughout geologic history became apparent. What controls the timing of glaciations is now largely understood, but how they grow or retreat—and why—continue to be important subjects of research.

GLOBAL FACTORS OF GLACIATIONS

The temperature of the surface of Earth is a function of how much energy it receives from the Sun and how much of this energy is radiated back into space. The atmosphere plays an important role, as its clouds and aerosols limit how much light energy reaches the surface and its greenhouse gases (GHGs) limit how much infrared energy escapes. Geologic evidence suggests that average global surface temperatures have remained within a limited range over the past two billion years, and GHGs, particularly carbon dioxide (CO_2), may have been responsible for this consistency.

CO_2 is produced by volcanism and removed by weathering. Weathering is temperature dependent, so it represents a negative feedback system: Higher temperature results in more weathering, which removes more CO_2, resulting in lower temperatures, or vice versa. Although over periods measured in hundreds of millions of years this feedback loop seems to have maintained a relatively constant temperature, its effects are not instantaneous, and perturbations have occurred. When these perturbations produce colder conditions, ice ages can result.

Evidence of a number of ice ages has been identified in rocks more than one billion years old. Little is known about these ice ages, as data are sparse and difficult to interpret. Between 750 and 550 million years ago, there were several major ice ages, usually lumped together and called "Snowball Earth," as there is evidence that glaciers then existed at sea level near the equator. Although they are of academic interest, at the time these ice ages occurred, there were no land plants, and the atmosphere had far less oxygen, so it is not clear that efforts to understand them will help in the study of contemporary climate change.

Other ice ages, including a short one 440 million years ago and the Permo-Carboniferous ice age between 325 and 240 million years ago, provide in-

How glaciations grow and retreat continue to be important research subjects. Photo via iStock.com/:sjoeman. [Used under license.]

sight into what conditions are required for ice ages. These occurred when the land on which the continental glaciers formed was over the South Pole. As continents were also over the South Pole in the time period between these two ice ages and Antarctica sat over the South Pole for 90 million years before its current glaciers formed, this location seems to be a necessary but not a sufficient condition for ice age formation.

PLEISTOCENE GLACIATIONS

About 50 million years ago, Earth began to cool. Deep-ocean temperatures gradually dropped from 13° Celsius to the present 1° Celsius. Study of ocean sediment cores has identified some minor ice ages around 40 million years ago, another ice age formed the East Antarctica Ice Sheet 34 million years ago, and a third ice age around 13 million years ago left evidence in Alaska. Orbital factors probably influenced these advances, perhaps by affecting the carbon cycle, but the details remain obscure.

Starting about three million years ago, the climate developed two different states. Since then, it has oscillated between them, causing glaciations—with continental glaciers extending down to the 40° north parallel of latitude—and interglacials such as the current period, some with temperatures even higher than those of the present. Geologists identified glacial deposits, determined that some were on top of others, and gave names to each, but the names were not standardized internationally. The name of the most recent glaciation is Valdaian on the Russian Plains, Devensian in Britain, Weichselian in Scandinavia, Würm in the Alps, and Wisconsinan in North America. The deposits of one glacier are easily removed by subsequent glaciers, so there might be some for which no evidence remains on the continents.

Oxygen isotope ratios of marine sediments, which are not removed by later glaciations, show that there have actually been around fifty glaciations. Because these ratios indicate how much water is tied up in ice, this record is now seen as the best way of delineating when an advance ended and a retreat began. Each marine isotope stage (MIS) is given a number (odd ones for interglacials, even ones for glaciations) starting with the most recent. Interglacial MIS 103 occurred 2,580,000 years ago.

Once all these additional glaciations were known, analysis showed that astronomical cycles controlled their timing, as had been suggested by James Croll in 1864 and Milutin Milanković in 1941. These results are so robust that geologists now use these cycles to calibrate the more recent part of the geologic time scale.

That Milankovitch cycles control the timing of glaciations is beyond dispute. How they do so, however, is not well understood. Feedback systems in the oceans, the atmosphere, the biosphere, or perhaps elsewhere are needed to amplify the tiny signal produced astronomically. Identifying and understanding these systems will help scientists understand current climate change.

CONTEXT OF GLACIATIONS

Skeptics who wonder if the climate is changing should consider glaciations. Doomsayers, who fear the human race is threatened by global warming, should also consider glaciations. Much of geologic history concerns things that happened so long ago that it is easy to dismiss or ignore them, but glaciations are recent history. *Homo erectus* was on the planet and using fire twenty or so glaciations ago. All of human evolution has taken place as glaciers ebbed and flowed. Human migration to North America occurred 14,600 years ago, just after the peak of the last glaciation. Ice sheets then, although they were getting smaller, still covered most of Maine and northern parts of New York, Vermont, and New Hampshire. They are not there now. Climate changes.

—Otto H. Muller

Further Reading

Ehlers, Jürgen, Philip Leonard Gibbard, and Philip D. Hughes. *Quaternary Glaciations—Extent and Chronology: A Closer Look.* Elsevier, 2011.

Ehlers, Jürgen, Philip D. Hughes, and Philip Leonard Gibbard. *The Ice Age.* Wiley-Blackwell, 2016.

Macdougall, J. D. *Frozen Earth: The Once and Future Story of Ice Ages.* U of California P, 2004.

Miller, Frederic P., Agnes F. Vandome, and John McBrewster. *Milankovitch Cycles.* Alphascript Publishing, 2009.

Ruddiman, William F. *Earth's Climate Past and Future.* 2nd ed. W. H. Freeman, 2008.

Woodward, Jamie. *The Ice Age: A Very Short Introduction.* Oxford UP, 2014.

GLACIERS

Fields of Study: Geology; Cryology; Glaciology

ABSTRACT

Glaciers are very sensitive to global temperature changes, retreating and growing as the Earth warms and cools. Because of this sensitivity, geologists have made careful measurements of the changes in benchmark glaciers in order to monitor the effects of global climate change and have studied the history of glaciers as a means of reconstructing the sequence of temperature fluctuations in Earth's past.

KEY CONCEPTS

alpine glaciers: large masses of ice found in valleys, on plateaus, and attached to mountains

cryosphere: the portion of the Earth's surface that is composed of frozen water

firn: the intermediary stage between snow and ice

glacial ice: ice created by the compression of snow, sometimes saturated with meltwater that is refrozen

ice caps: masses of ice covering areas smaller than 50,000 square kilometers

ice sheets: masses of ice covering large landmasses

SUMMARY OF GLACIERS

Of the freshwater ice of the world, 99 percent is located in Antarctica and in Greenland: 90 percent of the ice is located in Antarctica (an area of 14 million square kilometers containing 27.6 million cubic kilo-

Glaciers lose mass (a process called "ablation") by melting, sublimation, and calving. Photo via iStock.com/slowmotiongli. [Used under license.]

meters of ice); 9 percent is in Greenland (an area of 1.726 million square kilometers containing 2.85 million cubic kilometers of ice), and 1 percent exists in the glaciers and ice caps scattered throughout the world. The two ice sheets of Antarctica and Greenland represent about 10 percent of Earth's land area and contain more than three-quarters of its freshwater.

GLACIAL LANDSCAPE

Geologists can provide useful information about the history of glaciation by studying the geographic features left by glaciers that have long since melted away. When alpine glaciers form, they produce an amphitheater-like depression called a "cirque" at their highest point. When the compression of snow produces ice in a depth of about 20 meters, the ice begins to flow downward along the valleys of former streams. The ice deepens and widens the stream valley to produce a glacial trough. Valleys of tributary streams are filled with ice as well, but the shorter length of the tributary glaciers and the lesser discharge produce less deepening of the channel. Typically, these tributary glaciers erode their channels only down to the current ice surface of the main glacier, resulting in distinctive "hanging valleys" that end in a precipice at their juncture with the valley produced by the main body of the glacier. The height of these hanging valleys indicates the depth of the ice in the main glacier.

Materials the glacier erodes are pushed to its side, creating lateral moraines, or to its lower end, creating a frontal or end moraine, or accumulate under the glacier, creating a ground moraine, or till. These features identify the greatest breadth (lateral moraines) and furthest extent (terminal moraine) of ancient glaciers. The sediments deposited in glacial lakes—called "varves"—can be used to determine the length of the deposition process. Distinctive small hills called "drumlins" are formed under an advancing glacier and, because of their teardrop shape, indicate the direction of the ice and water flow. Meltwater accumulates in lakes found in the cirques; these lakes are called "tarns." Other lakes occupy the depressions in the glacial troughs. When the glacier retreats, it uncovers a U-shaped valley instead of the V-shaped valley it had prior to the glaciation.

Continental ice sheets transform the landscape differently. When they retreat, they leave a multitude of lakes, many of which are round (kettle lakes). Others are elongated in the direction of the ice flow—such as the Finger Lakes in New York. By reconstructing the history of glaciation from these alterations made to the landscape, glaciologists and geologists have been able to contribute greatly to the understanding of the nature of Earth's climatic changes.

ICE FORMATION AND LOCATION

Glaciers are masses of ice that are produced where the summer temperatures fail to melt the snow that fell during the preceding winter. Over time, this snow is compressed by overlying layers of more recent snow, forcing out some of the air that exists around snow crystals. As this air escapes toward the surface; the density of this older snow steadily increases, until the snow turns into ice. At that point, the density of the ice varies from 0.85 to 0.91. Ice therefore floats on liquid water, whose density is 1. The transformation of snow into ice is slower in polar areas, where compression is the major mechanism at work, because air temperatures remain low all year round, producing very little, if any, meltwater to be refrozen. In temperate climates, ice forms more rapidly than in the polar areas because there are periods of melting when the temperature is above the freezing point of water. The meltwater soaks the snow and refreezes during the next colder period. This process is much faster than compression to achieve the density of ice.

Glaciers can exist at any latitude, even along the equator. They are present on every continent except Australia. The necessary condition for glaciers is that the air temperature remains low enough to prevent melting of the last winter snow. Because temperature decreases with increasing elevation, glaciers are found on high mountains or volcanoes. In Africa, Mounts Kenya, Kilimanjaro, and Ruwenzori have small glaciers (though they are retreating rapidly). In South America, there are many small tropical glaciers located in the Andes as well. In North America, Europe, and Asia, small glaciers dot the summits of high mountain ranges and volcanoes. Glaciers gain mass—called "accumulation"—by the deposition of snow in the highest elevation, called the "cirque"—a large amphitheater at the summit of glaciers.

Glaciers lose mass (a process called "ablation") by melting, sublimation, and calving. Melting takes place in a glacier where the temperature is greater than 0° Celsius. This can occur where the air temperature reaches this value or at the underside of the glacier, where friction of the ice on the ground beneath the glacier causes the temperature to increase, producing meltwater. This water lubricates the underside of the glacier, resulting in faster movement.

The second component of ablation is sublimation. Sublimation is the transformation of ice directly into water vapor, without an intermediate liquid stage. In Antarctica, sublimation is a major contributor to the ablation of ice because the ice sheet is affected by very strong winds which enhance the process.

The third form of ablation is calving. It involves the breaking of the end of the glacier when it reaches an ocean or a lake. calving produces icebergs—masses of ice that float because of the lower density of ice (0.85-0.91) compared to that of water. Calving is responsible for about 40 percent of ablation in Greenland and 80 to 90 percent in Antarctica.

The term "mass balance" refers to the difference in mass between accumulation and ablation. A glacier will grow longer if accumulation is greater than ablation over a period of time. This is called a "glacial advance." A glacier will get shorter if the amount of ice removed by ablation exceeds the amount of snow that accumulates in the coldest part, the cirque. When glaciers become shorter they are said to retreat. This does not mean that the ice stops moving downward from the cirque to the lower end of the glacier, called the "front." Glacial retreat indicates only that the front of the glacier will be found closer to the cirque.

Today, most glaciers in the world are retreating. Although there are a few exceptions where glaciers are advancing, the worldwide trend is a steady retreat in response to the general raising of Earth's temperatures.

GLACIOLOGY AND PALEOCLIMATES

The science that studies the cryosphere is called "glaciology." One of the first important glaciologists was Louis Agassiz, a native of Switzerland, who studied the glaciers of the Alps in the nineteenth century. Building on Agassiz's work, modern glaciologists discovered that glaciers not only are a good source of information about the global impact of recent environmental changes but also provide valuable data about the long history of climate change on this planet. Modern study of ice cores, cylinders of ice retrieved from glaciers, has shown that ice records the temperatures of Earth's atmosphere at the time the snow fell. The paleotemperatures are inferred from the composition of the water that makes up the ice.

In nature, water is made of two atoms of hydrogen and one atom of oxygen; however, oxygen has three isotopes (elements that have the same number of protons but different numbers of neutrons). The lightest and most abundant of the three is ^{16}O. ^{18}O is the heaviest but exists in much smaller quantities. Higher Earth temperatures make it easier for the heavier ^{18}O-containing water molecules to evaporate, resulting in snow—and therefore glacial ice—that has an increased proportion of it. During glaciations, the lower temperatures lead to a depletion of ^{18}O in the ice of glaciers. The same principle is applied to ice when two isotopes of hydrogen are measured. Because the ice sheets of Antarctica and Greenland are very thick, ice cores obtained by drilling into these ice sheets are very long. They therefore provide a very long record of the climates of the past (known as "paleoclimates"). Based on the cores retrieved so far, scientists have been able to identify a sequence of glacial and interglacial cycles that covers the last 800,000 years.

Earth has had many ice ages in its 4.5-billion-year history. Most recently, beginning about one million years ago, the "Great Ice Age" occurred during a time period called the "Pleistocene." This glacial period formed an ice sheet in North America, Northern Europe, Northern Asia, and Antarctica that expanded until it reached its maximum extent about twenty thousand years ago.

The Pleistocene Glaciation was not uniformly cold; short interglacial periods of warming occurred several times. Finally, about ten thousand years ago, the warming trend continued, melting the ice sheets and uncovering the northern continents. The northern part of Canada became free of ice about six thousand years ago. The mountain glaciers attached to the high mountain ranges of the American West are remnants of the Pleistocene period. In more recent centuries, Earth experienced a shorter period of cold temperatures, called the Little Ice Age, beginning about 1650 and ending approximately in 1850, during which the Earth cooled by about 1° Celsius. (The term "Little Ice Age" is used differently by different writers. Many use it to refer to the climate cooling from about 1300 to 1850, while others use it for the latter half of that interval, when cooling was greatest, beginning around 1550 or 1600.) The mountain glaciers that advanced during this period are currently retreating in response to today's higher temperatures.

GLACIERS IN CONTEXT

At their present rate of melting and retreating, glaciers are having a major impact on the populations living in their vicinity. The increase in meltwater can increase the production of hydroelectricity for a short while but at the same time may impact the population's well-being in the very near future by decreasing the amount of available water for irrigation or electricity production when the glaciers will have completely disappeared. Monitoring glacier changes and drawing scientific conclusions about their retreat is not something new. As early as 1894 scientists began cataloging glaciers and their changes. These findings were published by the World Glacier Monitoring Service. Maximum extents of glaciers were computed by using the position of their terminal moraines, and their volumes were estimated by measuring the height of their lower end since it corresponds to the height of ice that used to occupy the glacial trough.

In the 1970s, during the International Hydrological Decade declared by United Nations Educational, Scientific, and Cultural Organization (UNESCO), the Temporal Technical Secretariat for the World Glacier Inventory was created and began making a comprehensive inventory of more than 100,000 glaciers worldwide. Since then, with the help of satellite instruments, it has been determined that there are almost 200,000 glaciers, thousands of whose outlines, retreats, and advances are readily mappable. These measurements allow scientists to rapidly assess the impact of the warming of Earth on the cryosphere, and their study has proven a valuable tool in monitoring their reaction to the warming of Earth's atmosphere.

The work of the Intergovernmental Panel on Climate Change (IPCC) and the research conducted for the Fourth International Polar Year (2007-2008) organized by the International Council for Science (ICS) in conjunction with the World Meteorological Organization focused on understanding the extremely complex relationships between glaciers and climates. One of the goals of the United States National Committee for the International Polar Year was the creation of a network of observation platforms to monitor glaciers in order to provide reliable data by which scientists can be able to assess the impact of global warming both on the glaciers themselves and upon the global ecosystem of which they are an essential part.

—Denyse Lemaire and David Kasserman

Further Reading

Davies, Bethan. "Mapping the World's Glaciers." *Antarcticglaciers.org*, 25 Nov. 2014. Accessed 23 Mar. 2014.

Gornitz, Vivien. *Vanishing Ice: Glaciers, Ice Sheets and Rising Seas.* Columbia UP, 2019.

Hambrey, Michael, and Jurg Alean. *Glaciers.* Cambridge UP, 2006.

Hubbard, Bryn, and Neil Glasser. *Field Techniques in Glaciology and Glacial Geomorphology.* John Wiley & Sons, 2005.

Krüger, Tobias. *Discovering the Ice Ages: International Reception and Consequences for a Historical Understanding of Climate.* Brill, 2013.

Pfeffer, W. T., A. A. Arendt, A. Bliss, et al. "The Randolph Glacier Inventory: A Globally Complete Inventory of Glaciers." *Journal of Glaciology*, vol. 60, no. 221, 2014, pp. 537-52.

Trewby, Mary. *Antarctica: An Encyclopedia from Abbott Ice Shelf to Zooplankton.* Firefly, 2002.

GONDWANALAND AND LAURASIA

Fields of Study: Geology; Morphology; Petrology; Paleontology

ABSTRACT

Earth scientists theorize that the present-day continents were produced from the division of a supercontinent, called "Pangaea," into two gigantic landmasses, "Gondwanaland" and "Laurasia," which continued to fragment. The continents moved into their present locations in accord with a phenomenon known as continental drift, which scientists believe is still taking place.

KEY CONCEPTS

basalt: a hard, dense volcanic rock

Carboniferous period: the fifth of the six periods in the Paleozoic era; it preceded the Permian period

craton: the crystalline portion of a continent; it may have a sedimentary rock veneer

fathom: a unit of length equal to 1.8 meters, used principally in the measurement of marine depths

geomorphology: the study of the origin of landscapes

isobath: the contour lines of continental slopes

Paleozoic era: the era immediately before the Mesozoic era; it included the Cambrian, Ordovician, Silurian, Devonian, Carboniferous, and Permian periods

placer: a deposit of sand or gravel containing eroded particles of valuable minerals

trough: a long, narrow depression, as between waves or ridges

FORMATION OF THE CONTINENTS

The existence of an ancient supercontinent was first postulated by Austrian geologist Edward Suess in 1885. He called this great continent "Gondwanaland," after a region in India inhabited by the Gonds, an aboriginal tribe. Whereas Suess proposed that much of Gondwanaland had sunk, in 1929 German geophysicist Alfred Lothar Wegener suggested that a supercontinent that he called "Pangaea" had broken up and that the individual continents had drifted to their present positions. In 1937, South African geologist Alexander Du Toit envisioned two primordial continents: Laurasia in the north and

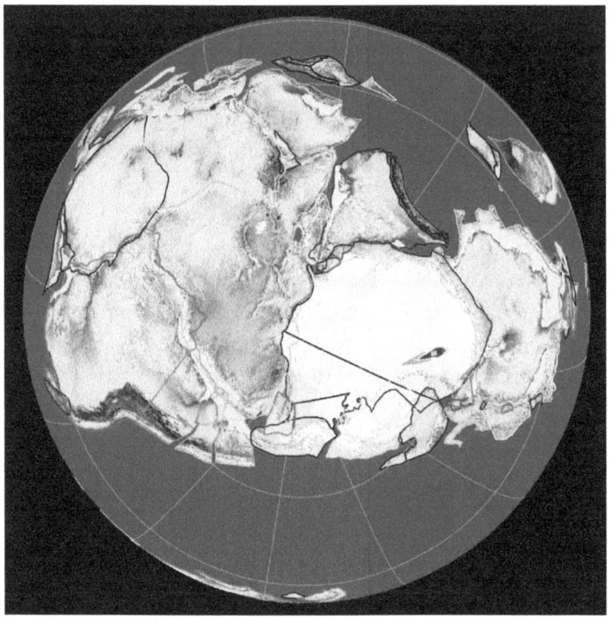

Gondwanaland as it was 420 million years ago; the view centered on the south pole. Image by Fama Clamosa, CC BY-SA 4.0, via Wikimedia Commons.

Gondwanaland in the south; subsequently, both parts broke again to form the present continents. English geologists Edward Bullard, J. E. Everett, and A. G. Smith's computerized assembly of the ancient continents in 1965 seems to verify Du Toit's continental arrangement.

The standard explanation of the formation of the continents begins 300 million years ago, at the end of the Carboniferous and start of the Permian geological periods, with the formation of Pangaea, a V-shaped supercontinent. The northern extension of Pangaea, called "Laurasia," straddled the equator and included North America, Europe, and Asia; the southern arm of Pangaea, called Gondwanaland and comprising South America, Africa, India, Australia, and Antarctica, was positioned so that its northern parts were in the tropical latitudes and its southern parts were beneath the polar ice cap. Although the Tethys Sea lay between Laurasia and Gondwanaland, they were linked from northwest Africa to North America and southern Europe.

The total area of Pangaea, when measured down to the 1,000-fathom isobath, seems to have been 200,000 square kilometers, or roughly 40 percent of Earth's surface. When the future continents were still part of Pangaea, they were to the south and east of their present locations. If New York had been in existence at the time, it would have been on the equator and at longitude 10 degrees east instead of 74 degrees west. Spain would have been near its present longitude, but it, too, would have been on the equator. Japan would have been in the Arctic, and India and Australia would have bordered the Antarctic.

This loose grouping of partly linked continents remained constant for the next 150 million years, although a few changes were slowly occurring. During this period, Gondwanaland gradually drifted northward. In addition, Gondwanaland, which had formed a huge basin-like area, was slowly being eroded, and the debris was deposited into the basin itself or into the long troughs that surrounded Pangaea. Toward the end of this period, this enormous basin was transformed into a collection of separate basins.

This scene was violently disrupted approximately 160 million years ago when gigantic floods of basaltic lava spread out on all of the southern continents except for South America, where lava flows did not occur until 40 million years later. As a result of the lava flow, the cracks from which the lava was released sank, forming basins. When the water in these basins evaporated, it left behind salt deposits that now border most of the southern continents.

The predecessors of the present deep oceans were formed between India and Somaliland as the areas sank even lower. Similar seas spread along the coast of India some 10 million years later, separating it from Australia. The seas reached southwest Africa roughly 120 million years ago, the Congo 110 million years ago, and Nigeria 105 million years ago. While the seas were spreading, the Benue Trough, which cuts across the bulge of Africa from Nigeria through Algeria, was slowly being filled. Instead of driving a great distance down the trough, however, the seas moved westward, where they merged with marginal seas that had been moving eastward between the bulge of Africa and northern Brazil. Thus, when the seas finally met, 92 million years ago, Africa became separated from South America. The Indian Ocean opened approximately 160 million years ago, and the South Atlantic Ocean opened roughly

120 million years ago, but neither ocean really began to widen until about 100 million years ago.

BREAKUP OF GONDWANALAND

Between 100 and 80 million years ago, Gondwanaland began to break up. South America rotated away from Africa before it drifted westward to its present position. At approximately the same time, India rotated away from Africa, moved northward, and collided with Asia. The Himalayan Mountains are a direct result of this collision. Before finally separating into their present positions during the past 50 to 60 million years, Australia and Antarctica drifted away from Africa. After rotating slightly, Africa moved northward to encroach upon Europe.

Laurasia had been lying lower than Gondwanaland at the beginning of this 300-million-year period, with the exception of the Appalachian-Caledonian mountains, the Russian Urals, and the eastern Siberian mountains, which all gradually wore down during the next 50 million years. During the next 150 million years, the seas that covered most of North America slowly receded to the south and west. Shallow seas also covered Europe west of the Urals. These seas, which spread northward from the Tethys, shifted their location periodically during the next 200 million years. A large portion of northeast Asia was covered with marine sedimentary troughs. Some of these troughs cut China off from the rest of Asia at the beginning of this period. Shallow seas spread out from these troughs over parts of China and western Siberia. The first strong geological activity occurred in Siberia 200 million years ago, when flood basalts up to 2.5 kilometers thick spread out over 500,000 square kilometers.

Laurasia and Gondwanaland were formed after the division of the supercontinent Pangaea. North America separated from Africa more than 200 million years ago, after the eruption of some volcanoes in eastern North America and Morocco. The proto-Central Atlantic was formed as a result of this rupture.

The main framework of fractures that would eventually form the North Atlantic was already in place 300 million years ago. Actual separation, however, began farther south. Even though the northern continents were linked to Gondwanaland at the beginning of this period, North America separated from Africa more than 200 million years ago, after the eruption of some volcanoes in eastern North America and Morocco. The proto-Central Atlantic was formed as a result of this rupture. Although Europe was still connected to North America after this movement occurred, a shallow sea was formed in which a layer of sediments between 3- and 4-kilometers thick was formed; traces of this sediment can still be found along the Atlantic continental shelf of North America. The Atlantic Ocean widened to one-quarter of its present width and connected to the Labrador Sea some 120 million years ago, separating Greenland and Canada. While the Atlantic continued to expand some 70 to 80 million years ago, the opening began between Europe and Greenland after the eruption of flood basalts.

STUDY OF CONTINENT FORMATION

Scientists have employed a wide variety of methods to study the formation of the continents. Glaciation has long been used both to determine the original fracturing of the great landmasses and to estimate the rate of their separation. The close grouping of the glaciated areas provides some of the most convincing evidence for continental drift. Glacial scratches on rocks found in Gondwanaland point to the movement of ice from areas that are now submerged under the ocean. In addition, some materials deposited in the glacial drift, such as Brazilian diamonds, are completely foreign in the lands where they now repose and, therefore, provide proof for continental drift.

The study of terrestrial vertebrates also provides clues to the existence of Pangaea and Gondwanaland. After the land bridge theory fell into disrepute, contiguity of the continents early in Earth's history seemed to be the only way to explain the fact that fauna is comparatively homogeneous throughout the world. The most convincing evidence for continental drift from terrestrial vertebrates is provided by an early Permian reptile called "Mesosaurus," whose remains are found only in South America and South Africa. Since this reptile was equipped for swimming only in shallow freshwater, it probably could not have swum the distance between the two continents.

The evolution of plants also supplies evidence for the existence of a supercontinent. The presence of araucarian pines in South America, the Falklands, Australia, and the South Pacific has led scientists to the inference of former land connections. The distribution of the *Glossopteris* land plants in the Lower Gondwana Formations is larger than that of the glacial deposits. Evidently, the cold climate stimulated the growth of this unique plant. These plants could not have developed independently of one another in widely separated continents but could have thrived in a supercontinent covered by enormous ice sheets.

Modern technology has provided scientists with more advanced methods of exploring the possibility of a supercontinent. Gathering paleomagnetic data is one such method. Magnetic particles in the seafloor rocks recorded the direction of Earth's magnetic field when the rock hardened. As the ocean floor formed, the direction of the field reversed itself from time to time. By paleomagnetically determining the pole positions for sedimentary and igneous rock in different continents, scientists have been able to determine the relative positions of the continents for any period of time. Some of the data that they have acquired in this fashion support the existence of Pangaea.

Examination of Precambrian cratons also provides invaluable information regarding continental drift. Drift can be demonstrated by matching pieces of cratons across facing continents. The cratons between Guyana and West Africa, for example, fit very well together and are well documented.

The complicated process of reassembling the continents has been greatly facilitated by computers. In 1965, following a 500-fathom isobath, Bullard, Everett, and Smith used a computer to fit the continents together into their original configuration. Even though computerized models have brought an air of mathematical precision to the task of reconstructing Pangaea, there is still considerable disagreement about certain features. For example, did Madagascar lie against east or southeast Africa when they were part of Pangaea?

SIGNIFICANCE

The study of ancient landscapes is of great value to the economy. The same techniques that geomorphologists devised to comprehend fossil landforms and deposits can also be brought to bear to locate ore deposits. Geomorphologists have also applied their methods to detect placers and to discover petroleum. Through the use of structure contour maps, the search for an "oil pool" almost always involves geomorphology applied to problems of fluid entrapment.

Knowledge of ancient landscapes also aids scientists in predicting the occurrence of natural catastrophes. Scientists have mapped out a worldwide system of ocean trenches and ocean ridges, which were probably made when the oceans were formed and the continents moved to their present locations. They have observed that many deep earthquakes occur beneath oceanic trenches and that volcanic activity is concentrated along the ocean ridges.

Finally, scientists have used the theory of continental drift to predict what Earth's surface will look like millions of years from now. They predict that the California coast may tear away from the mainland and drift north toward Alaska, that Africa and South America may move even farther apart, and that Australia may collide with Asia.

—*Alan Brown*

Further Reading

Ceuleneer, Georges. "Earth Science: Mantle Mapped in the Desert." *Nature*, vol. 432, 2004, pp. 156-57.

Condie, Kent C. *Earth as an Evolving Planetary System.* 2nd ed., Academic Press, 2011.

Nield, Ted. *Supercontinent: Ten Billion Years in the Life of Our Planet.* Granta Books, 2012.

Prothero, D. R., and Robert H. Dott. *Evolution of the Earth.* 8th ed., McGraw-Hill, 2009.

Rogers, John J. W., and M. Santosh. *Continents and Supercontinents.* Oxford UP, 2004.

Sengör, A.M. Celâl, and Saniye Atayman. *The Permian Extinction and the Tethys: An Exercise in Global Geology.* Geological Society of America, 2009.

Wegener, Alfred. *The Origins of Continents and Oceans.* Trans. By John Biram from 4th rev. German ed., Dover, 1966.

Wicander, Reed, and James S. Monroe. *Historical Geology.* Brooks/Cole, Cengage Learning, 2010.

Granitic Rocks

Fields of Study: Geology; Geochemistry; Geomorphology; Analytical Chemistry

ABSTRACT

Granitic rocks are coarse-grained igneous rocks consisting mainly of quartz, sodic plagioclase, and alkali feldspar, with various accessory minerals. These rock types occur primarily as large intrusive bodies that have solidified from magma at great depths. Granitic rocks can also occur to a lesser degree as a result of metamorphism, a process referred to as "granitization."

KEY CONCEPTS

aphanitic: a textural term that applies to an igneous rock composed of crystals that are microscopic in size

crystal: a solid made up of a regular periodic arrangement of atoms

crystallization: the formation and growth of a crystalline solid from a liquid or gas

granitization: the process of converting rock into granite; it is thought to occur when hot, ion-rich fluids migrate through a rock and chemically alter its composition

isotopes: atoms of the same element with identical numbers of protons but different numbers of neutrons, thus giving them a different mass

magma: a body of molten rock typically found at great depths, including any dissolved gases and crystals

migmatite: a rock exhibiting both igneous and metamorphic characteristics, which forms when light-colored silicate minerals melt and crystallize, while the dark silicate minerals remain solid

phaneritic: a textural term that applies to an igneous rock composed of crystals that are macroscopic in size, ranging from about 1 to over 5 millimeters in diameter

pluton: a structure that results from the emplacement and crystallization of magma beneath the surface of the earth

porphyritic: a texture characteristic of an igneous rock in which macroscopic crystals are embedded in a fine phaneritic or aphanitic matrix

GRANITIC ROCK OCCURRENCE

The term "granitic rocks" generally refers to the whole range of plutonic rocks that contain at least 10 percent quartz. They are the main component of continental shields and also occur as great compound batholiths in folded geosynclinal belts. Granitic rocks are so widespread, and their occurrence and relation to the tectonic environment are so varied, that generalizations often obscure their complexity. Major granitic complexes are, in general, a continental phenomenon occurring in the form of batholiths and migmatite complexes.

When large masses of magma solidify deep below the ground surface, they form igneous rocks that exhibit a coarse-grained texture described as phaneritic. These rocks have the appearance of being composed of a mass of intergrown crystals large enough to be identified with the unaided eye. A large mass of magma situated at depth may require tens of thousands or even millions of years to solidify. Because phaneritic rocks form deep within the crust, their exposure at ground surface reflects regional uplift and erosion, which has removed the overlying rocks that once surrounded the now-solidified magma chamber.

Along continental margins, belts of granitic rocks developed as batholiths composed of hundreds of individual plutons. The formation of batholithic volumes of granitic magma generally appears to require continental settings. Some of the more prominent batholiths in North America are the Coast Range, Boulder-Idaho, Sierra Nevada, and Baja California. The largest are more than 1,500 kilometers long and 200 kilometers wide, and have a composite structure. The Sierra Nevada, for example, is composed of about 200 plutons separated by many smaller plutons, some only a few kilometers wide.

GRANITIC ROCK CLASSIFICATION

As with other rock types, granitic rocks are classified on the basis of both mineral composition and fabric or texture. The mineral makeup of an igneous rock is ultimately determined by the chemical composition of the magma from which it crystallized. Feldspar-bearing phaneritic rocks containing conspicuous quartz (greater than 10 percent in total volume) in addition to large amounts of feldspar can be designated as granitic rocks. This nonspecific term is useful where the type of feldspar is not recognizable because of alteration or weathering, for purposes of quick reconnaissance field studies, or for general discussion.

Granitic rocks consist of two general groups of minerals: essential minerals and accessory minerals.

Essential minerals are those required for the rock to be assigned a specific name on a classification scheme. Essential minerals in most granitic rocks are quartz, sodic plagioclase with 30 to 50 percent calcic plagioclase, and potassium-rich alkali feldspars (either orthoclase or microcline). Accessory minerals include biotite, muscovite, hornblende, and pyroxene.

When an initial phase of slow cooling and crystallization at great depths is followed by more rapid cooling at shallower depths or at the surface, porphyritic texture develops, as is evident in the presence of large crystals enveloped in a finer-grained matrix or groundmass. The presence of porphyritic texture is evidence that crystallization occurs over a range of temperatures, and magmas are commonly emplaced or erupted as mixtures of liquid and early formed crystals.

Classification of granitic rocks can be based either on the bulk chemical composition or on the mineral composition. Because oxygen is ubiquitous in rocks, chemical compositions are given as the weight percentages of oxides rather than individual elements, whereas mineral composition units are in approximate percentages in total volume. The mineral composition of granitic rocks provides a reliable basis for

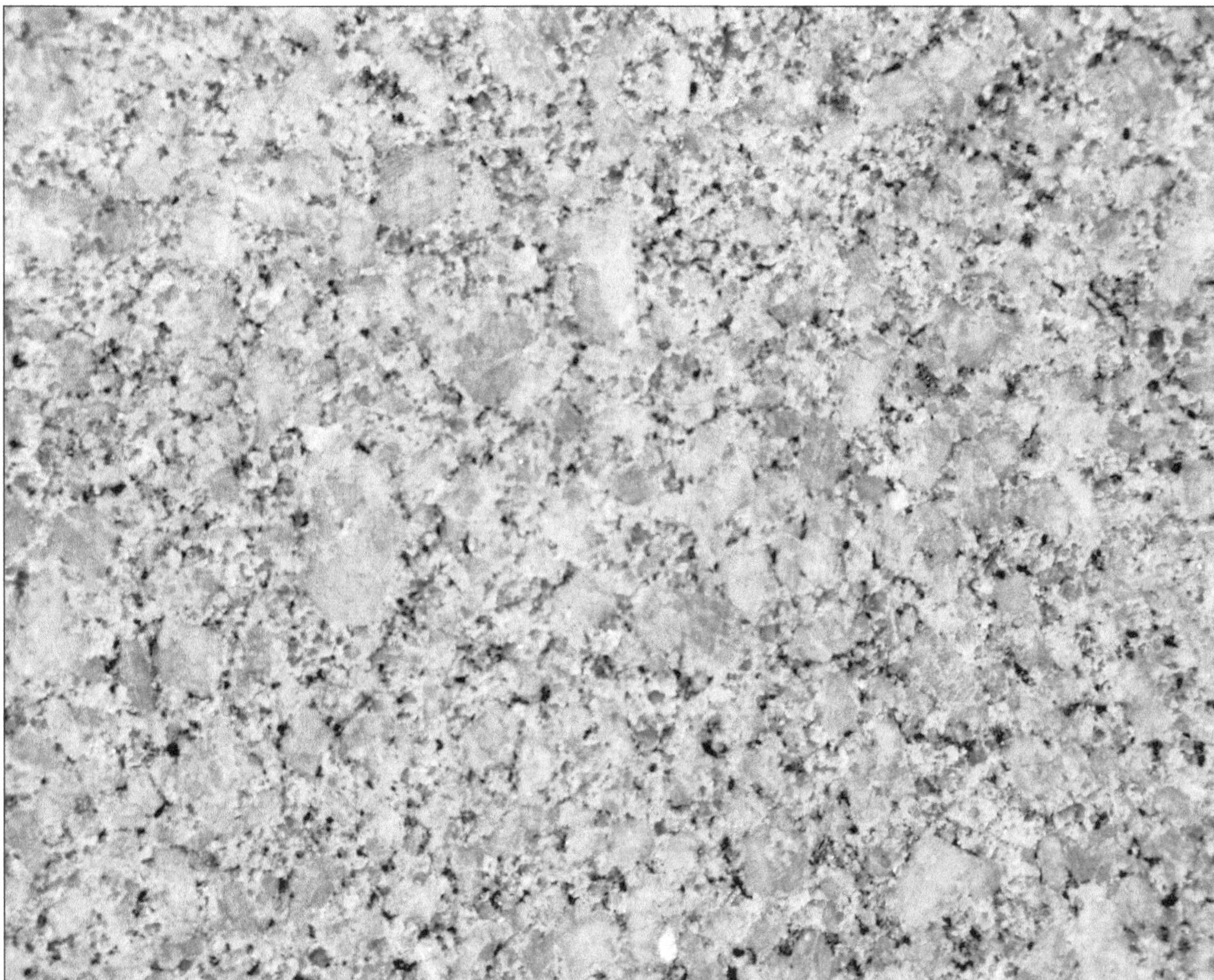

Granite containing potassium feldspar, plagioclase feldspar, quartz, and biotite and/or amphibole. Photo by I. Friman, CC BY-SA 3.0, via Wikimedia Commons.

classification. Specific granitic rock types are defined on the basis of the presence of quartz and the ratio of potassium feldspar and plagioclase. The presence of other accessory minerals may be indicated in the form of a modifier (such as biotite granite).

GRANITIC ROCK GROUPS

Granitic rocks include true granites, tonalite, granodiorite, quartz monzonite granite, soda granite, and vein rocks of pegmatite and aplite. The mineralogy of these varies, and the distinction between different granitic rocks can be gradational, but all are defined by the presence of quartz. Tonalite has mostly plagioclase feldspar and very little potassium feldspar. Granodiorite is composed predominantly of plagioclase feldspar, with subordinate potassium feldspar and biotite, hornblende, or both as accessory minerals. Quartz monzonite is composed of approximately equal amounts of potassium and plagioclase feldspars, with biotite, hornblende, or both as accessory minerals. Granite is composed predominantly of potassium feldspar with subordinate plagioclase feldspar and biotite alone or with hornblende or muscovite as accessory minerals. Soda granite is composed predominantly of very sodium-rich plagioclase feldspar, with small amounts of sodium-bearing pyroxene or amphibole.

Pegmatite is a very coarse-grained complex rock in terms of mineralogy. Structurally, pegmatites occur as dikes associated with large plutonic rock masses. Dikes are tabular-shaped intrusive features that cut through the surrounding rock. The large crystals are inferred to reflect crystallization in a water-rich environment. Most pegmatites are simple in terms of mineralogy, consisting primarily of quartz and alkali feldspar, with lesser amounts of muscovite, tourmaline, and garnet; they are referred to as simple pegmatites. Other pegmatites can be very complex and contain other elements that slowly crystallize from residual, deeply seated magma bodies. High concentrations of these residual elements can result in the formation of minerals such as topaz, beryl, and minerals rich in rare-earth elements, as well as quartz and feldspar. Aplite is a very fine-grained, light-colored granitic rock that also occurs as a dike and consists of quartz, albite, potassium feldspar, and muscovite, with almandine garnet as an occasional accessory mineral.

Other rocks that are occasionally grouped with granitic rocks are migmatites. Migmatites, or mixed rocks, are heterogeneous granitic rocks that, on a large scale, occur within regions of high-grade metamorphism or as broad migmatite-rich zones bordering major plutons. Migmatites appear as alternating light and dark bands. The light-colored bands are broadly granitic in mineralogy and chemistry, while the darker bands are clearly metamorphic. Migmatites are believed to reflect the partial melting of metamorphic rocks in which the portions rich in quartz and feldspar melt first.

Granitic rocks are classified by the relative proportions of quartz, plagioclase, and potassium feldspar, and can also be classified by their aluminum content. Peraluminous granites have excess aluminum present after the formation of feldspar, and thus contain aluminum-rich minerals like sillimanite or garnet. Peralkaline granites run out of aluminum before using up their sodium and potassium and have alkali-rich minerals like sodium amphibole. Granitic rocks can further be divided into two groups, S-types and I-types, according to whether they were derived from predominantly sedimentary or igneous sources, respectively. Enormous volumes of I-type granites, which constitute most plutons, occur along continental margins overriding subducting oceanic material.

INCLUSIONS

A common feature of granitic rocks, notably in granodiorite to dioritic plutons, is the presence of inclusions or rock fragments that differ in fabric and/or composition from the main pluton itself. The term "inclusions" indicates that they originated in different ways. Foreign rock inclusions, called "xenoliths," include blocks of wall rocks that have been mechanically incorporated into the magma body. This process is referred to as stoping. Some mafic inclusions are early-formed crystals precipitated from the magma itself after segregation along the margin of the pluton, which cools first. These inclusions are called "antoliths." Other inclusions may reflect clots of solidified mantle-derived magma that ascended into the granitic source region, or residual material that accompanied the magma during its ascent.

PETROGRAPHIC STUDY

Granitic magmas can be derived from a number of sources, notably the melting of continental crust, the melting of subducted oceanic crust or mantle, and differentiation. The problem facing the geologist is to decide which tectonic environment led to the formation of the granite. To accomplish this objective, both petrographic and geochemical information is used.

The standard method of mineral identification and study of textural features and crystal relationships is the use of rock thin sections. A thin section is an oriented wafer-thin portion of rock 0.03 millimeter in thickness that is mounted on a glass slide. The rock thin section is so thin that even normally opaque minerals transmit light and reveal mineral content, abundance and association, grain size, structure, and texture. The thin section also provides a permanent record of a given rock that may be filed for future reference.

Thin sections are studied with the use of a petrographic microscope, which is a modification of the conventional compound microscope commonly used in laboratories. The modifications that render the petrographic microscope suitable to study the optical behavior of transparent crystalline substances are a rotating stage, an upper polarizer or lower polarizer, and a Bertrand lens, which together are used to survey the optical properties of the mineral over a wide range of angles at once. With a magnification that ranges from about 30 to 500 times, the petrographic microscope allows one to examine the optical behavior of transparent crystalline substances or, in this case, crystals that make up granitic rocks.

The study of granitic rocks may be greatly facilitated by various staining techniques of both hand specimens and thin sections. Staining is employed occasionally to distinguish potassium feldspar from plagioclase and quartz, the three main mineral constituents of granitic rocks. A flat surface on the rock is produced by sawing and then polishing. The rock surface is etched using hydrofluoric acid, an extremely hazardous material that requires great care in handling. This step is followed by a water rinse and immersion in a solution of sodium cobaltinitrate. The potassium feldspars will then turn bright yellow. After rinsing with water and covering the surface with rhodizonate reagent, the plagioclase becomes brick red in color. Staining techniques are available for other minerals, including certain accessory minerals such as cordierite, anorthoclase, and feldspathoids.

Measurement of the relative amounts of various mineral components of a rock is called "modal analysis." The relative area occupied by the individual minerals is estimated or measured on a flat surface (on a flat-sawed surface, on a flat outcrop surface, or in thin section) and then related to the relative volume. Caution must be used, because the relative area occupied by any mineral species on a particular planar surface is not always equal to the modal (volume) percentage of that mineral on the rock mass, especially if mineral grains are clumped or strongly aligned.

GEOCHEMICAL STUDY OF GRANITIC ROCKS

When a rock specimen is crushed to a homogeneous powder and chemically analyzed, the bulk chemical composition of the rock is derived. Chemical analyses are normally expressed as oxides of the respective elements because of the overwhelming abundance of oxygen. Analysis of granitic rocks shows them to be typically rich in silica potassium and sodium, with lesser amounts of basic oxides such as magnesium, iron, and calcium oxides. Magnesium, iron, and calcium oxides are present in higher abundance in basalts, which contain plagioclase feldspar, pyroxene, and olivine.

Isotopes such as those of strontium, oxygen, and lead can also be used as tools in evaluating granitic magma sources, such as a mantle origin and crustal melting of certain rock types. Some accessory minerals reflect the trace element content of the magma and thus the possible nature of their source. Much research is focused on chemical tracers. Tracers help distinguish the source region of a granitic magma, such as lower-crustal igneous or sedimentary rock or mantle material.

INDUSTRIAL APPLICATIONS

Granitic rocks have been used as dimension stones for many years. Dimension stones are blocks of rock with roughly even surfaces of specified shape and size used for the foundation and facing of expensive buildings. When crushed, granitic rocks can be used as aggregate in the cement and lime industry. In addition to these uses, granitic rocks are valued because of their geographic association with gold. Gold ores are found in close proximity to the contacts of the granitic bodies within both granitic rocks and surrounding rocks.

Pegmatites can also be very valuable. Simple pegmatites are exploited for large volumes of quartz and feldspar, used in the glass and ceramic industries. Complex pegmatites can also be a source of gem minerals, including tourmaline, beryl, topaz, and chrysoberyl. In spite of varied mineral composition, relatively small size, and unpredictable occurrence, pegmatites constitute the world's main source of high-grade feldspar, electrical-grade mica (used for supporting heater elements in toasters), certain metals (including beryllium, lithium, niobium, and tantalum), and some piezoelectric quartz, although most industrial quartz is now grown synthetically rather than mined.

—Stephen M. Testa

Further Reading

Chappell, B. W., and A. J. R. White. "Two Contrasting Granite Types: 25 Years Later." *Australian Journal of Earth Sciences*, vol. 48, 2001, pp. 489-99.

Decrée, Sophie, and Laurence Robb, editors. *Ore Deposits: Origin, Exploration, and Exploitation.* Wiley, 2019.

Grotzinger, John, and Thomas H. Jordan. *Understanding Earth.* 6th ed., W. H. Freeman, 2010.

Frost, B. Ronald, et al. "A Geochemical Classification for Granitic Rocks." *Journal of Petrology*, vol. 42, 2001, pp. 2033-48.

Klein, Cornelis, and Cornelius S. Hurlbut, Jr. *Manual of Mineralogy.* 23rd ed., John Wiley & Sons, 2008.

Ramo, O. T. *Granitic Systems.* Elsevier, 2005.

Stevenson, David S. *Granite Skyscrapers: How Rock Shaped Earth and Other Worlds.* Springer/Praxis, 2018.

Van Kranendonk, Martin, J. R. Hugh Smithies, and Vickie C. Bennett, editors. *Earth's Oldest Rocks.* Elsevier, 2007.

GUTENBERG DISCOVERS EARTH'S MANTLE-OUTER CORE BOUNDARY

Fields of Study: Geophysics; Seismology; Geology

ABSTRACT

Beno Gutenberg expanded the use of seismographs to the global scale when he discovered the boundary between the earth's outer core and the lower mantle.

KEY CONCEPTS

mantle: the internal layer in the structure of Earth between the outermost layer (the crust) and the outer layer of the core

P-waves: compression waves, like sound waves, presumed to be produced by direct impingement of rock masses against each other at the focus of the earthquake

S-waves: sheer waves, like side-to-side rocking, presumed to be produced by lateral slippage of rock masses against each other at the focus of the earthquake

seismograph: a device designed to record the physical waves produced by an earth tremor

surface waves: the effective movement of rock at the focus of the earthquake as transmitted to the surface

DISCOVERING THE MANTLE AND OUTER CORE BOUNDARY

Very little is known about the interior of Earth because of its inaccessibility, even to modern technology. Based on evidence gained through indirect and secondary research and experiments, scientists have been able to determine that the structure of Earth consists of layers. The uppermost layer, on or in which most life exists, is the crust. Scientists have determined the structure of the remainder of Earth's interior and the size and composition of the deeper layers through various methods. The original source of information that founded modern understanding of Earth's structure was seismology, the study of earthquakes and vibrations of Earth.

The pioneers in the field of seismology used needle-and-paper seismographs—instruments that measure and record vibrations—to study wave patterns. In 1893, one of these pioneers, John Milne, perfected a clockwork-powered seismograph that produced a record of vibrations on light-sensitive film. The Milne seismograph was a self-recording instrument capable of preserving a record of ground movement. Milne discovered that the vibrations from a distant earthquake arrived in a series of separate vibrations traveling at different speeds. The greater the separation between the waves of different types, the farther away the earthquake. The method was identical to that used to calculate the distance of a storm from the time between the arrival of the lightning and the sound of the thunder.

The Milne seismograph allowed for the identification of three different kinds of waves. The fastest waves are those of compression, or P-waves, which move through the rock in a manner similar to sound waves. Another type are shear waves, or S-waves, which move by the sideways motion of the particle. The third type, surface waves, move along boundaries, such as the boundary between the rock and the air or the water. Surface waves are a varied group that behave in ways similar to waves in the sea.

Milne's early seismograph allowed for the base recording of earthquake waves. Later improvements on his model were computerized, with the waves recorded in binary code rather than on paper. Although the later models were more precise, recording new waves that aroused new investigations, Milne's early seismograph allowed researchers to make great discoveries. Later research in the field of seismology used the records from the seismograph—called "seismograms"—to learn the composition of Earth by comparing density curves to wave curves. Their research revealed that the innermost section of Earth is composed possibly of pure iron. Surrounding the inner core is the molten outer core, with a radius of approximately 3,500 kilometers (2,175 miles) and composed of an iron alloy with a metal lighter than nickel. The mantle of Earth is the area between the outer core and the crust. Probably the most active area in the layers of Earth, the mantle contains large amounts of the mineral olivine. The mantle is approximately 3,000 kilometers (1,864 miles) thick and consists of two parts, with the lower mantle forming the boundary between the upper mantle and the core. This boundary between the mantle and the outer core was discovered by Beno Gutenberg.

Gutenberg's method of research was first used by Andrija Mohorovičić, a Yugoslavian meteorologist who was one of the pioneers in the science of seismology. The key event that spurred Mohorovičić's inspiration was a minor earthquake in 1909 in Zagreb, Yugoslavia (now in Croatia). When Mohorovičić examined the seismogram of the earthquake, he found two primary (P) and two secondary (S) waves recorded for each tremor. The separate groups of P and S waves appeared to be traveling at different velocities. Because the type of material as well as its density changes the rate, and sometimes the direction, of the wave's speed, the seismogram indicated the presence of a layer of material under Earth's outer crust that was dense enough to alter the velocity of the second group of P- and S-waves. Because the second wave group reached the recording stations before the first wave group, Mohorovičić deduced that the boundary—now called the "Mohorovičić Discontinuity"—was denser than the crust. Along with the discovery of the boundary between Earth's outer crust and the upper mantle, Mohorovičić contributed a new application of seismology in exploring the interior structure of the planet.

Gutenberg, a graduate student in Germany, was encouraged to begin his own studies in seismology two years after the discovery of the Mohorovičić Discontinuity. Gutenberg's area of concentration was the mysterious "shadow zone" of seismology. For some unknown reason, P-waves disappeared when passing through an area 4,400 kilometers (2,734 miles) wide on the side of the planet opposite to the focus, or epicenter, of the earthquake. Gutenberg began a mathematical investigation to explain the temporary disappearance of P-waves. The most likely answer was suggested by the work of Richard Dixon Oldham, a geologist, and Emil Wiechert, a geophysicist. Their independently formed theories proposed that the center of the planet is composed of a large, dense, and perhaps partially molten core.

Assuming that Oldham and Wiechert were correct, Gutenberg made mathematical models of the effects a dense core would have on P-waves. He positioned his hypothetical core at various depths in the planet and then calculated the course and behavior of P-waves in each. After comparing his models to actual seismograph readings, he discovered one that confirmed his work. The model that eventually matched the real graphs was based on a core 2,900 kilometers (1,802 miles) below the surface of the planet. The shadow zone was the boundary between the lower mantle and the outer core. Later research that continued with Gutenberg's original outline altered the accuracy of his placement only slightly and revealed that the outer edge of the core must be molten. Gutenberg's discovery of the boundary between the mantle and the outer core is called the "Gutenberg Discontinuity".

SIGNIFICANCE OF GUTENBERG'S DISCOVERY

Gutenberg's confirmation of the presence of Earth's core and his discovery of the boundary between the outer core and the lower mantle in 1913 solved the old mystery of the shadow zone and raised new questions regarding the composition of these areas. In addition to the significance of the discovery of the boundary between Earth's lower mantle and outer core, the research techniques that Gutenberg applied opened new avenues for tackling old problems and also provided a built-in continuation of the original research.

As a sideline to Gutenberg's major discovery, later research found that the Gutenberg Discontinuity also marks the lowest edges of the continental plates. Through the use of seismographs, scientists can apply the speed of earthquake waves to learn the composition of a material through which a random wave will pass. Because he found that the waves traveled more slowly in the lower part of the mantle, Gutenberg postulated that the lower mantle is softer than the upper mantle. This theory shed new light on the study of plate tectonics, for if the lower mantle is the layer in which the plates extend the deepest and it is partially molten, the movement of plates is more easily explained.

With the use of seismic waves, Gutenberg and Mohorovičić revealed the interior structure of Earth. The improved sensitivity of later technology revealed a previously unrecorded set of P-waves in the shadow zone of the planet. Based on Gutenberg's findings, the understanding that seismologists had of the core did not coincide with this energy rebound. Using the later, more sensitive records for her research, Inge Lehmann solved the mystery by proposing the existence of an inner core that could reflect the waves. Through the use of seismic velocities, K. E. Bullen proved that the Earth's inner core is solid. Despite Bullen's ingenious argument, however, one important wave pattern is missing from the records. If the inner core is solid, there should be an extra wave recorded, but it has not appeared in seismographs. Although the interior of Earth is solid, the outer core is molten. The use of densities, as indicated by wave patterns, is essential in determining the composition of Earth's layers. The easier method of comparing infrared densities in the planet with known densities is inapplicable in this case because of the temperatures and pressures found in Earth that cannot be duplicated in laboratories. Thus scientists must compare observed seismic wave speeds and estimated densities.

Gutenberg later used the seismograph again while working with his colleague Charles Francis Richter at the California Institute of Technology. A universal way of judging the absolute size of earthquakes was needed, as well as a scale on which to measure earthquakes. Richter developed a way to define the amplitudes of the magnitudes of earthquakes. One forms magnitudes by plotting a curve based on the amplitudes of ground motion against the distance of seismograph recording stations from the epicenter, or heart, of the earthquake. Richter used the logarithms of the amplitudes of the earthquakes as they could be determined to the nearest tenth magnitude. Along with Gutenberg, he developed three scales by which to judge earthquakes: one for near earthquakes, one for distant shallow earthquakes, and one for deep-focus earthquakes.

—Earl G. Hoover

Further Reading

Havskov, Jens, and Gerardo Alguacil. *Instrumentation in Earthquakes Seismology.* 2nd ed., Springer, 2016.

Igel, Heiner. *Computational Seismology: A Practical Introduction.* Oxford UP, 2017.

Lowrie, William. *Geophysics: A Very Short Introduction.* Oxford UP, 2018.

Stein, Seth, and Michael Wyesession. *An Introduction to Seismology, Earthquakes, and Earth Structure.* Wiley, 2013.

Turcotte, Donald, and Gerald Schubert. *Geodynamics.* 3rd ed., Cambridge UP, 2014.

Yang, Wencai. *Reflection Seismology: Theory, Data Processing and Interpretation.* Elsevier, 2014.

H

Heezen and Ewing Discover the Midoceanic Ridge

Fields of Study: Geology; Oceanography; Marine Geology; Sonography

ABSTRACT

After leading an expedition to gather echograms of the ocean's floor, Maurice Ewing discovered the existence of the Midoceanic Ridge and postulated, with Bruce C. Heezen, the existence of a midoceanic rift.

KEY CONCEPTS

convection: a cyclic rise-and-fall movement of fluids caused by density changes with temperature, as material becomes less dense as it becomes hotter and rises from cooler, less dense material, then becomes denser as it loses heat energy and sinks through hotter, less dense material

rift valley: an elongated depression between two segments of continental mass that are being driven apart by convection currents in the underlying magma of the mantle layer

sonar: an underwater range-finding method that emits sharp sound pulses, or "pings," and uses the elapsed time for the echo to return as the means of calculating the distance to the reflecting surface; the principle on which the modern "fish finder" operates

subduction: downward movement of dense basaltic seafloor crust as it encounters a less dense continental mass and is forced to go beneath it and reenter the magma of the mantle layer

FINDING THE MIDOCEANIC RIDGE

From the blind creatures of sea, land, and air to the development of sonar and from a war that encompassed the world to seafloor mapping, modern science received the research necessary to prove the theory of continental drift and in the process to make great strides in filling in gaps and adding new information to the world of ideas. Alfred Wegener, the multifaceted scientist responsible for proposing the theory of continental drift, first began formulating his notion when he noticed that the coastlines of Africa and South America fit together as if they had once been parts of one huge continent. Preceded by at least five others, including Francis Bacon, in his observations on the jigsaw-like quality of continents, Wegener was the first to investigate the possibility fully. His research spanned the sciences and culminated in the publication *Die Entstehung der Kontinente und Ozeane* (1912; *The Origin of Continents and Oceans*, 1924). His hypothesis explained the mysteries of similar fossils found on opposite sides of the ocean and the similarity of geographic detail in parts of the world. Unfortunately, his hypothesis was unable to explain how the continents could be in motion, constantly breaking and rejoining.

The explanation Wegener lacked was found several years later during World War II. In 1912, spurred by the *Titanic* disaster, Reginald Aubrey Fessenden completed an off-again, on-again project aimed at creating an echo-detection device for seagoing vessels to be alerted to the presence of icebergs. About 1914, the former assistant to Thomas Alva Edison produced his invention that created sound waves under water. By listening to the echoes picked up by the underwater microphone, Fessenden could detect objects by changes in the echo in the same way many ocean creatures, and even some land creatures, had used blind vision for millennia. Fessenden's early device became known later as "sonar" when Paul Langevin improved on his concept and introduced it to the seafaring public for general use.

Sonar is an abbreviation for "sound navigation ranging"; it was used by ships to seek, hunt, or guard against the new terror of the seas, the Ger-

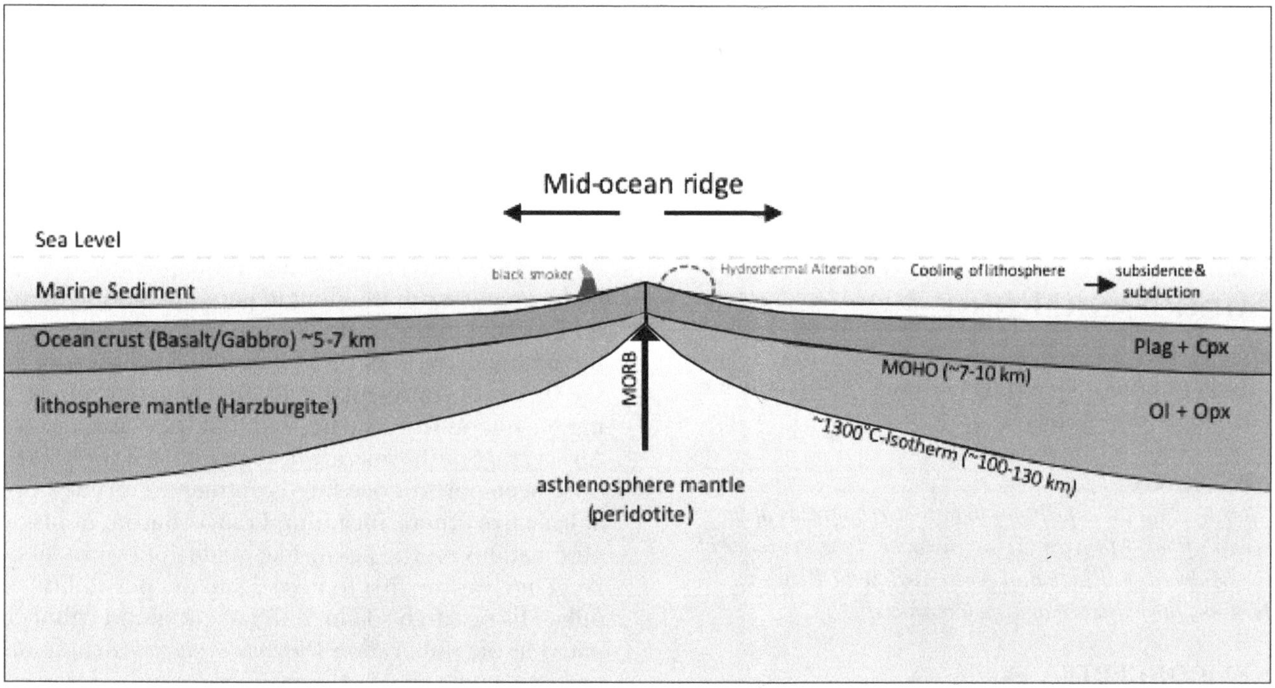

Cutaway view of midocean ridge. Image by 37ophiuchi BrucePL, CC BY-SA 4.0, via Wikimedia Commons.

man "u-boat" submarines. Sonar operates by the emission of high-pitched signals, followed by the interpretation of the echoes to indicate the presence of submarines, sea life, or anything else noteworthy below the surface. During the war, the Navy used a device similar in working principles to sonar, called the "Fathometer," to map the ocean's floor around Iwo Jima. Aboard the USS *Cape Johnson* was Harry Hammond Hess, a geologist from Princeton University. He realized that with the technology of echo sounding, the seafloor could be mapped at last. While Hess's eventual contribution to the exploration of the seafloor was more along the lines of data interpretation rather than actual physical data gathering, the task he made possible was carried out by a colleague and friend from Columbia University, Maurice Ewing.

Leading a group of scientists from Columbia University's Lamont Geological Observatory in Palisades, New York, Ewing began his journey in 1956 with the intent to map a string of undersea mountains called the "Mid-Atlantic Ridge." He found in the process, however, that simple mapping was next to impossible in the light of the discoveries made by his team. Sounding by echo with sonar and the Fathometer and utilizing data from oceanic earthquakes, they charted their progress, noting with interest that the underwater earthquakes reported by the seismologists Beno Gutenberg and Charles Richter occurred with increasing frequency along the ridge. Another exciting and inexplicable discovery was made during the process of dredging rock from the seafloor in the hope of finding ancient crustal rock from the formation of the earth. Instead of churning up chronological layers of rock with the newest at the top, the expedition found only relatively new rock formations; the oldest among the rocks found was approximately 150 million years old, and the others were even younger. Perhaps the greatest discovery of all—that of the Midoceanic Ridge—occurred after the actual data collection and during the interpretation of the echograms.

Columbia University graduate student Bruce Heezen, in collaboration with Marie Tharp, also from Columbia University, began mapping the ocean floor based on data from previous expeditions. Working in the Lamont Geological Observatory, they examined the echo-sounding passes, revealing more topography. Much of the topography was very rugged terrain pierced by land jutting upward in the form of cliffs, crags, and, more important, mountains. The increasing detail of their mapping soon revealed that the ridge was not a regular mountain range or system. The ridge was 1,829 to 3,048 meters high and 74,028 kilometers long, with a continuous V-shaped, groove-like valley, later known as a rift valley. Often compared to the laces of a baseball, the superlative mountain range crisscrossed the world.

Heezen expanded on Tharp's suggestion that the groove-shaped valley recorded by the echograms was a rift when it was compared to the median rift and the similar earthquake pattern found there. Although the then-current echograms—records of echo soundings—did not reveal the presence of a rift along the crest of the ridge, Heezen remained certain of his theory. He and Ewing developed a hypothesis on the existence of the rift and later suggested where the research team of the International Geophysical Year might look. By 1958, the Heezen-Ewing theory that the ridge contained a rift was proven correct. The work of Ewing and Heezen, which officially proposed the existence of the rift, sparked interest in further study of midoceanic ridges.

SIGNIFICANCE OF FINDING THE MIDOCEANIC RIDGE

Beginning in 1872, HMS *Challenger* began an expedition of discovery across the oceans. The first large-scale attempt to learn more about Earth's oceans resulted in massive amounts of data collected; it led to the precedent of large team efforts, combining the knowledge of several sciences to work toward one end. For various reasons, further group expeditions to the sea were put on hold for roughly half a century until, facilitated by war and technology, another group set out to discover the secrets beneath the sea. What had been previously impossible for HMS *Challenger* became feasible for the newest seagoing scientists—the actual mapping of the seafloor. Whereas the expedition of 1872 used a human-operated dredge and specimen containers, the 1956 expedition led by Maurice Ewing used the latest in sonar and dredging equipment to complete a goal first visualized by Hess.

Leading a group of scientists from the Lamont Geological Observatory, Ewing began his journey with the intent of mapping a string of undersea mountains called the "Mid-Atlantic Ridge." He found in the process, however, that simple mapping was next to impossible in view of the discoveries made by his team. Hess tackled the strange information they gathered in a piece he called "an essay in geopoetry," so named because of the extraordinary lack of supporting data for his proposition. Using the same information from which Ewing and Heezen developed and eventually proved their theory of the existence and location of a midoceanic rift, Hess went a step further in his speculations.

Hess proposed that hot rock rose from Earth's interior through the frequently recorded earthquakes surrounding the midoceanic ridges. The continuous upwelling of hot rock continually forces the rift apart and the seafloor to spread as the new material from each eruption cools and is pushed away at a rate of a few centimeters per year. When the spreading seafloor encounters an obstruction such as a continent base, it moves under the obstruction, or subducts. Dubbed "seafloor spreading" by Robert Dietz, Hess's theory helped validate Wegener's idea of continental drift. In addition to validating a decades-old theory, the work of Heezen and Ewing opened a new venue of research to scientists as the world entered the 1957-1958 International Geophysical Year.

—Earl G. Hoover

Further Reading

Felt, Hali Soundings. *The Story of the Remarkable Woman Who Mapped the Ocean Floor.* Henry Holt, 2012.

Frankel, Henry R. *The Continental Drift Controversy: Introduction of Seafloor Spreading.* Vol. 3. Cambridge UP, 2012.

Lawrence, David M. *Upheaval from the Abyss: Ocean Floor Mapping and the Earth Science Revolution.* Rutgers UP, 2002.

Neighbors, Thomas, and Leif Bjørnø. *Applied Underwater Acoustics.* Elsevier, 2017.

Searle, Roger. *Mid-Ocean Ridges.* Cambridge UP, 2013.

Seibold, Eugen, and Wolfgang Berger. *The Sea Floor: An Introduction to Marine Geology.* 4th ed., Springer, 2017.

Thurman, Harold V., and Alan P. Trujillo. *Introductory Oceanography.* 10th ed., Pearson Prentice Hall, 2004.

Hess, Harry Hammond

Fields of Study: Geology; Geochemistry; Marine Geology

ABSTRACT
American geologist Harry Hammond Hess developed the theory of seafloor spreading. This helped scientists come to a consensus about plate tectonic theory and changed the way people understood the formation of Earth.

Harry Hammond Hess. Photo via Wikimedia Commons. [Public domain.]

KEY CONCEPTS

guyot: an undersea mesa of volcanic origin that is believed to have once protruded above the surface near midocean ridges, where wave action eroded the upper portion flat before they were transported into the abyssal plains by seafloor spreading

Mohole Project: a deep drilling project to penetrate Earth's crust and reach the mantle at the Mohorovičić Discontinuity

Mohorovičić Discontinuity: the transition region between the crust and the mantle layer of Earth's inner structure

petrology: study of the formation and composition of rocks

seafloor spreading: the divergent motion of oceanic tectonic plates, driven by the force of upwelling convection currents in the molten magma of the mantle layer of the planet; new basaltic seafloor rock is produced continually by this process

EARLY LIFE OF HARRY HESS

Harry Hammond Hess was born in New York City on May 24, 1906. In 1923, he enrolled at Yale University in New Haven, Connecticut. He intended to pursue a degree in electrical engineering, but his interests shifted to geology. Hess earned his bachelor's degree in 1927 and moved to Rhodesia (now Zimbabwe), where he worked as a mineral prospector for two years. Upon returning, Hess enrolled in graduate courses at Princeton University in New Jersey, where he studied under petrologist Arthur Buddington.

Petrology is the study of the formation and composition of rocks. Much of Hess's research at Princeton concerned the structure of peridotite, a dense igneous rock that contains the mineral olivine. Peridotite, of which Earth's mantle is primarily composed, was the focus of Hess's doctoral thesis. In addition to his thesis, Hess wrote several papers that became influential geological documents. While working toward his doctoral degree, Hess traveled to the Caribbean to conduct underwater study. His mentor on the excursion was the renowned Dutch geophysicist F. A. Vening Meinesz. Using naval submarines as laboratories, Hess and Meinesz conducted experiments in the region of the Lesser Antilles islands.

Upon finishing his PhD at Princeton in 1932, Hess taught at Rutgers University in Newark, New Jersey, and worked as a research associate at the Geophysical Laboratory of the Carnegie Institution of Washington, DC. Hess began teaching at Princeton in 1934 and earned tenure as a professor of geology in 1950.

HESS'S LIFE'S WORK

During his years on the Princeton faculty, Hess also taught as a visiting professor at the University of Cape Town in South Africa from 1949 to 1950 and at the University of Cambridge in England in 1965. He was elected to a number of institutions, including the National Academy of Sciences (1952), the American Philosophical Society (1960), and the American Academy of Arts and Sciences (1968). Additionally, Hess was elected president of the Mineralogy Society of America (1955), the Geological Society of America (1963), and the Geodesy and Tectonophysics branches of the American Geophysical Union (1951 and 1956). In 1950, Hess became chair of the geology department at Princeton, a position he held until 1966. He succeeded Buddington as Blair Professor, a distinction awarded to outstanding professors, in 1964.

Although Hess had already conducted research with peridotite and had published several noteworthy papers, his career breakthrough came during a period of service in the United States Navy. Hess enlisted in the Navy Reserve and was called upon in 1941 to assist in the detection of German U-boats in the northeast Atlantic. Afterward, he requested a transfer to the USS *Big Horn*, which was built as a decoy to help determine whether the navy's detection systems were effective. Hess served on the *Big Horn* during the early part of the war. He was then appointed captain of the USS *Cape Johnson*, a large transport vessel assigned to tour the South Pacific.

During the tour, Hess used the ship's sonar technology, a new echo-sounding device, to gather information about the ocean floor. With the data he collected, Hess was able to map strategic routes to the South Pacific. In the process of surveying thousands of miles of the ocean floor, Hess noticed its distinct geological features. Hess detected the presence of raised, flat-topped platforms at the bottom of the ocean. He hypothesized that the platforms were the remnants of islands that sank thousands of years earlier. Hess named the platforms "guyots," in honor of Swiss geographer Arnold Guyot, the founder of the geology department at Princeton.

From his recorded observations, Hess formulated the theory of seafloor spreading to explain the features and arrangement of the rocks that make up the ocean floor. British geologist Arthur Holmes had conducted similar research in the 1930s. Hess's theory included both Holmes's original ideas and conclusions based on his own observations. Hess outlined the theory of seafloor spreading in a paper in 1960. Because he had very little concrete evidence or time-lapse observations of the ocean's floor, he did not publish the paper in a scholarly journal, and thus the theory received little recognition. However, Hess decided to expand his work despite the lack of initial support.

In 1962, Hess published "History of Ocean Basins." The paper discusses the theory of plate tectonics—the movement of continental plates—with regard to Hess's theory of seafloor spreading. Hess theorized that molten rock escapes Earth's interior through ridges in the oceanic crust. The escaped rock pushes the seafloor apart as it cools, and falls into deep trenches. Hess maintained that although the oceans had existed for four billion years, the sediment and other rock formations presently on the ocean floor had only been around for about 300 million years. He also suggested that seafloor spreading is primarily caused by mantle convection cells, heating and cooling currents that cause the movement of matter within Earth's mantle. "History of Ocean Basins," later considered a seminal work in plate tectonics, resulted in the widespread acceptance of the theory of seafloor spreading. The paper answered a number of geological questions regarding the nature of the ocean floor.

In 1957, American oceanographer Walter Munk initiated a campaign to drill through the ocean's floor and into the underlying mantle. This effort, called the "Mohole Project," became a matter of intrigue and controversy for geologists and politicians. A proponent of the Mohole Project, Hess served as its scientific advisor. Drilling began near Guadalupe, Mexico, in 1958. The Mohole Project was successful in providing mantle samples and information about Earth's crust. Divided into three phases, the project continued for nearly a decade. However, due to the increasing expense of the project, the U.S. Congress suspended the Mohole Project indefinitely in 1966.

U.S. President John F. Kennedy took note of Hess's accomplishments and appointed him chairman of the Space Science Board of the National Academy of Sciences. In this position, Hess assisted in the design and development of the U.S. space program.

In recognition of his theory of seafloor spreading, Hess received the Geological Society of America's Penrose Medal in 1966, the geologists' equivalent of the Nobel Prize. Yale University awarded Hess an honorary PhD in 1969. The Space Science Board later appointed Hess to chair a meeting at Woods Hole, Massachusetts. During the meeting, on August 25, 1969, Hess suffered a fatal heart attack.

IMPACT OF HESS'S WORK

As more scientists studied the ocean floor, they began to accept Hess's theory of seafloor spreading as the mechanism behind continental drift. By 1967, seafloor spreading was the dominant theory in the field, and virtually all Earth scientists were reinterpreting their data in light of it. The idea of seafloor spreading led to an overall theory of plate tectonics, which was developed in the 1960s and soon became accepted as fact. The theory of plate tectonics explains the formation and distribution of Earth's continents, including details such as why earthquakes occur.

Hess's theory of seafloor spreading had an impact on geology comparable to that of Darwin's theory of evolution on the field of biology. Upon his death, the National Academy of Sciences wrote that Hess was one of the most significant Earth scientists of the twentieth century. The National Aeronautics and Space Administration (NASA) posthumously awarded Hess the Distinguished Public Service Award. Because Hess provided invaluable leadership, research, and funding to a range of scientific fields, the American Geophysical Union named an award of distinction in Hess's honor in 1984. The Harry H. Hess Medal is awarded to people who demonstrate "outstanding achievement in research in the constitution and evolution of Earth and sister planets."

—Richard Means

Further Reading

Blatt, Henry, Robert Tracy, and Brent Owens. *Petrology: Igneous, Sedimentary, and Metamorphic.* 3rd ed., Freeman, 2006.

Fletcher, Charles. *Physical Geology: The Science of Earth.* 3rd ed., Wiley, 2017.

———, Dan Gibson, and Kevin Ansdell. *Introduction to Physical Geology: The Science of Earth.* Canadian ed., Wiley, 2014.

Frankel, Henry R. *The Continental Drift Controversy: Introduction of Seafloor Spreading.* Vol. 3. Cambridge UP, 2012.

Lawrence, David M. *Upheaval from the Abyss: Ocean Floor Mapping and the Earth Science Revolution.* Rutgers UP, 2002.

Oreskes, Naomi, editor. *Plate Tectonics: An Insider's History of the Modern Theory of the Earth.* Westview Press, 2001.

Seibold, Eugen, and Wolfgang Berger. *The Sea Floor: An Introduction to Marine Geology.* 4th ed., Springer, 2017.

HESS IDENTIFIES THE CAUSE OF CONTINENTAL DRIFT

Fields of Study: Oceanography; Marine Geology; Geophysics

ABSTRACT

Harry Hammond Hess's identification of seafloor spreading as the mechanism for continental displacement revolutionized geologic paradigms and had the same degree of impact on the Earth sciences that Charles Darwin's theory of evolution had had on biology.

KEY CONCEPTS

continental drift: the theory proposed by Alfred Wegener to explain the apparent complimentary outline of continental land masses on opposite sides of the Atlantic Ocean

paleomagnetism: the "signature" of Earth's magnetic field frozen in susceptible rock minerals as they cool from the molten state

plate tectonics: the evolved form of the continental drift theory, positing the movement of large "plates" of Earth's crust being moved about the planet by the force of magmatic convection currents in Earth's mantle layer

tensional rift: an elongated depression that forms between two segments of a tectonic plate that are being driven apart from each other by magmatic convection currents

DISCOVERING THE CAUSE OF CONTINENTAL DRIFT

Princeton University professor Harry Hammond Hess was noted for his scientific contributions to the field of geology, specifically his groundbreaking "History of the Ocean Basins" (published in 1962 but widely circulated beginning in 1960), in which he proposed seafloor spreading as the long-sought-after mechanism for Alfred Wegener's continental drift theory. The elements of seafloor spreading, the rifting of the Pangaea continent into several continent-size plates, and lateral displacement of those plates to their present latitudes remain the dominant paradigm in the earth sciences and are collectively known as "plate tectonics." The idea of drifting continents was not original with Hess, but the hostility the geologic community demonstrated toward the idea inhibited Hess initially from publishing his theory.

Wegener's 1915 theory of drifting continents lay dormant after his death in 1930. Several problems kept it from universal acceptance. First, the jigsaw-puzzle fit was not exact. It did not account very well for Central America, and Ireland and Newfoundland did not correlate geologically. Second, Sir Harold Jeffreys, leading geophysicist of the time, presented evidence in the 1923-1924 meeting of the British Association for the Advancement of Science that Wegener's forces were too small by three orders of magnitude to drive the drift mechanism.

Other geophysicists questioned Wegener's objectivity and accused him of generalizing on generalizations. A third problem was the repeated failure of Wegener's own test to prove his hypothesis. The timing of radio signals across the Atlantic in 1922, 1927, 1936, 1938, and 1948 did not reveal a widening of the Atlantic through progressively longer time intervals. Finally, in the eyes of the geologic community, Wegener was not a geologist but a meteorologist—an outsider—attempting to restructure their science.

Wegener's supporters tried to mollify the scientific community. Harvard University geologist Reginald A. Daly, an ardent believer in continental drift, wrote in his 1926 book, *Our Mobile Earth*: "So obvious are his logical inconsequences and his failure properly to weigh ascertained facts that there is danger of a too speedy rejection of the main idea involved." Geologist Alexander L. du Toit of South Africa's Johannesburg University and S. Warren Carey of the University of Tasmania, Hobart, defended Wegener's theory in their 1937 publications, and the theory rallied again in England with the 1944 publication of Arthur Holmes's textbook *Principles of Physical Geology*.

For the most part, resistance to Wegener's theory continued. World War II delayed research into the question, and from the 1930s to the mid-1950s, continental drift remained a paradigm held with great passion by a minority of geologists but deficient of evidence according to the majority. Bailey Willis, of Stanford University, criticized it in his 1944 article, "Continental Drift, Ein Märchen (a Fairy Tale)," published in the *American Journal of Science*. Even as supporting evidence was accumulating in the mid-1950s, *Crust of the Earth*, a 1955 landmark publication of the Geological Society of America, ignored continental drift, although it mentions both Wegener and du Toit in unrelated discussions. As late as 1966, University of Hamburg physicist Pascual Jordan described the theory as the geophysicists' "favorite fairy tale."

Postwar advances in technology, methodology, and the new science of paleomagnetism lent support to Wegener's theory. Paleomagnetism enabled scientists for the first time to reconstruct, from igneous rocks, a fossil imprint of Earth's earlier magnetic fields. Geologic research in these new areas continued to focus on data gathering, but now particularly in the oceans, where the U.S. Navy had directed its interest to the ocean floor. Other seagoing nations also initiated active research programs that culminated in the International Geophysical Year (IGY, July, 1957, to December, 1958), the first multinational research effort. The result of this effort was that in almost every area of geologic research, scientists found Earth and particularly the ocean very different from what they had imagined originally. One of the most curious features was the Mid-Atlantic Ridge. Understanding it led to an understanding of plate tectonics.

Matthew F. Maury, director of the U.S. Navy's Department of Charts and Instruments, first recognized the Mid-Atlantic Ridge in 1850 while collecting bathometric measurements aboard the USS *Dolphin*. Maury named it the "Dolphin Rise" and published a map of it in his *The Physical Geography of the Sea* (1855). Data from HMS *Challenger* expedition (1872 to 1876) supplemented Maury's map, but the details of the Mid-Atlantic Ridge remained vague

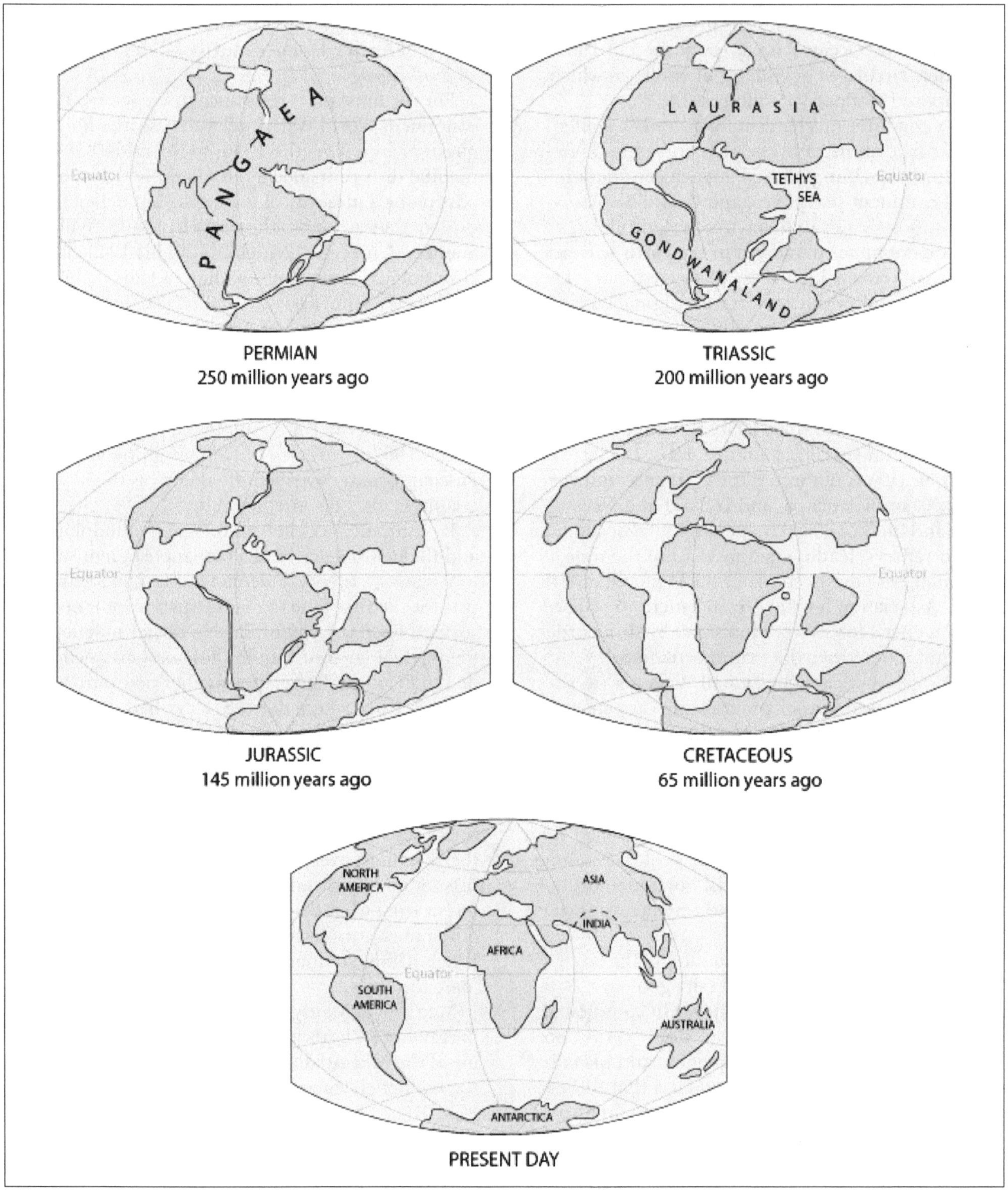

History of continents; land shifting over time. Image via USGS/Wikimedia Commons. [Public domain.]

until the German *Meteor* expedition of 1925 to 1927 led by Nobel laureate Fritz Haber. The expedition collected much oceanographic data and utilized an echo sounder to map the ocean floor. In 1933, German oceanographers Theodor Stocks and Georg Wust produced the first detailed map of the ridge noting a valley that seemed to be bisecting it. Later, in 1935, geophysicist Nicholas H. Heck found a strong correlation between earthquakes and the Mid-Atlantic Ridge.

Interrupted by World War II, oceanic exploration resumed in the late 1940s as a predominantly American venture. These studies produced a host of seemingly divergent data. In 1950, Maurice Ewing of the Lamont-Geological Observatory established that no continental crust exists beneath the ocean basins. In 1952, Roger Revelle, director of Scripps Institute of Oceanography at La Jolla, California, and his student Arthur E. Maxwell determined the rate of heat flow from Earth's interior in the Pacific Ocean. They found it was similar to continental heat flow rates that Edward Bullard of the University of Cambridge first described in 1935, except over the oceanic ridge, where the rate was substantially higher.

The idea of a structurally active ridge received further support in 1954 when Jean P. Rothé, director of the International Bureau of Seismology in Strasbourg, mapped a continuous belt of earthquake epicenters from Iceland through the mid-Atlantic around South Africa, through the Indian Ocean, and on to the African Rifts and the Red Sea. In 1956, Maurice Ewing and Bruce Heezen continued the German technique of echo soundings at Lamont-Geological Observatory and found that the Mid-Atlantic Ridge was more than 64,000 kilometers long and, more important, that it had a rift (tensional) valley along the entire crest. In 1961, Ewing and Mark Landisman discovered that this ridge system extends throughout all the world's oceans, is seismically and volcanically active, continues to exhibit a tensional valley, and is mostly devoid of sediment cover.

Ivan Tolstoy and Ewing first characterized the geologically important question of sediment cover in 1949, describing a main ridge of thin sediment and flanks of thick sediment. The age of the sediments increased from the ridge toward the continents, the oldest being only Cretaceous in age (65 million years). In 1959, Victor Vacquier discovered that enormously long faults crossed the oceans' floor in many places. The faults have offsets of hundreds of kilometers but do not extend into the continents.

Central to the interpretation of this ocean mountain range were the early 1950s paleomagnetic researches of Patrick M. S. Blackett and his student, Keith Runcorn, at the University of Manchester. Their studies of fossil magnetism suggested that in the geologic past, the inclination, declination, and even the polarity of Earth's magnetic field had been very different from current orientations. The seemingly chaotic data formed a consistent pattern only upon assuming that the continents had moved relative to the magnetic poles and to one another. Magnetic studies of the seafloor by oceanographers R. G. Mason, A. D. Raff, and Vacquier revealed a symmetrical, zebra-like pattern about the midoceanic ridge in 1957.

By the end of the 1950s, it was clear that existing geologic theories clearly had failed to predict or explain these seemingly unrelated phenomena; the new data required a new paradigm. In 1960, Hess circulated among his colleagues his theory of seafloor spreading to illuminate the working of Wegener's continental displacement hypothesis.

SIGNIFICANCE OF HESS'S DISCOVERY

The period between 1960 and 1965 was one of great uncertainty and multiple directions for geologists. In 1960, Hess synthesized the oceanic data of the 1950s into a bold new theory. Hess's paradigm was so novel and radical that he did not attempt to publish it in the usual professional journals, but included it in a 1960 report to the Office of Naval Research. Hess also widely circulated reprints among his colleagues.

In his 1960 report, Hess proposed that the midoceanic ridges were the loci of upwelling mantle convection cells that progressively moved the seafloor laterally and eventually under the continents. (Nearly thirty years earlier, in 1931, Arthur Holmes, of the University of Edinburgh, proposed that convection currents may exist in the mantle.) This driving mechanism brought together the divergent data of post-World War II research into one coherent theory. This creation of new seafloor from the upwelling of the mantle explained the high heat flow above the midoceanic ridge, the tensional rift-valley in the middle of the ridge, the correlation of the ridge with earthquake epicenters, the transform faults associated with the ocean floor, the continuation of the ridge throughout the oceans, the absence of continental-type rocks in the oceans, the thin sediment in the middle of the ridge and its thickening toward the edges, the symmetrical paleomagnetic zebra patterns, and the energy of the convection currents that was sufficient to drive the continents.

Robert S. Dietz, working for the Navy Electronic Laboratory in San Diego, published virtually identical ideas and coined the phrase "seafloor spreading" in a 1961 article "Continent and Ocean Basin Evolution by Spreading of the Sea Floor," published in *Nature*. Hess's reluctance to publish in mainstream scientific journals muddles the question of intellectual priority. He had scheduled his theory for publication in a multivolume set *The Sea: Ideas and Observations*, but delays kept it from the presses until 1963. Meanwhile, Hess was aware of Dietz's 1961 article, and in November, 1962, he published his ideas through a more expedient route, a collection of essays dedicated to his former teacher, *Petrologic Studies: A Volume in Honor of A. F. Buddington*. Dietz acknowledged Hess's priority of the mobile seafloor theory, and Hess credited Dietz for coining the term "seafloor spreading."

With the theory of seafloor spreading fully illuminated by both Hess and Dietz, the geologic community began reevaluating data and redesigning field studies to test it. One of the first tests occurred in 1963, following a prediction by Drummond H. Mathews and his student Frederick J. Vine at the University of Cambridge. They hypothesized that the magnetic pattern of polar reversals should be symmetrical along the axis of the oceanic ridge and correlate with its morphology. Their hypothesis proved correct and initiated a revolution in the earth sciences. In 1966, in recognition of his scientific breakthrough, Hess received the Geological Society of America's Penrose Medal, the geologists' equivalent of the Nobel Prize. By 1967, seafloor spreading was the dominant paradigm, and virtually all earth science specialists began to reinterpret their data in the light of the new paradigm.

—Richard C. Jones and Anthony N. Stranges

Further Reading

Fletcher, Charles. *Physical Geology: The Science of Earth.* 3rd ed., Wiley, 2017.

Frankel, Henry R. *The Continental Drift Controversy: Introduction of Seafloor Spreading*. Vol. 3. Cambridge UP, 2012.

Lawrence, David M. *Upheaval from the Abyss: Ocean Floor Mapping and the Earth Science Revolution.* Rutgers UP, 2002.

McCoy, Roger M. *Ending in Ice: The Revolutionary Idea and Tragic Expedition of Alfred Wegener.* Oxford UP, 2006.

Searle, Roger. *Mid-Ocean Ridges.* Cambridge UP, 2013.

Seibold, Eugen, and Wolfgang Berger. *The Sea Floor: An Introduction to Marine Geology.* 4th ed., Springer, 2017.

Vrielynck, Bruno, and Philippe Bouysse. *The Changing Face of the Earth: The Break-Up of Pangaea and Continental Drift over the Past 250 Million Years in Ten Steps.* UNESCO, 2003.

HUTTON PROPOSES THE GEOLOGICAL THEORY OF UNIFORMITARIANISM

Fields of Study: Historical Geology; Philosophy; Geomorphology

ABSTRACT

James Hutton's theory that Earth's formation was the result of a cyclic process of erosion and uplift—which, in turn, was the result of the compounding of the ordinary action of water and heat in deep time—was the intellectual precursor of Charles Lyell's uniformitarian geology and a fundamental premise of Charles Darwin's evolutionary theory.

KEY CONCEPTS

dynamic equilibrium: the condition of a system in which changes in parts of the system tend to be countered by changes in other parts of the system, producing an overall steady state; e.g., the formation of sediment by erosion of rocks is countered by formation of rocks from sediment

Neptunism: the constructive philosophical consensus that views geological history as a dynamic equilibrium

Plutonism: the destructive philosophical consensus that views geological history as a series of catastrophic events

uniformitarianism: the theory that the processes of Earth's geological history existed and will continue to exist in a state of dynamic equilibrium rather than as a series of catastrophic events

DEVELOPMENT OF HUTTON'S THEORY

On March 7, 1785, members of the Philosophical Society of Edinburgh assembled to hear a much anticipated paper. Its author, James Hutton, was ill that day and had chosen his closest friend, the renowned philosophical chemist Joseph Black, to deliver the first part of a new theory in a work titled "Concerning the System of the Earth, Its Durability and Stability." Four weeks later, on April 4, Hutton had recovered sufficiently to present the second part of his theory. In July, he privately printed and circulated an abstract of the paper. Eventually, in 1788, the full ninety-five-page manuscript from which Hutton's papers had been drawn was published in the first volume of the *Philosophical Transactions* of the society as "Theory of the Earth: Or, An Investigation of the Laws Observable in the Composition, Dissolution, and Restoration of Land upon the Globe."

Hutton's paper was written in the context of a well-established consensus in eighteenth-century geological science. The French naturalist the Comte de Buffon had provided a general framework for the consensus in the initial volumes of his *Histoire naturelle, générale et particulière* (1749-1789; *Natural History, General and Particular*, 1781-1812) and again in his *Époques de la nature* (1778; Epochs of Nature). According to Buffon, the Earth originated as solar matter. As Earth cooled, a universal ocean covered its surface. Sedimentation of materials suspended in this primitive ocean produced rock strata, which were exposed as the ocean receded.

In the 1780s, Abraham Gottlob Werner, a professor at the Freiburg Mining Academy in Saxony, supplemented Buffon's cosmogony with a stratigraphy that distinguished four mineralogical groups according to the order in which they had settled out of the universal ocean. Most primitive were chemical precipitates such as granite. Settling out next at lower elevations were heavier materials such as limestone, followed by basalts at still lower elevations and sand and other alluvial deposits. Werner's stratigraphy would potentially accommodate a great variety of geological phenomena and gain widespread acceptance. Indeed, when professor of natural history John Walker offered the first series of lectures in geology at the University of Edinburgh in 1781, it was Werner's stratigraphy that he introduced.

What the audience for Hutton's society paper heard was an audacious departure from the Wernerian consensus. To be sure, Hutton's theory acknowledged that water was one of the primary agents in geological change. During the 1750s and 1760s, Hutton had been a highly innovative agricultural improver on his Berwickshire estate and knew well the power of erosion. However, for Hutton, water's geological effect was destructive. The action of water could not explain dramatic features of the Earth's topography such as mountain ranges, nor

James Hutton, portrait by Henry Raeburn, 1776. Image via Wikimedia Commons. [Public domain.]

could it adequately explain the presence of unconformities in rock strata.

For these phenomena, Hutton required a more constructive force. A graduate of the University of Edinburgh in 1743, Hutton had had a longtime interest in chemistry and, during the 1740s, he had even invented an improved method of producing the ammonium chloride used in soldering metals. In 1768, Hutton leased his farm and returned to Edinburgh. It was at this time that Hutton developed his friendship with Joseph Black, a pioneer in the study of heat, and Black's former student, James Watt, the inventor of an improved steam engine. These friendships soon suggested to Hutton a second agency in geological change. Subterranean heat, he began to argue, drove the terrestrial machine.

Hutton was combining the so-called Neptunian consensus with a new Plutonism: Like the sea, the underworld could rise and fall. The origins of this conception of dynamic equilibrium again lay in Hutton's own past. As an undergraduate at the University of Edinburgh, Hutton had studied with Colin Maclaurin, one of the eighteenth century's most effective popularizers of Sir Isaac Newton's ideas. In the autumn of 1744, Hutton had begun to study medicine at Edinburgh, then in 1747 he transferred to the University of Paris, and in 1748 he enrolled in the center of medical Newtonianism at the University of Leiden, where he completed his medical degree with *Dissertatio physico-medica inauguralis de sanguine et circulatione microcosmi* (1749; *James Hutton's Medical Dissertation*, 1980). Just as in Newtonian astronomy, in which a balance of centrifugal and gravitational forces produced the solar system, and just as systolic and diastolic forces circulate the blood (or just as the piston pushes and pulls in Watt's double-acting steam engine), uplift complemented erosion in Hutton's system of the Earth. As water dissolved rock, the products of sedimentation accumulated, generating intense heat and pressure. The intense heat and pressure, in turn, caused rapid expansion and uplift. Even as old continents become new oceans, old oceans become new continents.

Hutton's argument had profound implications for eighteenth century geology. Whereas Buffon conceived the history of Earth proceeding in one direction of greater cooling and erosion, Hutton did for matter in time what Newton had done for matter in space and conceived the laws of motion in terms of reciprocal action and reaction. Whereas Neptunism implied catastrophism, the assumption that the geological processes of the past were of a qualitatively greater magnitude than the geological processes of the present, Hutton's dynamic equilibrium implied a uniformity, in which the observed geological processes of the present were the key to understanding the geological processes of the past. Also, whereas catastrophism offered a way to reconcile biblical accounts of the Creation and the Flood, Hutton entirely ignored Genesis as a source of geological knowledge.

Finally, Hutton undercut Werner's stratigraphy. In the years after 1785, Hutton traveled to sites across Scotland—most famously, to the unconformity at Siccar Point—to gather evidence for his theory. This evidence indicated that granite had not originated as an aqueous precipitate but had crystallized from molten magmas. It also revealed that granite, which Werner had made the oldest rock, was in some cases intruded upward into sedimentary strata. Conversely, the evidence also showed the ancient volcanic origin of basalts. By recognizing the igneous origin of many rocks, Hutton reconfirmed that the forces of geological change acted uniformly in past and present. He also reconfirmed that geological processes did not work in one direction and, therefore, no inherent limit could be set regarding Earth's geological history. In short, Hutton had discovered deep time. "We find," Hutton famously concluded in his 1788 paper, "no vestige of a beginning—no prospect of an end."

SIGNIFICANCE OF HUTTON'S THEORY

James Hutton extended the Scottish tradition of conjectural history to the economy of nature. From the perspective of deep time, cycles of dissolution and composition succeeded one another in a spontaneous and, ultimately, benevolent order. As if by an invisible hand, the nearly imperceptible action of heat and water shaped the grandest features of the physical environment. Yet, for all the assurance of durability and stability, Hutton's theory faced intense criticism in the years after 1788. The specter of the French Revolution rendered Hutton's defense of design an invitation to atheism, his understanding of nature's uniformity an incitement to political turbulence.

In 1795, a very ill Hutton attempted to rebut these charges in the two sprawling volumes of "Theory of the Earth." Most responsible for keeping Huttonian geology before a scientific public, however, were his two companions on the trip to Siccar Point in 1788, the chemist and geologist James Hall and the mathematician John Playfair. Under the pressure of criticism, Hutton's geology underwent its own reconstruction. In a series of ingenious experiments between 1798 and 1805, Hall proved aspects of Hutton's theory, demonstrating, for example, the thermal metamorphosis of limestone into marble. In 1802, geologist John Playfair published *Illustrations of the Huttonian Theory of the Earth*. Beautifully presented in compelling prose, Playfair's illustrations were no longer organized as mere proofs of an a priori theory but as the empirical foundation for a historical narrative. It was Hall's experimental and Playfair's empirical Huttonianism that, in 1830, provided the precedent for the scientific uniformitarianism of British geologist Charles Lyell's *Principles of Geology* (1830-1833). It was the same Huttonianism that, in 1859, provided two crucial elements to the variational evolution of Charles Darwin's *On the Origin of Species by Means of Natural Selection: Or, The Preservation of Favoured Races in the Struggle for Life* (1859): the notion that evolutionary change could take place by an incremental accumulation of small variations and the discovery of time that allowed this slow agency to work.

—Charles R. Sullivan

Further Reading

Craig, Gordon Younger, and John Hewett Hull. *James Hutton: Past and Future.* Geological Society of London, 1999.

Dean, Dennis R. *James Hutton and the History of Geology.* Cornell UP, 1992.

Hutton, James. *System of the Earth, 1785, Theory of the Earth, 1788, Observations on Granite, 1794, Together with Playfair's Biography of Hutton.* Hafner, 1970.

Playfair, John. *Illustrations of the Huttonian Theory of the Earth (1802).* Cambridge UP, 2011.

Repcheck, Jack. *The Man Who Found Time: James Hutton and the Discovery of the Earth's Antiquity.* Perseus Books, 2003.

I

Igneous Processes, Rocks, and Mineral Deposits

Fields of Study: Geology; Geomorphology; Geochemistry

ABSTRACT

Igneous rocks and mineral deposits, created by the crystallization and solidification of magma, are found all over the world. Many of the world's most economically important mineral deposits result, directly or indirectly, from igneous activity.

KEY CONCEPTS

cumulates: layers of large crystalline material, separated by gravity, that form at the bottom of a magma chamber as it begins to cool

dike: a magmatic intrusion that cuts vertically through layers of sedimentary rock

igneous: rock formed from the solidification or crystallization of magma

plutonic: refers to magma that solidifies within the crust rather than on the surface

pyroclastic: refers to particles of solidified volcanic magma ranging in size from fine dust, or "ash" to house-sized boulders, as well as to rock that is formed by consolidation of such particulate matter

sill: a magmatic intrusion that drives horizontally between layers of sedimentary rock

WHERE IGNEOUS ROCKS ARE BORN

Igneous rocks are created by the crystallization and solidification of hot, molten silicate magma. Magma consists of silicate liquid (the major component is the silicate ion SiO_4^{-4}), solid crystals, rock fragments, dissolved gases such as carbon dioxide, water, and various sulfurous oxides. Familiar examples of igneous rocks are granite (an "intrusive" or "plutonic" rock that is crystallized at depth) and basalt (as in dark "extrusive" lava flows, such as those in Hawaii). Igneous rocks are found worldwide on all continents, on oceanic islands, and on the ocean floors. They are particularly common in mountain ranges or other areas where Earth experiences tectonic activity. Oceanic islands, such as Hawaii and Iceland, are nearly exclusively igneous in origin, and the world's oceans are floored by basalt lava flows.

Metallic ores produced by igneous activity may be formed directly from the igneous rocks or through the injection of hydrothermal (hot water) solutions into adjacent rocks. Some of the most important commodities obtained from igneous sources include copper, nickel, gold, silver, platinum, iron, titanium, tungsten, and tin. Nonmetallic products include crushed stone, construction stones for buildings and monuments, and some precious and semiprecious gemstones.

Igneous (from the Latin word *ignis*, meaning fire) rocks form by the crystallization of hot, molten magma produced by the heat of Earth's interior. Surface exposures of igneous rock bodies are widespread through out the globe. On continents they mostly occur in mountainous areas or ancient "Precambrian shield" areas where billions of years of erosion have revealed the roots of old mountain ranges. The Canadian Shield region of northern Ontario and eastern Canada is an excellent example. In the oceans, igneous rocks cover the floors of ocean basins below a thin layer of sediment. Most oceanic islands owe their very existence to ocean floor volcanic eruptions that produce volcanoes of sufficient stature to project above the waves. Familiar examples are the Hawaiian chain, the Galápagos Islands, and Iceland.

TYPES OF IGNEOUS ROCKS

Igneous rocks are divided into two major categories defined by their mode of emplacement in or on

Earth's crust. If molten magma cools and solidifies below the surface, the rocks are called "intrusive" or "plutonic." Because these rocks generally take a long time to cool and solidify (a process called "crystallization"), their component minerals grow large enough to see with the naked eye (coarse-grained rocks). On the other hand, if magma flows out onto Earth's surface, it forms "extrusive" or "volcanic" rock. These rocks lose heat rapidly to air or water, and the resulting rapid crystallization produces tiny, nearly in visible crystals (fine-grained rocks). Some volcanic rocks cool so quickly that few crystals have time to form; these are glassy rocks such as obsidian. Two kinds of volcanic rock exist: lava flows and "pyroclastic" deposits formed by explosive volcanism. Pyroclastic materials (volcanic ash) are deposited as layers of particles that have been violently ejected into the air.

Igneous rocks are also classified according to chemical composition. At one extreme are the light-colored "felsic" rocks that contain high concentrations of silica (up to about 75 percent silicon dioxide, SiO_2) and relatively little iron, magnesium, and calcium. Examples of felsic rocks are granite, a plutonic rock, and its volcanic equivalent, rhyolite (obsidian glass is rapidly cooled rhyolite).

At the other extreme are the dark "mafic" rocks with relatively low silica (as low as about 46 percent SiO_2) but with higher concentrations of iron, magnesium, and calcium. Examples of mafic rocks are gabbro (plutonic) and its volcanic equivalent, basalt. Rocks of intermediate composition also exist, for example plutonic diorite and its volcanic equivalent, andesite. It is andesite (and a more silicic variety called "dacite") that is expelled from the potentially explosive volcanoes of the Cascade range in the American Pacific Northwest (Mount St. Helens, Mount Rainier, Mount Hood, and others).

INTRUSIVE (PLUTONIC) STRUCTURES

Intrusive igneous rock bodies come in many shapes and sizes. The term "pluton" applies to all intrusive bodies but mainly to granitic rocks (granites, diorites, and related rocks). Specific terms applied to plutons mostly describe the size of the body. "Stocks" are exposed over areas less than 100 square kilometers, whereas "batholiths" are giant, commonly lens-shaped, bodies that exceed 100 square kilometers in exposed area. The Sierra Nevada range in eastern California is a good example of a batholith.

Some specialized pluton varieties are "laccoliths," commonly mountainous areas (e.g., the Henry and La Salle mountains in Utah) in which intrusive granitic magma has invaded horizontal sedimentary layers and has bowed them up into a broad arch. A "phacolith" is similar to a laccolith only the magma has invaded folded sedimentary rocks so that the pluton itself appears to have been folded.

Minor intrusive bodies include "sills," tabular bodies intruded parallel to rock layers (a laccolith can be considered a "fat sill"), and "dikes," tabular bodies that cut across rock layers. Sills and dikes are common features around the margins of plutons where they contact "country rock" (older, preintrusion materials).

Another intrusive body, mostly produced by mafic (gabbroic) magmas, is the "lopolith." Lopoliths are relatively large funnel-shaped bodies (on the order of large stocks or small batholiths) in some cases created where magma fills the down-warped part (syncline) of a fold structure. An excellent example is the Muskox intrusion of northern Canada; another possible one (one limb is unexposed under Lake Superior) is the Duluth gabbro intrusion of northeastern Minnesota.

EXTRUSIVE (VOLCANIC) STRUCTURES

The nature of volcanoes and volcanic rock deposits in general is greatly influenced by the composition of their parent magmas. Basalt magma is a low-viscosity liquid (it is thin and flows easily) and thus produces topographically low, broad volcanic features. Typical of these are the "fissure flows" (also known as plateau basalts) in which basalt lava issues from fractures in the crust and spreads out almost like water in all directions. Examples are the Columbia River basalt plateau in Oregon and Washington, the Deccan plateau in India, and the Piraná basalt plateau in Brazil. The basalt flows that floor the oceans are underwater versions of fissure flows.

Basaltic volcanoes tend to have low profiles but laterally extensive bases typified by the "shield" volcanoes of Hawaii and other areas. These volcanoes resemble giant ancient shields lying on the ground. Pyroclastic eruptions of basalt, powered mostly by the violent release of dissolved carbon dioxide, produce cinder-cone volcanoes, otherwise known as "Strombolian" volcanoes, after the Italian volcano Stromboli.

In contrast to mafic magmas, the more silica-rich felsic and intermediate magmas are more viscous, and thus flow less readily. This magma tends to pile up in one place, producing towering volcanoes of mountainous proportions. Because felsic-intermediate magmas also tend to contain significant dissolved water, steam trapped during eruption may explode violently, producing thick blankets of volcanic ash near the volcano. The best North American example of these potentially violent volcanoes, called "stratovolcanoes" or "composite" volcanoes, is the Cascade Range in the Pacific Northwest. The terms for these volcanoes reflect their tendency to have layers of mud and lava flows (generally andesite or dacite) that alternate with pyroclastic ash deposits. Stratovolcanoes occur worldwide, particularly at continental margins and in the oceans near continents where "lithospheric plates" (thick horizontal slabs of crust and upper mantle) collide, with one plate moving under the other (subduction zones). Volcanism associated with subduction zones has produced the Andes of South America as well as islands such as Japan, the Philippines, New Zealand, the Aleutian Islands of Alaska, and the islands of Indonesia.

Another important volcanic feature is the "rhyolite complex," or "caldera complex," exemplified by Yellowstone National Park in Wyoming and the Valles Caldera (Jemez Mountains), New Mexico. When fully active, these areas produce violently explosive volcanism and rhyolite lava flows that blanket many square kilometers. The most violent activity occurs when the roof of a large underground magma chamber collapses into the shallow void created by expulsion of magma during previous eruptions. The crater formed during this process is called a "caldera." Roof collapse during caldera formation has the effect of ramming a large piston into the heart of the magma body, violently expelling gas-charged, sticky rhyolite into the atmosphere, from which it may cascade along the surface as a *nuée ardente* (French for "glowing cloud"). These roiling infernos of hot noxious gases, bubbling lava fragments, and mineral crystals are capable of speeds in excess of 300 kilometers per hour and temperatures in excess of 400° Celsius. They deposit ash blankets (welded ashflow tuffs) over wide regions, as in the case of Yellowstone. Stratovolcanoes (described above) can also form calderas and ashflow deposits, as exemplified at Crater Lake, Oregon.

ORE DEPOSITS OF FELSIC-INTERMEDIATE ROCK

Granite and related rocks are the source of many metals and other products that are the foundation of an industrial society. Quartz veins intruding granite may contain gold and other precious metals, as in the "mother lode" areas of the Sierra Nevada Range in California. These veins originate as hydrothermal deposits, minerals precipitated from hot-water solutions flowing through fractures in cooling granitic bodies. Felsic and intermediate composition igneous rocks contain significant dissolved water in their magmas (called "juvenile" water), which is finally expelled as hydrothermal fluids in the late stages of plutonic crystallization. Hydrothermal veins occur in the parent granite itself or are injected into cracks in the surrounding rocks. Many important metallic ore bodies formed as hydrothermal deposits.

So-called porphyry copper deposits such as those of the American southwest (Arizona, New Mexico, Colorado, and Utah) are low-grade deposits of widely scattered small grains of chalcopyrite ($CuFeS_2$) and other copper minerals in felsic plutonic and volcanic rocks, mostly residing in a multitude of extremely thin hydrothermal veins. Some porphyry copper deposits also have considerable deposits of molybdenite (in the sulfide molybdenite, used in high-temperature alloys), especially at the Questa mine in New Mexico and at Climax, Colorado.

By far the greatest concentration of valuable minerals associated with granitic rocks comes from pegmatite deposits. Like hydrothermal deposits, pegmatites form in the late stages of granite crystallization after most of the other rock-forming minerals have already crystallized. Another similarity to hydrothermal fluids is their high volatile content—materials that tend to melt or form gases at relatively low temperatures, such as water, carbon dioxide, and the halogens fluorine and chlorine. Elements with large atomic sizes (ionic radii) and valence charges also tend to concentrate in pegmatitic fluids because the majority of minerals in granites (mostly quartz and feldspars) cannot accommodate these giant atoms within their mineral structures. Thus, pegmatite deposits may contain relatively high concentrations of uranium, thorium, lithium, beryllium, boron, niobium, tin, tantalum, and other rare metals. The high water content of pegmatite fluids, some of it occurring as vapor, allows minerals such as quartz, feldspar, and mica to grow to enormous sizes, the largest of which are on the order of railway boxcars. Pegmatites are generally fairly small bodies; some deposits are no larger than a small house. They may also occur as veins or dikes. Excellent North American examples containing rare and exotic minerals are located in the Black Hills of South Dakota, Maine, New Hampshire, North Carolina, the Adirondacks of New York state, Pala and Ramona in California, and Bancroft and Wilberforce, Canada. Notable international occurrences are in Brazil (Minas Gerais), Russia (the Urals and Siberia), Greenland, Italy, Australia, Germany (Saxony), Madagascar, and Sri Lanka.

ORE DEPOSITS IN MAFIC AND RELATED ROCK

Owing to their low viscosity, mafic magmas produce some unique mineral deposits compared with thicker felsic magmas. In plutonic settings formed early, heavy mineral crystals can easily sink through the magma to form crystal-rich layers on the bottom of the magma chamber. These gravitationally deposited layers are called "cumulates" (from the word accumulate) and, depending on their mineralogical makeup, may constitute important ore bodies. Because cumulates are generally enriched in iron and depleted in silica compared with their mafic parent magma, they are termed "ultramafic," the common rock type being "peridotite," a rock rich in olivine $[(Fe,Mg)_2SiO_4]$. Most of the world's chromium that is used in high-temperature, corrosion-resistant alloys comes from cumulate layers of the mineral chromite ($FeCr_2O_4$), mostly mined in South Africa. The other major commodities recovered from cumulates are the precious metals platinum and palladium, mined in South Africa and Russia.

Intrusive mafic magmas may also form layers of sulfide-rich minerals called "late-stage immiscible segregations" that constitute some of the richest copper and nickel ore bodies in the world. As some mafic magmas cool and change chemically, sulfur and metal-rich fluids may separate from the silicate liquid, just as oil would from water. These "immiscible" (incapable of mixing) sulfide droplets then sink through the lower density silicate magma to form thick layers of "massive sulfide" deposits on the magma chamber floor. The major minerals in massive sulfide copper-nickel mines are chalcopyrite, bornite (Cu_5FeS_4), pyrrhotite ($Fe_{1-x}S$), and pentlandite $[(Fe,Ni)_9S_8]$. Platinum, gold, and silver, among minerals, are commonly recovered as by-products. Major magmatic segregation sulfide mines are located in South Africa (Messina and Bushveld districts, Transvaal) and Norway, and at Sudbury, Ontario, Canada, which has ore rich in nickel.

Titanium and iron ores may also form as magmatic segregations. Massive titanium ores, mostly the oxide ilmenite ($FeTiO_3$), are mined from anorthosite rock, a plagioclase $[(Ca,Na)AlSiO_8]$ feldspar-rich variation of gabbro. Typical examples of these deposits occur in the titanium mines in the Adirondacks of New York state and at Allard Lake, Quebec. Iron deposits of this type, mostly the mineral magnetite (Fe_3O_4), are located at Kiruna, Sweden; the Ozarks of Missouri; Durango, Mexico; and Algarrobo, Chile.

OTHER IMPORTANT IGNEOUS COMMODITIES

Some valuable mineral commodities are recovered from igneous rocks that do not lend themselves to simple classification. For example, diamonds occur in deposits called "kimberlites," a type of general deposit called "diatremes," explosively injected mixtures of mantle (mostly serpentine) and crustal materials that in rare localities contain diamonds. The diamonds form deep in the upper mantle, where pressures are sufficiently high to produce them by the reduction (removal of oxygen) of carbon dioxide. They are then injected into more shallow crustal levels upon the carbon dioxide-powered eruption of kimberlite. Diamonds are mostly mined in South Africa, Ghana, the Democratic Republic of the Congo, Russia, Canada, Brazil, India, and the United States (Murfreesboro, Arkansas).

Two other deposits with chemical affinities to kimberlites are "nepheline syenites" and "carbonatites." Like kimberlites, these bodies are rare, and their magmas probably originate deep in Earth's mantle. Nepheline syenites contain mostly the mineral nepheline ($NaAlSiO_4$) and are sources of apatite (phosphate mineral) and corundum (Al_2O_3), used as an abrasive. Nepheline itself is used to make ceramics. Carbonatites are unusual igneous deposits in that they are composed mostly of the carbonate mineral calcite ($CaCO_3$). They have become increasingly important as sources of the rare elements niobium and tantalum, used in the electronics industry.

—John L. Berkley

Further Reading

Best, Myron G. *Igneous and Metamorphic Petrology.* 2nd ed., Blackwell, 2003.

Blatt, Harvey, Robert J. Tracy, and Brent E. Owens. *Petrology: Igneous, Sedimentary, and Metamorphic.* 3rd ed., W. H. Freeman, 2006.

Decrée, Sophie, and Laurence Robb, editors. *Ore Deposits: Origin, Exploration, and Exploitation.* Wiley, 2019.

Klein, Cornelis, and Anthony Philpotts. *Earth Materials: Introduction to Minerals and Petrology.* 2nd ed., Cambridge UP, 2016.

Li, Z. X., D. A. D. Evans, and J. B. Murphy, editors. *Supercontinent Cycles Through Earth History.* The Geological Society, 2016.

Philpotts, Anthony R., and Jay J. Ague. *Principles of Igneous and Metamorphic Petrology.* 2nd ed., Cambridge UP, 2009.

Wilson, Marjorie. *Igneous Petrogenesis: A Global Tectonic Approach.* Springer, 2007.

Winter, John D. *An Introduction to Igneous and Metamorphic Petrology.* Prentice Hall, 2001.

Young, Davis A. *Mind over Magma: The Story of Igneous Petrology.* Princeton UP, 2003.

INTERNAL GEOLOGIC PROCESSES

Fields of Study: Geology; Geomorphology; Geochemistry; Crystallography

ABSTRACT

The topographic features at Earth's surface, such as mountains, rift valleys, ocean basins, islands, and ocean trenches, are produced by extension or compression forces that act along divergent, convergent, or transform plate boundaries. The extension and compression forces at Earth's surface are powered by convection within the asthenosphere.

KEY CONCEPTS

asthenosphere: a relatively thin layer of molten rock at the interface of the upper mantle and the crust

lithosphere: the rocky outer layer of Earth's internal structure

orogeny: the process of mountain formation

rift valley: an elongated depression that forms where two plates or plate segments are being pushed in opposite directions by the force of the underlying magma convection currents

THE STRUCTURE OF EARTH IN BRIEF

The structure of Earth is essentially a layered spheroid with a central core wrapped in a mantle and enclosed crust. The topmost mantle and the crust make up the lithosphere. Beneath this is a layer called the "asthenosphere," which is composed of moldable and partly liquid materials. Heat transference within the asthenosphere sets up convection currents that ascend from hot regions and sink to less hot regions. Consequently, the overlying lithosphere is segmented into ridged plates that are moved by the convection process. The hot asthenosphere does not rise along a line. This causes the development of a structure called a "transform plate boundary," which is perpendicular to and offsetting the divergent boundary.

The topographic features at Earth's surface, such as mountains, rift valleys, ocean basins, islands, and ocean trenches, are produced by extension or compression forces that act along divergent, convergent, or transform plate boundaries. The extension and compression forces at Earth's surface are powered by convection within the asthenosphere. At divergent plate boundaries, plates are being driven apart from each other, typically allowing molten magma from the mantle to rise into the growing space between them. At convergent plate boundaries, plates are being driven toward each other, creating compressive stress that forces one plate to subduct to, or move under, the other plate, and the other plate to rise and buckle in orogeny. At transverse plate boundaries, the plates experience sheer stress that forces them to move laterally relative to each other, and creates fault lines in the plates as they are torn apart by the stress.

MOUNTAINS AND DEPRESSIONS IN ZONES OF COMPRESSION

Compression along convergent plate boundaries yields three types of mountain system: island arcs that are partly under water; mountains along a continental edge, such as the Andes; and mountains at continental interiors, such as the Alps. At convergent plate boundaries, the denser of the two colliding plates slides down into the asthenosphere and causes volcanic activity to form on the leading edge of the upper plate. Island arcs such as the Aleutians and the Caribbean are formed when an oceanic plate descends beneath another oceanic plate.

Volcanic mountain chains such as the Andes of South America are formed when an oceanic plate descends beneath a continental plate. In both the island arc type and Andean type collisions, a deep depression in the oceans, called a "trench," marks the place where neighboring plates are colliding and where the denser plates are forced downward into the asthensophere. If the colliding plates are of similar density, neither plate will go into the asthenosphere. Instead, the edges of the neighboring plates will be folded and faulted and excess material will be pushed upward to form a block mountain, such as the mountain chain that stretches from the Alps through to the Himalayas. This type of mountain chain is not associated with a trench.

The Appalachians of the eastern United States are an example of the alpine type of mountain belt. When the Appalachians were forming 300 million

The main internal geological agent is the movement of the lithospheric plates, which can cause earthquakes and volcanic eruptions. Photo via iStock.com/guenterguni. [Used under license.]

years ago, rock layers were deformed. The deformation included folding to form ridges and valleys; fracturing along joint sets, with one joint set being parallel to ridges, while the other set is perpendicular; and thrust faulting, in which rock blocks were detached and shoved upward and northwestward.

Millions of years of erosion have reduced the height of the mountains and have produced topographic inversion in the foothills. Topographic inversion occurs because joints create wider fractures at upfolded ridges and narrower fractures at downfolded valleys. Erosion is then accelerated at upfolded ridges, converting ancient ridges into valleys, while ancient valleys stand as ridges. The Valley and Ridge Province of the Appalachians is noted for such topographic inversion.

West of the Valley and Ridge Province of the Appalachians is the Allegheny Plateau, which is bounded by a cliff on its eastern side. In general, plateaus are flat topped because the rock layer that covers the surface is resistant to weathering. The cliff side is formed by erosion along joint or fault surfaces.

The Sierra Nevada range, which formed about 70 million years ago, is an example of an Andean type of mountain belt. Millions of years of erosion there has exposed igneous rocks that formed at depth. Over the years, the force of compression that formed the Sierras has evolved to form a zone of extension between the Sierras and the Colorado Plateau.

MOUNTAINS AND DEPRESSIONS IN ZONES OF EXTENSION

Extension is a strain that involves an increase in length and causes crustal thinning and faulting. Extension is associated with convergent boundaries, divergent boundaries, and transform boundaries.

Extension associated with a convergent boundary. During the formation of the Sierra Nevada, an oceanic plate that was subducted beneath California declined at a shallow angle eastward toward the Colorado Plateau. Later, the subducted plate peeled off and molten asthenosphere took its place. From the asthenosphere, lava ascended through fractures to form volcanic mountains in Arizona and Utah; lava flowed and volcanic ash fell as far west as California. The lithosphere has been heated up and has become buoyant, so the Colorado Plateau rises to higher elevations, and rock layers slide westward from it in a zone of extension that characterizes the Basin and Range Province.

In the extension zone, the top rock layers move westward on curved displacement planes that are steep at the surface and nearly horizontal at depth. When rock layers move westward over a curved detachment surface, the trailing edge of the rock layers roll over and are tilted toward the east so they do not leave space in buried rocks. On the other hand, a west-facing slope is left behind on a mountain from which the rock layers were detached. Therefore, movement along one curved detachment surface creates a valley, and movement along several such detachment surfaces forms a series of valleys separated by ridges, as in the Basin and Range Province. The amount of the displacement along the curved surfaces is not uniform. For example, more displacement has created wide zones of valleys such as the Las Vegas valley in Nevada, and Death Valley in California.

Extension associated with a divergent boundary. The longest mountain chain on Earth lies under the Pacific Ocean. It is about 60,000 kilometers (37,500 miles) long, 50 kilometers (31.3 miles) wide, and 3 kilometers (2 miles) high. The central part of this midoceanic ridge is marked by a depression, about 1,000 meters (3,000 feet) deep, called a "rift valley." A part of the submarine ridge, called the "East Pacific Rise," forms the seafloor sector in the Gulf of California and reappears off the coast of northern California, Oregon, and Washington as the Juan de Fuca Ridge. Another part forms the seafloor sector in the Gulf of Aden and Red Sea seafloor, part of which is exposed in the Afar of Ethiopia. From the Afar southward to the southern part of Mozambique is the longest exposed rift valley on land, the East African Great Rift Valley.

A rift valley is the place where old rock is pushed aside and new rock is created. Blocks of rock that are detached from the rift walls slide down by a series of normal fault displacements. The ridge adjacent to the central rift is present because hot rocks are less dense and therefore buoyant. If the process of divergences continues from the rifting stage to a drifting stage, as the rocks move farther away from the central rift, the rocks become older, colder, and denser, and push on the underlying asthenosphere to create basins. These basins will be flooded by oceanic water as neighboring continents drift away. However, not all processes of divergence advance from the rifting to the drifting stage.

Extension associated with transform boundary. Perhaps the best-known example of a transform boundary is the San Andreas Fault that offsets the East Pacific Rise from the Juan de Fuca Ridge, and is exposed on land from the Gulf of California to San Francisco. Along transform boundaries, there are pull-apart basins that may be filled to form lakes, such as the Salton Sea in Southern California. Another example is the Aqaba transform of the Middle East, along which the Sea of Galilee and the Dead Sea are located.

—*H. G. Churnet*

Further Reading

Dawson, J. B. *The Gregory Rift Valley and Neogene-Recent Volcanoes of Northern Tanzania*. The Geological Society, 2008.
Eyles, Nick, and Andrew D. Miall. *Canada Rocks: The Geologic Journey*. Fitzhenry and Whiteside, 2018.
Lowrie, William. *Geophysics: A Very Short Introduction*. Oxford UP, 2018.
Searle, Roger. *Mid-Ocean Ridges*. Cambridge UP, 2013.
Sharma, Vijay K. *Introduction to Process Geomorphology*. CRC Press, 2010.
Treloar, P. J., and M. P. Searle. *Himalayan Tectonics: A Modern Synthesis*. Geological Society, 2019.
Van der Pluijm, Ben A., and Stephen Marshak. *Earth Structure: An Introduction to Structural Geology and Tectonics*. W.W. Norton, 2010.

INTERNATIONAL SYSTEM OF UNITS

Fields of Study: All Sciences; Metrology

ABSTRACT

The initial development of the metric system, which served as the basis of the International System of Units (SI), occurred in the mid-eighteenth century, coinciding with the beginning of the age of modern science. The common nature of science demanded an equally consistent system of units and measurements that would apply across all fields and branches of science-based endeavor. The units in the SI have been defined by international accord to provide that consistency.

KEY CONCEPTS

base unit: unit of measurement that is used for one of the seven fundamental properties; also known as defined unit

conversion factor: ratio of one unit of measure in a particular system to the unit of measure of the same property in a different system of measurements

derived unit: unit of measurement that is derived from the relationship of base units

dimensional analysis: consideration and manipulation of the units involved in a measurement and their relationship to one another

fundamental property: property ascribed to a phenomenon, substance, or body that can be quantified: length, mass, time, electric current, thermodynamic temperature, the amount of a substance, and luminous intensity

International Standard: unit of measurement whose consistent value is recognized as a common unit of measurement by international agreement

measurement: ascribing a value to the determination of the quantity of a physical property

Standard Temperature and Pressure (STP): value of the mean atmospheric pressure and mean annual temperature at sea level, chosen by international agreement to be designated as 1 standard atmosphere or 760 torr and 0° Celsius/32° Fahrenheit or 273.15° Kelvin

unit: single definitive basic magnitude of portions used to measure various properties

BEGINNINGS OF THE METRIC SYSTEM

The initial development of the metric system, which served as the basis of the International System of Units (i.e., Système International d'Unités; known as SI), occurred during the French Revolution in the mid-eighteenth century. This coincided with the beginning of the age of modern science, especially chemistry, physics and engineering, as the value of consistent physical measurements in the conduct of those pursuits became apparent. As scientific activities became more precise and founded on sound theory, the common nature of science demanded an equally consistent system of units and measurements that would apply across all fields and branches of science-based endeavor. The units in the SI have been defined by international accord to provide that consistency. The basic units are defined for only seven fundamental properties of matter. All other consistent units are derived as functions of these seven fundamental units.

DEFINITION AND BASIC PRINCIPLES OF SI

The International System of Units (SI) is the internationally accepted standard system of measurement in use throughout the world. The units of the SI are ascribed to seven fundamental physical properties and two supplementary properties. These are length, mass, time, electric current, thermodynamic temperature, the amount of a substance, luminous intensity, and the magnitude of plane and solid angles.

Length is the extent of some physical structure or boundary in two dimensions, such as the distance from one point to another, how tall someone is, and the distance between nodes of a sinusoidal wave. The SI base unit associated with length is the meter.

"Mass" refers to the amount of material in a bulk quantity. The term is used interchangeably with weight, although the mass of any object remains constant while its weight varies according to the strength of the gravitational field to which it is subject. The SI base unit for mass is the gram.

Time is a rather more difficult property to define outside of itself, as it relates to the continuous progression of existence of some state through past, present, and future stages. The SI base unit of time is the second.

Electric current is the movement of electronic charge from one point to another, and it is ascribed by the SI base unit of the ampere. One ampere is defined as the movement of one coulomb of charge, as one mole of electrons, for a period of one second.

The amount of a substance is defined by an SI base unit called the "mole." The term is essentially never used outside of the context of atoms, molecules, and certain subatomic particles such as electrons. A mole of any substance is the quantity of that substance containing the number of atoms or molecules as there are atoms in 12 grams of carbon-12; this number is referred to as Avogadro's number (or the Avogadro constant), which is 6.02214×10^{23} of atoms, molecules, or particles.

The term "luminous intensity" refers to the brightness, or the quantity, of light or other electromagnetic energy being emitted from a source. Initially luminous intensity was referred to by comparison to the light of a candle flame, but the variability of such a source is not conducive to standardization. The SI base unit for luminous intensity is the candela. This is about the amount of light emitted by a candle flame, but it has been standardized to mean an electromagnetic field strength of 0.00146 watts.

The derived unit known as the plane angle refers to the angular separation of two lines from a common point in a two-dimensional plane. The SI derived unit for plane angles is the radian. A complete rotation about a point origin is an angular displacement of 2π radians. The extension of this into three dimensions is known as the solid angle and is measured by the SI derived unit called the "steradian."

Measurement of any property or quantity is accomplished by comparing the particular amount of the property or quantity to the amount represented by the standard unit. For example, an object that is 3.62 times as long as the distance defined as 1 meter is said to be 3.62 meters long. Similarly, an object that is proportionately 6,486 times as massive as the quantity defined as 1 gram is said to weigh or to have a mass of 6,486 grams. The use of standard reference units such as the SI units ensures that measurements and quantities have the same meaning in all places where that system is used.

BACKGROUND AND HISTORY OF SI

Traditionally, systems of measurement have employed units that related to various parts or proportions of the human body. Ready and convenient measuring tools in earlier times included the hand, the foot, the thumb, and the pace. Because these human measures tend to vary somewhat from person to person, no work that depended on measurements could be repeated exactly, even if the same person carried it out. Specialization in tasks eventually led to the realization that standard units of measurement would be beneficial, and determination of certain units were established by royal decree as long ago as in the signing of the Magna Carta.

The Industrial Revolution in Europe and the growth of science as a common international pursuit drove the need for a unified system of measurement that would be independent of human variability and consistent from place to place. The International System of Units was developed and first put to use in about 1799. It represents the first real standardized system of measurements. Prior to this time, a broad variety of measurement systems were in use because many countries had developed their own measuring standards for use internally and in any territories that it held.

A need for a standardized system of measurement had been recognized by various luminaries and was proposed in 1670 by French scientist Gabriel Mouton. The incompatibility and variability of the many measurement systems in use at that time often resulted in unfair trade practices and power struggles. In 1790, the French Academy of Sciences was charged with developing a system of measurement that would be fixed and independent of any inconsistencies arising from human intervention. On April 7, 1795, the French National Assembly decreed the use of the meter as the standard unit of length. It was defined as being equal to one ten-millionth part of one quarter of the terrestrial meridian, specified by measurements undertaken between Dunkirk and Barcelona. The liter was later defined as being a unit of volume equal to the volume of a cube that is one-tenth of a meter on a side, and the kilogram as being equal to the mass of one liter of pure water. Larger and smaller quantities than these are indicated by the use of increasing and decreasing prefixes, all indicating units that were some power of ten times greater or lesser than the base units.

Standard reference models of the metric units were made and kept at the Palais-Royale, and it was then required that measuring devices being used for trade and commerce be regularly checked for accuracy against the official standard versions.

HOW SI WORKS

All measurement practices relate the actual properties and proportions of something to a defined standard that is relevant to that property or proportion. For example, two-dimensional linear quantities are related to a standard of length, and the mass of an object is proportionately related to a standard unit of mass. In the measurement of length, for example, the standard unit of the SI is the meter. An object that is determined to be, for example, 6.3 times longer than the base unit of one meter is therefore said to be 6.3 meters long. Similarly, an object that is, say, 5.5 times heavier than the base unit of one kilogram is said to weigh 5.5 kilograms, and something that has a volume of 22.4 times that of the base unit of one liter is said to have a volume of 22.4 liters.

The modern metric system is simple to use, especially when compared with its most common predecessor systems, the British Imperial and the U.S. customary systems. Whereas these predecessor systems use units such as inches, feet, yards, ounces, and pounds that relate to one another irregularly, the metric system has always used the simple relation of factors of ten for all of its units. For example, the decameter is equal to ten meters (10 × 1 meter), while the decimeter is equal to one-tenth of a meter (0.1 × 1 meter). A complete series of prefixes indicates the size of the smaller and larger units that are being used in a measurement. Thus, a decimeter is ten times smaller than a meter, a centimeter is a hundred times smaller than a meter, a millimeter is a thousand times smaller, and so on. Correspondingly, there are ten millimeters in a centimeter, ten centimeters in a decimeter, and ten decimeters in a meter. There are one thousand milliliters in a liter, one hundred liters in a hectoliter, one thousand grams in a kilogram, and one thousand milligrams in a gram. The uniformity of the system readily allows one to visualize and estimate relative sizes and quantities.

The basic units of SI have always been defined with the intention of relating to some universal and unalterable standard. In some cases, the definitions have been changed to relate the unit to a more permanent universal feature or a more accurately known property. For example, with the high accuracy of time measurement that has become available, the definition of the meter has been changed from being a specific fraction of the distance between two fixed terrestrial points to "the length of the path traveled by light in a vacuum during a time interval of 1/299792458 of a second."

The units of measurement of other properties and quantities reflect the fact that those properties and quantities can be viewed as combinations of the seven fundamental properties. That is, for any dimensional property Q, the dimensions of Q are derived from the expression:

$$\dim Q = L^{\alpha}M^{\beta}T^{\gamma}I^{\delta}\theta^{\varepsilon}N^{\zeta}J^{\eta}$$

The exponents a, b, c, d, e, f, and g are integers representing the degree of involvement of the corresponding property of length (L), mass (M), time (T), electric current (I), thermodynamic temperature (θ), amount of material (N), and luminous intensity (J). (Remember that any quantity raised to the power 0 has the numerical value of 1.)

APPLICATIONS AND PRODUCTS

Apart from its everyday fundamental use in all science-based fields, applications and products related to the metric system essentially fall into two categories: educational devices and training to promote familiarity with the system and metric versions of existing products and the tools required for their maintenance.

Educational devices and training. The long history of independent measurement systems has served to entrench those systems in common usage. Therefore, a very large body of materials, products, and devices have been constructed using nonmetric measurements. More significantly, those independent systems have been so deeply entrenched in the education systems of many nations that several generations of people have grown up using no other system. The familiarity gained through a lifetime of using a particular system generally gives individuals the ability to visualize and estimate quantities in terms of that measurement system. Making the change to a different measurement system such as the metric system represents a paradigm shift that leaves many individuals unable to associate the new measurement units with even the most familiar dimensions. Such a change, however, can be accommodated in a number of ways as the world continues to adopt the metric system of measurement as its universal standard.

The most basic method of replacing one system with another is to incorporate the new system into the basic education system, teaching it in such a manner that it becomes the entrenched measurement system as children progress through school. It is, therefore, important not only that teachers are educated in the use of the metric system but also that they actively replace their own reliance on any former system that they have used. This guarantees that, in time, the older system is completely displaced from the public lexicon and the metric system becomes the primary measurement system over a period of essentially one generation.

Those who have already left the school system can learn the metric system through training programs. Training in the metric system can be incorporated into professional development programs at the workplace or take the shape of formal training programs offered by local educational institutions or third-party providers.

Metric versions of existing devices. The vast majority of goods and devices produced in countries that have not adopted the metric standard must be maintained using their original component dimensions because metric values and nonmetric values are not generally interchangeable. The fundamental difference between the two bases of measurement ensures that any coincidence of size from one system to the other is exactly that: coincidental. Therefore, switching to metric goods and devices requires that complete new lines of products be made with metric rather than nonmetric dimensions. This requirement places an odd sort of constraint on the situation because nonmetric parts and tools must still be produced in order to maintain existing nonmetric devices. At the same time, new devices to replace those that fail are produced in metric dimensions. This means that tradespeople, such as automobile mechanics, maintenance workers, and engineers, must obtain double sets of tools, and supply stores must maintain double sets of components and supplies. In addition, tools for taking measurements must be capable of using both metric and nonmetric dimensions, although the incorporation of electronic capabilities into many tools has greatly minimized the difficulties that arise from this dual requirement.

IMPACT ON INDUSTRY

In some countries that have officially adopted the metric system to replace a system such as the Imperial system, the change has been sudden and dramatic. For example, when Canada adopted the metric system in the 1970s, the changeover was quick and, at times, dramatic. The Imperial system of measurement was phased out and replaced by the metric system in a relatively short time, with the key feature being that after a certain date, Imperial units would no longer be used for any quantities that were subject to government regulation. On that date, gas stations changed from price per gallon to price per liter, automobile speedometers were read in kilometers per hour, and Canadian schools taught using only the metric system. Average citizens were basically left on their own to learn and adapt to the new system. In many respects, it posed few hardships, because many goods were purchased by dollar amounts rather than unit quantities. Regulatory control of how goods were measured and delivered ensured that the consumer received the correct amount of what was purchased.

In other countries, such as the United States, the American democratic process has stalled conversion to the metric system. In 1800, the United States could have become the second nation in the world to adopt SI because of its strong ties with France, where SI was developed. It chose instead to adopt the U.S. customary system of measurement, which used units similar to the British Imperial system but sometimes varied in terms of quantities (an Imperial gallon held 160 fluid ounces, for example, while an American gallon held only 128 fluid ounces). The debate over adoption of the metric system in the United States continues, as many Americans oppose any official move to replace the customary system. As a result, the United States remains essentially the last major country in the world to maintain a system of measurement other than the metric system. In practical terms, the failure to completely adopt the metric system, even in the scientific and technical fields, has cost the country billions of dollars in lost trade and costly errors. For example, in 1999, a multimillion-dollar space probe crashed on Mars because the technical contractors had failed to notice that some of the data being used were in metric units while other data were in U.S. customary units.

CAREERS AND COURSEWORK

The International System of Units is not broad enough to provide a field of study or advanced coursework and, by itself, is unlikely to form the basis of a career. However, during the process of converting from a nonmetric measurement system to a metric system, opportunities will arise for those familiar with SI to create and provide training to facilitate understanding and use of the metric system.

Anyone entering a technical or scientific field must become familiar with the metric system, as this has been the standard system of measurement in those areas since the SI was developed. As the metric system becomes more and more widely adopted and accepted, students and others should expect to carry out measurements and calculations using the appropriate metric system units.

Careers that depend specifically on measurement include quality-control engineering and most branches of scientific research and physical engineering. Specific examples of measurement-based careers include civil engineering, medical and biochemical analysis, analytical chemistry, metrology, mechanical engineering, and industrial chemical engineering.

SOCIAL CONTEXT AND FUTURE PROSPECTS

It seems likely that the metric system will ultimately become the standard system of measurement in the United States, as it already is in the rest of the world. As international trade and offshore manufacturing increase, the necessity for American industry to adopt SI increases. In addition, educational systems have increased their focus on science and technology, and metric systems will become more familiar to children in American schools.

In the meantime, development of other measurement systems and hybrids of measuring systems progresses. For example, the manufacturing community has adopted a standard unit called the "metric inch," which is the standard inch divided into hundredths, to be used on visual measuring devices. There is some logic to this adoption, as this measurement apparently corresponds with the limits of differentiation of which the human eye is capable to a better degree than does the millimeter division of the metric system. However, such developments are likely to delay and otherwise interfere with the adoption of the metric system.

—Richard M. Renneboog

Further Reading

Butcher, Kenneth S., Linda D. Crown, and Elizabeth J. Gentry. *The International System of Units (SI): Conversion Factors for General Use*. National Institute of Standards and Technology, 2006.

Cardarelli, Francois. *Encyclopedia of Scientific Units, Weights and Measures: Their SI Equivalences and Origins*. Springer, 2003.

Himbert, M. E. "A Brief History of Measurement." *European Physical Journal Special Topics*, vol. 172, 2009, pp. 25-35.

K

Kimberlites

Fields of Study: Geology; Geomorphology; Geochemistry; Gemology; Crystallography

ABSTRACT

Kimberlite is a variety of ultramafic rock that is fine- to medium-grained, with a dull gray-green to bluish color. Often referred to as a mica peridotite, kimberlite originates in the upper mantle under high temperature and pressure conditions. It frequently occurs at Earth's surface as old volcanic diatremes and dikes. Kimberlite can be the source rock for diamonds.

KEY CONCEPTS

blue ground: the slaty blue or blue-green kimberlite breccia of the South African diamond pipes

diamond: a high-pressure, high-temperature mineral consisting of the element carbon; it is the hardest naturally occurring substance and is valued for its brilliant luster

diatreme: a volcanic vent or pipe formed as the explosive energy of gas-charged magmas breaks through crustal rocks

dike: a tabular body of igneous rock that intrudes vertically through the structure of the existing rock layers above

magma: a semiliquid, semisolid rock material that exists at high temperatures and pressures and is mobile and capable of intrusion and extrusion; igneous rocks are formed from magma as it cools

mantle: the layer of Earth's interior that lies between the crust and the core; it is believed to consist of ultramafic material and is the source of magma

peridotite: any of a group of plutonic rocks that essentially consist of olivine and other mafic minerals, such as pyroxenes and amphiboles

ultramafic: a term used to describe certain igneous rocks and most meteorites that contain less than 45 percent silica; they contain virtually no quartz or feldspar and are mainly of ferromagnesium silicates, metallic oxides and sulfides, and native metals

xenocrysts: minerals found as either crystals or fragments in some volcanic rocks; they are foreign to the body of the rock in which they occur

xenoliths: various rock fragments that are foreign to the igneous body in which they are present

KIMBERLITE COMPOSITION AND EMPLACEMENT

The rock known as kimberlite is a variety of mica-bearing peridotite and is characterized by the minerals olivine, phlogopite (mica), and the accessory minerals pyroxene, apatite, perovskite, and various opaque oxides such as chromite and ilmenite. Chemically, kimberlite is recognized by its extraordinarily low silicon dioxide content (25 to 30 percent), high magnesia content (30 to 35 percent), high titanium dioxide content (3 to 4 percent), and the presence of up to 10 percent carbon dioxide. It is a dark, heavy rock that often exhibits numerous crystals of olivine within a serpentinized groundmass. Upon weathering, kimberlite is commonly altered to a mixture of chlorite, talc, and various carbonates and is known as blue ground by diamond miners. Occasionally, kimberlites contain large quantities of diamonds, which makes them important economically.

"Kimberlite" was first used in 1887 by Carvill Lewis to describe the diamond-bearing rock found at Kimberly, South Africa. There, it primarily occurs as a breccia found in several deeply eroded volcanic pipes and also in an occasional dike. Unfortunately, the kimberlite is so thoroughly brecciated and chemically altered that it does not lend itself to detailed petrographic study. Instead, the kimberlite of Kimberly, like kimberlite elsewhere, is more often noted for the varied assortment of exotic xenoliths (inclusions of other rocks) and megacrysts (large crystals) it contains. The explosive nature of a kimberlite pipe leads to the removal of country rock (rock surrounding an igneous intrusion) as the magma passes through the crust and thus can provide scientists with samples of material that originated at great depths. Among the many xenoliths found in the kimberlite breccia pipes are ultramafic rocks such as garnet-peridotites and eclogites, along with a variety of high-grade metamorphic rocks. These specimens provide scientists with an excellent vertical profile of the rock strata at various locations, serve to construct a model of Earth's crust at various depths, and provide information on the chemical variations in magma.

The emplacement of kimberlite is believed to be due to the eruption of a gas-rich magma that has intruded rapidly up a network of deep-seated fractures and probably breached the surface as a volcanic eruption. The magma's rate of upward mobility must have been rapid, as evidenced by the occurrence of high-density xenoliths such as eclogite and peridotite within the kimberlite pipes. The final breakthrough of the kimberlite magma may have taken place at a depth of 2 to 3 kilometers, where contact with groundwater contributed to its propulsion and explosive nature. Brecciation rapidly followed, along with vent enlargement by hydrologic fracturing of the country rock. Fragments of deep-seated rock, along with other country rocks,

Picture of a diamond-bearing kimberlite rock. Photo by Woudloper, CC BY-SA 1.0, via Wikimedia Commons.

were then incorporated within the kimberlite magma.

KIMBERLITE MINERALOGY AND TEXTURE

As a rock, kimberlite is very complex. Not only does it contain its own principal mineral phases, but it also has multicrystalline fragments or single crystals derived from the various fragmented xenoliths that it collected along the way. These fragments represent upper mantle and deep crustal origins and are further complicated by the intermixing with the mineralogy of a highly volatile fluid. As a result, no two kimberlite pipes have the same mineralogy. The continued alteration of the high-temperature phases after crystallization can produce a third mineralogy that can affect the interpretation of a particular kimberlite's occurrence.

The characteristic texture of kimberlite is inequigranular because of the presence of xenoliths and megacrysts within an otherwise fine-grained matrix. In relation to kimberlite, the term "megacryst" refers to both large xenocrysts and phenocrysts, with no genetic distinctions. Among the more common megacrysts present are olivine (often altered to serpentine), picroilmenite, mica (commonly phlogopite), pyroxene, and garnet. These megacrysts are usually contained in a finer-grained matrix of carbonate and serpentine-group minerals that crystallized at considerably lower temperatures. Among the more common matrix minerals are phlogopite, perovskite, apatite, calcite, and a very characteristic spinel. Found within the textures of these matrix minerals are examples of both rapid and protracted cooling, with the latter evidenced by zoning. Zoning indicates that the matrix liquid cooled after emplacement and that there was sufficient time for crystals present to react with the remaining liquid, which is common to the megacrysts as well. In addition, the megacrysts exhibit an unusual, generally rounded shape that is believed to be a result of their rapid transport to the surface during the eruptive phase.

KIMBERLITE OCCURRENCE

The geological occurrence of kimberlite, clearly volcanic in origin, takes the form of diatremes, dikes, and sills of relatively small size. In shape, kimberlite diatremes usually have a rounded or oval appearance but can occur in a variety of forms. Quite often, diatremes occur in clusters or as individuals scattered along an elongated zone. They rarely attain a surface area greater than a square kilometer but in profile will resemble an inverted cone that descends to a great depth. When it occurs as a dike, kimberlite is quite small—often not more than a few meters wide. It may occur as a simple ring dike or in swarms of parallel dikes. Kimberlite's occurrence as a sill is quite rare and will have a wide variability in thickness.

As compared with other types of igneous rocks, the occurrence of kimberlite is considered to be quite rare. Based on factors such as specific matrix color, density, mineral content, and xenolith size, shape, and number, connections between a certain kimberlite and a specific occurrence can be made. Aside from their scientific value, most kimberlites are economically worthless, except when they contain diamonds.

DIAMOND-BEARING KIMBERLITE

Of the many minerals that constitute kimberlite, diamond is the most noteworthy because of its great economic value. Not all kimberlite contains diamonds, and when diamonds are present, they are not always of gem quality. Even in the most diamondiferous kimberlites, the crystals and cleavage fragments are rare and highly dispersed. Diamond mining from kimberlite requires the removal and processing of enormous amounts of rock to produce relatively few diamonds. This method is expensive and can be dangerous. A much better rate of return can be gained from the placer mining of riverbeds and shorelines where nature has weathered out the diamonds and concentrated them in more readily accessible areas. The discovery of a large diamond in a riverbed in 1866 led to the eventual search for the source rock that produced diamonds. Up to that time, diamonds were recovered only from alluvial deposits and not from the host rock. In 1872, as miners were removing a diamond-bearing gravel, they uncovered a hard bluish-green rock that also contained diamonds. Further mining revealed the now familiar circular structure of the diatreme that continued downward to an undetermined depth. In 1887, the first petrographic description was made of the rock now known as kimberlite, and it was then recognized to be similar to other volcanic breccias from around the world.

Diamond-bearing kimberlite locations can be found around the world, the most famous being in South Africa. The diamond pipes of South Africa have been the leading producer of diamonds for well over a hundred years and are still unrivaled in terms of world production. Other locations of kimberlites that produce diamonds include the pipes of Siberia, northern Canada, western Australia, and Wyoming. For many years, Arkansas had the only commercial diamond mine in the United States, but the site is no longer active and is now preserved as a state park where visitors can hunt for diamonds.

KIMBERLITE ORIGIN HYPOTHESES

The chemistry and mineralogy characteristic of kimberlite indicate a very complex set of conditions that existed during their emplacement and subsequent crystallization. This fact, combined with kimberlite's limited occurrence and the altered nature of its matrix, makes kimberlite a puzzle to scientists. Three hypotheses have been proposed to explain its origin. A zone-refining hypothesis describes a liquid generated at great depth (600 kilometers) that is dynamically unstable and, as a result, rises toward the surface. As it reaches specific lower pressures, its composition is altered through partial crystallization and fractionation. A second hypothesis envisions a residual process: Partial melting of a garnet-peridotite parent, at depths of 80 to 100 kilometers, produces fractional crystallization of a picritic basalt (olivine-rich, or more than 50 percent olivine) at high pressure, which could lead to the formation of eclogite and a residual liquid of kimberlite composition. The third hypothesis describes kimberlite as either the residual end product of a long fractionation process or the product of a limited amount of partial melting. Each of the three hypotheses has merit but falls short of providing a definitive answer, partly because the near-surface environment where kimberlites are found is one of complex chemical reactions, which makes interpretation difficult. Adding to the problem is that no kimberlite eruption has ever been observed, so the exact processes are still somewhat speculative.

Although kimberlites tantalize scientists with their complexity, the xenoliths and megacrysts that arise with the kimberlites provide substantial data on the relationship between the lower crust and the upper mantle. Kimberlites show Earth's upper mantle to be very complex in terms of petrography, and they define both large- and small-scale areas of chemical and textural heterogeneities. Pressure and temperature conditions have been accurately established for depths down to 200 kilometers through studies based on the specimens brought up during kimberlite eruptions.

KIMBERLITE SPECIMEN ANALYSIS

Analysis of a rock such as kimberlite begins with the collection of a fresh specimen (as unaltered by weathering as possible) and continues with the preparation of a series of thin sections and microprobe sections. In this process, a very thin slice of rock is cut with a diamond saw and glued to a glass plate. A second cut is then made to reduce the specimen's thickness to nearly 0.03 millimeter. A final polishing will achieve this thickness, which will permit light to pass through the specimen and provide the opportunity for microscopic examination. A similar procedure will produce a microprobe-thin section, but it requires extra polishing to ensure a uniform surface.

Once the thin sections have been prepared, a geologist uses a petrographic microscope, which employs polarized light, to make proper identification of the mineral phases present, along with their specific optical parameters. Opaque minerals require reflective light study. Microscopic examination is usually the first step in classifying a rock, and it may include an actual point count of the minerals present to determine their individual ratios. Afterward, an electron microprobe or electron microscope is used to give specific chemical compositions for individual mineral phases. With these devices, the scientist can analyze mineral grains as small as a few microns with a high degree of accuracy. Mineral grains or crystals that are smaller require an analyzing electron microscope to reveal their composition and fine detail. If the individual minerals are large enough and can be extracted, then X-ray diffraction can be used for a definitive identification of a particular mineral phase.

It is also important to understand the bulk chemistry of a rock specimen. Here, the scientist can select several different methods, including neutron activation analysis (NAA) and atomic absorption spectrometry (AAS). Both techniques provide excellent sensitivity and precision for a wide range of elements. A third method, X-ray fluorescence (XRF), is also commonly used and provides an accurate and quick means for analysis.

In NAA, a specimen is activated by thermal neutrons generated in a nuclear reactor. In this process, radioisotopes formed by neutron bombardment decay with characteristic half-lives and are measured by gamma-ray spectrometry. In AAS, a specimen is first dissolved in a solution and then identified element by element. XRF, which provides sensitivity and precision for quantitative determination of a wide range of elements (best for those above atomic number 12), is the principal method used by most geologists to gather element amounts in both rocks and minerals.

After all the bulk chemical analyses have been gathered and run through various computer programs to determine mineral compositions, the data are compared to the microscopy results for evaluation and classification of the rock. This evaluation is compared to similar data from other locations. A final check and comparison to other rock types and strata in the field collection area complete the analysis.

SCIENTIFIC AND INDUSTRIAL APPLICATIONS

Although some kimberlites have economic value in their diamond content, most do not. No dramatic breakthroughs from the study of kimberlites will make diamonds more abundant or cheaper in price. The result of these studies is a better understanding of the processes and conditions under which diamonds and semiprecious minerals such as garnet are formed. This knowledge has led to the development of synthetic gems that can be used for industrial applications.

The study of kimberlites also has a direct bearing on scientists' understanding of volcanic activity. The prediction of eruptions is an aim of geology, as it affects the hundreds of thousands of people who live near potentially dangerous volcanoes. Kimberlite magma, by the nature of its chemical composition, is a very explosive volcanic material and can produce violent eruptions. Even so, it is true that most kimberlite magmas do not reach the surface, and when they do, they affect only a small area. Kimberlite magmas are more effective at depth, where rock is being fractured by the rising magma and by the hydraulic pressures exerted by the various gases moving through the magma. By examining the results of the movement of a kimberlite magma, scientists can gain a better understanding of volcanic behavior.

—Paul P. Sipiera

Further Reading

Dawson, J. B. *Kimberlites and Their Xenoliths*. Springer, 2012.

Decker, Robert, and Barbara Decker. *Volcanoes*. 4th ed., W. H. Freeman, 2005.

Le Roex, Anton P., David R. Bell, and Peter Davis. "Petrogenesis of Group I Kimberlites from Kimberley, South Africa: Evidence from Bulk-Rock Geochemistry." *Journal of Petrology*, vol. 44, 2003, pp. 2261-86.

Miller, Roger H. *Kimberlites: Mineralogy, Geochemistry and Petrology*. Springer, 2013.

Nemeth, Karoly, editor. *Updates in Volcanology: From Volcano Modelling to Volcano Geology*. IntechOpen, 2016.

Sage, R. P., and T. Gareau. *A Compilation of References for Kimberlite, Diamond and Related Topics: Ontario Geological Survey, Open File Report 6067*. Queen's Printer for Ontario, 2001.

Tappert, Ralf, and Michelle C. Tappert. *Diamonds in Nature: A Guide to Rough Diamonds*. Springer, 2011.

L

Lehmann Discovers Earth's Inner Core

Fields of Study: Geology; Geophysics; Seismology

ABSTRACT
Inge Lehmann's hypothesis that Earth has an inner and outer core led to investigations that confirmed her theory in 1936, following the development of the seismograph.

KEY CONCEPTS
P-wave: a longitudinal compression wave produced by the impact of plate segments during an earthquake
PKP-wave: a third type of P-wave that passes through the core
PP-wave: a second type of P-wave that is refracted by a boundary within the core
S-wave: a lateral (side-to-side) wave produced by the relative movement of plate segments during an earthquake
seismograph: a device that detects seismic vibrations produced by earthquakes, bomb blasts, heavy vehicles, etc.
seismogram: a recording of the vibrations detected by a seismograph

BASIC SEISMOLOGY
Prior to the development of the seismograph, an instrument that records the vibrations of Earth, or earthquakes, very little was known about the composition of the inner parts of the planet. In the late 1800s, a basic knowledge of vibrations generated by a seismic source evolved. It was found that these vibrations, or waves, travel outward through Earth at measurable speeds.

The science of geophysics recognizes two major types of "elastic" waves, which were defined by the British geologist Richard Dixon Oldham. The first type is called a "P-wave" because it causes a "primary" disturbance that deforms Earth by means of alternately lengthening and shortening in the direction of the source of the wave. P-waves are also called "compressional waves" because the volume of the rock that is affected is alternately compressed and expanded. The second type of elastic wave is the S-wave, which produces a "secondary" distur-

Inge Lehmann in 1932. Photo by The Royal Library, National Libary of Denmark and University of Copenhagen University Library, CC BY-SA 4.0, via Wikimedia Commons.

bance. The S-wave is a transverse body wave that travels through the interior of an elastic medium. S-waves do not change the volume of the medium, but they do change its shape; for this reason, they are also called "distortional waves" or "shear waves." Both P-waves and S-waves pass through the interior of Earth; for this reason, they are called "body waves."

When an earthquake occurs, its waves travel through the body of Earth and are recorded by seismographs at earthquake observatories. These seismic waves carry to the surface information about the material through which they have passed. In 1883, the English engineer John Milne surmised that every large earthquake at any point of the globe could be recorded if there were an instrument designed for that purpose; however, it was not until 1893 that he perfected the first clockwork-powered seismograph.

In later years, when extremely sensitive seismographs had been developed, it was found that some weak P-waves actually were penetrating a shadow zone, an area opposite the projected core of Earth. The shadow zone, discovered in the research of the Croatian geophysicist Andrija Mohorovičić, was left unexplained in earlier research conducted by pioneers in the use of seismographs to map the interior. Inge Lehmann postulated the existence of an inner core that could reflect the waves back into the shadow zone.

At the Copenhagen Seismological Observatory, Lehmann had for a number of years been clearly observing, through the core, seismic waves caused by earthquakes in the Pacific Ocean. Among these were the shocks that occurred at Murchison and Hawke's Bay in New Zealand in 1928 and 1931, respectively. It was evident from these records that a P-type wave that should have been within the shadow zone was arriving at seismological stations. This phenomenon could be explained only by the existence of an inner core that was about 1,250 kilometers (roughly 777 miles) in radius and was denser than the outer core.

Lehmann believed that core waves could be classified into three separate types of P-wave. The standard explanation for the first two of these wave types was that they were refracted at the boundary between the mantle and core and focused toward the antipodes, placed opposite each other on the globe. She explained that waves of the third type were reflections from another sharp discontinuity within the core itself. This family of waves is made up of the core refractions. Beyond an angle of about 103°, the direct P-wave cannot be recorded because of the shadow effect of the core. Beyond this distance, the first wave to appear on long-period instruments is often a PP-wave, which does not penetrate so deeply and therefore is able to avoid the obstacle. Short-period instruments show a refracted wave arising from complexities within the core, but it is not quite as prominent as P when it makes its reappearance at an angle of 142°. Because it is deflected from its path and disappears altogether for nearly 40°, it is called a "PKP-wave" (K stands for *Kern*, the German word for "core").

DISCOVERING THE CORE

In 1936, after ten years of interpreting seismograms (records made by a seismograph) and using a well-established scientific method, Lehmann was prepared to discover the inner core. Her first step was to calculate a direct problem. She assumed an Earth model that was particularly simple. It had constant velocities in the mantle (10 kilometers, or 6.2 miles, per second) and in the core (8 kilometers, or 5 miles, per second). These were reasonable average values for both regions. She then introduced a small central core, which again had a constant velocity. These simplifications enabled her to view the seismic waves as straight lines; therefore, their travel times could be calculated by using elementary trigonometry. She then showed by making successive adjustments that a reasonable velocity and radius of the inner core could be found that predicted a travel-time curve close to the observations of the third type of P-wave. In effect, she proved an existence theorem: A plausible three-shell earth structure could be defined that explained the features of the observed waves.

Lehmann's discovery of the inner core was very complicated, but it convinced Beno Gutenberg in the United States and Harold Jeffreys in England that her hypothesis was a viable one. Within two years, they had independently carried out more detailed calculations involving many observed travel times of P-waves and calculated by means of an inverse method both the radius of the inner core and the P-velocity distribution in it.

SIGNIFICANCE OF LEHMANN'S DISCOVERY

After the discovery of Earth's inner core, the measured travel times could be transformed, using inverse theory, into plausible P-velocities in the mantle and the outer core. In late 1938 and 1939, Gutenberg and Jeffreys computed, independently, the average velocity based on thousands of observed travel times of P- and S-waves. Their agreement was extremely close; in fact, their calculations were so well developed that they have not been seriously altered since.

As a result of the development of sensitive seismographs, an increase in the number of seismographic stations around the world, and the availability of large-capacity computers, a better understanding of the inner structure of Earth has become possible. The core has a role in many geophysical studies, and the way it is affected during great earthquakes is being probed actively. If the physical properties inside Earth were better known, the frequencies and amplitude patterns of the resonant vibrations could be calculated, thereby making it possible to prevent loss of life.

—Earl G. Hoover

Further Reading

Lowrie, William. *Fundamentals of Geophysics.* 2nd ed., Cambridge UP, 2007.

Philips, Jon M. *The Earth's Core: Structure, Properties and Dynamics.* NOVA Science Publishers, 2012.

Stein, Seth, and Michael Wyesession. *An Introduction to Seismology, Earthquakes, and Earth Structure.* Wiley, 2013.

Swaby, Rachel. *Headstrong: 52 Women Who Changed Science-And the World.* Broadway Books, 2015.

Tarbuck, Edward J., Frederick K. Lutgens, and Dennis Tasa. *Earth: An Introduction to Physical Geology.* 13th ed., Pearson Education, 2019.

Limestone

Fields of Study: Geology; Geomorphology; Geochemistry

ABSTRACT

Limestone is one of the most widely used rock materials. It is used as road metal, as aggregate for macadam and concrete, and as a building stone.

KEY CONCEPTS

calcite: a mineral form of calcium carbonate, $CaCO_3$

coquina: a fossil-bearing form of limestone

diagenesis: the temperature and pressure-driven process that alters the structure of sedimentary rock such as limestone, and acts to cement the particles together by the formation of fine crystals in the spaces between the particles

dolostone: a mineral form of calcium carbonate, $CaCO_3$, that is dense and fine-grained

karst topography: a landform characteristic of limestone that has been eroded chemically by rain, named for the Karst region of Eastern Europe

marble: a mineral form of calcium carbonate, $CaCO_3$, that has undergone diagenesis through temperature and pressure

WHAT IS LIMESTONE?

Limestone is a widespread marine sedimentary rock found wherever shallow seas once encroached onto continents. Limestone accounts for 10 to 15 percent of all sedimentary rocks. Some limestones are formed in lakes, around springs, at geysers, and in caves. The term "limestone" encompasses many rocks of diverse appearance that have calcite as their essential component. They differ considerably in texture, color, structure, and origin, however.

OVERVIEW OF LIMESTONES

Although limestones may form by inorganic precipitation of calcite in lakes, springs, or caves, the most widespread limestones are of marine origin. Most limestones are formed by organic processes and consist largely of the shells and shell fragments of marine invertebrates. Because calcite is susceptible to solution and recrystallization, diagenetic processes may completely alter the texture of the original rock.

Limestone is a sedimentary rock composed largely of the mineral calcite (calcium carbonate). This relatively soft stone in its pure form is white, but it may be buff, pink, red, gray, or black, depending upon minor materials present. The texture ranges from fine- to coarse-grained and from highly porous to highly compact. Many limestones contain abundant fossils. Dolostone is a closely related rock composed primarily of dolomite (calcium-magnesium carbonate).

Coquina is a limestone of comparatively recent formation consisting of loosely cemented shell fragments. Compact rocks with abundant shell material are known as fossiliferous limestone. They may be described more specifically by adding the dominant fossil genera to the rock name. Chalk is a fine-grained, porous, white rock made up of minute tests of foraminifera. Lithographic limestone is a compact, fine-grained rock that is used in the printing process from which it derives its name. Travertine is an inorganic deposit usually formed in caves as coarse, crystalline dripstone. Tufa is a porous, spongy material deposited around springs and geysers. Oolitic limestone is composed of small, spherical bodies of concentrically layered calcite formed in shallow water with moderate agitation. Coarse crystalline limestone forms by recrystallization of primary, fine-grained limestones.

Limestone and other soluble rocks in warm, humid regions are susceptible to solution by meteoric water at the surface and in the subsurface. The resulting landscapes, characterized by abundant sinkholes and caverns, are known as "karst topography." Because water moves rapidly into the subsurface in karst regions, rapid spreading of contamination in groundwater is of special concern.

Some limestones that take a good polish are marketed as marble. Limestone is used as a flux in open-hearth iron smelters. It is a basic raw material in the manufacture of Portland cement. It is also used as an inert ingredient in pharmaceutical preparations. Limestone is the chief source of chemical and agricultural lime. It is also ground and pressed to make blackboard chalk. Limestone serves as a significant aquifer, and it constitutes about 50 percent of reservoir rocks for oil and gas. Prior to the introduction of electric lighting, carved chunks of limestone were fed into a gas flame to produce a fairly bright light used as stage lighting—hence the term "limelight."

—René A. De Hon

Further Reading

Boggs, Sam. "Limestones." *Petrology of Sedimentary Rocks*. 2nd ed., Cambridge UP, 2009.

Flügel, Erik. *Microfacies of Carbonate Rocks: Analysis, Interpretation and Application*. 2nd ed., Springer, 2010.

James, Noel P., and Brian Jones. *Origin of Carbonate Sedimentary Rocks*. Wiley, 2016.

Johnson, David. *Lime Kilns: History and Heritage*. Amberley Publishing, 2018.

Kogel, Jessica Elzea, et al., editors. "Lime." *Industrial Minerals and Rocks: Commodities, Markets, and Uses*. 7th ed., Society for Mining, Metallurgy, and Exploration, 2006.

Oates, J. A. H. *Lime and Limestone: Chemistry and Technology, Production and Uses*. Wiley-VCH, 1998.

LITHOSPHERE

Fields of Study: Geology; Geomorphology; Geophysics; Geochemistry

ABSTRACT

The lithosphere ("stone sphere," from Greek lithos) consists of the outer, brittle portions of the Earth, including the upper mantle and crust. The usable mineral resources of Earth are all within the lithosphere, and knowledge of its properties is particularly important in the search for mineral resources, as well as gas and oil.

KEY CONCEPTS

asthenosphere: the plastic, semifluid rock layer immediately underlying the crust

felsic rock: rock having a higher silica content than mafic rock

fractionation: the separation of mineral components in magma according to their density under the force of gravity

mafic rock: rock having a higher magnesium and iron content than felsic rock

ultramafic: rock that is greatly enriched in magnesium and iron content in comparison to mafic rock

EARTH'S LAYERS

The interior of the planet Earth has a number of layers, as concentric spheres. At the center of the planet is the inner core. Then, moving outward, come the outer core, the lower mantle, the upper mantle, and the crust. Scientists subdivide the upper mantle into the asthenosphere, a partially molten zone, and, above that, the lithosphere. The lithosphere, then, is the rigid (or brittle) outer shell of Earth, which extends to a depth of between 70 and 100 kilometers and rests on the asthenosphere. It includes Earth's crust and part of the upper mantle.

The upper mantle is approximately 700 kilometers thick. The asthenosphere begins at a depth of approximately 70 to 100 kilometers and shows a rapid increase in density and a temperature in excess of 1,000° Celsius. The asthenosphere is partially molten ultramafic material. Because of its partially molten properties, the asthenosphere probably exhibits plastic flow. Above the asthenosphere, the upper brittle portion of the upper mantle that is part of the lithosphere is a dense ultramafic material that directly underlies Earth's crust. The lithosphere comprises several major lithospheric "plates" that move slowly as they rest on the asthenosphere. Plate tectonics refers to the movement of these plates and the land and ocean forms that are created as a result.

Within the lithosphere, the boundary between the upper mantle and the crust is called the "Mohorovičić Discontinuity," or "Moho," which marks a compositional change in the rock. Earth's

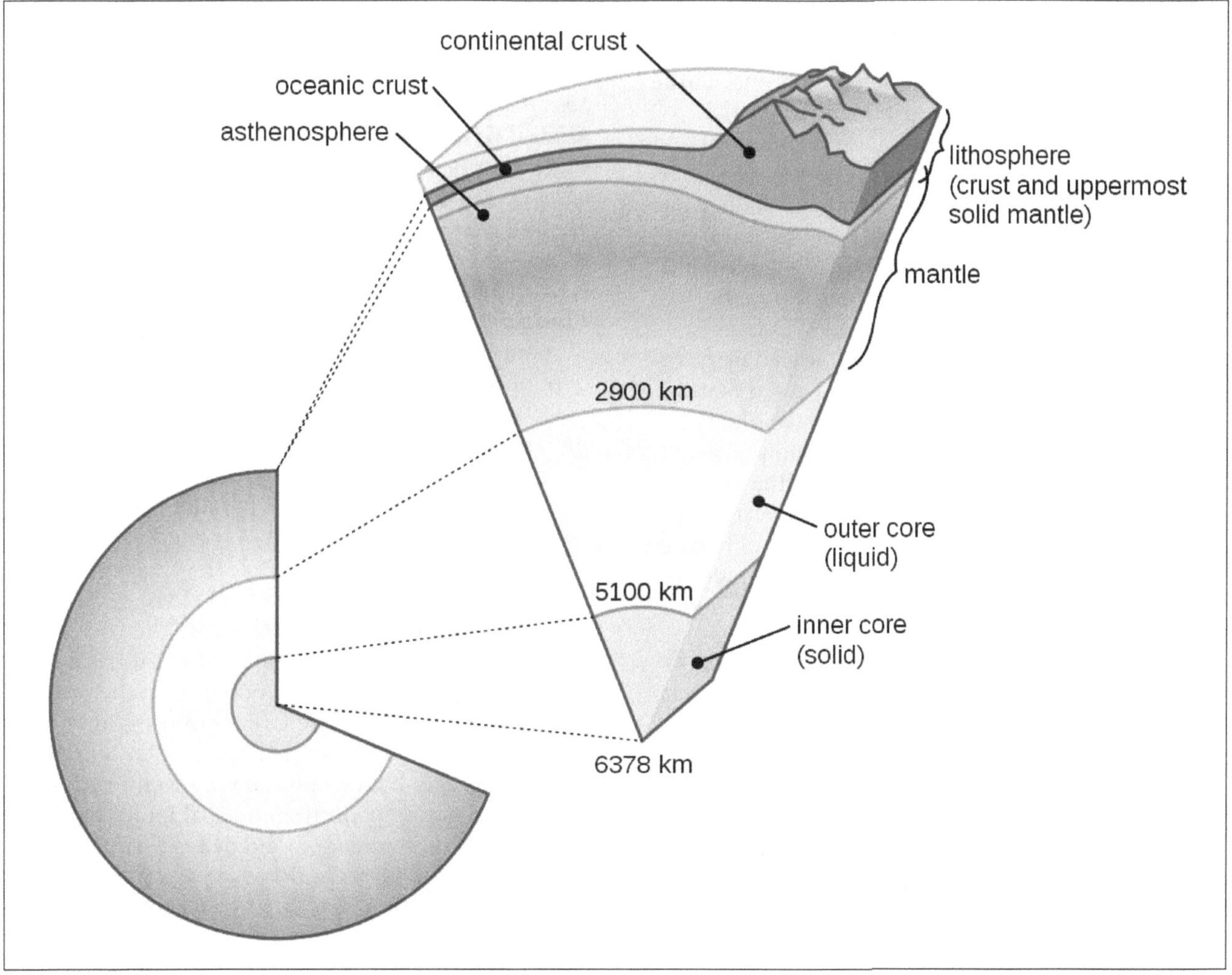

Earth cutaway from center to surface, the lithosphere comprising the crust and lithospheric mantle. Image via USGS/Wikimedia Commons. [Public domain.]

crust contains two basic types of crustal material, oceanic and continental, with an average density of 2.9 and 2.6, respectively. Oceanic crust ranges from 5- to 10-kilometers thick and is thinnest over seafloor-spreading areas. Oceanic crust is primarily composed of dense basaltic rock with a thin veneer of silt and carbonate precipitates; however, a variety of minerals have been observed at seafloor vents. Continental crust is primarily composed of felsic granitic rock, which is less dense than oceanic crust; however, continental crust also includes sedimentary and metamorphic rock and even uplifted oceanic basalt. A variety of minerals of varying economic importance occur in the continental crust. The continental crust averages 30 to 40 kilometers in thickness, but it may be more than 70-kilometers thick in some mountain areas.

Oceanic crust is less dense than the parent mantle material. This is probably attributable to partial melting and crystal fractionation. Felsic minerals have a lower melting temperature than mafic minerals, and mafic minerals are the first to crystallize out of a melt. As oceanic crust subducts below continental crust, the subducting plate eventually melts, and its upwelling liquid fraction produces less mafic intermediates.

The lithosphere is highly variable, according to regional studies. In parts of the middle United States and in the Gulf of Mexico region, for example, the crust has thick sedimentary layers. Oil companies were able to measure the seismic wave patterns generated by many controlled explosions and discover petroleum and natural gas within these layers. The later discovery of oil in northern Alaska was prompted by the similarity of the crust there to the crust of these regions. As the study of the characteristics of the lithosphere—including plate tectonics—continues, scientists will increasingly be able to use their knowledge to discover sites of mineral resources.

—Raymond U. Roberts

Further Reading

Fowler, C. M. R. *The Solid Earth. An Introduction to Global Geophysics.* 2nd ed., Cambridge UP, 2005.

Li, Z. X., D. A. D. Evans, and J. B. Murphy, editors. *Supercontinent Cycles Through Earth History.* The Geological Society, 2016.

Stein, Seth, and Michael Wyesession. *An Introduction to Seismology, Earthquakes, and Earth Structure.* Wiley, 2013.

Stüwe, Kurt. *Geodynamics of the Lithosphere: An Introduction.* Springer, 2007.

Svalova, Valentina, editor. *Heat-Mass Transfer and Geodynamics of the Lithosphere.* Springer, 2021.

Turcotte, Donald L., and Gerald Schubert. *Geodynamics.* 3rd ed., Cambridge UP, 2014.

Lomonosov Issues the First Catalog of Minerals

Fields of Study: Geology; Mineralogy; Geomorphology; Geochemistry; Gemology; Crystallography

ABSTRACT

Soon after returning from scientific studies in Germany, Mikhail Vasilyevich Lomonosov began sorting and cataloging the mineral cabinet of the Kunstkammer in St. Petersburg. His resulting catalog of more than thirty-five hundred mineral specimens was published in 1745.

KEY CONCEPTS

Kunstkammer: German word meaning "oddities room," the name of Czar Peter the Great's first public museum, which grew to include a vast collection of rock and mineral samples

pallasites: a type of meteor named for Peter Simon Pallas

paragenesis: a contraction of parallel genesis, referring to the common origin of related minerals

REGAL BEGINNINGS OF MINERAL COLLECTIONS IN RUSSIA

The history of the A. E. Fersman Mineralogical Museum, one of the finest such museums in the world, begins with Czar Peter the Great's interest in science and his desire to bring science to the attention of the Russian people. By the close of the seventeenth century, Peter started collecting unusual biological and anatomical specimens, sparked by his first trip to Europe in 1679. He kept these specimens in what was then known as a *Kunstkammer*, German for "cabinet of curiosities."

In 1703, as part of Peter's desire to modernize and westernize Russia, he began construction of his new capital of St. Petersburg on the Baltic Sea. He was eager to show his *Kunstkammer* to ordinary Russians, to educate them, to stimulate their interest in Western science, and to impress them with the knowledge, resources, and wealth of their monarch. The cabinet thus began to be exhibited to visitors in Kikin's Palace, St. Petersburg, in 1714. In 1716, a collection of 1,195 mineral specimens was purchased from Dr. I. Gotvald, a physician in Gdansk, as the foundation of a mineralogical collection called the "Mineral Cabinet." Items of geological interest that Peter received as gifts from foreign leaders were added to the Mineral Cabinet. These included a 16-centimeter-long specimen of native silver wire known as the Silver Horn, given to him by the king of Sweden in 1718.

Peter the Great founded the St. Petersburg Academy of Sciences in 1724. Members of the academy were actively recruited from various European countries, especially Germany. The *Kunstkammer* was formally put in the control of the Academy of Sciences. Mining companies throughout Russia sent samples to the Kunstkammer's Mineral Cabinet, and by the early 1740s, the collection included more than three thousand minerals, fossils, gems, and rocks. The responsibility for curating the Mineral Cabinet had fallen to Johann Gmelin, a German scientist, who had left St. Petersburg to participate in a Siberian expedition lasting from 1733-1743.

LOMONOSOV'S INFLUENCE

The Russian scientist Mikhail Vasilyevich Lomonosov returned to his native country in 1741. He had spent several years studying abroad and had received degrees in metallurgy at the University of Marburg and in mining engineering at the University of Freiburg, Austria. Lomonosov had returned to teach chemistry at the University of St. Petersburg, while Gmelin was still away on his Siberian expedition. Lomonosov was quickly made a member of St. Petersburg Academy of Sciences and appointed director of the Mineral Cabinet.

In 1742, Lomonosov was imprisoned for views hostile to the aristocracy and objections to certain academy policies. He continued to work while imprisoned and wrote two poems dedicated to Empress Elizabeth Petrovna, who decided to free him early in 1743. After his release from prison,

Lomonosov, portrait by Georg Caspar Prenner, 1787. Image via Wikimedia Commons. [Public domain.]

Lomonosov resumed his duties as director of the Mineral Cabinet, cataloged the entire thirty-five-hundred-specimen collection, and published the results in 1745 under the auspices of the Russian Academy of Sciences. This was a precedent-setting accomplishment, in that it was the first research paper in mineralogy to be published by the St. Petersburg Academy of Sciences.

Lomonosov's catalog helped establish the reputation of St. Petersburg's Kunstkammer as one of a very few public museums in Europe. (Oxford's Ashmolean Museum was founded in 1683; the British Museum would not be founded until 1753.) The catalog played an important role in establishing the Kunstkammer's reputation as more than a repository of bizarre anatomical specimens. It also demonstrated Lomonosov's credentials as a learned Russian scientist affiliated with the Kunstkammer.

In addition to producing the catalog of the Mineral Cabinet, Lomonosov directed construction of Russia's first chemical laboratory, where he kept records of the measure, weight, and composition of various substances. Lomonosov recognized the importance of studying pure metals and salts. He believed that minerals were mixtures of primary particles (now referred to as atoms) and that their structure, color, and weight were important. Lomonosov was the first scientist to establish that amber, peat, and coal arose from plant remains. His chemical analysis of amber showed that its specific gravity was identical to that of pine resin. He also separated water from amber, which produced an odor associated with growing plants.

Lomonosov was interested in the common genesis of related minerals, a concept now known as paragenesis. Over the next 250 years, considerable work on paragenesis would be conducted under the auspices of the academy. He was also interested in the crystallographic properties of minerals and recorded crystal interfacial angles. He wrote a paper on the geometric arrangements of ideal spheres in crystal lattices.

In 1747, a damaging fire swept through the Kunstkammer, and only the Mineral Cabinet's most valuable specimens (including the Silver Horn) were saved. Specimens continued to arrive, however. The first meteorite, the 687-kilogram, stony iron *Pallasovo Zheleso* (Pallas's Iron) was sent from Medvedevo (Siberia) in 1762 and arrived in St. Petersburg in 1772. This meteorite was acquired on a Siberian expedition by Peter Simon Pallas and, after its arrival at the Mineral Cabinet, was studied by academician Ernst Florens Friedrich Chladni, who realized that its unique composition of olivine and nickel iron was extraterrestrial. All meteorites of this compositional type are now referred to as "pallasites."

After the Mineral Cabinet was established in its restored building, Pallas, then its director, had a new catalog of the collection prepared in 1789. The Mineral Cabinet contained about twenty thousand specimens in 1836. Early in the twentieth century, specimens in the Mineral Cabinet were divided into five collections: the Systematic Collection (more than ninety thousand specimens, showing crystallographic and mineralogical features), the Crystal Collection (more than thirty-one thousand specimens, representing mineral associations and ore types), the Location Collection, the Pseudomorph Collection (containing examples of crystallogenesis under various conditions), and the Gems and Stone Art Collection (including thirty items designed by Peter Carl Fabergé and handcrafted from semiprecious gemstones). The Meteorite Collection includes more than 447 specimens from various sites in Russia and all over the world, including a large collection of tektites, natural glass objects sometimes associated with meteorites.

SIGNIFICANCE OF LOMONOSOV'S CATALOG

Mikhail Vasilyevich Lomonosov was the first Russian-born scientist to become a member of the St. Petersburg Academy of Sciences. By publishing a catalog of the Mineral Cabinet, detailing the geological contents of the Kunstkammer, he was able to place Russian mineralogy at the forefront of European science. As museums outside Russia were later established, mineral collections were displayed following the Mineral Cabinet's precedent. Lomonosov worked for many years at the Kunstkammer, and the Lomonosov Museum was founded there in 1947. This museum was dedicated to the early years of the Academy of Sciences, and it included a display recreating Lomonosov's office and laboratory.

Over the two centuries following the fire that almost destroyed it, the Mineral Cabinet grew steadily. Following the 1919 Russian Revolution, the Mineral Cabinet was placed under the direction of A. E. Fersman. It was at this time that the museum acquired Fabergé's collection, as well as several other private collections. The Mineral Cabinet moved from St. Petersburg to Moscow in 1934; its contents, then estimated to include eighty thousand specimens, filled thirty railroad cars. Relocation was completed to host the Seventeenth International Geological Congress in Moscow in 1937. The museum was renamed to honor Fersman, its longtime director, in 1955. The museum building was reconstructed in preparation for the Twenty-Seventh International Geological Congress, held in Moscow in 1984.

Between 1772 and 1996, one meteorite type (pallasite) and thirty-two mineral species were named in honor of staff members of the Mineral

Cabinet. The minerals lomosovite and lomosovite-beta were named in honor of Lomonosov. In 2000, an oxalate mineral discovered in the Chelkar salt dome in Kazakhstan was named "novgorodovaite" after the current director, Margarita Ivanovna Novgorodova.

At the start of the twenty-first century, the A. E. Fersman Mineralogical Museum housed one of the most complete mineral collections in the world, including more than twenty-three hundred of about four thousand known mineral species. Its collection has been curated using Dana's Mineralogical Classification. About ten thousand of its specimens are viewable on the internet. Its scientists continue in Mikhail Lomonosov's tradition of mineralogical research and publication.

—Anita Baker-Blocker

Further Reading

Fersman, A. E. *Geochemistry for Everyone.* Foreign Languages, 1958.

Generalov, Mikhail. "A. E. Fersman Mineralogical Museum: An Embassy of the Mineral Kingdom in Russia" *Rocks and Minerals,* vol. 76, no. 1, Jan.-Feb. 2001, pp. 17-21.

Pavlova, Galina Evgen'evna, and Aleksandr Sergeevich. Fedorov *Mikhail Vail'evich Lomonosov: His Life and Work.* Mir Publishers, 1980.

Popova, V. I., et al. "Murzinka: Alabasha Pegmatite Field." *Mineralogical Almanac,* vol. 5, 2002.

Rundqvist, Dmitrii V. *Mineral Collections of Russia, Part II: Twelve Museum Collections.* Vol 3. in *Mineralogical Almanac.* Ocean Pictures, 2000.

LUNAR EXPLORATION

Fields of Study: Geology; Astrogeology; Geomorphology; Geochemistry

ABSTRACT

The "Space Race" between the United States and the Soviet Union began in 1957, when the Soviets launched two satellites. In response, the U.S. Department of Defense (DoD) was authorized to begin lunar exploration. In April, 1961, Russian cosmonaut Yuri A. Gagarin became the first human being in space. In May, 1961—shortly after Alan Shepard's suborbital flight—President John F. Kennedy proposed that the United States send a man to the Moon. Both political desire and a quest for scientific knowledge motivated many lunar missions in the following decades.

KEY CONCEPTS

booster: the rocket used to propel a capsule into space

impact ejecta: material ranging in size from house-sized rocks to fine dust that is ejected from the site of impact of a meteor

mascon: an area of the lunar surface that significantly greater density than the rest of the surface, indicating a mass concentration (hence, mas-con)

resolution: the minimum size of an area that can be differentiated from its surroundings photographically

soft-land: to land a spacecraft on a planetary surface intact and in fully functioning condition, as opposed to crash landing and loss of the machine

SUMMARY OF LUNAR EXPLORATION MISSIONS

The American lunar exploration program began in 1957, before the creation of National Aeronautics and Space Administration (NASA), as a response to the Cold War political challenge issued by the launching of the first two Earth satellites by the Soviet Union. The U.S. Department of Defense (DoD), through its Advanced Research Projects Agency (ARPA), authorized the Air Force to launch three 45-kilogram Pioneer probes toward the Moon using the Thor-Able booster. The Army was also authorized to prepare two smaller 6-kilogram Pioneer probes for launch by the Juno 2 booster.

The first Air Force Pioneer met with disaster on August 17, 1958, when its Thor booster exploded in flight. A subsequent attempt on October 11, 1958, successfully sent Pioneer 1 into space. However, trajectory inaccuracies caused the probe to fall back to Earth after reaching 115,000 kilometers in altitude. Pioneer 2 also suffered launch failure in November, 1958. The Army rocket team successfully lofted Pioneer 3 into space on December 6, 1958, but again, a faulty aim doomed the probe to fall back to Earth after reaching an altitude of 107,000 kilometers.

Before the next Pioneer could be readied, the Soviet Union succeeded in launching the first craft to break free of Earth's gravity. Russian lunar efforts had previously suffered three secret launch failures in 1958, but on January 2, 1959, Luna 1 left Earth. The 361-kilogram spacecraft, popularly called *Mechta*, or "dream," was intended to hit the Moon, but a guidance error caused the craft to pass 5,955 kilometers from the Moon and enter solar orbit.

An American probe followed the Russian craft on March 3, 1959, when the Army succeeded in launching Pioneer 4 into solar orbit. However, the probe passed 56,350 kilometers from the Moon, too far away to return any lunar data.

Russian lunar efforts achieved two more successes in 1959, when Luna 2 impacted the Moon on September 14, and Luna 3, launched October 4, looped around the Moon and returned crude photographs of its previously unseen far side.

Following Pioneer 4, American lunar efforts were directed by the newly created NASA. A heavier series of Pioneer spacecraft, originally designed to study the planet Venus, were diverted toward lunar studies in response to the Russian Luna 3 success. All three launch attempts in 1960 failed when their more powerful Atlas-Able boosters failed in flight. NASA also created two new robotic lunar exploration programs: the Ranger series, which were both to rough-land instruments and to take close-up photographs of the lunar surface before crashing and the Surveyor soft landers, which were to deposit scientific instruments on the Moon.

Initially, the Ranger program had no better luck reaching the Moon than did the Pioneers. Rangers 3, 4, and 5 were rough-land seismic instruments on the Moon, but after successful launches in 1962, all failed on the way to the Moon. Rangers 6 through 9 were lunar impactors, which were to photograph the surface during terminal descent. After a successful launch on January 30, 1964, Ranger 6 struck the Moon but failed to return any pictures. The long-delayed American lunar success finally came on July 31, 1964, when Ranger 7 returned close-up photographs of Mare Cognitum. Rangers 8 and 9 also returned close-up images in February and March of 1965.

Russian's Luna 9 spacecraft, launched January 31, 1966, achieved a soft landing on the Moon. The craft's 100-kilogram landing spacecraft returned the first images from the lunar surface and recorded radiation levels. Luna 13, launched December 21, 1966, repeated the landing at a different site.

American efforts to soft-land on the Moon achieved success on the first attempt when Surveyor 1 landed on the Moon's Oceanus Procellarum on June 1, 1966. The 995-kilogram lander was considered a test flight and carried only a 7-kilogram television camera. Six additional Surveyor craft were successfully launched toward the Moon in 1966 and 1967. Of these, Surveyors 2 and 4 were lost before landing. The five successful landings returned a total of 87,700 images from the Moon's surface.

The Lunar Orbiter program was conceived from the need for detailed reconnaissance of the lunar surface in order to select suitable Apollo landing sites. The resulting 387-kilogram spacecraft carried a photographic package adapted from a military spy satellite and could map portions of the Moon at 1-meter resolution. All five Lunar Orbiter missions launched in 1966 and 1967 were successful. Of the 2,110 photographic frames available, the Lunar Orbiters returned 1,650 successful images; 78 percent of the photography was scientifically useful.

Between 1969 and 1972, six Apollo piloted landings were carried out. Beginning with the historic first piloted landing on July 20, 1969, Apollos 11 through 17 placed twelve American astronauts on the Moon's surface. Neil A. Armstrong became the first human to set foot on the Moon, while Eugene Cernan was the last human on the Moon in the twentieth century. Apollo 13 did not achieve a landing after an oxygen tank explosion forced the crew to make an emergency return to Earth in April, 1970.

Although the Soviet piloted lunar-landing program was not carried out following the success of the American landings, Russia continued the robotic exploration of the Moon. Luna 16, launched September 12, 1970, soft-landed and robotically returned 101 grams of lunar material to Earth. This feat was repeated by Luna 20, launched February 14, 1972, and again by Luna 24, launched August 9, 1976. The latter used a specially designed drilling mechanism to retrieve a 1.6-meter core sample. Additional Soviet lunar launches on November 10, 1970, and January 8, 1973, deposited two remote-controlled rovers, Lunokhods 1 and 2, on the Moon. These eight-wheeled robots traversed a combined 47 kilometers of the lunar surface photographing and chemically analyzing the lunar soil.

Apollo 12 Commander Charles Conrad Jr. examines the unmanned Surveyor III spacecraft during the second extravehicular activity (EVA-2). The Lunar Module (LM) "Intrepid" is in the right background. November 20, 1969. Photo via NASA/Wikimedia Commons. [Public domain.]

Since the 1976 Luna 24 landing, no further lunar missions of any kind were carried out by either the United States or Russia until January 25, 1994, when the U.S. DoD again entered the lunar exploration field. The Clementine spacecraft was placed into orbit around the Moon to gather lunar mineralogical data during a test of military missile detection sensors. On January 6, 1998, NASA also returned to the Moon when it placed the Lunar Prospector spacecraft in lunar orbit to continue lunar mineralogy studies.

The SMART-1 (Small Missions for Advanced Research in Technology 1) is a Lunar Orbiter designed to test spacecraft technologies for future missions. The primary technology being tested is a solar-powered ion drive. It carries an experimental deep-space telecommunications system and an instrument payload to monitor the ion drive and study the Moon. The primary scientific objectives of the mission are to return data on the geology, morphology, topography, mineralogy, geochemistry, and exospheric environment of the Moon. SMART-1 was launched as a secondary payload on September 27, 2003, by the Ariane-5 Cyclad booster.

Lunar Reconnaissance Orbiter (LRO) was sent into orbit about the Moon in 2009 for prolonged lunar research and imaging. LRO had been envisioned to support the planned Vision for Space Exploration which had called for astronauts to return to the Moon as part of the Constellation Program. LRO continued to provide useful information about the Moon long after the Obama Administration cancelled the Constellation Program. In 2019 NASA was challenged to send the first astronauts of the twenty-first century back to the Moon within five years. LRO would continue to provide support to lunar exploration begun over fifty years earlier at the dawn of the Space Age.

In the intervening years, a number of other nations of Earth, including China, Japan, India, Saudi Arabia and Iran, have sent or have plans to send rover-type missions to the Moon.

LUNAR MISSIONS CONTRIBUTIONS TO SCIENCE AND SOCIETY

Although they failed to reach the Moon, Pioneers 1 and 3 made fundamental discoveries by confirming that two zones of high radiation, known as the Van Allen radiation belts, were trapped by Earth's magnetic field and centered at 4,800 kilometers and 11,250 kilometers in altitude. Pioneer 4 demonstrated that above the Van Allen radiation belts out to lunar distances, the normal space radiation environment was low enough to allow humans to safely travel to the Moon.

Prior to the first landing, estimates of the conditions on the lunar surface ranged from hard rock to soft, nonsupportive, loose dust, which would swallow a landing craft like quicksand. The successful Luna 9 landing established that the lunar surface would indeed support a spacecraft. Subsequent television pictures from the first Surveyor showed that one of its own footpads had indented the lunar surface only a few centimeters, further demonstrating that the Moon could support a piloted lander. The mechanical and chemical soil analysis performed by the final three Surveyors suggested a general model for lunar composition, which was later confirmed by examination of samples returned by Apollo astronauts.

The most significant result of the space-based photographic reconnaissance of the Moon by both U.S. and Russian probes was the determination that the vast majority of lunar surface features were sculpted by the impacts of meteors, asteroids, and comets, and not by volcanic activity, as had been previously supposed. Photography also established that chains of craters, even those not radially aligned with nearby large craters, were in fact created by impact ejecta and not volcanism along faults or fractures. It was further established that many volcanic-appearing features associated with individual large craters were the result of impact melting, not true volcanism.

As seen from Earth, the Moon's surface is divided into two geologic types: the light-colored, heavily cratered highlands called "terrae," and the dark-colored smoother regions called "maria" (plural for *mare*, the Latin word for sea). Over its global surface, maria cover 17 percent of the Moon. Orbital tracking of the Lunar Orbiter spacecraft revealed that their initial orbits did not follow that expected for a craft circling a homogeneous body. Instead, areas of increasing gravity centered over the circular lunar maria perturbed the craft. These dark, relatively flat areas were assumed to be volcanic flows across the surface. It was realized that the circular lunar maria were giant impact basins, which had slowly filled with basalt over a period of time. As more basalt flowed into the basins, it slumped and became denser than the surrounding lunar crust. These gravitationally stronger areas are called "mascons," for mass concentrations.

Of all the terrestrial planets and their satellites, Earth and the Moon are most like a double-planet system. However, comparing Earth's mean density of 5.5 grams per cubic centimeter and the Moon's mean density of 3.3 grams per cubic centimeter shows marked differences in the overall composition of both worlds. The Moon possesses less volatile substances such as water, sodium, and lead.

The mission of Lunar Prospector ended on July 31, 1999, when it was deliberately targeted to impact in a permanently shadowed area of a crater near the lunar south pole. It was hoped that the impact would liberate water vapor from the suspected ice deposits in the crater and that the plume would be detectable from Earth; however, no plume was observed.

SMART-1 began to monitor the illumination of the lunar poles even before arriving at its final science orbit in March, 2005. The Moon's rotational axis is only 1.5° inclined to the ecliptic plane. This orientation allows areas to exist at the bottom of near-polar craters that do not see direct sunlight and where ice might be trapped. On the other hand, at higher elevations on the rims of polar craters, there are areas that see the Sun for more than half of the time. Following successful lunar orbital insertion, SMART-1's mission was extended a year. It was expected to continue collecting data through August, 2006. The engine would then be used to lower the orbit pericenter to 300-kilometer altitude. It was planned to nominally spend six months in lunar orbit mapping the Moon's surface and evaluating the new technologies onboard.

LRO identified and imaged Apollo program landing sites, and found evidence of water ice at the Moon's south polar region, enough to support a Moon base which was unofficially envisioned as part of the return to the Moon eventually supported by the Trump administration.

A relatively inexpensive spacecraft identified as LADEE or Lunar Atmosphere and Dust Environment Explorer launched from the Wallops Island facility in southeastern Virginia on September 7, 2013, and performed a seven-month-long investigation of the Moon's tenuous atmosphere and the nature of dust in low-lunar orbit. LADEE found evidence of helium, neon, and argon in the lunar exosphere. After concluding its mission, LADEE was purposely crashed into the far side of the Moon to remove it from orbit.

CONTEXT OF LUNAR EXPLORATION MISSIONS

In the 1950s, spaceflight visionary Wernher von Braun proposed an orderly progression of space exploration events that would ultimately result in humans landing on Mars. An Earth-orbiting space station and piloted flights to the Moon would be stepping stones to the Red Planet. Scientists realized that robotic spacecraft would precede humans to the Moon and determine the general lunar characteristics as the precursors for eventual piloted expeditions. However, this blueprint for piloted space exploration was discarded after the Soviet Union succeeded in launching Sputnik, the first artificial Earth satellite. The political shock of this event was seen by the Western world as a challenge to demonstrate which political system, the free West or the Communist-controlled East, possessed the best industrial and technological base.

The "Space Race" between the United States and the Soviet Union, born from the response to Sputnik, escalated on April 12, 1961, when the Russian cosmonaut Yuri A. Gagarin became the first human being in space. A long string of Soviet technological space "firsts," and American political setbacks on Earth, such as the failed Bay of Pigs invasion of Communist-controlled Cuba, led President John F. Kennedy to seek a national goal in space that offered a reasonable chance of exceeding Soviet efforts. In May, 1961, President Kennedy boldly proposed that the United States send a man to the Moon and safely return him to Earth by the end of the decade. Thus, the U.S. space program was spurred heavily by a political desire to restore American international prestige, and a political timetable was established to conduct the first piloted landing on the lunar surface.

An unfortunate side effect of President Kennedy's mandate was the abandonment of most lunar science. The rush to achieve piloted lunar success before the Soviet Union reduced the Ranger and Surveyor lunar probes to engineering support missions for the piloted landing effort. Any lunar science gained was of secondary importance to the politically driven goal of landing humans on the Moon by the end of the 1960s. However, during the time of active lunar exploration in the 1960s and 1970s, the geology of Earth was not well understood either. This lent validity to the idea that an understanding of the Moon's geology was essential to understanding the geologic composition of the Earth.

Technical problems in perfecting the complex Ranger spacecraft delayed the results from that program. By the time Ranger succeeded in reaching the Moon in July, 1964, the program was no longer of any direct benefit to the Apollo Program because the design of the piloted lunar lander had already been frozen. Similarly, the Surveyor Program suffered delays when its Centaur booster encountered technical problems and lacked the power needed to loft a full complement of exploratory instruments to the Moon. Conceived a year before Apollo as a science program, Surveyor was thus turned primarily into an engineering test to see whether a soft landing on the Moon was possible.

The early piloted landings of Apollo missions 11 through 14 were scientifically conservative. Yet, even as the initial piloted lunar landing was conducted by Apollo 11, subsequent missions devoted to lunar science were canceled by Congress. Only Apollo missions 15 through 17 were upgraded to true lunar scientific efforts.

After four decades of lunar exploration scientists have still not conclusively answered how the Moon was formed almost 4.5 billion years ago. However, post-Apollo studies of the chemical makeup of the Moon now lend strong support to the idea that the Moon was created from the debris flung into space by the impact on Earth of a Mars-sized body. This theory is supported by the similarity between the minerals in Earth's crust and the composition of the Moon, and the fact that the Moon has no large iron core like that of Earth.

Rock samples returned by the Apollo expeditions show the Moon is lifeless and contains no fossils or organic compounds. The lunar regolith, or surface layer of dust, rubble, and rock fragments created and stirred by billions of years of meteor bombardment, is composed primarily of three basic rock types; volcanic basalts, anorthosites, which are crystalline slow-cooled volcanic rock, and breccias, made up of previously shattered rock fragments welded back together by meteor impact.

The Apollo seismic experiments showed the Moon possesses a crust, mantle, and small core, and were thus differentiated by global melting early in its history. At that time, the Moon was covered with a magma ocean. Today, the lunar highlands are the remains of lower-density rocks that floated to the surface of the early global magma cover. The crust is about three times thicker than Earth's, and the mantle surrounds an iron core that is less than 25 percent of the Moon's radius. Below a 1,000-kilometer depth, the lunar interior is still semimolten.

The repercussions of the singular national space goal of landing humans on the Moon before 1970 still affect the space programs of both the United States and Russia. Once the political goal of a piloted lunar landing was achieved, funding for future lunar exploration was severely cut. By the beginning of the twenty-first century, both the United States and Russian space programs had abandoned the technological means for sending humans to the Moon.

However, since 2019 there is greatly renewed interest in establishing a human presence on the Moon as the gateway for missions to Mars. This renewed interest seems to have coincided with the 50th anniversary of the first human landing on the Moon in 1969. Several nations, including the United States, Canada, Russia, China, India, Japan, Israel, the European Space Agency and Saudi Arabia, as well as Space-X and other private companies, have concrete plans to send missions, both manned and unmanned, to the Moon in the mid-2020s. NASA is preparing to launch its "Artemis 2" program in 2022, with plans to land humans on the Moon each year thereafter. The plans are ambitious, and their value will depend as much, or more, on geology as on technological development.

—Robert V. Reeves

Further Reading

Ayrey, Bill. *Lunar Outfitters: Making the Apollo Space Suit*. UP of Florida, 2020.

Beattie, Donald. *Taking Science to the Moon: Lunar Experiments and the Apollo Program*. Johns Hopkins UP, 2001.

Burrows, William E. *This New Ocean: The Story of the First Space Age*. Random House, 1998.

David, Leonard. *Moon Rush: The New Space Race*. National Geographic, 2019.

Harvey, Brian. *Soviet and Russian Lunar Exploration*. Praxis Publishing, 2007.

Kelly, Thomas J. *Moon Lander: How We Developed the Apollo Lunar Module*. Smithsonian Institution Press, 2001.

McDougall, Walter A. *The Heavens and the Earth: A Political History of the Space Age*. 2nd ed., Johns Hopkins UP, 1997.

Reeves, Robert. *The Superpower Space Race: An Explosive Rivalry Across the Solar System*. Plenum Publishing, 1994.

Spudis, Paul D. *The Value of the Moon. How to Explore, Live, and Prosper in Space Using the Moon's Resources*. Smithsonian Books, 2016.

Ulivi, Paolo, and David M. Harland. *Lunar Exploration: Human Pioneers and Robotic Surveyors*. Praxis Publishing, 2004.

Wilhelms, Don E. *To a Rocky Moon: A Geologist's History of Lunar Exploration*. U of Arizona P, 1993.

M

Magma Crystallization

Fields of Study: Geology; Geochemistry; Mineralogy; Crystallography

ABSTRACT

Magma crystallization is a geologic process in which molten magma in Earth's interior cools and subsequently crystallizes to form an igneous rock. The crystallization process produces many different types of minerals, some of which are valuable natural resources.

KEY CONCEPTS

aphanitic: igneous rock that is so fine-grained as to be glassy, such as obsidian, formed from magma that has cooled and crystallized very rapidly

gravity layering: distinct layers of igneous rock in an igneous rock body, produced by fractionation of minerals according to density under the influence of gravity as the magma body slowly cools

intrusion: a body of igneous rock that results from the intrusion of magma into a preexisting rock body

kimberlite pipes: the throats of ancient volcanoes that have erupted explosively, allowing kimberlite magma to fill the void channel

phaneritic: igneous rock that contains large crystals formed as magma cools slowly

MAGMA

Magma is molten rock material consisting of liquid, gas, and early formed crystals. It is hot (900° to 1,200° Celsius), mobile, and capable of penetrating into or through Earth's crust from the mantle, deep in Earth's interior. Most magma cools within Earth's crust; in a process similar to ice crystallizing from water as the temperature drops below the freezing point, minerals crystallize from molten magma to form a type of rock generally called "igneous rock." Once completely crystallized, the body of igneous rock is called an "intrusion." Some magma, however, works its way to the surface and is extruded as lava from volcanoes.

MINERAL GROWTH

Magma that remains below the surface cools at a slow rate. Ions have time to collect and organize into orderly, crystalline structures to form minerals. These minerals grow larger with time and, if the cooling rate is slow enough, may grow to several centimeters in diameter or larger. Igneous rocks with minerals of this size are said to have a phaneritic texture. Magma that reaches the surface, on the other hand, cools very rapidly and forms rocks that consist of extremely fine-grained minerals or quenched glass. These rocks have an aphanitic or glassy texture. Consequently, it is those minerals that grow beneath the surface that reach sizes large enough to be considered economically feasible resources.

CONCENTRATION OF VALUABLE ELEMENTS

Minerals do not crystallize from magma all at once. Instead, they follow a sequence of crystallization as the temperature decreases. In general, silicate minerals (substances with silicon-oxygen compounds) with high contents of calcium, iron, and magnesium crystallize early, followed by silicate minerals with high contents of aluminum, potassium, and sodium. Excess silica crystallizes last as the mineral quartz. Bonding factors such as ionic size and charge prevent some elements from being incorporated into early crystallizing minerals. Thus, they are more highly concentrated in the residual magma and become incorporated into the last minerals to crystallize, forming rocks called "granites" and "pegmatites". These rocks may contain minerals such as beryl, spodumene, lepidolite, and uraninite, which include important elements such as beryllium,

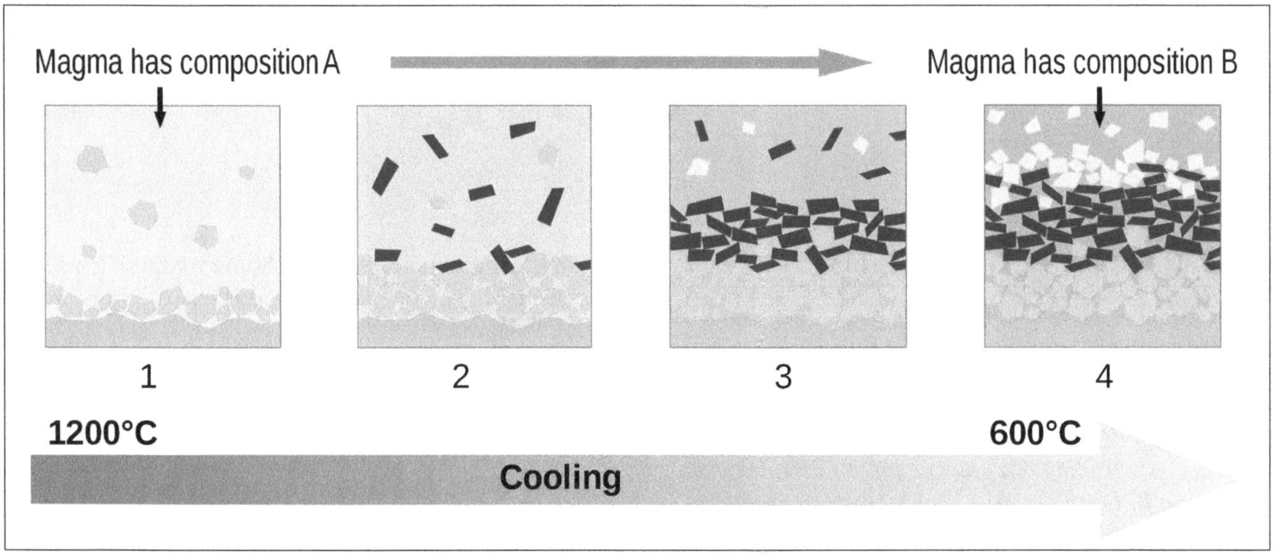

Schematic diagrams showing the principles behind fractional crystallisation in a magma. While cooling, the magma evolves in composition because different minerals crystallize from the melt. 1: olivine crystallizes; 2: olivine and pyroxene crystallize; 3: pyroxene and plagioclase crystallize; 4: plagioclase crystallizes. At the bottom of the magma reservoir, a cumulate rock forms. Image by Woudloper, CC BY-SA 3.0, via Wikimedia Commons.

lithium, and uranium. Granites and pegmatites are also important sources for feldspar and sheet mica.

DIAMONDS AND KIMBERLITES

Perhaps the best-known magmatic minerals are diamonds. Formed deep in the mantle at extremely high temperatures and pressures, diamonds are carried by a certain type of magma as it violently intrudes upward through the crust, sometimes reaching the surface. Upon cooling and crystallizing, this magma forms a "pipe" of igneous rock known as kimberlite. It is in kimberlites that most diamonds are found. Most kimberlite pipes are less than 1 square kilometer in horizontal area, and they are often grouped in clusters. Most of the known diamond-bearing kimberlite pipes are found in southern Africa, western Australia, Siberia, and Canada.

MAGMATIC SULFIDE DEPOSITS

Most major metals used in industry (copper, iron, lead, nickel, zinc, and platinum) are found in sulfide minerals, which are substances that contain metal-sulfur compounds. When magma is in the early stages of cooling and crystallizing underground, certain processes can cause droplets of liquid sulfur to form within it. These sulfur droplets attract metallic cations and concentrate them by factors ranging from 100 to 100,000 over their normal levels in the host magma. The droplets eventually cool and solidify to form sulfide minerals such as pyrite ("fool's gold"), galena (lead sulfide), and sphalerite (zinc sulfide). Sulfide minerals such as these become important targets for mining because of their high concentration of metals.

LAYERED MAGMATIC INTRUSIONS

Some magmas give rise to layered intrusions in which a specific sequence of minerals is repeated many times from bottom to top in a process called "gravity layering" (also called "rhythmic layering"). Dark-colored, heavier minerals such as pyroxene, olivine, and chromite concentrate near the base of each layer, grading to predominantly light-colored minerals such as plagioclase at the top. Each mineral sequence is a separate layer, averaging several meters thick and ranging from less than 2 centimeters to more than 30 meters. It has been suggested that the origin of gravity layering involves multiple injections of fresh magma into a crystallizing magma chamber, effectively replenishing the magma and allowing the same minerals to crystallize repeatedly.

The Bushveld intrusion in South Africa, one of the largest layered intrusions, contains multiple gravity layers and is more than 7,000 meters in total thickness. Layered intrusions contain Earth's main reserves for chromium and platinum. In the Bushveld intrusion, chromium occurs in the mineral chromite, and platinum in platinum-iron alloys, braggite, and other platinum-metal compounds. The main source for platinum minerals in the Bushveld intrusion, and the source for approximately half of Earth's supply of platinum, is the Merensky Reef, a layer of chromite and platinum minerals 1 meter thick and more than 200 kilometers long. Also present in the Bushveld intrusion is the mineral magnetite, which yields important elements used in steel manufacturing such as iron and vanadium. The Bushveld intrusion accounts for almost 50 percent of the world's production of vanadium.

—*Stephen C. Hildreth Jr.*

Further Reading

Best, Myron G. *Igneous and Metamorphic Petrology.* 2nd ed., Blackwell, 2003.

Brown, Michael, and Tracy Rushmer, editors. *Evolution and Differentiation of the Continental Crust.* Cambridge UP, 2006.

Hefferan, Kevin, and John O'Brien. *Earth Materials.* John Wiley & Sons, 2010.

Naldrett, Anthony J. *Magmatic Sulfide Deposits: Geology, Geochemistry, and Exploration.* Springer, 2004.

Robb, Laurence. *Introduction to Ore-Forming Processes.* Wiley-Blackwell, 2021.

Schmincke, Hans-Ulrich. "Magma." *Volcanism.* Springer, 2004.

Young, Davis A. *Mind over Magma: The Story of Igneous Petrology.* Princeton UP, 2003.

Vetere, Francesco, editor. *Dynamic Magma Evolution.* Wiley, 2021.

MARBLE

Fields of Study: Geology; Geomorphology; Geochemistry; Geophysics; Art History

ABSTRACT

Marble is used in architecture as both an ornamental and a structural stone. It is also used as an artistic medium for three-dimensional art such as sculpture, interior furnishings, and mortuary and historical monuments.

KEY CONCEPTS

bas-relief: a method of stonework in which the background layer of stone is removed from an area of rock, so that the remaining unaffected portion can be sculpted into a scene that stands out from the background

calcareous rock: rock composed primarily of calcium carbonate

contact metamorphism: metamorphic processes that occur when rock is forced into contact or close proximity to a source of high temperatures and pressures

regional metamorphism: metamorphic processes that occur slowly due to long periods of exposure to pressure and temperature due to tectonic plate movement

WHERE MARBLE IS FOUND

Marbles, geologically defined as metamorphically altered calcareous rocks, are found in the core areas of younger mountain chains formed by the collision of tectonic plates and the consequent uplift and distortion of carbonate sedimentary strata. They are also found in the exposed roots of ancient, very eroded mountain chains of continental shield areas. Important marble-producing areas include the Carrara area in the Italian Apennines and Vermont, Georgia, and Alabama in the United States.

TECHNICAL DEFINITION OF MARBLE

Geologists define marble as a type of rock produced by metamorphic processes acting on either limestone or dolomite (dolostone), causing recrystallization through heat and pressure to produce a coarser-grained, harder rock. Stonemasons and quarriers have a more generic definition, which calls almost any hard rock that accepts a fine polish marble.

DESCRIPTION, DISTRIBUTION, AND FORMS

As defined geologically, marble is a type of rock composed primarily of calcite. It can be, like limestone, monomineralic in nature—that is, a rock composed of only one, or nearly one, mineral. Thus, it can be up to 99 percent calcite (calcium carbonate). True marble can be derived from either limestone or dolomite (sometimes called "dolostone"). Dolomite (calcium magnesium carbonate) is a carbonate rock in which much, if not most, of the original calcium has been replaced by magnesium. True marbles are formed by two types of metamorphism: regional and contact. Regional metamorphism is usually tectonic in nature and involves the slow compression and heating of rocks by large-scale crustal movements of Earth over long periods of time. Contact metamorphism is caused by rocks coming into contact, or near contact, with sources of great geologic heat, such as intruding bodies of magma; in these cases, change can be affected within a short period of time.

HISTORY OF MARBLE USE

Marble in its various forms has been known and admired since remote antiquity as a stone of choice for many applications. Some of the earliest known works of true architecture that have survived from ancient Mesopotamia, Egypt, and Greece featured marble as either decorative or structural elements. Sculptures, bas-reliefs, dedicatory columns, and triumphal arches have frequently featured various marbles. Thus, marble has been in use at least five thousand years, dating back to the first civilizations, and its use continues up to the present. Many sculp-

The Taj Mahal is entirely clad in marble. Photo by Dhirad, CC BY-SA 3.0, via Wikimedia Commons.

tors through the ages—among them such giants as Michelangelo, working in the fifteenth and sixteenth centuries in Italy—have preferred marble, especially the pure white varieties.

PRIMARY USES OF MARBLE

Marble is used in architecture as both an ornamental and a structural stone. It is also used as an artistic medium for three-dimensional art such as sculpture, interior furnishings, and mortuary and historical monuments. The primary importance of marble is its use in architectural columns, floorings, wall coverings, sculpture, vases and other receptacles, and monuments of all sorts. Beginning in the twentieth century, new minor uses were found for marble, including electrical outlet baseplates and other electrical insulators, as it is a good natural insulator.

OBTAINING MARBLE

Marble deposits are quarried in large operations that may involve hundreds of workers. In Europe marble is often obtained from quarries that have been worked continuously since antiquity. Until the last century or so, work was laboriously performed with age-old traditional tools and methods, but with the advent of power equipment the methodology and speed of extraction have greatly improved. Some constants have remained, such as the general strategy regarding extraction of large blocks of marble: removing the overburden (overlying sediments and rubble, if any), defining a quarry floor and front by quarrying monolithic blocks of marble parallel to their natural jointing planes, cutting away large blocks on all sides and removing the marble to the quarry floor, trimming, removing the marble from the quarry, and transporting it to the purchaser (often by use of specially built railroad systems).

Marble extraction has never had significant environmental effects, as the true marbles are chemically inert for all practical purposes. The metamorphism they underwent in their natural development stabilized their constituent minerals, including the trace minerals such as iron and magnesium from which colored marbles derive their patterns and hues.

—Frederick M. Surowiec

Further Reading

Boggs, Sam, Jr. *Petrology of Sedimentary Rocks.* 2nd ed., Cambridge UP, 2009.
Kogel, Jessica Elzea, et al., editors. "Decorative Stone" and "Dimension Stone." *Industrial Minerals and Rocks: Commodities, Markets, and Uses.* 7th ed., Society for Mining, Metallurgy, and Exploration, 2006.
Miglorie, Catherine. *Vermont's Marble Industry.* Arcadia Publishing, 2013.
Napoli, J. Nicholas, and William Tronzo, editors. *Radical Marble: Architectural Innovation from Antiquity to the Present.* Routledge, 2018.
Pellant, Chris. *Smithsonian Handbook: Rocks and Minerals.* Dorling Kindersley Books, 2002.
Price, Monica T. *The Sourcebook of Decorative Stone: An Illustrated Identification Guide.* Firefly Books, 2007.
Tucker, E. R., and R. G. C. Bathurst. *Carbonate Diagenesis.* Reprint. John Wiley & Sons, 2009.

Mars Exploration Rovers

Fields of Study: Geology; Geomorphology; Geochemistry; Astrogeology

ABSTRACT

The Mars Exploration Rovers (MERs), twin robotic expeditions exploring the surface of Mars, landed at different locations on opposite sides of Mars to maximize the scientific return. With the ability to trek the length of a football field each day, they can explore a larger portion of the Martian surface than previous missions. Each rover contains a package of scientific instruments designed to analyze Mars's surface geology to help trace the history of liquid water on Mars.

KEY CONCEPTS

alpha proton X-ray spectrometer: a device that irradiates rock samples with X-rays of a specific wavelength that are absorbed by atoms in the sample and reemitted as X-rays of characteristic wavelengths. The reemitted X-rays are then captured by the spectrometer for identification of the particular minerals that are present in the sample

Mössbauer spectrometer: a device that irradiates rock samples with gamma rays of a known wavelength and measures the wavelength of gamma rays that are reemitted from the sample as a means of identifying the atoms present

thermal emission spectrometer: a device that identifies the wavelengths and relative intensities of infrared emissions from a sample; the wavelengths and patterns of intensities are characteristic of particular minerals

SUMMARY OF THE MARS EXPLORATION ROVER MISSIONS

The Mars Exploration Rovers (MERs) were launched about a month apart in the summer of 2003, and reached Mars in early 2004. The first rover, Spirit or MER-A, was launched on June 10 on a Delta II 7925 rocket. The second, Opportunity or MER-B, was launched on July 7 atop a Delta 7925H rocket. The "H" indicates Heavy. This con figuration was used because the second mission launched when the relative positions of Earth and Mars required more energy to get to Mars. Both were launched from Cape Canaveral, Florida. Both MER missions were timed to take advantage of the closest Mars approach to Earth in approximately sixty thousand years.

The probes took the fastest possible trajectories to Mars, and arrived in January, 2004. After a seven-month journey, Spirit landed at Gusev Crater on January 3, 2004. Opportunity soon followed by landing on Mars at Meridiani Planum on January 25, 2004. Unfortunately, two other missions, the Japanese Nozomi probe and the European Beagle 2 mission, which were planned to arrive at Mars nearly the same time did not succeed.

The MER mission used a landing system pioneered by the 1997 Pathfinder mission, which incorporated a combination of rockets, parachutes, and airbags to cushion the blow of striking the Martian surface. Because the MERs were heavier than the Mars Pathfinder, parachutes 40 percent larger were used. After the parachute was opened to provide initial slowing, rockets fired to bring the landers to a stop about 10 to 15 meters above the surface. Finally, airbags inflated to cushion the remaining fall. The landers bounced several times before finally coming to rest on the Martian surface and deflating the airbags. The lander's petals were then free to open, allowing the rovers to deploy their solar arrays and begin exploring Mars.

Shortly after landing, the computer on board Spirit experienced some memory difficulties. Mission controllers were able to resolve the problems in a little over a week by purging files and reformatting the flash memory. After this initial difficulty, both rovers performed well for their originally scheduled ninety-day mission and beyond. Each 170-kilogram rover contained a suite of instruments designed to allow it to explore the geology of Mars. These included cameras and instruments designed for detailed studies of the chemical compositions of selected rocks.

Nine cameras on each rover included six engineering and three scientific cameras. Four Hazard Avoidance Cameras (Hazcams) were mounted on the front and back of the rovers. Their 120° fields of view were pointed downward to see obstacles that might interfere with the rover's motion. Having two in each direction gave three-dimensional (3D) images, just as two eyes give human beings 3D depth perception. Two 45° field of view Navigation Cameras (Navcams) provided a 3D forward view higher than the Hazcams and allowed ground controllers to see where the rover was going.

Two Panoramic Cameras (Pancams) had the same resolution as the human eye, so they could mimic what a geologist would see standing on the surface of Mars. Because there were two Pancams, they could also provide 3D imaging. In addition, these cameras had a number of color filters to provide basic spectral information and color pictures. A solar filter allowed the Pancams to locate the Sun's position in the sky. When combined with the time of Mars's day, the Sun's position provided navigational information.

In addition to the large view cameras, the Microscopic Imager (MI) took extreme close-up pictures. This camera was mounted on a robotic arm that could swing into position very close to the rock of interest. The camera was also mounted behind a microscope, so it provided the equivalent of a human geologist's observations of a rock sample through a microscope.

As on most space probes, cameras provide dramatic images, but the information provided by other scientific instruments is at least as important. The Miniature Thermal Emission Spectrometer (Mini-TES) provided infrared spectra of the Martian

Artist's digital rendering of the MER rover on Mars. Photo by NASA/JPL/Cornell University, Maas Digital LLC, via Wikimedia Commons. [Public domain.]

rocks, soil, and atmosphere. These spectra revealed much about the chemical composition and mineral structure of the rocks and soil. The Mini-TES was particularly important in searching for minerals that were produced in the presence of water. Atmospheric spectra taken with the Mini-TES told mission scientists about the temperature and the presence of dust and water vapor in the atmosphere.

Two spectrometers studied the chemical composition of the Martian surface. The Mössbauer Spectrometer (MB) was designed specifically to study the iron abundance in the mineral content of Martian rock and soil samples. Because iron is magnetic, the MB also helped scientists study the sample's magnetic properties. Magnet arrays mounted at various positions on the MERs collected magnetic samples for analysis by the MB. Chemical compositions other than iron used the Alpha Proton X-Ray Spectrometer (APXS). Alpha particles emitted during radioactive decay told mission scientists about the presence and abundance of various radioactive isotopes in the mineral structure of rock samples. The X-ray spectra also provided information about the chemical samples' composition.

The Rock Abrasion Tool (RAT), located on the end of the robotic arm, could grind a 4.5-centimeter diameter hole into a selected rock sample. A half-centimeter hole took about two hours to grind. Mission scientists used the RAT to expose and study the interior structure of rocks.

In late October, 2004, the MER mission passed the 50,000-picture milestone. This number represents more than twice as many pictures as taken by all previous Mars landers. In early November, 2004, the Opportunity solar panels received a mysterious power boost that may have been caused by wind blowing accumulated dust off the solar panels. The MER missions, originally planned for about ninety days, performed so well that on October 1, 2004, the National Aeronautics and Space Administration (NASA) extended the missions for a second time. Some rover components were beginning to show signs of age, but they were still exploring Mars's surface.

On April 6, 2005, with both rovers suffering from only minor problems, and each covering greater and greater distances during translations across the Martian surface, NASA announced that the MER program was being extended for the third time. This latest extension would run for the coming eighteen months. Few program officials and scientists would have been so optimistic before launch to have expected the rovers to last this long in the severity of the Martian environment.

Spirit got stuck in soft sand on May 1, 2009. It took eight months to analyze the situation and devise an extraction procedure. Unfortunately, Spirit could not break free, and on January 26, 2010, NASA revised the rover's mission to be a stationary science platform. Communications with Spirit ceased on March 22, 2010. The Jet Propulsion Laboratory repeatedly attempted to reestablish a link to Spirit, but on May 24, 2011, Spirit's mission was officially declared complete.

Mars developed a wide-ranging dust storm in 2018. Inside Endeavour crater Opportunity was configured so that it had a decent chance of weathering the dust storm. Alas on June 10, 2018, Opportunity stopped communicating with Earth. Two days later with its power level dangerously low, Opportunity entered hibernation mode. Opportunity never rebooted itself or called Earth after the dust storm abated. Controllers on Earth attempted to reestablish contact with Opportunity without success. Either the spacecraft had succumbed to the cold or had drained its battery since dust covered its solar panels. In any event, after over a thousand signals had been sent to the rover, NASA declared Opportunity's mission completed on February 13, 2019.

CONTRIBUTIONS OF THE MER MISSIONS TO SCIENCE AND SOCIETY

The primary scientific objective of the MER missions was to find evidence of past water on Mars by looking for signs of past water activity in Martian rocks and soils. Other important scientific goals included understanding the environmental conditions when liquid water was present on Mars; understanding the composition and distribution of minerals, particularly those containing iron, near the landing site; and understanding the geologic processes that formed the minerals and terrain at the landing sites. Finally, studies on the ground could check the conclusions reached about the local Martian geology from orbiter studies, and thereby provide confidence that conclusions reached from orbital studies elsewhere on Mars were correct.

The Gusev Crater and Meridiani Planum landing sites were chosen for their potential as sites where water once existed. Gusev Crater, the landing site for Spirit, is an impact basin about the size of the state of Connecticut and was once thought to have been a giant lake. However, during the three-month primary mission Spirit failed to find any lake-related deposits at this site. If they had ever existed, they have apparently been disrupted by subsequent impacts. Rocks at the Spirit landing site are olivine-bearing basaltic rocks, a type of volcanic rock not previously seen on Mars. The RAT ground off surface layers on rocks at this site, revealing subsurface veins possibly altered by the presence of water. Rock coatings are also consistent with brief periods of moisture in the past, even if not indicative of a large, long-term lake. Spirit also found evidence that subsurface water percolated to the surface at Gusev Crater. After exploring Gusev Crater, Spirit started the trek for the nearby Columbia Hills, named after the crew of the ill-fated STS-107 space shuttle flight (destroyed upon reentry on February 1, 2003), to search for clues to the early history of Gusev Crater. Initial indications are that the hills are layers of volcanic ash.

Meanwhile, Opportunity found evidence of water at Meridiani Planum inside a crater named Eagle. Rocks containing sediments from evaporation appear to have formed in a body of slow-moving saltwater, perhaps at the shoreline of a salty sea. Rocks at this site contain high concentrations of hematite which forms in wet conditions. These rocks provide definite direct evidence of liquid water in Mars's past.

Samples collected by the Rovers' magnets indicate that most of the rocks on Mars contain iron. Oxidized iron gives Mars its rusty red color. Studies of patterns on the rocks are also helping mission scientists understand wind erosion on Mars, currently the most significant source of Martian erosion. Some rocks found on Mars resemble meteorites found on Earth that are thought to originate from Mars, bolstering evidence that they truly originated on Mars.

CONTEXT OF THE MER MISSIONS

The MER missions are part of NASA's long-term program to explore Mars. The program has four main scientific goals:

1. to determine if there was ever life on Mars
2. to understand the Martian climate
3. to understand the Martian geology
4. to pave the way for human exploration of Mars

In 1976, the first Viking landers on Mars found no direct evidence for life on Mars but did find evidence suggestive of large amounts of liquid water in the past in the form of dry riverbeds. The MER missions found more direct evidence of geologic processes requiring liquid water on the surface of Mars. The first Viking landers on Mars had no mobility. The 1997 Pathfinder mission was the first rover to land on Mars. The MER rover had greater mobility compared with the Pathfinder, and were able to trek several kilometers from their impact points.

—Paul A. Heckert

Further Reading

Carr, Michael H. *The Surface of Mars.* Cambridge UP, 2018.

Chaisson, Eric J., and Steve McMillan. *Astronomy Today.* 5th ed., Pearson Prentice Hall, 2004.

Coles, Kenneth S., Kenneth L. Tanaka, and Philip R. Christensen. *The Atlas of Mars: Mapping Its Geography and Geology.* Cambridge UP, 2019.

Hartmann, William K. *Moons and Planets.* 5th ed., Thomson Brooks/Cole, 2005.

Howell, Elizabeth, and Nicholas Booth. *The Search for Life on Mars.* Arcade Publishing, 2020.

Johnson, Sarah Stewart. *The Sirens of Mars: Searching for Life on Another World.* Crown/Random House, 2020.

Moomaw, B. "Spirit Lands at Gusev." *Astronomy*, vol. 32, no. 4, 2004, p. 32.

Naeye, R. "Red-Letter Days" *Sky and Telescope*, vol. 107, no. 5, 2004, p. 44.

National Aeronautics and Space Administration, Jet Propulsion Laboratory. "Mars Exploration Rovers." http://marsrovers.jpl.nasa.gov. Accessed June, 2005.

Vertesi, Janet. *Seeing Like a Rover: How Robots, Teams, and Images Craft Knowledge of Mars.* U of Chicago P, 2015.

MEAN SEA LEVEL

Fields of Study: Geology; Marine Geology; Geomorphology; Geochemistry

ABSTRACT

Sea level is the position of the surface of the ocean at any given location. Sea level is highly variable and can undergo very rapid changes. Mean sea level is the average position of the sea surface around the entire globe, independent of local short-term changes.

KEY CONCEPTS

isostatic rebound: the relaxing and return to its original level by continental crust material following the loss of the weight of a glacial mass

satellite-based radar altimetry: measurement of the altitude of a satellite above the surface of the ocean; because the altitude of the satellite is fixed, any differences of altitude measured are due to changes in the level of the ocean surface

shore processes: principally, erosion and characteristic intertidal regions, but also including sediment formations created by wave action

DEFINITION OF MEAN SEA LEVEL

Sea level is the position of the surface of the ocean at any given location. Sea level is highly variable and can undergo very rapid changes due to such events as tides, tsunamis, changes in barometric pressure, wind-generated waves, and even freshwater floods.

While these events can produce changes in sea level of several meters, they are local in scale and of a very short duration, generally lasting only for hours. Mean sea level is the average, position of the sea surface around the entire globe, independent of these local short-term changes. Changes in mean sea level are on the order of a few millimeters per year. Although small in the short term, over many years mean sea level can, and has, changed by as much as two hundred meters.

Mean sea level at specific locations can be calculated using tide gauge records and subtracting the effects of annual changes in atmospheric pressure and long-term changes in tidal ranges, which are driven by astronomical factors. Changes in global mean sea level can be calculated using satellite-based radar altimetry, such as with the TOPEX/Poseidon satellite. The radar altimeter measures the height of the satellite above the ocean, based on the time it takes for a radio signal to travel from the satellite to the sea surface and back. Since the actual altitude of the satellite is known, any changes in the altimeter measurement reflect changes in the height of the sea surface itself.

SIGNIFICANCE OF MEAN SEA LEVEL IN CLIMATE CHANGE

On short timescales (decades to centuries), mean sea level is a function of the amount of water stored as ice in glaciers and ice sheets. As global temperature rises, less water is stored as ice, contributing to a rise in mean sea level. A rise in mean sea level in response to global warming has important consequences for both human society and the natural world. First, such a rise contributes to a loss of land, as coastal areas are slowly inundated by water. This is a concern for low-lying island nations such as Maldives or Tuvalu. Maldives is a nation made up of

This marker indicating sea level is situated between Jerusalem and the Dead Sea. Photo via Wikimedia Commons. [Public domain.]

twelve hundred islands in the Indian Ocean, which has a maximum elevation of only 2.5 meters above current sea level. Thus, the Maldivian population of approximately 380,000 people is highly vulnerable to even a slow rise in mean sea level. For other nations, a rise in mean sea level is also a concern because of increased hazards from flooding during high tides—especially spring tides—and storms. A rise in mean sea level provides a higher baseline upon which tidal fluctuations build. This effect has been observed locally many times already, when hurricanes in the Gulf of Mexico or the Atlantic Ocean make landfall. The city of New Orleans, most of which sits on land that is significantly below sea level already, and the city of New York have both experienced the type of flooding from hurricane landfall that would be experienced in a corresponding rise of mean sea level. Vast areas of coastal lands would become submerged and uninhabitable. In addition, any agricultural lands that are affected would become "poisoned" by the saltwater in addition to being lost to inundation.

According to the Fourth Assessment Report released by the Intergovernmental Panel on Climate Change, from 1993 to 2003, mean sea level rose approximately 3.1 millimeters per year. According to the National Oceanic and Atmospheric Agency (NOAA), as of February 26, 2021, global mean sea level continues to rise at that rate. This means that in a span of just eight years the global mean sea level will have risen by about 2.5 centimeters, the equivalent of one inch, and in less than a hundred years it will have risen the one foot that would cause serious difficulties for many coastal cities around the world.

SIGNIFICANCE OF MEAN SEA LEVEL IN GEOLOGY

Global mean sea level has changed dramatically several times in the past, and undoubtedly will do so in the future as well. Both lowered mean sea level and increased mean sea level have important ramifications for geology, and both leave traces from shore processes in the geological record that can be discerned and interpreted by geologists. Because shorelines change their locations according to mean sea level, the associated shore processes and their signatures also change location accordingly.

Lowering of mean sea level is associated with cool periods in Earth's history, particularly the Ice Ages. During an Ice Age, great shields of ice develop and cover land areas as they expand outward from the polar regions. As water evaporates from the relatively warmer oceans and enters the atmosphere. It eventually falls once more as rain or snow and becomes incorporated into the water ice of ice shields and glaciers rather than returning to the oceans as surface runoff. The overall effect of this aqueous migration is lowering of the mean sea level. The geological record reveals that global mean sea level has been lowered by as much as two hundred meters during past Ice Ages, exposing vast areas of the continental shelves to the atmosphere and its associated erosion processes. This also exposes the marine sediments to air, and by effectively removing them from their watery environment alters the geochemistry and petrology they would otherwise undergo.

The presence of large ice shields on continental land surfaces also has seismic effects. As the mass of ice increases, it presses down with increasing force against the underlying crust. The crustal material slowly subsides beneath the weight of the ice as the underlying plastic layer of mantle material is squeezed away. The effect is similar to what would be observed by placing a good-sized ice cube on a balloon filled with water. The skin of the balloon would be seen to sag beneath the ice cube as the weight of the ice cube displaces the water below the balloon's skin. Similarly, when the weight is removed the water returns to its normal distribution within the balloon, causing the depressed area to rebound to its predepressed state. The same thing happens to Earth's crust when the weight if the ice shield is removed. The land surface experiences isostatic rebound as the underlying plastic layer of the mantle returns to its normal distribution. This process continues in the present day in northern Canada as the continent recovers geophysically from the last Ice Age. The land area about the region of Hudson's Bay is currently rising in isostatic rebound at a rate of about 1.3 meters per hundred years. Accordingly, the stress that was formed within the rock during the period of its depression under the ice is now being released, and although the movement is slow and the effects are typically not large enough to be felt there are occasionally rare earthquakes of sufficient magnitude to be felt in the Great Lakes basin, a region that is not seismically active, as the stress is released during isostatic rebound.

Raising of mean sea level, on the other hand, is associated with warm periods in Earth's history. As the global average temperature increases, so too does the rate of melting of polar ice caps and ice shields such as that of Greenland. The water that was imprisoned in those ice forms becomes free to enter the oceans once more, causing the elevation of mean sea level. The depth of the ice shields on Greenland and Antarctica is measured in kilometers, and represents a massive amount of water, enough to raise the current mean sea level by about 70 meters according to the United States Geological Survey (USGS). This would effectively flood all coastal land areas and a significant amount of interior lands that could be reached by the waters. In past warm periods, a large inland sea existed in the North American midwest, as proven by the fossils of ancient marine creatures, and the large mineral deposits of marine origin that are found in the region today. The Bonneville Salt Flats and the Death Valley borax deposits are examples of the latter. Raising of mean sea level alters the geochemical effects of groundwater movement, as well as the environment in which mineral and ore formation takes place.

—Anna M. Cruse and Richard M. Renneboog

Further Reading

Douglas, Bruce C., Michael S. Kearney, and Stephen P. Leatherman. *Sea Level Rise: History and Consequences.* Academic Press, 2001.

Everhart, Michael J. *Oceans of Kansas. A Natural History of the Western Interior Sea.* 2nd ed., Indiana UP, 2017.

Fu, Lee-Leung, and Anny Cazenave. *Satellite Altimetry and Earth Sciences: A Handbook of Techniques and Applications.* Academic Press, 2001.

Gaffney, Vincent, Kenneth Thomson, and Simon Fitch, editors. *Mapping Doggerland: The Mesolithic Landscapes of the Southern North Sea.* Archaeopress, 2007.

Meyer, David L., and Richard Arnold Davis. *A Sea Without Fish: Life in the Ordovician Sea of the Cincinnati Region.* Indiana UP, 2009.

METAMICTIZATION

Fields of Study: Geology; Geochemistry; Gemology; Crystallography

ABSTRACT

Metamictization is the process of rendering crystalline minerals partly or wholly amorphous (glasslike) as a consequence of radioactive decay. Metamict minerals such as zircon are important as gemstones, and metamict minerals that do not lose their radioactive components during the process of metamictization may possibly be used for the disposal of high-level nuclear wastes.

KEY CONCEPTS

anisotropic: having properties that differ with orientation

isotropic: having identical properties regardless of orientation

metamict: of minerals that have been altered by naturally occurring alpha radiation

zircon: a mineral form of zirconium silicate, $ZrSiO_4$

DEFINITION OF METAMICTIZATION

The term "metamict" (meaning "mixed otherwise") was proposed in 1893 by W. C. Broegger when he recognized that some minerals, although they show crystal form, are nevertheless structurally very similar to glass. Metamict minerals fracture like glass, are optically isotropic (have the same properties in all directions) to visible and infrared light, and to all appearances are noncrystalline.

OVERVIEW OF METAMICTIZATION

The discovery that all metamict minerals are at least slightly radioactive and that metamict grains contain uranium and thorium led to the conclusion that the process of metamictization results from radiation damage caused by the decay of uranium and thorium. Although all metamict minerals are radioactive, not all radioactive minerals are metamict. Many metamict minerals have nonmetamict equivalents with the same form and essentially the same composition.

Isotopes of uranium and of thorium decay, through a series of emissions of alpha particles (helium nuclei), into a stable isotope of lead. The alpha particle is emitted from the decaying nucleus with

An example of a metamict mineral is zircon. Photo by Rob Lavinsky, iRocks.com, CC BY-SA 3.0, via Wikimedia Commons.

great energy, causing the emitting nucleus to recoil simultaneously in the opposite direction. In the final part of its trajectory, the alpha particle is slowed enough to collide with hundreds of atoms in the mineral, but since the larger recoil nucleus travels a much shorter path, it collides with ten times as many atoms. Consequently, the majority of radiation damage is caused by the recoiling nucleus. The immense amount of heat generated by both particles in a small region of the mineral structure produces damage, but some of the energy also serves to self-repair some of the damage spontaneously. Radioactive minerals that remain crystalline have high rates of self-repair, while metamict minerals do not.

Metamict minerals are not common in nature, and they are generally found in pegmatites associated with granites. Showing little resistance to metamictization, the largest group of metamict minerals includes the thorium-, uranium-, and yttrium-bearing oxides of niobium, tantalum, and titanium. The second-largest group of metamict minerals are silicates, with zircon (a zirconium-silicate mineral) occurring most frequently. The smallest group of metamict minerals are the phosphates, including xenotime (yttrium phosphate), which has the same crystal structure as zircon.

Since metamict gemstones, such as zircon, are isotropic and look clear inside, they are often of greater value than the crystalline varieties, because the anisotropic properties of crystalline gems make them look cloudy inside. In addition, radiation damage often imparts attractive color to the metamict gemstones. Metamict minerals may possibly have another important use in the future: Since some of them retain their radioactive elements over millions of years despite metamictization, they may provide the key for safe disposal of high-level nuclear wastes. Many geochemists believe that synthetic versions of these metamict minerals could be "grown" to produce rocks that would be able to contain hazardous nuclear wastes safely for tens of thousands of years.

—*Alvin K. Benson*

Further Reading

Philpotts, Anthony, and Jay Aque. *Principles of Igneous and Metamorphic Petrology*. 2nd ed., Cambridge UP, 2009.
Read, Peter G. *Gemmology*. 3rd ed., Elsevier, 2005.

Metamorphic Processes, Rocks, and Mineral Deposits

Fields of Study: Geology; Geomorphology; Geochemistry

ABSTRACT

The word "metamorphism," based on Greek roots, translates as the "process of changing form." Existing sedimentary or igneous rocks are transformed in the solid state to metamorphic rocks as the temperature and pressure of their environment increase at various depths within the Earth. The numerous transformations that occur are collectively termed "metamorphic processes."

KEY CONCEPTS

foliation: formation of rock in sheet-like layers, analogous to the sheets of paper in a book or "folio"

metasomatism: the infusion of additional water-borne elements into protolithic rock that is undergoing metamorphism

monomineralic zone: a band or region between a limestone body and an igneous pluton, in which only one distinct mineral type is found

protolith: the original form of a metamorphic rock before it underwent metamorphosis

skarns: coarse-grained rock produced through metasomatism, differing from tactites with regard to metal content

tactites: coarse-grained rock produced through metasomatism, differing from skarns with respect to metal content

METAMORPHOSIS OF MINERALS

Every metamorphic process relates either to the formation of new minerals, called "neocrystallization," or to the formation of a new texture in the metamorphic rock. The new texture may simply be an increase in size and change in shape of existing minerals (recrystallization). The new texture may also involve the development of a "foliation," in which the elongate and platy minerals assume a parallel orientation. These general processes are further divided depending upon the specific chemical and mechanical changes occurring during the metamorphic transformation. Long periods of erosion can expose metamorphic rocks on the surface of Earth; surface metamorphic rocks are often valuable resources, either because of their new minerals or because of the physical properties that the rocks themselves have as a result of their new textures.

NEOCRYSTALLIZATION

New minerals form at the expense of old minerals. As the pressure and temperature increase on an existing igneous or sedimentary rock (called the "protolith"), the old minerals become unstable and break down into chemical components that recombine to form new minerals. Some of the chemicals, for example, H_2O and CO_2, occur as gases at metamorphic temperatures. These gases mix to form a vapor that exists in the cracks and along the boundaries between the individual grains of the minerals. The gain and loss of gases from the vapor are part of the overall chemical reconstruction that takes place during metamorphism. The vapor inevitably escapes from the rock during the long period of cooling and erosion that exposes such rocks on Earth's surface.

The neocrystallization process is usually expressed as a chemical reaction. The minerals of the protolith (existing rock) are the reactants, shown on the left side of the reaction, and the new metamorphic minerals that form are the products, listed on the right side. The reactions often will generate and/or consume chemicals residing in the vapor. As with any chemical reaction, there are specific temperature and pressure conditions that must exist before the reaction can occur. Each metamorphic mineral of interest forms within a specific temperature and pressure region within Earth. The exact temperature and pressure conditions under which a metamorphic mineral or group of minerals will form can be determined by laboratory experiments; geologists then deduce that similar conditions must have existed whenever these minerals are found in the geological environment. The geological environment required for the development of a given metamorphic mineral is usually controlled by tectonic plate movements. Explorations for metamorphic resources are targeted to specific tectonic regions that correspond to the proper temperature-pressure environments for their formation.

TYPES OF METAMORPHISM

There are three tectonic environments with specific pressure and temperature conditions that control the location for the development of metamorphic minerals. Burial metamorphism results from a high-pressure and low-temperature environment that occurs where two plates converge and one plate is actively subducted. During the recent geological past, the coastline along Oregon and Northern California experienced this tectonic environment. Contact metamorphism is a high-temperature, low-pressure environment occurring slightly farther inland from the region of burial metamorphism. Contact metamorphism results when magma generated during the subduction of a plate rises into the overriding plate and solidifies as shallow igneous plutons. Contact metamorphism has occurred along the margins of the Sierra Nevada batholiths of eastern California. The third tectonic environment is regional metamorphism, often called "dynothermal metamorphism," which corresponds to moderately high pressures and temperatures. Regional metamorphism is seen after extensive erosion of a contact metamorphism area has exposed deeper regions within Earth's crust.

ISOCHEMICAL PROCESSES

Neocrystallization that occurs without any influx of new chemicals (other than the water and carbon dioxide from the vapor) is called "isochemical metamorphism." Isochemical metamorphism produces about a dozen minerals that are considered valuable resources. The isochemical-neocrystallization processes responsible for the formation of some of these minerals are described below, with a brief indication of the tectonic environments that favor their formation.

Serpentine. When serpentine ($Mg_3Si_2O_5(OH)_4$) is the major mineral formed during the low-temperature, low-pressure metamorphism associated with the beginning of regional metamorphism, the resulting metamorphic rock is called a "serpentinite." Polished serpentinites are used widely as a facing stone in both interior and exterior applications. When the serpentinites contain some carbonate minerals they are marketed as "verde antique marble." Serpentine can occur in any one of three forms. The form called "chrysotile" is the most common asbestos mineral. Asbestos veins are common in serpentinites, and in many locations in eastern Canada and northern New England serpentinites have been mined for their asbestos.

Serpentine generally forms by metamorphism of ultramafic igneous rocks by one of two reactions. One type of serpentine reaction involves a mixed vapor phase of carbon dioxide and water, which produces some carbonate minerals. A second serpentine-forming reaction requires that some silica be dissolved in the water vapor.

Talc. Talc ($Mg_3Si_4O_{10}(OH)_2$) can form large masses of randomly oriented interlocking small flakes to make a rock called "soapstone," used extensively for carving and as a source of talcum powder for health and beauty applications. The term "steatite" refers to talc-rich rocks that are used because of talc's lack of chemical reactivity or its high heat capacity. Talc forms by regional metamorphism at low to moderate temperatures and low to moderate pressures.

A second common reaction that produces major talc deposits is the continuing metamorphism of a peridotite protolith. Talc forms by this reaction at temperatures slightly above 300°C; however, the temperatures must remain below 700° Celsius to prevent the breakdown of talc.

Graphite. Graphite (a form of carbon, C) is used in a wide variety of applications from lubrication to high-temperature crucibles. Deposits of amorphous graphite form by contact metamorphism of coal beds, whereas deposits of flake graphite form by regional metamorphism of sedimentary rocks with the graphite being disseminated in mica schist and micaceous quartzite. Extensive weathering of these rocks assists in the release of the graphite. The graphite content of such metamorphic ores is usually 5 to 6 percent.

Clinker is a common term used by English miners for the graphite ore created by the contact metamorphism of coal beds. The reaction involves the breakdown of a wide variety of organic molecules. Continued high-temperature metamorphism of coal beds can transform the graphite into a natural coke, which has been mined in Wyoming and Utah.

Kyanite. Kyanite (Al_2SiO_5) and the related minerals andalusite and sillimanite are used in the production of refractory ceramics, such as those used in spark plugs. Kyanite forms from aluminum-rich clay-shale protoliths during regional metamorphism at moderate to high temperatures.

Wollastonite. Wollastonite ($CaSiO_3$) is used extensively in the manufacturing of tiles. It forms by high-temperature contact metamorphism of silica-bearing limestones. An example may be found in Willsboro, New York, where the wollastonite mine is in a metamorphosed limestone on the margin of the igneous intrusion that forms the Adirondack Mountains. This reaction normally occurs at temperatures around 650° Celsius.

Jadeite. The pure form of the mineral jadeite ($NaAlSi_2O_6$) is the best quality of all materials called "jade." Jade has been a valued material for sculpture and other art-and-craft applications for more than twenty-five centuries. It forms during burial metamorphism of alkali-rich igneous rocks that have been subjected to very high pressures and low temperatures. Such conditions are found in the mountains of the Coast Range in California, where jade has been mined.

Corundum. Corundum (Al_2O_3) is used extensively as an abrasive, and its pure colored variants known as ruby and sapphire are valued as gemstones. Corundum forms during regional metamorphism of aluminum-rich shale protoliths. The progressing metamorphism of the shale makes an intermediate mineral called "staurolite," which commonly is sold in mineral shops and displayed in museums as "fairy crosses" because of its well-developed cruciform twining. Corundum forms when the staurolite breaks down at very high temperatures.

METASOMATISM

A special type of metamorphism occurs whenever a major influx of new dissolved chemical components is added to the chemistry of the protolith. A water-rich fluid or vapor is the means of transport for this added chemistry. The process of adding chemistry to the rock through the vapor is called "metasomatism." Metasomatism occurs chiefly in regions of contact metamorphism where highly volatile elements such as boron, fluorine, or chlorine are released into a water-rich fluid associated with the igneous pluton. The igneous-based fluid also carries dissolved silicon, aluminum, iron, magnesium, manganese, minor sodium, potassium, and often some tin, copper, tungsten, lead, and zinc. This saline fluid invades the adjacent limestone and reacts with calcium to form pronounced monomineralic zones at the contact between the pluton and the limestone.

The rocks produced by metasomatism are called "skarns" or "tactites," and they are the coarsest grained of all metamorphic rocks. The garnet zone of a skarn may have individual grains of garnet that are as large as 20 centimeters in diameter. Skarns are mined throughout the world. Scheelite ($CaWO_4$), a major ore of tungsten, is mined from numerous metasomatized contact zones in California, Nevada, Idaho, and British Columbia. Other minerals that are mined from skarns are wollastonite, galena (an ore of lead), sphalerite (an ore of zinc), magnetite (an ore of iron), and chalcopyrite (an ore of copper).

TEXTURE CHANGES AND RECRYSTALLIZATION IN METAMORPHISM

During metamorphism changes may occur in the size, the shape, and often the orientation of the mineral grains within the rock. There are at least six different processes related to texture changes; the exact process is dependent upon which of the texture variables are changed and the mechanics of the change.

A change in size and shape of an existing mineral without the formation of any new minerals is a process called "recrystallization." Certain sedimentary protoliths may be monomineralic rocks; two common examples are a limestone that is made entirely of the mineral calcite and a silica-cemented sandstone that is made entirely of the mineral quartz. Such single-mineral rocks are unable to promote any form of neocrystallization, and recrystallization is the only result of metamorphism.

Marble. The transformation from a sedimentary limestone to a metamorphic rock called "marble" often results in more than a thousandfold increase in the size of the calcite grains. The grains in the limestone protolith are commonly round in shape, whereas the grains in the marble interlock like a jigsaw puzzle to give a mosaic texture. The interlocking texture in marble imparts a high coherence to the rock, yet its calcite mineralogy gives it a low hardness, allowing marble to be easily cut and polished. Pure white marble is used extensively for sculpting to form statues, as in the Lincoln Memorial; for building stone, as in the Greek Parthenon; and for ornamental carvings. Many marbles may contain an impurity that imparts a striking color pattern allowing their use in architecture as facings, tabletops, and flooring. Italy has more marble quarries than any other country. In the United States marble is quarried from both the Rocky and Appalachian Mountain chains, with major quarries in Vermont and Colorado.

FOLIATION

Slate. A metamorphic rock in which the platy and elongate-shaped minerals are parallel in their orientation is said to be foliated. A foliated texture can be seen in the rock by a tendency for the rock to break along parallel planes.

Slate is a foliated metamorphic rock in which the individual mineral flakes are so small that they can be seen only under the highest magnifications of a microscope. The foliation imparts to the slate the ability to break in near perfect planes. Slate is used as flagstones, roofing, floor tiles, hearthstones, and tabletops, especially billiard tables. A few slates are used not because of their foliation but because of their composition. Very clay-rich slates are ground because the smaller pieces will bloat when heated to form a material used as a lightweight aggregate.

METAMORPHIC DIFFERENTIATION

Gneiss. At relatively high temperatures a metamorphic process occurs in which minerals segregate. The light-colored minerals such as quartz and feldspar move into zones parallel to the rock's foliation, leaving behind alternate zones of dark minerals such as biotite and amphibole. Metamorphic differentiations cause a marked dark versus light layering in the rock. Such rock is commonly called "gneiss." Gneiss is quarried locally in many places as dimension stone.

ANATEXIS

Migmatites. At the more extreme temperatures for regional metamorphism, partial melting will begin to occur within the light-colored layers of a gneiss. The process of partially melting a rock is called "anatexis," and this process begins the transformation from metamorphic to igneous rocks. Migmatite is the name for such a mixed rock. Migmatites occur in regions that have experienced a great amount of erosion to reveal the highest levels of metamorphism. Migmatites are common in the shield regions of the major continents. The shield for the North American continent is exposed in the upper peninsula of Michigan, northern Wisconsin and Minnesota, and throughout most of Canada.

Migmatites are commonly used as monument stone. The contortions of pattern generated by the partial melting make each stone unique and generally quite handsome. Migmatites are quarried in Minnesota, New York, and Michigan and are used as building stone throughout the United States.

Cataclastite. A special texture develops in rocks when the metamorphic pressure involves tectonic forces having a distinctly linear or planar orientation on the rock. Such opposing forces result in shear stress, and they cause mechanical breakage of the mineral grains in the rock. The name "cataclastite" refers to a metamorphic rock that exhibits a sheared texture containing many fragmented and distorted mineral grains that are often cemented together by a calcite matrix. Cataclastites are formed in tectonic regions that are experiencing active crustal movements. Some cataclastites are quarried and polished for use as a decorative "marble." A famous cataclastite, the "Fantastica di Lasa," is quarried from the northern Alps in Italy because of its attractive and unique appearance.

—Dion C. Stewart

Further Reading

Best, Myron G. *Igneous and Metamorphic Petrology.* 2nd ed., Blackwell, 2003.

Blatt, Harvey, Robert J. Tracy, and Brent E. Owens. *Petrology: Igneous, Sedimentary, and Metamorphic.* 3rd ed., W. H. Freeman, 2006.

Bucher, Kurt, and Martin Frey. *Petrogenesis of Metamorphic Rocks.* 7th Rev. and Updated ed., Springer, 2002.

Philpotts, Anthony R., and Jay J. Ague. *Principles of Igneous and Metamorphic Petrology.* 2nd ed., Cambridge UP, 2009.

Raymond, L. A. *Petrology: The Study of Igneous, Sedimentary, and Metamorphic Rocks.* 2nd ed., Waveland Press, 2007.

Rollinson, Hugh, and Victoria Pease. *Using Geochemical Data to Understand Geological Processes.* Cambridge UP, 2021.

Winter, John D. *An Introduction to Igneous and Metamorphic Petrology.* Prentice Hall, 2001.

Yardley, Bruce, and Clare Warren. *An Introduction to Metamorphic Petrology.* 2nd ed., Cambridge UP, 2021.

Mineralization and Hydrothermal Solutions

Fields of Study: Geochemistry; Geomorphology; Geology; Geophysics

ABSTRACT

Hydrothermal solutions are "hot-water" solutions rich in base metals and other ions that create deposits of minerals. Hydrothermal mineralization occurs when mineral particulates precipitate from water. This process is closely tied to tectonics and volcanic activity, which create the heat and convection necessary for the transportation and deposition of minerals. Most hydrothermal solutions are exhalations from magmas, but some hydrothermal deposits have no identifiable magma source. Hydrothermal mineralization can occur in submarine or terrestrial environments. Hydrothermally created mineral deposits are economically important, and they provide data that help geologists to understand Earth processes.

KEY CONCEPTS

black smoker: a columnar structure formed deep in the ocean when sulfides precipitate from hydrothermal fluids

crust: the outermost layer of the earth; crust may be continental or oceanic

divergent boundary: an area where two tectonic plates are spreading apart instead of crashing together or sliding against each other

fumarole: a gas- and steam-producing vent at volcanic craters

hot spot: a column of magma that rises from the mantle, remaining in one place as the lithospheric plate moves over it; also known as a mantle plume

hydrothermal vent: a fissure through which magmatically heated fluids escape to Earth's surface

hypothermal: refers to mineral solutions formed at great depth and high pressures within Earth's crust, the associated minerals being precipitated from solution as it travels through fissures and cracks, cooling and becoming unable to maintain its mineral content as it does so

isotopes: atoms of an element that contain the same number of protons but different numbers of neutrons

magma: hot fluid or semifluid from beneath the earth's crust; later forms igneous rock

mantle: a hot layer of the earth below the crust, asthenosphere, and lithosphere; surrounds the inner core of the planet

silicate: a compound containing silica; includes mica, feldspar, and quartz

sulfide: a compound made with sulfur; an iron sulfide is a compound of iron and sulfur

BACKGROUND

Essential conditions for the formation of hydrothermal mineral deposits include metal-bearing mineralizing solutions, openings in rocks through which the solutions are channeled, sites for deposition, and chemical reaction resulting in deposition. The term "ore" is used for any assemblage of minerals that can be mined for a profit. "Gangue" is the nonvaluable mineral that occurs with the ore.

During the crystallization of igneous rocks, water and other volatile fluids concentrate in the upper part of the magma. These volatiles carry with them varying amounts of the ions from the melt, including high concentrations of ions that are not readily incorporated into silicate rock-forming minerals. If the vapor pressure in the magma exceeds the confining pressure of the enclosing rocks, the fluids are expelled to migrate though surrounding country rock. These solutions travel along natural pathways in the rock such as faults, fissures, or bedding planes in stratified rocks. As the solutions migrate away from their source region, they lose their mineral content through deposition in natural openings in the host rock (forming open space-filling deposits) or by chemical reaction with the host rock (forming metasomatic replacement deposits). A part of these solutions may make it to the surface to form fumaroles (gas emanations) or hot springs. In addition, some hydrothermal solutions may be derived from water trapped in ancient sediments or by dehydration of water-bearing minerals during metamorphism.

The observed volatiles from magmas, as seen during volcanic eruptions and at fumaroles, are 80 percent water. Carbon dioxide, hydrogen sulfide, sulfur, and sulfur dioxide are also abundant. Nitrogen, chlorine, fluorine, boron, and other elements are present in smaller amounts. In addition, metal ions are carried in this residual fluid. Especially abundant are the base metals—iron, tungsten, copper, lead, zinc, molybdenum, silver, and gold. Quartz is the most common nonore, or gangue, mineral deposited. Calcite, fluorite, and barite are also common as gangue minerals. Base metals combined with sulfur as sulfide minerals, with arsenic as arsenides, or with tellurium as tellurides form the most common ore minerals. Gold often occurs as a native mineral.

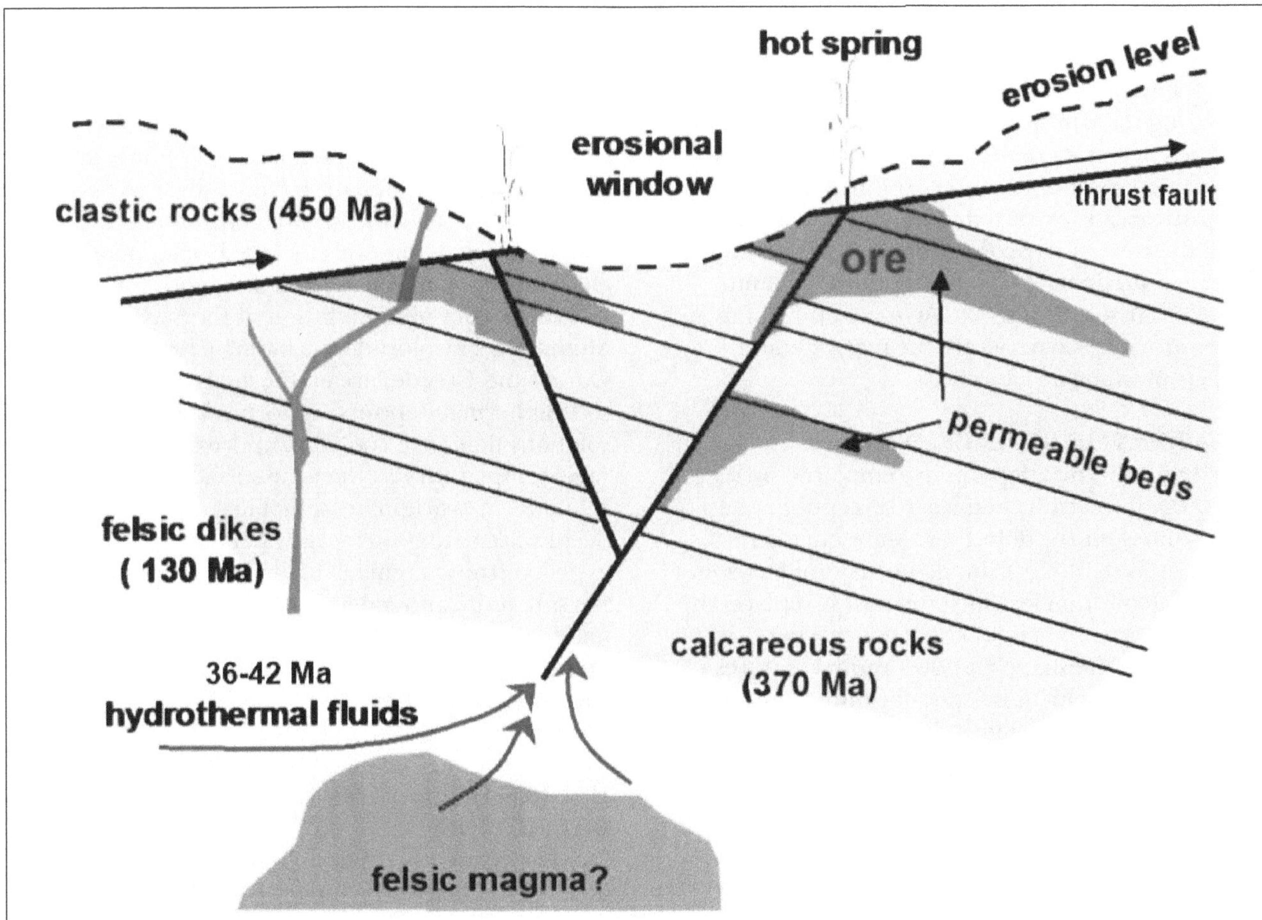

Ascending hydrothermal solutions rich in gold, sulfur and metals were channelled upward along major fracture and fault zones. Fluid that made it to the surface would have vented as hot springs and geysers. Adapted from Edwards and Atkinson (1985). Image by Alexgiovi, CC BY-SA 4.0, via Wikimedia Commons.

NATURE OF OPEN SPACES

Hydrothermal solutions find ready-made escape routes through the surrounding country rock in the form of faults and fissures. Ore and gangue minerals of cavity-filling deposits are found in faults or fissures (veins), in open spaces in fault breccias, in solution openings of soluble rocks, in pore spaces between the grain of sedimentary rocks, in vesicles of buried lava flows, and along permeable bedding planes of sedimentary strata. The shape of the mineral deposit is controlled by the configuration of structures controlling porosity and permeability. Fracture patterns, and therefore veins, may take on a wide variety of geometric patterns, ranging from tabular to rod-shaped or blanket-like deposits.

Some deposits are characterized by ore minerals that are widely disseminated in small amounts through out a large body of rock such as an igneous stock. These igneous bodies undergo intense fracturing during the late stage of consolidation, and residual fluids permeate the fractured rock to produce massive deposits of low-grade ores. In such deposits, the entire rock is extracted in mining operations. The famous porphyry copper deposits of the south western United States—including those of Santa Rita, New Mexico; Morence, Arizona; and Bingham, Utah—are of this type, as are the molybdenum deposits of Climax, Colorado.

METASOMATIC REPLACEMENT

Some hydrothermal deposits are emplaced by reaction of the fluids with chemically susceptible rocks such as limestone or dolostone. Metasomatic replacement is defined as simultaneous capillary solution and deposition by which the host is replaced by ore and gangue minerals. These massive deposits or lodes take on the shape and the original textures of the host. Replacement is especially important in deep-seated deposits where open spaces are scarce. Replacement deposits of lead-zinc are common in limestones surrounding the porphyry copper of Santa Rita, New Mexico, and at Pioche, Nevada.

CLASSIFICATION BY TEMPERATURE AND DEPTH

Veins are zoned, with higher-temperature minerals deposited near the source and lower-temperature minerals farther away. Hypothermal or high-temperature and high-pressure mineral assemblages include the minerals cassiterite (tin), scheelite and wolframite (tungsten), millerite (nickel), and molybdenite (molybdenum), associated with gangue minerals quartz, tourmaline, topaz, and other silicates. The mineral deposits of Broken Hill, Australia, the tin deposits of Cornwall, England, and Potosí, Bolivia, and the gold of the Homestake Mine, South Dakota, are hypothermal.

Mesothermal, or moderate-temperature and moderate-pressure deposits consist of pyrite (iron sulfide), bornite, chalocite, chalcopyrite and enargite (copper), galena (lead), sphalerite (zinc), and cobaltite or smaltite (cobalt). Gangue minerals include calcite, quartz, siderite, and rhodochrosite. The zinc-lead-silver replacement deposits of Leadville, Park City, and Aspen, Colorado, and the Coeur d'Alene, Idaho, lead veins are mesothermal.

Epithermal or low-temperature, near-surface deposits are often associated with regions of recent volcanism. The ore is characterized by stibnite (antimony), cinnabar (mercury), native silver and silver sulfides, gold telluride, native gold, sphalerite, and galena. Gangue minerals include barite, fluorite, chalcedony, opal, calcite, and aragonite. The extensive silver-gold mineralization of the San Juan Mountains of Colorado, including Cripple Creek, Ouray, and Creede, are epithermal deposits.

Telethermal deposits are formed by hydrothermal solutions that have cooled to approximately the same temperature as the near-surface rocks. These solutions may originate as mobilized connate and deeply circulating meteoric waters rather than fluids expelled from magma. The principal ore minerals are sphalerite and galena, with gangue minerals marcasite, fluorite, calcite, and chalcopyrite. The Mississippi Valley-type deposits of the tristate district of Missouri, Kansas, and Oklahoma exemplify this low-temperature mineralization.

HYDROTHERMAL MINERALIZATION: THE BIG PICTURE

Liquids frequently contain minerals, and when the liquids are heated, the minerals tend to fall from solution (precipitate) as insoluble combinations form. An everyday example of this is the mineral layer that sometimes forms on the bottom of tea kettles or in coffee makers.

In geologic terms, hydrothermal mineralization means that water is heated by magmas or hot rocks, and as the water travels through rock structures, it cools and leaves behind the minerals it was holding in solution. Other factors that affect precipitation are chemical reactions, pressure, oxygen levels, acidity, and rate of cooling.

To understand how water becomes heated and chemically active, it is necessary to understand wider geologic forces. Humans are familiar with Earth's crust because it is the layer of inhabitation, but beneath the crust are the lithosphere (composed of tectonic plates), the asthenosphere, and the mantle, which is hot. Convection in the mantle causes the plates to move. Sites where two tectonic plates crash together are called "convergent boundaries." Sites where they are moving apart are called "divergent boundaries."

Movement of tectonic plates creates seismic activity and facilitates the release of heat and hot magma from Earth's interior. Evidence of this process comes in the form of volcanoes. Hydrothermal activity occurs at divergent boundaries (often midocean ridges) because new, hot material comes to the surface as the plates spread apart. Hydrothermal processes also occur at subduction zones (associated with convergent boundaries), where surface crust slides under another plate as they collide and descends to an area of increased heat and pressure. The rock in the crust then melts into the surrounding material, allowing the rock to again rise to the surface.

Because hydrothermal activity is associated with plate boundaries and hot spots (localized areas where a column of magma rises from the mantle), one can infer that hydrothermal mineralization occurs at divergent boundaries such as the East African Rift, the Mid-Atlantic Ridge, the East Pacific Rise, and the Red Sea Rift. Hydrothermal mineralization also occurs at subduction zones, such as the ones along the west coast of the Americas. Hydrothermal activity also occurs at oceanic hot spots such as the Hawaiian Islands and at continental hot spots such as the Yellowstone National Park region of the United States.

HYDROTHERMAL PROCESSES AND MINERAL DEPOSITION

Hydrothermal vents—fissures in which water heated by magmatic activity rises and escapes to the surface—can occur in submarine or terrestrial environments. The types of mineral deposits associated with these vents depend upon the composition of the nearby rock and magma. Magma from divergent plate boundaries is made of a high percentage of iron and magnesium, but at convergent plate boundaries, the magmas tend to be higher in silicon and oxygen because they are mixing with continental sediments rich in these elements.

At submarine vents, ocean water heated to temperatures nearing 350° Celsius (662° Fahrenheit) rises, leaching metals from the basaltic rocks it passes through. The metals later precipitate when the hot water escapes and encounters the cold ocean water, creating mineral deposits. These deposits can occur in the form of black smokers, which are tower-like structures built from iron-rich sulfides. White smokers, which have lower temperatures, contain calcium sulfate and silica deposits. Some silver, lead, zinc, and copper sulfides are produced near submarine hydrothermal vents, too.

Deep ocean deposits are neither large enough nor accessible enough to mine economically. However, in the future, resource scarcity and improved mining technology may make these deposits more attractive for exploitation. Some exploitable ore deposits can be traced to ancient seafloor spreading centers that have moved great distances in time and have been lifted by plate collision. An example can be seen in the island nation Cyprus, where copper and other minerals, formed eons ago, are now near the surface.

Silicates, including feldspar, quartz, and mica, are not as economically valuable as metal deposits, but they are also formed by hydrothermal processes. Continental hydrothermal deposits are often rich in silica because igneous rocks that make up continental crust are high in silica. Silica-rich igneous rock that cools and solidifies underground is called "granite," but silica-rich igneous rock that cools at the surface is known as "rhyolite." These rock compositions are important because hydrothermal fluids pass through the rock and pick up minerals from them.

Hydrothermal fluids seep into existing spaces and precipitate minerals called "cavity-filling deposits." Sometimes, however, the hydrothermal fluids can react with the rocks they pass through and alter the rock to form a deposit made from both hydrothermal precipitates and host-rock material. This process is called "hydrothermal replacement."

Terrestrial hydrothermal vents occur in several forms associated with volcanic activity. Examples can be seen in and around Yellowstone National Park, which is located at a hot spot. A fumarole is a type of gas- and steam-producing vent that occurs in volcanic craters or along their sides. Deposits form when the vent's gases cool. The Yellowstone area is rich in rhyolite, so the hydrothermal fluids there are silica-rich. Hot springs or geysers frequently contain silicon dioxide and calcium carbonate. Precipitation of these minerals can be aided by the biologic activity of organisms that live in and around hot springs.

Other continental hydrothermal deposits were formed in ancient subduction zones where metals were liquefied as they made their way down toward the mantle then rose into fractures in the crust. As the fluids flowed through, such metals as copper, molybdenum, silver, gold, lead, and zinc were left behind. These metals are called "porphyry metal deposits" because they occur in porphyritic igneous rock associated with hydrothermal activity. Deposits of this type have been found in North America, and are being mined.

READING EARTH

Even when hydrothermal deposits have no economic value, they have scientific value. Because of the association of seafloor spreading centers with hydrothermal activity and the creation of mineral deposits, scientists can use deposits and rock samples from and near oceanic ridges to learn about cycles of spreading, subduction, and sediment transport. In addition, because humans cannot descend into the mantle to gather data about it, the material brought to the crustal layer through geologic events is a valuable extrapolative tool.

Hot hydrothermal fluids emerge from point sources (vents) and begin to spread out and cool as they mix with cold ocean water. Scientists can track the movement of these fluids by using a method called "isotopic tracing," which uses isotopes (forms of an element with the same number of protons but differing numbers of neutrons) to provide information about the vertical and horizontal movement of water. Isotopes also can be used to determine the age of particular rock samples or the conditions under which they formed. This is helpful in establishing patterns in tectonic plate movement or the rate of seafloor spreading.

Because many of the hydrothermal deposits and magmas at seafloor spreading centers contain iron, which can be magnetized, they also provide a record of Earth's magnetism. The needle of a compass will invariably indicate north because that is where Earth's north magnetic pole is located. This has not always been true. Earth's magnetic poles reverse approximately every few hundred thousand years. The process occurs gradually and does not follow strict cycles, but because seafloor spreading occurs with magmatic activity, and because magma fixes a permanent magnetic signature at a specific point during cooling, scientists can construct a relatively precise timeline of Earth's magnetism based on the magnetism of rocks and deposits near seafloor spreading centers.

One of the other benefits of studying mineral deposits at seafloor spreading centers is to compare magnetic and other qualities of the deposits on each side of the ridge. Minerals on one side of the ridge should have a matching counterpart on the other side of the ridge. By looking for similarities, scientists can see plate movement outward from a spreading center, but they also can track horizontal movement, such as the type of movement at transform fault boundaries.

—J. D. Ho and René A. De Hon

Further Reading

Chernicoff, Stanley, and Donna Whitney. *Geology: An Introduction to Physical Geology.* Prentice Hall, 2007.

Decrée, Sophie, and Laurence Robb, editors. *Ore Deposits: Origin, Exploration, and Exploitation.* Wiley, 2019.

Golani, Prakash R. *Assessment of Ore Deposit Settings, Structures and Proximity Indicator Minerals in Geological Exploration.* Springer, 2021.

Guilbert, John M., and Charles F. Park, Jr. *The Geology of* Ore *Deposits.* Waveland Press, 2007.

Jerram, Dougal, and Nick Petford. *The Field Description of Igneous Rocks.* Wiley-Blackwell, 2011.

Keller, Edward A. *Introduction to Environmental Geology.* Prentice Hall, 2012.

Libes, Susan M. *Introduction to Marine Biogeochemistry.* Academic Press, 2009.

Morgan, Lisa A. *Integrated Geoscience Studies in the Greater Yellowstone Area: Volcanic, Tectonic, and Hydrothermal*

Processes in the Yellowstone Geoecosystem. U.S. Geological Survey, 2007.
Pirajno, Franco. *Hydrothermal Processes and Mineral Systems.* Springer/Geological Survey of Western Australia, 2009.

Mineralogy

Fields of Study: Geology; Mineralogy; Geochemistry; Crystallography; Geomorphology; Gemology

ABSTRACT
Mineralogy is the study of the chemical composition and physical property of minerals, the arrangement of atoms in the minerals, and the use of the minerals. Minerals are naturally occurring elements or compounds. The composition and arrangement of the atoms that make up minerals is reflected in their physical characteristics. Sometimes mineral resources are broadened to refer to oil, natural gas, and coal, although those materials are not technically minerals.

KEY CONCEPTS
clay mineral: any of a group of tiny silicate minerals (less than 2 micrometers in size) that form by varied degrees of weathering of other silicate minerals
cleavage: breaking of a crystallized mineral along a plane, leaving a smooth rather than an irregular surface
hardness: resistance of a mineral to being scratched by another material
hydrothermal deposit: mineral deposit precipitated from a hot water or gas solution
igneous rock: rock formed from molten rock material
ion: atom or group of atoms with a positive or negative charge
lava: molten rock material at Earth's surface
magma: molten rock material and suspended mineral crystals below Earth's surface
major elements: elements that make up the bulk of the chemical composition of a mineral or rock
petrology: study of the origin, composition, structure, and properties of rocks and the processes that formed them
sedimentary rock: rock formed from accumulated particles such as sand, silt, mud, etc.
silicate mineral: mineral that contains silicon bonded with oxygen to form silicate groups bonded with positive ions; silicates make up 90 percent of Earth's crust
trace element: element present only in tiny amounts (in quantities of parts per million or less) in a mineral or rock

BASIC PRINCIPLES OF MINERALS AND MINERALOGY
Minerals are solid elements or compounds that have a definite but often not fixed chemical composition and a definite arrangement of atoms or ions. For instance, the mineral olivine is magnesium iron silicate, with the formula of $(Mg,Fe)_2SiO_4$, meaning that it has oxygen (O) and silicon (Si) atoms in a ratio of 4:1 and a total of two ions of magnesium (Mg) and iron (Fe) in any ratio. The magnesium and iron component can vary from 100 percent iron with no magnesium to 100 percent magnesium with no iron, and all variations in between. Other minerals, however, such as gold, have nearly 100 percent gold atoms. Minerals are usually formed by inorganic means, but some organisms can form minerals such as calcite (calcium carbonate, with the formula $CaCO_3$).

Minerals make up most of the rocks of Earth, so they are studied in many fields. In geochemistry, for example, scientists determine the chemical composition of the minerals and rocks to develop hypotheses about how various rocks may have formed. Geochemists might study what rocks melt to form magmas or lavas of a certain composition. Environmental geologists attempt to solve problems regarding minerals that pollute the environment. For instance, environmental geologists might try to minimize the effects of the mineral pyrite (iron sulfide, with the formula FeS_2) in natural bodies of water. Pyrite, which is found in coal, dissolves in water to form sulphuric acid and high-iron water, which can kill some organisms. Forensic mineralogists may determine the origin of minerals left at a crime scene. Economic geologists discover and determine the distribution of minerals that can be mined, such as lead minerals (galena), salt, gypsum (for wall board), or granite (for kitchen countertops, monuments and

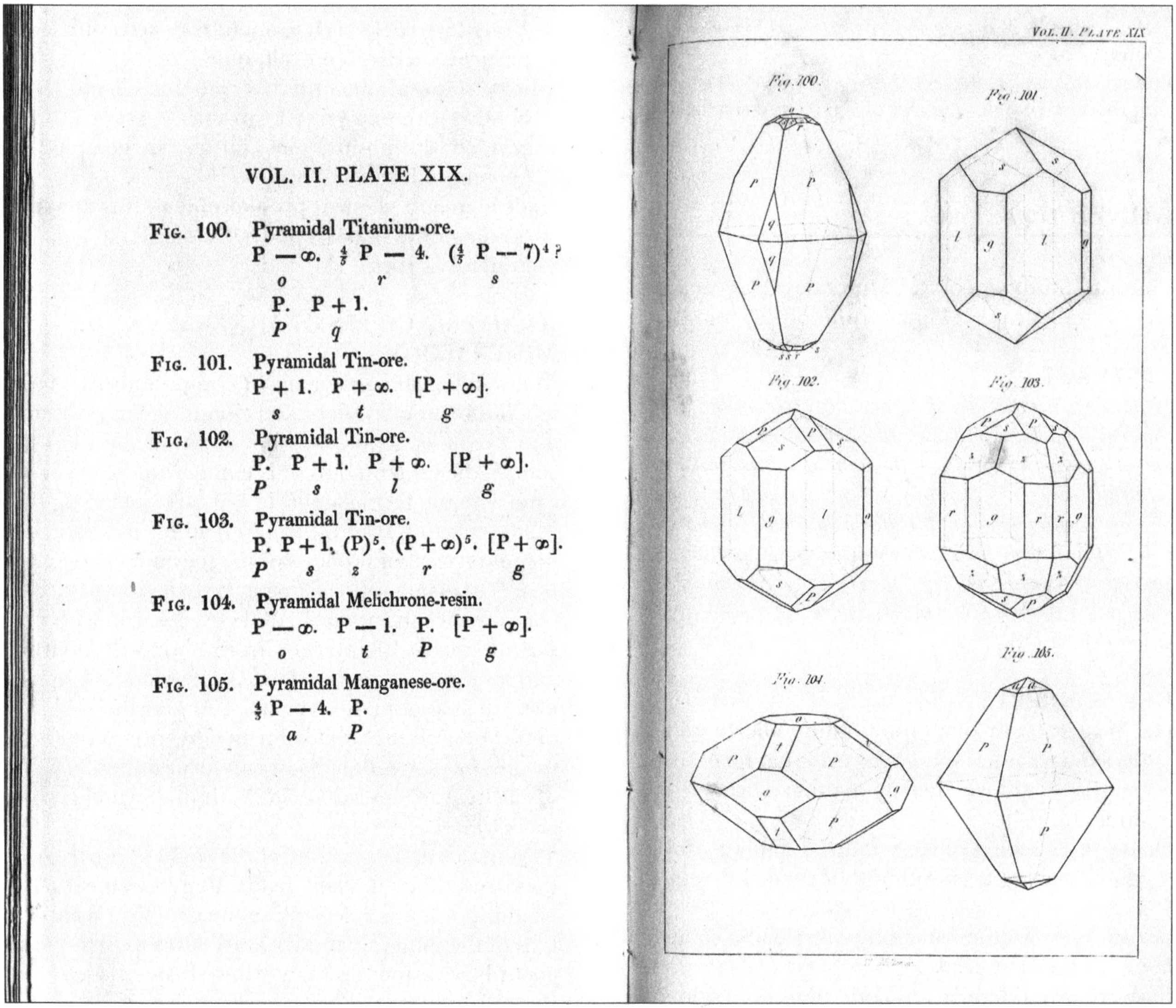
VOL. II. PLATE XIX.

FIG. 100. Pyramidal Titanium-ore.
$P - \infty.\ \frac{4}{3}P - 4.\ (\frac{4}{3}P - 7)^4 ?$
o r s
$P.\ P + 1.$
P q

FIG. 101. Pyramidal Tin-ore.
$P + 1.\ P + \infty.\ [P + \infty].$
s t g

FIG. 102. Pyramidal Tin-ore.
$P.\ P + 1.\ P + \infty.\ [P + \infty].$
P s l g

FIG. 103. Pyramidal Tin-ore.
$P.\ P + 1.\ (P)^5.\ (P + \infty)^5.\ [P + \infty].$
P s z r g

FIG. 104. Pyramidal Melichrone-resin.
$P - \infty.\ P - 1.\ P.\ [P + \infty].$
o t P g

FIG. 105. Pyramidal Manganese-ore.
$\frac{4}{3}P - 4.\ P.$
a P

Page from Treatise on Mineralogy by Friedrich Mohs (1825). Photo via Wikimedia Commons. [Public domain.]

cemetery headstones). Geophysicists may study the minerals below Earth's surface that cannot be directly sampled. They may, for example, study how the seismic waves given off by earthquakes pass through the ground to estimate the kinds of rocks present.

BACKGROUND AND HISTORY OF MINERALOGY

Archaeological evidence suggests that humans have used minerals in a number of ways for tens of thousands of years. For instance, the rich possessed jewels, red and black minerals were used in cave drawings in France, and minerals such as gold were used for barter. Metals were apparently extracted from ores for many years, but the methods of extraction were conveyed from person to person without being written down. In 1556, Georgius Agricola's *De Re Metallica* (English translation, 1912) described many of these mineral processing methods. From the late seventeenth century into the nineteenth century, many people studied the minerals that occur in definite crystal forms such as cubes, often measuring the angles between faces.

In the nineteenth century, the fields of chemistry and physics developed rapidly. Jöns Jacob Berzelius developed a chemical classification system for minerals. Another important development was the polarizing microscope, which was used to study the optical properties of minerals to aid in their identification.

During the late nineteenth century, scientists had theorized that the external crystal forms of minerals reflected the ordered internal arrangement of their atoms. In the early twentieth century, this theory was confirmed through the use of X-rays. Also, it became possible to chemically analyze minerals and rocks so that chemical mineral classifications could be further developed. Finally, in the 1960s, the use of many instruments such as the electron microprobe allowed geologists to determine variations in chemical composition of minerals across small portions of the minerals so that models for the formation of minerals could be further developed.

MINERAL AND ROCK IDENTIFICATION

A geologist may tentatively identify the minerals in a rock using characteristics such as crystal form, hardness, color, cleavage, luster (metallic or nonmetallic), magnetic properties, and mineral association. The rock granite, for instance, is composed of quartz (often colorless, harder than many other minerals, with rounded crystals, no cleavage, and nonmetallic luster) and feldspars (often tan, softer than quartz, with well-developed crystals, two good nearly right-angle cleavages, and nonmetallic luster), with lesser amounts of black minerals such as biotite (softer than quartz, with flat crystals, one excellent cleavage direction, and shiny nonmetallic luster).

The geologist then slices the rock into a section about 0.03 millimeters thick (most minerals are transparent and will transmit light when in such thin slices). The section is examined under a polarizing microscope to confirm the presence of the tentatively identified minerals and perhaps to find other minerals that could not be detected by eye because they were too small or present in very low quantities. The geologist can determine other relationships among the minerals, such as the sequence of crystallization of the minerals within an igneous rock.

INSTRUMENTAL IDENTIFICATION OF MINERALS

The minerals in a rock can be analyzed in a variety of other ways, depending on the goals of a given study. For instance, X-ray diffraction may be used to identify some minerals. One of the most useful applications of X-ray diffraction is to identify tiny minerals such as clay minerals that are hard to identify using a microscope. The wavelengths of the X-rays are similar to the spacing between atoms in the clay minerals, so when X-rays of a single wavelength are passed through a mineral, they are diffracted from the minerals at angles that are characteristic of a particular mineral. Thus, the mixture of clay minerals in a rock or soil may be identified.

The electron microprobe has enabled the analysis of tiny portions of minerals so that changes in composition across the mineral may be determined. The instrument accelerates masses of electrons into a mineral that releases X-rays with energies that are characteristic of a given element so that the elements present can be identified. The amount of energy given off is proportional to the amount of the element in the sample; therefore, the concentration of the element in the mineral can be determined when the results are compared with a standard of known concentration. The electron microprobe can also be used to analyze other materials such as alloys and ceramics.

The scanning electron microscope uses an electron beam that is scanned over a small portion of tiny minerals and essentially takes a photograph of the mineral grains in the sample. Some electron microscopes are set up to determine qualitatively what elements are present in the sample. This information is often enough to identify the mineral.

OTHER ANALYTICAL TECHNIQUES

Many instruments and techniques are used to analyze the major elements, trace elements, and the isotopic composition of minerals and rocks. Commonly used methods are X-ray fluorescence (XRF), inductively coupled plasma mass spectrometer (ICP-MS), and thermal ionization mass spectrometry. X-ray fluorescence is used to analyze bulk samples of minerals, rocks, and ceramics for major elements and many trace elements. A powdered sample or a glass of the sample is compressed and is bombarded by

X-rays so that an energy spectrum that is distinctive for each element is emitted. The amount of radiation given off by the sample is compared with a standard of known concentration to determine how much of each element is present in the sample.

The inductively coupled plasma mass spectrometer is used to analyze many elements in concentrations as low as parts per trillion by passing vaporized samples into high-temperature plasma so that all elements have positive charges. A mass spectrometer sorts out the ions by their differing sizes and charges in a magnetic field, which permits a determination of the elemental concentrations when the results are compared with a standard of known concentration. Up to seventy-five elements can be rapidly analyzed in a sample at precisions of 2 percent or better.

Thermal ionization mass spectrometers can be used to analyze the isotopic ratios of higher mass elements such as rubidium, strontium, uranium, lead, samarium, and neodymium, which may be used to interpret the geologic age of a rock. The mineral or rock is placed on a heated filament so that the isotopes are ejected into a magnetic field in which the ions are deflected by varied amounts depending on the mass and charge of the isotope. The data may then be used to calculate the amount of a certain isotope in the sample and eventually the isotopic age of the sample.

Other specialized instruments are also available. For instance, gemologists use specialized instruments to study and cut gemstones.

ABUNDANT METALS AND USES

The most abundant metals are iron, aluminum, magnesium, silicon, and titanium. Iron, which is mostly obtained from several minerals composed of iron and oxygen (hematite and magnetite), accounts for 95 percent by weight of the metals used in the United States. Much of the hematite and magnetite is obtained from large sedimentary rock deposits called "banded-iron formations" that are up to 700 meters thick and extend for up to thousands of square kilometers. The banded-iron formations formed 1.8 billion to 2.6 billion years ago. They are abundant, for instance, around the Lake Superior region in northern Minnesota, northern Wisconsin, and northwestern Michigan. The banded-iron formations have produced billions of tons of iron ore deposits. The ores are made into pellets and are mixed with limestone and coke to burn at 1,600°C in a blast furnace. The iron produced in this process is molten and can be mixed with small amounts of scarce metals (called "ferroalloy metals") to produce steel with useful properties. For instance, the addition of chromium prevents corrosion and gives the resulting "stainless steel" strength at high temperatures. The addition of niobium produces strength, and the addition of copper increases the resistance of the steel to corrosion.

Much of the aluminum occurs in clay minerals in which the aluminum cannot be economically removed from the other elements. Thus, most of the aluminum used comes from bauxite, which is a mixture of several aluminum-rich minerals such as gibbsite ($Al(OH)_3$) and boehmite ($AlO(OH)$). Bauxite forms in the tropics to subtropics by intense chemical weathering from other aluminum-rich minerals such as the clay minerals. The production of aluminum from bauxite is very expensive because the ore is dissolved in molten material at 950°C and the aluminum is concentrated by an electric current. Electricity represents about 20 percent of the total cost of producing aluminum. Recycling of aluminum is economical because the energy expended in producing new aluminum from old aluminum is about 5 percent of the energy required to refine aluminum from bauxite.

SCARCE METALS AND USES

At least thirty important trace metals occur in Earth's crust at concentrations of less than 0.1 percent. Therefore, these elements are trace elements in most minerals and rocks until specific geologic conditions concentrate them enough so that they become significant portions of some minerals. For instance, gold is present in Earth's crust at only 0.0000002 percent by weight. Gold often occurs as gold metal uncombined with other elements and precipitated from hydrothermal solutions as veins. Gold may also combine with the element tellurium to form several kinds of gold-tellurium minerals. Also, gold does not react very well chemically during weathering, so it may concentrate in certain streams to form placer deposits. Gold has been used for thousands of years in jewelry, dental work, and coins. In the past, much world exploration has been motivated by the drive to find gold.

Gold is a precious metal. Other precious metals are sliver and the platinum group elements (e.g., platinum, palladium, and rhodium). Silver, like gold, has been used for thousands of years in jewelry and coins. In modern times, silver also finds uses in batteries and in photographic film and papers because some silver compounds are sensitive to light. Silver occurs in nature as silver sulfide, and it substitutes for copper in copper minerals so that much of the silver produced is a side product from copper mining. The platinum metals occur together as native elements or combined with sulfur and tellurium. They concentrate in some dark-colored igneous rocks or as placers. The platinum group metals are useful as catalysts in chemical reactions, so they are, for example, used in catalytic converters in vehicles.

Another group of scarce metals are base metals. The base metals are of relatively low economic value compared with the precious metals. The base metals include copper, lead, zinc, tin, mercury, and cadmium. Copper minerals are native copper and various copper minerals combined with sulfate, carbonate, and oxygen. Copper minerals occur in some igneous rocks, including some hydrothermal deposits, mostly as dilute copper-sulfur minerals. These minerals may be concentrated during weathering, often forming copper minerals combined with oxygen and carbonate. Copper conducts electricity very well, and much of it is used in electric appliances and wires. Lead and zinc tend to occur together in hydrothermal deposits in combination with sulfur. Lead is used in automobile batteries, ammunition, and solder. Lead is biochemically very harmful to organisms, which restricts its potential use. Zinc is used as a coating on steel to prevent it from rusting, in brass, and in paint.

FERTILIZER AND MINERALS FOR INDUSTRY

Fertilizers contain minerals that have nitrogen, phosphorus, potassium, and a few other elements necessary for the growth of plants. In Peru, deposits of guano (seabird excrement), which is rich in these elements, have been mined and used as fertilizer. Saltpeter (potassium nitrate) has also been mined in Peru. A calcium phosphate mineral, apatite, occurs in some sedimentary rocks, so some of these have been mined for the phosphorus.

Some minerals are a source for sodium and chlorine. Halite, commonly known as rock salt, is used as a flavoring, to melt ice on roads, and in many other applications. Halite is produced in some sedimentary rocks from the slow evaporation of water in closed basins over long periods of time, a process that is occurring in the Great Salt Lake in Utah.

A variety of rocks or sediment—including granite, limestone, marble, sands, and gravels—are used for the exteriors of buildings, construction materials, or for countertops. Sands and gravels may be mixed with cement to make concrete.

MINERALOGY AND INDUSTRY

The materials of Earth interact, which means that all societies depend on mineral resources. For example, crops grow in soil containing clay minerals formed from the weathering of rocks. Fertilizers composed of nitrogen, phosphorus, and potassium minerals must be added to the soil to grow enough food to support the growing population of the world. Machinery such as tractors, trains, ships, and trucks made of steel and other materials obtained from processing metal ores must be used to fertilize fields and to move food to markets. Electricity, a significant portion of which is still obtained by burning coal, must be used to process food, whether in factories or in homes. Most of these mineral resources cannot be renewed quickly, so many are being rapidly used up.

MINERALOGY IN INDUSTRY AND BUSINESS

The largest industries directly related to mineralogy are public or private corporations engaged in the exploration, production, and development of minerals. As of 2009, the United States was the world's third largest producer of copper, after Chile and Peru, and of gold, after China and Australia. In 2009, net exports of mineral raw materials and old scrap materials was $10 billion. In 2009, Freeport-McMoRan Copper and Gold's Morenci open-pit mine in Arizona produced about 460,000 pounds of copper (about 40 percent of its total North American production), Barrick Gold Corporation's Goldstrike Complex in Nevada produced 1.36 million ounces of gold, and Teck's Red Dog Mine in Alaska produced 79 percent of the total U.S. production of zinc.

GOVERNMENT REGULATION

Although most governments do not directly control the mineral resources in their countries, they often regulate mining to ensure the workers' safety and support mining through subsidies and funding. In the United States, government agencies regulating mining include the Mine Safety and Health Administration of the U.S. Department of Labor and the mining division of the National Institute for Occupational Safety and Health. The U.S. Geological Survey maintains statistics and information on the worldwide supply of and demand for minerals.

ACADEMIC RESEARCH

Engineering departments in colleges and universities engage in mineralogical research such as improving the safety and efficiency of mining operations and searching for better, cleaner ways to process ores. For instance, some engineers analyze ways to break up ores and separate the ore minerals or develop ways to use more of the metals in the ores.

Anyone interested in pursuing a career in the minerals industry should study geology, chemistry, physics, biology, chemical engineering, and mining engineering. A bachelor's degree in one of these subjects is the minimum requirement for working in oil exploration or as a mining technician or water-quality technician. A master's degree or a doctorate is required for more challenging and responsible jobs in geochemistry, geophysics, environmental geology, geochronology, forensic mineralogy, and academics. Geochemists should have a background in geology and chemistry. Geophysicists use physics, mathematics, and geology to remotely tell what kinds of rocks are below the surface of Earth. Environmental geologists should have backgrounds in geology, chemistry, and biology. Geochronologists use natural radioactive isotopes to give estimates of the time when some kinds of rocks formed. They should have a background in geology, physics, and chemistry. Forensic mineralogists should have a good background in mineralogy and criminology. Economic geologists should major in geology, with a concentration in courses concerned with ore mineralization. Those interested in an academic career should have a PhD in geology, mining engineering, or chemical engineering. They will be expected to teach and do research in their subject area.

SOCIAL CONTEXT AND FUTURE PROSPECTS IN MINERALOGY

Mineral resources are not evenly distributed throughout the world. In the nineteenth century, the United States did not import many mineral resources, but it has increasingly become an importer of mineral resources. The need to import is driven by many factors, including an increase in the minerals used. U.S. production of final products, including cars and houses, using mineral materials accounted for about 13 percent of the 2009 gross domestic product. Other factors include the depletion of some U.S. sources of minerals and the increased use of minerals that are not naturally available in the United States. In 2009, foreign sources supplied more than 50 percent of thirty-eight mineral commodities consumed in the United States. Nineteen of those minerals were 100 percent foreign sourced. The United States must import much of the manganese, bauxite, platinum group minerals, tantalum, tungsten, cobalt, and petroleum that it uses. In contrast, the United States still has abundant resources of gold, copper, lead, iron, and salt. Japan, however, has few mineral resources, so it must import most of them.

This need to import and export mineral resources has forced countries to cooperate with one another to achieve their needs. The mineral resources traded in the largest quantities, in descending order, are iron and steel, gemstones, copper, coal, and aluminum. The total annual world trade in mineral resources is approaching $700 billion and is likely to rapidly increase as countries such as China, India and Brazil use and produce more mineral resources.

As the supply of minerals decreases worldwide and environmental and safety concerns raised by mining gain in importance, it is likely that mineralogy research may turn to improving extraction and processing methods and looking for ways to recycle materials, capture minerals remaining in wastes, and restore mining areas after the minerals have been depleted.

—Robert L. Cullers

Further Reading

Aydinalp, Cumhur. *An Introduction to the Study of Mineralogy*. IntechOpen, 2012.

Craig, James R., David J. Vaughan, and Brian Skinner. *Earth Resources and the Environment*. Prentice Hall, 2010.

Dyar, Melinda Darby, and Mickey E. Gunter. *Mineralogy and Optical Mineralogy*. Mineralogical Society of America, 2019.
Guastoni, Alessandro, and Roberto Appiani. *Minerals*. Firefly Books, 2005.
Klein, Cornelis, and Barbara Dutrow. *The Twenty-Third Edition of the Manual of Mineral Science*. Wiley, 2008.
Pellant, Chris. *Smithsonian Handbook: Rocks and Minerals*. Dorling Kindersley Books, 2002.
Ross, Cyaden. *Introduction to Mineralogy*. Callisto Reference, 2019.
Sinding-Larsen, Richard, and Friedrich W. Wellmer, editors. *Non-renewable Resource Issues: Geoscientific and Societal Challenges*. Springer, 2012.
Turner, David, and Lee A. Groat. *Geology and Mineralogy of Gemstones*. Wiley, 2021.

Mohs Hardness Scale

Fields of Study: Geology; Mineralogy; Geochemistry; Gemology; Crystallography

ABSTRACT

The resistance of minerals to abrasion or scratch is a valuable diagnostic physical property used in mineral identification. The Mohs hardness scale, proposed by Friedrich Mohs in 1822, provides a method of ranking minerals according to their relative hardness and thus is a way to help identify them.

KEY CONCEPTS

boron carbide: a compound of boron and carbon with the chemical formula BC, not a simple alloy
hardness pencils: a set of mineralogist's or gemologist's tools prepared with a standard assortment of mineral tips in accord with the Mohs hardness scale, used to carry out scratch tests
scratch: a test of abrasiveness carried out by scratching a 5- to 6-millimeter mark on one material with another material to determine which is the harder of the two materials
silicon carbide: a compound of silicon and carbon with the chemical formula SiC, not simply an alloy

DEFINITION OF THE MOHS SCALE

The resistance of minerals to abrasion or scratch is a valuable diagnostic physical property used in mineral identification. In 1822, Friedrich Mohs, an Austrian mineralogist, developed a relative scale of mineral hardness. This scale consisted of ten common minerals that were ranked from 1 (softest) to 10 (hardest). The values assigned to each member of the scale indicate the relative hardness of the minerals. Intervals between minerals in the scale are approximately equal, except between nine and ten.

USING THE MOHS HARDNESS SCALE

The resistance of a mineral to scratch is tested by sliding a pointed corner of one mineral across the smooth surface of another mineral. If the mineral with the point is harder, it will cut or scratch the other mineral. The scratch should be as short as possible, not more than 5 or 6 millimeters. If the pointed mineral is softer, a smear or powdered residue is left on the flat surface of the other mineral.

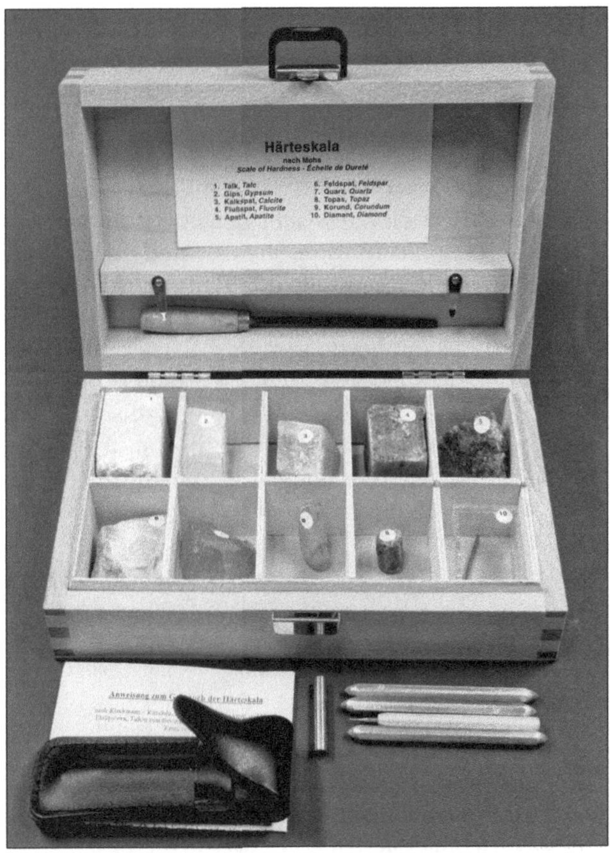

Mohs hardness kit, containing one specimen of each mineral on the ten-point hardness scale. Photo by Hannes Grobe, CC BY 3.0, via Wikimedia Commons.

This residue could be mistaken for a scratch; however, the smear can be easily rubbed off. A mineral from the high end of the scale will usually produce a significant "bite" on the softer mineral. Two minerals that have the same hardness will scratch each other equally well. Common objects are sometimes used as aids in hardness determination.

Brass rods set with conical-shaped fragments of test minerals on the ends are sometimes used to determine the hardness of small specimens and gemstones; these rods are known as "hardness pencils." Most gems, with the exception of pearls, have a hardness of 6 or above. In testing rough and uncut gems, some jewelers use these pencils to determine the specific hardness of the stones. Other minerals, such as chrysoberyl, epidote, olivine, and zircon, are included with the set of instruments. Six test pencils are sometimes conveniently arranged in a hardness wheel.

THE MODIFIED MOHS HARDNESS SCALE

With the advent of extremely hard manufactured abrasives in the second half of the twentieth century, a new sequence of index minerals was proposed for the upper part of the Mohs scale. This modified Mohs scale has found some use in industry. In this scale, quartz was elevated to 8, garnet was introduced as 10, and corundum was elevated to 12. Diamond, the hardest naturally occurring substance on Earth, topped the scale at 15. The artificial abrasives silicon carbide and boron carbide were designated as 13 and 14, respectively. Silicon carbide is produced by heating a mixture of carbon and sand in a specially designed electric furnace. Boron carbide, the hardest known substance next to diamond, is manufactured in an electric furnace from coke and dehydrated boric acid.

—Donald F. Reaser

Further Reading

The CRC Handbook of Chemistry and Physics. 93rd ed., CRC Press, 2012.

Deline, Bradley, Randa Harris, and Karen Tefend. *Laboratory Manual for Introductory Geology*. U of North Georgia, 2016.

Klein, Cornelis, and Barbara Dutrow. *The Twenty-Third Edition of the Manual of Mineral Science*. Wiley, 2008.

Nesse, William D. *Introduction to Mineralogy*. 3rd ed., Oxford UP, 2017.

MOUNTAIN BUILDING

Fields of Study: Geology; Geomorphology; Seismology

ABSTRACT

A diversity of geological and geodynamic factors contributes to the formation of mountains. Plate tectonics is central to this process. As these plates interact with each other, the outer crust is pushed outward, forming mountains. In some cases this push is dramatic, while in others it is more gradual and "gentle." Many mountains are remnants of much larger, ancient mountains that have since been eroded by geodynamic forces.

KEY CONCEPTS

convergent boundary: the geodynamic process in which two tectonic plates directly collide

divergent boundary: the geodynamic process whereby two tectonic plates separate and create a space for magma to fill

folding: the effect of the collision of two plates pushing Earth's outer crust outward, causing the rock to bend slowly over time

lithosphere: the outermost geological layer of Earth consisting of massive plates of rock

orogenesis: the process of mountain building

plate tectonics: the theoretical concept in which Earth's lithosphere, comprising continental and smaller plates, is in constant motion atop the plastic layer beneath it

transform boundary: the geodynamic process in which two tectonic plates collide moving laterally relative to each other

BASIC PRINCIPLES OF MOUNTAIN BUILDING

Mountains are formed in a variety of ways, but most scientists believe that it is primarily done through plate tectonics. The theory of plate tectonics, which has been overwhelmingly accepted in the scientific community, states that Earth's outer crust is composed of massive rock plates called the "lithosphere." These continental plates (the North American, African, South American, Eurasian, Antarctic, Australian, and Pacific), along with other smaller plates, are constantly in motion and moving in different directions, driven by the movement of convection currents in the magma of the underlying mantle layer.

The tectonic plates are in contact with one another at plate boundaries. In the case of convergent boundaries, the force of this direct contact causes the adjacent rock to fold slowly over time; the upward fold pushes outward in the form of mountains and mountain ranges. When plates move away from one another (to create divergent boundaries), an opening is created between them, allowing magma to flow into the empty spaces. The rising magma pushes upward, creating cone- and dome-shaped mountains. The mountain becomes an active volcano if magma is pushed through the surface of the domes. When plates move laterally relative to each other, as at transform boundaries, the shearing effect may produce mountainous deformations on the surface.

Mountainous structures such as buttes and plateaus are also formed on Earth's surface through erosion rather than plate tectonics. In this manner, wind, ice floe, and running water carve away the erodable mass of existing mountains, sculpting the shape of mountains through millions of years.

BACKGROUND AND HISTORY OF MOUNTAIN BUILDING

Throughout history, humans have been awed by the grandeur and powerful visage of mountains. Ancient Tibetans, from 20,000 BCE, believed that gods resided atop the Himalayan mountains. Ancient Greeks later shared that sentiment, believing that their gods resided atop Mount Olympus.

In the early nineteenth century, scientists began to explore the nature of higher elevations and mountain ranges. In 1829, French geologist Jean-Baptiste Élie de Beaumont theorized that mountains formed as Earth cooled and contracted. In 1842, American scientist Henry Darwin Rogers argued that mountains were the result of wavelike undulations caused by molten rock flowing beneath Earth's crust.

In 1912, German climatologist Alfred Wegener moved the field of geology to an entirely new level. Wegener noticed that fossils of certain species of plants and animals were unearthed thousands of miles apart, on opposite sides of the Atlantic Ocean. He expanded upon the observations of others, who had noticed that each continent seemed to fit together with other continents like pieces of a puzzle. Wegener argued, in developing his theory of continental drift, that the continents were once connected and that they had later separated, and were now drifting away from one another over Earth's crust. Continental drift theory lacked a valid "driver" for the movement of the continents but remained an intriguing theory until the 1960s, when advances in seismological and volcanic studies seemed to show the outlines of tectonic plate boundaries. The discovery and identification of the Mid-Atlantic Ridge provided the necessary driver of continental drift, and the theory evolved into plate tectonics. Plate tectonics has since been overwhelmingly embraced by the scientific community, with clear applications in orogenesis, the formation of mountains on Earth's surface.

CONVERGENT BOUNDARIES AND OROGENESIS

The concept of plate tectonics is central to the process of orogenesis. The shape and configuration of mountains and mountain ranges depend on how tectonic plates converge. For example, the Himalayan range, which includes the world's tallest surfacial mountain (Mount Everest), was formed at a convergent boundary of the Eurasian and Indian plates, which first collided about 25 million years ago. The force with which the two plates are being driven into each other causes a crumpling effect, pushing rock outward in the form of mountain peaks. The collision is ongoing, which means that the Himalayas continue to form and grow.

Scientists frequently use the Himalayan example to study other mountain ranges. In one study, scientists attempting to understand the orogenesis of the Caledonide mountain range in Scandinavia have noticed similarities between that range and the composition of the Himalayas. This comparison has led them to argue that the Caledonides were formed by a convergent boundary plate collision. Such information is useful for scientists who are surveying this range for geothermal energy and certain minerals within its bedrock.

DIVERGENT BOUNDARIES AND OROGENESIS

In the second form of orogenesis, known as divergent boundaries, plates move away from one another and leave an opening between them. This opening subsequently fills with magma from Earth's mantle; the magma pushes outward, raising mountains. One of the best-documented examples of divergent boundary orogenesis is the Mid-Atlantic Ridge, which rests on the floor of the Atlantic Ocean, spreading from the Arctic regions to beyond the Cape of Good Hope off the southern coast of South Africa.

The divergent boundary of the Mid-Atlantic Ridge provides scientists with a rare ability to monitor the spread of the two plates in question (the North American and Eurasian plates). Scientists have placed a number of observation stations in Iceland, which lies along the boundary of the two plates. In addition to the volcanic activity that occurs frequently from this divergence, seismic activity enables scientists to use a wide range of scientific technologies (including global positioning systems [GPS]) to monitor the continuing rate of divergence and the continuing formation of mountains.

TRANSFORM BOUNDARIES AND OROGENESIS

The third form of orogenesis involves transform boundaries. This manifestation of mountain building occurs when two plates "sideswipe" each other rather than collide head-on. The San Andreas fault is one of the most famous of these transform boundaries. In this case, the North American and Pacific plates are grinding alongside each other in opposite directions (the Pacific plate in a northwesterly direction and the North American plate on a southeasterly heading). The Sierra Nevada range in California provides an example of the mountains that form at transform boundaries. The Sierra Nevada was formed (and continues to be formed) by the shearing effects of the plate motions. The mountains are fragments of the plates, pushed upward and outward by the encounter between the two tectonic plates.

COMPUTER MODELS

Because the subterranean movement of tectonic plates cannot be directly observed, the study of these rocks relies heavily on computer models. In a similar vein, scientists studying orogeny (which for even the youngest mountain ranges spans millions of years) rely on computer models to help create a profile of a region's mountain-building history. For example, scientists estimate that the creation of Wyoming's Laramie range, which extends from Wyoming to the Black Hills of South Dakota, took place about 70 to 80 million years ago through subduction, a process whereby the leading edge of one tectonic plate is forced to go under, or subduct to, the leading edge of the other tectonic plate. The pressure exerted by the subducting plate works to buckle and raise the rock in the leading edge of the other plate into mountains. The deformation typically allows magma to rise to the surface and add to the mountainous structure. In the early twenty-first century, scientists compiled thermal data taken from various points in the Laramie range and entered them into a computer database. Based on the models they generated, researchers created a profile of the degree of magma flow that contributed to the range's formation.

REMOTE-SENSING SATELLITES

One of the most invaluable tools for geologists and geophysicists studying orogeny is satellite technology. Remote-sensing satellites can compile large volumes of photographs and geodynamic data each time they pass over a target. Researchers have used data from satellite flyovers to assess the elevations of certain major mountain ranges. The satellite images showed scientists how natural elements such as water and wind erode mountains and contribute to a reduction in mountain elevation. Satellites have also been called into service to observe the many different features of the Alps. The information collected by the ERTS-1 satellite revealed a complex system of fault lines, folding, and shearing.

GROUND-BASED SENSORS

Geologists and geophysicists commonly use ground-based sensory equipment to better understand the forces involved in building mountains. For example, the rock that is brought to the surface in a divergent boundary scenario maintains a certain magnetic imprint. Magnetometers detected the magnetic rocks introduced along the Mid-Atlantic Ridge's divergent boundary. Closer study of magnetic imprints has revealed that Earth's magnetic field has inverted many times in the planet's history. Today, magnetometers continue to prove invaluable for surveying the orogeny of mountain ranges.

Scientists also use thermal equipment to observe the different minerals that help form mountains.

Such equipment can help differentiate between the elements that foster orogenesis. For example, researchers used thermal imaging equipment to reveal the heat radiating from the minerals found in different locations in the Adirondack Mountains. The use of this thermal equipment helped researchers create a chronology of the orogenesis of mountains in this area. Thermal detection equipment has also helped scientists determine the effects of glaciation on prehistoric mountain ranges. By examining the temperatures of certain components of the bedrock, researchers can determine the rates of glacial development and erosion in a given mountain range.

RELEVANT NETWORKS AND ORGANIZATIONS FOR OROGENY RESEARCH

Orogeny involves looking at the creation of mountains to understand what is occurring (or has occurred) within Earth's surface. Among the different types of groups involved in orogeny research are governments, universities, and nonprofit organizations and networks.

The U.S. government is typically involved in the study of mountain building in two ways. First, it dedicates resources and personnel directly to such research. For example, the U.S. Geological Survey has geologists, geophysicists, and other professionals to analyze plate tectonics, erosion, and volcanic activity, all of which are closely linked to the study of mountain building. Second, the government provides grants and other financial support to independent researchers. The National Science Foundation, for example, provides a wide range of financial support for independent research.

Universities are some of the most important institutions involved in orogeny research. The scientists working at these institutions have two roles to play: conduct field research on relevant topics and teach undergraduate and graduate students the principles of geodynamics, plate tectonics, and orogenesis. The California Institute of Technology, the University of Cambridge, and the University of Geneva are among the institutions with active programs in the study of mountain building, mostly within geology departments.

Many nonprofit organizations and scientific networks and societies have an interest in the field of orogeny studies. The Geological Society of America, for example, offers publications, conferences, and other venues for experts in this field to share information and research. Other geological societies with similar interests (including orogeny and plate tectonics) are found in Scandinavia, Great Britain, Japan, South Africa, and other areas of the world.

IMPLICATIONS AND FUTURE PROSPECTS IN OROGENY RESEARCH

The study of how mountains are formed continues to evolve, especially as available technologies improve. One of the most significant of these technological advances is the use of satellites to observe orogenic changes. The use of the extensive GPS makes it possible to carefully monitor the movement of the continents along their boundaries. Furthermore, satellites can track orogenic changes using other shipboard equipment, including thermal imaging and radar systems.

In addition to the ever-evolving use of satellites and state-of-the-art sensory equipment, the study of mountain building is greatly aided by the Internet. Researchers around the world can share, in real time, large volumes of the data they have acquired in the field. As human civilization learns more about plate tectonics using cutting-edge technology, it will also learn more about the mountains and ranges plate tectonics produce.

—Michael P. Auerbach

Further Reading

Egholm, D. L., et al. "Glacial Effects Limiting Mountain Height." *Nature,* vol. 460, 2009, pp. 884-87.

Fossen, Haakon. *Structural Geology.* 2nd ed., Cambridge UP, 2016.

Frisch, Wolfgang, Martin Meschede, and Ronald C. Blakey. *Plate Tectonics: Continental Drift and Mountain Building.* Springer, 2010.

Johnson, Michael, and Simon Harley. *Orogenesis: The Making of Mountains.* Cambridge UP, 2012.

Lamb, Simon, and Anthony Watts. "The Origin of Mountains: Implications for the Behaviour of Earth's Lithosphere." *Current Science,* vol. 99, no. 12, 2010, pp. 1699-1718.

Schellart, W. P. "Overriding Plate Shortening and Extension Above Subduction Zones: A Parametric Study to Explain Formation of the Andes Mountains." *Geological Society of America Bulletin,* vol. 120, no. 11, 2008, pp. 1441-54.

N

Nesosilicates and Sorosilicates

Fields of Study: Geology; Geochemistry; Gemology; Crystallography

ABSTRACT

Nesosilicates are a large and diverse group of rock-forming minerals. The group contains a number of minerals of geologic importance, among them olivine, which may be the most abundant mineral of the inner solar system. Nesosilicates have a few special but limited industrial uses and include a number of important gemstones. Sorosilicates are a small group of silicates with related structures.

KEY CONCEPTS

anion: an atom that has gained electrons to become a negatively charged ion

cation: an atom that has lost electrons to become a positively charged ion

crystal structure: the regular arrangement of atoms in a crystalline solid

igneous rock: a type of rock formed by the solidification of magma

metamorphic rock: a type of rock formed when another type of rock undergoes changes in mineralogy, chemistry, or structure owing to changes in temperature, pressure, or chemical environment

mineral: a naturally occurring solid, inorganic compound with a definite composition and an orderly internal arrangement of atoms

silicates: minerals containing both silicon and oxygen, usually in combination with one or more other elements

solid solution: a solid that shows a continuous variation in composition in which two or more elements substitute for each other on the same position in the crystal structure

tetrahedron: a four-sided pyramid made out of equilateral triangles

CHEMICAL STRUCTURE OF NESO- AND SOROSILICATES

Nesosilicates, or orthosilicates, are one of the major groups of silicate minerals. Oxygen and silicon are the two most common elements in the rocky outer layers of earth-like inner planets. Thus, it is not surprising that silicates are the most abundant rock-forming minerals on these planets. They are not only very abundant but also very diverse: It has been estimated that almost one-third of the roughly 3,000 known minerals are silicates.

In all but a few of these minerals, each silicon atom is surrounded by a cluster of four oxygen atoms, which are distributed around the silicon in the same way that the corners of a tetrahedron are distributed around its center. Silica tetrahedra can join together by sharing one oxygen at a corner. In this way, two or more silica tetrahedra can link together to form pairs, rings, chains, sheets, and three-dimensional frameworks. This linking provides the basis for the classification of silica structures. Sorosilicates contain pairs of tetrahedra, cyclosilicates contain rings, inosilicates contain chains, phyllosilicates contain sheets, and tectosilicates contain three-dimensional frameworks. In nesosilicates, however, silica tetrahedra do not link to each other. Each of the silica tetrahedra is isolated from the others as if it were an island, and hence these minerals are often known as nesosilicates, from the Greek word for "island." In the structure of some minerals, isolated silica tetrahedra are mixed with silicate pairs formed when two tetrahedra share an oxygen at a common corner. Minerals with pairs of silica tetrahedra are called "sorosilicates."

CHEMICAL BONDING IN NESO- AND SOROSILICATES

The electrostatic attraction between negatively charged ions (anions) and positively charged ions (cations) is the basis of the ionic chemical bond. The chemical bonding in nesosilicates is predominantly ionic. Under normal geological conditions, silicon loses four electrons to become a cation with a charge of +4, while oxygen gains two electrons to become an anion with a charge of -2. The ionic bond between oxygen and silicon is usually the strongest bond in nesosilicate structures. For a mineral to be stable it must be electrically neutral; in other words, the total number of negative charges on the anions must be equal to the total number of positive charges on the cations. It would take two oxygen anions to balance the charge on one silicon cation. In the silica tetrahedra, however, the silicon cation is surrounded by four oxygen anions, so in the group as a whole there are four excess negative charges. In the nesosilicate structures this excess negative charge is balanced by the presence of cations outside the silica tetrahedra. It is the bond between the oxygen in the tetrahedra and these other cations that holds the tetrahedra together to form a coherent, three-dimensional structure. The most common cations to play this role are aluminum with a charge of +3, iron with a charge of +3 or +2, calcium with a charge of +2, and magnesium with a charge of +2. Aluminum, iron, and calcium are, respectively, the third, fourth, and fifth most abundant elements in Earth's crust, while magnesium is the seventh most abundant. Conspicuously absent from the common nesosilicate structures are the alkali cations sodium and potassium, which are the sixth and eighth most abundant elements in the earth's crust.

Of all the different kinds of silicates, nesosilicates have the lowest ratio of silicon to oxygen. As a result they often form in environments relatively low in silicon. The atoms in nesosilicates tend to be packed closer together than the atoms in many other silicates, causing them to be somewhat denser. Greater densities are favored at higher pressure; hence a number of the more important minerals that prove stable at high pressures are nesosilicates. Nesosilicates also tend to be harder (resistant to being scratched) than the average silicate. This property helps give many nesosilicates high durability (resistance to wear), which contributes to their use as gemstones. The nesosilicates are a large and diverse group, and a description of more than the most common of them is well beyond the scope of this article.

OLIVINE

The mineral olivine is the most common and widespread of the nesosilicates. Indeed, it is probably the most abundant mineral in the inner solar system. Olivine contains magnesium (Mg) and iron (Fe), in addition to silicon (Si) and oxygen (O). The chemical formula of olivine is generally written as $(Mg,Fe)_2SiO_4$. The parentheses in this formula indicate that olivine is a solid solution; in other words, magnesium and iron can substitute for each other in the olivine structure. Pure magnesium olivines (Mg_2SiO_4) are known as forsterite, and pure iron olivines (Fe_2SiO_4) are known as fayalite. Most olivines have both magnesium and iron, and hence have compositions that lie between these two extremes. Each iron and magnesium atom is surrounded by six oxygen atoms, while each oxygen atom is bonded to three iron or magnesium atoms and one silicon atom, thereby creating an extended three-dimensional structure. Olivine is usually a green mineral with a glassy luster and a granular shape.

Rocks made up mostly of olivine are known as "peridotites." Although peridotites are relatively rare in Earth's crust (that layer that begins at the surface and extends to depths of between 5 and 80 kilometers), this rock makes up most of Earth's uppermost mantle. The olivine-rich layer begins at depths ranging between 5 and 80 kilometers, and extends downward to depths of roughly 400 kilometers. Evidence indicates that the moon and the other inner planets (Mercury, Venus, and Mars) also have similar olivine-rich layers. Olivine can also be an important mineral in basalts and gabbros, which are the most abundant igneous rocks in the crusts of the inner planets. Additionally, it is an abundant mineral in different kinds of meteorites and some kinds of metamorphic rocks.

GARNETS

The garnets are a group of closely related nesosilicate minerals, each of which is a solid solution. The chemistry of common garnets can be fairly well represented by the somewhat idealized general formula $A_3B_2Si_3O_{12}$, where the A stands for either magnesium, iron (with a charge of +2), manganese (Mn), or calcium (Ca), and the B stands for either aluminum (Al), iron (with a charge of +3), or chromium (Cr). In the crystal structures of these minerals, the cations in the A site are surrounded by eight oxygen anions, while the cations in the B site are surrounded by six oxygens. Most garnets can be described as a mixture of two or more of the following molecules: pyrope ($Mg_3Al_2Si_3O_{12}$), almandine ($Fe_3Al_2Si_3O_{12}$), spessartine ($Mn_3Al_2Si_3O_{12}$), grossular ($Ca_3Al_2Si_3O_{12}$), andradite ($Ca_3Fe_2Si_3O_{12}$), and uvarovite ($Ca_3Cr_2Si_3O_{12}$).

The most abundant and widespread garnets are almandine-rich garnets, which form during the metamorphism of some igneous rocks and sediments rich in clay minerals. Grossular-rich and andradite-rich garnets are found in marbles formed through the metamorphism of limestone. The formation of spessartine-rich or uvarovite-rich garnets occurs during the metamorphism of rocks with high concentrations of manganese or chrome. Rocks with these compositions are relatively unusual, and hence these garnets are fairly rare. Pyrope-rich garnets are widespread in Earth's mantle, although they typically do not occur in abundance (more than 5 or 10 percent of the rock). Garnets that have weathered out of other rocks are sometimes found in sands and sandstones. Garnets may also occur in small amounts in some igneous rocks.

ALUMINOSILICATES

Minerals are called "polymorphs" if they are made up of the same kinds of atoms in the same proportions but in different arrangements. Polymorphs are minerals with the same compositions but different crystal structures. Aluminosilicates are a group of nesosilicate minerals containing three polymorphs: kyanite, sillimanite, and andalusite. Each of these minerals has the chemical formula Al_2SiO_5. The differences in the structures of these minerals are best illustrated by considering the aluminum atoms; in kyanite, all the aluminum atoms are surrounded by six oxygen atoms; in sillimanite, half of the aluminum atoms are surrounded by six oxygens, while the other half are surrounded by four oxygens; in andalusite, half the aluminum atoms are surrounded by four oxygens, and half are surrounded by five oxygens. Kyanite usually forms elongated rectangular crystals with a blue color. Sillimanite typically occurs as white, thin, often fibrous crystals. Andalusite is most commonly found in elongated crystals with a square cross-section and a red to brown color.

Aluminosilicates typically form during the metamorphism of clay-rich sediments; such rocks have the relatively high ratios of aluminum to silicon necessary for the formation of these minerals. The identity of the aluminosilicate formed depends upon the temperature and pressure of metamorphism; kyanite forms at relatively high pressures, sillimanite forms at relatively high temperatures, and andalusite forms at low to moderate temperatures and pressures.

TOPAZ, ZIRCON, TITANITE, AND EPIDOTE

Topaz is another aluminum-rich nesosilicate; however, this mineral also contains fluorine (F) and/or the hydroxyl ion (OH^-). The chemical formula of topaz is $Al_2SiO_4(F,OH)_2$. As the parentheses indicate, this mineral is a solid solution in which fluorine and hydroxyl can substitute for each other. Topaz is formed during the late stages in the solidification of a granitic liquid.

The mineral zircon contains the relatively rare element zirconium (Zr). Zircon has the chemical formula $ZrSiO_4$; it also generally contains small amounts of uranium and thorium. The decay of these radioactive elements can be used to determine the age of a rock, making zircon particularly important to geologists. It is most commonly found as brown rectangular crystals with a pyramid on either end. It is a widespread mineral in igneous rocks, although it generally occurs in relatively minor amounts.

Titanite ($CaTiSiO_5$), sometimes known as "sphene," is one of the most common minerals bearing titanium (Ti). It is a fairly widespread mineral, occurring in many different kinds of igneous and metamorphic rocks, but it is rarely present in abundance.

Epidote ($Ca_2(Al,Fe)Al_2Si_3O_{12}(OH)$) contains both isolated silica tetrahedra and tetrahedra pairs. Epidote, a fairly common sorosilicate mineral, most typically forms during low-temperature metamorphism in the presence of water. It most often occurs as masses of fine-grained, pistachio-green crystals. A pistachio-green mineral in granite is almost certainly epidote.

ANALYTICAL TECHNIQUES FOR NESO- AND SOROSILICATES

To characterize a mineral requires both its chemical composition and its crystal structure. There are many different analytical techniques that will give chemical compositions; probably the most popular for silicates is electron microprobe analysis. In this technique, part of the mineral is bombarded by a high-energy beam of electrons, which causes it to give off X-rays. Different elements in the mineral give off X-rays of different wavelengths, and the intensities of these different X-rays depend on the abundance of their elements. By measuring the wavelengths and intensities of the X-rays, the composition of the mineral can be obtained. This technique has the advantage of being nondestructive (the mineral is still available and undamaged after the analysis) and applicable to very small spots on a mineral: The typical electron microprobe analysis gives the composition of a volume of mineral only a few tens to hundreds of cubic microns (millionths of a meter) in size.

The crystal structures of silicates are generally obtained using the technique of X-ray diffraction. In this technique an X-ray beam is passed through the mineral. As the beam interacts with the atoms in the mineral, it breaks up into many smaller, diffracted beams traveling in different directions. The intensity and direction of these diffracted beams depend on the positions of atoms in the mineral. By analyzing the diffraction pattern, a scientist can discover the crystal structure of a mineral.

SCIENTIFIC AND ECONOMIC VALUE OF NESO- AND SOROSILICATES

Nesosilicates are one of the important building blocks of the planets of the inner solar system, and this reason alone provides an important scientific rationale for studying them. Garnet and aluminosilicates also provide important clues about the temperatures and pressures at which rocks formed. Despite their importance in nature, these minerals have had only a limited technological use. The relatively high hardness of garnet makes it suitable as an abrasive, and it is used in some sandpapers and abrasive-coated cloths. Aluminosilicates are used in the manufacture of a variety of porcelain, which is noted for its high melting point, resistance to shock, and low electrical conductivity. This material is used in spark plugs and brick for high-temperature furnaces and kilns. Zircon is mined in order to obtain zirconium oxide and zirconium metal. Zirconium oxide has one of the highest known melting points and is used in the manufacture of items that have to withstand exceptionally high temperatures. Zirconium metal is used extensively in the construction of nuclear reactors. Titanite is mined as a source of titanium oxide. Titanium oxide has a number of uses but is most familiar as a white pigment in paint.

Probably the most widespread use of nesosilicates is gemstones. Especially fine, transparent crystals of olivine make a beautiful green gem known generally as peridot, although the names chrysolite and evening emerald are sometimes used instead. Relatively transparent garnets also make very beautiful gems. The most common garnets are a deep red, and this is the color usually associated with the stone. Yet gem-quality garnets can also be yellow, yellow-brown, orange-brown, orange-yellow, rose, purple, or green. Garnet is a relatively common mineral and is typically among the least valuable gemstones. The major exception is the green variety of andradite garnet. Known in the gem trade as demantoid, this relatively rare material is one of the more valuable gems. When properly cut, zircon, a popular gemstone, has a brilliancy (ability to reflect light) and fire (the ability to break white light up into different colors) second only to diamond. Topaz is also widely used as a gem. The most valuable topaz is orange-yellow to orange-brown in color. Unfortunately, all yellow gems are sometimes incorrectly referred to as topaz: When this practice is followed, true topaz is generally known as precious topaz or oriental topaz. Gem-quality topaz may also be colorless, faintly green, pink, red, blue, and brown.

—*Edward C. Hansen*

Further Reading

Deer, W. A., R. A. Howie, and J. Zussman. *Rock-Forming Minerals.* 2nd ed., 5 vols. Pearson Education Limited, 1992.

Hammond, Christopher. *The Basics of Crystallography and Diffraction.* 2nd ed., Oxford UP, 2001.

Klein, Cornelis, and Cornelius S. Hurlbut, Jr. *Manual of Mineralogy.* 23rd ed., John Wiley & Sons, 2008.

Lauf, Robert. *The Collector's Guide to Silicate Crystal Structures.* Schiffer Publishing, Ltd., 2010.

Liebau, Friendrich. *Structural Chemistry of Silicates: Structure, Bonding, and Classification.* Springer-Verlag, 1985.

Misra, Kula C. *Introduction to Geochemistry: Principles and Applications.* Wiley-Blackwell, 2012.
Wenk, Hans-Rudolf, and Andrei Bulakh. *Minerals: Their Constitution and Origin.* Cambridge UP, 2004.

NEW HORIZONS PLUTO-KUIPER BELT MISSION

Fields of Study: Geology; Geomorphology; Geochemistry; Astrogeology

ABSTRACT

The New Horizons spacecraft was the first human-made probe to explore Pluto, its moon Charon, and the distant Kuiper Belt Object (KBO) 2014 MU69 (also known as Ultima Thule). The mission provided insights into the origin and evolution of the outer solar system, including the geology and geochemistry of Pluto, its atmosphere, Charon, and the Kuiper Belt in general. This first reconnaissance of the Kuiper Belt by the New Horizons spacecraft has led to an increased understanding of how stars, planets, and solar systems form and evolve.

KEY CONCEPTS

Kuiper Belt: a ring of asteroidal bodies in and beyond the orbit of Pluto, broader and more diverse than the asteroid belt between Mars and Jupiter

magnetosphere: the region of influence in space of a planet's magnetic field

SUMMARY OF THE NEW HORIZONS MISSION

Shortly after the discovery of the first Kuiper Belt object in 1992, the Pluto Fast Flyby mission was outlined by the National Aeronautics and Space Administration (NASA). Subsequently named the Pluto-Kuiper Express mission, its goal was to send a robotic spacecraft to study Pluto and its moon, Charon, in 2012 and 2013 (after a December, 2004, launch), then move outward to investigate the Kuiper Belt. One of the primary objectives was to characterize the geology, surface composition, and atmosphere of Pluto. Initially, the mission was to be a joint effort of the United States and Russia. Zond probes, which would separate from the flyby spacecraft and relay data prior to their impact with Pluto, were to be included in the mission. However, in September, 2000, budgetary considerations forced NASA to cancel the Pluto-Kuiper Express mission.

In early 2001, NASA proposed another mission to go to Pluto and beyond. After competitive bidding, Lockheed Martin's New Horizons proposal was chosen on November 29, 2001, as the one that would provide the design of the spacecraft, scientific instruments, and ground systems for the mission. On February 26, 2003, the New Horizons Pluto-Kuiper Belt Mission was approved as part of NASA's budget. Total mission cost was estimated at approximately $550 million. The Johns Hopkins University's Applied Physics Laboratory in Laurel, Maryland, was chosen to manage the mission, with Thomas B. Coughlin serving as the project manager. Dr. S. Alan Stern of the Southwest Research Institute in Boulder, Colorado, was selected as the principal investigator for the project.

Scheduled to be launched in January 2006 on a Lockheed Martin Atlas V551 booster, the New Horizons Pluto- Kuiper Belt spacecraft was planned to reach Jupiter in March 2007. For a time in 2005, concerns over the availability of nuclear fuel for the probe's power generation plants threatened to delay launch another year, which would alter the probe's trajectory to the Pluto-Charon system. Fortunately, that delay did not materialize. Spacecraft assembly was completed during April 2005. During that time a comprehensive performance test verified functionality of major spacecraft systems and New Horizons' scientific instruments. High-fidelity flight simulations began on April 30. Engineers ran another comprehensive performance test during the first half of May. Intensive environmental testing and analysis followed that test. In late May through early June 2005, NASA Headquarters managers conducted a mission readiness review. The spacecraft's preparation remained on schedule and was launched on January 19, 2006.

During the spacecraft's passage through the Jupiter system, the spacecraft carried out four months of investigations that included imaging Jupiter and some of its moons. The New Horizons spacecraft had a trajectory that took it outward through Jupiter's magnetic field "tail" which allowed the first such study of a planetary magnetosphere. Passage close to Jupiter allowed the spacecraft to perform a gravity-assisted slingshot maneuver that greatly increased its velocity and gave it a trajectory toward the Pluto-Charon system for a close approach to Pluto on July 15, 2015 at a distance of 9,600 kilometers.

Beginning about three months prior to its closest encounter with Pluto, New Horizons cameras acquired images of Pluto and its moons and its compliment of instruments began an intensive campaign of observing Pluto's environment. The New Horizons spacecraft, with hydrazine fuel included, had an initial mass of 465 kilograms upon launch. Upon its arrival near Pluto, approximately 228 watts of power was available to the spacecraft from its radioisotope thermoelectric generator (RTG). A 2.5-meter dish antenna on top of the triangular-shaped spacecraft provided the necessary Radio Science communications with the Earth. Five scientific instruments were aboard the spacecraft. The Long-Range Reconnaissance Imager (LORRI) was a high-resolution charge-coupled device (CCD) that produced images in visible light. The Pluto Exploration Remote Sensing Investigation (PERSI) instrument consisted of a visible CCD imager, as well as near-infrared and ultraviolet imaging spectrometers, which were used to determine the atomic and molec-

New Horizons space probe. Photo via NASA/Wikimedia Commons. [Public domain.]

ular species that make up Pluto's atmosphere. The main component of the third instrument, the Radio Science Experiment (REX), is an ultrastable oscillator that was used to investigate surface composition and temperatures. The spectrometer for analyzing plasmas and high-energy particles (PAM) was composed of the Solar Wind Around Pluto (SWAP) detector and a time-of-flight ion and electron sensor. The fifth separate scientific instrument, a student-built dust counter (SDC) constructed at the University of Colorado, made dust measurements in the outer regions of the solar system.

The first major goal of the Pluto-Kuiper Belt Mission was to characterize the geology, geomorphology, surface composition, temperatures, atmospheres, and internal structure of Pluto and Charon from data gathered using the visible wavelength cameras, infrared spectral mapping, radio waves, ultraviolet spectroscopy, and *in situ* plasma sensors. High-resolution images of Pluto and Charon, far superior to those taken by the Hubble Space Telescope, were collected for a period of over seventy-five days. Global maps of Pluto's surface were generated with a 40-kilometer resolution, and hemispheric maps were made with a 1-kilometer resolution. These maps allowed scientists to characterize the diverse geologic processes occurring on Pluto and Charon. Localized visible color maps with up to a 3-kilometer resolution and infrared spectral maps with a 7-kilometer resolution allowed characterization of the composition and physical state of surface frost, composed of methane, carbon monoxide, water and nitrogen ices, on Pluto and Charon. The Radio Science Experiment mapped surface temperatures with a 50-kilometer resolution. Particle detectors sampled escaping material from Pluto's atmosphere.

After gathering data for approximately 150 days in the Pluto-Charon system, the spacecraft traveled outward through the Kuiper Belt. On January 1, 2019, the spacecraft performed a flyby of the Kuiper Belt Object (KBO) 2014 MU69. The surface of 2014 MU69 was imaged and mapped at infrared and visible wavelengths in order to analyze its composition. Due to the primordial nature of KBOs, insights into their origin and evolution provide constraints on the origin and evolution of our solar system.

NEW HORIZON'S CONTRIBUTIONS TO SCIENCE AND SOCIETY

The Pluto-Charon system and the Kuiper Belt had never been explored by spacecraft before New Horizons, and thus a great deal of new knowledge was gained from the mission. Using the experience, observations, and analogies derived from missions to other bodies in our solar system, as well as information gathered from ground-based, airborne, and Earth-orbiting telescopes, scientists crafted the New Horizons mission to optimize the knowledge gained from the Pluto flyby. From the data returned from the New Horizons mission, fundamental questions about Pluto's atmosphere, surface properties, geology, and the interior makeup of Pluto, Charon, and KBOs were answered, and many new more specific questions were formulated. Additionally, spacecraft observations provided basic information on what happens to the solar wind as it interacts with bodies in the outer regions of the solar system. The images and data gathered while New Horizons was near Jupiter provided new knowledge about the moons and Jupiter's giant magnetosphere.

While approaching Pluto, the New Horizons spacecraft searched unsuccessfully for rings that might exist around Pluto as well as for other satellites of Pluto. During the time of the spacecraft's closest approach to Pluto, a sequence of high-resolution images were taken that allowed scientists to search for time-variable phenomena associated with Pluto, including meteorology, clouds, atmospheric waves, and active volatile transport over its surface and through its atmosphere. Instrumentation on the spacecraft provided a deeper understanding of Pluto's atmospheric composition, its weather, winds, and atmospheric escape into space from the top of the atmosphere. The complex geology determined from the features imaged by New Horizons showed that Pluto has one of the most diverse surfaces among planetary bodies in the solar system.

The known frost on the surface of Pluto is composed of nitrogen, methane, carbon monoxide, and water, and the frost on Charon is composed primarily of water and ammonia. These frosts were analyzed from high-resolution images and spectral analysis data gathered by the spacecraft. These observations and subsequent studies have provided insights into the possibilities of simple microbial life associated with the liquid water possibly beneath the surface of Pluto. Many of the known properties of Pluto, particularly its red color, resemble those of Neptune's moon Triton. The New Horizons mission was able to reveal other similarities and differences between these two bodies that reflected their separate origins and independent evolution.

By mapping the surfaces of Pluto, its five moons Charon, Styx, Nix, Kerberos, Hydra, and the distant KBO Ultima Thule, the New Horizons mission has helped answer questions about their contrasting diversity. From the New Horizons measurements of cosmic dust, insights were gained into the source of water in the Kuiper Belt. Planetologists believe that KBOs represent primordial remnants from the formation of the outer planets around 4.5 billion years ago. This makes KBOs among the oldest known objects in the solar system, and thus they contain clues as to the nature of the primordial nebula from which the Sun and planets formed. The New Horizons' data on KBOs has provided the best information to date on these primordial planetary building blocks.

CONTEXT OF THE NEW HORIZONS MISSION

By providing the first close reconnaissance of Pluto and its moons as well as Ultima Thule, New Horizons mission has led to a dramatic increase in our understanding of the outer solar system and its history. Since the center of mass of the Pluto-Charon system lies between Pluto and Charon, the Pluto-Kuiper Belt Mission has given us the first close-up inspection of a true double-planet system. Because processes similar to what are believed to be occurring in the Kuiper Belt have been observed in apparent "Kuiper Belts" surrounding other stars in the Milky Way galaxy—particularly Vega and Beta Pictoris—the New Horizons mission is helping scientists better understand the development and evolution of other solar systems. Since the KBOs have probably not changed since the beginning of the solar system, their properties provide important constraints for theoretical models of the formation of stars and planets in general and of their evolution. The New Horizons mission has also provided valuable knowledge about spacecraft and instrumentation that are best suited for possible future missions into the outer realms of the solar system, as well as providing technology transfer to other branches of science.

There is still controversy as to whether or not Pluto merits planetary designation, rather than a designation as a Kuiper Belt Object or as a dwarf planet. From what has been learned from the New Horizons mission, Pluto is a complex and diverse world with a rich and equally complex history. Given the discovery of amazingly diverse planetary type bodies elsewhere in the galaxy, the study of Pluto and its categorization among these bodies is an important part of understanding the full range of planetary possibilities.

—Alvin K. Benson Michael and E. Summers

Further Reading

Cruikshank, Dale P., and William Sheehan. *Discovering Pluto: Exploration at the Edge of the Solar System.* The University of Arizona Press, 2018.

Davies, John Keith. *Beyond Pluto: Exploring the Outer Limits of the Solar System.* Cambridge UP, 2001.

Hartmann, William K. *Moons and Planets.* 5th ed., Thomson Brooks/Cole, 2005.

McBride, Neil, and Iain Gilmour, editors. *An Introduction to the Solar System.* Cambridge UP, 2004.

Moche, Dinah L. *Astronomy: A Self-Teaching Guide.* 6th ed., John Wiley & Sons, 2004.

Petersen, Carolyn Collins, and John C. Brandt. *Visions of the Cosmos.* Cambridge UP, 2003.

Stern, S. Alan. *Pluto and Charon: Ice Worlds on the Ragged Edge of the Solar System.* John Wiley & Sons, 1998.

———, editor. *Worlds Beyond: The Thrill of Planetary Exploration as Told by Leading Experts.* Cambridge UP, 2003.

———, and David Grinspook. *Chasing New Horizons: Inside the First Epic Mission to Pluto.* Picador, 2018.

NONSILICATES

Fields of Study: Geology; Seismology; Geophysics

ABSTRACT

Nonsilicate minerals, although not as abundant as the silicates in the part of the earth that is accessible to humankind, are important because they are the major sources of many of the critical elements and compounds upon which civilized society is based.

KEY CONCEPTS

crystalline: a property of a chemical compound with an orderly internal atomic arrangement that may or may not have well-developed external faces

ion: an atom that has a positive or negative charge

metal: an element with a metallic luster, high electrical and thermal conductivity, ductility, and malleability

mineral: a naturally occurring, solid chemical compound with a definite composition and an orderly internal atomic arrangement

ore: mineral or minerals present in large enough amounts in a given deposit to be profitably mined for the metal(s)

rock-forming mineral: the common minerals that compose the bulk of the earth's crust (outer layer)

semimetal: elements that have some properties of metals but are distinct because they are not malleable or ductile

CLASSIFICATION OF MINERALS

The earth is divided into several distinct layers (crust, mantle, and core), each with unique physical and chemical properties. Since only the crust is readily accessible to scientists, this discussion is largely restricted to this outermost layer. The crust is dominated by only eight elements (oxygen, silicon, aluminum, iron, magnesium, calcium, sodium, and potassium).

One of the most important elements is silicon. Geologists commonly divide minerals into two broad categories, silicates and nonsilicates, partly because of the sheer abundance of silicon and oxygen, but also because of the immense variety of atomic structures in the silicates. Silicates are those minerals that contain silicon as an essential part of the composition; nonsilicates lack silicon in their formulas. The silicates make up the vast majority of minerals in the crust, with all the other classes accounting for only 3 percent of the total.

A common classification scheme for minerals divides them into chemical families, or suites. One suite is silicates (usually subdivided into several groups).

The other principal suites (or nonsilicates) include the native elements sulfides and sulfosalts, oxides and hydroxides, carbonates, halides, nitrates, borates, phosphates, arsenates, vanadates, sulfates and chromates, and tungstates and molybdates. Generally the ending "ide" denotes an anion element by itself (sulfide, halide, oxide), and "ate" denotes a complex anion made of an element combined with oxygen.

NATIVE ELEMENTS

Of all the elements, about twenty, the native elements, are known to occur on Earth in the free state. These elements can be separated into metals, semimetals, and nonmetals. Within the metals, based on atomic structure, three groups are recognized: the gold group (gold, silver, copper, and lead), the platinum group (platinum, palladium, iridium, and osmium), and the iron group (iron and nickel). Mercury, tantalum, tin, and zinc are metals that have also been identified. Within the semimetals, two groups are commonly recognized: the arsenic group (arsenic, antimony, and bismuth) and the tellurium group (selenium and tellurium). Sulfur and three forms of carbon (fullerenes, graphite and diamond) are the nonmetallic native element minerals. On the microscopic level, other elements have been found as well; for example, native silicon can be formed in tiny amounts by lightning striking quartz sand.

SULFIDES AND SULFOSALTS

The sulfides include a great number of minerals, many of which are important economically as sources of metals. Although sulfur is the dominant anion (negatively charged ion), this group also includes compounds with arsenic (arsenides), tellurium (tellurides), selenium (selenides), antimony, and bismuth as the anion. Commonly included with the sulfides are the sulfosalts. These sulfosalts are generally distinct because they contain the semimetals arsenic and antimony in place of sulfur. Only four sulfides are considered to be rock-forming minerals: pyrite, marcasite, chalcopyrite, and pyrrhotite. The most common sulfosalt is arsenopyrite, which, as the name implies, is much like pyrite but contains arsenic as well as sulfur.

HALIDES, NITRATES, AND BORATES

The halides are minerals that contain one of four anions—fluorine, chlorine, bromine, and iodine. The chlorides are the most abundant halides, with the fluorides second in abundance, but the bromides and iodides are very rare. The only two halides that are considered to be rock-forming minerals are halite and fluorite.

The nitrates include minerals made up of one nitrogen ion surrounded by three oxygen ions as NO_3^-. None of the nitrates is considered common, and they are relatively few in number. Most occur only in deposits in very arid regions because they are extremely soluble.

The borates are minerals that have boron strongly bound to either three or four oxygen ions. All the borates are restricted to dry lake deposits in extremely arid regions. None is considered a common rock-forming mineral, but borax is probably the most readily recognized.

PHOSPHATES AND SULFATES

In phosphates, the phosphorus cation (positively charged ion) is surrounded by four oxygens. Although this class includes many minerals, most are extremely rare; only one, apatite, is common. Apatite is also the principal ingredient of bone. Arsenates and vanadates have similar atomic structures but contain arsenic or vanadium in place of phosphorus.

The sulfate minerals are another large class in which a sulfur cation is surrounded by four oxygens. Two main subgroups of sulfates are recognized: the anhydrous sulfates and the hydrous sulfates. Although this class includes many minerals, very few are considered to be common. Examples of anhydrous sulfates are barite and anhydrite. The most common hydrous sulfate is gypsum.

TUNGSTATE AND MOLYBDATE

The tungstate, chromate, and molybdate minerals are very similar to the sulfates, with either chromium, tungsten, or molybdenum cations surrounded by four oxygens in a pattern slightly different from that in the sulfates. None of the tungstates or molybdates is a common rock-forming mineral, and all are relatively rare but significant ore minerals.

STUDY OF PHYSICAL PROPERTIES

A number of physical properties are studied not only because they help identify particular minerals but also because the physical properties dictate whether minerals have any commercial use. Measurements of the angular relationships between faces with the optical goniometer are helpful in describing minerals. The systematic study of the external form of minerals is commonly called "morphological crystallography." Other properties that are studied include the tendency of minerals to fracture (break along irregular surfaces) or cleave (break along straight planar surfaces that represent planes of internal weakness), as well as tenacity (resistance or response to attempts to break, bend, or cut). Hardness, another essential property, refers to the resistance of a substance to abrasion. Hardness is commonly evaluated on a scale of relative hardness, called the "Mohs hardness scale," which uses a set of common minerals of different hardness for comparison. Other properties that may have industrial applications include magnetism and electrical properties. Minerals containing uranium and thorium exhibit another property, radioactivity, which is measurable with a Geiger counter or scintillation counter.

A variety of thermal properties are commonly determined. Differential thermal analysis (DTA) measures the temperatures at which compositional and structural changes take place in minerals. The DTA curve, which is a graphic recording of the thermal changes in a mineral, is a characteristic of many minerals. Density (mass per unit of volume) is another key property. Density is normally determined as specific gravity (the ratio of the density of the substance to that of an equal volume of water at 4° Celsius and 1 atmosphere of pressure). Depending on the amount and size of the mineral sample available, three methods may be used to determine density: Jolly balance, Berman balance, or pycnometer.

The petrographic microscope is an important analytical tool used to study the optical properties of minerals in polarized light transmitted through the specimen. Both crushed samples and thin sections (0.030 millimeter thick) are commonly utilized. Many of the metal-bearing minerals, particularly the native metals and sulfides, do not readily allow polarized light to pass through them, so microscopic analysis is conducted on highly polished samples in reflected light.

STRUCTURAL AND CHEMICAL ANALYSIS

The study of the orderly internal atomic arrangement within minerals is commonly called "structural crystallography." The most common method used to study this internal morphology is the X-ray diffraction powder method, in which a small, finely powdered sample is bombarded with X-rays. Like the human fingerprint, every mineral has unique diffraction patterns (caused by X-rays interacting with atomic planes). Several other important X-ray diffraction methods involve single crystals (oscillations, rotation, Weissenberg, and precession methods) and are primarily used not for identification purposes, but for refining the complex internal geometries of minerals.

In addition to the physical properties of minerals, a variety of chemical methods can be used to identify and study nonsilicate minerals. Until the 1960s, quantitative chemical analysis methods such as colorimetric tests, X-ray fluorescence spectrography, and atomic absorption spectrography were commonly used for mineral analysis. Since that time, however, most mineral analyses have been produced by electron microprobe analysis.

ECONOMIC IMPORTANCE OF NATIVE ELEMENTS

A great number of the nonsilicates under consideration touch people's lives in a multitude of ways. Of the native elements, one only needs to look over the commodity reports in the daily newspaper to understand the importance of the likes of gold, silver, and platinum to world economies. Most of the gold in the world is owned by national governments, which commonly use it to settle international monetary accounts. Like silver and platinum, gold is also becoming increasingly popular as a form of investment. In addition, gold is used in jewelry, scientific instruments, and electroplating. Silver is used in the photography, electronics, refrigeration, jewelry, and tableware industries. Most copper is used in the production of electrical wire and in a variety of alloys, brass (copper and zinc) and bronze (copper, tin, and zinc) being the most common. Platinum is a very important metal because of its high melting point, hardness, and chemical inertness. It is primarily used by the chemical industry as a catalyst in the production of chemicals but is also used in jewelry and surgical and dental tools. Sulfur has a wide variety of uses in the chemical industry (production of insecticides, fertilizers, fabrics, paper, and soaps). In addition to their use as gems, diamonds are used in a variety of ways as cutting, grinding, and drilling agents.

USES OF OTHER NONSILICATES

Like the native elements, the sulfides are also the source of a number of metals. The most important ores of silver, copper, lead, zinc, nickel, mercury, arsenic, antimony, molybdenum, and arsenic are sulfides. The government of the United States considers most of these commodities to be critical in time of war and has huge stockpiles of them in reserve throughout the country, since most are mined in other countries.

The two most important halides, halite and fluorite, are widely used. Halite, or ordinary table salt, is a source of sodium and sodium compounds, chlorine, and hydrochloric acid, which are all important in the chemical industry. It is also used as a deicing compound, in fertilizers, in livestock feeds, and in the processing of hides. Most fluorite is used to make hydrofluoric acid for the chemical industry or as a flux in the production of steel and aluminum.

Of the borates, borax is used to produce glass, insulation, and fabrics. It is also used in medicines, high explosives and propellants, detergents and soaps, and as a preservative. Of the phosphates, only apatite is a common mineral, but at least three others are important for the elements that they contain. Monazite is the primary source of thorium, which is a radioactive element with considerable potential as a source of nuclear energy. Autunite, a complex phosphate, and carnotite, a complex vanadate, are both important sources of uranium. A fine-grained form of apatite called "collophane" (a source of phosphorus) is most widely used in fertilizers and detergents.

Of the sulfates, barite, anhydrite, and gypsum are the three most commonly used. The bulk of the barite is used as a drilling mud in the mineral and energy exploration industry. Barite is unusually dense for a nonmetallic mineral and forms a dense pasty fluid that helps keep oil and gas contained in wells. Anhydrite and gypsum occur in similar geological conditions and have similar compositions. Both are used as soil conditioners. Gypsum is mainly used for the manufacture of plaster of Paris, which is used in wallboard.

In the tungstates, wolframite and scheelite are the main ores of tungsten, which is used in lamp filaments, in tungsten carbide for cutting tools, and as a hardening alloy.

—*Ronald D. Tyler*

Further Reading

Bowles, J. F. W., et al. *Rock-Forming Minerals: Non-Silicates; Oxides, Hydroxides and Sulfides.* 2nd ed., Geological Society Publishing House, 2011.

Hammond, Christopher. *The Basics of Crystallography and Diffraction.* 2nd ed., Oxford UP, 2001.

Klein, Cornelis, and Barbara Dutrow. *The Twenty-third Edition of the Manual of Mineral Science.* J. Wiley, 2008.

Nespolo, Massimo, and Giovanni Ferraris. "A Survey of Hybrid Twins in Non-Silicate Minerals." *European Journal of Mineralogy,* vol. 21, 2009, pp. 673-90.

Nesse, William D. *Introduction to Mineralogy.* 3rd ed., Oxford UP, 2017.

Perkins, Dexter. *Mineralogy.* 2nd ed., Prentice Hall, 2002.

Schumann, Walter. *Minerals of the World.* 2nd ed., Sterling, 2008.

Wenk, Hans-Rudolf, and Andrei Bulakh. *Minerals: Their Constitution and Origin.* Cambridge UP, 2004.

NUCLEOSYNTHESIS

Fields of Study: Cosmology; Theoretical Astronomy; Astrogeology

ABSTRACT

Nucleosynthesis is the process by which new atomic nuclei, and thus new elements, are created. The two main types of nucleosynthesis are Big Bang nucleosynthesis (BBN) and stellar nucleosynthesis. BBN occurred first, during the formation of the universe, and created the lightest elements. These elements later underwent stellar nucleosynthesis to create heavier elements. Nucleosynthesis accounts for all of the elements in the universe and is the reason that life on Earth was able to develop.

KEY CONCEPTS

Big Bang nucleosynthesis (BBN): the coming together of free neutrons and protons to create hydrogen isotopes, and the subsequent fusion of hydrogen nuclei to form other light elements, within the first three minutes of the Big Bang

Big Bang theory: the theory that the universe began in an extremely hot, dense state and then began a process of expansion that continues today

fusion: the joining together of two or more atomic nuclei to form a new, heavier nucleus of a different element

stellar nucleosynthesis: the fusion of light elements into heavier elements, such as carbon, oxygen, and iron, within the interiors of stars

BASIC PRINCIPLES OF NUCLEOSYNTHESIS

Nucleosynthesis is the process in which protons and neutrons fuse together to form new atomic nuclei. These nuclei then capture free electrons to form neutral atoms of new elements. The two principal types of nucleosynthesis are Big Bang nucleosynthesis (BBN) and stellar nucleosynthesis. BBN refers to the initial fusion of protons and neutrons to create nuclei of light elements. Stellar nucleosynthesis refers to the fusion of lighter nuclei within stars to create nuclei of heavier elements.

Big Bang nucleosynthesis is so called because it took place during the first few minutes of the Big Bang. In contrast, stellar nucleosynthesis is an ongoing process that takes place within every star and continues for as long as that star has fuel for the process. The elements created during BBN were mainly hydrogen and helium, plus trace amounts of lithium, beryllium, and possibly boron. Stellar nucleosynthesis is responsible for the creation of all elements between carbon (atomic number 6) and iron (atomic number 26) in the periodic table. Naturally occurring elements heavier than iron are formed during supernova explosions via a process called "neutron capture." This phenomenon is sometimes called "supernova nucleosynthesis."

For nucleosynthesis to take place, nuclei must collide with one another at very high speeds. Nucleosynthesis also requires extremely high temperatures —millions or even billions of degrees, depending on the element. For example, the fusion of hydrogen into helium requires a temperature of at least 3 million kelvins (5.4 million degrees Fahrenheit). But that temperature is low compared to carbon fusion, which requires a temperature of at least 600 million kelvins (1.1 billion degrees Fahrenheit). The more protons an element has in its nucleus, the higher the minimum temperature required for fusion.

THE BIG BANG THEORY

Although scientists may never know for certain how the universe was created, the Big Bang theory is widely accepted and well supported by the available evidence. According to this theory, the universe began as an extremely small, hot mass of near-infinite density. Then, approximately 13.8 billion years ago, that mass rapidly expanded, becoming cooler and less dense as it grew. One of the biggest misconceptions about the Big Bang is that it was an explosion in space. Rather, the term usually refers to the initial moment of expansion. It did not take place "in space" because before the Big Bang, there was no such thing as "space." Everything that would become the universe was contained within that infinitesimal mass.

The modern understanding of the Big Bang theory was developed and refined based on two main scientific concepts: the general theory of relativity and the cosmological principle. The general theory of relativity was published in 1915 by German-born physicist Albert Einstein (1879-1955). It builds on the law of universal gravitation, which Isaac Newton (1643-1727) developed in the late seventeenth century. While Newton's theory identifies mass as the source of gravity, Einstein proposed that gravity is generated by a combination of energy and momentum, while mass is just a different way of measuring energy. Put simply, wherever matter exists in the universe, its combined energy and momentum causes space-time to curve around it. The effects of gravity, such as planetary orbits, are the result of objects traveling along these curves.

In 1917, Einstein and several others started trying to use this idea to explain the entire universe. They found that the equations for general relativity did not permit the universe to exist in a static, unchanging state, as was previously thought to be the case. Instead, the universe had to be either expanding or shrinking. Einstein thought this was a flaw in his equations and tried to correct it by adding a constant value, known as the cosmological constant. However, in 1929, American astronomer Edwin Hubble (1889-1953) used his observations of the light from distant galaxies to prove that the universe is in fact expanding. Einstein accepted Hubble's proof and renounced the cosmological constant.

The fact that the universe is expanding implies that it had a definite beginning. This was the first step toward a theory of the origin of the universe. However, this implication raised another problem: it seemed to contradict a long-standing principle of cosmology. The cosmological principle states that the matter in the universe is uniformly distributed in all directions when viewed on a large enough scale. Originally, this was thought to be true with respect to time as well; in other words, scientists believed that the universe as a whole had not changed significantly over time. If this were not the case, then the relative uniformity of the universe would have to be accounted for in some way.

The problem was that the universe is too large for more distant regions to have a causal connection to one another. If the universe had changed over time, one would expect different parts to have evolved in different ways, because it would take too long for information to travel from one region to another. The fact that this is not the case implies that at some point in the past, all parts of the universe must have been close enough together to all be affected at once. Resolving the conflict with the cosmological principle provided yet more evidence in support of the Big Bang theory.

The cosmological principle is still being tested today, particularly with regard to the distribution of galaxies. Together, the theory of general relativity and the cosmological principle allow for precise predictions and observations regarding the universe. For example, they allow scientists to conclude that the universe must have one of three shapes: it may be flat and infinite; it may be negatively curved and infinite, resembling a saddle; or it may be positively curved and finite, resembling a ball.

BIG BANG NUCLEOSYNTHESIS

Nucleosynthesis first occurred shortly after the Big Bang, when the expansion of the universe allowed it to cool enough to produce light elements. Russian physicist George Gamow (1904-68) and his doctoral student Ralph Alpher (1921-2007) first proposed the idea in Alpher's doctoral dissertation in 1948. They claimed that the process was responsible for creating all of the elements in the universe. It was later determined that Big Bang nucleosynthesis accounts only for the light elements, and heavy elements were produced through a different process.

Big Bang nucleosynthesis occurred within just a few minutes of the Big Bang. Scientists believe that immediately after the Big Bang, the universe was so hot and dense that no individual particles could exist. All matter was condensed into a form of matter called "quark-gluon plasma." This plasma, which can exist only at extremely hot temperatures—roughly 1.8×10^{12} kelvins, or one hundred thousand times hotter than the core of the sun—consists of elementary particles called "quarks" and "gluons." Less than one second after the Big Bang, the quark-gluon plasma lost enough energy that the quarks and gluons could join together to form protons and neutrons. Because the nucleus of a hydrogen atom is just a single proton, hydrogen was the first element created in the universe.

As quarks continued to combine, the ratio of neutrons to protons in the universe reached approximately 1:6, or one neutron for every six protons. However, free neutrons decay, which brought the ratio closer to 1:7. About one hundred seconds after the Big Bang, the temperature of the universe had cooled enough for deuterons to form. A deuteron is the nucleus of a heavy isotope of hydrogen called "deuterium" (hydrogen-2). It consists of one proton and one neutron.

The deuterons began to react to form other nuclei. The most common product of these reactions was the helium-4 nucleus (two protons and two neutrons). Other products included "light" helium (helium-3, two protons and one neutron) and tritium (hydrogen-3, one proton and two neutrons). Additionally, a small number of helium nuclei joined together to create lithium-7 (three protons and four neutrons). The entire process of Big Bang nucleosynthesis lasted less than twenty minutes. At the end, the matter in the universe consisted of approximately 75 percent hydrogen, slightly less than 25 percent helium-4, about 0.01 percent deuterium and helium-3, and infinitesimal amounts of lithium-7 and other isotopes.

STELLAR NUCLEOSYNTHESIS

Protons and neutrons fuse together to form atomic nuclei. A hydrogen ion (proton) fuses with a neutron to form the nucleus of a deuterium (hydrogen-2) atom. If the deuterium nucleus fuses with another neutron, it becomes tritium (hydrogen-3). When this tritium nucleus fuses with another proton, it becomes a helium nucleus.

Stellar nucleosynthesis is the other main type of nucleosynthesis. It is the process by which heavy elements such as carbon, oxygen, and iron are created. Stellar nucleosynthesis takes place in stars, and unlike Big Bang nucleosynthesis, it continues to occur. English astronomer and physicist Arthur Eddington (1882-1944), who was an adamant supporter of Einstein, first proposed the idea in 1920. However, English astronomer Fred Hoyle (1915-2001) is credited with developing the concept into a theory following World War II. Notably, Hoyle did not believe in the Big Bang theory, because he thought that the idea of the universe having a beginning was too irrational. Instead, he helped formulate an alternative theory known as the "steady state theory." This theory proposed that matter was being constantly created in the space between galaxies, resulting in the universe's ongoing expansion. While the steady state theory was ultimately rejected as new evidence supported the Big Bang theory instead, Hoyle's theory of stellar nucleosynthesis is well supported and remains widely accepted.

All stars are fueled by nuclear fusion. When a star first forms, it consists primarily of hydrogen. For the majority of its life cycle—a period known as the "main sequence"—the star will fuse the hydrogen in its core into helium. Once the majority of its hydrogen has been burned, the star will evolve away from the main sequence. What happens next depends on the mass of the star. If it is greater than 0.5 solar masses (one-half the mass of the sun), the core will become hot enough for helium to fuse into carbon. If the star is more than about 8 solar masses, it will then go on to fuse carbon into a succession of heavier elements until it reaches iron. Iron is the heaviest element that can be produced through stellar nucleosynthesis, because to fuse iron would consume more energy than it would release. Elements heavier than iron can be produced only in supernovas.

SIGNIFICANCE OF NUCLEOSYNTHESIS

Much like the Big Bang theory, nucleosynthesis is an important part of cosmology. It is the process that created all of the elements in the universe, thus making human life—and, in fact, all forms of life—possible. Stellar nucleosynthesis created the elements that are essential for life to exist, particularly carbon, nitrogen, oxygen, and iron. When the earliest stars ejected some of their mass, or when they reached the ends of their lives and went supernova, these elements were dispersed into space for the first time. They became part of the interstellar medium, out of which planets would later form. As a result, when Earth formed about 4.6 billion years ago, it already contained the elements necessary for life to eventually evolve. In a sense, then, it can be said that human beings are made of stardust.

Scientists at the Relativistic Heavy Ion Collider (RHIC) in New York and the Large Hadron Collider (LHC) in Switzerland, the two most powerful particle accelerators in the world, continue to investigate nucleosynthesis and the conditions of the Big Bang. In 2005, researchers at the RHIC reported that for the first time ever, they had created a plasma similar to the quark-gluon plasma that was present at the beginning of the universe. It was such a remarkable achievement that the results were only verified in 2010. The following year, LHC researchers announced the production of a plasma even hotter than the one produced by the RHIC. Scientists hope that studying these plasmas will help them better understand the origin of the universe.

—Michael Mattei and Randa Tantawi

Further Reading

Boyd, Richard N. *An Introduction to Nuclear Astrophysics.* U of Chicago P, 2007.

Chaisson, Eric J. "Origin of Heavy Elements." *Cosmic Evolution.* Harvard-Smithsonian Center for Astrophysics, 2013. Accessed 2 June 2015.

"Cosmology: The Study of the Universe." *Wilkinson Microwave Anisotropy Probe.* NASA, 21 Dec. 2012. Accessed 2 June 2015.

Moskowitz, Clara. "Hottest Particle Soup May Reveal Secrets of Primordial Universe." *LiveScience.* Purch, 13 Aug. 2012. Accessed 2 June 2015.

"Nucleosynthesis." *Cosmos: The SAO Encyclopedia of Astronomy.* Swinburne University of Technology, n.d. Accessed 2 June 2015.

Oakes, Kelly. "On the Origin of Chemical Elements" *Basic Space.* Scientific American, 2 Aug. 2011. Accessed 16 July 2015.

Perlov, Delia, and Alex Vilenkin. *Cosmology for the Curious.* Springer, 2017.

Prantzos, Nikos, and Sylvia Ekström. "Stellar Nucleosynthesis." *Encyclopedia of Astrobiology*, edited by Muriel Gargaud, et al., Vol. 3, pp. 1584-92, Springer, 2011.

Ryan, Sean G., and Andrew J. Norton. *Stellar Evolution and Nucleosynthesis.* Cambridge UP, 2010.

Than, Ker. "Densest Matter Created in Big-Bang Machine." *National Geographic.* National Geographic Society, 26 May 2011. Accessed 16 July 2015.

"Tests of Big Bang: The Light Elements." *Wilkinson Microwave Anisotropy Probe.* NASA, 16 Apr. 2010. Accessed 2 June 2015.

Weiss, Achim. "Big Bang Nucleosynthesis: Cooking Up the First Light Elements." *Einstein Online.* Max Planck Institute for Gravitational Physics, 2006. Accessed 2 June 2015.

Wright, Edward L. "Big Bang Nucleosynthesis." *Ned Wright's Cosmology Tutorial.* UCLA Astronomy & Astrophysics, 26 Sept. 2012. Accessed 2 June 2015.

O

Ocean Acidification

Fields of Study: Geology; Marine Geology; Geochemistry; Oceanography

ABSTRACT

Ocean acidification describes the decrease in pH of the oceans due to increased concentration of dissolved carbon dioxide (CO_2). CO_2 is a trace gas found in the atmosphere. The concentration of dissolved oceanic CO_2 is in equilibrium with the gas's atmospheric concentration, such that an increase in the atmospheric concentration leads to an increase in the dissolved concentration, and vice versa.

KEY CONCEPTS

calcification: the process by which marine organisms secrete calcium carbonate to produce shells, tests and casings as body parts

equilibrium: in chemical systems, the state in which the reactants and products of different reactions involving related materials function in concert to maintain the system in an overall steady state

partial pressure: in any gaseous mixture, the total pressure is the sum of the proportions of the total pressure due to each individual gas; that is, total pressure = sum of partial pressures

pH: a measure of the acidity of a solution in water based on the concentration of free hydrogen ions; specifically, pH = log[H^+] where [H^+] is the free hydrogen ion concentration

primary producers: in biological systems, the organisms that produce environmental components from inorganic raw materials

DEFINITION OF OCEAN ACIDIFICATION

Ocean acidification describes the decrease in pH of the oceans due to increased concentration of dissolved carbon dioxide (CO_2). CO_2 is a trace gas found in the atmosphere. The concentration of dissolved oceanic CO_2 is in equilibrium with the gas's atmospheric concentration, such that an increase in the atmospheric concentration leads to an increase in the dissolved concentration, and vice versa.

Throughout geologic time, atmospheric CO_2 reflected a balance between various sources and sinks found on land and in the oceans—such as photosynthesis, respiration, chemical weathering of rocks and burial of organic and inorganic carbon. Since the onset of the industrial age, human activities, such as fossil fuel combustion and intensification of agriculture, represent a new source of atmospheric CO_2.

These activities are thought to cause the observed increases in atmospheric CO_2, which in turn lead to increased dissolved CO_2 concentrations in the oceans. Dissolved CO_2 combines with water to form carbonic acid (H_2CO_3), which subsequently dissociates to produce bicarbonate (HCO_3^-) and carbonate ions (CO_3^{2-}) and protons (H^+). The balance among these different reactions is such that increased concentrations of dissolved CO_2 lead to increased proton concentrations (acidity). Since pH is an inverse logarithmic scale of the concentration of dissolved H^+, the higher the proton concentrations, the lower the pH and the greater the acidity. Modeling experiments using different scenarios to project future atmospheric CO_2 concentrations indicate that by 2100, the ocean's pH will drop by up to 0.45 unit.

SIGNIFICANCE OF OCEAN ACIDIFICATION FOR CLIMATE CHANGE

The reactions by which CO_2 and water form H_2CO_3, HCO_3^- and CO_3^{2-} control the pH of the world's oceans. Oceanic pH has important ecological consequences, because many plankton (including the *Coccolithophoridae*, which are marine algae, and the *Foraminiferida*, which are planktonic protists) and macrofauna (corals, mollusks, brachiopods, and so on) precipitate the minerals calcite or aragonite (which both have the chemical formula $CaCO_3$) to make exoskeletons, shells, and tests in a process

A normally protective shell made thin, fragile and transparent by acidification. Photo by NOAA's National Ocean Service, CC BY 2.0, via Wikimedia Commons.

called "calcification." Under normal conditions, calcite and aragonite are stable minerals in surface waters, because CO_3^{2-} concentrations are naturally maintained at levels that prevent their dissolution. However, increasing atmospheric CO_2 concentrations lead to decreased oceanic pH values and a concomitant decrease in carbonate ion concentration.

This effect is currently observed in a vertical profile of the oceans, where carbonate ion concentrations decrease with increasing depth. At a depth known as the lysocline, the rate of carbonate mineral dissolution increases rapidly with similarly rapid decreasing carbonate ion concentrations. Thus, in surface and intermediate waters, carbonate minerals are not dissolved. Below the lysocline, carbonate minerals are readily dissolved. Many laboratory experiments seem to indicate that ocean acidification will prove detrimental to many marine ecosystems. This finding has been recently validated by a study of a benthic ecosystem located near a naturally occurring volcanic vent that delivers CO_2 to the surrounding waters. The full-scale consequences of acidification of the global ocean are likely to be numerous, and could include extinctions as food webs collapse due to a loss of calcareous planktonic primary producers and consequences for the strength of the biological pump. The biological pump is the process by which CO_2 is actively removed from the atmosphere by primary producing phytoplankton who convert the CO_2 to organic matter via photosynthesis. When these organisms die, their hard parts act as ballast to help the organic matter sink to the ocean floor, where it is buried in sediments. Ultimately, this sequesters the CO_2 in rocks over geologic time scales. However, if ocean acidification leads to dissolution of calcareous tests, organic matter will not be effectively buried because it is not dense enough to settle to the ocean floor.

Not all scientists agree with the predicted negative consequences of ocean acidification. They argue that carbonate minerals present in sedimentary rocks on the ocean floor should, over time, consume the excess H^+ produced from increased atmospheric CO_2 concentrations, causing ocean acidification to slow or even stop. The increase in atmospheric CO_2 concentrations appears to be linked to an increase in global temperatures. Temperature is a strong control on the conversion of atmospheric CO_2 to dissolved CO_2 in the ocean. Henry's Law states that the dissolved concentration of a gas is proportional to the partial pressure of that gas in the atmosphere in contact with that liquid. Gases are characterized by different Henry's Law constants that give the proportionality of dissolved gas that will be in equilibrium with the overlying atmosphere. For most gases, including CO_2, the value of the Henry's Law constant decreases with increasing temperature—that is, at higher temperatures, the amount of CO_2 that can be dissolved in the ocean will decrease. This temperature dependence provides a negative feedback on the amount of ocean acidification that could occur. However, this also means that a greater proportion of anthropogenic CO_2 would remain in the atmosphere as global temperature rises.

SIGNIFICANCE OF OCEAN ACIDIFICATION FOR GEOLOGY

The equilibrium balance between atmospheric carbon dioxide, ocean pH, and the formation and decomposition of carbonate rock has functioned for many millions of years, and evolved to its present state over much of that time. All fossil-bearing limestone, which consists primarily of calcium carbonate, and much of the nonfossil-bearing limestone in the world was formed from the deposition of calcareous particles precipitated as the spent remains of trillions upon trillions of tiny aquatic s creatures or the precipitation of insoluble carbonate salts as they formed.

Carbon dioxide in the atmosphere also reacts with atmospheric water, or meteoric water, to produce carbonic acid in atmospheric precipitation. Exposed limestone subjected to such precipitation experiences the reaction of calcium carbonate with acid, releasing carbon dioxide and soluble calcium salts. Over time, this has produced the well-known karst landscapes and carved out vast caverns in the susceptible underlying rock. This acidulated water eventually makes its way into the oceans, where it compounds the effects of ocean acidification, or into the underground where it becomes part of the geochemical processes occurring there, such as metasomatism and hydration.

The level of the oceans has changed drastically several times over the geological time scale. Each change has either exposed marine carbonate sediments to the atmosphere, or buried terrestrial carbonate sediments under the ocean medium. The effects of those changes are captured in the geological record, to be discovered and interpreted by geologists in learning and understanding the full range of geochemical processes occurring on Earth and presumably on all other seismically active rocky planets with a water-bearing atmosphere.

—Anna M. Cruse and Richard M. Renneboog

Further Reading

Doney, Scott C. "The Dangers of Ocean Acidification." *Scientific American,* vol. 294, no. 3, Mar. 2006, pp. 58-65.

Gattuso, Jean-Pierre, and Lina Hansson, editors. *Ocean Acidification.* Oxford UP, 2011.

Hall-Spencer, Jason M., et al. "Volcanic Carbon Dioxide Vents Show Ecosystem Effects of Ocean Acidification." *Nature,* vol. 454, no. 7200, 3 July 2008, pp. 96-99.

Loáiciga, Hugo. "Modern-Age Buildup of CO_2 and Its Effects on Seawater Acidity and Salinity." *Geophysical Research Letters,* vol. 33, no. 10, 6 May 2006.

Millero, Frank J. *Chemical Oceanography.* 4th ed., CRC Press, 2013.

Pandolfi, John M., Sean R. Connolly, Dustin J. Marshall, and Anne L. Cohen. "Projecting Coral Reef Futures under Global Warming and Ocean Acidification." *Science,* vol. 333, no. 6041, 2011, pp. 418-22.

Royal Society. *Ocean Acidification Due to Increasing Atmospheric Carbon Dioxide.* Author, 2005.

Zeebe, Richard E., and Dieter Wolf-Gladrow. *CO_2 in Seawater: Equilibrium, Kinetics, Isotopes.* Elsevier, 2001.

OCEAN MARGINS

Fields of Study: Geology; Geomorphology; Marine Geology; Oceanography

ABSTRACT

Ocean margins are the areas where land borders the sea. Although often referred to as coastlines or beaches, ocean margins cover far greater territory than beaches. An ocean margin extends from the coastal plain—the fertile farming belt of land along the seacoast—to the edge of the gently sloping land submerged in water, called the "continental shelf."

KEY CONCEPTS

barrier island: a low-lying type of island that may form from current-borne sediments along a coastline; such islands act as a barrier against incoming oceanic waves

delta: a formation that forms when sediment carried by a river is deposited where the river enters an ocean or other large body of water, similar in manner to the formation of barrier islands but remaining connected to the coastline

primary coast: coastal features that are formed as a result of processes occurring on land

secondary coast: coastal features that exist as a fundamental part of the land structure

NATURE OF OCEAN MARGINS

Ocean margin constitutes 8 percent of the world's surface. It is rich in minerals, both above and below water, and is home to 25 percent of Earth's people, along with 90 percent of the marine life. This fringe of land at the border of the ocean is ever changing. Tides wash sediment in and leave it behind, just below sea level. This process, called "deposition," builds up land in some areas of the coastline. At the same time, ocean waves, winds, and storms wear away or erode parts of the shoreline. As land is worn away or built up, the amount of land above sea level changes. Factors such as climate, erosion, deposition, changes in sea level, and the effects of humans constantly change the shape of the ocean margin on Earth.

BEACH DYNAMICS

The two types of coasts or land formations at the ocean margin are primary coasts and secondary coasts. Primary coasts are formed by systems on land, such as the melting of glaciers, wind or water erosion, and sediment deposited by rivers. Deltas and fjords are examples of primary coasts. Secondary coasts are formed by ocean patterns, such as erosion by waves or currents, sediment deposition by waves or currents, or changes by marine plants or animals. Beaches, coral reefs, salt marshes, and mangrove swamps are examples of secondary coasts.

Sediment carried by rivers to the sea is deposited to form deltas at the mouths of the rivers. Some of the sediment can wash out to sea, causing formations to build up at a distance from the shore. These formations eventually become barrier islands, which are often little more than 3 meters, or about 10 feet, above sea level. As a consequence, heavy storms, such as hurricanes, can cause great damage to barrier islands. Barrier islands naturally protect the coastline from erosion, however, especially during heavy coastal storms.

Sea level changes also affect the shape of the coastline. As oceans slowly rise, land is slowly consumed by the ocean. Barrier islands, having low elevations above sea levels, may slowly be covered with water. The melting of continental glaciers increased the sea level 0.15 centimeter (0.06 inch) per year during the twentieth century. As ocean waters warm, they expand, eating away at sea levels. Global warming caused by carbon dioxide levels in the atmosphere could cause sea levels to rise as much as 0.6 centimeter (0.24 inch) per year as a result of the warming of the water and glacial melting.

HUMAN INFLUENCE

The shape of the ocean margin also changes radically as a result of human influence. According to the U.S. Geological Survey, 39 percent of the people living in the United States live directly on the coasts. According to United Nations Atlas of Oceans, about 44 percent of the world's population lives within 150 kilometers (93 miles) of the coast. Pollution from toxins, dredging, recreational boating, and waste disposal kills plants and animals along the ocean margin. This changes the coastal shape, as the influence of mangrove forests, coral reefs, and other coastal lifeforms disappears.

A greater concern along the coastal fringe, however, is human development. Not only are people drawn to the fertile soil along the coastal zone of the continent, but they also develop islands and coves into resort communities. To protect homes and hotels along the coastal zone from coastal erosion, people build breakwalls, jetties, and sand and stone bars

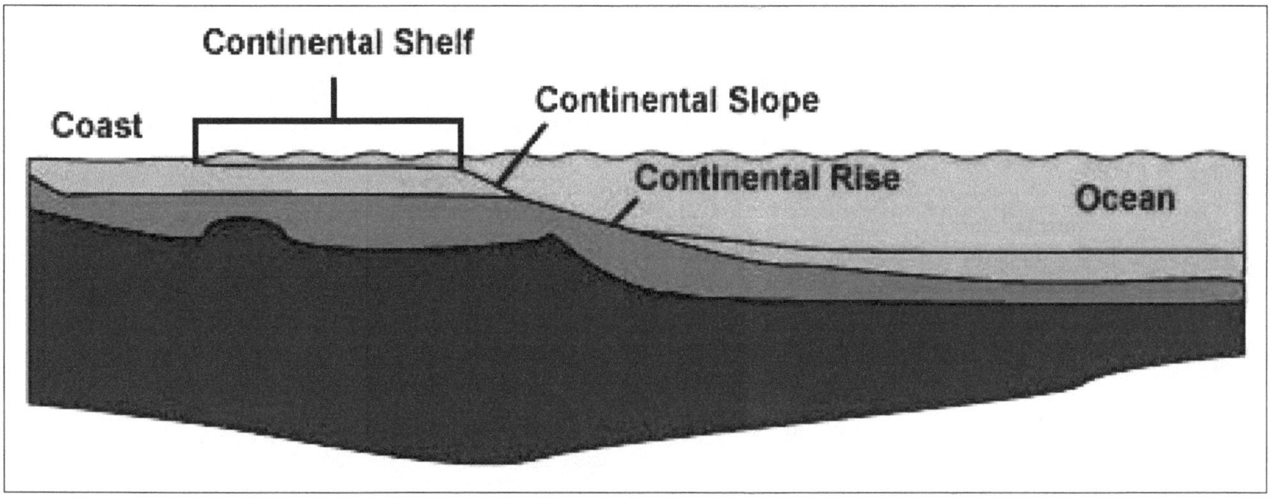

Profile illustrating shelf, slope, and rise. Image via Wikimedia Commons. [Public domain.]

called "groins." These human-made barriers disrupt the natural method by which the ocean carries material along the coast. Longshore drift, a zigzag movement, deposits sediment from one area of the beach farther along the shoreline. Breakwalls, jetties, and groins disrupt this flow. As the ocean smashes against a breakwall, the property behind it may be safe for the present, but the coastline neighboring the breakwall takes a greater beating. The silt and sediment from upshore, which would replace that carried downshore, never arrives. Eventually, the breakwall will break down under the impact of the ocean force. Areas with breakwalls and jetties often suffer greater damage in coastal storms than areas that remain naturally open to the changing forces of the ocean.

To compensate for the destructive nature of human-made barriers, many recreational beaches replace lost sand with dredgings or deposit truckloads of sand from inland sources. For example, Virginia Beach in the United States spends between US$2 million and US$3 millon annually to restore beaches for the tourist season in this way.

Despite the changes in the shape of the ocean margin, it continues to provide a stable supply of resources—fish, seafood, minerals, sponges, and other marine plants and animals. Offshore drilling of oil and natural gas often takes place within 200 miles (322 km.) of shorelines.

—Lisa A. Roble

Further Reading

Linnemann, Ulf, R. Damian Nance, Petr Kraft, and Gernold Zulauf, editors. *The Evolution of the Rheic Ocean: From Avalonian-Cadomian Active Margin to Alleghenian-Variscan Collision.* Geological Society of America, 2007.

Simpson, John H., and Jonathan Sharples. *Introduction to the Physical and Biological Oceanography of Shelf Seas.* Cambridge UP, 2012.

Wefer, Gerold, David Billett, Dierk Hebbeln, et al., editors. *Ocean Margin Systems.* Springer, 2003.

OCEANS' ORIGIN

Fields of Study: Geology; Astrogeology; Geochemistry; Geophysics

ABSTRACT

Oceanic waters were derived from the outgassing of hydrated minerals bound up during Earth's formation. Subsequent evolution of the waters primarily involved ions dissolving in the fluid medium by interactions with the continental and oceanic bottom sediments to give the basic salinity of Earth's oceans.

KEY CONCEPTS

carbonaceous chondrites: a class of meteoritic bodies found to contain large amounts of carbon in conjunction with other elements; used to date the solar system and provide chemical composition assessments of the original solar nebula

geochemical sinks: the processes by which elements and compounds are removed from the crust and oceans to be recycled in active chemical cycles

outgassing: the process by which volatile materials trapped within rock formations are released into the atmosphere and the environment

primordial solar nebula: the original collection of dust and gases that constituted the basic cloud from which the solar system formed

smokers: undersea vents on the active rift areas that emit large amounts of superheated water and dissolved minerals from within the crust

solar wind: the stream of highly charged particles emitted into space from the surface of the sun

volatiles: chemical elements and compounds that become gaseous at fairly low temperatures

water of hydration: water that is bound in the crystal structure of minerals without being part of their actual molecular structure

EARTH'S OCEANS

Of all the planets in the solar system, Earth stands out as a watery world, distinguished from the others by great quantities of liquid water. Encompassing more than 1.35 billion cubic kilometers, Earth's seas contain enough salt to cover all of Europe to a depth of 5 kilometers. The salt content of the oceans is composed primarily of sodium chloride, some 86 percent of the ions by weight, in association with other ions of magnesium, calcium, potassium, sulfate, and carbonate groups, providing a salinity of 35,000 parts per million, with pH 8 for the hydrogen ion concentration (slightly alkaline in nature). Of all the water on Earth, 97 percent of it is salty,

The ocean formed billions of years ago. Photo via National Oceanic and Atmospheric Association (NOAA). [Public domain.]

the remainder being proportioned among ice (77 percent of total freshwater) and continental and atmospheric waters. The ice itself, principally in the Arctic-Greenland area (1.72 million square kilometers, 3,200 meters thick) and the Antarctic area (12 million square kilometers, 4,000 meters thick), provides effects ranging from climatic control to habitats for living organisms and sources of new seawater that, in the past, have caused sea levels to rise more than 100 meters.

Consideration of the oceans' origin is twofold: the primordial origin of the water itself and the origin and rate of addition of the salt ions present in the past and contemporary waters. The data sources for consideration include the chemistry of water, the amounts and types of runoff delivered by rivers into the sea, and the composition of volcanic gases, geysers, and other vents opening to the surface. In addition, most researchers find the oceans and atmosphere to be linked in origin, providing even more data for analysis.

Numerous sources for the world's water have been proposed, although a definitive resolution has not been achieved. Possible sources include the primordial solar nebula, the solar wind acting over time, bodies colliding with Earth, impact degassing, and outgassing from the planetary interior. A complete consideration of the origin of Earth's oceans necessarily involves the investigation of factors controlling water on Earth, particularly the rates, amounts, and types of outgassing, modes of planetary formation, possible chemical reactions providing water, loss rates of gases to space, and, finally, internal feedback mechanisms such as changes in Earth's albedo (reflective power), temperature, alteration of mass, and other factors not clearly understood.

SOLAR WIND, COLLIDING BODIES, AND IMPACT DEGASSING

The solar wind as a primary source of terrestrial water can be eliminated for several reasons. Its basic constituents, charged protons, can formulate water in the atmosphere by reactions with oxygen, but all evidence indicates that there was no free oxygen in the primordial atmosphere. The geologic record shows the presence of liquid water at least 4 billion years ago that was substantially devoid of free oxygen. Astrophysical evidence suggests that the solar flux of energy, associated with the solar wind, was such that, early in Earth's history, water on Earth should have been frozen, not liquid, if that was the primary source. The presence of liquid water through at least 80 percent of Earth's history, however, has been established.

Colliding bodies would include two primary sources: meteorites and comets. The occurrence of cometary bodies striking Earth has yet to be documented. However, the basic chemical makeup of comets consisting of various ions, metals, organic molecules, and dust grains in a mass of water ice, could theoretically have supplied enough water to account for the amount of water present on the planet, providing that large numbers of cometary objects struck the early Earth during the first half billion years of its history. No evidence for such happenings is available at present, although a theory of Earth still being bombarded incessantly by small comets composed primarily of water ice has been fiercely debated.

Meteoritic impact, particularly during the earlier stages after final planetary accretion, would also have added water to the crust via two mechanisms. Through the study of carbonaceous chondrites (the oldest and most primitive meteorites), abundant volatiles, such as water, are found to be bound chemically to minerals such as serpentine. Additional water, trapped in crustal and mantle rocks, would have been released during impact, particularly from large meteoritic rocks. It has been calculated that such impact degassing could have released 10^{22} kilograms of volatiles, quite close to the currently estimated value of 4×10^{21} kilograms for Earth as a whole. Remnants of such ancient astroblemes are lacking, however, because of subsequent erosion, filling in by molten magma, or shifting of the continental masses over 4 billion years.

OUTGASSING

The most widely accepted origin for the oceans and atmosphere combines the features of the primordial solar nebula and slow outgassing from within the solidifying Earth. Original water would have been combined, under gravitational collapse, with silicates and metallic materials during the planetary accretion process, with the hydration of minerals assisted by the heating of Earth due to infalling bodies and radioactive elemental decay. Such "wet silicates" appear to hold large quantities of bound water for indefinite periods of time. The primordial Earth is believed to have formed by cold accretion, and would have trapped the water molecules. If it had been too hot during the accretion, all the minerals would have been dehydrated, and if too cold, no water would have been released. A delicate balance of temperature must therefore have been achieved. Further, the volatiles forming the atmosphere must have out gassed first, because water must be insulated from solar radiation in order to form a liquid phase.

A secondary problem examines how swiftly the fluids would have outgassed, whether all at once, as individual events, or in a continuous fashion. Most data suggest the continuous mode of emission, with the greatest reliance on data from still active sources—mainly volcanoes, undersea vents, and associated structures. Fumaroles, at temperatures of 500° to 600°C, emit copious quantities of water, sulfur gases, and other molecules. These bodies grade gradually into hot spots and geysers, areas where water is moved upward through the crust from great depths. Magmatic melts rising in volcanoes release water and other gases directly to the surface.

In Hawaii, for example, the Halemaumau Pit, the volcano Kilauea's most active vent, emits—in terms of material—68 percent water, 13 percent carbon dioxide, and 8 percent nitrogen, with the rest mostly sulfurous gases. Similar types of values are found in ridge axis black and white smokers, where hydrothermal accretions result in spectacular deposits of minerals falling out of solution from the emerging hot waters. Detailed studies show water trapped in the structures of altered minerals within the basaltic crustal rock of the oceanic plates, with 5 percent of the rocks, by weight, in the upper 2 to 3 kilometers being water and hydroxide ions. Free water is known to be extremely buoyant, rising in the crust along shallow dipping faults. Bound water, subducted to great depths, would be expected to cook, moving upward as the rock density lessens, then acting as a further catalyst for melting the surrounding rocks.

In Earth's earliest stages, the primordial atmosphere probably escaped from Earth's gravitational pull because of overheating. In the second phase, gases released from molten rocks, with a surface temperature of 300°C, provided 70 percent water and large quantities of carbon dioxide and nitrogen. In stage three, the atmosphere and oceans gradually changed, with gases and liquid water ejected from volcanoes resulting in more and more water deposited as liquid as the temperature fell. Oxygen was added to the atmosphere through thermal dissociation of water molecules, photochemical breakdown of high altitude water, or photosynthetic alteration of carbon dioxide to oxygen in plants.

OCEAN SALT

The saltiness of the oceans can be accounted for by the inordinately large dielectric constant of water, a property that essentially ensures that it does not remain chemically pure. Geologic evidence shows the general composition to be similar over time, the stability of water content attributable to the continuous seawater sediment interface. Geophysicist John Verhoogen determined that only 0.7 percent of the present ocean has been added since the Paleozoic era, primarily from lava materials. The salty quality is a product of acidic gases from volcanoes (forming hydrochloric, sulfuric, and carbonic acids) that act to leach out the common silicate rocks. Paleontological evidence indicates that the change in ions must have been extremely slow, as demonstrated by the narrow tolerance of organisms then alive, such as corals, echinoderms, brachiopods, and radiolarians. Ion concentrations in present day river waters differ drastically from the ocean's values, however, indicating a different atmospheric environment in the past. Geochemists Robert M. Garrels and Fred T. Mackenzie have divided the oceans into three historical periods. In the earliest, water and volcanic acidic emissions actively attacked the crustal rock, leaching out ions and leaving residues of alumina and silicates. During the next period, from 3.5 to 1.5 billion years ago, slow continuous chemical action continued to attack sedimentary rocks, adding silica and ferrous ions. During the third period, from 1.5 billion years ago to the present, ions have accumulated to the point of modern concentrations, such that the composition is in apparent equilibrium with a mixture of calcite, potassium feldspars, illite montmorillonite clays, and chlorite.

Because it is known that, in an equilibrium state, the output of ions is equal to the input of ions, a new problem, that of geochemical "sinks," has been identified. Calcium carbonate (limestone) is removed from solution by living organisms to form skeletons, as is silica for opaline skeletons. Metals are dropped from seawater as newly formed mineral clays, oxides, sulfides, and zeolites, and as alteration products at the hot water basaltic ridges. Sulfur is removed as heavy metal sulfides precipitating in anaerobic environments, while salts are moved in pore waters trapped in sediments. Residence times for many of the ions have been determined. For example, sodium cycles in 210 million years, magnesium in 22 million years, calcium in 1 million years, and silicon in 40,000 years. With such effective removal systems, it is truly a measure of the geochemical resistance of Earth's ocean waters to change that has allowed their composition to achieve stability over 4 billion years.

STUDY OF THE OCEAN

Numerous avenues of approach have been used to investigate the ocean and its ions, including geological, chemical, and physical means. Geology has supplied basic data on the types and makeup of rocks from the earliest solidified materials to present depositional formations. Using a petrographic microscope to view thin sections of rock under polarized light enables the identification of minerals and provides quantitative measurements of the water attached to the minerals themselves. Paleontological studies of fossil organisms and paleosols indicate the range of ions in the sea at diverse geologic periods, both by the ions themselves left in the deposited soils and rocks and through studies of the ion tolerance ranges for similar, twentieth-century organisms. Such studies, along with sedimentology investigations of the rates and types of river depositions, dissolved ion concentrations, and runoff rates for falling rain, provide determinants for comparing present day ion concentrations with those of the past for continentally derived materials.

Chemical analysis reveals the various ions present in seawater and rocks via two principal methods. Use of the mass spectrometer identifies the types and quantities of ions present by use of a magnetic field to accelerate the charged ions along a curved path with a radius that is strictly determined by the weight and charge of the ions. Although the material present at the end of the path is far too small to be physically captured, its collection would provide a pure sample of the different ions present. What is captured instead is a permanent record of the charge-to-mass ratios of the particles that were detected. For solid samples, electron beam probe studies provide analysis from an area only one micron in diameter. The electrons, fired at the sample, cause characteristic X-rays to arise from the point. The energy of each type of X-ray is characteristic of a specific element or compound. By use of various optics to focus the X-rays, identification of even minute variations in concentration is possible.

Solubility studies provide residence times as a means of geochemical analysis for cyclical research. Similar laboratory projects, testing the ability of water to dissolve and hold ions in solution, argue for a primordial atmosphere that was essentially neutral, or mildly reducing in nature. Such reduction characteristics are based on the study of planetary composition for Earth and for other solar planets as supplemented by the various "lander missions" to Venus and Mars, and by modern astrophysical satellite observations. Chemical analysis from such missions in interplanetary space has also determined compositions for meteoritic gases, cometary tails and nuclei, and the mixing ratios for noble gases, all important for determining the origin of the solar system. The latter study, involving analysis of radioactive isotopes such as helium-3 (3He), an isotope of helium that is formed in the mantle, has prompted geophysicists to consider the mantle as a major source and sink for elements in various geochemical cycles.

Laboratory analysis reaches two other areas. Petrographic studies of returned lunar rocks suggest that the moon is essentially devoid of water, lacking even hydroxyl ions, at least on the surface. This discovery initially helps eliminate the solar wind and meteoritic impact as major factors in the initial formation of oceans. However, the recent identification of bound subsurface water on the moon is cause for reconsideration of that conclusion. Furthermore, high temperature/high pressure metallurgical and chemical studies indicate molten granite, at temperatures of 900°C and under 1,000 atmospheres of pressure, will hold 6 percent water by weight, while basalt holds 4 percent. Based on geochemical calculations of the amounts of magma in the planet and lavas extruded over the first billion years, all the oceans' waters can be accounted for, particularly if parts of the fluid, as steam under pressure, are a result of oxidation of deep seated hydrogen deposits trapped within or combined with mantle rocks. This supposition is thus considered an excellent likelihood from evidence gathered on radioactive decay in Earth's interior.

—Arthur L. Alt

Further Reading

Catling, David C., and James F. Kasting. *Atmospheric Evolution on Inhabited and Lifeless Worlds.* Cambridge UP, 2017.

Hamblin, Kenneth W., and Eric H. Christiansen. *Earth's Dynamic Systems.* 10th ed., Prentice Hall, 2003.

Kandel, Robert. *Water from Heaven: The Story of Water from the Big Bang to the Rise of Civilization and Beyond.* Columbia UP, 2003.

Plummer, Charles C., Diane H. Carlson, and Lisa Hammersley. *Physical Geology.* 13th ed., McGraw Hill, 2009.

Redfern, Ron. *Origins: The Evolution of Continents, Oceans and Life.* U of Oklahoma P, 2001.

Schwartz, M. *Encyclopedia of Coastal Science.* Springer, 2005.

Zalasiewicz, Jan, and Mark Williams. *Ocean Worlds: The Story of Seas on Earth and Other Planets.* Oxford UP, 2014.

Oil Shale and Tar Sands

Fields of Study: Geology; Petroleum Geology; Geochemistry; Geophysics; Environmental Science

ABSTRACT

Oil shale and tar sands are sources of oil and gas fuel, lubricants, and chemical feedstock. Tar sands are dense soil with pore spaces filled by solid or semisolid bitumen. Oil shale is any fine-grained sedimentary rock containing kerogen and yielding petroleum when heated in the absence of oxygen.

KEY CONCEPTS

cracking: a petroleum production process in which crude oil is heated over catalytic materials to "crack" the large molecular components of the crude product into smaller molecules that are then separated and recovered

fracking: a shale oil extraction method in which the oil shale bedrock is physically fractured in place, then subjected to heat to facilitate the flow of oil to recovery wells

kerogen: a waxy, insoluble organic compound with a large molecular structure as an organic residue permeating the pore spaces of sedimentary rock, particularly shale

oil shale: fine-grained sedimentary rock that incorporates a solid hydrocarbon substance in its structure

tar sands: sands that are permeated with bitumen, a tarlike petroleum that is too thick to flow; the term is interchangeable with "oil sands"

PETROLEUM RESERVOIRS

Oil is not found in vast oil-filled underground caverns; rather, it is found in the spaces between grains that make up rock. Permeable rock has more space between grains than does impermeable rock. If the oil is thin enough, pressure will force it to slowly flow through permeable rock and into the borehole of a well, from which it can be brought to the surface.

The hydrocarbons in oil shale cannot flow because they are in a solid form called "kerogen." Kerogen is a waxy, insoluble organic compound with a large molecular structure. Almost all sedimentary rocks contain some kerogen; those that both contain kerogen and yield a few liters of oil per metric ton are considered oil shale. About 45 liters per tonne generally is the minimum figure used for calculating reserves. Ninety to 135 liters are required for development. Some shales contain more than 225 liters per metric ton. Worldwide shale oil resources have been estimated at 3 trillion barrels, of which the United States has 2 trillion.

When some types of kerogen are heated by natural processes (60° to 150° Celsius, or 140° to 300° Fahrenheit) they slowly release crude oil and natural gas. In place of nature's slow process, oil shale may be mined and then burned as a low-grade fuel, or it can be crushed and then heated in a retort to produce crude oil and natural gas. Retorting above ground is usually carried out between 300° and 520°C (570° and 970°F). It is also possible to process shale underground (*in situ*, or in place) by fracturing the rock underground and then using fire or an-

Mining of oil shale. Photo by Kaupo Kikkas, CC BY-SA 4.0, via Wikimedia Commons.

other heat source underground to produce crude oil, which can be pumped to the surface. In place of fire, steam or other hot fluid can be circulated underground. After retorting, the oil is usually upgraded through a process in which it is passed over a catalyst and treated with hydrogen gas to increase the hydrogen content and remove impurities such as sulfur, nitrogen, iron, and arsenic.

ORIGIN OF OIL SHALE

Oil shale forms in oxygen-deficient environments where organic debris accumulates more rapidly than it is destroyed by oxidation, scavengers, or decay. Deep, confined ocean basins with stagnant water or restricted water circulation may preserve organic debris. Baltic and Manchurian oil shales are of this origin. Swamp lakes, with slow circulation and rapid accumulation of plant debris, also may produce oil shale. Oil shale accompanying coal in Scotland and North America are examples. Lakes with noncirculating water at the bottom also may accumulate oil shale. The gigantic Green River oil shale deposit of Wyoming and Colorado is of this type.

PRODUCING OIL FROM OIL SHALE

hale oil is very expensive to produce, although expected increases in production and improvements in the process are projected to bring its cost down significantly over time. The process also has environmental costs. Processing oil shale releases carbon dioxide, which has been linked to global warming. Very great amounts of shale are needed for economically significant petroleum production. Open-pit mining is much more economical than underground mining, although there are some large underground mines. Near-surface deposits of oil shale are open-pit mined or strip-mined, both processes that damage surrounding ecosystems; underground mining uses the less efficient room-and-pillar method. Pits 300 meters deep and 3 kilometers across have costs equal to those of underground mining in the Green River deposit. Further expansion reduces expense, making gigantic pits the most economical mining option. Heating shale in the absence of oxygen (retorting) converts kerogen (solid organic material that is insoluble in petroleum solvents) to liquids. The liquid then requires hydrogenation to make petroleum. In any kind of mining of oil shale, one to five barrels of water are used for each barrel of shale oil produced; in addition, the process produces mountains of crushed rocks that require disposal and management.

Retorted shale is saline and/or alkaline powder. Open pits are ready disposal sites for waste material, and underground mines may be backfilled. The remaining 10 percent or more of retort waste, however, requires disposal elsewhere. Finally, the waste must be isolated from surface water and groundwater to prevent contamination. *In situ* processing solves some disposal problems. In this method large blocks of oil shale are undermined and collapsed, creating an underground porous rubble. Gas introduced to the top of the rubble is ignited, after which the shale burns on its own. Gas and oil "cooked" from the rock are withdrawn from the base of the rubble, leaving the spent shale underground.

SHALE OIL HISTORY

Shale oil for medicinal purposes was produced in 1350 at Seefeld, Austria. The manufacture of illuminating oil and lubricants from oil shale began in France around 1830 and quickly spread through Europe and North America. Petroleum almost entirely supplanted shale oil during the late nineteenth century. Afterward, shale oil production was largely limited to periods of oil shortage or of military or economic blockade. Flammable oil shale, rich in kerogen, has been burned to generate steam in Latvia. Scotland, the former Soviet Union, Manchuria, Sweden, France, Germany, South Africa, the United States, Brazil, and Australia all have produced shale oil, but total world production through 1961 was only about 400 million barrels.

The world's use of shale oil peaked in 1980, several years after the 1973 OPEC oil embargo. In 2008 Estonia was the top producer at 5,500 barrels per day, followed by Brazil at 3,100 barrels per day. It has been estimated that worldwide deposits of oil shale could produce 2.8 to 3.3 trillion barrels of shale oil. The United States has the largest share, with enough oil shale to produce 1.5 to 2.6 trillion barrels in the Green River Formation, located in parts of Utah, Wyoming, and Colorado. In comparison, the world's conventional oil reserves are estimated at 1.317 trillion barrels.

Different grades of oil shale yield different amounts of oil. About 16 percent of the Green River Formation yields 95 to 380 liters (25 to 100 gallons) of oil per ton of shale, about 33 percent yields 38 to 95 liters (10 to 25 U.S. gallons), and the final 51 percent yields 19 to 38 liters (5 to 10 U.S. gallons) per ton of shale. Unable to compete with cheap conventional oil, oil shale mining ceased in the United States in 1982, but in 2003 an oil shale development program was restarted, now incorporating a recovery process commonly called "fracking." In fracking, the oil shale structure is fractured in place and the subsequent injection of high-pressure steam "cracks" the oil from within the structure of the shale rock (fracturing + cracking = fracking), allowing it to flow out for recovery. The method is contentious, and the destabilization of the underground shale structure has been associated with a number of negative effects, including groundwater contamination and increased earthquake activity in areas where fracking has taken place.

TAR SANDS

Tar sands (also called "oil sands") are mixtures of sand, clay, and bitumen, a tarlike form of petroleum. The largest deposits of tar sands are in Canada and Venezuela, each country having the equivalent of the world's reserves of crude oil. Canada's production of oil from tar sands reached 1.25 million barrels per day in 2009 (at a cost of about $27 per barrel) and provided 47 percent of the nation's oil production. This allowed Canada to become the biggest supplier of products to the United States. In Canada about two tons of tar sand are processed to produce one barrel of oil. Producing oil in this way has the same environmental problems as producing oil from oil shale: Strip mining degrades ecosystems, and the process releases more carbon dioxide into the atmosphere than does the production of conventional oil, uses large amounts of water, and leaves tons of spent sand that require disposal and management.

Tar sandstone. Photo by James St. John (jsj1771), CC BY 2.0, via Wikimedia Commons.

Bitumen may be extracted *in situ* through the injection of steam, hot air, or solvents to allow the bitumen to flow. Alternatively, a deposit may be strip-mined and the materials then hauled to a processing facility, where the bitumen is separated from water, sand, and other waste. Like shale oil, the heavy crude extracted from bitumen must be upgraded through a process in which it is passed over a catalyst and treated with hydrogen gas to increase the hydrogen content and remove impurities such as sulfur, nitrogen, and arsenic. It can then be used as feedstock for conventional refining.

The only significant tar sands in the United States are in eastern Utah. The Utah tar sand deposit is far smaller than Canada's, and the Utah sands are hydrocarbon-wetted rather than water-wetted as the Canadian sands are, so their use would require different processing techniques.

TAR SAND OCCURRENCE

Tar sand bitumens are larger, heavier, and more complex hydrocarbon compounds than those in liquid petroleum, and they include substantial amounts of nitrogen and sulfur. ("Bitumen" is a term for a very thick, natural semisolid material such as asphalt or tar.) Deposits are most abundant in association with sandstone or limestone. Most large deposits are in deltaic, estuarine, or freshwater sandstone. The largest occur at depths of less than 1,000 meters on sedimentary basin margins where inclined layers of petroliferous rocks approach the surface. Here, upward migrating petroleum could lose volatiles and, with oxygenation and biodegradation, leave asphalt-impregnated rock. Some solid bitumens may be hydrocarbons not yet sufficiently altered to form liquids rather than residues of once liquid material.

United States reserves, although large, are insignificant compared to those of Canada and Venezuela. Additional deposits are known in Albania, Siberia, Madagascar, Azerbaijan, the Philippines, and Bulgaria.

TAR SAND HISTORY

Tar sands have been used since ancient times for surfacing roads, laying masonry, and waterproofing. The Athabasca tar sand deposit of northern Alberta, Canada, was discovered in 1778 when Peter Pond, a fur trader, waterproofed his canoes with tar. Geologic exploration began in the 1890s, and by 1915 tar sand was being shipped to Edmonton, Alberta, to pave streets. Pilot plant extraction of oil began in 1927. Afterward, exploitation continued with provincial and federal subsidies and support. By the end of the twentieth century, operations were self-supporting.

TAR SAND EXPLOITATION

Canadian tar sand is mined in large open pits and transported to processing plants where steam treatment produces bitumen froth and sand slurry. Naphtha steam removes the remaining sand, leaving viscous bitumen. Raw bitumen then is "cracked," a chemical process by which the large organic molecules in the bitumen are broken into smaller, more liquid molecules, gas, and coke. Finally, cracked oil is hydrogenated to produce synthetic crude oil. Sulfur is a salable byproduct. Sand ultimately is returned to the pit, overburden is replaced, and the site is reforested.

Open-pit production, however, is feasible only at the shallow periphery of the deposit, so *in situ* extraction will be required for about 90 percent of the Canadian deposit. In one system, wells drilled into the deposit are injected with steam to liquefy the bitumen. Bitumen then is pumped until flow ceases, after which the well is again steamed. In another system wells sunk into the tar sand are ignited. Heat then cracks the bitumen, producing liquid and gas that flow to production wells.

—Charles W. Rogers and Ralph L. Langenheim Jr.

Further Reading

Bartis, James T., et al. *Oil Shale Development in the United States: Prospects and Policy Issues*. RAND Corporation, 2005.

Chastko, Paul. *Developing Alberta's Oil Sands: From Karl Clark to Kyoto*. U of Calgary P, 2004.

Clarke, Tony. *Tar Sands Showdown: Canada and the New Politics of Oil in an Age of Climate Change*. J. Lorimer, 2008.

Marsden, William. *Stupid to the Last Drop: How Alberta Is Bringing Environmental Armageddon to Canada (and Doesn't Seem to Care)*. Vintage Canada, 2008.

Nikiforuk, Andrew. *The Tar Sands: Dirty Oil and the Future of a Continent*. Greystone, 2009.

Russell, Paul L. *Oil Shales of the World: Their Origin, Occurrence, and Exploitation*. Pergamon Press, 1990.

Selley, Richard, and Stephen A. Sonnenberg. *Elements of Petroleum Geology*. 3rd ed., Elsevier, 2015.

Speight, James G. *Enhanced Recovery Methods for Heavy Oil and Tar Sands*. Gulf, 2009.

OLDHAM AND MOHOROVIČIĆ DISCOVER EARTH'S INTERIOR STRUCTURE

Fields of Study: Geology; Seismology; Geophysics

ABSTRACT

Richard Dixon Oldham, Andrija Mohorovi Mohorovičić, and other seismologists, using data from earthquakes, revealed the layered internal structure of Earth.

KEY CONCEPTS

continental drift: the theory proposed by Alfred Wegener to account for the apparent "fit" of the continental masses with each other

contraction theory: the erroneous theory that Earth's mountains and valleys are the result of the outer crust buckling and folding due to shrinkage as the planet cooled from some primordial molten state

focus: the exact location within the crust at which an earthquake originates

Mohorovičić Discontinuity: the boundary layer in which the hard rock of the crust becomes the plastic layer of the upper mantle

P-wave: the compression wave produced in an earthquake as tectonic segments impact each other at the focus
S-wave: the laterally oscillating wave produced in an earthquake as tectonic segments rebound to adjust to their new equilibrium positions
surface wave: the wave produced in an earthquake as a reflection of the S-waves, felt only at the surface

HISTORICAL VIEWS OF EARTH'S INTERIOR

Scientists must rely on indirect evidence to create a picture of the deep interior of Earth. This picture has thus changed over time as new evidence has become available. Only in the twentieth century, with the development of new measurement techniques, did a clear picture of Earth's interior structure begin to emerge.

The preeminent geological text of the latter half of the seventeenth century, German Jesuit and scholar Athanasius Kircher's *Mundus Subterraneus* (1664; *The Subterranean World*), described Earth with a fiery central core from which emerged a web of channels that carried molten material into fire chambers or "glory holes" throughout the "bowels" of Earth. The "contraction" theory, popular through the first half of the nineteenth century, held that Earth had formed from material from the Sun and had since been slowly cooling, creating a solid crust. As the fluid cooled, it would contract, and the crust would collapse around the now smaller core, causing earthquakes and producing wrinkles that formed the surface features of mountains and valleys.

Physicists in the 1860s, however, pointed out several physical consequences of the contraction model that appeared to be violated. For example, the gravitational force of the Moon, which creates tides in the ocean, would produce tides also in a fluid interior. The resulting effect should either cause the postulated thin crust to crack and produce earthquakes whenever the Moon was overhead or, if the crust were sufficiently elastic, cause tides in the crust as well. Instruments were developed to detect crustal tides, but no such tides were discovered.

By the beginning of the twentieth century, geologists were forced to abandon the contraction theory and conclude that Earth probably was completely solid and "as rigid as steel." Thus, when the German geophysicist and meteorologist Alfred Wegener suggested the idea of "continental drift" in 1912, it was dismissed on this basis as being physically impossible. Ironically, the very measuring instruments that overturned the simple liquid interior contraction theory would, in turn, disprove the solid-earth view that had challenged and replaced it. Seismographs—such as the inverted pendulum seismograph invented in 1900 by German seismologist Emil Wiechert—would provide a completely new source of information about the internal structure of Earth.

EARTHQUAKE WAVES

In 1900, Richard Dixon Oldham published a paper that established that earthquakes give rise to three separate forms of wave motion that travel through Earth at different rates and along different paths. When an earthquake occurs, it causes waves throughout Earth that fan out from the earthquake's point of origin or "focus." Primary (P) waves emanate like sound waves from the focus, successively compressing and expanding the surrounding material. Whereas P-waves can travel through gases, liquids, and solids, secondary (S) waves can travel only through solids. S-waves travel at about two-thirds the speed of P-waves. Surface waves, the third type of seismic wave, travel only near Earth's surface.

Seismologists now speak of seismic data as being able to provide an X-ray picture of Earth. In his groundbreaking 1906 article "Earth's Interior as Revealed by Earthquakes," Oldham analyzed worldwide data on fourteen earthquakes. He also observed that within certain interior earth zones, P-waves behaved differently from what he expected. In 1912, Beno Gutenberg was able to establish that at a certain depth the velocity at which the seismic waves traveled changed sharply. He estimated this depth to be about 2,900 kilometers (roughly 1,802 miles—almost half of Earth's radius).

Oldham had also recognized in his own data the suggestion of a thin outer crust, but his data were insufficient to determine its depth. This estimation was to be made by Andrija Mohorovičić, a professor at the University of Zagreb, from an analysis (published in 1910) of an earthquake that had hit Croatia's Kulpa Valley in late 1909. Mohorovičić had noticed that at any one seismic station, both P- and S-waves from the earthquake appeared in two sets, but the time between the sets varied according to how far away the station was from the earthquake's focus.

Based on these different arrival times, Mohorovičić reasoned that waves from a single shock were taking two paths, one of which traveled for a time through a "faster" type of rock. He calculated the depth of this change in material. His estimate fell within the now-accepted figure of 20 to 70 kilometers (12.4 to 43.5 miles)—the crustal thickness varies under the continents and shrinks to 6 to 8 kilometers (about 3.7 to 5 miles) under the ocean. This boundary between crust and mantle is now called the "Mohorovičić Discontinuity," or the "Moho."

SIGNIFICANCE OF DISCOVERING EARTH'S INTERIOR STRUCTURE

The use of seismic data to plumb Earth's interior produced an outpouring of research, both theoretical and experimental. By the mid-1920s, seismologists had learned to make subtler interpretations of their data and realized that no S-waves had been observed to penetrate through the core. Given that S-waves cannot pass through liquid, there was reason to think that the core was liquid.

Independent support for this hypothesis came from rigidity studies. Seismic waves could be used to determine the rigidity of the mantle; it was revealed to be much greater than the average rigidity of Earth as a whole (the existence of a low-density fluid region would account for this discrepancy). In 1936, however, Inge Lehmann was able to show that P-waves passing close to Earth's center changed velocity slightly, and she correctly inferred the existence of an inner core to account for this phenomenon. The current picture of Earth now includes a liquid outer core surrounding a solid inner core about 1,200 kilometers (745.6 miles) in radius, surrounded by the lower and upper mantle layers, the asthenosphere, and the crust.

—Robert T. Pennock

Further Reading

Fowler, C. M. R. *The Solid Earth: An Introduction to Global Geophysics*. 2nd ed., Cambridge UP, 2005.

Lowrie, William. *Geophysics: A Very Short Introduction*. Oxford UP, 2018.

Stein, Seth, and Michael Wyesession. *An Introduction to Seismology, Earthquakes, and Earth Structure*. Wiley, 2013.

Tarbuck, Edward J., Frederick K. Lutgens, and Dennis Tasa. *Earth: An Introduction to Physical Geology*. 13th ed., Pearson Education, 2019.

Turcotte, Donald, and Gerald Schubert. *Geodynamics*. 3rd ed., Cambridge UP, 2014.

Yang, Wencai. *Reflection Seismology: Theory, Data Processing and Interpretation*. Elsevier, 2014.

OPHIOLITES

Fields of Study: Geology; Geophysics; Structural Geology

ABSTRACT

Ophiolites are pieces of oceanic crust and upper mantle that have been thrust up above continental crust. They contain a wide range of minerals.

KEY CONCEPTS

convergence: the condition in which tectonic plates are driven toward each other where their margins are in contact with each other

divergence: the condition in which tectonic plates are driven apart from each other from the locations where their margins meet

obduction: the process occurring in convergence when a tectonic plate that would normally be subducted is instead folded and buckled so that a significant portion of it is driven upward rather than downward

sheeted dyke complex: a series of essentially parallel vertically-oriented magma intrusions into fractured overlying rock

subduction: the condition in which a tectonic plate, typically of dense seafloor basalt, is forced to go downward into the mantle beneath a less dense continental mass due to seafloor spreading

DEFINITION OF OPHIOLITES

The process that forms ophiolites is not well understood, but is thought to occur where continental crust is bent down and subtends oceanic crust. There is debate as to whether this occurs generally in a subduction zone or as an artifact of tectonic plate convergence. Ophiolites consist of a vertical sequence of (from bottom to top) mantle rocks, gab-

Ordovician ophiolite in Gros Morne National Park, Newfoundland. Photo by Jcmurphy, CC BY 3.0, via Wikimedia Commons.

bro, sheeted dykes, and pillowed lavas. Ophiolites are remnants of ancient ocean basins, demonstrating that an ocean basin once existed in the area and that plate convergence has destroyed the ocean basin. Ophiolites generally define and decorate suture zones, places where once-separated continental blocks have collided. Ophiolites host a wide range of minerals, including chromite, platinum-group elements, and gold. Ophiolites are also important from a natural resources' perspective because the tectonic forces that have put them in place often also form sedimentary basins that can contain fossil fuel deposits (oil, gas, and coal).

OVERVIEW OF OPHIOLITES

The term "ophiolite" comes from the Greek word *ophis*, meaning snake or serpent; the term's origin is similar to that of the rock called "serpentinite." Both terms refer to the mottled green (reptilian) appearance of these rocks. Ophiolites are sequences of oceanic crust and upper mantle that have been emplaced on continental crust by a process known as obduction. Obduction often faults, folds, and otherwise disrupts the original sequence of rocks. Nevertheless, ophiolites provide the best-known example of the structure of oceanic crust.

The typical ophiolite sequence is on the order of 5 kilometers thick, and it records the products of seafloor spreading. The top kilometer is composed of pillowed basalt. (Basalt is a dense black lava containing about 50 percent by weight silicon dioxide formed as a result of melting of the mantle.) A sequence of these basalts looks like a huge pile of black to green pillows. Pillows form when basalt lava erupts and is quenched underwater, perhaps 3 kilometers or more beneath the surface. These basalt flows are fed from a system of vertical magma channels known as a sheeted dyke complex (about 1 kilometer thick). The dykes are supplied with magma from an underlying magma chamber, which upon cooling forms gabbro (about 3 kilometers thick). Gabbro is the base of the crustal section and is underlain by mantle rocks called "peridotite."

Although there is wide agreement that ophiolites form by seafloor spreading, scientists do not agree on precisely where these form. Early ideas, that ophiolites formed at true mid-ocean ridges such as the East Pacific Rise or Mid-Atlantic Ridge, have been superseded by the idea that they form at convergent margins undergoing extension. A more recent idea, that ophiolites for the most part represent back-arc basin crust, has been challenged by the hypothesis that ophiolites form in forearc environments during the initiation of subduction zones.

Ophiolites prove that processes of seafloor spreading and plate tectonics operated at the time the ophiolites formed. Therefore, the record of ophiolites can be interpreted as a record of when plate tectonics occurred on Earth. Ophiolites are common in rock sequences of the following time ages: 1,800 million years ago, 800 to 400 million years ago, and younger than 180 million years ago.

—Robert J. Stern

Further Reading

Dilek, Yildrim, and Paul T. Robinson, editors. *Ophiolites in Earth History.* The Geological Society, 2003.

Ghosh, S. K. *Structural Geology: Fundamentals and Modern Developments.* Elsevier, 2013.

Moores, Eldridge, and Robert J. Twiss. *Tectonics.* Waveland Press, 2014.

Twiss, Robert J., and Eldridge M. Moores. *Structural Geology.* W.H. Freeman, 2006.

Wright, James E., and John W. Shervais. *Ophiolites, Arcs, and Batholiths: A Tribute to Cliff Hopson.* The Geological Society of America, 2008.

ORBICULAR ROCKS

Fields of Study: Geology; Geochemistry; Geophysics

ABSTRACT

Orbicular rocks are a type of intrusive igneous rock deposit, occurring primarily at the outer margin of plutonic dikes, which are cooled from magma. Orbicular rocks contain inclusions called "orbicules" that are composed of concentrated minerals layered in concentric spheres around a central core. Most orbicular deposits have been uncovered from geologic strata in Finland, though they have been found on every continent.

KEY CONCEPTS

biotite: type of layered, silicate mineral composed of silica molecules bonded with magnesium and iron and appearing in dark brown and black varieties

carbonatite: silicate mineral consisting of reduced levels of silica and rich in carbonate molecules formed from carbonic acid

dike: type of plutonic rock formation consisting of cooled magma oriented perpendicular to the earth's surface and generally proceeding at a diagonal through the layers of surrounding parent rock

felsic: mineral group characterized by high proportions of silicate minerals, including quartz, feldspar, plagioclase, and micas

hornblende: type of silicate rock characterized by its dark color, which is the result of inclusions of iron and magnesium within the rock matrix

mafic: igneous deposits of rocks consisting of silica enriched with high levels of iron, magnesium, sodium, and calcium and with low levels of potassium-rich feldspar

orbicule: formation within orbicular igneous rocks consisting of similar materials to the parent rock; organized in concentric layers of crystals surrounding a core comprising smaller crystal clusters, larger individual crystals, or fragments of other rock types

petrology: branch of geology that studies the formation and classification of rock types

plagioclase: silicate mineral formed from molecules of silica bonded with sodium, aluminum, and calcium

plutonic: rock formed from magma that cools in pockets beneath Earth's surface

OCCURRENCE OF ORBICULAR ROCKS

Orbicular rocks are igneous rocks that contain orb-like formations as inclusions within the rock body. Igneous rocks form when magma generated within the mantle layer of Earth rises toward the surface and cools to form solid structures. Orbicular rocks form within intrusive or plutonic igneous rock bodies, which are deposits that form when magma solidifies beneath Earth's surface rather than breaking through the crust to become lava.

Orbicular rocks can form within a variety of different types of intrusive rock deposits and most commonly form from magma that is high in silica, a molecule composed of silicon and oxygen that makes up the vast majority of most igneous rock. Orbicular rocks have been found in dikes, which are intrusive rock formations that form when magma rises toward the surface perpendicular to the existing rock, forming a horizontal or vertical wedge within the parent rock.

The vast majority of orbicular rocks come from small dikes, and orbicular sections of the deposit are generally found to occur at the margin of the igneous deposit. Orbicular rocks generally occur only within portions of an igneous deposit that are 3 to 5 meters (10 to 16.5 feet) in width and less than 5 m in depth. Many orbicular rocks have been discovered in isolated boulders that are the remains of ancient igneous dikes that have eroded and split into individual rocks.

In rare situations, orbicular rocks have been found in sills, which are horizontal deposits of igneous rock that form within layers of the parent rock and are generally oriented parallel to Earth's crust. Alternatively, orbicular rocks may be part of larger intrusive deposits called "batholiths," which result from the combination of subterranean magma pockets and that may extend for many kilometers beneath Earth's surface.

Orbicular rocks have been found in more than one hundred locations worldwide, but they are a relatively rare type of igneous formation; the exact geologic setting required for orbicular rock formation is still poorly understood. The vast majority of orbicular rocks have been uncovered in igneous deposits from Europe, specifically in Finland. This is not believed to represent a unique feature of Fin-

Orbicular granite sourced from a stream in New Zealand. Photo by brewbooks, Seattle, CC BY-SA 2.0, via Wikimedia Commons.

land's geologic environment but rather is the result of dedicated efforts by Finnish geologists to complete detailed surveys of that country's mineralogical terrain. Significant quantities of orbicular rocks also have been uncovered from igneous deposits in the United States and in Indonesia.

CHEMICAL COMPOSITION OF ORBICULAR ROCKS

In general, the chemical petrology of orbicular rocks is similar to that of the igneous formations in which the rocks have been discovered. Morphological variations in the orbicular deposits are therefore related to the concentration and depositional orientation of the molecules within the formation, rather than to the presence of unique molecules not found in the surrounding parent rock.

Igneous rocks are classified according to both the chemical structure of the molecules found within the rock and the overall size of grains within the rock. Granitic rock is one of the most common types of igneous rock that contains orbicular inclusions. Granitic rock has visible grains and is composed largely of minerals derived from the molecule silica. Granitic rocks are classified as felsic, which are silicate rocks containing large quantities of feldspar, a common type of silicate mineral. Felsic rocks also contain other types of silica-based minerals, such as quartz and micas. Granitic orbicular rocks are the most common type of orbicular inclusion.

Orbicular deposits also can be found in mafic rocks, which are igneous rocks that contain a higher proportion of heavier elements, such as magnesium and iron. Unlike felsic rocks, mafic rocks contain lower levels of potassium but have higher levels of calcium and sodium along with iron and other metal deposits. Some igneous rocks are ultramafic, meaning they contain less than 45 percent silica, with the rest of the rock consisting largely of heavier elements like iron. In one rare occurrence, orbicular inclusions formed within a sample of carbonatite found in Finland. Carbonatite is a type of igneous rock containing high proportions of carbonate minerals, a molecular salt formed from carbonic acid.

Most orbicular rock deposits that are found are made of feldspar, one of the most common types of rock-forming minerals, accounting for more than 60 percent of Earth's crust. Feldspar is composed largely of silica molecules, with such elements as potassium, calcium, and sodium mixed into their matrix. Feldspar samples tend to be either rich in potassium, called "K-feldspars," or rich in sodium or calcium.

Plagioclase is a type of feldspar rich in either calcium or sodium, with low potassium content. Plagioclase is one of the most common types of rock and is found in a variety of igneous deposits. Orbicular rocks also may contain quartz, which is a crystalline variety of silica that generally forms in superheated aqueous environments or within magma pockets.

Orbicular rocks contain inclusions of dark minerals within their matrices, which may represent different mineral types. Among the most common mineral inclusions in orbicular rocks is hornblende, which is a dark form of amphibole, a type of mineral consisting of silica molecules joined to either iron or magnesium atoms. Hornblende occurs in two basic varieties—one that is rich in iron, called "ferrohornblende," and one that is rich in magnesium, called "magnesiohornblende."

Alternatively, orbicular rocks may contain biotite, which is a type of silicate rock from the group of minerals known as micas. The dark color of biotite is derived from iron atoms bonded to the silicate molecules within the rock. Biotite also may contain sheets of magnesium and aluminum atoms within the rock, further contributing to the mineral's dark color.

STRUCTURE OF ORBICULAR ROCKS

Orbicular plutonic rocks contain orb-like nodules of mineral called "orbicules," which contain layers of minerals in a radial array surrounding an internal core, sometimes called the "nucleus." The outer portion of an orbicule consists of concentric circles of minerals called "shells."

The number of shells within an orbicule varies, and some orbicules may have only a single shell while others may have hundreds of shells surrounding the core. Between the shells are layers of minerals called "subshells," which contain molecules similar to those of the surrounding rock but that may be organized differently.

The central core of an orbicule may consist of a cluster of small crystals, usually composed of quartz or a similar silica-rich mineral. Alternatively, some orbicules have cores consisting of one or a few large crystals, again generally derived from the quartz family of minerals. Some orbicular cores consist of fragments of parent rock, called "xenoliths," which may differ from the type of rock contained within the remaining part of the orbicule, including the minerals of the shells and subshells.

Outer shells are usually composed of feldspar or plagioclase and therefore appear as lighter segments of rock. The portion between shells consists of plagioclase or granite with some darker minerals mixed into the matrix, making these areas somewhat darker than the outer shells. Subshells generally consist of plagioclase bonded to hornblende or biotite, making them darker than the outer shells or the area between shells. The typical orbicule therefore consists of alternating bands of light- and dark-colored rock surrounding the inner core, which tends to be light in color and to consist of quartz crystals or other silicate minerals. Many orbicule cores contain small xenoliths that generally appear as dark flecks or nodules within the crystals of the core.

The size of orbicules is related to the type of igneous rocks in the surrounding deposit. Large orbicules tend to develop within felsic rocks containing high proportions of quartz and feldspar. Orbicules in this type of environment can range from 10 to 40 centimeters (4 to 16 inches) in diameter. Mafic igneous deposits generally contain smaller orbicules; similarly, ultramafic rocks contain the smallest and least developed orbicules.

FORMATION OF ORBICULES

Several hypotheses explain the origin of orbicules within igneous formations. Initially, the bands of various minerals were thought to be related to properties of liquid diffusion that affect the formation of mineral layers during the solidification process.

An alternative theory, developed in the 1970s, sees orbicules as the result of xenoliths of other rock types or crystals becoming part of (included in) a flow of magma. When this occurs, the magma surrounding the included segments cools such that "comb layering" occurs, in which crystals form like the teeth of a comb perpendicular to the surface of the inner core. Alternating layers within the orbicules are then seen as the result of successive periods of cooling, leading to comb layering of minerals around the forming nodule.

While comb layering appears to play a role in the formation of the shells and subshells contained within orbicules, chemical investigation has found that not all orbicules contain xenoliths within their core. Theories of orbicule genesis focus on the phenomenon of undercooling, which occurs when a sample of magma rapidly cools to temperatures significantly below those needed for solidification. In undercooling, crystals may form within the body of the magma and serve as attractors for layers of cooling material that form comb layers around the crystal structure. As this process continues, chemical properties within the mineral components cause them to layer into discrete segments of different types of minerals.

The orientation of orbicules within the matrix of an igneous deposit indicates that orbicules form while the surrounding rock is still cooling and that orbicules can shift within the rock matrix after formation. In some cases orbicules have been found where the outer shell has broken apart, perhaps after encountering disruptive contact with other orbicules as the inclusions shifted within the pliable rock matrix.

In general, the processes underlying the formation of orbicules is still an area of concerted research. The precise physical and chemical conditions required for the formation of orbicules appear to be relatively narrow in scope, as orbicular rock formations are rare among igneous deposits. There appears to be a narrow range of temperatures and pressures during which orbicule formation is possible, which explains why orbicules are relatively rare and why they tend to form only in certain portions of a developing dike, where temperatures fall within this range.

USES OF ORBICULAR ROCKS

Orbicular rocks contain attractive patterns of light and dark concentric circles, so they have been prized by rock collectors and are often used as decorative stones. Some orbicular rocks have been used as an alternative to marble for making counters and tabletops. A variety of collectible minerals can be made from polished and cut orbicular rocks.

Because of their rarity, orbicular rocks cannot be collected by amateurs in most locations in which they are found. In Finland, for instance, where the largest variety of orbicular rocks have been found, provisions prohibit removing orbicular specimens from rock quarries, thereby preserving the remaining orbicular deposits for geologic research.

—*Micah L. Issitt*

Further Reading

Dahl, P. S., and D. F. Palmer. "The Petrology and Origin of Orbicular Tonalite from Western Taylor Valley, Southern Victoria Land, Antarctica." *Antarctic Earth Science*, edited by R. L. Oliver, P. R. James, and J. B. Jago. Cambridge UP, 2011.

Fenton, Carol Lane, and Mildred Adams Fenton. *The Rock Book.* Dover, 2003.

Gill, Robin. *Igneous Rocks and Processes: A Practical Guide.* Wiley-Blackwell, 2011.

Lahti, Seppo L., Paula Raivio, and Ilka Laitakari, editors. *Orbicular Rocks in Finland.* Geological Survey of Finland, 2005.

Vernon, Ron H. *A Practical Guide to Rock Microstructure.* Cambridge UP, 2004.

P

Paleontology

Fields of Study: Geology; Paleogeology; Paleontology; Geochemistry; Geomorphism; Paleobiology

ABSTRACT

Paleontology is an interdisciplinary field that is concerned with the study of the record of life through time and the application of that information to solve scientific problems. Fossils are particularly used in the relative dating of rocks, which is important information for geologists. Paleontology also provides valuable data toward understanding past life and therefore of extinction events and the effects of climate change. Modern technology is increasingly used to help solve problems, including computed tomography (CT) scanning to reconstruct the anatomy of extinct organisms.

KEY CONCEPTS

biostratigraphy: use of fossil animals and plants in the dating of rocks

cladistics: computer-based method for delineating relationships among groups of organisms

evolution: change through time in successive generations of organisms

extinction: elimination of a taxon of organisms

fossil: remains of a once-living organism or a trace of its behavior

fossilization: process of preserving the remains of organisms within sediment

ichnology: study of trace fossils

paleobiology: approach to paleontology that stresses the understanding of processes

paleoecology: study of the interaction between organisms and their environment through time

systematics: study of the relationships among organisms

taphonomy: study of the processes that affect organisms between death and burial

zone fossil: organism that characterizes a particular period of time

DEFINITION AND BASIC PRINCIPLES OF PALEONTOLOGY

Paleontology is the study of fossils, or once-living organisms whose remains have been preserved within sediment. The term "paleontology" is derived from the Greek *palaios* for "ancient," *ontos* for "being," and *logos*, for "study." Therefore, paleontology does not encompass the allied field of archaeology and overlaps only slightly with anthropology. Traditionally, the study of fossils has been aimed at reconstructing the organisms and analyzing them to develop an understanding of their evolutionary relationships. This has been enhanced by the advent of "cladistics," a computer-based analytical method that greatly increases the speed and rigor of such studies. In addition, the careful tabulation of fossils over many years of study has enabled the development of a relative geological timescale that is the basic tool in all study of Earth's crust. Paleontology is an important industrial tool in that it helps extractive industries such as the petroleum and minerals industries understand where to find their respective resources.

Increasingly, paleontology is being viewed as a potentially valuable predictive tool. The results of earlier climate changes are preserved in the fossil record and may help scientists understand the effects of global warming or other future climate changes.

BACKGROUND AND HISTORY OF PALEONTOLOGY

Fossils were probably picked up and admired long before they were studied, but the Greeks recognized that some of these objects represented marine organisms and could be used to determine where oceans had once been. In medieval times, fossils were often viewed as magical objects. For example, fossil shark teeth were termed *glossopetrae*, or "tongue stones," and were thought to provide pro-

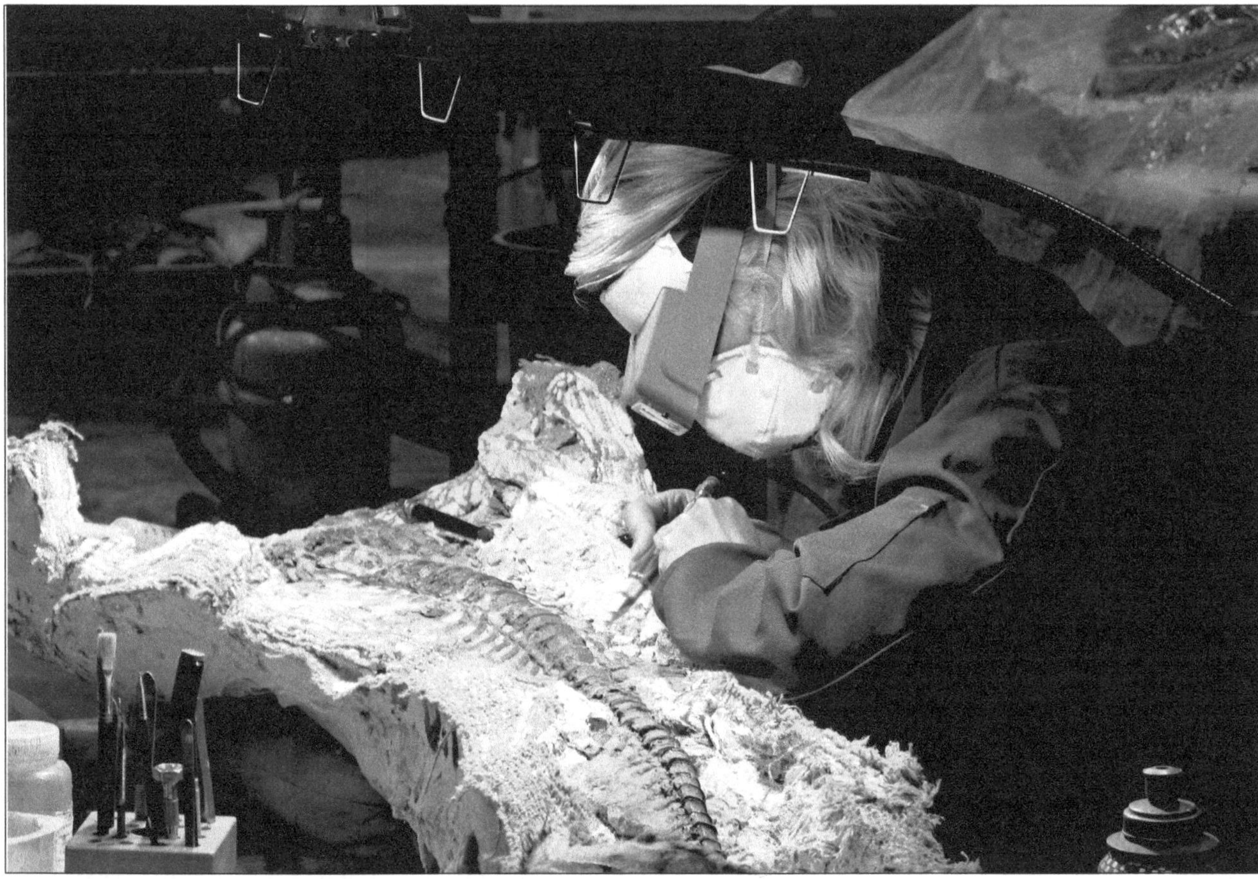
A paleontologist working on a dinosaur fossil. Photo via iStock.com/benedek [Used under license.]

tection from snakebite. They were viewed as structures that had "grown" within the rocks rather than the remains of actual organisms. In the seventeenth century, Nicolaus Steno demonstrated that *glossopetrae* were shark teeth by dissecting a modern shark and showing the similarity of its teeth to the fossils. At about the same time, Robert Hooke used early microscope technology to show that fossils and modern wood and shells had the same internal structure.

Once fossils were accepted as representing past life, they could be studied as such. In some sequences, scientists found clear evidence of extinction events in which whole faunas were replaced over relatively short periods of time. In the early nineteenth century, work by French naturalist Georges Cuvier in the Paris basin showed that numerous mammals that lived when those rocks were laid down no longer existed. Studies of this type showed that long periods of time had been necessary for the evolution and extinction of groups of fossil organisms. At about the same time, English canal surveyor William Smith determined that fossils could be used as time markers, representing the period they were present on Earth. His findings were developed into the geological timescale, which is still being refined but provides a basis for dating rocks in a relative sense.

RELATIVE GEOLOGIC DATING

Understanding the spatial relationships of rocks developed from the need to predict where materials of economic value could be found. Initially, distinct rock beds were used to correlate between places, but this method was imprecise over long distances or in areas where the rocks were folded and faulted. Rock units could not be dated effectively until Smith realized that the geological succession corresponded to a regular, nonrepeating succession of fossils. Fundamental to this process of biostratigraphy is the use of zone fossils that represent a period of time equal to the duration of the species before it became extinct and that are widespread enough to enable useful correlations to be made. Increasingly, microscopic fossils (such as the marine, planktonic, and unicellular foraminifera) are used as they have the advantage of being very abundant in sediments and of having short time ranges, which allows for accurate dating.

TAPHONOMY

The quality of the fossil record must be understood for it to be useful in applications such as dating or understanding mass extinctions or biodiversity loss. Taphonomy, which is the study of the processes that affect organisms from death to burial in sediment, is integral to such an understanding as it allows an evaluation of the extent to which the fossil record mirrors the original faunal situation. Most organisms do not become fossils, and those that do will first undergo decay of the soft tissues, then transport and breakage of the hard tissues, and finally their burial and modification. For example, soft-bodied organisms can make up 50 to 60 percent, or more, of the organisms in present-day marine settings, and essentially none of those organisms will ever become fossilized. Experimentation on modern forms shows that the presence of oxygen is vital to the initial breakdown of soft tissues, and its absence can lead to conditions of special preservation that result in unusually well-preserved faunas such as the Burgess Shale fauna (a Middle Cambrian site in Canada). The breakage and disarticulation of organisms can be simulated in the laboratory or observed in nature. These studies show that organisms go through a number of stages as they progress toward fossilization. Factors that affect whether they are preserved include the presence of hard parts, how common the organisms are, and their contemporary environment, as organisms living in areas of sedimentation (particularly aquatic environments) have a better chance of preservation. Processes within the rock, such as water filtering through, may result in replacement of the fossil or its total removal. Of course, before the information contained in the fossils can be used, the fossils must be found by a paleontologist and studied.

TAXONOMY AND PHYLOGENY

An understanding of the relationships of organisms enables scientists to appreciate and evaluate extinction events and the effects of climatic change, as well as to use fossils in dating. Naturalist Charles Darwin's views on natural selection led to an understanding not only of the evolutionary processes that result in the development of new species but also of phylogeny, the pattern of relationships of organisms, often shown as a branching diagram. Phylogeny has been greatly aided by the method called cladistics, or phylogenetic analysis, developed by the German entomologist Willi Hennig in the mid-1900s. Hennig's method recognizes and orders characters in an objective fashion to show the most likely evolution of organisms through time. As it involves the processing of large amounts of data, this system lends itself well to the use of computer programs such as PAUP (Phylogenetic Analysis Using Parsimony). Results are presented as regularly branching diagrams, or cladograms, that show the relative sequence of speciation events.

FOSSIL-BASED GEOLOGICAL DATING

The ability of geological industries to accurately locate the materials they seek depends largely on an understanding of the structure of Earth's crust. It is important initially to realize that rocks are normally laid down horizontally and that they become younger toward the top of any sequence. This is the principle of original horizontality. Detailed understanding, however, requires the ability to relate rocks of the same time period to one another, and this was not possible until fossils were recognized as providing unique temporal markers. Once this was understood, the sequences of rocks that had been recognized in discrete outcrops could be related to one another. During the first half of the nineteenth century, most of the relative timescale was assembled, showing the age relationships of rocks and detailing the sequence of fossils that could be used. The oldest fossil-bearing rocks were described as from the Paleozoic (ancient life) era, followed by the Mesozoic (middle life) and Cenozoic (new life) eras. Within these eras, the rocks were divided into periods, such as the Cambrian, Ordovician, and Silurian, which are part of the early Paleozoic era, and the Cretaceous period, the last in the Mesozoic era.

As the study of dating became more laboratory based and in response to the requirements of the oil industry, whose samples were typically drilling chips, which were too small to contain recognizable fossils, paleontologists moved toward using microfossils. These can be the remains of very small organisms or small parts of organisms, as long as they are common and widespread and show evidence of having evolved rapidly.

The field of applied micropaleontology started in the late 1800s in Poland when paleontologist Józef Grzybowski described the foraminifera from the Carpathians in a series of monographs and then applied his findings to foraminifera from the Galician oil fields. He was soon able to demonstrate that it was possible to correlate subsurface strata in wells drilled for petroleum exploration using these fossils. The methods pioneered by Grzybowski continue to be used, and the study of microorganisms has been very much centered on the requirements of the petroleum industry.

THERMAL MATURATION

Organic microfossils such as spores and pollen can be used to indicate the temperature reached by the rocks (thermal maturity) that contain them because temperature increases as the rock is buried deeper. Organic microfossils are progressively altered by the

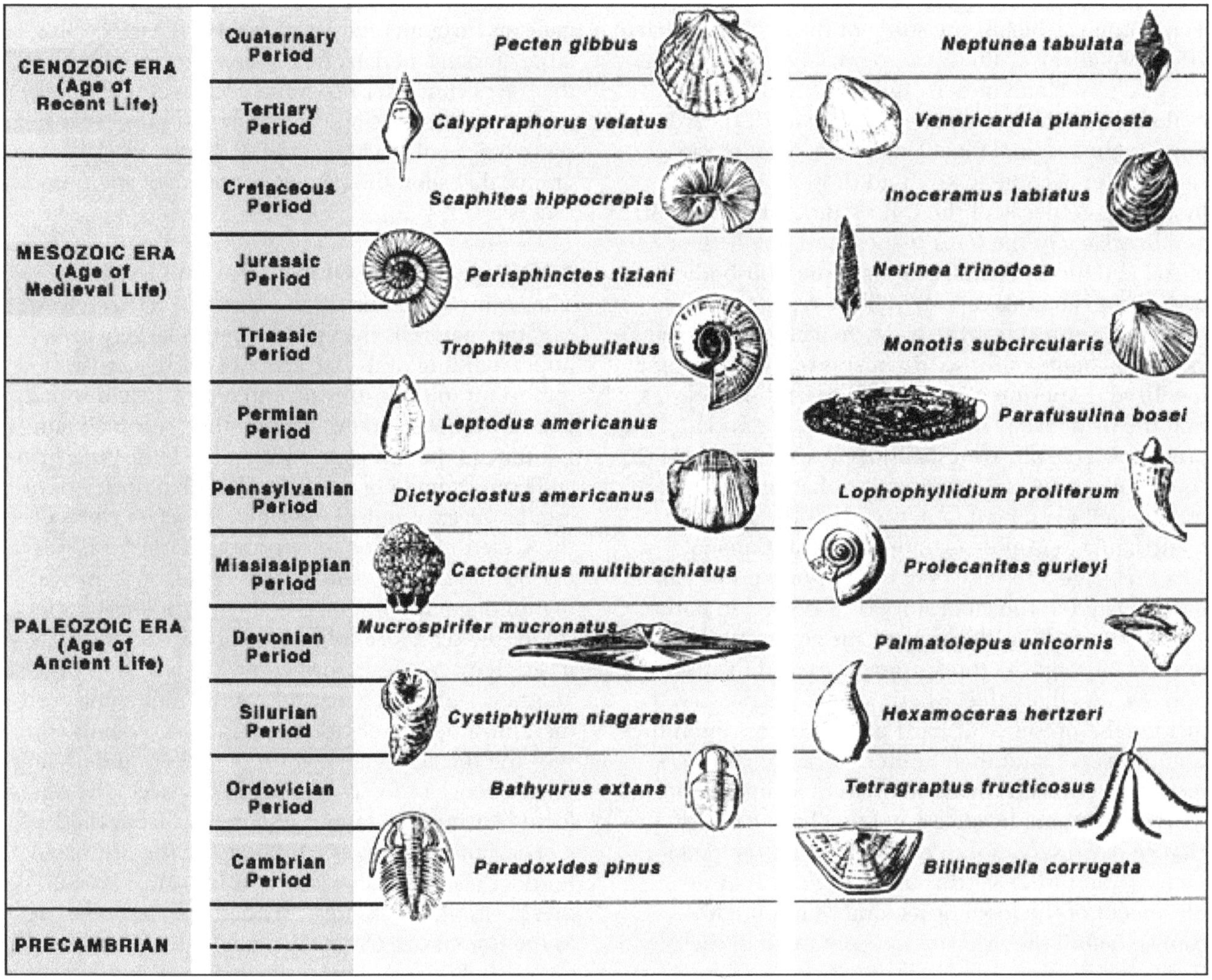

A fossil index. USGS. [Public domain.]

loss of hydrogen and oxygen, and the resultant changes to physical properties such as color, reflectivity, and fluorescence can then be measured.

Reflectivity studies using vitrinite, an organic component, were initially carried out on coal to determine its rank, or thermal maturity. These studies were then applied to hydrocarbon generation, as hydrocarbons such as oil and gas are generated over time by the action of heat on fossil organic material. The reflectivity of vitrinite in the hydrocarbon source rocks reveals maturity and the likelihood of the presence of oil and gas in the sediments.

Spores and pollen change color from pale yellow to orange to dark brown as the temperature increases, and these changes can be used to show the level of organic maturity. A similar change is seen in conodonts, tiny jaw elements from primitive craniates that were abundant in the Paleozoic era, and an eight-point scale has been developed detailing a color change from pale yellow through black to colorless or clear. Below 60° Celsius, organic material in rock is converted to kerogens and bitumen. Rock generates and expels most of its oil at temperatures between 60° and 160° Celsius, and above 160° Celsius, natural gas is formed. It is important for industry to understand the temperature level the rock has reached, and the application of thermal indices based on color or reflectivity provides that information.

ENVIRONMENT AND CLIMATE CHANGE

Organisms are intimately connected to environments and therefore can be used to deduce past environments. This is particularly true of plants, which, unlike marine organisms, are directly exposed to the atmosphere. The size of leaves is related to temperature as well as to humidity and light levels; leaves decrease in size as the temperature or humidity drops. Similarly, the ratio of serrated leaf margins to smooth margins varies according to temperature, and a 3 percent change in this ratio is equivalent to a change of 1° Celsius in the mean annual temperature. In addition, the numbers of stomata (small openings on the surface through which plants lose oxygen and water and gain carbon dioxide) vary according to the carbon dioxide level. When carbon dioxide levels are high, plants need only a small number of stomata, but when the levels are low, plants will commonly have large numbers of stomata to enable them to get all the carbon dioxide they need.

Because carbon dioxide is a greenhouse gas and traps radiation from the sun, the higher the atmospheric carbon dioxide levels, the higher the temperature. Therefore, a correlation can be made between high global temperatures and low numbers of stomata on fossil leaves. If this technique is applied to the end of the Triassic period extinction, the large drop in the number of stomata on leaves that occurred at this time indicates an increase in atmospheric carbon dioxide and a consequent increase in temperature. Calculations suggest that the temperature would have increased by 5° Celsius globally and possibly up to 16° Celsius locally. In the twenty-first century, carbon dioxide levels are rising because of the burning of fossil fuels, and these studies of carbon dioxide in past eras can provide information that will help scientists understand what changes in biodiversity are likely to occur because of higher carbon dioxide levels.

To obtain information about the temperature of ancient ocean waters, it is possible to use oxygen isotopes extracted from calcareous marine fossils. The ratios of oxygen-16 (^{16}O) and oxygen-18 (^{18}O) vary, with oxygen-18 increasing as the temperature decreases. Using isotope ratios from foraminifera from deep-sea sediments, it has been possible to show that over the last 100 million years, there has been a steady increase in the oxygen-18 level, indicating a decrease in ocean temperatures of up to 15° Celsius over that period.

EVOLUTION AND EXTINCTION

Studies on modern taxa have provided much information on how change in organisms takes place, but the fossil record is still the best evidence of evolution through time and also of its correlative, periods of extinction. Both of these areas are of interest because of their relevance to the changes in biodiversity that are taking place in the twenty-first century. To detail changes through time, it is necessary to develop a database of information on taxa and their longevity. That database is constantly gaining new entries, and therefore never complete; however, it has reached the point where information can be derived from it using statistical techniques aided by computers. Extinctions are always occurring, of course, but these studies show when the level of extinction exceeds that of the development of new species by a significant amount, a mass extinction has taken place. Although numerous extinction events have been recorded, there are seven major ones. The first occurred in the Middle Vendian period about 650 million years ago, followed by events at the end of the Cambrian, Ordovician, Devonian, Permian, Triassic, and Cretaceous periods, of which the end-Permian event is by far the most severe.

Analysis of extinction rates of shallow marine organisms at the family level by University of Chicago paleontologists David M. Raup and Jack Sepkoski showed a 26-million-year periodicity to extinction events through the Mesozoic and Ceonozoic eras that was interpreted to indicate an extraterrestrial cause for the extinction events. Corroborative evidence is generally lacking for all but the Late Cretaceous event, however, leaving uncertainty as to the underlying cause of the periodicity.

IMPACT ON INDUSTRY

Fossil fuels will continue to be important to the development of economies in the foreseeable future, and the need for these fuels will increase as nations strive to increase their standards of living. Paleontology is important in the search for oil and natural gas, particularly in the search for coal. In addition, paleontologists are important in the development of a more complete understanding of the history of organisms on Earth, work that is carried on in museums and universities. The main centers for research in paleontology are in North America and Europe, with paleontologists being involved in projects worldwide.

GOVERNMENT AND UNIVERSITY RESEARCH

A considerable amount of paleontological research is carried out by the U.S. Geological Survey. Although some of this is related to the dating of rocks as support in mapping projects, a significant amount of the work involves the study of microorganisms, particularly foraminifera, coccoliths, and nannofossils, in analysis of pollution levels and modern and historical environmental changes. For example, pollen levels in the Florida Everglades have been used to document the response of vegetation, beginning in the early twentieth century, to environmental changes related to nutrients from agriculture and disturbances caused by construction and fires.

Most of the basic research in paleontology takes place in universities, principally in biology and geology departments. Much of this research is related to the understanding of the evolution of organisms through time and the development of databases and collections that are available to answer broad evolutionary questions. A modern trend is for paleontologists to work as part of integrated teams that include members of other disciplines, such as geochemists or paleoclimatologists, to answer highly significant problems.

INDUSTRY AND BUSINESS

The petroleum industry has been one of the largest employers of paleontologists and uses their expertise in deciding where to locate wells and in identifying the strata reached during drilling. However, the number of paleontologists employed at U.S.-based major oil companies has decreased by 90 percent since 1985, as other techniques, such as magnetostratigraphy and seismic information have partially supplanted microfossil dating. Generally, the major oil companies contract out their paleontology requirements to firms that exist to provide these services. Because of this, the demand for paleontologists in the petroleum industry will probably remain low.

The environmental industry has developed extensively since the 1980s as interest in the preservation of natural resources and the environment has become more focused. Paleontologists are employed in this area and are involved in assessment of construction projects and the preparation of environmental impact statements as well as the study of pollution levels using microfossils.

Paleontologists may be trained initially as biologists or as geologists, but in either case, a solid foundation of basic science courses is required. For jobs in industry, a master's degree or a doctorate is required, and the field of study would normally have to be in some aspect of micropaleontology, as that is the area of interest to the petroleum industry. However, as positions in the petroleum industry decreased, universities eliminated or reduced the size of their training programs. In the mid-1980s, more than thirty universities in North America had programs in stratigraphic palynology (the use of spores in dating), but by the late 1990s, only two universities had such programs. Partnerships between industry and university programs are being touted as the way to maintain a flow of adequately qualified paleontologists. As of 2010, only one in ten graduating paleontologists was likely to get a job in industry.

Paleontologists are employed by universities and museums and by the U.S. Geological Survey and the corresponding organizations of other countries. However, the number of university paleontologists in North America has not grown since the 1980s, and most of them are approaching retirement age. The trend is for these paleontologists to be replaced by faculty in other fields, particularly environmental geology, which became the top geology degree program in the 1980s and 1990s. The U.S. Geological Survey has actually reduced the number of its paleontology staff since the 1990s, and although museums have responded to the popular interest in dinosaurs, this has not resulted in a significant increase in jobs for paleontologists.

SOCIAL CONTEXT AND FUTURE PROSPECTS OF PALEONTOLOGY

In 2007, 86.4 percent of world energy consumption came from fossil fuels, which includes oil, coal, and natural gas. Although there is a trend toward greater use of renewable resources, it will be many years before renewable resources provide a significant proportion of energy use. Therefore, paleontologists will still be needed to provide some of the basic data required in exploration to find fossil fuels. However, the numbers of paleontologists needed are likely to decline rather than to rise. Similarly, the need for paleontologists in academia and governmental agencies is declining.

One area in which paleontology could well have an effect in the future is in environmental studies and the study of climate change. As more emphasis is placed on the protection of natural resources, an understanding of past changes and an ability to monitor present-day changes using microorganisms will become increasingly important.

—David Elliott

Further Reading

Benton, Michael J. *Vertebrate Palaeontology.* 3rd ed., Blackwell, 2009.

———., and David A. T. Harper. *Introduction to Paleobiology and the Fossil Record.* Wiley, 2020.

Benton, Rachel C., Dennis O. Terry Jr., Emmett Evanoff, and H. Gregory McDonald. *The White River Badlands, Geology and Paleontology.* Indiana UP, 2015.

Foote, Michael, and Arnold L. Miller. *Principles of Paleontology.* 3rd ed., W. H. Freeman, 2007.

Lieberman, Bruce S., and Roger Kaesler. *Prehistoric Life: Evolution and the Fossil Record.* Wiley-Blackwell, 2010.

Prothero, Donald R. *Bringing Fossils to Life: An Introduction to Paleobiology.* 2nd ed., McGraw-Hill Higher Education, 2004.

Saraswati, Pratul Kumar, and M. S. Srinavasan. *Micropaleontology Principles and Applications.* Springer, 2016.

PHYSICAL PROPERTIES OF ROCKS

Fields of Study: Geology; Geochemistry; Geophysics; Mineralogy

ABSTRACT

Rocks and rock products are used so widely in everyday life that physical properties of rocks affect everyone. Major properties of rocks fall into two categories: those properties that compose the exterior nature of the rock, such as color and texture, and those properties that make up the rock's internal nature, such as strength, resistance to waves, and toughness.

KEY CONCEPTS

coefficient of thermal expansion: the linear expansion ratio (per unit length) for any particular material as the temperature is increased

compressive strength: the ability to withstand a pushing stress or pressure, usually given in pascals or pounds per square inch

density: mass per unit volume

elasticity: the maximum stress that can be sustained without suffering permanent deformation

hardness: the resistance to abrasion or surface deformation

shear, or shearing, strength: the ability to withstand a lateral or tangential stress

toughness: the degree of resistance to fragmentation or resistance to plastic deformation

COLOR

Physical properties of rocks are important not only to geologists and geophysicists but also to construction engineers, technicians, architects, builders, and highway planners. In fact, rock properties affect everyone. Major physical properties include color, hardness, toughness, density, compressive strength, shear strength, tensional and bending (transverse) strength, elasticity, coefficient of thermal expansion, electrical resistivity, absorption rate of fluids, rates of weathering, chemical activity, response to freeze-thaw tests, texture, spacing between fractures, ability to propagate waves, radioactivity, and melting point.

Color is one of the most interesting physical properties of rocks. It is determined by the colors of the component minerals, arrangement of these minerals, and weathering pigments. In minerals, particularly those of gem quality, color is often determined by the electronic state or the configuration of the electrons in atoms of an impurity entrapped within the crystal structure of the mineral. Color and patterns of colors are important in assessing the aesthetic qualities of ornamental and building stone. The unweathered colors must be pleasing to the eye, and staining that is detrimental to the appearance must not occur in appreciable amounts. The quality and beauty of many marbles are renowned when used in ornamental stone and sculpture. Building limestone must have a pleasing gray or tan color and not contain iron or manganese minerals, which would stain the rock brown or black during weathering. The pink-and-black or white-and-black mottling of attractive granites, diorites, and syenites is a familiar sight in business, school, university, and other public buildings. It is not, therefore, necessary or even always desirable that a color be uniform throughout.

HARDNESS, TOUGHNESS, AND DENSITY

Hardness, or resistance to abrasion, can be found by using the Mohs scale, which was derived by Friedrich Mohs in 1822 to measure relative resistance to abrasion in minerals and rocks. Any substance will be able to scratch any substance softer than or as hard as itself. Ten standard natural minerals are used: Talc is the softest and designated as having a hardness of 1, and diamond is the hardest and designated as having a hardness of 10. The modified Mohs scale, however, includes thirteen natural minerals and two synthetic carbide materials. Rock hardness is of particular importance to those people who use rocks or rock products in building façades, monuments, tombstones, patios, and other structures. A variety of tests are used to determine resistance to abrasion over a time interval.

Toughness differs from hardness. This property is defined as the degree of resistance to brittle fracture or plastic deformation of a particular substance. Toughness is determined by a test using repeated impacts of a heavy object. A standard weight called a "hammer" is dropped from a specified height upon a sample, and this height is continually increased. The height of the fall in centimeters upon breakage is then defined as toughness. An example of a very tough material is jade, whereas brittle substances include rock salt. The so-called French coefficient of wear is also used to measure toughness. This test measures the amount of material worn off a sample by tumbling in a drum under standard conditions. Toughness is a very important property in rocks that are used as crushed rock in building roads and airstrips, which are subject to repeated stresses.

The density of a rock is its mass or weight divided by its volume, expressed in standard units of grams per cubic centimeter. More frequently, the rock is compared with an equal volume of water that has a density of almost exactly 1 gram per cubic centimeter. This number, which is dimensionless, is known as specific gravity. There is a considerable variation in specific gravity among rock types and even within them. Most limestone and dolomite rocks range from about 2.2 to 2.7 in specific gravity, whereas basalt and traprock are considerably denser, at 2.8 to 3.0. Density is not synonymous with strength and toughness. Some low-density materials are strong, and some very dense materials cut by fractures are weak.

COMPRESSIVE, SHEARING, AND BENDING STRENGTHS

Compressive strength, or bulk modulus, is measured on the basis of the highest pressure or stress that a rock can withstand per square unit, usually measured in pascals or pounds per square inch. One pound per square inch, or psi, is about 7,000 pascals, and 1,000 psi is about 7 million pascals or 7 megapascals, abbreviated MPa. This strength is generally greater than transverse, tensional, or shearing strength. Limestones average about 100 MPa (15,000 psi), granites about 175 MPa (25,000 psi), quartzite about 200 MPa (30,000 psi), and basalts and traprock about 350 MPa (48,000 psi). This strength will vary considerably from rock to rock of the same composition because of variation in structural properties, as most rocks have fractures and voids and differ in grain size and shape.

Shear strength, or modulus of rigidity, is measured on the basis of the highest lateral stress in pounds a rock can withstand per square inch. An example of shear stress is the stress exerted by a car sideswiping another car. Values for limestones average about 14 MPa (2,000 psi) and granites about 20 MPa (3,000 psi). Shear strength is generally much less than compressive strength.

Bending, or transverse, strength is defined as the strength of a slab loaded at the center and supported only by adjustable knife edges. This strength is determined by the "modulus of rupture," which is a function of the rupture load in pounds, the length of a slab, the width or breadth of a slab, and the thickness of a slab.

ELASTICITY AND LINEAR EXPANSION

The so-called modulus of torsional rigidity, or elasticity (Young's modulus), is also measured in pascals or pounds per square inch. This property is a measurement of how difficult it is to deform a material. Young's modulus is the force that would have to be applied to deform a material by 100 percent of its original size. The amount that a material deforms under pressure is the ratio of the applied force to Young's modulus. This modulus is extremely variable, ranging typically from about 200,000 to 400,000 MPa (3×10^6 psi to 6×10^6 psi) for limestone, and about 400,000 to 560,000 MPa (6×10^6 psi to 8×10^6 psi) for granite; that is, a stress of 20 MPa applied to limestone will deform it by about 1/10,000 of its length. Elasticity refers to the ability of a material to spring back to its original shape without plastic deformation or rupture. Beyond a certain value of stress, rupture or flow will occur.

The coefficient of thermal expansion measures expansion along a line through the rock with increase in temperature. It is also termed "linear expansion." This property is expressed as a ratio of change in length divided by unit length times change in temperature. Typical values for common rocks used in crushed stone or building stone range from about 4×10^{-6} (four millionths) to 12×10^{-6} (twelve millionths) per degree Celsius. Knowledge of the coefficient of thermal expansion is very important to engineers who design structures such as highways and bridges. On bridges, expansion-contraction joints are commonly used in consideration of this property.

RESISTIVITY AND FLUID ABSORPTION RATE

A measurement important to petroleum geologists, geophysicists, and engineers is electrical resistivity of rocks. The resistivity may be defined as the reciprocal of electrical conductivity. Resistivity is measured in ohm-centimeters—that is, electrical resistance along a centimeter's length. It will vary greatly depending on whether a rock is dry or contains saline water, organic compounds such as petroleum or natural gas, or fresh water in its pores or cracks. Igneous rocks such as basalts and traprock will vary from about 1×10^4 to 4×10^5 ohm-centimeters, and granites about 10^7 to 10^9 ohm-centimeters. Basalts have less resistivity and greater conductivity than granite because they generally contain more abundant amounts of dark, nonmetallic minerals than granites. Limestones and sandstones commonly range from 10^3 to 10^5 ohm-centimeters, but those containing much saline water will have much smaller values, because water with dissolved salts is a good conductor of electricity. Metallic ore deposits will show very low resistivities because of the high conductivity of metals.

The absorption rate of fluids by rocks is another property often measured. A dry rock sample of known weight (or dry aggregate of rock chips) is soaked in water for twenty-four hours, dried under surface conditions, and weighed. The weight of water absorbed is given as a percentage of the dry weight of the rock. This property is extremely variable in rocks. For example, limestones may vary from 0.03 percent to 12 percent absorption rate. Very fine-grained limestones will commonly have a higher absorption rate than coarse-grained limestones.

SOUNDNESS, TEXTURE, AND SPACING BETWEEN FRACTURES

Soundness, or resistance to weathering, is important because rock materials are exposed to outside conditions, where temperature and moisture conditions may vary considerably. Chemicals may also attack the rock, and in arctic or temperate climates, freezing and thawing may occur. In a climate where much freezing and thawing occur, the breakage and deterioration of cement and rock materials caused by this phenomenon will be noticed. Engine-driven vehicles and equipment would not last very long in winter if antifreeze were not added to the engine block and radiator. Examples of chemical weathering can be found in an old cemetery, where gravestone inscriptions of the same age on chemically resistant rocks such as quartzite are much clearer than those on chemically reactive rocks such as marble. Texture relates to grain size and grain shape, as well as fabric, which is the relationship of the grains to one another. Appearance, resistance to weathering, absorption rate characteristics, and strength are all related to texture.

Spacing between fractures is a significant structural property of rocks. For building stone, it is usually desirable to have massive rock or rock separated at intervals of meters by layering or bedding planes or by vertical cracks or joints. Preferably, the distance between planes will be fairly constant and the cracks flat and even. Other faces may be sawed out by means of wire saws and silica grit. For some uses of building stone, such as roofing or facing slates and paving stones, smooth, closely spaced fractures are desirable, provided the stone is relatively strong and free from fractures in between.

WAVE TRANSMISSION ABILITY

The ability to transmit waves is an important rock property to petroleum geologists and engineers, as well as construction engineers regarding the design of buildings, roads, bridges, and dams in earthquake-prone areas. This property depends on the density and the strength of the rock, especially its compressibility and its rigidity (shear) modulus. The property is useful to engineers in predicting how strongly rocks will shake in an earthquake, and to petroleum geologists for interpreting the results of seismic surveys. Basically, there are two types of waves that result from earthquakes or human-made explosions: longitudinal and transverse waves. Longitudinal waves, similar to sound waves, create pushing and pulling effects on molecules while traveling along a straight line. Transverse waves oscillate sideways as they travel. Compressive strength (pushing strength) is important in determining the speed of longitudinal waves, whereas modulus of rigidity or shear strength is more important in determining the speed of transverse waves. In general, the greater the strength, the greater the wave speed through a material; the greater the density, the less the wave speed through a material or rock under a given temperature and pressure. The density of darker igneous rocks is generally greater than that of lighter-colored igneous and sedimentary rocks. Also, the density of rocks increases toward Earth's center, but velocity is usually greater, because strength tends also to increase. This factor generally outweighs the density factor and produces a net increase in velocity or speed.

RADIOACTIVITY AND MELTING POINT

Radioactivity, another physical-chemical property, is demonstrated by those rocks that spontaneously emit radiant energy (which results from the disintegration of unstable atomic nuclei). Scientists use the heat generated by absorption of the radiation emitted by these rocks to establish the temperature of Earth's interior and its thermal evolution. Radioactivity has been of particular interest since the early 1980s because of the health hazard posed by radon. Radon is an intermediate product of radioactive decay, and it can seep into basements and cause lung cancer. Proper designing or repair of basement interiors may be necessary to reduce rates of radon infiltration.

Rock melting point is a rock property that is generally of interest to petrologists, engineers on deep-drilling projects, and engineers designing furnaces or kilns that require rock or brick that has a melting point higher than that of the materials being processed in the kiln. Except in the case of dry, completely one-mineral rocks, melting will occur over a range of temperatures rather than at merely one temperature. This type of melting occurs for three reasons. First of all, several types of minerals, each with a different melting point when pure, usually occur in rocks. Second, each mineral mutually affects the melting points of the other minerals, usually lowering the melting points and extending them over a range of temperatures. Third, water and other fluids may mutually lower the melting point of a rock.

PHYSICAL PROPERTY TESTS

A great variety of tests are used to determine the degree or types of physical properties of rocks. Hardness may be determined by comparing a rock with known hardness points made from minerals of known hardness or from ceramic cones of various known hardnesses, either on the Mohs ten-point scale or on a more elaborate 1,000-point industrial scale. Another test is called the "Dorry hardness test." Rock cores 2.5 centimeters in diameter are loaded with a total weight of about 1,000 or 1,250 grams and then subjected to the abrasive action of a fine quartz of known size fed upon a rotating cast-iron disk. The loss in weight of the core after 1,000 revolutions is used to compute value of hardness. One of the most widely used machines to test hardness is the Los Angeles Rattler (abrasion testing machine). It has a steel drum 70 centimeters in diameter and 50 centimeters long. The sample is inserted with a certain number of steel balls, and the drum is rotated five hundred revolutions at thirty revolutions per minute. The material is then sized into further grades, and the coarser sets of remaining fragments are subjected to one thousand further revolutions using various numbers of steel balls weighing about 5,000 grams. Another test, the Brinell test, determines hardness by pressing a small ball of hard material on the sample and measuring the size of the depression.

Toughness is tested on samples with a diameter of about 2.5 centimeters. The sample is held in a test cylinder on an anvil and subjected to the fall of a steel hammer or plunger weighing 2 kilograms. The first fall is from a 1-centimeter height; this height is progressively increased by 1 centimeter until breakage occurs. Height of fall is then expressed as toughness.

Specific gravity is measured by placing the sample in water. It is measured, dried to obtain dry weight, then immersed in water for twenty-four hours, surface dried, and weighed. Finally, it is weighed again in water. Loss of weight from dried weight initially is true specific gravity. The second step, involving the twenty-four-hour immersion, is necessary to ascertain that pore spaces in the sample are filled with water so that a true measure of buoyancy can be taken in the third step.

Compressive strength is measured on test rock cylinders 5 centimeters in diameter and 5.6 centimeters long. The sample is placed in a cylindrical casing in a press, and the press is lowered at a specified rate. Amount of compression in pounds per square inch at failure is the compressive strength.

The so-called soundness test measures resistance to chemical weathering or, alternatively, resistance to a major type of mechanical weathering, called "ice" or "frost wedging." In the first case, the sample (usually about 1,000 grams) is covered with a saturated solution of sodium or magnesium sulfate for eighteen hours. Then it is oven-dried. The test is repeated usually five times. If the sample decomposes, it is called "unsound"; if it shows signs of decomposition but is still intact, it is called "questionable." If it shows no signs or extremely minor signs of decomposition, it is called "sound." In place of ice, the test uses expansion of the sulfate to measure resistance to wedging. In the case of freeze-thaw tests, different-sized materials are tested by freezing in water to an air temperature of -22° Celsius, then thawed at room temperature (about 20° Celsius). A first cycle lasts twenty-four hours and may be repeated. Damage to the sample is then noted.

SEISMOGRAPHIC STUDY

Wave velocities, or speeds through rocks, are dependent on density and strength. These velocities can be determined theoretically or in the laboratory if density and strength can be measured. Conversely, compressive and shearing strength may be measured knowing rock density and velocity of wave transfer.

Geophysicists and engineering geologists may determine wave speeds through Earth's crust by setting up a seismograph station in the field. Explosives are set off, and travel time and velocity of waves are monitored by a system of geophones that relay the wave energies to the seismograph, which then measures the waves on a recording drum with a stylus. Characteristic wave reflections occur at certain depths. By knowing the times of explosive detonation and the time of arrival of the two kinds of earth body waves (primary and secondary waves), and by understanding that distance traveled for the two waves is the same, scientists can determine the wave speeds. This information is vital when drilling for oil or gas and digging deep foundations, especially in limestone sinkhole areas or in areas prone to earthquakes.

ENGINEERING AND CONSTRUCTION APPLICATIONS

Knowledge of particular rock physical properties is important to geologists, engineers, geophysicists, hydrologists, builders, architects, industrial city and regional planners, and the general public. Rock properties are especially important in regard to the construction of buildings, dams, roads, airstrips, human-made lakes, and tunnels; drilling for petroleum and natural gas; mining and quarrying; monitoring earthquake hazards; measuring properties of aggregate and construction materials; determining rates of heat flow and mechanical strain underground; monitoring radioactive hazards; studying chemical makeup and physical structure of Earth's interior; and evaluating aesthetic properties of ornamental construction materials.

Within earthquake-prone areas such as California or areas affected by wind shear like the tops of high buildings, the utmost importance is granted to the nature and strength of materials surrounding foundations, the foundations themselves, groundwater behavior and its effects on building strength, and the ability of building materials to withstand vibrations. Engineering codes have been enacted within some of these areas so that earthquake-resistant or windshear-resistant buildings, dams, tunnels, and highways may be constructed.

—David F. Hess

Further Reading

Fettes, Douglas, and Jacqueline Desmons, editors. *Metamorphic Rocks: A Classification and Glossary of Terms*. Cambridge UP, 2007.

Haussühl, Siegfried. *Physical Properties of Crystals: An Introduction*. Wiley-VCH, 2007.

Jerram, Dougal, and Nick Petford. *The Field Description of Igneous Rocks*. 2nd ed., Wiley-Blackwell, 2011.

Pellant, Chris. *Smithsonian Handbooks: Rocks and Minerals*. Dorling Kindersley, 2002.

Schön, Jürgen H. *Physical Properties of Rocks: Fundamentals and Principles of Petrophysics*. 2nd ed., Elsevier, 2015.

Vernon, Ron H. *A Practical Guide to Rock Microstructure*. Cambridge UP, 2004.

Wenk, Hans-Rudolf, and Andrei Bulakh. *Minerals: Their Constitution and Origin*. Cambridge UP, 2004.

Placer Deposits

Fields of Study: Geology; Mineralogy; Geochemistry; Geophysics

ABSTRACT

Placer deposits are mechanical concentrations of debris weathered out of rocks. Commonly, economically important minerals have higher densities, so they are concentrated as the lighter-density minerals are winnowed out by the action of water or wind.

KEY CONCEPTS

aeolian: due to transport by wind

density: the ratio of an element or objects mass to its volume, expressed for elements and minerals as grams per cubic centimeter

equant grains: grains of similar size and shape

panning: a method of separating particles according to density by swirling the source material with water in a particular type of pan

sluicing: a method of separating particles according to density by entrapping the densest particles in transverse grooves as they are carried down a sluice, a specially constructed channel, by flowing water

spiral classifier: essentially a spiral sluice that operates vertically rather than as a lengthy run of channel

THE NATURE OF PLACER DEPOSITS

Placer deposits are found throughout the world wherever the mechanisms of concentration, water and wind, have been active and the resulting concentrates have not been dispersed by later processes.

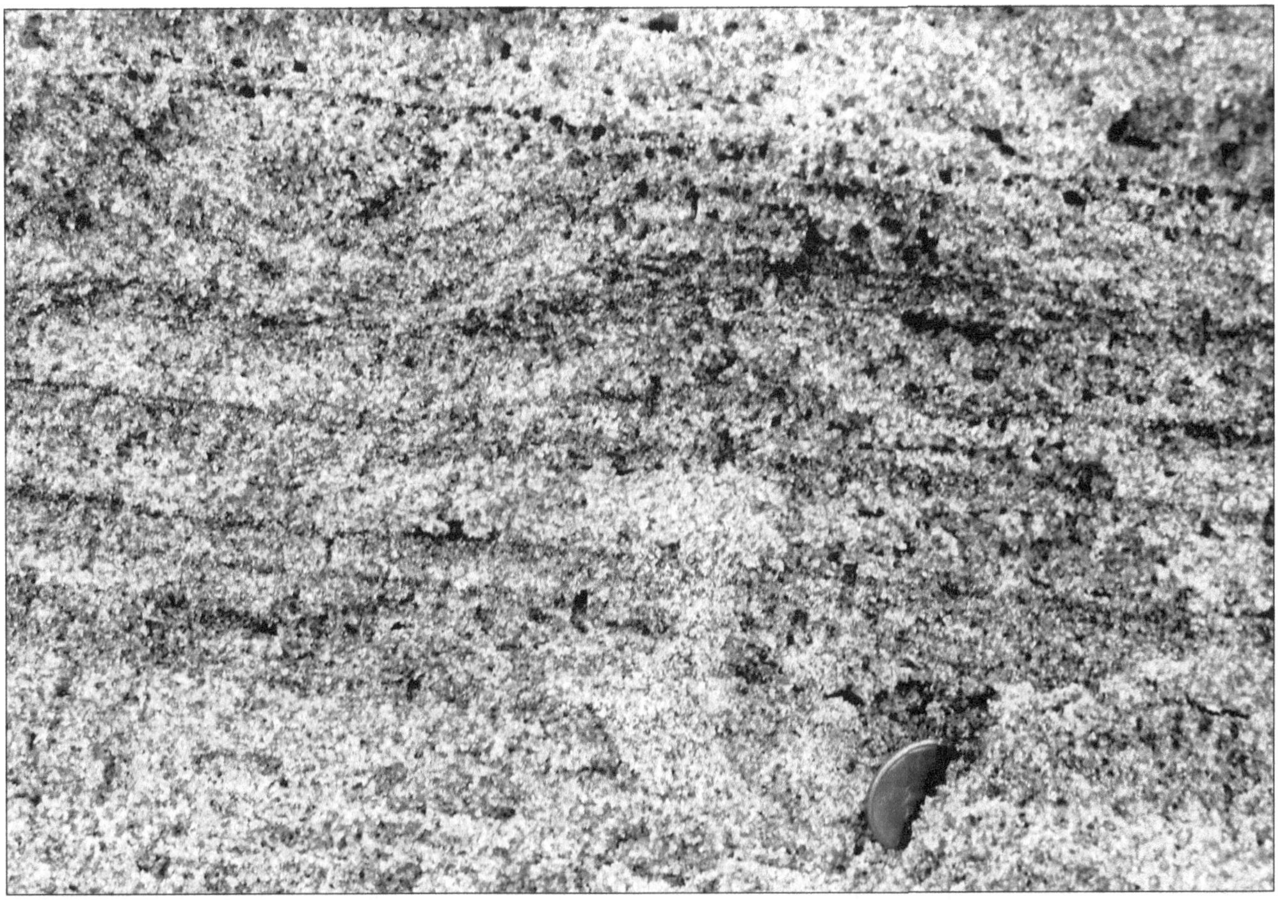

Beach placer deposit of heavy minerals (dark) in a quartz sand (Chennai, India). Photo via Wikimedia Commons. [Public domain.]

The best-known types of placers occur in river channels and in beach sediments. The weathering and erosion of rocks release particles of varying size, shape, and density. Soluble materials are dissolved and removed in surface water or groundwater. Some minerals, such as feldspars, are hydrated and converted into clay minerals, which, being soft, small, and of low density, are relatively readily removed in suspension. Quartz (SiO_2), common in many kinds of rocks, generally weathers out as roughly equant grains that, because of their hardness and insolubility, wash into streams and rivers, where they are moved by rolling, bouncing, and pushing along toward the oceans. The abundance of quartz and its resistance to mechanical and chemical weathering results in its being the most abundant placer mineral and the principal constituent of temperate and cold climate beaches throughout the world.

During the weathering and erosion processes, other minor or trace minerals, which are resistant to breakdown, are also transported along with the quartz grains and pebbles in river channels to the ocean margins. If the mineral particles possess high densities, they may be selectively concentrated as the transporting agent (usually water) more readily removes the lighter-density minerals. Thus, gold nuggets, with densities of 15-19 grams per cubic centimeter, are commonly concentrated in residual materials as the quartz grains, with a density of about 2.65 grams per cubic centimeter, are removed. The densities of several other valuable and resistant minerals are sufficiently higher than that of quartz to allow them to also be concentrated in placer deposits (examples include ilmenite, $FeTiO_3$, 4.8; rutile, TiO_2, 4.25; zircon, $ZrSiO_4$, 4.7; cassiterite, SnO_2, 7.0; and diamond, C, 3.5).

ECONOMICALLY IMPORTANT PLACER MINERALS

Many types of minerals and rock materials can occur in placer deposits; among the most important are gold, titanium minerals, zircon, tin oxide, diamonds, platinum, and sand and gravel. Throughout history, gold has no doubt been the most important placer mineral. Gold is soft and malleable but is otherwise nearly inert in the weathering realm. Hence, once gold is weathered out of the lode deposits where it initially formed, the grains may survive transport in streams and rivers over long distances. It was the discovery of such gold grains that led to nearly all of the world's major gold rushes, including the California gold rush in 1849. Gold placers have formed throughout geologic time; the world's largest gold reserves in South Africa occur in placers formed 2 billion years ago.

The titanium minerals, ilmenite and rutile, occur in minor amounts as small grains in many types of igneous and metamorphic rocks. These minerals are highly resistant to weathering and hence are liberated intact from their host rocks. Although their densities are less than twice that of quartz, they are quite effectively concentrated by flowing water in rivers and by the agitation of waves along beaches as the lower density quartz grains are winnowed out. The zirconium silicate zircon is a common accessory mineral in alkaline igneous rocks. It weathers out as the titanium minerals do and is generally found with those minerals in river and beach deposits. The tin oxide cassiterite, like the titanium minerals, is heavy, hard, and extremely resistant to weathering. Consequently, where there are cassiterite-bearing lode deposits, the cassiterite weathers out and may be concentrated into economic placers. The hardness of diamonds and their resistance to normal weathering agents has allowed them to wash down rivers that drain from the areas of exposed diamond pipes and to occur in placer river and beach deposits. Because the density of the diamonds is only about one third greater than that of quartz, the diamonds are not as selectively concentrated in these placers as are the heavier minerals noted above. Platinum, like gold, is chemically inert and has a high density, 15 to 19 grams per cubic centimeter. Platinum lode deposits are much more restricted geologically than are gold deposits; hence, placer platinum deposits occur in only a few places in the world. The sand and gravel deposits found in rivers and lakes and on beaches constitute the largest placer deposits in that they represent mechanically concentrated residual materials. Although they have a much lower per unit value than many other placer materials, the large volumes of sand and gravel mined from these deposits actually make them the most important placer deposits economically.

TYPES OF PLACER DEPOSITS

Placer deposits have been classified into several different types on the basis of location of formation. The major types include residual placers, alluvial placers, stream or river placers, riverbank and flood placers, aeolian placers, and beach placers.

Residual placers are occurrences of minerals at or near their point of release from the original source rocks. There has been some degree of enrichment of the placer minerals as the result of the removal of other portions of the host rocks by weathering.

Alluvial placers are transitional placers in which concentrations of placer minerals occur downslope from the source rocks but where the valued minerals have not yet washed into streams and rivers that would transport them for long distances.

Stream or river placers are the best-known placers and are the types responsible for most famous gold discoveries. The movement of the running water, especially where there is turbulence, is effective in sorting rock fragments and mineral grains according to size and density. Because of their higher densities, gold grains and several other placer minerals settle out. They are readily trapped in crevices and irregularities on the streambed or among larger boulders, as the lower density materials are more easily washed away. This type of placer sometimes grades into deltaic beds where a river drains into a lake or the ocean.

Riverbank and flood placers are deposits adjacent to streams and rivers that have been left as the rivers meander, cut downward, or overflow their banks in flood conditions. During the natural development of rivers, they commonly shift laterally across their floodplains, eroding banks on one side while depositing materials on the other side. In some areas, changes in base levels result in rivers cutting downward though the sediments they had previously deposited in their floodplains. In both these circumstances, valuable placer deposits may be left in the riverbank sediments that are adjacent to the present rivers.

Flood placers also occur in the sediments adjacent to rivers. They form during episodic flooding when water flow is sufficiently rapid and turbulent to transport gold or other valued placer mineral grains up and out of the channels onto the adjacent floodplains. As the water spreads laterally and its velocity decreases, the gold grains are left as placer deposits along the adjacent floodplains.

Aeolian placers are wind-formed placers that occur locally in desert regions where high winds have removed lighter mineral grains, thereby enriching the heavy minerals in the residuum.

Beach placers are generally formed by the combined effects of river transport of weathered materials to coastal margins and the action of tides and storm waves along beaches. The ebb and flow of the waves and the generation of longshore currents, especially under storm conditions, can effectively winnow and sort beach materials such that certain areas are highly enriched in heavy minerals. Gold-bearing beach placers are known in many localities, but the most famous are probably those at Nome, Alaska, where the beaches were actively mined for many years. Diamond-bearing beach placers have been extensively mined along the west coast of central and southern Africa, where rivers draining the interior have transported diamonds into the beach sands. Beach placers containing ilmenite, rutile, and zircon are the world's major sources of these minerals.

MINING OF PLACER DEPOSITS

Gold panning is probably the best-known method of exploiting placer deposits. A circular motion of water in a pan containing gold along with other sediments effectively separates the minerals on the basis of their densities. The same general principle is used in sluices, channel-like boxes with barriers to create turbulence in the water so that sorting can take place. On a large scale, modern placers are mined by the scooping up of the unconsolidated materials and the use of either spiral classifiers or heavy media to separate the heavy materials from the light materials. The differences in the densities of the minerals allow for effective separation.

—James R. Craig

Further Reading

Bell, Fred J., and Laurance J. Donnelly. "Placer Deposits and Mining." *Mining and Its Impact on the Environment.* Taylor & Francis, 2006.

Boggs, Sam, Jr. *Principles of Sedimentology and Stratigraphy.* 5th ed., Prentice Hall, 2011.

Guilbert, John M., and Charles F. Park, Jr. *The Geology of Ore Deposits.* Waveland Press, 2007.

Hartman, Howard L., and Jan M. Mutmansky. *Introductory Mining Engineering.* 2nd ed., J. Wiley, 2002.

Koch, Adam Gregory. *Gold Prospecting and Placer Deposits: Finding Gold Made Simpler.* Author, 2014.

McCulloch, Robin, et al. *Applied Gold Placer Exploration and Evaluation Techniques.* Montana Bureau of Mines and Geology, 2003.

Macdonald, Eoin H. *Alluvial Mining: The Geology, Technology, and Economics of Placers.* Chapman and Hall, 1983.

Valentine, David. "Chinese Placer Mining in the United States: An Example from American Canyon, Nevada." *The Chinese in America: A History from Gold Mountain to the New Millennium,* edited by Susie Lan Cassel, AltaMira Press, 2002.

Wenk, Hans-Rudolf, and Andrei Bulakh. *Minerals: Their Constitution and Origin.* Cambridge UP, 2004.

Planet Formation

Fields of Study: Geology; Astrogeology; Geochemistry; Geophysics

ABSTRACT

The processes that influence planetary formation are extremely complicated but reasonably well understood. In the densest portion of a contracting solar nebula, a star forms, and, assuming that conditions are suitable, a family of planets can also form around that star. Planet formation may be a very common occurrence throughout the galaxy.

KEY CONCEPTS

accretion: the accumulation of mass by the impact of small celestial bodies against large celestial bodies

condensation: the change of state of elements from hot gases to liquid or solid forms as temperature decreases

gravitational collapse: the infall of gaseous particles to form a more dense mass about the center of gravity of a large gaseous cloud of atoms

gravitationally induced wobble: planets and stars orbit each other about their common center of mass, like a very lopsided spinning dumbbell; the wobble is the minor but regular change in the observed position of the star as the orbital rotation proceeds

solar wind: the flow through space of particles ejected from a star

T-Tauri stage: a highly active period in the life of a very young star, in which large amounts of matter are ejected from it as solar wind

OVERVIEW OF PLANET FORMATION

"Planet formation" is the term used to describe how various planetary bodies formed from residual material surrounding a forming star. The two most discussed theories suggest either a catastrophic or an evolutionary origin for the planets. In either case, the planets would have formed in a manner similar to the way the star formed, but on a much smaller scale and with considerably less mass.

One theory suggests that the formation of a solar system begins with the explosion of a massive nearby star, called a "supernova". A supernova begins with a slowly contracting cloud of gas and dust that is capable of forming many individual stars, called a "cocoon nebula." Where the concentrations of matter are densest, giant stars will form. If they are massive enough, they will progress rapidly through their evolutionary stages and explode with gigantic force. Energy released may rival that of a galaxy for a short period of time. Compressional shock waves generated by the explosion, along with the excess matter thrown off, can initiate formation of smaller nearby stars such as the Sun. Evidence for such an event is suggested by certain isotopic ratios for various elements present in meteorite compositions and even the Sun itself.

The majority of mass in a solar nebula will go into the formation of individual stars. Some estimates indicate that a star will account for greater than 99 percent of the total mass. The remaining 1 percent will represent matter available for planetary formation. Increasing pressures and temperatures at the gravitational center of a solar nebula will eventually form a protostar. This is the initial stage that precedes the beginning of nuclear processes in a star. The protostar is more representative of a hot glowing ball of gas than a true star. It is believed that planets will commence their earliest formation processes at this time as well.

Planet formation is characterized by three distinct processes: gravitational collapse, condensation, and accretion. The initial process that took place in the solar nebula was gravitational collapse, the process in which atom attracts atom to form molecules of considerably greater mass. This greater mass increases gravitational pull and thereby attracts more atoms. As these localized regions became denser through the infall of matter, temperatures began to increase. Gases and solid dust grains that were originally cold now begin to heat up and glow as more and more matter becomes concentrated in a rapidly shrinking volume.

After the majority of gas has been incorporated into the star's formation, residual hot gases begin to cool and re-form as solids during the condensation process. Every element has a specific condensation temperature at which a phase change takes place. Metallic elements such as aluminum and magnesium condense from a gaseous state at very high temperatures, while gases such as methane or water vapor condense at low temperatures. Where concentrations of matter were greatest near Earth's proto-Sun, for example, only heavy elements condensed. As matter thinned out, temperatures dropped progressively and permitted lower-temperature elements to condense. Based on existing temperatures, distance from the protostar controlled, in part, the abundance ratios of the various elements found in those regions. This is one reason that the inner planets are high in iron content relative to the outer gaseous planets.

Material in the residual gas rings that surrounded the proto-Sun was pulled to the plane of the ecliptic by the force of gravity. This material tended to clump together as it condensed out of the cooling gases. Some scientists believe that in our solar system this condensation phase was rapid enough to trap some of the lighter gases within the forming clumps of heavier elements. This may account for the large amounts of volatiles present during the final stages of planetary formation.

As condensation progressed, only microscopic grains were in evidence. Later, these combined to form larger and larger particles. Growth continued through sand-sized to pebble-sized and then to boulder-sized objects. This describes the accretion process, where collision produced a larger mass rather than fragmentation. Accretion did not stop with boulders, but continued, eventually forming planetesimal-sized objects much like present-day asteroids. They had a size range from 30 meters up to about 1,000 kilometers in diameter. Finally, planetesimals collected one another and assumed the planetary dimensions familiar today.

Present-day planets are not uniform in their chemical compositions. The inner, so-called terrestrial planets (Mercury, Venus, Mars, and Earth) are rich in heavy metals, while the outer planets (Jupiter, Saturn, Uranus, and Neptune) are gaseous—hence, the so-called gas giants. One explanation for this may be that the solar nebula was not uniformly hot, because of nonhomogeneous distribution of material based on distance from the proto-Sun. Only the area from the Sun outward to the asteroid belt (a region of small rocky and metallic bodies between the orbits of Mars and Jupiter) was sufficiently hot to experience heating of the original nebular material and subsequent condensation. Material that was present at the distance of Jupiter and beyond re-

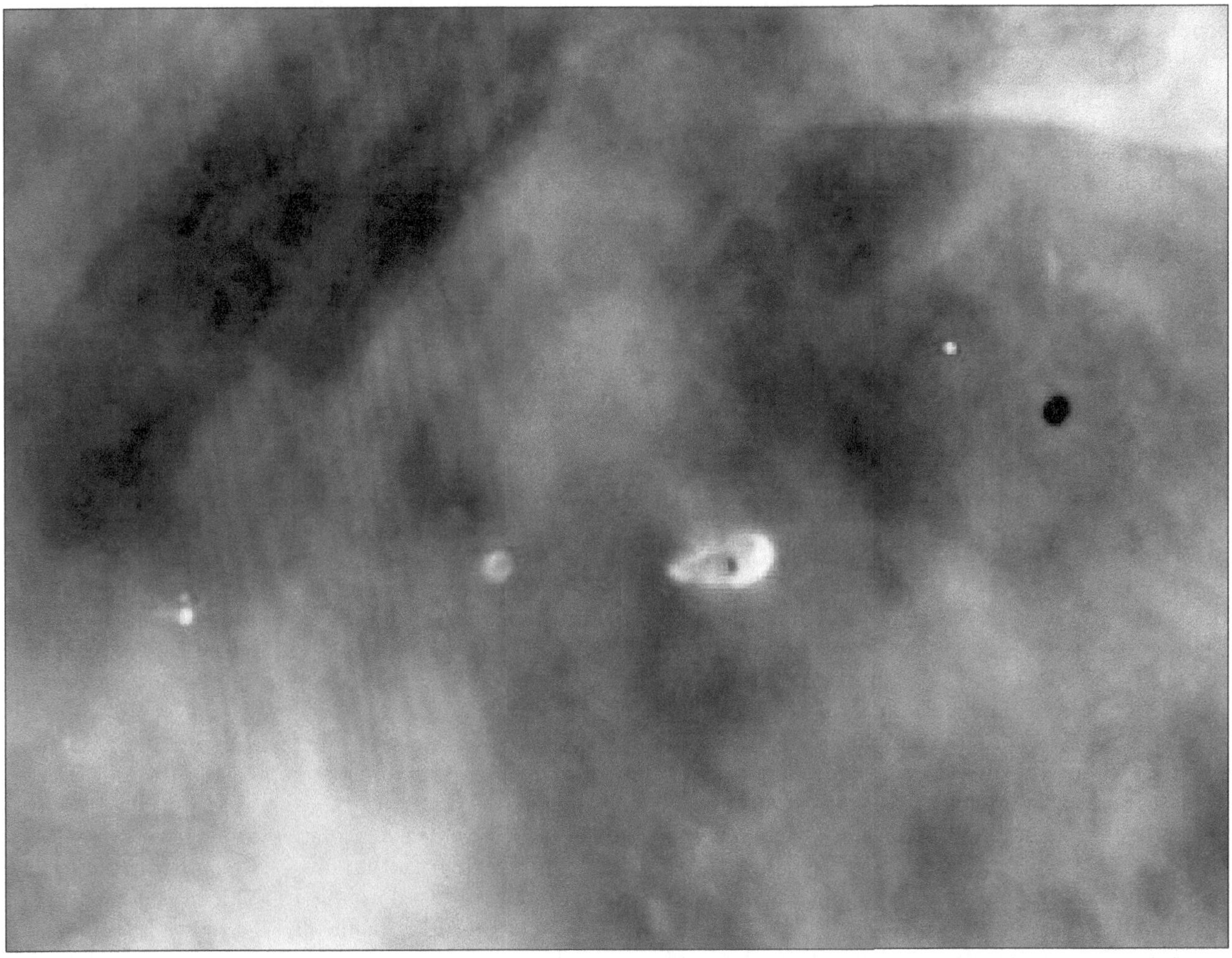

Hubble image of protoplanetary discs in the Orion Nebula, a light-years-wide "stellar nursery" probably very similar to the primordial nebula from which the Sun formed. Photo by C.R. O'Dell/Rice University via NASA/Wikimedia Commons. [Public domain.]

mained in its unaltered state until it became concentrated in the protoplanetary stage of formation.

An additional factor to be considered in planetary formation is the T-Tauri stage in the development of a main sequence star like the Sun. This is a hyperactive stage in a young star's life, as fusion reactions begin to dominate. As a result, large quantities of matter are ejected at a very rapid rate. It has been suggested that the early Sun may have lost as much as one-half its original mass in as short a period as 1 million years. This would be evident as an intense solar wind many times greater than it is today.

Distance would once again be an influencing factor in planet formation. As the solar wind came in contact with the inner protoplanets, it would have stripped away the lighter elements, thereby concentrating the heavier ones. Yet, even with its great intensity during the T-Tauri stage, the solar wind would have lost most of its effectiveness at the distance of the present asteroid belt. It may have been virtually ineffective by the vicinity of Jupiter, thereby leaving the outer planets with most of their original mass. In fact, the Jovian planets may have increased their mass as gravity collected additional matter flowing past from the inner solar system.

In the end, the planets as they are known today were formed. The processes at work were very complex, involving many variables. Collectively, the aforementioned theories are known as the nebular hypothesis. It offers reasonably logical explanations for a very difficult problem.

METHODS OF STUDY OF PLANET FORMATION

Experimentally, several aspects of the nebular hypothesis can be examined in the laboratory through high-temperature and pressure studies on various materials believed to have been present in the nebula. Nevertheless, in the final analysis, no experiment can reproduce an actual solar system or a planet. It is possible, however, to compile all available data into a computer model and develop a graphic representation of what may have taken place.

Scientists do not have to rely totally on computer modeling for answers; there are actual clues. Evidence for the condensation and accretion phases of planet formation can be found in the chemistry and mineralogy of stone meteorites. Chemical compositions of minerals found in meteoritic objects called "chondrules" bear witness to an early high-temperature condensation history. The physical condition of these chondrules—as they are found in the meteorite—also exhibits evidence of a very violent phase during which collisions were commonplace. These normally spherically shaped objects can be found as fragments or in a partially melted and deformed state. They were incorporated later into a re-forming object that eventually would reach planetesimal proportions. Although scientists do not totally agree about the mechanisms for the origin of chondrules, they are certain of their primitive nature.

The bulk chemistry of meteorites also provides additional information about planetary formation. One type of meteorite, called a "type one carbonaceous chondrite," contains the most primordial chemistry of all materials in the solar system. Its chemical composition most closely resembles the "condensable" part of the Sun's chemical makeup. Also, present are low-temperature, water-bearing minerals; carbonate and sulfate minerals; and 8 to 22 percent water. High-temperature minerals are rare, and chondrules are notably absent. This strongly suggests that this type of meteorite may be very close in chemical composition to the original unaltered material of the solar system.

The unique primordial nature of meteorites was destroyed as they were incorporated into the growing planetesimals. As the planetesimals accreted further into actual planet-sized masses, evolutionary processes took over, erasing all evidence of their protoplanetary stage. For the inner planets, intense bombardment from infalling planetesimals turned their surfaces into a crater-dominated world. Those planets that had sufficient mass would heat up internally, with volcanism, plate tectonics, and erosion eventually reshaping their surfaces. In contrast, the outer planets never got beyond their protoplanet stage. Their great masses kept them hot, and they never cooled off.

The first extrasolar planet was discovered in 1995 using a technique that involves measuring the gravitationally induced wobble in a star produced by a planet orbiting it. Ironically, the first planet found outside the solar system did not fit the typical model envisioned for a planetary system. It was a very large gas giant in an extremely tight orbit about a pulsar. Pulsars had not been considered likely candidates for solar systems with planets. A decade after that initial exoplanet discovery, more than one hundred other planets had been detected through a small number of indirect means. Most extrasolar planets were hot Jupiters, but some were found having masses down to a multiple of a Uranus-class mass. After the first decade of discoveries, the number of objects beyond the solar system that were accepted as planets rose dramatically. By 2008, there were more than three hundred recognized exoplanets. The tremendous variety of these planets and the nature of the solar systems in which they were found forced a serious reexamination of the theory of planetary system formation.

The nebular hypothesis for planetary formation provides good explanations for many of the dynamical and chemical properties found in the solar system, but it is not all-inclusive. Some inconsistencies do exist, such as the axial tilt and the rotational direction of Uranus. The possible formation of this planet is more representative of a catastrophic event that may have literally knocked it on its side.

Comets are another example. Although much is known about their physical and chemical characteristics, little is known about how they fit into the nebular hypothesis. They apparently come from the outer fringe of the solar system, but their highly elliptical orbits are not consistent with protoplanetary theory. A likely explanation would suggest that a catastrophic event is somehow also responsible for the origins of comets.

One place scientists can look for possible confirmation of the nebular hypothesis is in the satellite systems of the gaseous planets. At their formation these protoplanetary bodies resembled small stars rather than planets. It is possible that they acted in a manner similar to a forming star with respect to their satellite families. One can envision a protoplanetary Jupiter gravitationally gathering in the largest amount of mass and perhaps radiating a small amount of energy just as the early proto-Sun did. Its mass would not permit it to progress further, but it may have remained hot long enough to influence the chemical compositions of its inner higher-density satellites (Io, density 3.5 and Europa, density 3.0). In contrast, the two larger outer satellites, Ganymede and Callisto, have densities of 1.9 and 1.8, respectively. This comparison between the satellites of Jupiter and the arrangement of planets by density is inescapable. The inner two Jovian satellites are composed of rock, ice, and small amounts of metal, while the outer two are dominated by ice and dust. Astronomers are investigating if they truly formed according to protoplanetary theory. Only future spacecraft exploration will tell.

The ring systems of the outer planets may well answer some of the questions concerning accretion and planetesimal formation. For example, are the rings representative of a fragment satellite, or one just coming together? What about orbital dynamics and the influence of shepherding satellites on the orbital motion of the particles in the rings? These are all valid questions whose answers will undoubtedly shed light on planetary origin.

As one moves outward through the solar system, the density comparison between satellites in a given system is not so apparent as it is in the Jovian system. These satellites all appear to be relatively uniform mixtures of various ices and dust. Even large Titan, with a density of 1.9, does not match up to Io and Europa. It is then apparent that temperatures in these far-removed regions of the solar nebula were merely too cold to permit any large concentration of heavier elements, regardless of the influence of the planet they orbit.

Perhaps the most striking evidence for accretion can be found in the physical characteristics of the small Uranian moon called "Miranda." Apparent in the Voyager 2 photographs were two distinctly different surfaces. One was a densely cratered terrain typical of the low-density ice satellites of the outer planets. The other had no comparison. There were circular patterns of angular landforms surrounded by alternating bright and dark bands, along with fault scraps and ridges. These were clearly of a much younger age than the ancient cratered surface. Scientists are searching for a possible explanation for such a unique satellite. One theory suggests that Miranda may have experienced an early partial internal melting. This would have been followed by a low-velocity impact that broke it in several pieces. These large pieces remained in orbits close to one another and in a relatively short period of time reaccreted into a single large mass. The patchwork nature of Miranda's surface makes this a strong possibility and would thus provide astronomers with an example of an actual accretionary body.

Theory and the examples from the solar system can provide valuable evidence for planet formation, but scientists would like to see the actual event as it happens. Considering the vast number of young stars in the Milky Way, this should be possible, but distance prevents examination by earthbound telescopes. Methods of observation that make use of nonvisible radiation, however, may provide the information necessary to confirm the nebular hypothesis.

In 1983, the Infrared Astronomical Satellite (IRAS) began gathering data on various sources of infrared energy in space. This included energy emitted from newly forming planetary systems. One of the first candidate sources was the young star Vega. IRAS showed it to be surrounded by a spherical cloud of cool dust, just as theory predicted for a forming planetary system. A second observation showed the star Beta Pictoris to have a dust cloud surrounding it in a typical flattened plane. Optical observations confirmed this disk to have a radius ten times larger than that of Pluto's orbit and confirmed that the dust was orbiting within 5° of that star's plane of the elliptic. Evidence such as this does not necessarily confirm the nebular hypothesis, but it strongly supports it.

The Hubble Space telescope and Spitzer Space Telescope have since found many more protostellar objects and forming planetary systems. The indirect detection of more than three hundred extrasolar planets (a number that is continually increasing), the vast majority of which do not appear to have formed in a manner consistent with the model of our solar system condensing out of a planetary nebula, puts constraints on the acceptability of the nebular hypothesis, or at least its universality. "Hot Jupiters" close to a star, pulsars with planets, and a few other unusual star-planet relationships fall well outside the classic planetary nebula model.

Much anticipated is the first direct imaging of an Earth-class planet about another star. The Kepler Observatory, designed to detect planets while transiting their star, deployed in 2009 and was designed to survey as many as 100,000 star systems. It is estimated that the chances for finding a terrestrial planet are on the order of 1 in 210; if that low probability proves to be correct, then Kepler might detect as many as 476 Earth-class planets. Such continuing indirect and direct observations of existing and forming planetary systems will surely require major modifications to the theory of planet formation.

In mid-June, 2008, European astronomers announced remarkable results found using the European Southern Observatory (ESO) La Silla Observatory's High Accuracy Radial Velocity Planet (HARP) spectrograph. Surrounding the star HD40307 were found three super-Earth-scale planets. HARP was capable of detecting planets between two and ten Earth masses—thus, objects on a planetary scale, between Earth and Uranus or Neptune in mass. This technique was sensitive to stellar wobbles of only a few meters per second, something necessary to determine the tiny tug on a star by an Earth-sized extrasolar planet at significant distances from its star. These researchers also announced that they expected to refine data on forty-five candidate Earth-scale exoplanets with masses between 1 and 30 Earth masses as they continued to observe with HARP. However, an extrasolar planetary system with three Earth-class planets was unique thus far in the history of exoplanet identification.

CONTEXT OF PLANET FORMATION

Prior to the mid-seventeenth century, most theories for the origin of Earth and the solar system were based on either myth or religious interpretation. Later, science would tend to take either an evolutionary or a catastrophic approach. It was realized quickly that no comprehensive answer would be forthcoming, and the "best-fit" concept was developed to provide answers based on the best science available.

In 1644, French philosopher René Descartes presented the first scientific theory for the origin of the solar system. His model was based on an evolutionary process. Descartes envisioned space as filled with swirling gases. Large whirlpools developed that would form stars eventually, with smaller vortices forming planets and their satellites. In 1755, Immanuel Kant applied Sir Isaac Newton's laws of gravity to Descartes's model and concluded that the swirling gases would assume a disk shape. In 1796, the French mathematician Pierre-Simon Laplace proposed further that the disk would separate eventually into rings. The planets would condense from these rings. Collectively, the theories of Descartes, Kant, and Laplace became known as the nebular hypothesis. It offered reasonable explanations for the flattened appearance of the solar system, the nearly circular orbits of the planets, and their similar orbital directions around the Sun. Nevertheless, it left many questions unanswered and by the late nineteenth century had fallen from popularity.

In 1745, French naturalist Comte de Buffon proposed a catastrophic explanation that contrasted with the nebular hypothesis. His theory suggested that the solar system formed after a collision between the Sun and another massive object. The material thrown out by that collision eventually formed the planets. It was a relatively simple theory that provided answers to questions left unanswered by the nebular hypothesis. The Buffon theory was resurrected and modified by American geologist Thomas Chronder Chamberlin and American astronomer Forest Ray Moulton in 1900, and in 1917 it was revised by Sir James Hopwood Jeans. Jeans's revision had no collision but instead envisioned matter being pulled out of the Sun by the tidal forces of a passing star. This matter would then condense to form planets. Jeans's theory answered certain questions that the nebular hypothesis could not, but it was not a complete explanation.

Technological advances made during the twentieth and twenty-first centuries gave the nebular hypothesis new credibility. From the work of Carl von Weizsäcker, Gerard Peter Kuiper, and Hannes Alfvén, a revised model of the nebular hypothesis arose, envisioning a vast interstellar cloud of gas and dust that fragmented and contracted into smaller, dense regions from which stars would form. It is from within one of these smaller regions that the Sun and its planets formed.

In astronomical terms, the formation of the Sun and planets happened relatively quickly, perhaps in only 100 million years. It is suggested that zones of turbulent eddies once existed in the gas and dust cloud. Through a combination of processes of regional gravitational collapse, condensation, and subsequent accretion of solid particles, large protoplanetary bodies approximated the chemical compositions and were probably the precursors of the present-day planets.

Theories of planetary formation have given humankind a better understanding of the cosmic origins and the intricacies that led to Earth's existence in the solar system. In seeking answers to such questions, both mental and technological capabilities have been expanded, and this has overflowed into everyday lives. Experimentally, the biggest goal of the search for extrasolar planets is the detection and later the imaging of Earth-like planets.

—Paul P. Sipiera

Further Reading

Armitage, Philip J. *Astrophysics of Planet Formation.* 2nd ed., Cambridge UP, 2020.

Arny, Thomas T. *Explorations: An Introduction to Astronomy.* 3rd ed., McGraw-Hill, 2003.

Brandner, Wolfgang. *Planet Formation: Theory, Observations, and Experiments.* Cambridge UP, 2011.

De Pater, Imke, and Jack J. Lissauer. *Planetary Sciences.* Cambridge UP, 2001.

Elkins-Tanton, Linda T., and Benjamin P. Weiss, editors. *Planetesimals: Early Differentiation and Consequences for Planets.* Cambridge UP, 2017.

Fraknoi, Andrew, David Morrison, and Sidney C. Wolff. *Astronomy.* Samurai Media, 2017.

Freedman, Roger A., and William J. Kaufmann III. *Universe.* 8th ed., W. H. Freeman, 2008.

Hartmann, William K. *Moons and Planets.* 5th ed., Thomson Brooks/Cole, 2005.

Henning, Thomas. *Astromineralogy.* Springer, 2003.

Perryman, Michael. *The Exoplanet Handbook.* Cambridge UP, 2011.

PLANETARY EXPLORATION

Fields of Study: Geology; Astrogeology; Geochemistry; Geophysics

ABSTRACT

American, European, and Soviet robotic planetary space probes from 1962 to the present have explored all the known planets in the solar system. The greatest beneficiary of knowledge gained from those investigations has been the field of planetary geology. By comparing our planet to others, a greater understanding of Earth has been acquired.

This has also added to the understanding of planetary formation, and is of great interest to those searching for planets beyond our solar system.

KEY CONCEPTS

flyby: a trajectory through space that passes by a planet or other body without entering into orbit about that body

lander: a spacecraft designed to land in one location on a planet's surface and remain at that location while gathering data about the planet

orbit: a gravitationally bound path in space around a planet

orbital platform: a spacecraft purposed to remain in orbit about a planet and act as a relay station for transfer of data from surface-based devices back to Earth

rover: a vehicle designed to be delivered to a location on the surface of a planet and move from that location to obtain data about the planet

SUMMARY OF THE PLANETARY EXPLORATION MISSIONS

Nearly six decades of planetary exploration by spacecraft dispatched from Earth cannot be completely recounted in an article of this length. What follows is a synopsis of many of the highlights of robotic investigations throughout the solar system. Some of the story is presented sorted by solar system object, and part is chronological in nature. For more detailed descriptions see other articles devoted to individual missions. Successful robotic exploration began in 1962 when the American Mariner 2 probe reached Venus. The first robotic mission to draw major public attention was the Mariner 4 flyby of Mars in 1965. That spacecraft returned a small number of close-range photographs of the Martian surface showing heavy cratering, but also some evidence of flowing water in past geologic ages.

During the mid-1960s, the United States gained valuable experience in interplanetary exploration with the Ranger and Surveyor Programs, designed to prepare the way for the Apollo piloted lunar program. Ranger vehicles returned thousands of pictures before impacting the surface; the Surveyor series soft-landed to collect and analyze samples of the lunar surface. Extensive mapping was done by five Lunar Orbiters which helped scientists pick the Apollo landing sites.

The Soviet probe Venera 3 became the first human-made object to impact a planet, in 1966. Venera 4 and the American Mariner 5 both arrived at Venus in 1967 within hours of each other. Venera 4 released an instrument package that was designed to land on the planet, but stopped transmitting some 30 kilometers above the surface. Mariner 5 passed radio signals through the atmosphere to determine its composition.

In 1969, Mariners 6 and 7 returned much larger numbers of photographs of Mars of far higher quality than those produced by Mariner 4, revealing Mars to a be a planet of very complex geology. Mariner 9, which reached Mars in 1971, became the first probe to orbit the planet rather than just make a flyby, providing scientists with much time to use its instruments to gather data. In the same year, a Soviet probe attached to Mars 3—part of a sequence of mostly unsuccessful missions by the Soviet Union in the early 1970s—reached the surface of Mars but transmitted for only about ninety seconds. (Following these failures, the Soviet Union turned its attention for the next fifteen years to Venus with a series of more successful Venera orbiters and landers.)

Pioneers 10 and 11 were the first missions to the outer planets. Pioneer 10 passed Jupiter in late 1973 and headed on a trajectory that would take it beyond the bounds of the solar system; it transmitted until late 1997. Pioneer 11 visited both Jupiter (1974) and Saturn (1979) and continued transmitting through 1995. Mariner 10 made three flybys of Mercury in 1974-1975, photographing about half of the planet's surface. En route, it passed Venus and photographed the planet's cloud cover. The spacecraft was the first to use the gravitational field of one planet to assist in propelling it to the vicinity of another.

Two Viking missions to Mars in 1976 became the first successful landers on the planet. Viking 1 enthralled the public with the first photographs from the Martian surface and a dramatic set of experiments designed to detect evidence of possible Martian microbial life. The Viking landers continued to transmit meteorological reports for several years. The orbiters that delivered the Viking landers to Mars remained in orbit, adding to the depth and de-

Solar System depiction. Image by Beinahegut, CC BY-SA 4.0, via Wikimedia Commons.

tail of our knowledge of surface features. Several of the Mariner and Viking missions also returned photographs of the two tiny Martian satellites Deimos and Phobos.

Perhaps the highlights of robotic planetary exploration in the twentieth century were the epic missions of Voyagers 1 and 2 to the outer solar system's gas giant planets and beyond. Launched in 1977, Voyager 1 reached Jupiter early in 1979 and Saturn in late 1980. In order to bring it into position to examine Saturn's satellite Titan, scientists bent Voyager 1's trajectory out of the plane of the ecliptic, the plane containing the orbits of the major planets. Voyager 2 arrived at Jupiter in 1979 and reached Saturn in 1981. Timed for launch when planetary alignments made possible an almost incredible "Grand Tour" of the outer solar system, the mission continued on to Uranus in 1986, and Neptune in 1989. Before exiting the system, it returned some of the most spectacular photographs in history.

A pair of Pioneer Venus spacecraft reached their objective in 1978, the first of which entered orbit and transmitted information until 1992. The second carried probes that impacted on the planet's surface. Following a hiatus of more than a decade in robotic planetary exploration, activity resumed in 1989 with the launch of Magellan, which orbited Venus. Magellan used radar to map most of the surface of Venus, which is almost totally obscured by clouds.

Also in 1989, the complex Galileo expedition to Jupiter got under way. Originally intended to be launched out of Earth orbit by the *Atlantis* shuttle, Galileo had to be redesigned for surface launch following the *Challenger* accident. This required a complex series of gravitational boosts from Venus and Earth to acquire the necessary velocity and greatly lengthened the mission. Galileo did not reach Jupiter until the end of 1995. It released a probe into the Jovian upper atmosphere that survived nearly an hour. Galileo then went into orbit around Jupiter to study the planet and its major satellites, especially volcanic Io and enticing Europa. Some scientists believe Europa harbors an ocean beneath its frozen surface. After a greatly extended mission, Galileo was purposely plunged into Jupiter's atmosphere to prevent it from potentially colliding with and contaminating Europa or any other Jovian moon. Galileo was followed by the Juno spacecraft, which launched on August 5, 2011, and reached Jovian orbit on July 5, 2016. The seven-year mission was intended to include 37 orbits about Jupiter with its science mission to concentrate on magnetic and gravity fields, particle detections, plasma wave observations, and investigations using an ultraviolet imaging spectrometer. Also, Juno's imaging system was far superior than Galileo's, and did not suffer from a low transmission rate for data. As of 2019 Juno remains operational, providing surprises about Jupiter so intriguing that its mission was extended until at least 2021.

Robotic exploration of Mars resumed in 1988 with a number of extremely ambitious missions.

Spectacular failures punctuated attempts to return to the planet. In August, 1993, ground stations lost contact with the American Mars Observer shortly before its scheduled arrival, the victim of an explosion in its propulsion system plumbing; in 1988-1989, two complex Soviet missions, Phobos 1 and 2—loaded with international experiments and assisted by the American Deep Space Tracking Network—reached Mars orbit but failed in novel attempts to place landers on the Martian satellites; in 1996 a Russian probe named Mars '96 failed to reach escape velocity from Earth.

A huge success finally came in July, 1997, when Mars Pathfinder used a radical method of landing on Mars, encasing the entire vehicle in a cluster of large air bags to cushion the impact. After settling on the surface, Pathfinder deployed a small robot vehicle named Sojourner, which could receive commands from Earth. Sojourner spent nearly three months exploring the immediate vicinity of the landing site. In September, 1997, the Mars Global Surveyor arrived to begin making highly detailed physiographic maps of the Martian surface. Surveyor was the first vehicle to establish an orbit around another planet using atmospheric braking, which gradually transformed a highly elliptical orbit into a nearly circular one.

Catastrophe returned to haunt the American Mars exploration program in September, 1999, with the loss of the Mars Climate Orbiter, which evidently burned up in the Martian atmosphere as the result of confusion by ground controllers between U.S. and metric units of measurement in plotting the course. Perhaps even more of a setback was the loss, just three months later, of the Mars Polar Lander, a much-anticipated mission to explore the southern polar regions of the planet and search for water ice. Ground controllers lost contact with the lander—which also carried two microprobes designed to search for water or ice beneath the surface—shortly after it entered the Martian atmosphere. These much-publicized misfortunes prompted full-scale review of the Mars exploration program and missions scheduled for future launch. Tremendous success followed when a pair of Mars Exploration Rovers named Spirit and Opportunity landed independently at vastly separated sites in January, 2004. Both rovers were much larger and more capable than Mars Pathfinder, traversing distances on the surface measured in kilometers rather than meters. Each continued to produce important results well after its first year on Mars. Among the most exciting of these was the definitive identification of rocks and minerals that required the presence of water on Mars in the past. Results hinted at the potential for past life on Mars, but no direct evidence was found.

Mars studies enjoyed a great deal of success during the early portion of the twenty-first century. The National Aeronautics and Space Administration (NASA) sent the Mars Odyssey and Mars Reconnaissance Orbiter to act as scientific platforms, imaging systems, and relays to surface robots. Spirit continued to operate until it became stuck in soft sand and was lost in 2010. Opportunity lasted even longer roving across the Martian surface. It encountered a nearly complete planetary dust storm which left it without the ability to recharge its solar battery, and its mission was declared over in 2019. For a time, there were three rovers operating on the Red Planet. The Mars Science Laboratory or Curiosity rover was delivered to the Red Planet in 2012. It landed in Gale crater and moved toward the centrally located Mount Sharp to seek evidence of sedimentary layers. As of 2019 Curiosity was the only rover working on the Martian surface, but the launch of the Mars 2020 rover was already planned.

Surface operations on Mars early in the twenty-first century also involved stationary landers. On May 25, 2008 the Mars Phoenix spacecraft landed at a high latitude in the Martian Northern Hemisphere. Its robotic arm was able to reveal evidence of subsurface water ice. Then on November 26, 2018 the InSight lander touched down in Elysium Planitia to perform heat flow and seismology research on Mars. As of 2019 the seismometer had detected Marsquakes, but the heat flow experiment was having trouble digging itself deep enough to deploy its temperature sensors.

Cassini, intended to be the first orbital mission of the Saturn system, left Earth in October, 1997, carrying the Huygens probe, designed to land on Titan, the planet's largest satellite. A joint effort of NASA and the European Space Agency (ESA), the spacecraft entered Saturn's orbit in July, 2004, and began its studies of the planet, its system of moons, and the ring plane. Late in 2004, the Huygens probe was detached from the main spacecraft, and on January 14, 2005, it descended through the atmosphere of Titan and touched down safely on that large moon's surface. Cassini remained in orbit about Saturn for over twelve years. The orbiter flew past Titan a great many times at close range, discovered that the small moon Enceladus sported water geysers near its south pole and thereby developed an interest in this world for both planetary scientists and astrobiologists, studied the ring system of the planet in great detail, and basically rewrote the textbooks on the solar system's most beautiful planet. To avoid potential plutonium contamination of any world in the Saturn system, when the mission was drawing to a close, Cassini was maneuvered to fly through the ring plane, and on September 15, 2017, be destroyed in the planet's atmosphere.

Clementine, a joint venture of NASA and the United States Department of Defense, signified renewed interest in exploration of Earth's Moon and spent two months of 1994 producing highly detailed maps of the surface. Lunar Prospector, launched in 1998, spent eighteen months in orbit studying lunar geology. Controllers caused it to impact near the lunar south pole in July, 1999, in an unsuccessful attempt to reveal the presence of water ice.

The planet Mercury had not been revisited since the initial expedition, Mariner 10, in 1975. A Mercury orbiter named MESSENGER (Mercury Surface, Space Environment, Geochemistry, and Ranging) was launched on August 3, 2004. After a series of gravity assists, the spacecraft entered Mercury orbit in 2011, and remained operational until April 30, 2015 when it was crashed purposely into the planet's surface.

Expeditions to smaller objects in the solar system contribute significantly to our knowledge of planetary evolution. An international flotilla of probes encountered Comet Halley in 1986. The Near Earth Asteroid Rendezvous (NEAR) mission, launched by the United States in 1996, entered orbit around the asteroid Eros in 2000. Stardust, launched in 1999, rendezvoused with Comet Wild 2 on January 2, 2004, and captured cometary dust samples; the spacecraft returned those samples to Earth in 2006. The Genesis spacecraft had demonstrated the method of sample return to be used by Stardust when it came back to Earth over the salt flats of Utah on September 8, 2004, carrying solar wind particles trapped in aerogel-based collectors. Unfortunately, the spacecraft crashed into the ground and cracked open rather than having its parachute snagged safely by a recovery helicopter. Fortunately for the Genesis program, samples survived the crash, but the incident created consternation as to whether or not Stardust would suffer a similar fate upon return to Earth. The Dawn mission launched on September 27, 2007, and flew to the asteroid belt. It first entered orbit about Vesta, and remained there studying that large asteroid for nearly 14 months. Dawn then broke out of orbit and headed for the dwarf planet Ceres. During a period of over 3.5 years Dawn discovered a number of unexpected features on Ceres, not the least of which was a large crater with two very bright spots inside. Dawn was the only spacecraft to enter orbit about one solar system body only to go and orbit yet another outside the Earth-Moon system.

Throughout the last half of the twentieth century, vastly improved Earth-based and orbital telescopes, radar imaging, and radio astronomy supplemented the findings of robotic probes. For example, as early as 1956, radio astronomers confirmed the high temperatures of the surface and atmosphere of Venus. Telescopes detected rings around Uranus and a previously unknown satellite for Pluto in 1978. In 1992, radar mapping of Mercury suggested possible water ice deposits near the poles. Observation of the impacts of fragments of Comet Shoemaker-Levy on Jupiter in 1994 illustrated the potentially disastrous consequences of impacts even by small bodies, and in 1998-1999 radar signals returned conflicting data for a possible hydrocarbon ocean on Titan. Growing evidence of a vast number of planetesimals in orbit far beyond Pluto greatly complicated our image of the solar system. In the 1990s, astronomers began to detect evidence of planets, Jupiter-sized or much larger, orbiting other stars. As these discoveries continue, and improved technology begins to reveal smaller planets as well, there will be an invaluable comparative basis upon which to better understand our own solar system. Those studies continued in the early twenty-first century, and were joined by investigations performed by space-based telescopes such as the Hubble Space Telescope and Spitzer Space Telescope.

CONTRIBUTIONS FROM PLANETARY EXPLORATION MISSIONS TO SCIENCE AND SOCIETY

Robotic planetary exploration generated far more knowledge about the solar system in the last third of the twentieth century and the first two decades of the twenty-first century than had been accumulated in all previous periods of history combined. Photographic returns alone represented such a stunning advance over fuzzy images from Earth-bound telescopes that textbooks about the solar system had to be completely rewritten. Other planets and their satellites turned out to be far more varied and individually peculiar than could have been imagined. Major accomplishments include a nearly complete topographic map of Venus and working theories of how that planet, so similar to Earth in size, could have developed a runaway greenhouse effect leading to its thick atmosphere and extremely high temperatures and pressures. During the first billion or so years of its existence, Venus may have evolved along lines similar to Earth, perhaps even with oceans, but later developed a hellish environment resulting from this runaway greenhouse effect.

Mars also shows many signs of early evolution along terrestrial lines. Although heavily cratered and far colder today than even Antarctica, and lacking all but the thinnest atmosphere of carbon dioxide, the topography of Mars is rich with hints of huge amounts of flowing water during its first billion years. The comparative flatness of the planet's northern hemisphere suggests the existence of a primordial ocean, and for liquid water to have been present on the surface, the atmosphere must have been much denser and ambient temperatures much higher than at present. Despite such signs, both Venus and Mars are significantly different from Earth; both, for example, nearly lack magnetic fields, required to deflect dangerous radiation.

The outer planets—Jupiter, Saturn, Uranus, and Neptune, once thought to be more or less alike—revealed many differences. The atmospheric composition of Jupiter, for example, as analyzed by the Galileo probe, is not consonant with formation of the planet at its current distance from the Sun. Jupiter's major satellites have vastly different conditions. Europa appears covered with water ice, but the fractures in the ice indicate a liquid ocean beneath, perhaps kept above freezing by volcanism and tidal stresses as the satellite resonates with Jupiter and other moons. Ganymede and Callisto also may be covered with water ice and may have subsurface salty oceans. Io, caught in powerful gravitational fields within the system, is the most geologically active body in the solar system, completely resurfacing itself with lava flows about every hundred million years. Saturn revealed the immense complexity of its rings (though all the gas giants turned out to have thin ring systems); Voyager photographs discovered a multitude of previously unknown satellites. The number of satellites of Uranus increased to eighteen, and those of Neptune to eight; again, the larger satellites showed great individuality. Geologic activity appears to be present on Neptune's largest satellite, Triton. Postmission analysis of Voyager data, coupled with telescopic observations, found many more moons around the gas giants.

CONTEXT OF THE PLANETARY EXPLORATION MISSIONS

Although planetary probes have accounted for a tiny percentage of robotic missions, they generate disproportionate public interest through both their spectacular successes and their much-publicized failures. Earlier missions were relatively expensive, and the odds against success continue to be daunting. Many links exist between public attitudes toward piloted missions and these presumed robotic precursors. The captivating nature of deep-space photography, robotic systems, and other instrumentation—when they succeed—serve the argument that piloted missions are dangerous and unnecessary; the failures often do the same. Others argue that a human presence could deal with malfunctions and that planning and development time—especially added to flight time to the outer planets—is so long that when a mission arrives at its destination, its technology is two decades old.

Inevitably, much public attention has been invested in findings that might suggest life, or at least the conditions for life, elsewhere in the solar system. Witness the suspenseful coverage of test programs of Martian soil samples on Vikings 1 and 2. Increasingly, robotic missions explore for water or water ice (now known to be relatively abundant), in the hope that some form of life might also be present or that they might supply constituents for rocket propellants and sustain eventual human colonization.

Something else that spacecraft search for at Mars is methane. The Curiosity rover in 2019 detected a change in methane emission at the surface. Previous orbiting spacecraft had noted significant methane on Mars. The open question remains whether that methane originated geologically or by biological means.

The real beneficiaries of this kind of exploration thus far have been the planetary sciences. Our knowledge of volcanism, plate tectonics, and planetary evolution has expanded enormously with the data returned from other planets as a comparative base. Other models of atmospheric circulation add to our understanding of atmospheric physics and climatology. The complex gravitational resonances in the satellite systems of the outer planets show that the overall gravitational structure of the solar system is far more complex than previously believed. Compared to Mars, whose angle of axial inclination appears to have varied greatly over geologic time, with consequent upheavals in climate, scientists now value the presence of a large satellite such as the Moon in lending stability to a planetary system. Growing concerns about the number of small asteroids discovered in the vicinity of Earth orbit—punctuated by a consensus that major asteroid impacts may have led to the near-extinction of life on Earth on several occasions—have completely changed how scientists and the informed public view the significance of space exploration.

Many other planetary science missions were conducted since the dawn of the space age than the ones listed in this summary article. What constitutes a planet remained somewhat controversial even in 2019. At a meeting of the International Astronomical Union in Prague in 2006, four special criteria had been established for determining what constitutes a planet. That classification scheme resulted in our solar system having terrestrial planets (Mercury, Venus, Earth, and Mars), gas giants (Jupiter and Saturn), ice giants (Uranus and Neptune), and dwarf planets. Pluto had been demoted from planetary status and was referred to in 2019 as a dwarf planet like Ceres. Pluto was encountered by the New Horizons spacecraft, and the interested reader is directed to a separate article concerning that historic mission.

—Ronald W. Davis and David G. Fisher

Further Reading

Adamson, Thomas K. *The Secrets of Jupiter (Planets).* National Geographic, 2015.

———. *The Secrets of Neptune (Planets).* National Geographic, 2015.

———. *The Secrets of Uranus (Planets).* National Geographic, 2015.

Baines, Kevin H., F. Michael Flasar, Norbert Krupp, and Tom Stallard. *Saturn in the Twenty-First Century.* Cambridge UP, 2019.

Brandt, John C., and Robert D. Chapman. *Introduction to Comets.* Cambridge UP, 2004.

Carlson-Berne, Emma. *The Secrets of Venus (Planets).* National Geographic, 2015.

David, Leonard. *Mars: Our Future on the Red Planet.* National Geographic, 2016.

———. *Moon Rush: The New Space Race.* National Geographic, 2019.

Grinspoon, David Harry. *Venus Revealed: A New Look Below the Clouds of Our Mysterious Twin Planet.* Perseus Book Group, 1998.

Harland, David M. *Jupiter Odyssey: The Story of NASA's Galileo Mission.* Springer-Praxis, 2000.

———. *Mission to Saturn: Cassini and the Huygens Probe.* Springer-Praxis, 2002.

Irwin, Patrick G. J. *Giant Planets of Our Solar System: Atmospheres, Composition, and Structure.* Springer-Praxis, 2003.

Kaufman, Marc. *Mars Up Close: Inside the Curiosity Mission.* National Geographic, 2014.

Miner, Ellis D., and Randii R. Wessen. *Neptune: The Planet, Rings, and Satellites.* Springer-Praxis, 2002.

Mishkin, Andrew. *Sojourner: An Insider's View of the Mars Pathfinder. Mission* Berkley Books, 2003.

Morrison, David, and Tobias Owen. *The Planetary System.* 3rd ed., Addison-Wesley, 2003.

Radomski, Kassandra Kathleen. *The Secrets of Saturn (Planets).* National Geographic, 2015.

Sparrow, Giles. *Mars.* Quercus, 2015.

Strom, Robert G., and Ann L. Sprague. *Exploring Mercury: The Iron Planet.* Springer-Praxis, 2003.

PLANETARY INTERIORS

Fields of Study: Geology; Astrogeology; Geochemistry; Geophysics

ABSTRACT

Planetary interiors characteristically have a nested shell structure, with concentric layers of differing physical prop-

erties and chemical compositions. Often, planetary materials are compressed into unusual forms by the tremendous pressures deep within planets.

KEY CONCEPTS

bar: a metric unit of pressure equivalent to one standard atmosphere

convection: a circular movement within a fluid by which hotter, less dense material rises within the fluid medium and cooler, denser material sinks within the fluid medium

Mohorovičić Discontinuity: a narrow transition zone between the base of the crust and the upper mantle layer

moment of inertia: a physical quality of a spinning mass, analogous to the momentum of an object undergoing linear motion

radiogenic heating: heat generated by the decomposition of radioactive elements within Earth's interior

OVERVIEW OF PLANETARY INTERIORS

The interior of Earth is typical in many ways of the interiors of other terrestrial, or "rocky," planets and serves as a useful starting point for understanding those planetary interiors. Earth consists of three major layers, differing sharply in composition and physical properties: a thin outermost layer called the "crust," a thick intermediate layer called the "mantle," and a dense central core. Earth's crust consists of two quite different materials: granitic rocks averaging about 40 kilometers thick beneath the continents and basaltic rocks 5 to 10 kilometers thick beneath the ocean basins. Sea waters fill the ocean basins because the crust is thin there and its elevation low, but there is no direct connection between oceans and oceanic crust. The crust of Earth composes only about 0.3 percent of its mass and about 0.8 percent of its volume. Beneath Earth's crust is a dense mantle of iron and magnesium silicate rocks extending to a depth of about 2,885 kilometers. The mantle occupies about five-sixths of the volume of Earth and accounts for two-thirds of its mass.

At the center of Earth is a dense core consisting of two parts. The outer core, with a radius of 3,486 kilometers, is slightly more than half the diameter of Earth and about the size of the planet Mars. This outer core is most likely molten iron and nickel with other elements present in smaller amounts. Fluid movements in the liquid outer core are believed to be the source of Earth's magnetic field. The inner core, which is solid but similar to the outer core in composition, has a radius of 1,216 kilometers. The core occupies about one-sixth of the volume of Earth and accounts for a third of its mass. The inner core accounts for only 0.7 percent of the volume of Earth and 1 percent of its mass.

The composition of Earth's interior, as well as that of other planets, is dictated by the chemistry of the solar system. If the solar system formed from a cloud of material similar to the present Sun in composition, any solid grains that condensed should have a composition similar to the Sun, minus those elements that would have formed gases. In the inner solar system (above the freezing point of water), most hydrogen, nitrogen, and carbon would have formed gases. Oxygen would largely be tied up in water vapor, but much would form oxides with various metallic elements or combine with silicon to form silicate minerals. Some sulfur would be available for iron sulfide minerals such as troilite (FeS). Noble gases (helium, neon, argon, krypton, and xenon) combine rarely with other elements and would be nearly absent from any solid grains. The most abundant elements of the inner planets should be oxygen, magnesium, silicon, sulfur, and iron. Direct chemical analysis of solar system bodies has been largely limited to Earth, the Moon, meteorites, and more recently Mars, but these analyses and data from robotic missions carrying out spectroscopic examinations of Venus and Mercury are consistent with theoretical predictions.

Meteorites are particularly valuable because they provide samples of the materials that condensed in the early solar system. A class of meteorites called "chondrites" closely matches the theoretically predicted composition of the planets; they are believed to be samples of very primitive solar system meteorites. Chondrites are usually the starting point for studies of the chemical composition of the inner planets. The presumed composition of Earth's mantle, mostly iron and magnesium silicates, agrees well with the theory that Earth has an overall chondritic composition. Much of the iron in Earth has settled into the central core, together with nickel and perhaps much of Earth's sulfur.

Pressure deep within Earth is caused by the weight of overlying rocks. It is easy to calculate once the distribution of mass within Earth is known. The gravitational attraction within Earth is solely caused by the mass of Earth between the observer and the core; the gravitational effect of a shell surrounding an observer is zero. In a uniform Earth, gravitational attraction would decrease steadily toward the center, but because Earth's core is dense, gravitational attraction actually increases slightly with depth and is about 4 percent higher at the boundary of the core than at Earth's surface. Pressure at the base of Earth's continental crust is about ten thousand times greater than Earth's atmospheric pressure. Geologists commonly use a metric unit, the bar, to measure pressure. A bar is convenient because it is nearly equal to the pressure of Earth's atmosphere at sea level and thus is easily visualized. A bar is equal to 100,000 pascals. The pressure at the bottom of the mantle is about 1.4 million bars, and the pressure at Earth's center is about 3.6 million bars.

Temperatures deep inside the planet are not known as precisely as many other physical quantities. Earth becomes hotter at an average rate of 25 kelvins per kilometer near the surface, reaching about 1,250 kelvins at the base of the continental crust. Thereafter, temperatures increase much more slowly, reaching 3,000 to 5,000 kelvins in the core. Despite high temperatures in the mantle, it is still solid, because high pressure elevates the melting point of rocks. Portions of the mantle melt at depths of 50 to 100 kilometers only locally to generate magma for volcanic eruptions. Much of Earth's internal heat is caused by radiogenic heating, but

Physical processes active in Earth's interior may be at work in other planets as well. Photo via iStock.com/Maximusnd. [Used under license.]

some may be primordial remnants from the accretion of Earth or gravitational energy released when dense material sank to create the core.

Higher temperatures in Earth's deep interior create convection, the rise of hot, light material and the sinking of cooler, denser material. Although the rocks in the mantle are very rigid over short time spans, they can flow slowly over long timescales. It is still not known whether convection occurs throughout the mantle, whether there is one layer of convective motion or several, or precisely how the changes in crystal structure of mantle rocks at different depths might affect convection. Crustal plates consist of the crust, plus part of the mantle beneath. The material of the plates, collectively called the "lithosphere," is about 100 kilometers thick.

Movement of crustal plates on Earth results in recycling of oceanic crust over a period of about 150 to 200 million years. New oceanic crust is formed as plates spread apart at the midocean ridges, and old oceanic crust returns to Earth's interior at oceanic trenches, mostly around the Pacific Rim. As oceanic crust is drawn back into Earth's interior, it melts partially, and molten material invades Earth's crust. This molten material is enriched in elements of large ionic radius and large electric charge that do not fit easily into the crystal structures of mantle rocks. During Earth's lifetime, the crust has become enriched in potassium, rubidium, silicon, uranium, thorium, and the rare earth elements. As a result, the mantle has become depleted. Volatile materials such as water and carbon dioxide have also escaped to the surface.

Physical processes active in Earth's interior may be at work in other planets as well: internal heating, recrystallization of rocks and other materials to denser forms at great depths, gravitational separation of a dense metallic or metallic sulfide core, internal flow caused by convection, and segregation of particular materials into a crust. These processes are all driven, directly or indirectly, by internal heat. Small planetary bodies have less internal heat than large ones. As the planets accreted from the impact of smaller bodies, impacts released heat. Smaller bodies had less mass to retain heat and therefore cooled more quickly than larger ones. Gravitational separation of a dense core would also release heat, but separation of a core is less likely in a body with weak gravity and a cool interior. Small bodies have smaller amounts of radioactive material for radiogenic heating and smaller thicknesses of insulating mass to retain the heat that is produced. Larger bodies of the inner solar system show considerably more internal activity than smaller ones.

Spectroscopic evidence indicates that some large asteroids have basaltic surfaces, which means that bodies a few hundred kilometers in diameter have undergone enough internal heating at some time in their history to melt rocks. The Moon is believed to have a very small core, probably rich in iron or iron sulfide. Mercury is unusually dense for a small planet and probably has a core with a radius three-fourths that of the planet itself. Although both Mercury and the Moon are covered with vast volcanic plains, neither has experienced significant volcanic or tectonic activity in the last 3 billion years, perhaps because their interiors are too cool and rigid to permit much convection or melting.

Venus is only slightly smaller than Earth, and its interior is probably similar to Earth's. The surface of Venus shows many manifestations of intense internal activity such as fracturing and folding of the crust and widespread volcanic activity. Nevertheless, Venus lacks the sharp division between two kinds of crust that is so evident on Earth. Also, Venus lacks obvious topographic features typical of plate tectonics on Earth: ridges with crustal spreading, fracture zones, or trenches. One possible reason for its different geology may be that Venus has a thin, flexible crust that results from its very high surface temperature (650 kelvins), rather than thick, rigid crustal plates, as on Earth.

Mars is believed to have a small core of iron or iron sulfide and shows considerably more evidence of recent dynamic internal activity than Mercury or the Moon. A gigantic rift valley, Valles Marineris, extends more than 5,000 kilometers on Mars. The Tharsis Ridge is close by, crowned with several enormous volcanoes. The relative lack of craters on these features of Mars indicates that they formed much more recently than the ancient volcanic plains on Mercury or Venus, probably about one billion years ago. Nevertheless, the opposite side of Mars consists of ancient cratered terrain that has been inactive since its early history. Mars never developed a global system of crustal plates as had Earth, probably because the planet cooled enough for a very thick, rigid lithosphere to form.

In the outer solar system, water ice and other volatile materials (collectively termed "ices") are major constituents of planetary bodies. Water ice evaporates quickly in a vacuum even from the solid state. Only at 175 kelvins or a little sunward of Jupiter was it cold enough for ice to form in a vacuum and survive billions of years. At 150 kelvins, methane hydrate (a solid mixture of methane and water) condensed, followed by ammonia hydrate at 120 kelvins, and argon and pure methane at about 65 kelvins. These temperatures are comparable to the surface temperatures on the satellites of Jupiter (135 kelvins), Saturn (100 kelvins), Uranus (85 kelvins), and Neptune (50 kelvins).

Satellites of all the large outer planets, and perhaps the dwarf planet Pluto and its satellite Charon as well, consist of rocky cores surrounded by mantles of ices. One of the great surprises of planetary exploration has been the degree of internal activity that occurs on objects made largely of ices. A rocky core need not be very large to generate enough radiogenic heat to warm water ice to near or above the melting point, and some larger bodies, especially the large satellites of Jupiter, may have liquid water or icy slush within their mantles. A number of satellites in the outer solar system show evidence that ice has flowed upward from the interior to the surface. This process has been called "ice volcanism," but it is more nearly akin to vertical glacial flow. Water ice undergoes a remarkable series of changes in crystal structure at high pressure, and these high-pressure forms of ice are present undoubtedly in the interiors of large satellites in the outer solar system.

The icy satellites of the outer solar system circle very massive planets and experience a force that is only of minor importance on Earth: tidal forces. Tidal forces arise because the gravitational attraction between two bodies is greater on their near sides than their far sides. Tidal forces distort bodies into ellipsoidal shapes whose long axes point toward the other attracting body. The Moon exerts only a small effect on the solid Earth and a more noticeable effect on the oceans, but Earth's tidal force has locked the Moon's rotation so that it always shows the same face to Earth. Tidal stresses also produce very small "moonquakes" when the Moon is nearest to Earth in its orbit.

Almost all satellites in the solar system are similarly locked to their planets. If a planet has several large satellites, the satellites also exert tidal forces that tend to twist the satellites into alignment with one another. Under the right conditions, the conflicting tidal forces can generate a significant amount of internal frictional heat. The most spectacular example is Io, Jupiter's innermost satellite. Frictional heating from tidal forces makes Io's interior hot enough to cause active volcanism, despite the fact that Io is smaller than Earth's moon. The ultimate source of energy for this heat is the orbital motion of Jupiter's satellites and the rotation of Jupiter. Heating of Io is but a negligible energy drain on these bodies. Other satellites, notably Europa, another satellite of Jupiter, the Saturnian satellite Enceladus, and several satellites of Uranus have extensive fracture networks also believed to be related to tidal stresses.

Jupiter, Saturn, Uranus, and Neptune, the so-called gas giants, probably formed by accumulating rock and ice cores several times as massive as Earth. These large cores were massive enough to attract and hold large amounts of gases. Jupiter and Saturn, which were massive enough to retain essentially all of their gases, are similar to the Sun in composition. Uranus and Neptune did not become massive enough to attract or retain as much hydrogen and are made mostly of denser gases such as ammonia or methane. The deep interiors of all of these planets are very dense gases or fluids. In Jupiter and Saturn, pressures become great enough that hydrogen is believed to have metallic properties, a form of matter predicted in theory and only created in small systems in the laboratory. Fluid movements in the interiors of the gas giants are believed to account for their magnetic fields, in the same manner that fluid motions in Earth's core produce a magnetic field.

METHODS OF STUDYING PLANETARY INTERIORS

The deepest mine on Earth is 4 kilometers deep, and the deepest well ever drilled is 12 kilometers deep, about one one-thousandth the diameter of Earth. Direct sampling of Earth's interior has barely penetrated through the crust, to the Mohorovičić

Discontinuity, much less reached the core. Geologic processes have brought materials from the deep crust and upper mantle to the surface, from depths perhaps greater than 100 kilometers. Knowledge of Earth's deep interior is derived from a variety of indirect techniques, therefore, and knowledge of the interiors of other planets is based in part on knowledge of Earth's interior.

Some information about Earth's interior can be derived from the surface. Most rocks have densities of 2.7 to 3.0 grams per cubic centimeter. The size of Earth and hence its volume are known from mapping. Earth's mass is known from its gravitational attraction. The average density of Earth, 5.5 grams per cubic centimeter, implies that there is dense material deep in the interior. Another observation supports this conclusion. All rotating objects, including Earth, can be described by a physical quantity called "moment of inertia." Moment of inertia plays much the same role in equations of rotational motion that mass plays in equations of linear motion. Earth's moment of inertia can be found from observations of gravitational effects of the Moon on Earth. It does not match the moment of inertia of a uniform body with the size and mass of Earth; instead, Earth's moment of inertia indicates that much of Earth's mass is concentrated near its center. Similar studies applied to other planets also provide clues to their internal structure.

Samples of Earth's mantle are found in two geologic settings: as parts of ophiolites (fragments of oceanic crust incorporated into mountain ranges) and as inclusions in kimberlite pipes, volcano-like vents that appear to have brought rocks (and occasionally diamonds) from the mantle to the surface with great speed and violence. The composition of mantle rocks—predominantly iron and magnesium silicates—agrees with the theory that Earth is largely composed of chondritic raw material and their mechanical properties match the properties of the upper mantle as deduced from seismic studies.

One measure of solar-system bodies can provide great insight into their composition, even for bodies not sampled directly: bulk density, which is the mass of a body divided by its volume. The volume of a body can be computed readily once the diameter is known, and diameter can be obtained with high precision with spacecraft-based imagery. The same spacecraft that obtains imagery can also provide an accurate mass determination, through the gravitational effect of the body on the path of the spacecraft. Bodies with bulk densities of 3 to 4 grams per cubic centimeter, such as the Moon, Mars, and Io, are probably composed mostly of silicates of the type found in Earth's mantle.

The only solid that is abundant in the solar system and less dense than silicate rocks is water ice, with a density of 0.9 gram per cubic centimeter. Small solid bodies with densities of 1 to 3 grams per cubic centimeter are very likely made of varying proportions of silicate rock and water ice. Most of the satellites in the outer solar system are of this composition. The very large outer planets also have low bulk densities (Saturn has a bulk density of only 0.7 gram per cubic centimeter). These planets are known, from spectroscopic evidence as well as direct imaging by spacecraft, to be composed mostly of dense gases, probably with solid cores.

CONTEXT OF PLANETARY INTERIORS

Until seismic methods became available, the only way of knowing anything of Earth's interior was by analyzing its exterior properties, such as mass, shape, and gravitational attraction. During the eighteenth and early nineteenth centuries, accurate determination of the shape of Earth was at the foremost frontiers of research in science. Many now-indispensable advanced mathematical methods were stimulated by study of Earth's shape and gravitation.

Seismic studies of Earth's interior began in the late nineteenth century. John Milne, an English engineer working in Japan, refined the crude seismographs then in existence to new heights of sensitivity. By the 1880s, seismographs were capable of detecting earthquakes at distances of thousands of kilometers. In 1896, Richard Dixon Oldham argued from astronomical and geologic evidence that Earth must have a dense iron core. By 1906, he presented evidence for a core by showing that seismic waves passing through the center of Earth traveled more slowly than they did in the mantle. In 1909, the Yugoslavian seismologist Andrija Mohorovičić noted that seismic waves for nearby earthquakes showed an abrupt change in velocity beyond a certain distance, which he interpreted as evidence for a thin outer crust. By 1912, Beno Gutenberg had determined accurately the dimensions of the core. In 1936, the Danish seismologist Inge Lehmann presented evidence for the existence of the inner core. Until the 1920s, most seismographs were mechanical, curious blends of massive weights and delicate lever mechanisms to obtain enough sensitivity and magnification to record faint signals. Since that time, electronic signal amplification has vastly improved the sensitivity of seismographs.

One of the great conceptual advances in the understanding of Earth's interior was the hypothesis of continental drift, proposed by the German meteorologist Alfred Lothar Wegener in 1912. Beginning in the late 1950s, evidence began to accumulate that the active processes for crustal motion on Earth actually were concentrated in the ocean basins. Between 1960 and 1975, a modified concept of continental drift, or plate tectonics, became the generally accepted view among geologists.

Study of the interiors of other planets was not possible until spacecraft could provide accurate dimensions and masses for these objects, as well as views of their surfaces that might provide clues to internal processes. The only bodies besides Earth to be investigated seismically are the Moon and Mars. Apollo astronauts placed seismographs on the Moon between 1969 and 1972. Some spent third stages of the Saturn V rocket and some discarded the ascent stage of the Lunar Module were purposely impacted on the Moon to create seismic events of known magnitude for surface experiments to detect. These impacts helped to calibrate the experiments, and also to decipher the subsurface structure of the Moon. Seismic waves caused by natural meteor impacts have also helped probe the lunar interior. Seismographic instruments emplaced by robotic missions to Mars have detected seismic activity, or Marsquakes, there.

One of the most dramatic insights into planetary interiors was provided by American scientists Stanton Peale, Patrick Cassen, and Ray Reynolds. They calculated the amount of tidal heating that might occur on Io and found that Io could get hot enough to sustain volcanic activity. They published this hypothesis in the journal *Science* less than a month before the Voyager spacecraft returned the first photographs of volcanic eruptions on Io, the first active volcanism ever seen outside Earth.

—Steven I. Dutch

Further Reading

Blondel, Phillippe, and John Mason, editors. *Solar System Update*. Springer, 2006.

Esposito, Larry W., Ellen R. Stofan, and Thomas E. Cravens, editors. *Exploring Venus as a Terrestrial Planet*. American Geophysical Union, 2007.

Faure, Gunter, and Teresa M. Mensing. *Introduction to Planetary Science: The Geological Perspective*. Springer, 2007.

Freedman, Roger A., and William J. Kaufmann III. *Universe*. 8th ed., W. H. Freeman, 2008.

Hartmann, William K. *Moons and Planets*. 5th ed., Thomson Brooks/Cole, 2005.

Irwin, Patrick G. J. *Giant Planets of Our Solar System: An Introduction*. 2nd ed., Springer, 2006.

Lauretta, Dante S. and Harry Y. McSween Jr., editors. *Meteorites and the Early Solar System II*. U of Arizona P, 2006.

Leverington, David. *Babylon to Voyager and Beyond: A History of Planetary Astronomy*. Cambridge UP, 2003.

Lewis, John S. *Physics and Chemistry of the Solar System*. 2nd ed., Academic Press, 2004.

McSween, Harry Y. Jr., Jeffrey E. Moersch, Devon M. Burr, et al. *Planetary Geoscience*. Cambridge UP, 2019.

Morrison, David, and Tobias Owen. *The Planetary System*. 3rd ed., Pearson/Addison-Wesley, 2003.

Squyres, Steve. *Roving Mars: Spirit, Opportunity, and the Exploration of the Red Planet*. Hyperion, 2006.

Plate Tectonics

Fields of Study: Geology; Astrogeology; Geochemistry; Geophysics

ABSTRACT

Plate tectonics provides an explanation for the present-day structure of the outer part of Earth. It provides a framework for understanding the global distribution of mountain building, earthquake activity, and volcanism; the geology of ocean basins; various associations of igneous, metamorphic, and sedimentary rocks; and the formation and location of mineral resources.

KEY CONCEPTS

asthenosphere: a transition zone of molten rock, or magma, between the base of the lithosphere and the upper mantle

continental crust: the high silica-content rock of continental land masses, less dense than the basaltic rock of oceanic crust

lithosphere: the solid rock crust of the planet

mantle: the thick layer of hot, semi-molten rock surrounding the core of the planet between the core and the lithosphere

oceanic crust: the dense, iron-rich basaltic rock of the seafloor formed by solidified magma rising at divergent plate boundaries under the ocean

rifting: being separated by force applied in opposite directions

BACKGROUND OF PLATE TECTONICS

Plate tectonics theory developed from Alfred Wegener's theory of continental drift and is based on a concept of Earth in which a rigid, outer shell, the lithosphere, lies above a hotter, weaker, partially molten region of the mantle known as the asthenosphere. The thickness of the lithosphere varies between 50 and 150 kilometers, and it consists of crust and the underlying upper mantle. The asthenosphere extends from the base of the lithosphere to a depth of about 700 kilometers. The brittle lithosphere is broken into a pattern of internally rigid plates that move across Earth's surface relative to each other. Seven major plates and a number of smaller ones have been identified, and they grind and scrape against one another as they move independently, similar to chunks of ice on water. Most of Earth's dynamic activity, including earthquakes and volcanism, occurs along plate boundaries, and the global distribution of these tectonic phenomena delineates the boundaries of the plates.

PLATE BOUNDARIES AND MOTIONS

Geophysical data, geological observations, and theoretical deductions support the existence of three basic types of plate boundaries: divergent boundaries, where adjacent plates move apart (diverge) from each other; convergent boundaries, where adjacent plates move toward each other (converge); and transform boundaries, where plates slip past one another in a direction parallel to their common boundary. The velocity with which plates move varies from plate to plate and within portions of the same plate, ranging from two to twenty centimeters per year. This rate is determined from radioactive dating estimates of the age of the seafloor as a function of distance from midoceanic ridge crests (seafloor spreading ridges).

DIVERGENT PLATE BOUNDARIES

At midoceanic ridges, or divergent plate boundaries, new seafloor is created from molten basalt (magma) rising from the asthenosphere. A great deal of volcanic activity thus occurs at divergent boundaries. Because of the pulling apart (rifting) of the plates of lithosphere, earthquake activity will also occur along divergent boundaries, and since the rift is caused by magma rising from the mantle, the earthquakes will be frequent, shallow, and mild.

An example of continental rifting (divergence) in its embryonic stage is seen in the Red Sea, where the Arabian plate is separating from the African plate, creating a new oceanic ridge. Another modern-day example is the East African Rift system, which is the site of active rifting extending in a north-south direction through Africa, creating the Great Rift Valley structure of that continent. As it continues, it will eventually fragment Africa, and the ocean will extend between and separate the resulting pieces. Through divergence, or rifting, large plates are broken up into smaller ones.

CONVERGENT PLATE BOUNDARIES

Because Earth is neither expanding nor contracting, the increase in lithosphere created along divergent boundaries must be compensated for by the destruction of lithosphere elsewhere. Otherwise the radius of the planet would increase. At convergent plate boundaries, plates are moving together, and three scenarios are possible depending on whether the crust of the lithosphere is oceanic or continental.

If both converging plates are made of oceanic crust, one will inevitably be older, and thus cooler and denser than the other plate. The denser plate will plunge below (subduct to) the less-dense plate and descend down into the asthenosphere. This type of plate boundary is called a "subduction zone," and the boundary along the two interacting plates forms a trench. The subducted plate is heated by the hot asthenosphere and, in time, becomes hot enough to melt. Some of the melted material rises buoyantly through fissures and cracks to form volcanoes on the overlying plate, whereas other parts of the melted material will eventually migrate to and rise again at a divergent boundary (spreading ridge). Thus the oceanic lithosphere is constantly being recycled. The volcanoes along the overriding plate may form a string of islands called "island arcs." Japan, the Aleutians, and the Marianas are good examples of island arcs resulting from subduction of two plates consisting of oceanic lithosphere.

If the leading edge of one of the two convergent plates is oceanic crust but the other leading edge is continental crust, the subduction differs from the case above. Since continental crust is less dense than oceanic, the oceanic plate is always the one subducted. A classical example of this case is the western boundary of South America. On the oceanic side of the boundary, a trench is formed, where the oceanic plate drives underneath the continental plate. On the continental side, a fold mountain belt (the Andes) is formed as the oceanic lithosphere pushes against the continental lithosphere. As the oceanic plate descends into the mantle, some of the material melts and works its way up through the fold mountain belt to form violent volcanoes. The boundary between the plates is a region of earthquake activity, with the earthquakes ranging from shallow to relatively deep, and some are quite severe.

The last type of convergent plate boundary involves the collision of two continental masses of lithosphere. When the plates collide, neither is dense enough to be forced into the asthenosphere. Thus the collision compresses and thickens the continental edges, twisting and deforming the rocks and uplifting the land to form unusually high fold mountain belts. The prototype example is the collision of India with Asia that resulted in the formation of the Himalayas. In this case the earthquakes are typically shallow, but frequent and severe.

TRANSFORM PLATE BOUNDARIES

The actual structure of a seafloor spreading ridge is more complex than a single straight crack. Rather, ridges consist of many short segments slightly offset from one another. The offsets are a special kind of fault, or break in the lithosphere, known as a transform fault, and their function is to connect segments of a spreading ridge. The opposite sides of a transform fault belong to two different plates, and these are moving apart in opposite directions. The transform faults are just boundaries along which the plates move past one another. The classic transform boundary is the San Andreas fault that slices off a sliver of western California that rides on the Pacific plate from the rest of the state, which is on the North American plate. As the two plates scrape past each other, stress builds up and is released in earthquakes.

WHY TECTONIC PLATES MOVE

One mechanism that creates energy to move the huge plates is convection currents that are driven by heat from radioactive decay in the mantle. These convection currents in Earth's mantle carry magma up from the asthenosphere. Some of this magma escapes to form new lithosphere, but the rest spreads out sideways beneath the lithosphere, slowly cooling in the process. As it flows outward, it drags the overlying lithosphere along with it, thus continuing to open the ridges. When it cools, the flowing material becomes dense enough to sink back deeper into the mantle at convergent boundaries. A second plate-driving mechanism is the pull of the dense, cold, downward-moving slab of lithosphere in a subduction zone on the rest of the trailing plate, opening up the spreading ridges so magma can move upward.

PLATE TECTONICS AND MINERAL DEPOSITS

The theory of plate tectonics has greatly enhanced understanding of why many mineral deposits form where they do and has thus made mineral exploration more efficient. During the evolution of new oceanic plates and mountain belts by plate tectonics, a large number of mineral deposits form, particularly in association with the plate boundaries.

Hot fluids (hydrothermal fluids) circulate at spreading ridges (divergent boundaries) and deposit minerals. For example, niobium deposits are found in the intrusions in the East African Rift zone, and iron and manganese are found in the sediments of the Red Sea. Hydrothermal fluids also flow through the cracks and pores in rock along convergent boundaries and deposit metals along these boundaries as they cool. Good examples are the copper ore deposits associated with the collisional boundary of the Himalayas and tin ores in southwestern England. A general sequence of minerals found when passing inland from a trench associated with subduction is iron, gold, copper, molybdenum, gold, lead, zinc, tin, tungsten, antimony, and mercury.

—*Alvin K. Benson*

Further Reading

Erickson, Jon. *Plate Tectonics: Unraveling the Mysteries of the Earth*. Rev. ed., Facts On File, 2001.

Fossen, Haakon. *Structural Geology*. 2nd ed., Cambridge UP, 2016.

Frisch, Wolfgang, Martin Meschede, and Ronald C. Blakey. *Plate Tectonics: Continental Drift and Mountain Building*. Springer, 2010.

Hamblin, W. Kenneth, and Eric H. Christiansen. *Earth's Dynamic Systems*. 10th ed., Prentice Hall, 2003.

Kearey, Philip, Keith A. Klepeis, and Frederick J. Vine. *Global Tectonics*. 3rd ed., Wiley-Blackwell, 2009.

Keller, Edward A., and Nicholas Pinter. *Active Tectonics: Earthquakes, Uplift, and Landscape*. 2nd ed., Prentice Hall, 2002.

Kusky, Timothy. *Earthquakes: Plate Tectonics and Earthquake Hazards*. Facts On File, 2008.

Oreskes, Naomi, editor. *Plate Tectonics: An Insider's History of the Modern Theory of the Earth*. Westview Press, 2001.

Rafferty, John P., editor. *Plate Tectonics, Volcanoes, and Earthquakes*. Britannica Educational Publishing, 2011.

Van der Pluijm, Ben A., and Stephen Marshak. *Earth Structure: An Introduction to Structural Geology and Tectonics*. W. W. Norton, 2010.

Plutonic Rocks and Mineral Deposits

Fields of Study: Geology; Mineralogy; Geochemistry; Geophysics

ABSTRACT

Plutonic rocks are formed by slow magma crystallization below Earth's surface. Erosion can expose plutonic formations containing valuable mineral deposits, including metallic resources vital to industry.

KEY CONCEPTS

basaltic: formed from iron-rich magma from deep within the mantle

batholith: an exposed plutonic intrusion spanning hundreds to thousands of square kilometers in size

country rock: the preexisting rock into which magma or a mineral formation intrudes

felsic-intermediate: igneous rock having the highest range of silica content

hydrothermal vein: a mineral deposit formed by deposition and crystallization from hot, mineral-laden water solutions

mafic-ultramafic: igneous rock having the lowest range of silica content

segregation deposit: a mineral deposit formed when a component of liquid magma separates from the mixture and drifts down to the bottom of the magma chamber where mineralization is subsequently completed

IN THE PLUTONIC REALM

Plutonic rocks (from *Pluto*, Roman god of the underworld) form by slow crystallization of molten silicate magma intruded below Earth's surface. Subsequent erosion may expose "plutons" that have the area of a large house or huge granitic "batholiths" that encompass thousands of square kilometers. Many plutons are sources of valuable industrial and metallic mineral deposits, including granite building stone, gold, copper, molybdenum, chromium, and other metals. Plutonic bodies occur on all the major continents of Earth. They are common in Precambrian shield areas, the ancient cores of continents composed of rocks formed billions of years ago, such as the Canadian Shield of North America. They also occur in younger mountain ranges such as the Rockies and the Appalachian Mountains of North America.

Most of the major metallic resources consumed by modern industries—and thus many commercial products—have their origin in plutonic rocks. The rocks themselves may be cut as building stones or used for monuments and art objects. Plutons are also an important source of metal ores, including gold, silver, platinum, chromium, copper, molybdenum, lithium, beryllium, and nickel.

Plutonic rocks crystallize from igneous intrusions of molten magma that cool slowly beneath Earth's surface. Plutonic intrusions underlie many volcanic areas where magma has made its way to the surface. Later erosion by water or glacial scouring removes most or all traces of the overlying volcanic material and other overburden to expose the pluton at the surface. Plutonic igneous rocks can be distinguished from volcanic rocks by their grain size: The slowly cooled plutonic rocks allow for relatively large grain sizes (from a few millimeters to several centimeters in diameter) compared with very fine-grained, quickly cooled volcanic rocks.

TYPES OF PLUTONS AND PLUTONIC ROCKS

Major plutonic rock bodies in the world can be divided into two major compositions based mainly on silica (SiO_2) content: "felsic-intermediate" rocks, in which silica content ranges from about 52 percent to more than 70 percent (examples include granite and diorite), and "mafic-ultramafic" rocks, in which silica content is much lower, 52 percent down to about 45 percent SiO_2 (examples are gabbro and peridotite).

Light-colored felsic-intermediate, or "granitic," plutons are the most common type and include relatively small "stocks" (surface exposures of tens of square kilometers) up to immense "batholiths" that cover hundreds or thousands of square kilometers. The largest of these in North America are the Idaho, Boulder (Montana), Sierra Nevada (California), and Southern California batholiths. Many large stocks and batholiths also occur in the Canadian Shield of Canada and the northern states of Minnesota, Wisconsin, and New York. Granitic plutons also occur in every continental shield area (ancient core) of every continent and in most of the world's major mountain ranges.

Less common are the iron-rich, dark-colored mafic-ultramafic plutons. Most of these bodies consist of relatively iron-rich basaltic magma (like the dark lava flows in Hawaii) that, upon intrusion, crystallize dense minerals that settle to the bottom of the magma chamber to form mineralogically distinct layers. These layered rocks may contain economically important ore minerals. Prominent examples are the Muskox intrusion (Canada), the Skaergaard complex (Greenland), the Stillwater complex (Montana), and the metallic ore-rich intrusions in South Africa (Bushveld, Great Dyke).

ORE DEPOSITS IN FELSIC-INTERMEDIATE PLUTONS

Granitic plutons are the source of many valuable mineral commodities, either directly from the rock itself or indirectly as placer minerals (e.g., gold or cassiterite in stream sediments). Some important sedimentary ore deposits form by deposition of minerals leached from granitic plutons by groundwater (for example, many uranium ores). Ore minerals may also occur in hydrothermal (deposited by hot water) quartz veins as in the case of native gold. These veins may occur within the plutons themselves or penetrate adjacent "country rocks." Most North American copper mines (principally in New Mexico, Arizona, and Utah) obtain their ore from so-called porphyry copper deposits. These are low-grade deposits (about 0.65 percent copper) of widely disseminated copper sulfide grains within the granitic pluton or in adjacent rocks mined from large open-pit mines. Typical of these mines is the Kennecott Utah Copper mine at Bingham, Utah, the largest copper mine in the world. Perhaps the richest source of valuable minerals in granitic plutons is pegmatite deposits, generally small exposures of large crystals, the largest measuring many meters in diameter. Pegmatites are the source of many rare metals (beryllium, lithium, zirconium, boron, tantalum, and niobium) and some valuable gemstones.

ORE DEPOSITS IN MAFIC-ULTRAMAFIC PLUTONS

Because the magma associated with these plutons is low in viscosity ("thin") compared with felsic-intermediate bodies, many important ore deposits in mafic-ultramafic intrusions arise by gravity settling of minerals. South African chromite deposits, for example, form as thick-layered accumulations of crystallized chromite grains that have settled on the floor of the pluton like sand falling through thick motor oil. Such ore-rich layers are called "cumulates" (from "accumulation"), and include platinum-palladium deposits commonly associated with chromite layers.

Similar-layered ore bodies involve the formation of large blobs of dense sulfide liquids that separate from the silicate liquid and accumulate on the pluton floor. These "segregation" deposits include some of the richest nickel and copper mines in the world, including the nickel mine at Sudbury, Ontario, the Duluth complex in Minnesota, and the Bushveld of South Africa. Gold, silver, and other valuable metals are also mined from these rich deposits. Some iron and titanium deposits were created this way as well but involve the segregation of titanium or iron-rich fluids in mafic plutons, which eventually crystallize the mineral magnetite (Fe_3O_4; iron ore) or ilmenite ($FeTiO_3$; titanium ore). An example of this type of iron deposit is the Kiruna district in Sweden. Allard Lake, Quebec, is a good example of a segregation titanium mine.

—John L. Berkley

Further Reading

Best, Myron G. *Igneous and Metamorphic Petrology.* 2nd ed., Blackwell, 2003.

———, and Eric H. Christiansen. *Igneous Petrology.* Blackwell Science, 2001.

Gill, Robin. *Igneous Rocks and Processes: A Practical Guide.* Wiley-Blackwell, 2011.

Haldar, Swapan Kumar. *Introduction to Mineralogy and Petrology.* 2nd ed., Elsevier, 2020.

McBirney, Alexander R. *Igneous Petrology.* 3rd ed., Jones and Bartlett, 2007.

Philpotts, Anthony R., and Jay J. Ague. *Principles of Igneous and Metamorphic Petrology.* 2nd ed., Cambridge UP, 2009.

Young, Davis A. *Mind over Magma: The Story of Igneous Petrology.* Princeton UP, 2003.

PUMICE

Fields of Study: Geology; Volcanology; Geochemistry; Geophysics

ABSTRACT

Pumice is a vesiculated form of solidified lava, containing so many pockets of entrapped volcanic gases that its density is less than that of water. Floating masses of pumice are occasionally encountered on the open ocean where a volcano has erupted with frothy lava that has solidified quickly. Large deposits of pumice are also found on land as the result of pyroclastic flows.

KEY CONCEPTS

champlevé: a French word literally meaning "raised field" for the process of roughening the surface of, or raising, the field that is to be enameled

pyroclastic flow: typically, a large, superheated cloud of volcanic ejecta; may be described as a fast-moving, violent landslide borne by extremely hot gases

vesiculated rock: rock containing numerous gas pockets within it, typically as the result of eruption of lava with high gas content

WHERE PUMICE IS FOUND

Pumice is found where volcanic rock and pyroclastic flows have occurred. Major U.S. regions include California, Arizona, New Mexico, Idaho, Oregon, Nevada, and Kansas. Top world producers include Italy, Greece, Iran, Chile, and the United States.

PRIMARY USES OF PUMICE

More than half of the pumice mined used in construction building block. The remainder is used in horticulture, concrete, abrasives, absorbents, filtration media, landscaping materials, laundry stones, "sand soaps," and other miscellaneous uses such as beauty aids.

TECHNICAL DEFINITION OF PUMICE

When molten volcanic rock, or magma, under high pressure and charged with dissolved gases and vapors rises to the surface of Earth (terminologically becoming "lava"), the gases escape rapidly as the lava quickly cools to a solid state. If the rates of cooling and gas escape are properly matched, the lava hardens to a vesiculated glass foam called "pumice." The gas bubbles that form within the volcanic glass are called "vesicles." Vesicle size and vesicle wall thickness vary greatly.

DESCRIPTION, DISTRIBUTION, AND FORMS OF PUMICE

Pumice is a naturally occurring glass foam associated with volcanism. Block or fragmental pumice is porous and lightweight, commonly lighter than water, very abrasive because of the sharp edges of the thin

Kutkhiny Baty, a pumice rock formation in Russia. Photo by Igor Shpilenok, CC BY-SA 3.0, via Wikimedia Commons.

vesicle (gas bubble) walls, and of low compressive strength.

Because of the violent character of pumice formation, pumice deposits may consist of large masses but more commonly consist of pumice fragments ranging from gravel to sand size. If pumice vesicle walls are thin, the density of pumice may be lower than that of water. The pumice vesicles are not normally connected, so water will not flow through the pumice, and pumice will commonly float on water indefinitely. Very fine-grained granular pumice is essentially volcanic ash and is called "pumicite."

Pumice occurs in geologically young volcanic regions. In the United States, geologically young volcanic rocks occur only in and west of the Rocky Mountains, although volcanic ash deposits, blown eastward across the Rocky Mountains in the late Cenozoic era, occur in the Great Plains. Some important pumice-producing areas are Mono Craters and Siskiyou, California; Flagstaff, Arizona; and Bend, Oregon. In Europe, important deposits occur in Italy, Germany, and Greece. Pumice is a relatively low-priced and low-density product. As a raw material, therefore, it cannot be economically shipped to markets distant from the source, so international trade is limited. However, because of increased demand for 2- to 8-centimeter, low-density pumice for fabric treatment, sources of pumice have been developed in Turkey, Latin America, and Indonesia.

HISTORY OF PUMICE

The term "pumice" comes from *pumex*, Latin for foam. The resource has figured prominently in world history, specifically in the Mediterranean region. In fact, pumice has recently been studied as a historical place marker. Chemical tests allow scientists to trace pumice samples to their place of origin and therefore deduce whether or not different cultures of the Mediterranean were in contact with each other. For example, pumice used in Egypt can be traced to volcanoes of Greek island origin, linking the two civilizations as trading partners.

In 1883, the eruption of Krakatau, or Krakatoa as it is commonly called in the western hemisphere, one of the strongest and most devastating volcanic eruptions on record, occurred in modern-day Indonesia. The sea was blanketed in pumice and rose approximately 1.5 meters. Other prominent historical events involving pumice include the eruption of Mount Vesuvius, in which pumice buried the city of Pompeii, and the eruption of a volcano on the island of Thera, which occurred in 1500 or 1600 BCE.

Pumice was also used in *champlevé*, a finishing process for art pieces that involves enameling, first popularized by the Celts in Europe during the Roman period. Ferdinand von Hochstetter, a German geologist commissioned by the government of New Zealand to complete a geological survey of the country, first deduced that pumice was the result of volcanic expulsion of lava, based on his observations at Lake Taupo.

OBTAINING PUMICE

Pumice is essentially frozen or solidified volcanic lava bubbles and can therefore be extracted with relative ease. Open-pit mining is the primary method of extraction. Pumice can then be crushed into a powder or shaped into blocks, depending on its intended use.

SECONDARY USES OF PUMICE

By far, the largest volume use of pumice is for lightweight concrete aggregate. The use of cast concrete or concrete block incorporating significant quantities of pumice reduces structure weight and bearing-strength requirements for load-bearing elements of a structure. As compared with normal concrete blocks, lightweight blocks are easier to handle and are superior in terms of heat and sound insulation. Loose granular pumice, in fact, is used as insulation.

Small blocks of pumice, normally sawed to shape, are used as abrasives, and pumicite is a common ingredient of polishes, scouring powders, and other fine abrasives. Pumice fragments 2 to 8 centimeters in diameter and lighter than water are used to "stone wash" finished garments of stiff fabric such as heavyweight denim. The process softens the fabric. Other uses of pumice and pumicite relate to their high surface area per unit of volume. These uses include absorbents, fillers, and filter media.

—Robert E. Carver

Further Reading

Cas, Ray, Guido Giordano, and John V. Wright. *Volcanology: Processes, Deposits, Geology and Resources.* Springer International, 2021.

Gill, Robin. *Igneous Rocks and Processes: A Practical Guide.* Wiley-Blackwell, 2011.

Nemeth, Karoly, editor. *Updates in Volcanology: From Volcano Modelling to Volcano Geology*. IntechOpen, 2016.

White, James D. L., John L. Smellie, and David A. Clague. *Explosive Subaqueous Volcanism.* Wiley, 2003.

Wilkinson, Paul. *Pompeii: The Last Day.* BBC, 2003.

PYROCLASTIC FLOWS

Fields of Study: Geology; Astrogeology; Geochemistry; Geophysics

ABSTRACT

A pyroclastic flow is a fast-moving avalanche of hot gas, fluidized lava fragments, and ash that is ejected from an erupting volcano. In some cases, reaching speeds of 160 kilometers (100 miles) per hour or higher, pyroclastic flows can reach temperatures of more than 540° Celsius (1,000° Fahrenheit), are extremely destructive, and can cover large areas surrounding an eruption.

KEY CONCEPTS

caldera: an underground magma chamber

fluidization: process whereby the rocks and minerals contained in a volcanic eruption are vaporized within a pyroclastic flow

ignimbrite: a pumice-filled pyroclastic flow caused by the collapse of a volcanic column and accelerated by gravity

lahar: combination of water, ash, and mud that creates tremendous mudslides in a pyroclastic flow

nuée ardente: a type of dense pyroclastic flow that occurs after a volcanic dome collapse; also known as block and ash and Peleen flow

Plinian: volcanic eruption that sends ash and debris high above the volcano in a column

pumice: light, frothy volcanic rock formed when lava is cooled above ground following a violent eruption and pyroclastic flow

vesiculated: gas-infused rock within a pyroclastic flow

BASIC PRINCIPLES OF PYROCLASTIC FLOWS

A pyroclastic flow occurs when superheated gas and rock are ejected forcefully from a volcano during eruption. Pyroclastic flows may occur when, after a volcanic eruption sends a plume of ash and rock fragments into the sky, gravity causes the cloud to collapse and the superheated material to flow to the surface. In other cases, pyroclastic flows can occur when a volcano explodes outward, sending a pyroclastic flow along the side of the mountain and along the ground.

Pyroclastic flows comprise hot gases, rock, and ash. When such a flow is released from an active volcano, it travels downhill at extreme speed, causing enormous destruction along its path. Pyroclastic flows have been the most destructive components of some of the most violent eruptions in history, including the eruptions of Mount Vesuvius, Krakatoa, and Mount St. Helens.

Pyroclastic flows combine with a great deal of other material as they travel after an eruption. Flows are so destructive that they can, for example, disrupt or even obliterate lakes and ponds that are located on the volcano's surface. Additionally, the intense heat that accompanies the flows is enough to instantly melt the snow and ice packs at summits. Furthermore, volcanic eruptions release a tremendous amount of water vapor into the sky, along with ash and soot, causing intense rainstorms. All of this water combines with mud and falling ash to form a slurry called a "lahar." Lahars are fast-moving mudflows that often carry large debris including gigantic boulders that add to the pyroclastic flow's overall destructiveness.

There are two general components of a pyroclastic flow. One is basal flow, which comprises dense rock fragments and material. Basal flow moves along the contours of the land as it travels down and away from the volcano. Another component of a pyroclastic flow is the cloud of ash, gas, and other lighter material, which surges above the basal flow.

In addition to causing extreme damage to the area surrounding a volcano, pyroclastic flows also deposit a large volume of volcanic material. Pyroclastic flow deposits, which are sometimes found many kilometers from the eruption site, occur in three general layers. The bottom layer (the basal or sillar level) contains fine grains of pumice and glass. The middle level usually comprises welded pumice containing a combination of unsorted fragments of pumice and other volcanic minerals (such as silicon). Atop the welded middle level is a layer of ash. In many cases, the pyroclastic flow deposit is mixed with the water and debris of lahars. Scientists often look to flow deposits for information about the type of activity that has occurred in a volcanic area.

FLUIDIZATION AND FLOW VELOCITY

A central element of the velocity of a pyroclastic flow is its fluidization. A pyroclastic flow contains large quantities of rock and ash in addition to gases. However, substances therein are superheated to the point at which the particles experience very little friction. Combined with the water and gases from the eruption and with the vaporized water and snow that the flow encounters as it moves downslope, the flow travels rapidly and unobstructed, moving along the contours of the mountain.

Fluidization is, in turn, dependent on the types of material contained in the flow. The granular minerals and rocks are major contributors to how the flow moves, playing a role in the flow's density. The more dense the flow, the faster and farther it travels. Pyroclastic flows have been known to travel great distances (sometimes over large bodies of water). In Japan, for example, an eruption in 4000 BCE sent a pyroclastic flow more than 60 km (37 mi) from its source, including 9.5 km (6 mi) over water. Researchers found that the flow deposit was about 2 meters (6.5 feet) thick only, a testament to the flow's low density.

Scientists are still attempting to study the natural process of fluidization and how pyroclastic flows carry fluidized minerals from volcanic sites. Increased research on the natural process of fluidization will help researchers understand the nature of pyroclastic flows and how they transport minerals across Earth's surface.

IGNIMBRITES

There are two general classifications of pyroclastic flows. The first of these is the ignimbrite or "pumice flow." Ignimbrites are volcanic clouds that consist largely of pumice, a light, gaseous rock that is formed from frothy lava. These clouds are vesiculated (the rock contained within the cloud contains a large amount of gas), which means that they have a low density. The grains of pumice contained in the flow vary in size, particularly as they proceed from their volcanic source.

Because they are low in density, ignimbrites are expelled rapidly. During a Plinian eruption, wherein the eruption is vertical, a cloud is expelled high into the air above the volcano. However, as the eruption continues to send denser debris into the sky, the debris causes the cloud to collapse. The main component of ignimbrite flow, containing large concentrations of pumice, stays on the ground; the cloud of ash hovers above it. Meanwhile, the fluidized pumice flow rushes down the slope and outward, accelerated by its fall from high above the volcano.

Pumice flows emanating from another type of eruption—the caldera collapse—can be some of the most devastating types of pyroclastic flows. Here, a caldera (an underground magma chamber) is drained in a volcanic eruption. The collapse sends a sheet of pumice outward rather than upward, filling adjacent valleys and causing widespread destruction. In some cases, ignimbrites emanating from a caldera collapse have been known to wipe out an area hundreds of kilometers wide, leaving extremely thick deposits of pumice far from the volcano.

NUÉE ARDENTE FLOW

The second type of pyroclastic flow is called the *nuée ardente* ("glowing cloud"), so named by French geolo-

Pyroclastic flows sweep down the flanks of Mayon Volcano, Philippines, in 1984. Photo via Wikimedia Commons. [Public domain.]

gist Antoine Lacroix. Nuées ardentes are frequently associated with Peléen eruptions, after the deadly 1902 eruption of Mount Pelée in Martinique. That eruption was marked by two massive, fiery pyroclastic flow clouds. The vertical column's intimidating luminescence captured observers' attention, although the surface level of the flow was far more destructive.

Nuées ardentes are considerably denser in composition than are ignimbrites. Nuées ardentes occur when the dome (a mass of viscous lava that cools and forms a bulbous mass around the volcano's vents) of a volcano collapses explosively in an eruption. The release of fluidized ash and rock in the pyroclastic flow is too dense to travel into a volcanic column; it therefore spills rapidly from the vents and along the contours of the mountain at incredible speeds. Nuées ardentes are nonvesiculated because they contain blocks of fragments from the dome and ash and other particles (hence, the more accurate name provided to these occurrences: "block and ash" flows).

When a dome collapses, the flow surges from the vent, upward or outward, or both. The dense cloud then bends down as the pyroclastic flow begins to travel along the surface of the volcano. A second component of the nuée ardente, the ash cloud surge, moves above the flow, masking it. Because they are so dense, nuées ardentes are not as extensive as ignimbrites. Still, they move with tremendous speed and, as was the case of St. Pierre, Martinique, in 1902, can cause widespread damage to the nearby environment.

HISTORIC EXAMPLES OF PYROCLASTIC FLOWS

Pyroclastic flows have been the main force behind some of the most awesome and destructive volcanic eruptions in recorded history. For example, 7,700 years ago, a 3.6-km (12,000-foot) volcano dubbed Mount Mazama erupted in what is now central Oregon. The caldera feeding the volcano collapsed with such force that the resulting ignimbrite flow sent a sheet outward for nearly 32 km (20 mi) in every direction. The eruption destroyed a large portion of the mountain itself and created what is now Crater Lake, the deepest freshwater lake in North America.

The origin of the term *Plinian* (named for Pliny the Younger, who recorded his observations of the subsequent devastation) comes from the eruption of Mount Vesuvius in 79 CE, one of the most infamous volcanic eruptions in human history. This Plinian eruption created a devastating ignimbrite pyroclastic flow, which quickly engulfed the nearby Roman cities of Pompeii and Herculaneum (now the modern-day city of Ercolano), leaving both communities (and the people and animals who did not escape the rapidly advancing flow) buried in pumice.

On May 7, 1902, Mount Pelée erupted near St. Pierre in Martinique. Witnesses report having seen two clouds emitted from atop the 1,400-m (4,600-ft) mountain: A black, glowing column spewed vertically and another rushed down the sides of the mountain. Only one of the thirty thousand people living in St. Pierre, a prisoner locked in a small cell with only a small barred window, survived the nuée ardente flow.

One of the most famous examples of pyroclastic flows in modern history is the 1980 eruption of Mount St. Helens in Washington State. This event featured both types of pyroclastic flow. On May 18, the mountain erupted violently, sending a Plinian column more than 19 km (12 mi) into the sky. When the massive column collapsed, the resulting ignimbrite flows, coupled with a significant lahar, destroyed millions of trees and caused massive steam explosions in nearby lakes and rivers.

In the following few weeks, a lava dome formed at the mountain's vent. During subsequent eruptions in June, the dome's collapse produced several nuée ardente flows, triggering more avalanches and mudslides that contributed to the major alteration of the area around the mountain.

EFFECTS OF PYROCLASTIC FLOWS

Pyroclastic flows are major events that can substantially alter the landscape and environment in and around a volcano. Many of the effects of pyroclastic flows are destructive in nature, devastating to all forms of life in their paths.

Pyroclastic flows also can significantly alter the topography of the mountain and surrounding areas, creating new crevasses and surface features along the volcano's slopes and, therefore, new routes for descending lava and debris in future eruptions. In the long term, nuées ardentes and ignimbrites also can alter the shapes and courses of rivers, lakes, and streams.

STUDYING PYROCLASTIC FLOWS

One of the difficulties scientists encountered while attempting to study the volcanic activity at Mount St. Helens in 1980 was that so much of the mountain was obscured by the massive ash clouds that the eruption produced. Indeed, even when dormant a volcano presents a number of challenges for scientists, not the least of which includes traveling up steep, dangerous terrain to the cone.

Modern technology has greatly enhanced scientists' ability to study pyroclastic flows and other aspects of volcanism. For example, the use of computer models has enabled scientists to analyze potential flow courses and velocities. Such data enable them to isolate hazard zones and potentially save lives in the event of an eruption.

In other cases, scientists may utilize radar systems, based on the ground and mounted on aircraft and satellites, to study the paths of pyroclastic flow surges. For example, in a recent study, scientists successfully analyzed flow deposits and mapped flow paths on an active volcano by using synthetic aperture radar. The use of such technology precludes the need to climb up a potentially dangerous slope in unpredictable weather and atmospheric conditions. Other scientists have used ground-penetrating radar systems to map flow paths and layers in the topography of a volcano, making it easier to create models of volcanic activity.

In addition to radar and computer modeling, another invaluable technological tool in the study of pyroclastic flows is the global positioning system (GPS). Utilizing this satellite-based technology, scientists can analyze pyroclastic flow patterns, movement, and composition despite cloud cover and other visual obstructions.

—*Michael P. Auerbach*

Further Reading

Berry, Joanne. *The Complete Pompeii.* Thomas and Hudson, 2013.

Constantinescu, Robert, Jean-Claude Thouret, and Ioan-Aurel Irimus. "Computer Modeling as a Tool for Volcanic Hazards Assessment: An Example of Pyroclastic Flow Modeling at El Misti Volcano, Southern Peru." *Geographia Technia*, vol. 11, no. 2, 2011, pp. 1-14.

Dale, Virginia H., et al., editors. *Ecological Responses to the 1980 Eruption of Mount St. Helens.* Springer, 2005.

Morgan, Peter. *Fire Mountain: How One Man Survived the World's Worst Volcanic Disaster.* Bloomsbury, 2003.

Oppenheimer, Clive. *Eruptions That Shook the World.* Cambridge UP, 2011.

Saucedo, R., et al. "Modeling of Pyroclastic Flows of Colima Volcano, Mexico: Implications for Hazard Assessment." *Journal of Volcanology and Geothermal Research,* vol. 139, nos. 1/2, 2005, pp. 103-15.

Scarth, Alwyn. *La Catastrophe: The Eruption of Mount Pelée, the Worst Volcanic Disaster of the Twentieth Century.* Oxford UP, 2002.

Q

Quartz and Silica

Fields of Study: Geology; Geochemistry; Geophysics; Mineralogy

ABSTRACT

Silica is one of the most common substances in Earth's crust and mantle. It is composed of silicon dioxide tetrahedra that bond according to unique atomic characteristics to form crystalline structures. Quartz is a crystalline form of silica that forms under conditions of high temperatures and pressures in liquid environments. Silica and quartz are used for a variety of industrial applications, including chemistry and construction, and in the development of fiber-optic technologies.

KEY CONCEPTS

amethyst: a variety of quartz noted for its violet to purple color produced by irradiated molecules of iron contained within its crystalline structure

chalcedony: a form of cryptocrystalline quartz formed from microcrystals of quartz and other included minerals

conchoidal fracture: fracturing of the type exhibited by materials that do not break along planes of separation

feldspar: a silicate rock that forms in igneous environments and is the most common mineral in the earth's crust

fiber optics: flexible transparent fiber made of fused quartz or silica glass; used in computer and telecommunications applications

flint: a type of microcrystalline quartz that exists in large quantities within the earth's crust and is notable for its glass-like characteristics

igneous rock: rocks formed from magma that cools either on the earth's surface or in pockets within the mineral layers of the earth's crust

quartz: crystalline form of silica that develops when silica is heated in a liquid or semiliquid environment and then slowly cooled

silicates: compounds originating from tetrahedra of silicon dioxide

tetrahedron: a polyhedron composed of four triangular sides organized into pyramidal shape

SILICON DIOXIDE

Silicon dioxide (SiO_2), also called "silica," is an abundant chemical substance in hundreds of different types of minerals and rocks. It forms a majority of Earth's lithosphere. The upper layers of the planet consist of the rocky crust, which comprises the outer 30 to 90 kilometers (19 to 56 miles) of the terrestrial environment and 5 to 8 km (3 to 5 mi) of the ocean floor. Below the crust is the mantle, a layer of ductile rock that reaches from the outer core of Earth to the bottom of the crust and comprises more than two-thirds of Earth's mass. Silica accounts for more than 80 percent of the crust and approximately 50 percent of the minerals contained within the mantle, making it the most abundant single mineral compound on the planet.

Within Earth's crust, silica usually appears as quartz, a crystalline form of silicon dioxide. Quartz is the second most abundant mineral in Earth's crust, second only to feldspar, which is another mineral composed almost entirely of silica and included minerals. Quartz is composed of silica molecules organized into a tetrahedral pattern, which lends quartz its distinct six-sided crystalline shape, topped with a six-sided pyramid at the leading end of each crystal. Sand found on beaches is composed almost entirely of fragmented quartz crystals, blended with other elements in smaller quantities.

Both silica and quartz have been used in a variety of industrial and technological applications. Silica is a component of many building materials and has been used in the manufacture of microelectronic components. Quartz crystals are used in the manufacture of watches. Some of the more attractive varieties of quartz are prized for their aesthetic properties and are used in the manufacture of jewelry.

Silica is also melted and polished to make a type of glass called "fused quartz" or "fused silica," which is used in the manufacture of eyeglasses and other optical lenses. Fused silica is also used in the construction of optical fibers for telecommunications applications.

Optical fibers transmitting light pulses are superior to metal fibers at transmitting electrical signals because of the chemical properties of silica, which make the material largely immune to the effects of electromagnetic interference. Fiber optics have been used in a variety of applications, from lighting and telecommunications to computing. Quartz also is the ore used in the development of silicon computer chips. The silicon, a semiconductive element obtained from silica is melted and slowly drawn out into a single large crystal that is then sliced into wafers, to be etched with the patterns of transistor structures that are the heart of digital processors.

CHEMISTRY OF SILICA AND QUARTZ

Silicon dioxide forms spontaneously at a variety of temperatures and pressures when silicon atoms encounter atomic oxygen. Oxygen atoms form covalent bonds to a silicon atom such that four atoms of oxygen are organized around each silicon atom in a tetrahedron or pyramid-shaped array.

The connections between silicon and oxygen and between molecules of silicon dioxide create flexible bonds that allow silicon dioxide to organize into a variety of shapes. By controlling the temperature and pressures at which silicon and oxygen interact, it is possible to develop silicate forms with a variety of structures.

Each tetrahedral molecule of silica then bonds to other silica molecules to form a three-dimensional structure. The nature of the crystal structure depends on the angle at which the silica molecules bond. Typically, molecules bond at an angle of between 140 to 160 degrees. Quartz crystals form when silica molecules bond at angles between 143.6 and 153 degrees.

Once formed, silica is highly stable and will resist further chemical reactions because of the strength of the covalent silicon-oxygen bonds within the crystal structure. Silica glass, which is a pure form of quartz glass containing more than 99 percent silica, is therefore used for a number of chemical applications, providing an inert environment that can contain chemical substances without becoming reactive.

CRYSTAL STRUCTURE AND FORMATION

Quartz comes in two basic varieties, based on the size of the crystals within a deposit. Macrocrystalline quartz contains crystals that can be viewed with the naked eye, while crypto- or microcrystaline quartz are a form in which the constituent crystals can be viewed only with a microscopic viewing aid.

Microcrystaline quartz can further be broken down into two basic types, fibrous and grainy, based on the appearance of thin layers of material viewed through a polarizing microscope. The difference in structure between macro- and microcrystalline quartz is related to the environment in which the crystals form.

Most macrocrystalline quartz forms either in heated, silica-rich water solutions or within igneous rock deposits from silica-rich magma. Macrocrystalline quartz is a primary component of many rock types, meaning that it was one of the basic components present during the formation of the rock. Individual crystals of quartz are generally considered a secondary product, forming within quartz rocks or within other rocks that contain quartz as part of their overall structure.

Macrocrystalline quartz grows in hydrothermal environments, where water temperature is between 100° and 450° Celsius (212° to 842° Fahrenheit) with correspondingly high pressure. In this environment, rocks containing silica dissolve to produce orthosilicic acid (H_4SiO_4), which reacts to form layers of crystalline silica, giving off water in the process. If quartz grains are already present in the solution, these new layers of crystallized silica will add to existing layers, building into larger crystals in the process.

If no preformed crystals are present, then crystalline silica can bond together to form new crystals, though this occurs only in environments in which the temperature and pressure slowly decrease. This is largely because silica tetrahedra are continually being both added and removed from the aggregating crystals. For crystals to form, there must be sufficient dissolved silica in the environment, but macrocrystalline structures will not form if silica saturation rises above a certain critical level. At this point, crystal aggregation occurs so frequently throughout the entire medium that only tiny, microcrystalline structures will form.

In environments closer to the surface of the planet, the medium will cool more rapidly, and fewer crystals will form. In some situations, crystal-forming aqueous environments develop within a solid host rock, which may lead to large crystals as the heat gradually escapes, buffered by the surrounding rock layer. Extremely large quartz deposits result from tectonic collisions, which can drive sediment from deep within the mantle or crust toward the surface through millions of years. In this situation, temperatures and pressures slowly change, and pockets of heated water may be able to generate quartz crystals for thousands of years before crystal formation ceases.

In molten rock, silica does not exist as an individual tetrahedral molecule but forms instead into long chains of tetrahedra that move together through the molten rock. Silica chains of this type cause the viscosity of flowing magma and lava, and magma with higher proportions of silica will be more viscous. Whether cooling magma will form quartz crystals depends on the chemical composition of the magma and the conditions under which the magma cools.

Magma that erupts onto the surface and cools rapidly will not form crystals because insufficient time exists for the bonds between tetrahedra in the silica chains to break and then re-form into crystallized structures. Lava therefore gives rise to either porous volcanic rock or, when the lava cools extremely rapidly, glassy structures like obsidian. Similarly, high-viscosity magma, which is rich in silica chains, will not generally form into crystals and usually gives rise to other types of silicate mineral.

Alternatively, when molten magma cools beneath the surface, cooling occurs gradually, forming into a variety of silica-based rock types. In granite, for instance, mica, feldspar, and quartz are the primary types of silicate crystals that form as the granitic magma cools. Quartz is usually the last type of crystal to form in a sample of granite and therefore occurs in pockets within the stone.

Quartz crystal cluster. Photo by JJ Harrison, CC BY-SA 2.5, via Wikimedia Commons.

Microcrystalline structures, sometimes called "chalcedony," form in aqueous environments through a similar mechanism as macrocrystalline quartz, but they tend to form under different basic conditions. Higher saturation of silica tetrahedra favors the development of microcrystalline structures. Microcrystalline quartz also tends to form in environments where the overall temperature is below 150° Celsius (320° Fahrenheit); microcrystal growth is highly inhibited at temperatures above 200° Celsius (395° Fahrenheit).

In addition, microcrystalline structures tend to grow only in aqueous environments and are inhibited by environments in which there is insufficient water for the formation of dissolved silica tetrahedra. For this reason, microcrystalline structures are largely absent from magmatic rocks and other types of igneous rock deposits.

VARIETIES OF QUARTZ

There are many varieties of quartz, all based on the size and shape of the crystals and on the inclusion of other minerals and elements that cause variations in color and texture. Common types of macrocrytalline quartz include rock crystal, amethyst, milky quartz, and smoky quartz.

Rock crystal is the name for basic, colorless macro-crystalline quartz, consisting of 99.5 percent or more silica. Rock crystal quartz can be found around the world and in many types of geological environments.

A valuable and highly prized type of quartz is amethyst, a variety of macrocrystalline quartz with included pockets of purple, red, or violet formed by the presence of embedded iron atoms. When these iron atoms are exposed to radiation during formation, the oxygen that is bonded to silicon oxidizes in such a way to produce the distinctive color. Amethyst remains highly sensitive to radiation; the violet hues will fade in amethyst exposed to sunlight and other forms of radiation. The color in most amethyst is concentrated at the pyramidal heads of the crystals, though in some cases iron can form into waves of color called "phantoms" within the body of the crystals.

Iron present in quartz also can produce ametrine, a type that varies in colors and often includes both yellow and violet sections within a single crystal. The yellow areas reflect yellow light because of the higher concentration of iron atoms within the lattice.

Smoky quartz is named for its gray to brown color formed by the irradiation of aluminum atoms within the lattice. The color in smoky quartz can occur in isolated areas or can affect the entire crystal structure. The color in smoky quartz forms long after the initial formation of the crystal, after extended exposure to irradiation chemically transforms the included aluminum ions through oxidation, thereby altering their structure and interaction with light. Smoky quartz can be artificially created by exposing quartz with included aluminum to radiation.

Milky quartz has a translucent white color that forms from the inclusion of numerous microscopic chambers containing liquid or gas molecules that alter the way the crystal reacts to light. Milky quartz has been discovered in a variety of environments and ranges in color from slightly translucent to specimens that are nearly solid white and opaque. Because inclusions tend to distort crystal growth, milky quartz exhibits unusual morphological shapes, with crystals often shifting in their growth patterns as the crystal develops.

One of the most common types of microcrystalline quartz is agate, which is a translucent, multicolored rock that exhibits bands of different colors organized parallel to the surface of the rock. Agate does not appear to have a crystalline structure unless examined with a high-powered electron or other type of microscope. Agate is not composed of pure silica, and the colored bands result from inclusions of various types of minerals and other rock types. Agates can have a variety of colors, from greens and reds to shades of brown and gray.

Another of the most common types of microcrystalline quartz is flint, which is translucent to opaque and exhibits no obvious crystalline structure. Flint is usually dark brown, gray, or tan. Color varieties are caused by included minerals and oxides, which typically include iron, aluminum, and other metals. One of the most unusual properties of flint is that it exhibits conchoidal fracturing, which is fracturing that does not follow a standard plane of separation. Conchoidal fractures and flint fractures, which are formed similar to glass fractures, have made flint useful in a variety of applications. Because it is fine grained and can be sharpened due to its fracturing pattern, flint was used to manufacture arrowheads and other types of weaponry, making it one of the most anthropologically important rocks on Earth.

—Micah L. Issitt

Further Reading

Götze, Jens, editor. *Mineralogy of Quartz and Silica Minerals*. MDPI, 2018.

Grotzinger, John, and Thomas H. Jordan. *Understanding Earth*. 6th ed., W. H. Freeman, 2010.

Monroe, James S., Reed Wicander, and Richard Hazlett. *Physical Geology*. 6th ed., Thompson Higher Education, 2007.

Montgomery, Carla. W. *Environmental Geology*. 4th ed., McGraw-Hill, 2008.

Stanley, Steven M. *Earth System History*. W. H. Freeman, 2004.

R

Reefs

Fields of Study: Geology; Marine Geology; Marine Geochemistry; Oceanography; Geophysics

ABSTRACT

Reefs are among the oldest-known communities, existing at least 2 billion years ago. They exert considerable control on the surrounding physical environment, influencing turbulence levels and patterns of sedimentation. Ancient reefs are often important hydrocarbon reservoirs.

KEY CONCEPTS

calcareous algae: green algae that secrete needles or plates of aragonite as an internal skeleton; very important contributors to reef sediment

carbonate rocks: sedimentary rocks such as limestone, composed of calcium carbonate minerals

coralline algae: red algae that secrete crusts or branching skeletons of high magnesium calcite; important sediment contributors and binders on reefs

rugose corals: a Paleozoic coral group also known as "tetracorals," sometimes colonial, but more often solitary and horn shaped

scleractinian corals: modern corals or "hexacorals," different from their more ancient counterparts in details of the skeleton and the presence of a symbiosis with unicellular algae in most shallow water species

stromatolites: layered columnar or flattened structures in sedimentary rocks, produced by the binding of sediment by blue green algal (cyanobacterial) mats

stromatoporoids: spongelike organisms that produced layered, mound shaped, calcareous skeletons and were important reef builders during the Paleozoic era

tabulate "corals": colonial organisms with calcareous skeletons that were important Paleozoic reef builders; considered to be more closely related to sponges than to corals

"TRUE" REEFS VERSUS REEF-LIKE STRUCTURES

Reefs or reef-like structures are among the oldest-known communities, extending more than 2 billion years into Earth's history. These earliest reefs were vastly different in their biotic composition and physical structure from modern reefs, which are among the most diverse of biotic communities and display amazingly high rates of biotic productivity (carbon fixation) and calcium carbonate deposition, despite their existence in a virtual nutrient "desert." Reefs are among the few communities to rival the power of humankind as a shaper of the planet. The Great Barrier Reef of Australia, for example, forms a structure some 2,000 kilometers in length and up to 150 kilometers in width.

It is necessary to distinguish between "true," or structural, reefs and reef-like structures or banks. Reefs are carbonate structures that possess an internal framework. The framework traps sediment and provides resistance to wave action. Thus, reefs can exist in very shallow water and may grow to the surface of the oceans. Banks are also biogenically produced but lack an internal framework. Thus, banks are often restricted to low energy, deep-water settings. "Bioherm" refers to mound-like carbonate buildups, either reefs or banks, and "biostrome" to low, lens shaped buildups.

REEF CLASSIFICATION

Modern reefs are classified into several geomorphic types: atoll, barrier, fringing, and patch. Many of these may be further subdivided into reef crest or flat, back reef or lagoon, and fore reef zones. Atoll reefs are circular structures with a central lagoon, thought to form on subsiding volcanic islands. Barrier reefs are elongate structures that parallel coastlines and possess a significant lagoon between the exposed reef crest and shore. These often occur on the edges of shelves that are uplifted by faulting. Fringing reefs are elongate structures paralleling and extending seaward from the coastline that lack a lagoon between shore and exposed reef crest. Patch reefs are typically small, mound-like structures, occurring isolated on shelves or in lagoons. The majority of fossil reefs would be classified as patch reefs, although many examples of extensive, linear, shelf edge trends are also known from the geologic record.

Reefs form one of the most distinctive and easily recognized sedimentary facies (or environments). In addition to possessing a characteristic fauna consisting of corals, various algae, and stromatoporoids, they are distinguished by a massive (nonlayered) core that has abrupt contacts with adjacent facies. Associated facies include flat-lying lagoon and steeply inclined fore reef talus, the latter often consisting of large angular blocks derived from the core. The reef core is typically a thick unit relative to adjacent deposits. The core also consists of relatively pure calcium carbonate with little contained terrigenous material.

REEF ENVIRONMENTS

Modern reefs are restricted to certain environments. They occur abundantly only between 23 degrees north and south latitudes and tend to be restricted to the western side of ocean basins, which lack upwelling of cold bottom waters. This restriction is based on temperature, as reefs do not flourish where temperatures frequently fall below 18° Celsius. Reef growth is largely restricted to depths less than 60 meters, as there is insufficient penetration of sunlight below this

Coral reef at Nusa Lembongan, Bali, Indonesia. Photo by Ilse Reijs and Jan-Noud Hutten , CC BY 2.0, via Wikimedia Commons.

depth for symbiont bearing corals to flourish. Reefs also require clear waters lacking suspended terrigenous materials, as these interfere with the feeding activity of many reef organisms and also reduce the penetration of sunlight. Finally, most reef organisms require salinities that are in the normal oceanic range. It appears that many fossil reefs were similarly limited in their environmental requirements.

Some of the most striking features of modern reefs include their pronounced zonation, great diversity, and high productivity and growth rates. Reefs demonstrate a strong bathymetric (depth-related) zonation. This zonation is largely mediated through depth-related changes in turbulence intensity and in the quantity and spectral characteristics (reds are absorbed first, blues last) of available light. Shallow (1to 5 meter) fore reef environments are characterized by strong turbulence and high light intensity and possess low diversity assemblages of wave resistant corals, such as the elk horn coral, Acropora palmata, and crustose red algae.

With increasing depth (10 to 20 meters), turbulence levels decrease and coral species diversity increases, with the occurrence of mound and delicate branching colonies. At greater depths (30 to 60 meters), corals assume a flattened, platelike form in an attempt to maximize surface area for exposure to ambient light. Sponges and many green algae are also very important over this range. Finally, corals possessing zooxanthellae, which live in the coral tissues and provide food for the coral host, are rare or absent below 60 meters because of insufficient light. Surprisingly, green and red calcareous algae extend to much greater depths (100 to 200 meters), despite the very low light intensity (much less than 1 percent of surface irradiance). Sponges are also important members of these deep reef communities.

REEF COMMUNITIES

Coral reefs are among the most diverse communities on Earth; however, there is no consensus on the mechanisms behind the maintenance of this great diversity. At one time, it was believed that reefs existed in a low disturbance, highly stable environment, which allowed very fine subdivision of food and habitat resources and thus permitted the coexistence of a great number of different species. Upon closer inspection, however, many reef organisms appear to overlap greatly in food and habitat requirements. Also, it has become increasingly apparent that disturbance, in the form of disease, extreme temperatures, and hurricanes, is no stranger to reef communities.

Coral reefs exhibit very high rates of productivity (carbon fixation), which is a result of extremely tight recycling of existing nutrients. This is necessary, as coral reefs exist in virtual nutrient "deserts." Modern corals exhibit high skeletal growth rates, up to 10 centimeters per year for some branching species. Such high rates of skeletal production are intimately related to the symbiosis existing between the hermatypic or reef-building scleractinian corals (also gorgonians and many sponges) and unicellular algae or zooxanthellae. Corals that, for some reason, have lost their zooxanthellae or that are kept in dark rooms exhibit greatly reduced rates of skeleton production.

In addition to high individual growth rates for component taxa, the carbonate mass of the reefs may grow at a rate of some 2 meters per 1,000 years, a rate that is much higher than that of most other sedimentary deposits. This reflects the high productivity or growth rates of the component organisms and the efficient trapping of derived sediment by the reef frame. Although the framework organisms, most notably corals, are perhaps the most striking components of the reef system, the framework represents only 10 to 20 percent of most fossil reef masses. The remainder of the reef mass consists of sedimentary fill derived from the reef community through a combination of biosynthesis (secretion) and bioerosion (breaking down) of calcium carbonate. An example of the relative contributions of reef organisms to sediment can be found in Jamaica, where shallow water, back reef sediment consists of 41 percent coral, 24 percent green calcareous algae, 13 percent red calcareous algae, 6 percent foraminifera, 4 percent mollusks, and 12 percent other grains. The most important bioeroders are boring sponges, bivalves, and various "worms," which excavate living spaces within reef rock or skeletons, and parrot fish and sea urchins, which remove calcium carbonate as they feed upon surface films of algae.

TYPES OF REEF COMMUNITIES

A diversity of organisms has produced reef and reef-like structures throughout Earth's history. Several distinct reef community types have been noted, as well as four major "collapses" of reef communities. The oldest reefs or reef-like structures existed more than 2 billion years ago during the Precambrian eon. These consisted of low diversity communities dominated by soft, blue green algae, which trapped sediment to produce layered, often columnar structures known as stromatolites, similar if not identical to those being formed today. During the Early Cambrian period, blue green algae were joined by calcareous, conical, spongelike organisms known as archaeocyathids, which persisted until the end of the Middle Cambrian. Following the extinction of the archaeocyathids, reefs again consisted only of blue green algae until the advent of more modern reef communities in the Middle Ordovician period. These reefs consisted of corals (predominantly tabulate and, to a much lesser extent, rugose corals), red calcareous algae, bryozoans (moss animals), and the spongelike stromatoporoids. This community type persisted through the Devonian period, at which time a global collapse of reef communities occurred. The succeeding Carboniferous period largely lacked reefs, although algal and crinoidal (sea lily) mounds were common. Reefs again occurred in the Permian period, consisting mainly of red and green calcareous algae, stromatolites, bryozoans, and chambered calcareous sponges known as sphinctozoans, which resembled strings of beads. These reefs were very different from those of the earlier Paleozoic era; in particular, the tabulates and stromatoporoids no longer played an important role. The famous El Capitan reef complex of West Texas formed during this interval. The Paleozoic era ended with a sweeping extinction event that involved not only reef inhabitants but also other marine organisms.

After the Paleozoic extinctions, reefs were largely absent during the early part of the Mesozoic era. The advent of modern type reefs consisting of scleractinian corals and red and green algae occurred in the Late Triassic period. Stromatoporoids once again occurred abundantly on reefs during this interval; however, the role of the previously ubiquitous blue green algal stromatolites in reefs declined. Late Cretaceous reefs were often dominated by conical, rudistid bivalves that developed the ability to form frameworks and may have possessed symbiotic relationships with algae, as do many modern corals. Rudists, however, became extinct during the sweeping extinctions that occurred at the end of the Cretaceous period. The reefs that were reestablished in the Cenozoic era lacked stromatoporoids and rudists and consisted of scleractinian corals and red and green calcareous algae. This reef type has persisted, with fluctuations, until the present.

STUDY OF MODERN REEFS

Modern reefs are typically studied by divers, which enables observation and sampling to a depth of approximately 50 meters. Deeper environments have been made accessible through the availability of manned submersibles and unmanned, remotely operated vehicles that carry mechanical samplers as well as still and video cameras. The biological compositions of reef communities are determined by census (counting) methods commonly employed by plant ecologists. Studies of symbioses, such as that between corals and their zooxanthellae, employ radioactive tracers to determine the transfer of products between symbiont and host. Growth rates are measured by staining the calcareous skeletons of living organisms with a dye, such as Alizarin red, and then later collecting and sectioning the specimen and measuring the amount of skeleton added since the time of staining. Another method for determining growth is to X-ray a thin slice of skeleton and then measure and count the yearly growth bands that are revealed on the radiograph. Variations in growth banding reflect, among other factors, fluctuations in ocean temperature.

Reef sediments, which will potentially be transformed into reef limestones, are examined through sieving, X-ray diffraction, and epoxy impregnation and thin sectioning. Sieving enables the determination of sediment texture, the relationships of grain sizes and abundance (which will reflect environmental energy and the production), and erosion of grains through biotic processes. X-ray diffraction produces a pattern that is determined by the internal crystalline structure of the sediment grains. As each mineral possesses a unique structure, the mineralogical identity of the sediment may be determined. Thin sections of embedded sediment or lithified rock are examined with petrographic microscopes, which reveal the characteristic microstructures of the individual grains. Thus, even highly abraded fragments of coral or algae may be identified and their contributions to the reef sediment determined.

STUDY OF FOSSIL REEFS

Because of their typically massive nature, fossil reefs are usually studied by thin sectioning of lithified rock samples collected either from surface exposures or well cores. Reef limestones that have not undergone extensive alteration may be dated through carbon-14 dating, if relatively young, or through uranium series radiometric dating methods.

NATURAL LABORATORIES AND ECONOMIC RESOURCES

Modern reefs serve as natural laboratories, enabling the geoscientist to witness and study phenomena, such as carbonate sediment production, bioerosion, and early cementation, that have been responsible for forming major carbonate rock bodies in the past. The study of cores extracted from centuries old coral colonies shows promise for deciphering past climates and perhaps predicting future trends. This is made possible by the fact that the coral skeleton records variations in growth that are related to ocean temperature fluctuations. The highly diverse modern reefs also serve as ecological laboratories for testing models on the control of community structure. For example, the relative importance of stability versus disturbance and recruitment versus predation in determining community structure is being studied within the reef setting.

Modern reefs are economically significant resources, particularly for many developing nations in the tropics. Reefs and the associated lagoonal seagrass beds serve as important nurseries and habitats for many fish and invertebrates. The standing crop of fish immediately over reefs is much higher than that of adjacent open shelf areas. Reef organisms may one day provide an important source of pharmaceutical compounds, such as prostaglandins, which may be extracted from gorgonians (octocorals). In addition, research has focused upon the antifouling properties exhibited by certain reef encrusters. Reefs also provide recreational opportunities for snorkelers and for scuba divers, a fact that many developing countries are utilizing to promote their tourist industries. Finally, reefs serve to protect shorelines from wave erosion.

Because of the highly restrictive environmental tolerances of reef organisms, the occurrence of reefs in ancient strata enables fairly confident estimation of paleolatitude, temperature, depth, salinity, and water clarity. In addition, depth- or turbulence-related variation in growth form (mounds in very shallow water, branches at intermediate depths, and plates at greater depths) enables even more precise estimation of paleobathymetry or turbulence levels. Finally, buried ancient reefs are often important reservoir rock structures containing hydrocarbons and thus represent an important economic resource.

—John Brelsford Southard

Further Reading

Aronson, Richard B., editor. *Geological Approaches to Coral Reef Ecology*. Springer, 2007.

Dubinsky, Zvy, and Noga Stambler, editors. *Coral Reefs: An Ecosystem in Transition*. Springer, 2011.

Harris, Peter T., and Elaine K. Baker, editors. *Seafloor Geomorphology as Benthic Habitat*. Elsevier, 2011.

Hopley, David. *Encyclopedia of Modern Coral Reefs: Structure, Form and Process*. Springer, 2011.

Hovland, Martin. *Deep-Water Coral Reefs: Unique Biodiversity Hot-Spots*. Praxis, 2008.

Rohwer, Forest, Merry Youle, and Derek Vosten. *Coral Reefs in the Microbial Seas*. Plaid Press, 2010.

Sheppard, Charles. *Coral Reefs: A Very Short Introduction*. Oxford UP, 2014.

Sheppard, Charles R. C., Simon K. Davy, and Graham M. Pilling. *The Biology of Coral Reefs*. Oxford UP, 2009.

Spalding, Mark D., Edmund P. Green, and Corinna Ravilious. *World Atlas of Coral Reefs*. U of California P, 2001.

Vacher, Leonard, and Terrence M. Quinn. *Geology and Hydrogeology of Carbonate Islands*. Elsevier, 2004.

Wolanski, Eric. *Oceanographic Processes of Coral Reefs: Physical and Biological Links in the Great Barrier Reef*. CRC Press, 2001.

Regional Metamorphism

Fields of Study: Geology; Geochemistry; Geophysics

ABSTRACT

Regional metamorphism, which takes place in the roots of actively forming mountain belts, is a process in which in-

creased temperatures and pressures cause rock to undergo recrystallization, new mineral growth, and deformation. These changes diagnostically cause minerals to develop a preferred orientation, or foliation, in the rock.

KEY CONCEPTS

equilibrium: a situation in which a mineral is stable at a given set of temperature-pressure conditions

facies: a part of a rock or group of rocks that differs from the whole formation in one or more properties, such as composition, age, or fossil content

igneous rock: a rock formed from the cooling of molten material

mica: a platy silicate mineral (one silicon atom surrounded by four oxygen atoms) that readily splits into thin, flexible sheets

mineral: a naturally occurring chemical compound that has an orderly internal arrangement of atoms and a definite formula

sedimentary rock: a rock formed from the physical breakdown of preexisting rock material or from the precipitation, chemically or biologically, of minerals

shale: a sedimentary rock composed of fine-grained products derived from the physical breakdown of preexisting rock material

strain: change in volume or size in response to stress

stress: force per unit of area

texture: the size, shape, and arrangement of crystals or particles in a rock

zeolites: members of a mineral group with very complex compositions: aluminosilicates with variable amounts of calcium, sodium, and water and a very open atomic structure that permits atoms to move easily

ROCK TEXTURE AND COMPOSITION

Metamorphism is a process whereby igneous or sedimentary rocks undergo change in response to some combination of increased temperature, pressure, and chemically active fluids. Several types of metamorphism are recognized, but regional metamorphism is the most widespread. Regionally metamorphosed rocks in association with igneous rocks make up about 85 percent of the continents. In order to describe regionally metamorphosed rocks and to understand the processes by which they form, one must understand two properties of these rocks: texture and composition. Texture refers to the size, shape, and arrangement of crystals or particles in a rock. Composition refers to the minerals present in a rock.

The effect of temperature and pressure on texture and composition is aided by fluids. Fluids transport chemical constituents and facilitate chemical reactions that take place in the rocks. As metamorphic conditions increase, metamorphic reactions readily occur if fluids are present; however, fluids are gradually driven off as recrystallization reduces open spaces. Most of the chemical reactions that occur during metamorphism result in the loss of water or carbon dioxide. Once fluids are eliminated and temperatures and pressures decrease, the reactions generally occur very slowly. Therefore, in most metamorphic rocks, the minerals preserved are those that represent the highest temperatures and pressures attained, assuming equilibrium when adequate fluids were still available.

Regionally metamorphosed rocks characteristically develop distinctive textures, largely because of the role of two types of pressure in concert with elevated temperatures: confining pressure, or the force caused by the weight of the material overlying a rock, and a stronger directed pressure, or a horizontal force created by mountain-building processes.

FOLIATION, SCHISTOSITY, AND GNEISSIC TEXTURE

As metamorphism commences, rocks become deformed, and original features are strongly distorted or obliterated. Original minerals either recrystallize or react to form new minerals and develop a pronounced preferred orientation, or foliation, in response to the directed pressure. Preferentially aligned platy minerals (such as the micas), formed during the early stages of metamorphism, enhance the development of foliation. Foliation assumes different forms depending on the degree of metamorphism.

Although the minerals in most rock types characteristically develop preferred orientations, shales and other fine-grained rock types display more distinct changes of texture. Shales undergo textural changes as they pass from conditions of low-grade metamorphism (low temperatures and pressures) at shallow depths to high grades (higher temperatures and pressures) deeper in the planet. At low grades of regional metamorphism, microscopic platy minerals grow and align themselves perpendicular to the orientation of the directed pressure. This realignment allows this fine-grained rock, called a "slate," to split or cleave easily along this preferred direction. Under somewhat higher metamorphic grades deeper in the planet, the platy minerals continue to develop parallel arrangements in response to the directed pressures, but the resultant mineral grains, typically the micas, are now large enough to be identified with the unaided eye. This texture, schistosity, forms in a rock called a "schist." At higher grades, the platy minerals react to form new minerals, and a gneissic texture develops. This texture is distinguished by alternating layers of light and dark minerals that give the rock, called a "gneiss," a banded appearance.

INDEX MINERALS

Just as textural changes in metamorphic rocks are generally predictable as grade increases, minerals also appear or disappear in a systematic sequence and become even better gauges of the conditions of metamorphism. A progression of key minerals in shales, called "index minerals," forms as the grade of metamorphism increases. These index minerals, therefore, occur in rocks metamorphosed at a given range of temperature-pressure conditions. When mapped in the field, areas containing index minerals are called "metamorphic zones." As the metamorphic grades increase, the typical sequence of the index minerals is chlorite, biotite, garnet, staurolite, kyanite, and sillimanite.

Because metamorphic zonation based on index minerals utilizes minerals most likely to form from shales, zonation can be limited if shales are not abundant. A more precise definition of metamorphic change is recognized, which is not based on single minerals in one rock type but on sequences of mineral assemblages in groups of associated rocks within ranges of temperature-pressure conditions. These ranges of temperatures and pressures, in which diagnostic mineral assemblages in different rock types may exist, are called "metamorphic facies." The zeolite facies represent the lowest conditions of regional metamorphism. At the lowest extremes of this facies, conditions considered sedimentary merge with those that are metamorphic. These facies are named for the zeolite group of minerals that forms within it.

As metamorphic conditions increase, the assemblages of minerals distinctive of the zeolite facies break down, and new assemblages form that are distinctive of either the blueschist or the greenschist facies. As the names of these facies imply, the textures formed under these conditions are commonly schistose, and a number of the minerals impart either blue or green tints to the rocks in their respective facies. In a given area of regional metamorphism, it is typical to find rocks that represent a sequence of metamorphic facies that follow one of two pathways. The greenschist path has temperature increasing more rapidly than pressure, relatively speaking, so the combination of facies that it crosses is referred to as a low pressure/high temperature sequence. The less common blueschist path, in which temperature increases more slowly relative to pressure, represents a high pressure, low temperature sequence.

UNDERSTANDING REGIONAL METAMORPHISM PROCESSES

By understanding how mountains are formed, scientists can learn how regionally metamorphosed rocks are generated. Regional metamorphism occurs in the roots of actively forming mountain belts. Linear zones thousands of kilometers long and hundreds of kilometers wide can be involved. It will take millions of years before the products of regional metamorphism are exposed in the Andes or Cascades that are forming now. Among the youngest exposed metamorphic rocks in the world are those in the Olympic Mountains of Washington, some of which are only ten million years old. On all the continents, however, broad areas of exposed rocks, called "shields," now reveal the products of numerous ancient mountain-building events.

In the 1960s, the concept of plate tectonics revolutionized geology and scientists' understanding of the processes in regional metamorphism. For decades, geologists have known that Earth is divided into a number of concentric layers. The outermost zone, the crust, consists of relatively low-density rocks and averages 35 kilometers thick on the continents and 10 kilometers thick under the oceans. Beneath the crust is the mantle (nearly 3,000 kilometers thick), and then the core.

With the advent of plate tectonics, another important subdivision was recognized. The upper 60 to 100 kilometers, which includes the crust and a portion of the upper mantle, behaves in a relatively rigid fashion and is composed of about twenty pieces that move independently of one another. This rigid zone is called the "lithosphere," and the pieces are called "lithospheric," or "tectonic," plates. Because each plate is moving relative to its neighbors, three types of boundaries with other plates exist: those in collision, those pulling apart, and those sliding by one another. The rigid lithosphere apparently floats on a soft plastic layer called the "asthenosphere," which extends from 60 to 100 kilometers to at least 250 kilometers into the mantle. Plate tectonics allows us to understand the processes that occur during mountain building.

CONVERGENT BOUNDARIES

Regional metamorphism is restricted to margins in collision, called "convergent boundaries." The collision of two lithospheric plates produces several profound effects. One plate subducts, or sinks, under the other but manages to crumple the edge of the continent into a mountain belt. An oceanic trench forms where the descending plate sinks beneath a continent. As one plate moves downward, melting occurs where temperatures sufficiently rise. The molten material rises through the lithosphere to produce volcanoes at the surface and masses of igneous rock at depth. It is in this realm—extending from the zone of collision to well beyond the igneous bodies—that regional metamorphism occurs.

Within this area, two distinct sequences of regional metamorphism develop. The high pressure/low temperature sequence forms in the region nearest the trench. As sediments and volcanic rocks are carried rapidly (in geological terms) downward into the subduction zone, they attain relatively high pressures but remain relatively cold because they have not had the time to heat up to the temperature of the surroundings. The minerals that form in these high-pressure conditions are those of the zeolite and blueschist facies. If the downward movement continued uninterrupted, the rocks would eventually heat up. Before that can happen, however, thin, cold slabs with high-pressure mineral assemblages within them are chaotically pushed or thrust back toward the surface. Why this thrusting occurs is not entirely clear, but it is not uncommon to find a region 100 kilometers wide, adjacent to a trench, composed entirely of highly deformed high pressure/low temperature facies rocks. This region is called an "accretionary wedge" because of its inferred shape in a vertical slice through the earth. It is this material that allows the continents to grow larger through time. The California Coast Ranges are a good example of rocks that formed in such a setting.

STUDY OF REGIONALLY METAMORPHOSED ROCKS

Geologists study regionally metamorphosed rocks from several different perspectives. At the core of these studies are field observations. Descriptions of rock units and structures provide information on how these rocks have formed. Much of this fieldwork is synthesized into a geological map.

In the laboratory, samples from the field are studied by a variety of physical and chemical methods. Studies with the petrographic microscope yield information on mineral compositions and textural relationships that provide clues for classifying and determining modes of origin for these rocks. X-ray powder diffraction techniques are commonly used to identify metamorphic minerals not readily identified by visual inspection. Since metamorphic rocks are markedly changed texturally and mineralogically from their original state, chemical analysis provides data that can be compared to probable premetamorphic rock types. Standard classical chemical methods of analysis may be used, but a number of more sophisticated, faster, and simpler spectrographic methods are popular. Atomic absorption and X-ray fluorescence spectroscopy are two of the more widely applied techniques to determine elemental abundances, although emission spectrographic and neutron activation analyses are also used. Another important tool is the electron microprobe. Most published mineral analyses are generated by electron microprobe analyses. Mass spectroscopic analyses also provide information on the distribution of isotopes in metamorphic rocks and minerals. These data are useful in determining age and conditions of formation.

Rock deformation experiments provide information about the mechanical properties of metamorphic rocks. Rock samples subjected to stress and strain tests yield data on properties such as plasticity, strength, and viscosity. These data are particularly important in understanding how directed pressures influence textures. Another phase of the study of metamorphic rocks is experimental petrology. Here, metamorphic minerals are synthesized under controlled equilibrium conditions. From these studies, geologists gain knowledge about the actual ranges of stability of these minerals in naturally occurring environments. The study of the crystal chemistry and thermodynamics provides information regarding actual conditions of formation and potential reactions. The roles of variables—such as entropy, volume change, and heat of reaction—provide clues to the behavior of mineral assemblages at varying temperature and pressure conditions. Theoretical petrology—the treatment of data on metamorphic rocks by mathematical models or the principles of theoretical physics—also provides important information in the study of regional metamorphism. In addition to analyzing the samples collected, scientists analyze field data in the laboratory. Statistical analysis of orientations of foliations and other structural features is greatly facilitated by the computer. Laboratory analyses are recast to generate a variety of graphical treatments to check for chemical trends or metamorphic facies relationships.

APPLICATIONS OF REGIONAL METAMORPHISM IN INDUSTRY AND ENGINEERING

The products of regional metamorphism touch people's lives in many ways. Marbles, gneisses, and slates are used as building and cut ornamental stones, and some quartzites, marbles, and gneisses are used as aggregate. The minerals graphite, talc, vermiculite, and asbestos are four of the more commonly used mineral products of regional metamorphism, whereas wollastonite, garnet, kyanite, emery, and pyrophyllite are mineral products of limited use. Graphite is most commonly used in the metallurgical industry, with lesser amounts used as lubricants, paints, batteries, and pencil "leads," and in electrodes. Talc is also a product with diverse uses. The ceramics industry is the largest consumer, followed by paint and paper manufacturers. Vermiculite is a common product in thermal and acoustic insulation. Asbestos, despite concerns about its health risks, remains an important fire-retardant material for appropriate uses.

Regionally metamorphosed rocks provide certain advantages in construction but also present unique engineering problems. A scan of the skyline of the island of Manhattan in New York City reveals that the largest skyscrapers are restricted to several areas of the island, surrounded by expanses of much shorter buildings. The reason that buildings such as the Empire State Building occupy only certain areas is that they sit directly on structurally strong regionally metamorphosed rock that occurs at or near the surface. In other areas of the island, such rocks are too deeply buried, and the overlying sediments are too weak to support the larger structures. In other cases, inclined foliations paralleling slopes in hilly or mountainous regions represent potential planes of slippage that can give way and produce massive rock slides, particularly when road, mine, or dam construction modifies the landscape. In areas where foliations are not adequately taken into account, the cost can be millions of dollars in repairs and/or great loss of life. Special techniques and restrictive measures need to be applied when construction occurs in areas where foliations present potential problems.

—Ronald D. Tyler

Further Reading

Best, Myron G. *Igneous and Metamorphic Petrology*. 2nd ed., Blackwell Science Ltd., 2003.

Blatt, Harvey, Robert J. Tracy, and Brent Owens. *Petrology: Igneous, Sedimentary, and Metamorphic*. 3rd ed., W. H. Freeman, 2005.

Bucher, Kurt, and Martin Frey. *Petrogenesis of Metamorphic Rocks*. 7th ed., Springer-Verlag, 2002.

Chernicoff, Stanley. *Geology: An Introduction to Physical Geology*. 4th ed., Prentice Hall, 2006.

Fettes, Douglas, and Jacqueline Desmons, editors. *Metamorphic Rocks: A Classification and Glossary of Terms*. Cambridge UP, 2007.

Gillen, Cornelius. *Metamorphic Geology: An Introduction to Tectonic and Metamorphic Processes*. 2nd ed., Springer, 2011.

Newton, Robert C. "The Three Partners of Metamorphic Petrology." *American Mineralogist,* vol. 96. 2011, pp. 457-69.
Philpotts, Anthony, and Jay Aque. *Principles of Igneous and Metamorphic Petrology.* 2nd ed., Cambridge UP, 2009.
Vernon, Ron H. *A Practical Guide to Rock Microstructure.* Cambridge UP, 2004.
Winter, John D. *An Introduction to Igneous and Metamorphic Petrology.* Prentice Hall, 2001.

Richter, Charles Francis

Fields of Study: Geology; Geophysics; Seismology

ABSTRACT
Richter is best remembered for developing a scale that relates the energy released by an earthquake. He was a pioneer in seismology and was instrumental in developing a database of all earthquakes that occur worldwide every year.

KEY CONCEPTS

logarithmic: based on powers of ten, such that each unit increase in designation represents an amount ten times greater than the next lower designation and ten times less than the next higher designation

seismicity: the degree of seismic activity, as earthquakes, that occurs in a defined region over time

seismology: the study of vibratory and displacement perturbations of Earth's crust, typically as earthquakes and volcanic eruptions, although modern seismometers are able to detect the vibrations produced by such events as explosions and the traffic of heavy vehicles

seismometer: an electromechanical device capable of detecting and recording vibrations traveling through the planet such as those caused by earthquakes, volcanic eruptions, landslides, explosions, or even just traffic

EARLY LIFE

Charles Francis Kinsinger was born on a farm near Hamilton, Ohio, north of Cincinnati. His father, Frederick William Kinsinger, a farmer, and his mother, Lillian Anna Richter, a schoolteacher, were married on July 15, 1891, in Butler County, Ohio. Charles's sister, Margaret, was born in 1892, but their parents divorced shortly after her birth. Their parents remarried before Charles's birth in 1900, but, unfortunately, the second attempt at matrimony was no more successful than the first. After separating from her husband, Lillian resumed using her maiden name and changed her children's names to Charles Francis and Margaret Richter. Returning to live in her father's household, she raised her children in Ohio until the death of her mother in 1907, prompting her father to move his family to Los Angeles, California, where he purchased a house in the Wilshire district.

Charles Richter. Photo via CalTech/Wikimedia Commons. [Public domain.]

Accounts of Charles Richter's early childhood family life indicate that he and his sister were educated at home while they were in Ohio but were sent to Hobart Boulevard School when they arrived in Los Angeles. These accounts also depict a dysfunctional family in which his mother was extremely attached to her father. Charles, who had survived a bout of cholera when he was fifteen months old, was a small, thin boy, shy at school, lonely, and dominated by his extremely assertive sister at home. (Margaret's aggressive personality, in some ways, served her well; in an era in which women did not commonly pursue an advanced degree, she ultimately earned a PhD at the prestigious Stanford University.)

Academically, Charles did well in public school and was soon transferred to a preparatory school where he developed a taste for astronomy, chemistry, mathematics, and physics. He performed very well in high school, and after graduating at the age of sixteen in 1916, he was accepted at Stanford University, where he obtained his bachelor's degree in physics in 1920 and his PhD in 1928.

Although his intellectual assets served him well, Richter was so afflicted with depression, nervous breakdown, and mental confusion that he spent a year (1920-1921) in a private sanatorium under the care of a psychiatrist, Dr. Ross Moore. At Moore's suggestion, Richter put aside his academic work temporarily while he took a series of jobs (clerk in a hardware company, messenger for the Los Angeles County Museum) in order to gain a degree of financial (and social) independence from his mother. Moore also encouraged him to take up writing as a way to explore his emotions, sparking in Richter a love of poetry that lasted throughout his life. (The vast majority of the poems Richter wrote have been preserved in his personal papers, which he bequeathed to the California Institute of Technology library.)

LIFE'S WORK

In 1922, Richter returned to Stanford and tackled the research he needed to complete his doctorate in physics. After receiving his doctorate in 1928, Richter, who had been an isolated and lonely student, was invited to a friend's house where he met divorcée Lillian Brand. They were married on July 19, 1928. Ignoring the advice of his psychiatrist to separate himself from his mother's influence, Richter took his new family, which included Lillian's two-year-old son, Reginald "Butch" Floyer Saunders Jr., to live with his mother and sister. Predictably, the arrangement was a social disaster; Richter's wife, mother, and sister argued continuously.

Fortunately for Richter, his professional life was more successful. At the end of 1927, he had been hired as a part-time research assistant at the Caltech Seismological Laboratory near Pasadena, under the direction of Robert Andrews Millikan. As soon as he received his PhD, Richter was hired full-time, acquiring at once a stable position and access to the emerging science of seismology that became his lifelong passion. Rarely do people spend their entire lives working at the same place. Richter was one of these rare people, remaining at Caltech until he retired in 1970.

The founding of the Caltech Seismological Laboratory in Pasadena made good sense in light of the fact that many earthquakes had occurred on the San Andreas Fault in California. Under Millikan, a physicist, Harry Wood, an astronomer, Hugo Benioff, a former astronomer and newly appointed seismologist, and John Anderson, an engineer, the laboratory flourished. Creating and maintaining an almost perfect research atmosphere, the laboratory encouraged its members to achieve their best. All scientists would gather at 10:30 a.m. and 2:30 p.m. for coffee and at noon for a game of bridge. These daily meetings would keep everyone aware of the progress made in each scientist's research. Problems would be discussed and solutions offered.

Thanks to the new and precise seismometer that Wood and Anderson created, a network of seismic stations, each equipped with a seismometer, was put into place. These seismometers recorded the basic data necessary to create a catalog of earthquakes. The successes, technological advances, and energetic research projects of the Caltech Seismological Laboratory stimulated other research institutes, such as the Seismology Institute of the University of California (Berkeley), the Lick Observatory of San Jose, and St. Louis University, to a healthy competition in which the science of seismology flourished.

As the laboratory grew, new well-known geophysicists joined the staff. Among these new employees was Beno Gutenberg, a Jewish German seismologist who had received from Caltech a job offer he could not turn down; the salary and living conditions in California were excellent, while the situation in Germany was worsening for Jews. Upon his arrival in 1930, Gutenberg immediately teamed up with Richter in the difficult task of measuring earthquakes. They published their research in four scientific papers that studied the nature of the seismic waves, the way they reflect and refract, and how they move within Earth's interior. Acknowledging the partnership that both scientists developed over the years, many researchers believe that the Richter scale, used for measuring the magnitude of (the amount of energy released by) an earthquake, should be called the "Gutenberg" or "Gutenberg-Richter" scale. In an interview in 1979, Richter himself said that the term "somewhat underrates Gutenberg's part in developing it for further use." Nevertheless, Richter is credited with having produced the scale, which was developed and used beginning in 1932. The paper that introduced it to the scientific community was published in the January, 1935, issue of the *Bulletin of the Seismological Society of America*.

The Richter scale is logarithmic and has no upper limit (although no earthquake greater than 9.5 has ever been recorded). Richter most certainly did his share in the development of the scale, and the contributions of Richter and Gutenberg to the science of seismology in their book *Seismicity of the Earth* (1941) are such that their book is still valued today.

Richter was an avid reader of science fiction books and an excellent linguist. He also enjoyed hiking in the Sierra Nevada. Year after year, he would spend his vacation in Sequoia and Kings Canyon National Parks. This communion with nature was a perfect match for his shy personality. At the beginning of his marriage, his wife Lillian would accompany him, but within a few years, her health deteriorated when she developed acute colitis. Charles and Lillian lived very independent lives while married. She visited France and Africa without her husband, while Charles spent a year alone in Japan as a Fulbright scholar.

In 1969, Charles and Lillian moved to Altadena in a house with very large windows from which Charles could see the San Gabriel Mountains, a region where he loved to walk. Shortly after their move, Lillian developed intestinal cancer. She underwent surgery in July, 1970, but the cancer spread and she developed lymphoma. She had another surgery in October, 1972, but had complications leading to her death on November 6, 1972.

Richter's health declined after the death of his wife, but he continued to work in the field of seismology. Even after his retirement in 1970, Richter returned to the laboratory every time there was a major earthquake in the region. He progressively needed more and more nursing help but remained in his house in Altadena. In July, 1984, he suffered a heart attack that required him to move to a nursing home. He died on September 30, 1985, of congestive heart failure and was buried in the Mountain View cemetery in Altadena next to his wife and stepson.

Richter was an unconventional man, plagued throughout his life with neurological disorders that researchers today identify as probably being Asperger's syndrome, a condition that makes social interaction very difficult. Perhaps this helps to explain his loneliness and his lifelong preference for solitary pursuits, and may even relate to his commitment to nudism. However, while people often remarked on his isolation, they also thought that he was a "walking encyclopedia" because of his incredible capacity to memorize terms in physics and seismology, and many were impressed with his commitment to teaching.

IMPACT

An accomplished teacher and prolific researcher, Richter published extensively and mentored generations of graduate students from 1927 to 1970 while working at Caltech Seismological Laboratory. Richter is remembered as a pioneer in the science of seismology; his achievements are even more enormous when one considers that the theory of plate tectonics had not been advanced when he coauthored his treatise *Seismicity of the Earth* with Beno Gutenberg in 1941. Richter's legacy encompasses not only his magnitude scale that revolutionized the measurement of earthquakes but also the earthquake monitoring system. He produced his long-term catalogs of earthquakes that researchers still use today.

—Denyse Lemaire and David Kasserman

Further Reading

Gutenberg, Beno, and Charles Richter. *Seismicity of the Earth.* The Society, 1941.

Hough, Susan Elizabeth. *Predicting the Unpredictable: The Tumultuous Science of Earthquake Prediction.* Princeton UP, 2010.

———. *Richter's Scale: Measure of an Earthquake, Measure of a Man.* Princeton UP, 2007.

Richter, Charles. "An Instrumental Earthquake Magnitude Scale." *Bulletin of the Seismological Society of America,* vol. 25, no. 1, Jan. 1935, pp. 1-32.

Witze, Alexandra. "Charles F. Richter, 1900-1985, American Seismologist." *Notable Scientists: From 1900 to the Present,* edited by Brigham Narins, Gale, 2001.

Richter Develops a Scale for Measuring Earthquake Strength

Fields of Study: Geology; Seismology; Geophysics

ABSTRACT

Charles Francis Richter devised a scale for measuring the strength of earthquakes based on their seismograph recordings.

KEY CONCEPTS

intensity: of an earthquake, the combination of the energy released in the event and the duration of the event

logarithmic scale: a scale based on powers of ten rather than on unit steps such that 0=1, 1=10, 2=100, 3=1,000...etc.

magnitude: a scaled measurement of the amount of movement of rock at the epicenter of an earthquake

EARTHQUAKE MEASUREMENT BEFORE RICHTER

Earthquakes range in strength from imperceptible tremors to catastrophic shakings that devastate large regions and cause the loss of thousands of lives. The human impact of earthquakes is not an accurate measure of their power; minor earthquakes in heavily populated regions may cause great destruction, whereas powerful earthquakes in remote areas may go unnoticed. To study earthquakes, it is essential to have an accurate means of measuring their power.

The first attempts to measure the power of earthquakes involved the development of intensity scales that relied on damage effects and the reports of witnesses as measures of the force of vibration. The first such scale was devised by Michele Stefano de Rossi and François-Alphonse Forel in 1883. It ranked earthquakes on a scale of 1 to 10. The de Rossi-Forel scale proved to have two serious limitations: Its level 10 encompassed a great range of effects, and its description of effects on human-made and natural objects was so specifically European that the scale was difficult to apply elsewhere. To remedy these problems, Giuseppe Mercalli published a revised intensity scale in 1902. The Mercalli scale, as it came to be called, added two levels to the high end of the de Rossi-Forel scale, making its highest level 12. It also was rewritten to make it more globally applicable. With later modifications by Charles Francis Richter, the Mercalli scale is still in use.

Intensity measurements, even though they are somewhat subjective, are very useful in mapping the extent of earthquake effects. Nevertheless, intensity measurements are still not ideal measuring techniques. Intensity varies from place to place and is strongly influenced by local geologic factors, and different observers frequently report different intensities. There is a need for an objective method of describing the strength of earthquakes with a single measurement.

Richter devised an objective technique for determining the power of earthquakes in the early 1930s at the California Institute of Technology in Pasadena, California. The eventual usefulness of what became known as the Richter scale was completely unforeseen at first. In 1931, the California Institute of Technology was preparing to issue a catalog of all earthquakes detected by its seismographs in the preceding three years. Several hundred earthquakes were listed, most of which had not been felt by humans, only detected by instruments. Richter was concerned about possible misinterpretations of the listings. With no indication of the strength of the earthquakes, the public might overestimate the risk of earthquakes in areas where seismographs were numerous and underestimate the risk in areas where seismographs were few. To remedy the lack of a

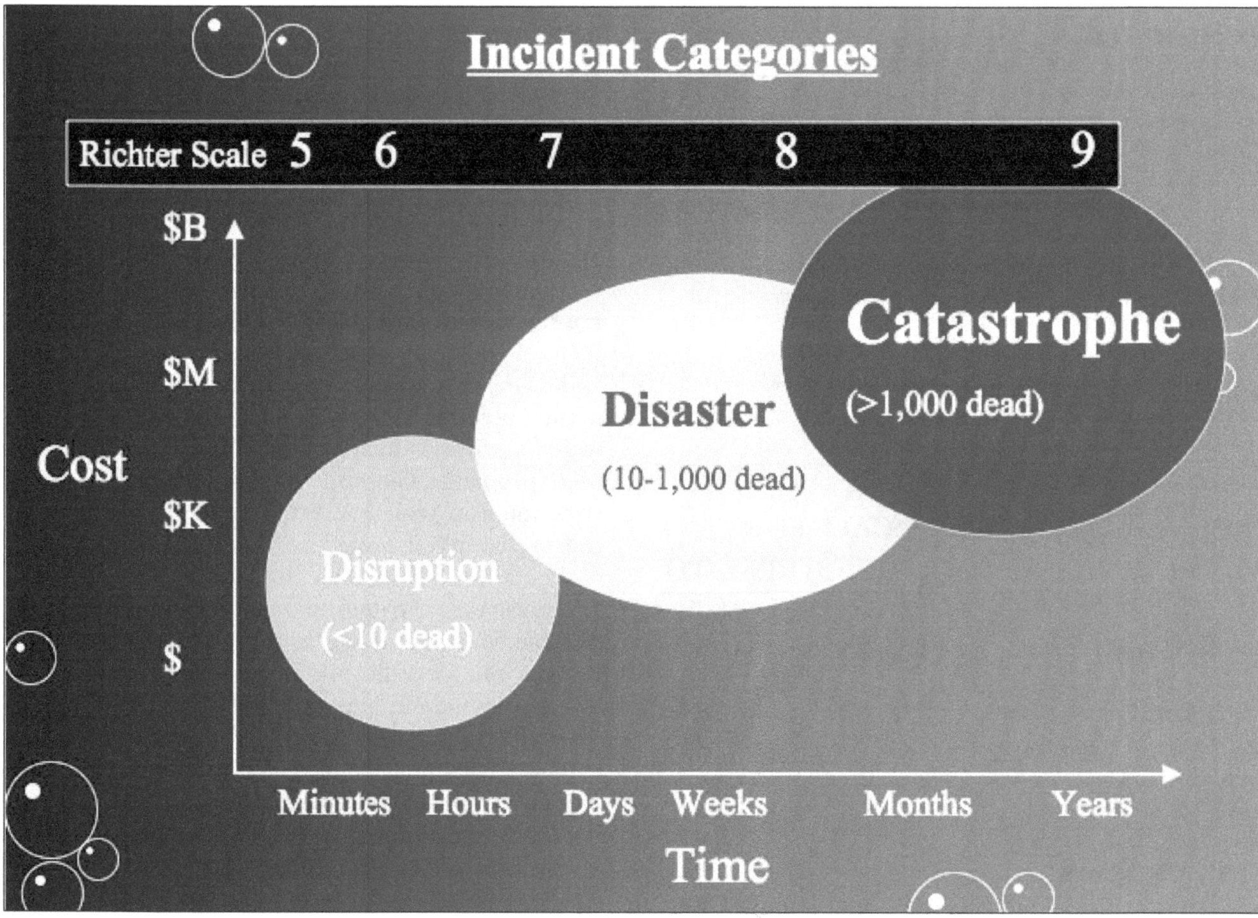

Increasing severity and costs along the Richter scale. Image via Wikimedia Commons. [Public domain.]

measuring method, Richter devised the scale that now bears his name.

Richter defined the magnitude of an earthquake as the logarithm of the height of its seismograph trace in microns (thousandths of a millimeter), as recorded on a standard instrument. Thus, an earthquake that produces a trace one millimeter (1,000 microns) high would be magnitude 3, one that produces a trace 1 centimeter high (10,000 microns) would be magnitude 4, and so on. These measurements were defined for a standard seismograph magnifying ground motion twenty-eight hundred times and located 100 kilometers (about 62 miles) from the earthquake. The magnification of the instrument means that the actual ground motion caused by a magnitude 3 earthquake 100 kilometers away is not 1 millimeter but only 1/2,800 millimeter, or only about three thousand times the diameter of an atom. By comparing records for earthquakes recorded on different devices at different distances, Richter was able to create conversion tables for measuring magnitudes for any instrument at any distance. He also set up the scale so that any event likely to be felt by humans would have a positive magnitude, because scales with zero and negative numbers tend to be confusing to most people.

Richter had hoped to create a rough means of separating small, medium, and large earthquakes, but he found that his scale was capable of making much finer distinctions. Most magnitude estimates made with a variety of instruments at various distances from earthquakes agreed to within a few tenths of a magnitude. Richter formally published a description of his scale in January, 1935, in the *Bulletin of the Seismological Society of America*. Other systems of estimating magnitude had been attempted, notably that of pioneering Japanese seismologist Kiyoo Wadati, published in 1931, but Richter's system proved to be the most workable scale yet devised and rapidly became the standard.

Over the next few years, the scale was refined. One critical refinement was in the way seismic recordings were converted into magnitude. Earthquakes produce many types of seismic waves, but it was not known which type should be the standard for magnitude. So-called surface waves travel along the surface of Earth. It is these waves that produce most of the damage in large earthquakes; therefore, it seems logical to let these waves be the standard. On the other hand, earthquakes deep within Earth produce few surface waves. Magnitudes based on surface waves would be too small for these earthquakes. Deep earthquakes produce mostly waves that travel through the solid body of Earth, or so-called body waves. Actually, two scales are needed: one based on surface waves and one on body waves. Richter and his colleague Beno Gutenberg developed scales for the two different types of waves that are still in use. Magnitudes estimated from surface waves are symbolized by a capital *M*, and those based on body waves are denoted by lowercase *m*.

From a knowledge of movements of Earth's crust associated with seismic waves, Richter and Gutenberg succeeded in relating the energy output of an earthquake with its magnitude. Each increase of one Richter magnitude corresponds to about a thirtyfold increase in energy. A magnitude 6 earthquake releases about as much energy as a one-megaton nuclear explosion; a magnitude 0 earthquake releases about as much energy as a small car dropped off a two-story building.

An additional refinement to the Richter scale was developed in the 1970s. Extremely large earthquakes release their energy over time spans as long as several minutes and over a fault break of hundreds of kilometers. The highest seismograph trace for the earthquake, however, measures the energy received at only one instant. It is also possible to estimate energy released from the length of the fault rupture and the amount of fault displacement, and by this measure, conventional magnitudes for the largest earthquakes are too small. A magnitude corrected for the long duration and great spatial extent of the largest earthquakes is called a "seismic-moment magnitude." Japanese seismologist Hiroo Kanamori devised a seismic-moment magnitude scale during the 1970s; on this scale, unlike the conventional Richter scale, some of the greatest earthquakes exceed magnitude 9.

SIGNIFICANCE OF THE RICHTER SCALE

The Richter scale is a good example of an accidental scientific discovery. Richter's original intent had been to develop a scale for very rough measurements, but the scale proved to be useful on a completely unforeseen scale. Richter scrupulously avoided using the term "Richter scale" in his writings and went to some length to give credit to other researchers who also had similar ideas. The Richter scale has now become firmly established in popular and professional usage.

The Richter scale had been in use among scientists for about fifteen years before Richter scale readings began to be widely quoted by the press in the 1950s. Just as scientists had, the press found the scale a useful tool for defining the strength of earthquakes. Initially, there was much confusion among laypersons over the difference between magnitude and intensity. Also, the logarithmic nature of the scale, clear enough to scientists, was widely misunderstood. It was not until about 1970 that the press generally stopped describing the Richter scale as running from 1 to 10.

The scientific utility of the Richter scale stems from its logarithmic nature. Wave phenomena occur over a great range of intensities and are best described by logarithmic scales; other examples include the speed and shutter settings on cameras (each step doubles or halves the original amount of light) and the decibel scale for sound. Virtually all the important physical quantities associated with waves are proportional to the amplitude of the wave raised to some power; such relationships are very simple to express in logarithmic terms. Because the magnitude of an earthquake is defined as the logarithm of its ground motion, many of the physical dimensions of earthquakes, such as their energy release, are directly and simply related to magnitude. Also, it is simple to develop equations to relate ground motion, magnitude, and distance from an earthquake.

The Richter scale made it possible for scientists to compare the energy outputs of earthquakes in a quantitative way and to derive new understandings of earthquake phenomena. For example, most of the energy released by earthquakes is released by the few most powerful ones. Also, earthquakes are more powerful in some geologic settings than in others. The greatest earthquakes (above magnitude 6.5) are associated with faults on or bordering the continents, whereas earthquakes along the midocean ridges rarely exceed magnitude 6.5. Deep earthquakes are rarely as strong as magnitude 7. These differences provide scientists with important insights into the forces at work within Earth.

—Steven I. Dutch

Further Reading

Bolt, Bruce A. *Earthquakes.* 5th ed., W. H. Freeman, 2003.

Gutenberg, Beno, and Charles Richter. *Seismicity of the Earth.* The Society, 1941.

Hough, Susan Elizabeth. *Earthshaking Science: What We Know (and Don't Know) About Earthquakes.* Princeton UP, 2002.

———. *Predicting the Unpredictable: The Tumultuous Science of Earthquake Prediction.* Princeton UP, 2010.

———. *Richter's Scale: Measure of an Earthquake, Measure of a Man.* Princeton UP, 2007.

Richter, Charles. "An Instrumental Earthquake Magnitude Scale." *Bulletin of the Seismological Society of America,* vol. 25, no. 1, Jan. 1935, pp. 1-32.

Stein, Seth, and Michael Wyesession. *An Introduction to Seismology, Earthquakes, and Earth Structure.* Wiley, 2013.

Riverbed Forms

Fields of Study: Geology; Sedimentology; Geomorphology; Geophysics

ABSTRACT

Bed forms, produced by flows of water or air in natural environments and in artificial channels, are a distinctive aspect of the transport of granular sediment. They have a strong effect on the magnitude of bottom friction felt by the flow. Because of their great variety as a function of flow conditions, bed forms are valuable in interpreting depositional conditions of ancient sedimentary deposits.

KEY CONCEPTS

aeolian deposits: material transported by wind

angle of repose: the natural angle formed between the horizontal plane and the side of a freestanding pile of sediment particles under the force of gravity in a fluid medium

antidune: an undulatory upstream moving bed form produced in free surface flow of water over a sand bed in a certain range of high flow speeds and shallow flow depths

bed configuration: the overall geometry of a sediment bed molded by sediment transport in a flowing fluid

bed form: an individual geometrical element of a bed configuration

combined flow: a flow of fluid with components of both unidirectional and oscillatory flow superposed on one another to produce a more complex pattern of fluid motion

current ripple: a small bed form, oriented predominantly transverse to the direction of flow, produced at low to moderate flow speeds in unidirectional water flows

dune: a large bed form, oriented predominantly transverse to the direction of flow, produced at moderate to high flow speeds

flume: a laboratory open channel in which water is passed over a sediment bed to study the nature of the sediment movement

oscillation ripple: a small to large bed form, oriented predominantly transverse to the direction of flow, produced at low to moderate flow speeds in oscillatory water flows

oscillatory flow: a flow of fluid with a regular back-and-forth pattern of motion

plane bed: a bed configuration without rugged bed forms produced in both unidirectional and oscillatory flows at high flow speeds

unidirectional flow: a flow of fluid oriented everywhere and at all times in the same direction

CHARACTERISTICS OF RIVERBEDS

A striking feature of the transport of loose granular sediment over a bed of the same material by a turbulent flow of fluid such as air or water is that in a wide range of conditions of flow and sediment size, the bed is molded into topographic features, called "bed forms," on a scale ranging from hundreds to millions of times larger than the grains themselves. Examples of such bed forms are sand ripples at the seashore or on a dry riverbed, sand dunes in the desert, and (less apparent to the casual observer) large underwater sand dunes in rivers and in the shallow ocean. The overall geometry of a sediment bed molded by a flow of fluid is called the "bed configuration." Bed forms are individual elements of this configuration. The term "bed form" includes both the overall geometry and the individual elements of that geometry.

The enormous range of bed form-producing flows, together with the complex dynamics of the response of the bed, makes for striking variety in the scale and geometry of bed forms. Scales typically span a range of five orders of magnitude, from a few centimeters to more than 1,000 meters in spacing. Bed forms may appear as long ridges or circumscribed mounds, and their crests may be rounded or sharp. Most bed forms are irregular in detail, but elongated bed forms tend to show a more or less strong element of regularity in their overall arrangement, some even being perfectly regular and straight crested. Elongated bed forms tend to be oriented transversely to the direction of flow, although flow parallel forms are produced under certain conditions, and forms with no strongly preferred orientation are produced in some flows. Most bed forms are approximately wave shaped and are often likened to waves, but they are waves only in a geometrical sense, not in a functional sense.

The most common bed forms are in sands (sediments having a mean particle size between about 0.1 millimeter and 2 millimeters), but bed forms are produced in silts (sediments having a mean particle size between about 4 micrometers and 0.1 millime-

Current ripples preserved in sandstone of the Moenkopi Formation, Capitol Reef National Park, Utah. Photo by Mav, CC BY-SA 2.5, via Wikimedia Commons.

ter) and gravels (sediment having a mean particle size greater than 2 millimeters) as well. Bed forms produced by flows of air or water over mineral sediments in natural flow environments are of greatest interest to geologists. A far wider range of bed forms can be produced by flows of fluids with other densities and viscosities over sediments that are more or less dense than the common mineral sediments, which have densities mostly in the range 2.5 to 3.0 grams per cubic centimeter.

TYPES OF FLUID FLOW

Bed forms are made by unidirectional flows of air or water, as in rivers and tidal currents and under sand-moving winds, and by oscillatory flows, as on the shallow seafloor beneath wind-generated surface waves, which cause the water at the bottom to move back and forth with a period the same as that of the waves and with horizontal excursion distances of a few centimeters to a few meters. Bed forms are also made by what are called "combined flows," which are superpositions of unidirectional and oscillatory flows. Such combined flow bed forms are not as well understood as those made by unidirectional and oscillatory flows, but they are common in the shallow ocean. Bed forms made under wind are called "subaerial" or "aeolian bed forms," and bed forms made under water are called "subaqueous bed forms."

It might seem that the natural mode of sediment transport would be over a planar bed surface. In certain ranges of flow, a planar transport surface is indeed the stable bed configuration, but technically, such a plane bed is a bed configuration with no bed forms. In reality, plane bed transport is the stable configuration only under a very limited set of conditions, whereas rugged bed forms cover the transport surface over a wide range of conditions in both oscillatory and unidirectional flows. Why such bed forms develop at all on transport surfaces is poorly understood. In certain ranges of flow, the planar transport surface is unstable in the sense that small bed irregularities of the kind that can be built at random by the plane bed sediment transport become amplified to grow eventually into bed forms rather than being smoothed out again. The physics behind this instability is chaotic and complex, and is still not clearly understood. An essential element of this complexity is that there is a strong interaction or feedback between the bed configuration and the flow. The flow molds the bed configuration, but the bed configuration in turn affects the nature of the flow.

In unidirectional flow, current ripples are formed in fine sands as soon as the current is strong enough to move the particles, and they persist to moderate currents of about 0.5 meter per second. In a vertical cross section parallel to the flow, these current ripples are triangular in shape, with downstream slopes about equal to the angle of repose of sand under water (about 30 degrees) and with gentler upstream slopes. In top or horizontal view, they are oriented mostly transverse to the direction of flow and are irregular in detail. Their spacings range from 10 to 20 centimeters, and their heights are a few centimeters. This characteristic size changes little with flow strength or sediment size. With increasing unidirectional flow speed, ripples give way to dunes, which are geometrically fairly similar to ripples but are of much larger scale, ranging from meters to thousands of meters in spacing and tens of centimeters to tens of meters in height, depending in a complex and poorly understood way on conditions of flow as well as sediment size. With further increase in flow speed to about 1 meter per second, dunes are replaced by a plane bed configuration. The sequence of bed configurations with increasing unidirectional flow speed over coarse sands is different from that over fine sands: Transport is over a plane bed at flow strengths just above the threshold for sediment movement, and then dunes develop with further increase in flow strength. In sediments coarser than about 0.6 millimeter, current ripples are not formed in any range of flow speeds.

Current ripples and dunes move upstream or downstream at a speed far lower than the flow speed. This movement proceeds by erosion of sediment on the upstream side of the bed form and deposition on the downstream side. When an individual ripple or dune can be watched carefully for a time, it is seen to change its size, shape, and speed irregularly, eventually to disappear by being absorbed into a neighboring bed form. Offsetting this loss of bed forms is the production of new ones by a kind of subdivision of one larger form into two smaller ones. In flows of water with a free upper surface, like rivers and tidal currents, undulatory bed forms called "antidunes" make their appearance at high flow speeds and shallow flow depths. Antidunes, so named because they tend to move slowly upstream by erosion of sediment on the downstream sides and deposition on the upstream sides, come about by a complex effect of standing surface waves (waves that move upstream about as fast as the flow is moving downstream) on the sediment bed.

In oscillatory flow, regular straight crested bed forms called "oscillation ripples," with sharp crests and broadly rounded troughs, are formed as soon as the sediment begins to be moved. At their smallest, their spacing is a few centimeters, but they grow in size to more than 1 meter in spacing as the period and amplitude of the oscillatory motion increase. In fine sands, these larger ripples become irregular and mound shaped, but observations are not yet adequate for a detailed description. When the maximum flow speed during a single oscillation reaches about 1 meter per second, the bed forms are washed out to a plane bed mode of transport. Combined flows produce a whole range of ripples intermediate between current ripples and oscillation ripples.

If the flow changes with time, as is the rule rather than the exception in natural flows, the bed configuration adjusts in response. Usually, the bed configuration lags behind the change in the flow, with the result that the bed configuration is more or less out of equilibrium. Dunes formed by reversing flow in a tidal channel are a good example of such disequilibrium; commonly the weaker of the two flows (whether ebb or flood) modifies the shape of the dunes but does not reverse the asymmetry. Bed forms on a riverbed during passage of a flood also tend to lag behind changes in the flow.

STUDY OF RIVERBED FORMS

Bed forms can be observed and studied in natural flow environments and in laboratory tanks and channels. Each of these approaches has its advantages and disadvantages. In nature, observations on bed configurations are limited by various practical and technical difficulties. In laboratory tanks and channels, the bed forms can be studied much more easily, but for the most part water depths are unnaturally shallow.

A laboratory open channel in which a flow of water is passed over a sediment bed is called a "flume." Flumes range from a few meters to more than 100 meters long, from about 10 centimeters to a few meters wide, and from several centimeters to about 1 meter deep. The water is usually recirculated from the downstream end to the upstream end to form a kind of endless river. The sediment may also be recirculated, or it may be fed at the upstream end and caught in a trap at the downstream end. In a flume, it is fairly easy to measure the profile of the bed configuration with a mechanical pointer gauge or with a sonic depth sounder, and the bed forms can be studied visually and photographed through the sidewalls and also from the top when the absence of surface turbulence permits. Time lapse motion pictures of bed form movement are also instructive. Oscillatory flow bed forms can be studied either in long open tanks, in which water waves are passed over a sand bed to produce oscillatory flow at the bed, or in closed horizontal ducts, in which water is pushed back and forth in a regular oscillation between tanks at the ends of the duct.

In some natural flow environments, such as rivers and tidal currents, the geometry of the bed forms can be studied when they are exposed at low water. Observations of the bed forms when they are being molded by the flow, however, are more difficult. If the flow is not too strong, divers can make direct observations. Profiling of the bed geometry along lines or even across wide areas is usually possible by means of various sonar techniques, whereby the travel times of sound pulses reflected from the bed are converted to water depths. If the water is clear enough, bottom photographs of small areas can be taken. Current velocities can be measured with current meters anchored on the bed. Movement of large bed forms is difficult to measure because bed form speeds are slow and usually no fixed reference points are available.

The status of observations on bed configurations leaves much to be desired. Even in laboratory channels and tanks, where the major outlines are by now fairly well known, there is much room for further work for two reasons: The narrowness of the flow tends to distort the three-dimensional aspects of the bed geometry, and little work has been done on bed forms in combined flows. There is a need for more observations on the geometry and movement of bed forms in natural flows as a function of flow strength, as well as on the effect of disequilibrium.

SIGNIFICANCE OF RIVERBED FORMS

Bed forms are ubiquitous in natural flow environments. They are most apparent to the casual eye in fields of sand dunes in deserts and in certain coastal environments where the wind molds available loose sand into dunes. They are widely present but less obvious in rivers and the shallow ocean. Apart from their intrinsic scientific interest as a widespread natural phenomenon, bed forms are of importance in both engineering and geology for various reasons. Large underwater bed forms many meters high in rivers and marine currents can be obstacles to navigation, and their movement can be a threat to submarine structures. Also, the inexorable movement of desert dunes or coastal dunes can bury roads and buildings. In other engineering applications, technology imitates nature and various processes to be carried out rely on the flow of a particle bearing fluid medium, as is the case in fluidized bed processes. A comprehensive understanding of the behavior of the particle load within the fluid is essential to enhancing and ensuring the efficiency of those processes.

The rugged topography of ripple and dune bed forms leads to a pattern of flow over each bed form in which the pressure on the upstream surface is relatively high and the pressure on the downstream surface is relatively low, much like an unstreamlined motor vehicle on the highway or a house in a strong wind. This pressure difference adds greatly to the force the flow exerts on the bed and, conversely, to the force the bed exerts on the moving flow. Hydraulic engineers have expended much effort on studying the effect that this resistance force has on the depth a river assumes when it is given a particular rate of flow to carry. A river with a planar bed can pass a given flow rate at a shallower depth and greater velocity than can a river with a bed roughened by large dunes, which exert a large resistance force on the flow and make the velocity smaller and the depth greater.

Geologists have given attention to bed forms partly because of their effect on the geometry of the stratification that develops as a sediment bed is deposited while bed forms are active on the sediment surface. It is often possible to tell the kind of bed form that was present just by examining the stratification in a sedimentary rock like a sandstone. If the flow conditions responsible for making that kind of bed form are already known, from laboratory experiments or from observations in modern natural flow environments, the depositional conditions of that sedimentary rock (which may be geologically very ancient) can be interpreted. Such interpretations are one of the tools used in mapping the geometry of subsurface petroleum reservoirs in sedimentary rocks.

—John Brelsford Southard

Further Reading

Allen, J. R. L. *Principles of Physical Sedimentology*. Blackburn Press, 2001.

Collier, Michael. *Over the Rivers: An Aerial View of Geology*. Mikaya Press, 2008.

Collinson, J. D., Nigel Mountney, and D. B. Thompson. *Sedimentary Structures*. 3rd ed., Dunedin Academic Press, 2006.

Darby, Stephen, and David Sear. *River Restoration: Managing the Uncertainty in Restoring Physical Habitat.* John Wiley & Sons, 2008.

Freitag, Bob, et al. *Floodplain Management: A New Approach for a New Era*. Island Press, 2009.

Kondolf, G. Mathias, and Hervé Piégay. *Tools in Fluvial Geomorphology*. Wiley, 2016.

Leeder, M. R. *Sedimentology and Sedimentary Basins: From Turbulence to Tectonics*. Wiley-Blackwell, 2011.

Newson, Malcolm. *Land, Water and Development: Sustainable and Adaptive Management of Rivers*. 3rd ed., Routledge, 2008.

Rhoads, Bruce L. *River Dynamics. Geomorphology to Support Management*. Cambridge UP, 2020.

Tsutsumi, Daizo, and Jonathan B. Laronne. *Gravel-Bed Rivers: Process and Disasters*. Wiley, 2017.

Vanoni, V.A., editor. *Sedimentation Engineering*. 2nd ed., American Society of Civil Engineers, 2006.

Wohl, Ellen. *A World of Rivers*. U of Chicago P, 2011.

S

Salt Domes

Fields of Study: Geology; Geochemistry; Geophysics

ABSTRACT

Salt domes are a major source of the world's salt. Their caprocks are major sources of gypsum and sulfur. Upturned sediments on the flanks of salt domes form stratigraphic traps for oil and natural gas.

KEY CONCEPTS

caprock: impermeable rock that overlies a particular geological stratum or deposit

stratigraphic trap: a formation in rock strata where a mobile material can collect but is subsequently prevented from migrating farther

stratum (strata): a layer (layers) of rock

updip: the upward direction of an inclined stratum

DEFINITION OF SALT DOME

Salt domes consist of roughly cylindrical to mushroom-shaped plugs of massive rock salt extending toward Earth's surface from depths as great as 6,000 meters. These salt pillars typically range in diameter from 1 to 3 kilometers; however, some cores reach 12 kilometers in diameter. The plug is usually topped by a limestone, gypsum, and anhydrite caprock.

HOW SALT DOMES ARE FORMED

Salt beds, ranging in thickness from a meter to a few hundred meters, are deposited in shallow, hypersaline, marine environments such as basins with restricted circulation in regions where evaporation exceeds precipitation. The salt is commonly pure white and is associated with gypsum, anhydrite, and shales. Some deeply buried salt beds form mobile salt columns that rise toward the surface. Salt domes occur in the Colorado-Utah area, the Gulf Coast of the United States and Mexico, Spain, France, Romania, Iran, Arabia, and India.

Salt domes are emplaced when beds of salt deform plastically under the pressure of overlying rocks and rise through overlying layered sediments. The rising salt forces the overlying rocks into domes and punches through them to leave rock layers upturned along its flanks. The depth of the salt core beneath the surface varies widely. Deep domes may be more than 1,750 meters beneath the surface, but others may expose salt at the surface. As salt approaches the surface, it encounters groundwater, which dissolves the rising salt. A caprock of less soluble minerals, mostly anhydrite, forms on top of the rising plug. Often, anaerobic bacteria in groundwater break down the anhydrite of the caprock, forming calcite and native sulfur in the process.

Commercial quantities of sulfur are dispersed within the caprock of a few domes. Sulfur is extracted from the caprock by the Frasch process. Water heated to 150° Celsius is discharged into the caprock to melt the sulfur, and hot air is used to drive it to the surface. The molten sulfur is then piped to storage, where it solidifies.

At shallow domes, anhydrite, gypsum, and limestone in the caprock may be quarried for road metal

Astronaut photography of Jashak salt dome in the Zagros Mountains in Bushehr Province, Iran (the white area in the middle). Photo via NASA/Wikimedia Commons. [Public domain.]

or building materials. Salt is recovered by underground mining techniques. Salt domes account for only 5 percent of the world's reserves of salt, but alone they could supply the world's demand for thirty thousand years. Upturned sedimentary beds around the flanks of the domes provide traps for oil and gas that migrate updip and are impounded against the impermeable salt. The limestone of the caprock also forms a petroleum reservoir in some salt domes.

Cavities may be excavated within salt domes either by standard mining techniques or by pumping fresh water into the dome to form a solution cavity. The resulting cavity may then be used to store oil or gas. Because salt is impermeable and self-healing when fractured, it has been used as a storage site for nuclear waste disposal.

—René A. De Hon

Further Reading

Belenitskaya, Galina. *Salt Systems of the Earth: Distribution, Tectonic and Kinematic History, Salt-Naphthids Interrelations, Discharge Foci, Recycling*. Wiley, 2018.

Fossen, Haakon. *Structural Geology*. 2nd ed., Cambridge UP, 2016.

Jackson, Martin P. A., and Michael R. Hudec. *Salt Tectonics: Principles and Practice*. Cambridge UP, 2017.

Palmer, Andrew. *Introduction to Petroleum Exploration and Engineering*. World Scientific, 2017.

Selley, Richard C., and Stephen A. Sonnenberg. *Elements of Petroleum Geology*. 3rd ed., Elsevier, 2015.

Sand and Gravel

Fields of Study: Geology; Geomorphology; Sedimentology; Geophysics; Economic Geology

ABSTRACT

Sand and gravel are the most common solid particulates on Earth. They are transported by fluid flows of air or water and act to change the outward surface of Earth through erosion processes that in turn result in the formation of more sand and gravel over time. Accretion of these sediments produces a variety of rock types, particularly sandstone. Both sand and gravel are used in vast quantities in construction and other human activities.

KEY CONCEPTS

alluvial fan: the deposit of sediments that forms as the sediments being carried in suspension by flowing water settle and spread out as the flow rate decreases

glacial deposit: accumulations of particulate matter ranging in size from fine silt to large boulders that are pushed ahead of glaciers and eventually dropped in place when the glaciers retreat

gravel: assorted rock particles whose dimensions are within a defined range of about 5 to 90 millimeters

road metal: the hard road surface that forms from a tightly packed mixture of gravel and sand

sand: assorted rock particles whose dimensions are within a defined range of 0.05 to 5 millimeters

WHERE SAND AND GRAVEL ARE FOUND

Sands and gravels are widely distributed on Earth's surface; all fifty of the United States have producing deposits. Sand and gravel deposits are not spatially ubiquitous, however. Sand and gravel are heavy or dense, high in bulk, and low in value, and they cannot be shipped economically for long distances. Most sand and gravel in the United States and Canada come from glacial deposits, stream terraces and channels, including alluvial fans, or from beach deposits of either current or relict shorelines. Some specialty or industrial sands are derived from bedrock when more rigid control over the character or particle size of the sand is required.

PRIMARY USES OF SAND AND GRAVEL

By far the greatest use of sand and gravel is in construction, where they may be employed as fill material or as the aggregate in concrete. Industrial sands are more specialized, and their uses demand higher quality. Most industrial sand is used to make glass or as molding sand in foundries. The United States produces more than 1 billion tonnes of construction sand and gravel and between 25 and 30 million tonnes of industrial sands.

TECHNICAL DEFINITION OF SAND AND GRAVEL

Sand particles are 0.05 millimeter to 4.76 millimeters in diameter. Gravel particles are larger, 4.76 to 80 or 90 millimeters in diameter. Sand fragments are composed almost entirely of single minerals, chiefly quartz, with significant fractions of feldspars and smaller proportions of mica, chert, and heavy minerals. Gravels, on the other hand, are usually fragments of rocks that are composed of several minerals. Gravels reflect the geology of the stream basin in which they are located, because this is the source of the gravel deposit. Most gravels are resistant, but if the source stream basin is underlain largely by soft sediments, the gravels are less valuable as a resource. Impurities in sand and gravel deposits consist of silts, clays, or excessive proportions of micas, soft sediments, or rock fragments that have an undesirable chemistry.

DESCRIPTION, DISTRIBUTION, AND FORMS

Sand and gravel are the most widely distributed of the construction aggregates, are the easiest to recover or mine, and require only simple beneficiation, usually washing and screening. Historically, in the United States, they have dominated the market for aggregates. However, for many purposes, even in rough construction, they are not as suitable as their closest competitor—crushed stone or rock—because sharp-edged broken stones can interlock, unlike gravels, which are rounded by stream transport.

Both the quality of a sand or gravel deposit and its location with respect to market determine the resource value of that deposit. Quality concerns include the lithology of the particles (their chemical and physical character), the size and shape of the particles, their resistance to abrasion and cracking, the potential for chemical reactivity, and the freedom of the deposit from organic matter, silt, and clay (in other words, the deposit's cleanness). Fortunately, most sand deposits are dominated by quartz particles, which are both resistant and inert.

Gravels can pose a greater problem because of the variety of rocks in different drainage basins. Soft rocks or those that weather relatively rapidly (shales, friable sandstones, some limestones, and certain metamorphic rocks, especially schists and slate) do

Gravel on a beach. Photo by Stan Zurek, CC BY-SA 3.0, via Wikimedia Commons.

not make valuable gravels. A variety of rocks react with the alkalies in Portland cement and must be avoided for that particular use. Iron impurities rust, and certain other minerals weather or decompose rapidly. These conditions lead to weakened construction and are avoided. Thus, all gravel deposits are not equally valuable as resources, even if they are favorably located with respect to markets. Just as high-quartz sands are more valuable, so are gravels with high proportions of resistant rocks of the proper chemical composition.

SAND AND GRAVEL ECONOMICS

Market, in the case of construction sand and gravel, is defined by population and appropriate construction, such as highways. Thus, a sparsely populated region serves as a significant market while interstate highway construction is under way but becomes a small market when the highway is completed. The fortunes of construction sand and gravel suppliers wax and wane with the general economy; boom times of expanded residential or office building construction provide an excellent market. Recessions with little construction activity result in a shrinkage in production.

The low-value, high-bulk character of sand and gravel dictates that only surface mining is economical (the exception is some higher-value industrial sands, which may be mined underground). Moreover, the mining must be close to metropolitan centers, where most new construction occurs. Inevitably, the urban centers grow and encounter the sand and gravel mining. Zoning may then displace the mining to more remote locations because of complaints about dust, noise, truck traffic, or the unsightliness of gravel pits. Restrictions on the use of wetlands are increasing, particularly in cases in which endangered species may be involved. There is also an increasing concern with silica dust, which may affect specialty industrial sands.

HISTORY OF SAND AND GRAVEL USE

A measure of the overall relationship between population numbers and construction may be seen in the history of sand and gravel production in the United States. During the Depression of the 1930s annual production was about 180 million tonnes. In 1946, before widespread construction began in the postwar era, production was about 230 million tonnes. By 1960 construction had expanded significantly and production stood at 641 million tonnes. In 1970, residential construction and the interstate highway program were active; production was about 856 million tonnes. By 1994, although population had grown, highway and commercial construction had declined, and total sand and gravel production was just more than 918 million tonnes, of which about 27 million tonnes were industrial sands. By 2008, U.S. production of sand and gravel for construction was about 1 million tonnes; for industrial use, about 30 million tonnes. Throughout the post-World War II period, metropolitan population concentrations were far more important as markets for construction sand and gravel than were rural regions, which generally failed to generate construction in proportion to their population numbers. The major exception to this generalization is the interstate highway construction program, which generated temporary markets for sand and gravel in even the most sparsely settled portions of the country.

OBTAINING SAND AND GRAVEL

Nearly all gravel deposits, and most sand deposits, are found in stream sediments. Present-day stream deposits include channels, low terraces, and active portions of alluvial fans. Under these circumstances, sands and gravels removed by dredging or open-pit mining may be renewed by recurrent flooding or high stream flows. Relict stream deposits are those created by glaciation, including outwash fans and valley train deposits (the latter extending to the oceans from the glacial source), as well as relatively minor sources such as eskers, kames, and moraines deposited close to the ice margin. Most alluvial fans in western North America are also relict or inactive in that they were formed during the Pleistocene era and are not renewed by current geologic processes. In either case, virtually all sand and gravel are found in surficial deposits, which are frequently wetlands. This fact has advantages in terms of mining costs, but it also results in environmental problems and land-use conflicts.

USES OF SAND AND GRAVEL

The overwhelming use of sand and gravel is in construction. Use of these materials for fill, base, or subgrade of highways is the least demanding of quality requirements, and sand and gravel may not even be washed or screened for these uses. Usage in concrete, however, is far more demanding, both in terms of size-of-particle requirements (sorting, screening, or crushing may be used to produce the desired size) and in terms of quality (avoiding easily weathered or alkali-reactant rocks). Substitutes for sand and gravel in construction are crushed stone or rock and lightweight aggregates. Lightweight aggregates, largely volcanic rocks, are increasingly employed in specialty concretes and building blocks. Crushed rock is utilized where more rigid specifications for concrete exist or in regions where sand and gravel are scarce (this high-bulk, low-value commodity is shipped largely by truck, and rarely for distances greater than 30 meters).

Industrial sand and gravel encompass a variety of uses, each with its own specifications as to desirable characteristics in the product and its own market—hence the resultant location of mining activities. Glassmaking and foundry or molding sands lead the list of uses by tonnage; the former requires more rigid specifications and is located where construction is active, and the latter is located where metalworking is significant. The petroleum industry uses significant quantities for hydraulic fracturing of oil and gas wells. Abrasives, especially for blast sands, also rank high. Each use or type of sand has competition from substitutes that may reduce the resource value of deposits or the profitability of an industry. Glass, for example, has largely been replaced by aluminum and plastics as the material for containers in the food and beverage industry, although the push to a so-called green economy will undoubtedly see a resurgence in the use of recyclable glass containers. Abrasives used for "sand blasting" have come under fire for reasons of health, such as the breathing of dust by workers, particularly microscopic silica dust that can lead to a condition called "silicosis." Increasingly, however, solid carbon dioxide (dry ice) particles are being used to replace traditional blasting media.

—Neil E. Salisbury

Further Reading

Bell, Fred J., and Laurance J. Donnelly. "Gravel, Sand, and Clay Pits." *Mining and Its Impact on the Environment.* Taylor & Francis, 2006.

Gyr, Albert, and Klaus Hoyer. *Sediment Transport: A Geophysical Phenomenon.* Springer, 2006.

Kogel, Jessica Elzea, et al., editors. "Industrial Sand and Sandstone." *Industrial Minerals and Rocks: Commodities, Markets, and Uses.* 7th ed., Society for Mining, Metallurgy, and Exploration, 2006.

Randazzo, Giovanni, Derek W. T. Jackson, and J. Andrew G. Cooper. *Sand and Gravel Spits.* Springer, 2015.

Smith, M. R., and L. Collis, editors. *Aggregates: Sand, Gravel, and Crushed Rock Aggregates for Construction Purposes.* 3rd ed., Revised by P. G. Fookes, et al., Geological Society, 2001.

Welland, Michael. *Sand: The Never Ending Story.* U of California P, 2009.

SANDSTONE

Fields of Study: Geology; Sedimentology; Geophysics

ABSTRACT

Sandstone is found throughout the world. It is probably the most familiar, but not the most abundant, of all sedimentary rock, that group of rocks composed of consolidated rock fragments of all sizes.

KEY CONCEPTS

calcrete: the calcium carbonate formations that cement grains together in calcareous sandstone

permeability: a measurement of the quantity of a fluid like water that can pass through a sandstone formation per unit of time

porosity: a measurement of the amount of space between grains in a particular type of sandstone

silcrete: the silica formations that cement grains together in siliceous sandstone

tuffaceous sandstone: a type of sandstone formed from grains of volcanic ash (tuffa)

PRIMARY USES OF SANDSTONE

Sandstone has numerous uses in the construction industry. It is used to make bricks, concrete, and plaster. It has been used for thousands of years as a basic

Cut slab of sandstone showing Liesegang banding. Photo via Wikimedia Commons. [Public domain.]

construction block material as it is fairly easily cut and shaped, yet is also very strong.

TECHNICAL DEFINITION OF SANDSTONE

Sandstone is a rock composed of abundant rounded or angular, sand-sized fragments produced by the erosion and disintegration of existing rock. The sediments are subsequently subjected to great pressure from overlying layers of sediment, rock and other materials. The sand fragments in the resulting rock are commonly cemented together by calcium carbonate, silica, or iron oxide.

DESCRIPTION, DISTRIBUTION, AND FORMS OF SANDSTONE

In chemical composition, the average sandstone is principally composed of approximately 80 percent silica (silicon dioxide), 6 percent aluminum oxide, and 3 percent each of calcium oxide and carbon dioxide, or calcium carbonate. Arkose is a sandstone that contains fairly large, angular granules of pink feldspar. In North America two economically significant sandstones are the Oriskany sandstone of New York State and the Saint Peter sandstone of Minnesota. Both these sandstones are important as glass sand and natural gas reservoir rock.

Sandstones of a wide variety of physical characteristics and mineral compositions are known. While many sandstone classification schemes exist, a common scheme lists four typical varieties. The chief constituent (90 percent or greater) of siliceous sandstone is the mineral quartz, whereas more than 25 percent of arkose is composed of the mineral feldspar. A third variety is graywacke, a heterogeneous mixture of quartz and feldspar surrounded by fine-grained clay material. In regions of volcanic activity, accumulations of sand-sized detritus ejected from active volcanoes form tuffaceous sandstone.

Field studies suggest that sandstone constitutes approximately 30 percent of the three principal types of sedimentary rock, exceeding the abundance of limestone (approximately 20 percent), but less commonly seen than shale (approximately 50 percent).

HISTORY OF SANDSTONE USE

Sandstone has been studied seriously for more than two hundred years, especially in terms of its distribution in riverine environments. It has been used historically as a building material in many famous structures, from ancient times to the present day.

OBTAINING SANDSTONE

Individual sandstones can be analyzed as to their particular environment of deposition. For instance, the siliceous sandstone variety is indicative of deposition in a low-lying region that experienced intense chemical weathering, while the presence of arkose is suggestive of rapid erosion and deposition in a region lacking severe chemical weathering.

In the geologic study of sandstone, particular emphasis is placed on analyses of physical and chemical properties. A typical review would include the degree of angularity, roundness, sphericity, size, and mineralogy of individual sand-sized grains. The manner of sand grain arrangement is helpful in determining the amount of interstices a sandstone may contain (known as porosity) and the amount of fluid that can pass through a volume of sandstone (known as permeability). Porosity and permeability are important in evaluating a sandstone's potential as an oil, natural gas, or freshwater reservoir.

USES OF SANDSTONE

As quarried material, sandstone has long been employed in the construction industry. In crushed form, sandstone finds common usage in the manufacture of bricks, concrete, mortar, and plaster and is a primary resource employed in the foundry, abrasive, filter-purification, pottery, metallurgical, and glass-making industries. Indirectly, sandstones are important to society as they commonly form underground reservoirs containing reserves of crude oil, natural gas, or potable groundwater. Tarry sandstone deposits, or tar sands, are found in large volumes in Alberta, Canada, and Venezuela. Bitumen and other solid organic material from tar sands can be refined to produce oil similar to liquid petroleum.

—Albert B. Dickas

Further Reading

Boggs, Sam, Jr. *Petrology of Sedimentary Rocks*. Cambridge UP, 2009.

———. *Principles of Sedimentology and Stratigraphy*. 5th ed., Prentice Hall, 2011.

Craig, James R., David J. Vaughan, and Brian Skinner. *Earth Resources and the Environment*. Prentice Hall, 2010.

Mackenzie, F. T., editor. *Sediments, Diagenesis, and Sedimentary Rocks*. Elsevier, 2005.

Roy, Raja Ram Mohan. *Taj Mahal Was a Red Sandstone Palace: The Saga of the Sikarwars*. Golden Egg Publishing, 2015.

Tucker, Maurice E. *Sedimentary Rocks in the Field*. 4th ed., John Wiley & Sons, 2011.

SEAFLOOR SPREADING

Fields of Study: Geology; Seismology; Geophysics; Geochemistry

ABSTRACT

Seafloor spreading is the process by which the seafloor splits and new oceanic crust is created. The ocean basins are renewed and enlarged by the movement of molten basalt (magma) through cracks in a centralized oceanic ridge system, such as the Mid-Atlantic Ridge.

KEY CONCEPTS

basalt: the general type of rock typical of oceanic seafloor, produced by upwelling magma from Earth's mantle

geomagnetic signature: the alignment of the magnetic moment of magnetically susceptible atoms with the magnetic field of Earth

lithosphere: the solid rock portion of Earth's structure, primarily the crust but including parts of the upper mantle layer

midoceanic ridges: a line of volcanic peaks roughly bisecting the ocean floor where convection currents in the magma of Earth's mantle tend to force separation of the tectonic plates as basaltic magma is forced upward into the rift

OVERVIEW OF SEAFLOOR SPREADING

During the 1950s and 1960s, oceanographic studies revealed that the seafloors of Earth were marked by a nearly continuous system of submarine ridges, more than 40,000 kilometers in total length. Detailed investigation revealed that this midoceanic ridge system has a central rift valley that runs along its length and that the ridge system is associated with volcanic and earthquake activity. Magnetic studies of the seafloor indicated that the oceanic lithosphere (Earth's oceanic crust and part of the upper mantle) was segmented into long strips paralleling the axis of the midoceanic ridges. Furthermore, on either side of the ridge, the floor of the ocean was found to consist of alternating bands of rock bearing a geomagnetic signature. The magnetic signatures parallel to the present-day direction of Earth's magnetic field alternate with rocks magnetized exactly opposite to Earth's present-day field direction.

Integrating these data, Henry Hess postulated in 1960 that the midoceanic ridges are tensional features representing zones of weakness within Earth's crust and that new seafloor is created by the welling up of molten mantle material into cracks along the ridges, followed by lateral spreading. Hess hypothesized that the mantle is capable of plastic flow and that the ridges represent the meeting place of two rising convection currents of molten rock material from within the mantle. As rifting proceeds, magma ascends to fill in the fissures, thus creating new oceanic crust. The oceanic floor that is thus created would slowly move away from the oceanic ridge to ward the deep-ocean trenches, where it would descend into the mantle to be melted and recycled.

THE GEOMAGNETIC RECORD

In 1963, Frederick John Vine and D. H. Matthews of Cambridge University, and independently two Canadian scientists, Lawrence Whitaker Morley and André Larochelle, suggested an explanation for the magnetic pattern on the seafloor that supported Hess's hypothesis. They reasoned that ocean seafloor behaves like a magnetic "tape" that registers changes in the polarity of Earth's magnetic field. Following the upwelling of molten basalt into the midoceanic rift, iron minerals within the material become aligned to the then-existing Earth polarity as the rock cools and crystallizes. As the seafloor spreads out ward from the center of these deposits, about half the material is carried to either side of the rift, which later is filled by another influx of molten basalt.

At times the polarity of Earth changes, and subsequent molten basalt results in rocks of the opposite polarity. The continuation of this process over geologic time would thus lead to the young geologic age of the seafloor and to the magnetic symmetry around the midoceanic ridges. Thus, the ocean crust contains a magnetic record of its own formation. Depending on the particular ridge involved, the rate of spreading away from the midoceanic ridges varies from 2 to 15 centimeters per year.

—Alvin K. Benson

Further Reading

Christiansen, Eric H., and W. Kenneth Hamblin. *Dynamic Earth: An Introduction to Physical Geology*. Jones & Bartlett, 2015.

Fletcher, Charles. *Physical Geology: The Science of Earth*. 3rd ed., Wiley, 2017.

Mathez, Edmond A., and James D. Webster. *The Earth Machine: The Science of a Dynamic Planet*. Columbia UP, 2004.

Molnar, Peter. *Plate Tectonics: A Very Short Introduction*. Oxford UP, 2015.

Searle, Roger. *Mid-Ocean Ridges*. Cambridge UP, 2013.

Seibold, Eugen, and Wolfgang Berger. *The Sea Floor: An Introduction to Marine Geology*. 4th ed., Springer, 2017.

SECONDARY ENRICHMENT OF MINERAL DEPOSITS

Fields of Study: Geology; Seismology; Geophysics

ABSTRACT

Secondary enrichment is an ore-deposition process that involves the dissolution and redeposition of an ore mineral. This process concentrates the mineral, creating an enriched deposit or zone that is often more valuable than the original deposit from which the mineral was derived.

KEY CONCEPTS

gossan: a cap layer of ferric hydroxide atop an ore deposit

metal ion displacement: exchange of one metal ion for a different metal ion in minerals

oxidation-reduction process: a reaction between two different atoms in which one atom loses one or more electrons (oxidation) as the other atom gains one or more electrons (reduction)

supergene enrichment: enrichment of the mineral in an ore deposit by miner-bearing water that migrates downward from near the surface

DEFINITION OF SECONDARY ENRICHMENT

Secondary enrichment is a naturally occurring, near-surface process whereby a valuable substance is dissolved, carried downward in solution, and redeposited, thus producing a concentrated state of the mineral at the lower depths. This ore-deposition process is also known as downward enrichment or supergene (literally, "origin from above") enrichment. Secondary enrichment does not create a new and separate mineral deposit; rather, it concentrates and makes more valuable an existing deposit. Some low-grade mineral deposits have been made ore-grade as a result of secondary enrichment.

OVERVIEW OF THE SECONDARY ENRICHMENT PROCESS

For secondary enrichment to occur, there must be an exposed or near-surface deposit of a metallic ore mineral, and the mineral must be present in a soluble form. Many metallic sulfides (minerals composed of sulfur and a metal or metals) combine with oxygen as they weather to become sulfates. The sulfates tend to dissolve more readily in water than the original sulfides. Pyrite (FeS_2), an iron sulfide mineral present in most metallic mineral deposits, contributes to the dissolution process. Pyrite weathers to form ferric sulfate ($Fe_2(SO_4)_3$) and sulfuric acid (H_2SO_4), powerful solvents that help to break down the other primary sulfide ore minerals. Weathering and solution take place in the oxidation zone, the portion of the mineral deposit that is above the groundwater table. At the surface, a cap of ferric hydroxide often forms a residual surface deposit. Called a "gossan" or "iron hat," this rusty-looking deposit may contain enough iron to be mined as an ore.

A METAL ION DISPLACEMENT AND OXIDATION-REDUCTION PROCESS

The mineral-rich solutions percolate downward through the oxidation zone. If these solutions are near saturation, they may precipitate metal oxides or carbonates above the groundwater table. Just below the groundwater table, the solutions enter an environment where oxidation cannot take place and the sulfide ore minerals of the primary deposit have remained essentially unchanged. There, in the reduction zone, these sulfide minerals interact with the dissolved metals to create what is known as a supergene enrichment zone. When a metal in solution encounters a mineral that contains a metal with a greater ionization potential (that is, one with a stronger tendency to enter into solution), the metals exchange places. Copper, for instance, has a lower ionization potential than iron; when a copper-bearing solution encounters a primary deposit of an iron sulfide mineral such as pyrite, iron begins to enter into solution while copper is deposited. The pyrite changes to chalcopyrite ($CuFeS_2$), which in turn becomes bornite (Cu_5FeS_4), then covellite (CuS), and finally chalcocite (Cu_2S), the most common supergene sulfide. Iron in primary sulfide deposits appears to be the controlling metal in enrichment reactions. Because copper and silver are the only common metals with lower ionization potentials than iron, supergene enrichment rarely occurs in deposits of other metals.

Secondary enrichment has occurred in most of the nonglaciated land areas of the world. The supergene enrichment zones found within the porphyry copper deposits of the western and southwestern United States are classic examples.

—*Karen N. Kähler*

Further Reading

Guilbert, John M., and Charles F. Park Jr. *The Geology of Ore Deposits.* Waveland Press, 2007.

Misra, Kula C. *Understanding Mineral Deposits.* Kluwer Academic, 2000.

Tarbuck, Edward J., Frederick K. Lutgens, and Dennis Tasa. *Earth: An Introduction to Physical Geology.* 13th ed., Pearson Education, 2019.

Wicander, Reed, and James S. Monroe. *Essentials of Physical Geology.* 5th ed., Cengage Learning, 2008.

SEDIMENTARY PROCESSES, ROCKS, AND MINERAL DEPOSITS

Fields of Study: Geology; Sedimentology; Geophysics; Geochemistry

ABSTRACT

Sedimentary processes occur only at the Earth's surface, because they are driven by various components of the hydrologic and biologic cycles. Sedimentary processes involve the breakdown, movement, and ultimate deposition of broken rock fragments and chemicals in solution. These processes create many of the important resources used by humans.

KEY CONCEPTS

abrasion: damage and removal of material from materials by friction against another material

cementation: the binding together of particles in a sedimentary deposit resulting in a single consolidated mass

clastic particles: particles formed by the mechanical breakdown of rocks

compaction: the pressing together of particles in a sedimentary deposit by the weight of overlying materials

erosion: removal of material from a location by a moving fluid

nonclastic particles: particles formed by chemical reactions or biosynthetically

sorting: the deposition of particulate material as it settles out from a moving fluid according to the ability of the fluid to transport the particles

SEDIMENTS AND SEDIMENTATION

Most sediments originate from the weathering of existing rocks. Weathering is the breakdown of earth material by physical and/or chemical processes. Physical weathering only breaks the original material into smaller sizes. This is accomplished through mechanical means such as freezing and thawing, plant-root wedging, differential heating of rocks, and crystal growth in rock cracks. Chemical weathering, on the other hand, actually changes the composition of the original material into completely different components through solution, oxidation, hydration, and/or hydrolysis. The byproducts of physical and chemical weathering provide the different types of sediment and ion-bearing solutions that create sedimentary rocks and minerals.

Biological activity can also create sediments. Many invertebrates and some algae utilize calcium carbonate in seawater to make shells. Organisms, such as coral, live in warm, shallow seas and construct reefs. In addition, both macro- and microscopic shells of organisms that do not live in reef communities become sediments that blanket the seafloor after the organisms' deaths. Also, some algae and bacteria, as well as swamp vegetation, can become organic sediments under certain conditions of oxygen deficiency, creating valuable fossil fuels.

TRANSPORTATION OF SEDIMENTS

Once particles of rock are loosened and broken free by weathering processes, those particles can be carried off by water, wind, or glaciers. Ions released through chemical activity are also free to travel in moving water, but they will never settle out of the water unless a specific chemical reaction occurs that causes the ions to precipitate as solid particles, or until shell-bearing organisms use the ions to build shells that later become sediments.

The amount of energy available in a transporting medium determines which rock particles are picked up (eroded) and moved along. For example, more energy is required to move larger or more dense grains than smaller or less dense grains. If a transporting medium loses energy, the sediments being carried will drop out in order of their transportability, typically by relative size, the largest ones first. Sediment sizes range from boulders to clay. Each has a specific size determination for the name: boulder (more than 256 millimeters), cobble (256 to 64 millimeters), pebble (64 to 4 millimeters), granules (4 to 2 millimeters), sand (2 to 0.062 millimeter), silt (0.062 to 0.0039 millimeter), and clay (less than 0.0039 millimeter). This separation of grains according to size by the transporting medium is called "sorting." A well-sorted sediment is one that contains nearly all the same size grains. The farther sediments travel, the better sorted they become. Thus, a poorly sorted sediment probably is still fairly close to its source area.

Transport of material also tends to make grains more rounded in shape. Rounding occurs as particles bump into one another or into objects in the environment, causing abrasion of their edges. The longer and farther a grain is moved, the smaller and rounder it becomes. Transportation tends to winnow material either by sorting or by continued weathering of the particles. Some minerals are more resistant to breakdown than other minerals. Quartz is more resistant, while minerals such as feldspar, mica, pyroxene, and amphibole are less resistant. The longer and farther the sediments are transported, the more likely the weaker minerals are to disappear, which would increase the percentage of quartz present. Therefore, rocks composed of nearly all quartz can be interpreted as having traveled a great distance from their source of weathering. These rocks also exhibit good sorting and are well-rounded.

WATER TRANSPORT

Most sediments are carried by running water. Running water begins as rainfall, flows into increasingly larger streams, and eventually runs into the oceans. Running water is present nearly everywhere on Earth, including in arid regions (where, although unusual, it is particularly effective as an agent of erosion because of the lack of vegetation for ground cover and because much of the rain in arid lands occurs in the form of downpours).

Waves and currents along shorelines also move large amounts of sediment, as can some currents out in deeper waters. Energy is usually greater nearer a shoreline, so, if grains accumulate, beaches form. Depending on the energy level, the beaches can be composed of sand, pebbles, cobbles, or even boulders. The energy usually decreases offshore, so finer-grained silt and clay (mud) often accumulate in the deeper, quieter water.

Groundwater can carry dissolved minerals. Sometimes when the rock is dissolved, minerals of economic importance, such as copper, aluminum, or iron, can be left behind. In some limestone areas, groundwater can dissolve enough limestone to form caves. If conditions are right in a cave, the groundwater can redeposit some minerals in the forms of dripstone and flowstone, features admired by tourists.

The stratification of rock layers can be clearly seen in this photo of the Devil's Punchbowl in Ontario, Canada. Image by Carolyn McKenzie. [Used with permission.]

WIND AND GLACIER TRANSPORT

Wind can move sediment grains as well, but neither in the quantity nor in sizes as large as does water. Fine sand is usually the upper size limit that wind can transport. Grains carried by wind do not have the cushioning effect that water provides, so grains are rapidly abraded as they hit each other during transport. Quartz sand grains become frosted in this manner.

Sediments are also transported by glaciers. Movement of grains in glacial ice will break the grains, making them smaller, but they remain fairly angular because they do not tumble around next to each other as they would in water or wind and, thus, do not undergo much rounding. Because ice will not let particles readily settle, as water or wind will, the particles do not become sorted at all. Glacial deposits are probably the most poorly sorted of all sediments. However, some sediments do not end their journeys under a glacier; they may move away from the ice in glacial meltwater. When this happens, the sediments are transported by running water and take on the characteristics of any sediment carried in running water.

DEPOSITION (SEDIMENTATION)

All particles eventually are carried to specific environments where the transporting energy decreases to the point of no longer being able to carry the material, and the grains are deposited. If not destroyed by further erosion, they may consolidate into various sedimentary rocks. There are three main depositional systems: marine, transitional, and terrestrial.

Marine environments are important mainly, but by no means exclusively, for deposition of chemical sediments. Marine environments include shallow-marine environments (from the shore to the edge of the continental shelf), reefs, and deep marine environments. Transitional and terrestrial depositional environments are important for deposition of mainly clastic sediments. Transitional environments include beaches, deltas, barrier islands, lagoons, and tidal marshes. Terrestrial environments include rivers, lakes, alluvial fans, glacial environments, and windy areas such as deserts.

DIAGENESIS

Diagenesis is any postdepositional alteration of sediments or sedimentary rocks. Diagenetic processes usually take place after sediments are buried by newer sediments. An important diagenetic process is lithification, the process by which loose sediments are hardened into rock. There are two main processes involved in lithifying sediments: cementation and compaction. Cementation occurs when pore spaces between the grains are large enough for mineral-rich water to seep around the grains, depositing either silica crystals (silcrete) or calcite crystals (calcrete) that grow and eventually coalesce and hold the grains together. Many sediments typically lithify by physical compaction. As more and more sediments collect, the weight of overlying sediments compresses the lower sediments, squeezing out much of the water and pressing the grains close enough together that they become a harder mass of material. Lithification processes can occur below water; an ocean basin does not have to dry up before the sediments it contains can lithify into rock.

SEDIMENTARY ROCKS AND MINERALS

Sedimentary rocks are the most abundant type of rocks found at the surface of Earth. All sedimentary rocks and minerals can be put into one of two categories: clastic, composed of rock and mineral fragments that were weathered from preexisting rock materials and lithified in new combinations to create a sedimentary rock, and nonclastic, made of precipitated chemicals or of organically derived material such as shells or plants and animals. Identification of sedimentary rocks begins with the determination of whether a rock is clastic or nonclastic. Once such textural characteristics have been determined, the grain sizes and composition of the rock are used to complete the identification process.

Clastic rocks are fairly easy to identify because individual particles in the rock can usually be seen. The rock is identified mainly on the sizes and shapes of the grains it contains. Rocks made of large particles such as boulders, cobbles, and/or pebbles in a matrix of sand are called "conglomerate" if the grains are mostly rounded or breccia if the grains are angular. The sandstone family contains many varieties of rocks depending on the composition of grains present. Sandstone feels like sandpaper, and the purer variety can be white, tan, or pink and is composed almost entirely of quartz grains. More commonly sandstone is gray, indicating that it contains particles other than quartz. Arkose is a sandstone that contains fairly large, angular granules of pink feldspar. Sedimentary rocks composed of silt are called "siltstone," and shale is composed of grains of clay that are so small they cannot be seen, even with a microscope. (Because the grains are so small, shale can easily be misidentified as a nonclastic rock.)

The nonclastic rocks are dominated by limestone and dolostone (or dolomite). Limestone is calcium carbonate that formed in warm, shallow seas. Most limestone originated as accumulations of shells either in reefs or in beds on the seafloor. Some limestone can also form by chemical precipitation. Dolostone, a close relative of limestone, is thought to form diagenetically from limestone deposits when magnesium ions in the environment replace some of the calcium in the limestone. Microscopic shell accumulations from calcareous algae have created thick layers of chalk, while similar accumulations from siliceous plankton create diatomaceous earth and chert (flint). Plants that die in swamps and do not completely rot away can evolve into peat and eventually into coal, and microscopic algae and bacteria are the source of petroleum deposits. Although petroleum is not a rock, it is a resource that originates in a sedimentary setting.

Some resource minerals form under evaporative conditions. Ions in solution travel with running water and eventually end up in the oceans or in lakes. In shallow embayments where seawater can flush into the bay and not be diluted with fresh stream water, salts can accumulate if the water evaporates. Likewise, in arid regions where streams enter landlocked lakes (playa lakes), evaporative conditions cause salts to accumulate. There are a great variety of salts (gypsum, halite, magnesium sulfate, and potassium salts, to name a few), and each one forms in turn as the chemical concentrations change as more and more of the water evaporates.

ECONOMIC IMPORTANCE OF SELECTED SEDIMENTARY ROCKS AND MINERALS

Sedimentary resources are classified into four main groups: sedimentary metallic ore deposits, sedimentary nonmetallic deposits, evaporites, and energy resources. The sedimentary ore deposits contain some of the world's most valuable mineral resources. Many of these deposits were formed in depositional environments where large amounts of dissolved metals collected. For example, the iron ores of the famous Mesabi Range in Minnesota originated when Earth's early atmosphere was poor in oxygen. This permitted an abundance of iron in its soluble (ferrous) form to be leached from large areas of the Earth's surface and transported in solution to vast, shallow marine environments, where it oxidized to its insoluble (ferric) form and precipitated in thin layers.

Although gold originates from igneous and hydrothermal processes, once it weathers out of its original rock setting, it becomes influenced by the sedimentary processes of transportation by running water and ultimate deposition in streambeds. This is called a "placer deposit." Placer deposits are not limited to gold. Diamonds, tin, chromite, platinum, and magnetite can undergo similar histories. Placers can sometimes be traced upstream to find the source rock, which can then be mined.

The nonmetallic deposits are also of great economic importance. Limestone is quite extensive and has a variety of uses. It is obtained by quarrying; the rock can be either cut into large blocks or blasted into fragments. The blocks, used for building stone, are taken to mills and cut to order for particular buildings; decorative carvings can also be made. Limestone that has been blasted is usually ground into lime for either agricultural purposes or the manufacture of cement. Larger blocks may be used as riprap along shorelines or rivers to protect those areas from excessive erosion.

Pure quartz sandstone, which usually originates from beach deposits, is quarried for use in glassmaking and fiber-optic cables. Coarser sand and gravel, which originates from glacial deposits or channel deposits in rapidly moving rivers, is quarried for construction purposes. Clays of high purity, often formed in coal swamps or in areas of prolonged weathering, are used for both craft and industrial ceramics. Phosphatic rocks, usually marine shales and limestones that have been chemically enriched in phosphate in deep marine environments, are an ingredient in agricultural fertilizers.

Evaporite deposits have created vast and varied salt resources. Halite is used both as table salt and as an ice melter for road clearance. Gypsum is used in the making of plaster as well as the writing utensils used on modern "chalk" boards, contrary to popular belief that these writing implements are really chalk. Potassium salts can be used as a table salt substitute and in fertilizer. Epsom salts (magnesium sulfate) have health benefits, and borates (such as borax and boron) have uses that range from manufacturing of enamel to additives in soap and gasoline.

The fossil fuels, the major energy resources in use in the twentieth century and early twenty-first century, all have biogenic origins in depositional environments. Coal forms from vegetation that grew in ancient swamps. Coal is obtained from both strip mining and underground mining. Petroleum products are buried microscopic planktonic life-forms that lived in seas of the past. Although these organics collect on small scales, the sedimentary rocks with which they are associated permit the migration and eventual accumulation of great enough volumes of oil and gas that they can be extracted for human use.

—Diann S. Kiesel

Further Reading

Boggs, Sam, Jr. *Petrology of Sedimentary Rocks.* 2nd ed., Cambridge UP, 2009.

———. *Principles of Sedimentology and Stratigraphy.* 5th ed., Prentice Hall, 2011.

Chernicoff, Stanley, and Donna Whitney. *Geology: An Introduction to Physical Geology.* 4th ed., Pearson Prentice Hall, 2007.

Grotzinger, John, and Thomas H. Jordan. *Understanding Earth.* 6th ed., W. H. Freeman, 2010.

Koch, Adam. Gregory *Gold Prospecting and Placer Deposits: Finding Gold Made Simpler.* Author, 2014.

Nichols, Gary. *Sedimentology and Stratigraphy.* 2nd ed., Wiley-Blackwell, 2009.

Tucker, Maurice E. *Sedimentary Petrology: An Introduction to the Origin of Sedimentary Rocks.* 3rd ed., Blackwell Science, 2001.

SHALE

Fields of Study: Geology; Sedimentology; Geophysics; Economic Geology

ABSTRACT

Shale is the most common of the three principal types of sedimentary rock, that category of rock formed by consolidation of rock fragments or by chemical precipitation. Shale is used in construction materials, in everyday products such as cosmetics and toothpaste, and as an energy source.

KEY CONCEPTS

bedding planes: because the sediments that make up shale are laid down in successive thin layers, a plane of separation exists between the layers and will become the planes along which the rock will separate after it has become compressed into shale

claystone: shale formed from fine clay particles that are flat or plate-like in structure

indurated: made significantly harder than its original form

mudstone: shale formed from clay particles that are generally rounded or irregular rather than flat in structure

shale: a sedimentary rock composed of fine-grained products derived from the physical breakdown of preexisting rock material

slate: shale that has undergone a period of intense pressure that has deformed the clay particles and expressed all organic matter that may have been present, resulting in a very hard and consolidated shale rock

WHERE SHALE IS FOUND

Shale is found throughout the world. It is the most common of the three principal types of sedimentary rock, that category of rock formed by consolidation

Shales in the Devonian of Kentucky (geology hammer for scale). Photo by James St. John, CC BY 2.0, via Wikimedia Commons.

of rock fragments or by chemical precipitation. In the geologic record, for every approximate five units of shale known, three units of sandstone and two units of limestone (the remaining two common categories of sedimentary rock) are also known.

PRIMARY USES OF SHALE

Shale is used as a filler in numerous construction materials. It is also used in everyday products such as cosmetics and toothpaste, and as an energy source. Because shale is formed from fine, plate-like particles of clay, it can be ground into a very fine powder consisting of fine, plate-like particles that present a high degree of sheen to such products.

TECHNICAL DEFINITION OF SHALE

Shale is a fine-grained consolidated rock principally composed of silt-size (particles between 0.0039 and 0.0625 millimeter in diameter) and clay-size (less than 0.0039 millimeter in diameter) rock detritus. Shale is generally characterized by a tendency to break along well-defined bedding planes.

DESCRIPTION, DISTRIBUTION, AND FORMS OF SHALE

The classification of shale is generally based on the presence or absence of well-defined bedding (lamination) planes. Fine-grained rock lacking this characteristic is termed "mudstone," while a similar rock composed entirely of clay-size material is known as "claystone." The ubiquity of shale is explained by its representing approximately 75 percent of all sedimentary rock produced throughout the entirety of geologic time.

Because of their fine-grained nature, shales cannot be conveniently examined mineralogically. Bulk chemistry and X-ray studies show, however, that the average shale is composed principally of the following oxides: silicon (approximately 58 percent), aluminum (approximately 15 percent), iron (approximately 7 percent), and calcium, potassium, and carbon (each approximately 3 percent).

HISTORY OF SHALE USE

Shale rich in organic material deposited by the Mississippi River over the past several tens of millions of years caused the Gulf of Mexico to be one of the richest hydrocarbon provinces in the world. Throughout ten of the eastern United States and three western states, the Chattanooga Shale and the Green River Shale are identified as significant oil shale resources.

OBTAINING SHALE

Shale is a prime geologic source of crude oil and natural gas (hydrocarbon). Hydrocarbon originates from organic matter that accumulates in varieties of shale generally deposited under marine conditions. The preserved organic matter is converted to petroleum and natural gas by burial and related postdepositional changes through the passage of geologic time. The general lack of permeability of shale will later form a barrier to the upper subsurface migration (and thus to possible loss by surface evaporation) of generated hydrocarbon.

USES OF SHALE

Kaolinite-rich shale supplies the basic material for a wide range of ceramic products, from pottery and fine porcelain to sewer pipe. Shale rich in barite is employed in the hydrocarbon industry to prevent oil and natural gas blowouts during the drilling of exploratory boreholes. Clay-rich shales are also employed in the cosmetics, insulator, printing ink, medicine, and toothpaste industries. The highly indurated form of shale known as slate is used in the construction industry as roofing and paving material. One major economic importance of shale is associated with the worldwide distribution of oil shale, a dark-colored rock containing 5 to more than 25 percent solid organic material, from which oil can be extracted by distillation. Shale rich in organic material also acts directly as a primary source of crude oil and natural gas.

—Albert B. Dickas

Further Reading

Boggs, Sam, Jr. *Petrology of Sedimentary Rocks.* 2nd ed., Cambridge UP, 2009.

———. *Principles of Sedimentology and Stratigraphy.* 5th ed., Prentice Hall, 2011.

Fettes, Douglas, and Jacqueline Desmons, editors. *Metamorphic Rocks: A Classification and Glossary of Terms.* Cambridge UP, 2007.

Tucker, Maurice E. *Sedimentary Petrology: An Introduction to the Origin of Sedimentary Rocks.* 3rd ed., Blackwell Science, 2001.

Vernon, Ron H. *A Practical Guide to Rock Microstructure.* Cambridge UP, 2004.

SILICATES

Fields of Study: Geology; Geochemistry; Geophysics; Gemology; Crystallography

ABSTRACT

The two most common elements in Earth's crust are silicon and oxygen. The prevalence of these two elements and their ability to combine as stable complex ions accounts for the fact that silicate minerals constitute a major portion of the minerals in Earth's crust. Their wide range of physical properties leads to many commercial uses.

KEY CONCEPTS

tetrahedral: arranged at the four apices of a tetrahedron about the tetrahedron's central focus; in silicates, four oxygen atoms are arranged tetrahedrally about a central silicon atom

semiprecious stones: mineral crystals, typically types of quartz or beryl, that are not classifiable as gemstones

silicate: chemically, the anion SiO_4^{4-}; geologically, any of the rock types that are based on silica (SiO_2) structural combinations

zeolites: hydrous aluminum silicates containing any of sodium, calcium, barium, strontium, or potassium as cations

DEFINITION OF SILICATES

Silicon and oxygen combine to form stable complex ions composed of one silicon ion surrounded by four oxygen ions. The resulting complex ion is a four-sided figure known as a tetrahedron. Silicate tetrahedra may exist independently separated by cations or link together by sharing oxygens to form a wide range of structural groupings. Tetrahedron groupings act as skeletons in which charge neutrality is maintained by the addition of cations between and within silicate tetrahedra. Silicate skeletons may exist as isolated tetrahedra; as two tetrahedra sharing one oxygen atom; as rings of three, four, or six tetrahedra; as single and double chains; as sheets; and as continuous three-dimensional frameworks. These skeletal arrangements impart many diverse physical characteristics to the various silicate minerals.

OVERVIEW OF SILICATES

No generalization describes all silicates, although most are translucent to transparent, have moderate specific gravity, and are chemically inert. Silicates range from extremely soft to hard. Some display excellent cleavage, but others are uniformly resistant to breaking in all directions.

Because silicate minerals exhibit such a wide variation in physical properties, they have variable commercial uses. Talc, a magnesium silicate, is used as a filler in paint, ceramics, rubber, insecticides, roofing, and paper. Its most familiar form is as talcum powder. Clay minerals (extremely small platy grains of hydrous aluminum silicates) are important industrial minerals. They are used in a variety of fired products, ranging from bricks to fine china and porcelain. Clay is used as a filler in many products, including paper. Montmorillonite, an expandable clay, is widely used as a sealant. Zeolites, which are hydrous silicates with open tunnels within a framework lattice, are widely used as molecular sieves and for ion exchange resins; they are valuable for oil-spill cleanup and wastewater treatment and are used in water softeners.

In many applications, natural silicate minerals have been replaced by the industrial manufacture

Copper silicate mineral chrysocolla. Photo by Lloyd.james0615, CC BY 3.0, via Wikimedia Commons.

of synthetic and substitute materials. Muscovite mica, a sheet structure, was used largely as an electrical insulator in capacitors and electronic tubes. Asbestos, a term describing flexible, fibrous silicate materials, was widely used for its heat resistance and its ability to be woven as a fabric in fire-retardant cloth, in heat-resistant sheets, in blown insulation, and in brake linings. Almost all asbestos use is outlawed in North America, as it has been identified as the causal carcinogen of the lung cancer that bears its name (asbestosis). Quartz (silicon dioxide) is used as an abrasive, in the manufacture of optical components, and as thin wafers to control the frequency of radio and radar transmission. Quartz crystals are now grown by commercial hydrothermal processes.

Many semiprecious stones are silicate minerals. Microcrystalline varieties of quartz that are used in jewelry include fibrous-appearing tiger's eye, red jasper, multicolored agate, and the red-spotted bloodstone. Crystalline varieties of quartz that serve as semiprecious stones include yellow citrine and violet amethyst. Topaz, jade, garnet, opal, and peridot are silicates, as is emerald, which is gem-quality beryl.

—René A. De Hon

Further Reading

Aydinalp, Cumhur. *An Introduction to the Study of Mineralogy*. IntechOpen, 2012.

Blatt, Harvey, Robert J. Tracy, and Brent E. Owens. *Petrology: Igneous, Sedimentary, and Metamorphic*. 3rd ed., W. H. Freeman, 2006.

Chernicoff, Stanley, and Donna Whitney. *Geology: An Introduction to Physical Geology*. Prentice Hall, 2007.

Fossen, Haakon. *Structural Geology*. 2nd ed., Cambridge UP, 2016.

Grotzinger, John, and Thomas H. Jordan. *Understanding Earth*. 6th ed., W. H. Freeman, 2010.

Klein, Cornelis, and Cornelius S. Hurlbut, Jr. *Manual of Mineralogy*. 23rd ed., John Wiley & Sons, 2008.

SILICICLASTIC ROCKS

Fields of Study: Geology; Geomorphology; Geochemistry; Geophysics

ABSTRACT

Siliciclastic rocks, which include siltstone, sandstone, and conglomerate, are second only to mudrocks in abundance among the world's sedimentary rock types. They form major reservoirs for water, oil, and natural gas and are the repositories of much of the world's diamonds, gold, and other precious minerals. In their composition and sedimentary structures, siliciclastic rocks reveal much about paleogeography and, consequently, Earth history.

KEY CONCEPTS

clastic rock: a sedimentary rock composed of particles without regard to their composition; this term is sometimes used, incorrectly, as a synonym for "detrital"

detrital rock: a sedimentary rock composed mainly of grains of silicate minerals as opposed to grains of calcite or clays

diagenesis: chemical and mineralogical changes that occur in a sediment after deposition and before metamorphism

lithic fragment: a grain composed of a particle of another rock; in other words, a rock fragment

mudrock: a rock composed of abundant clay minerals and extremely fine siliciclastic material

WEATHERING AND EROSION

Weathering and erosion constantly strip Earth's surface of its rocky exterior. Rocks experience fluctuating temperatures, humidity, and freeze-thaw cycles that physically disaggregate (break up) mineral grains. Simultaneously, these mineral grains are exposed to water, oxygen, carbon dioxide, and dissolved acids or bases that chemically attack them. As a result, a hard, rocky surface is eventually transformed into a collection of mineral grains, clay minerals, and ions dissolved in water.

Weathering and erosion are slow but continuous agents of destruction. Because mountains and other high regions are constantly being uplifted, weathering and erosion provide a more or less steady supply of sediment to streams and rivers and, from there, into the ocean for deposition. Deposited sediment is eventually compacted and cemented into rock. The rocks that are formed by this process are known collectively as siliciclastic rocks because they are made mostly of particles of silicate minerals.

Cross beds in siliciclastic shoreface sediment (Agadir-Essaouira Basin, Morocco). Photo by ManuRoquette, CC BY-SA 4.0, via Wikimedia Commons.

SILICICLASTIC ROCK COMPOSITION

Siliciclastic rocks differ in the sizes of their grains and their composition. Three main size classes are recognized: silt, sand, and gravel. Silt includes particles between 0.004 and 0.0625 millimeter (1/256 to 1/16 millimeter) in diameter; sand is composed of particles between 0.0625 (1/16) and 2 millimeters in diameter; and gravel generally includes particles larger than 2 millimeters. Most sedimentologists refer to sand and sandstone when discussing siliciclastic rocks.

Siliciclastic rocks are composed of three broad classes of material: framework grains, matrix, and cement. Framework grains are particles of minerals or small fragments of other rocks that usually make up the bulk of a siliciclastic rock. Matrix is extremely fine-grained material such as clay that is deposited at the same time as framework grains. Cement is any material precipitated within the spaces, or pores, between grains in a sediment or rock. Framework grains and matrix are primary deposits, whereas cement is a secondary deposit because it is precipitated in pores after the primary material.

QUARTZ AND FELDSPAR MINERALS

The main minerals that compose siliciclastic rocks are quartz, various types of feldspar, and micas. The ferromagnesian minerals, such as olivine, pyroxene, and amphibole, are not as common in sedimentary rocks as they are in igneous and metamorphic rocks. Ferromagnesian minerals are more easily dissolved during weathering, and they fracture more readily during erosion, than the more durable quartz and feldspar. Consequently, quartz and feldspar (and, to a lesser extent, the micas) are concentrated in sediments because the other minerals are selectively removed beforehand. Carbonate rocks dissolve very easily and seldom form much residue to contribute to sand.

Quartz is the most chemically and physically stable mineral that forms siliciclastic rocks. Although quartz is uncommon in some igneous and metamorphic rocks, it is sufficiently abundant in granite and gneiss to contribute a large amount of bulk to almost all sands. Subtle differences may provide useful clues as to the origin of some quartz grains. For example, quartz from plastically deformed metamorphic rocks (schist, gneiss) usually occurs as polycrystalline aggregates or displays a distorted crystal structure, revealed by the petrographic microscope. Quartz from volcanic rocks often possesses planar crystal faces or embayments (deep, rounded indentations) that were formed as the crystal grew in magma. Quartz grains eroded from older sedimentary rocks sometimes retain the previous rock's quartz cement as a rind, which is called an "inherited overgrowth."

Feldspar minerals are also common in most siliciclastic rocks. Two main groups of feldspar are recognized: plagioclase feldspar and the potassium feldspars (microcline, orthoclase, and sanidine). The plagioclase group is actually a collection of many similar minerals with different chemical compositions. Anorthite is plagioclase feldspar composed of calcium, silicon, and oxygen, whereas albite is plagioclase composed of sodium, silicon, and oxygen. Calcium and sodium substitute rather easily for each other, so most plagioclase feldspars have both calcium and sodium. In general, plagioclase with more calcium is more susceptible to chemical weathering than is sodium-rich plagioclase. Consequently, sodium-rich albite is more common than anorthite in most siliciclastic rocks.

Potassium feldspars all have the same chemical composition of potassium, silicon, and oxygen. They differ from one another in their chemical structure. Microcline is the most highly organized, crystallographically, followed by orthoclase and then sanidine. Microcline is formed in igneous and metamorphic rocks that crystallize very slowly, permitting the greatest amount of crystallographic ordering. At the other extreme, sanidine is formed in volcanic rocks where little time is available for ions to get into their "proper places." During chemical weathering of detrital potassium feldspars, sanidine dissolves much more readily than does microcline or orthoclase. Microcline is usually about equal in abundance to orthoclase in siliciclastic rocks, whereas sanidine is usually rare.

MICAS AND ACCESSORY MINERALS

The mica minerals biotite and muscovite are common in some siliciclastic rocks and rare in others. In general, muscovite (white mica) is more durable and, consequently, more abundant than biotite (brown mica). Chlorite is a mineral with a sheetlike crystal structure similar to the micas. It is common in a few sandstones but generally less abundant than the micas. A large variety of grains is often present in siliciclastic rocks as "accessory minerals." Altogether they seldom constitute more than 1 percent of any siliciclastic rock. Sometimes these accessory minerals are called "heavy minerals" because most of them have a much greater density than does quartz or feldspar. This group includes zircon, tourmaline, rutile, and garnet; the ferromagnesian minerals pyroxene, amphibole, and olivine; and iron oxides such as hematite, magnetite, chromite, and limonite. If one examines a handful of sand, the accessory minerals are usually the dark grains.

Accessory minerals are important in several ways. First, they may reveal information about the source rock from which the sand was eroded. For example, the accessory mineral chromite is generally formed in basalt, so its presence in a sand indicates that basalt (usually from the seafloor) had previously been uplifted and eroded nearby. Similarly, garnets of particular composition may be representative of particular metamorphic source rocks. Second, accessory minerals may tell geologists how "mature" a sand is—in other words, how great has been its exposure to chemical and physical weathering. Zircon, tourmaline, and rutile are far more durable than the other accessory minerals. If they are the only accessory minerals in a sand, then the sand probably experienced prolonged weathering in the period before its final deposition. Third, accessory minerals may be economically valuable. Diamonds and gold are accessory minerals in a few cases. Much of the United States' titanium comes from unusual concentrations of rutile and ilmenite in beach sands along the coast of South Carolina. Zircon in sands provides an important source of zirconium, used in high-temperature ceramics.

LITHIC FRAGMENTS AND MATRIX

Pieces of rocks (lithic fragments) may form an important part of coarse-grained sandstone and conglomerate. Lithic fragments may be composed of pieces of almost any igneous, metamorphic, or sedimentary rock. Their presence is a powerful indicator of the source rock from which the sand or gravel was weathered and eroded. As one might expect, not all rock fragments are equally durable in the sedimentary environment. Limestone and shale fragments disintegrate very rapidly, whereas granite and gneiss rock fragments are quite robust. Consequently, both the presence and the absence of certain rock fragments may reveal information about the origin of a siliciclastic sediment.

Matrix in siliciclastic rocks is usually clay or fine silt. This substance is the "mud" that often accompanies sand or gravel during deposition. It is commonly deformed between the more rigid framework grains as the sediment is compacted. Some types of matrices were not originally deposited as fine-grained matrix. This "pseudomatrix" is actually squashed fragments of soft grains of, for example, shale or schist. During compaction of the sediment, these fragments are more easily deformed than durable quartz and feldspar, and the fragments superficially resemble original matrix. Considerable skill is required to distinguish between true matrix and pseudomatrix.

SANDSTONES

Sand-sized siliciclastic rocks are often classified based on their mineral composition. Sandstones may be named with the term "arenite" (after the Latin *harena*, "sand"). Relative abundances of quartz, feldspar, and rock fragments determine the name. One of the major types of sandstone is quartz arenite, which is composed almost entirely of quartz, with less than 5 percent of other framework minerals and less than 15 percent clay matrix (usually much less). This rock is usually white when fresh. Another type is feldspathic arenite, which is sometimes called "arkose." This rock has abundant feldspar and up to 50 percent lithic fragments. The typical composition of arkose is perhaps 30 percent feldspar, 45 percent quartz, and 25 percent lithic fragments. Up to 15 percent detrital clay matrix may also be present. This rock is often pinkish-gray in color. A third type is lithic arenite, a kind of sand composed largely of lithic fragments and up to 50 percent feldspar. The typical composition of lithic arenite might be 60 percent lithic fragments, 30 percent feldspar, and 10 percent quartz. Up to 15 percent matrix may be present as well. Almost all conglomerates are lithic arenites. Lithic arenites are often dark gray and sometimes feature a salt-and-pepper appearance from the lithic fragments. A fourth type is wacke, sometimes called "greywacke," which has between 15 and 75 percent clay matrix at the time of deposition. The term "wacke" may be modified by prefixing with the name of the dominant nonmatrix component; examples might include "quartz wacke," "feldspathic wacke," or "lithic wacke." Some wackes may be "formed" diagenetically from lithic arenites by alteration of shale or schist fragments so that they appear similar to originally detrital clay matrix. This rock is usually gray in color (thus the name "greywacke").

The composition of sandstones holds important clues as to the kinds of rocks from which they were derived. These clues may help petrologists to reconstruct the plate tectonic setting of the sand's source region even though the source region has long since disappeared because of erosion or tectonic destruction. For example, quartz arenite often reflects sedimentation on or near a stable craton (a piece of Earth's crust), where physical and chemical weathering are permitted to eliminate unstable minerals for an extended period of time. At the other extreme, a lithic arenite with abundant basaltic rock fragments, calcic plagioclase, and rare quartz may represent detritus shed from an island arc during plate collision. Careful observation of the mineral composition of sandstones has become an important part of reconstructing the earth's history.

DIAGENESIS

The diagenesis of siliciclastic rocks is a complex collection of processes that include cementation, replacement, recrystallization, and dissolution. Diagenesis includes chemical reactions that begin at the time of deposition and may continue until metamorphism takes place. Some sedimentary petrologists consider low-temperature cementation as part of diagenesis, whereas others consider diagenesis to include only those reactions that occur at temperatures exceeding 100° Celsius. The upper range of diagenesis grades into metamorphism; for that reason, sedimentary petrologists and metamorphic petrologists share some common territory in diagenesis.

Siliciclastic sediments are transformed into rocks by compaction and cementation. Common cements in siliciclastic rocks are quartz, calcite, clays, and iron oxides such as limonite or hematite. It is common for a rock to have several cements, each deposited at different times or under different conditions of pore-water chemistry. Cement is introduced into sediment in the form of elements or compounds dissolved in water. For example, silica may be dissolved from quartz grains or given off during low-grade metamorphism and then may saturate water in the pores of a rock. One means of dissolving silica from quartz grains is called "pressure solution." Quartz grains exert very high pressures upon one another at their points of contact. Quartz is more soluble under high pressure; therefore, it is more easily dissolved where grains touch each other. Well-compacted sands often display flattened grain-to-grain contacts as a consequence of pressure solution. The quartz lost from these grains is carried away, dissolved in pore water, to be precipitated elsewhere.

As the water moves slowly through the pores of the rock, it may encounter different temperatures, pressures, acidity, or gas concentrations. Changes in these conditions may cause the water to lose some of its ability to dissolve silica, and the silica may therefore precipitate as crystals. Similar factors influence the dissolution and precipitation of other cements in siliciclastic rocks. Cement not only holds sediment grains together, forming a rock, but it also fills the pores between grains, which may inhibit further flow of fluids. As a consequence, the porosity and permeability of most rocks are reduced by cementation. Rocks with abundant cement generally make poor reservoirs for petroleum or water.

In some cases, the pore-water chemistry conducive to precipitation of one cement is capable of dissolving another cement. A common example is the generally inverse relationship between quartz and calcite cement. Pore water that precipitates quartz often dissolves calcite at the same time, or the other way around. Sometimes dissolution occurs without immediate precipitation of another mineral type in the void that is formed. Pores formed by dissolution are called "secondary porosity"; such porosity may be the main cause of porosity in many sandstones. Minerals commonly dissolved include carbonates such as calcite and aragonite, silicates such as feldspars and ferromagnesian minerals, and, though rarely, accessory minerals such as garnet. In some cases, earlier cements may be dissolved to form secondary porosity.

RECRYSTALLIZATION AND REPLACEMENT

Recrystallization is a common feature in diagenesis. In recrystallization, one mineral is transformed into another mineral with the same (or nearly the same) chemical composition or into differently sized (usually larger) crystals of the original mineral. The most common example is the transformation of aragonite into calcite in limestones. In siliciclastic rocks, microcrystalline quartz may recrystallize into more coarsely crystalline quartz, or rutile may recrystallize into anatase.

A more prevalent form of diagenesis among ancient sandstones is replacement. In this process, one mineral is replaced volume-for-volume by another mineral. In some cases, it appears that the replacing mineral has combined material from its host grain with dissolved ions from pore water. The most common replacement examples involve clay minerals. For example, kaolinite may replace potassium feldspar, forming a grain of kaolinite that looks superficially like a potassium feldspar grain.

EXAMINING SILICLASTIC ROCKS

The principal means of studying siliciclastic rocks is with the petrographic microscope. This microscope is similar to other microscopes except that light passes through the specimen rather than illuminates its surface. Rocks are normally not transparent, however, and they must be cut into very thin slices in order for light to pass through them. Magnifications of up to four hundred times are routinely used. In thin section, it is possible to view closely individual grains and their relations to other grains, identify the mineral grains in the rock, and view matrices and cements. Thin sections are routinely used to estimate the amount of pore space in rocks, revealing their potential as reservoirs for fluids such as water, oil, and natural gas.

When illuminated by polarized light, different minerals display characteristic interference colors that reveal the identity of a mineral to the experienced observer. Other clues are the existence and pattern of cleavages in the minerals, alteration products resulting from diagenesis, and overall grain shape. In addition, it is possible to view cements that surround grains and partially or completely fill pores. The sequence of cementation can be interpreted by the relative positions of cements filling pores. Knowing the sequence of cements can help geologists understand how deeply buried the rock was when it was cemented and how the chemical composition of the pore water changed through time.

Another means of examining siliciclastic rocks is with the scanning electron microscope (SEM). The SEM permits examination of clastic grains with much greater magnification than with the conventional petrographic microscope. Magnifications of up to ten thousand times are easily accomplished, with details as small as 0.001 micrometer being easily resolved. The SEM is very useful in examining the surface texture of minerals. The presence of pits, grooves, and cracks can reveal clues as to the environment under which the sand grains were deposited. Were the sand grains blown by wind, carried by water, or transported by glaciers? In some cases, this question may be answered by examination under the extreme magnification of the SEM. The SEM is also very useful for studying the cements that bind sandstones.

ELECTRON MICROPROBE AND X-RAY STUDY

When accurate data concerning chemical composition of minerals are needed, the electron microprobe is used. The microprobe is actually an elaborate version of an SEM. It differs in that its beam does not normally scan but instead remains fixed in one spot, and its detectors are carefully calibrated to give accurate readings of the elemental composition of the material being examined. Microprobes collect X-rays and can determine from them the amounts of different elements in almost any mineral. In some instances, the abundance of trace elements in minerals such as zircon, tourmaline, garnet, or rutile may provide clues as to the source region from which the grains were derived, giving an indication of the source for a sedimentary formation. The electron microprobe is a very useful tool in these studies.

There are limitations to the microprobe. It cannot analyze for elements lighter than sodium, so the amount of oxygen or carbon cannot be determined. Furthermore, rock material must be very carefully prepared for readings to be accurate. Also, electron microprobes are rather expensive and require considerable maintenance; however, X-rays are also employed in the study of rocks. X-ray diffraction of powdered or whole specimens can identify minerals and their degree of crystal ordering. X-ray fluorescence identifies the chemical composition of minerals. Both techniques are commonly used in the study of clays in the matrix of siliciclastic rocks. Sample preparation is relatively easy, and beginners can operate most modern X-ray machines. X-ray analysis usually requires destruction of the sample by grinding, so the sizes, shapes, and relationships between grains are lost.

ECONOMIC SIGNIFICANCE OF SILICICLASTIC ROCKS

Sandstone and its cousins—siltstone and conglomerate—are common sedimentary deposits. These rocks, collectively known as siliciclastic rocks, are important for a number of reasons. For example, they are porous—that is, they contain small spaces between grains that may be filled with valuable fluids. Siliciclastic rocks are the major type of aquifer in most of the world. Also, they form reservoirs for oil and natural gas. Sand and gravel are essential parts of modern construction because they form the "bulk" in concrete. In this role, they represent the most economically significant mineral resource in the United States, ahead of petroleum, coal, and precious metals.

—Michael R. Owen

Further Reading

Blatt, Harvey, Robert J. Tracy, and Brent Owens. *Petrology: Igneous, Sedimentary, and Metamorphic.* 3rd ed., W. H. Freeman, 2005.

Boggs, Sam, Jr. *Petrology of Sedimentary Rocks.* Cambridge UP, 2009.

Haldar, Swapan Kumar. *Introduction to Mineralogy and Petrology.* 2nd ed., Elsevier, 2020.

Lauf, Robert. *The Collector's Guide to Silicate Crystal Structures.* Schiffer Publishing, 2010.
Lutgens, Frederick K., Edward J. Tarbuck, and Dennis G. Tasa. *Essentials of Geology.* Pearson Education, 2017.
Mackenzie, F. T., editor. *Sediments, Diagenesis, and Sedimentary Rocks.* Elsevier, 2005.
Middleton, Gerard V., editor. *Encyclopedia of Sediments and Sedimentary Rocks.* Springer, 2003.
Prothero, Donald R., and Fred Schwab, editors. *Sedimentary Geology.* W.H. Freeman, 2013.
Tucker, Maurice E. *Sedimentary Rocks in the Field.* 4th ed., John Wiley & Sons, 2011.

Slate

Fields of Study: Geology; Geochemistry; Geophysics

ABSTRACT

Slate is found worldwide, generally in areas where older rocks from beneath Earth's surface have been exposed or in areas where younger rocks have been subjected to the pressure of mountain building. Slate is formed when clay or smooth, fine-grained rock such as shale or basalt is subjected to relatively low temperature and pressure beneath the Earth's surface.

KEY CONCEPTS

bedding planes: the silt sediments that will eventually become slate are deposited at regular intervals as thin layers, or beds. The slate that they form will separate readily along the plane of those beds, or bedding planes

color: a distinctive color tone apparent in and unique to particular formation of slate due to minor mineral components in the slate

quarrying: obtaining slate blocks by cutting and splitting them from exposed slate beds; other types of rock and geological products are obtained by similar methods called "quarrying"

WHERE SLATE IS FOUND

Slate is found worldwide, generally in areas where older rocks from beneath Earth's surface have been exposed or in areas where younger rocks have been subjected to the pressure of mountain building. The most important areas of slate production in the United States are Pennsylvania, Vermont, and Wales.

PRIMARY USES OF SLATE

Cut slate is used for paving, roofing, electrical panels, blackboards, and tabletops. While most modern-day chalkboards in classrooms are large sheets of glass that have been etched and tempered, the "blackboards" they replaced were large sheets of cut slate. Slate remains the material of choice for the bed of quality pool tables. Crushed slate is used in aggregates (materials mixed with cement to make concrete).

TECHNICAL DEFINITION OF SLATE

Slate is formed when clay or smooth, fine-grained rock such as shale or basalt is subjected to relatively low temperature and pressure beneath the Earth's surface. This low-grade metamorphic process results in the formation of very small, often microscopic, crystals. The pressure inside the Earth causes these crystals to be flattened and elongated, resulting in the ability of slate to be split into thin slabs.

DESCRIPTION, DISTRIBUTION, AND FORMS OF SLATE

Slate is a smooth, hard, metamorphic rock that easily splits into thin, strong slabs. Slate is usually red, green, gray, or black, but some slate is blue or purple. The color of slate is often mottled, streaked, or spotted. A few large grains of various minerals may be found within the otherwise homogeneous mate-

Slate. Photo via Wikimedia Commons. [Public domain.]

rial. Thin lines may appear, which reveal the structure of the rock from which the slate was formed.

Slate varies widely in color because of the presence of small particles of various minerals within it. Red and purple slates contain hematite (iron oxide). Green slates contain chlorites (various green materials composed of aluminum silicates). Gray and black slates may contain carbon in the form of graphite, organic matter, or iron sulfide.

Other rocks that can be split into thin slabs are often loosely called "slates." True slates can be distinguished by the fact that the angle at which they split is different from the angle at which they lie. Other rocks known as "slate" usually split into slabs parallel to the way they lie.

HISTORY OF SLATE USE

Slate has been used as a roofing material in the United States since early colonial times. In other places where slate is common, particularly in parts of Britain, Wales and Scotland, slate has been used as a primary building material of walls and roofs of houses, as well as fences around fields. Individual pieces of slate about the size of a modern-day notebook were also used as pupils' "workbooks" in schools until the 1930s before being replaced in favor of paper notebooks. Ironically, the trend in present-day classrooms has been to replace paper notebooks with the modern "slate" called the "tablet."

OBTAINING SLATE

Slate is obtained by quarrying from shallow open-pit mines. When an area containing slate is discovered, the material covering the slate is removed with power shovels and bulldozers. Large blocks of slate are then removed by drilling and by using the natural tendency of the rock to split along well-defined bedding planes. The blocks of slate, which are about 7.5 centimeters thick, are split with chisels and mallets into slabs about 0.5 centimeter thick. The slabs are then cut to the desired size by rotating blades.

USES OF SLATE

Cut slate, known as dimension slate, is used to form panels for roofs, floors, and pavement. It is also used to provide a smooth, strong surface for electrical panels, blackboards, billiard tables, and laboratory tabletops. Slate that is not smooth enough to be used in the form of slabs may be mined using drills and explosives. The blasted slate is crushed and used in filler, in aggregate, or in roofing.

—Rose Secrest

Further Reading

Boggs, Sam, Jr. *Petrology of Sedimentary Rocks.* 2nd ed., Cambridge UP, 2009.

Fettes, Douglas, and Jacqueline Desmons, editors. *Metamorphic Rocks: A Classification and Glossary of Terms.* Cambridge UP, 2007.

Gwyn, David. *Welsh Slate: Archaeology and History of an Industry.* Royal Commission on the Ancient and Historical Monuments of Wales, 2015.

Jones, Reg Chambers. *Dinorwic: The Llanberis Slate Quarry 1780-1969.* Bridge Books, 2006.

Lyons, Arthur R. *Materials for Architects and Builders.* Elsevier Butterworth Heinemann, 2004.

Vernon, Ron H. *A Practical Guide to Rock Microstructure.* Cambridge UP, 2004.

Wichert, Jörn. *Slate as Dimension Stone: Origin, Standards, Properties, Mining and Deposits.* Springer, 2020.

SOILS

Fields of Study: Geology; Geochemistry; Biogeology; Geophysics

ABSTRACT

A soil is formed when decomposed organic material is encompassed into weathered mineral material at Earth's surface. The climate, the organisms living in and on the soil, the type of parent material, the local topography, and the amount of time the soil has been developing all influence the resulting soil characteristics. Changes in the soil arise from the main processes of soil formation: addition, removal, transfer, and transformation of parts of the soil.

KEY CONCEPTS

horizon: a layer of soil material approximately parallel to the surface of the land that differs from adjacent related layers in physical, chemical, and biological properties

leaching: the dissolving out or removal of soluble materials from a soil horizon by percolating water

sediment: rock fragments such as clay, silt, sand, gravel, and cobbles

soil profile: a vertical section of a soil, extending through its horizons into the unweathered parent material

weathering: the mechanical disintegration and chemical decomposition of rocks and sediments

A BRIEF DESCRIPTION OF SOIL

Soils are the loose masses of broken and chemically weathered rock mixed with organic matter that cover much of the world's land surface, except in polar regions and most deserts. The two major solid components of soil—minerals and organic matter—occupy about half the volume of a soil. Pore spaces filled with air and water account for the other half. A soil's organic material comes from the remains of dead plants and animals, its minerals from weathered fragments of rock. Soil is also an active, dynamic, ever-changing environment. Tiny pores in soil fill with air, water, bacteria, algae, and fungi working to alter the soil's chemistry and speed up the decay of organic material, making the soil a better living environment for larger plants and animals.

SOIL FORMATION

Many processes contribute to the two stages of soil formation, which may grade indistinctly into each other. The first stage is the accumulation of rock fragments known as parent material. These sediments may have formed in place from the physical and chemical weathering of hard rocks below or by the accumulation of sediments moved by wind, water, gravity, or glaciers. The second stage is the formation of soil horizons, which may either follow or occur simultaneously with the first stage. The near-surface horizon, where materials are removed, is called the "A horizon." Some of the removed material accumulates in the subsurface, in a zone called the "B horizon." If the soil has an organic-rich top layer, that horizon is called the "O horizon." In some cases, the A horizon has so much iron removed that it looks bleached. A very light zone at the base of the A horizon is called the "E horizon". Unmodified parent material is termed the "C horizon."

The characteristics of the horizons and their degree of differentiation depend upon the relative strengths of four major processes in the soil: weathering, transfers, gains, and losses of the soil constituents. These processes have been observed directly in laboratory experiments but also are implied by relating physical, chemical, and morphological properties in different parts of the soil profile.

Climate is the principal factor in determining the type and rate of soil formation. Temperature and precipitation are the two main climatic factors that influence soil formation, and they vary with elevation and latitude. Water is the main agent of weathering, and the amount of water available depends on how much falls and how much runs off. The amount of precipitation and its distribution during the year influence the kind of soil formed and the rate at which it is formed. Increased precipitation usually results in increased rates of soil formation and deep soils. Temperature and precipitation also determine the kind and amount of vegetation in a region, which determines the amount of available organics.

Topography is a characteristic of the landscape involving slope angle and slope length. Topographic relief governs the amount of water that runs

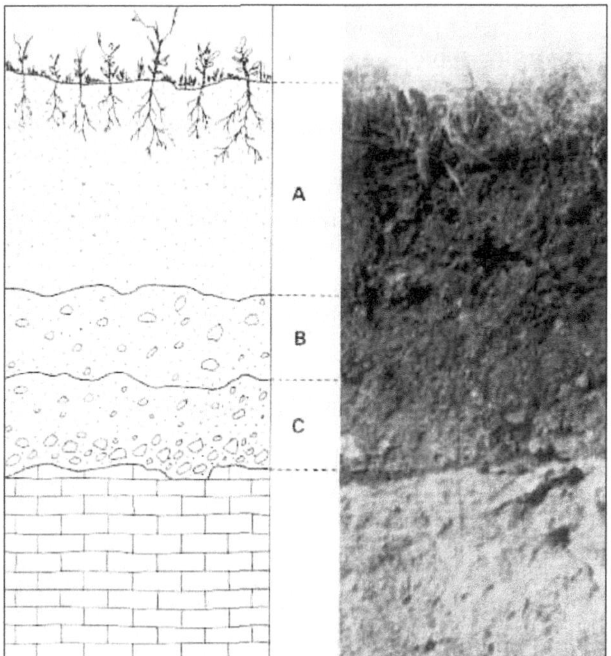

A, B, and C represent the soil profile, a notation firstly coined by Vasily Dokuchaev (1846–1903), the father of pedology; A is the topsoil; B is a regolith; C is a saprolite (a less-weathered regolith); the bottom-most layer represents the bedrock. Photo by Carlosblh, CC BY-SA 3.0, via Wikimedia Commons.

off or enters a soil. On flat or gently sloping land, soil tends to stay in place and may become thick, but as the slope increases so does the potential for erosion. On steep slopes, soil cover may be very thin, possibly only a few centimeters, because precipitation washes it downhill; on level plains, soil profiles may be several meters thick.

TRANSFORMATIONS AND TRANSFERS

Weathering is the fundamental process of soil formation, as it transforms soil constituents into new chemicals. The primary minerals found in the initial sediments and rocks are decomposed by physical and chemical agents into salts that are soluble in water and new minerals. Clay is a very important new mineral formed from the breakdown of primary minerals by the agents of water and air. Calcite, or calcium carbonate, is another mineral formed through weathering, especially of rocks such as basalt. Iron and aluminum hydroxides form a third major group of weathering-produced minerals. Organic matter is added to the soil when roots die and vegetation residues fall onto the surface. These are transformed by microorganisms into a mixture of chemically stable, organic substances called "humus."

Typically, bedrock first weathers to form regolith, a protosoil devoid of organic material. Rain, wind, snow, roots growing into cracks, freezing and thawing, uneven heating, abrasion, and shrinking and swelling break large rock particles into smaller ones. Weathered rock particles may range in size from clay to silt, sand, and gravel, with the texture and particle size depending largely on the type of bedrock. For example, shale yields finer-textured soils than sandstone. Soils formed from eroded limestone are rich in base minerals; others tend to be acidic. Generally, rates of soil formation are largely determined by the rates at which silicate minerals in the bedrock weather: the more silicates, the longer the formation time.

The transfer of soil constituents is generally in a downward direction. Rainwater is the major transporting agent. As it moves through the soil, it picks up salts and humus in solution and clays and humus in suspension. These materials are usually deposited in another horizon when the water is withdrawn by roots or evaporation, or when the materials are precipitated as a result of differences in pH (degree of acidity) or salt concentration. The clays normally coat the exteriors of sediment grains and line the interiors of pores. This transfer of materials through solution is commonly called "leaching" and requires good drainage. The soil profile often becomes vertically zoned according to the solubility of the dissolved substances. In forest soils, leaching generally removes iron and aluminum, leaving behind a bleached-looking zone. This process is called "podzolization." In arid soils, calcium minerals precipitate in the subsoil to form calcite or gypsum, a deposit called "caliche."

GAINS AND LOSSES

Gains consist of additions of new materials to the soil. Wind can deposit fine sands and silts if there is a source nearby. Chemical reactions during weathering add oxygen and water to the soil minerals. Groundwater brings new minerals into the soil pores. Floods can add new sediment to the soils in a floodplain. The largest addition to soils, however, comes from organic matter from the decomposition of plant material. Annual additions of new organic material between 255 and 22,900 kilograms per hectare have been estimated for soils in different environments ranging from deserts to tropical rain forests.

In regions where organic materials, such as plant and animal remains, may be deposited on top of regolith, rudimentary soils can begin to form. When waste material is excreted, or a plant or animal dies, the material usually ends up on Earth's surface. Organisms that cause decomposition, such as bacteria and fungi, begin breaking down the remains into a beneficial substance known as humus. Humus restores minerals and nutrients to the soil. It also improves the soil's structure, helping it to retain water. Over time, a skeletal soil of coarse, sandy material with trace amounts of organics gradually forms. Even in a region with good weathering rates and adequate organic material, it can take as long as fifty years to form 30 centimeters of soil. When new soil is formed from weathering bedrock, it can take from 100 to 1,000 years for less than two to three centimeters of soil to accumulate.

Losses from the soil profile generally occur when there is so much water moving through the system that the dissolved and suspended materials exit at the bottom of the profile and move downslope and eventually enter a stream. Soil erosion also occurs on the surface by water runoff, especially where there is minimal vegetation to control the sediment. Soil erosion is one of the major environmental problems facing the world today. Oxidation of plant matter balances some of the additions of new organic material.

Because Earth's surface is made of many different rock types exposed at differing amounts and weathering at different rates at different locations, and because the availability of organic matter varies greatly around the planet due to climatic and seasonal conditions, soil is very diverse and fertile soil is unevenly distributed. Structure and composition are key factors in determining soil fertility. In a fertile soil, plant roots are able to penetrate easily to obtain water and dissolved nutrients. A loam is a naturally fertile soil, consisting of masses of particles from clays (less than 0.002 mm across), through silts (ten times larger) to sands (100 times larger), interspersed with pores, cracks, and crevices.

CLIMATE

After describing many soil profiles around the world and classifying them, scientists concluded that this weathered material at Earth's surface was quite variable and differed from environment to environment. In 1980, Hans Jenny summarized the work of early researchers and listed the five main soil-forming factors that created this variation: climate, organisms, parent material, topography, and time. The rates and degrees of these processes in different soils depend upon these five factors. These factors are interdependent, and, therefore, very different soils may be formed from the same parent material.

Of the five factors, climate has probably the greatest impact on soil development. This dominance is best shown on a large-scale geographic basis, in which the distribution of soils on soil maps is related to vegetation type, which is in turn related to climatic zones. Precipitation provides the water for the transfer processes in the soils. In dry climates, where little water is moving through the system, calcium carbonate and salts build up in the soils. In humid climates, the abundance of water moving through the systems allows leaching and clay movement processes to dominate in the soils. Temperature is the second important element of climate, as it controls the rates of chemical weathering reactions in the soil. In colder environments, soil weathering and development are slower because weathering reactions are not active all year round. Some extreme climatic conditions also can control soil development, such as permafrost, which limits soil depth and drainage. Overall, climate controls vegetation, which also controls soil development.

ORGANISMS AND PARENT MATERIALS

Living organisms, primarily vegetation, control the distribution and types of humus that are formed in soils. In turn, humus controls some of the important soil-forming processes. Grassland vegetation produces humus that is highest in mineral matter and not very acid, so resulting soils have thick A horizons that are high in organics. Conifer forests produce humus that is low in mineral constituents and highly acidic. These forest soils are dominated by leaching and the production of O, E, and B horizons (B horizons are rich in iron and aluminum). Deciduous forest soils are located somewhere between the conifer forest and the grassland soils in characteristics. Therefore, it is rare to find white, bleached E horizons under grassland vegetation, and vice versa—it is rare to find thick A horizons under conifer forest vegetation. Through its leaves and root systems, vegetation also protects the soil from erosion and, therefore, promotes soil development.

The parent material is the rock or sediment deposit from which the soil has developed. The composition of the original minerals controls the complexity of the soil processes by determining the weathering byproducts, and controls the rate of these processes by porosity. A parent material of pure quartz sand would undergo minimal weathering, producing a simple soil profile. A parent material of wind-deposited silt, or loess, produces a more complex soil, as it has many minerals initially that can weather to byproducts such as clays, salts, carbonates, and oxides. A soil produced on granite would yield a fairly deep, sandy soil, whereas a soil produced on basalt would yield a fairly thin, clay-rich soil. A soil developed in shale would be clay-rich, whereas one developed on sandstone would be sandy. Soils formed on rocks rich in iron tend to produce very red soils. Soils formed on limestone are difficult to acidify and podzolize because the parent material neutralizes those chemical processes. The porosity, or the amount of pore space, of the parent material gives ready access of gases and liquids to weathering. In porous, or permeable, soils, weathering and transfers of soil constituents are more rapid than in soils with low permeability.

TOPOGRAPHY AND TIME

Topography is the geometric configuration of the soil surface and is defined by the slope gradient, orientation, and elevation. This factor essentially controls the microclimate. A slope modifies rainfall by allowing rapid runoff on a steep slope or ponding in low places. Slope orientation modifies soil temperatures by the differences at which the sun's rays strike the soil. In the northern latitudes, south-facing slopes are warmer and drier than those on north-facing slopes. In mountain ranges, rainfall increases and temperature decreases with increases in elevation.

Time is the factor of the duration of the processes working in the soil. All the characteristics in the soil take time to form. The A horizon forms most rapidly, taking as little as 10 years with the aid of humankind to more than 1,000 years in a cold and dry arctic or alpine environment. B horizons take longer to form because minerals weather slowly. In the dry western parts of the United States, discoloration from the weathering of iron-rich minerals appears at about 1,000 years, but the strongest red colors require 100,000 years of development in the same environment. In the same part of the United States, initial detection of clay movement into the B horizon in a sandy parent material occurs after about 10,000 years of development, with maximum development taking more than 100,000 years. Calcium carbonate deposition in arid regions of the world takes between 100,000 and 500,000 years to produce a strongly cemented horizon. It is estimated that it takes more than 1 million years to produce a highly weathered lateritic soil (decayed iron-rich rock that is red in color) in the tropics.

Soils change with time just as humans change as they grow. Young or poorly developed soils in a grassland environment have only A and C horizons. With time and the movement of iron oxides and clay into the lower parts of the profile, a faint B horizon develops between the A and C horizons. An old or well-developed soil in this environment would have a thick A horizon and a thick B horizon with abundant clay moved into it, all overlying a C horizon.

THE ROLES OF SOIL

In any ecosystem, soils play six key roles. First, soil serves as a medium for plant growth by mechanically supporting plant roots and supplying the eighteen nutrients essential for plants to survive. Different types of soil contain differing amounts of these eighteen nutrients; their combination often determines the types of vegetation present in a region, and as a result, influences the number and types of animals the vegetation can support, including humans. Humans rely on soil for crops necessary for food and fiber.

Second, the property of a particular soil is the controlling factor in how the hydrologic system in a region retains and transports water, how contaminants are stored or flushed, and at what rate water is naturally purified. Water enters the soil in the form of precipitation, irrigation, or snowmelt that falls or runs off soil. When it reaches the soil, it will either be surface water, which evaporates or runs into streams, or subsurface water, which soaks into the soil where it is either taken up by plant roots or percolates downward to enter the groundwater system. Passing through soil, organic and inorganic pollutants are filtered out, producing pure groundwater.

Soil also functions as an air-storage facility. Air is pushed into and drawn out of the soil by changes in barometric pressure, high winds, percolating water, and diffusion. Pore spaces within soil provide access to oxygen to organisms living underground as well as to plant roots. Soil pore spaces also contain carbon dioxide, which many bacteria use as a source of carbon.

Soil is nature's recycling system, through which organic waste products and decaying plants and animals are assimilated and their elements made available for reuse. The production and assimilation of humus within soil converts mineral nutrients into forms that can be used by plants and animals, who return carbon to the atmosphere as carbon dioxide. While dead organic matter amounts to only about 1 percent of the soil by weight, it is a vital component as a source of minerals.

Soil provides a habitat for many living things, from insects to burrowing animals, from single microscopic organisms to massive colonies of subterranean fungi. Soils contain much of the earth's genetic diversity, and a handful of soil may contain billions of organisms, belonging to thousands of species. Although living organisms only account for about 0.1 percent of soil by weight, one hectare of good-quality soil can contain at least 300 million small invertebrates—mites, millipedes, insects, and worms. Just 1 ounce (30 grams) of fertile soil can contain 1 million bacteria of a single type, 100 million yeast cells, and 50,000 fungus mycelium. Without these, soil could not convert nitrogen, phosphorus, and sulphur to forms available to plants.

Finally, soil is an important factor in human culture and civilization. Soil is a building material used to make bricks, adobe, plaster, and pottery, and often provides the foundation for roads and buildings. Most important, soil resources are the basis for agriculture, providing people with their dietary needs.

LAND-USE PLANNING

Scientists are finding that soils can be useful tools in the interpretation of past environments of an area and also in the formulation of land-use plans for that region. First, the soil scientists must determine the exact soils that are in the area through a mapping program. During this project, they determine the regional developmental sequence of soils—that is, the poorly developed and the well-developed soils of the different places on the landscape. They assign ages to each soil in the sequence using radiocarbon-dated samples from the soils. As a result, when a particular soil is located in that region and compared with the sequence, an age can be assigned to the land surface upon which the soil developed. Geologists commonly use soils to estimate the ages of deposits, especially on glacial landscapes. The idea of multiple glaciations has been developed through the use of soils. Poorly developed soil on glacial sediments represents a recent glaciation compared with a site that features well-developed soil lying on sediment deposited by an older glacial advance.

The frequency of geological hazard events such as landslides, rockfalls, and flooding can be estimated by the study of buried soils in sediment deposits. An ancient, buried A horizon within a soil profile means that some geological event covered an old land surface with sediment. If the organic matter of that buried A horizon is dated using radiocarbon dating, an approximate time of burial can be determined. If a valley bottom has many buried soils, it probably means that the nearby stream floods quite frequently. These decisions are important for land-use planning, for one does not want to build in an area where hazards occur frequently. For example, nuclear power plants must not be located in areas where there are active faults (ground breaks). Soils are commonly used to determine if faults are active in an area and, if so, the frequency of movement on the fault.

The U.S. Department of Agriculture produces maps of soils across the country. The maps describe the productivity of crop growth and manageability related to land use of each soil. These data help farmers grow crops that are best suited for their soils; they tell the farmers which soils are most susceptible to erosion so that extra precautions can be taken with the fragile soils. The data also help land-use planners keep the most productive soils in agriculture instead of paving them. In addition, the maps help to predict how much fertilizer to put on the fields. Natural soil fertility is related to geological activity and weathering, as the oldest soils are leached of most of their nutrients and are least fertile. Young soils have had little weathering and leaching and are most fertile, especially those on floodplains, young glacial deposits, and volcanic deposits. Soil maps are also used to recognize where shrink-swell clays occur in the soils. The presence of shrink-swell clays in the soil is important for home owners. If present, these clays can crack the foundation of a house. To prevent these damages, special foundations must be constructed and the clays stabilized with the use of lime.

Because the human use of soils has been haphazard and unchecked for millennia, soil resources in many parts of the world have been harmed severely. Human activities, such as overcultivation, inexpert irrigation, overgrazing of livestock, elimination of tree cover, and cultivating steep slopes, have caused natural erosion rates to increase many times over. As a result of mismanaged farm and forest lands, escalated erosional processes wash off or blow away an estimated 75 billion tons of soil annually, eroding away one of civilization's crucial resources.

—Randall L. Milstein and Scott F. Burns

Further Reading

Birkeland, P. W. *Soils and Geomorphology.* 3rd ed., Oxford UP, 1999.

Buol, S. W., F. D. Hole, and R. J. McCracken. *Soil Genesis and Classification.* 5th ed., Iowa State UP, 2003.

Retallack, G. J. *Soils of the Past.* 2nd ed., Blackwell Science, 2001.

Schaetzl, Randall J., and Sharon Anderson. *Soils: Genesis and Geomorphology.* Cambridge UP, 2005.

Singer, M. J., and D. N. Munns. *Soils.* 6th ed., Prentice Hall, 2005.

Soil Science Society of America. *Glossary of Soil Science Terms.* Author, 2001.

U.S. Department of Agriculture, Soil Survey Staff. *Keys to Soil Taxonomy.* UP of the Pacific, 2005.

Van Breemen, Nico, and Peter Buurman. *Soil Formation.* 2nd ed., Kluwer Academic Publishers, 2002.

Walther, John Victor. *Essentials of Geochemistry.* 2nd ed., Jones & Bartlett Publishers, 2008.

STENO PRESENTS HIS THEORIES OF FOSSILS AND DYNAMIC GEOLOGY

Fields of Study: Historical Geology; Paleontology; Geophysics; Geochemistry

ABSTRACT

By establishing that physical Earth indeed has a history, with geologic strata deposited by sedimentation and mountains and valleys formed by the collapse of plains, Nicolaus Steno explained the nature of marine fossils found far from the sea. His work may be considered the first modern form of paleontology and the first presentation of a geology of changes over time.

KEY CONCEPTS

fossil: an old term that referred to anything that came from beneath Earth's surface, regardless of its actual nature; now refers to the preserved remains of ancient living things

principle of superposition: the principle that geological and historical strata are laid down in successive layers such that the lower stratum is the older one, with the oldest at the bottom of the strata layers and the youngest at the top

SUMMARY OF STENO'S IDEATION

Though the Greeks had observed the presence of petrified seashells inland and correctly concluded that the sea had once covered the land and deposited the shells, fossils posed several difficulties for the seventeenth century mind. First, no clear idea existed to explain fossils as the remains of once-living beings. The concept of "fossil" included anything buried in the earth or lying on its surface, and comprised a continuum from things clearly inorganic—such as gemstones—to things clearly resembling living organisms.

Interpreting fossils as organic did not follow for several reasons. Some of these fossils had formed crystals and hence seemed to be inorganic. Another conclusion held that the resemblance of such fossils to organisms was fortuitous: They were "sports"

Nicolas Steno. Image via Wikimedia Commons. [Public domain.]

(jokes) of nature. Moreover, that many organic fossils were of extinct, unknown species further obscured their nature. A second issue involved how to explain the formation of fossils. According to the Aristotelian view, fossils grew from seeds in the earth; the neo-Platonic idea held that a molding force created the fossils in a web of affinities that united all parts of the universe. Neither view required the notion that the fossils resembling life forms were organic. A final difficulty arose from the location of the fossils, especially seashell forms located on mountains far from the sea. The biblical flood provided a reasonable explanation.

Whereas several authors of the sixteenth and early seventeenth centuries, including Leonardo da Vinci, argued that fossils were of organic origin, an understanding of the nature of fossils ultimately rested upon a theory of geological change. In the seventeenth century, Nicolaus Steno provided the first such theory.

Born Niels Stensen, Steno received training at the University of Copenhagen from Thomas Bartholin, perhaps the most famous anatomist of the day, and from his younger brother Erasmus, a mathematician, physician, and Cartesian scholar. Steno received additional training in Amsterdam, at the University of Leiden, and in Paris. Having gained a reputation for his anatomical studies, Steno arrived in Florence in March, 1666, where Ferdinand II de' Medici, the grand duke of Tuscany, received him and appointed him to a post that provided a living. He was also elected to the Academia del Cimento (Academy of Experiments), founded in 1657 by Leopold de' Medici, brother of the grand duke. The academy's members included Francesco Redi, a physician who performed experiments to discredit spontaneous generation.

With the help of Ferdinand, Steno published in 1667 a work on the action of muscular contraction, to which he attached his findings from the dissection of the head of a shark caught in October of 1666. His examination of its teeth convinced Steno of the organic origin of fossilized shark teeth known as "tongue stones," which he had observed in rocks during sojourns with Ferdinand. To support his theory, Steno had to argue convincingly that tongue stones had been produced in the sea and that they could not have grown in or on the solid Earth. One of his arguments rested on the observation that some of them had eroded and thus were quite old. He then presented an explanation that the fossils had formed in soil that was not firm and was covered with water. The soils containing the fossils thus were sediments, an image suggested perhaps by Steno's early experience with chemical precipitates. Despite these arguments, Steno asserted that he drew no conclusions concerning the origins of the fossils.

Two years later, Steno published *De solido intra solidum naturaliter contento dissertationis prodromus* (1669; *The Prodromus to a Dissertation Concerning Solids Naturally Contained Within Solids*, 1671; better known as *Prodromus*), a work significant for its presentation of principles governing geologic history. A major goal was to explain how a solid—including a fossil or a crystal—could have formed within rocks. In the first part of this work, Steno argued definitively that fossilized animal and plant parts were formed in exactly the same way as the parts of living animals. He explained that the hard organic parts impressed their form on soft sediments. Furthermore, Steno noted that because fossil shells as well as fossil plants and the fossil remains of other animals resembled living species, it follows that fossils were organic. To account for the hardness of fossils, Steno suggested that the original matter had lost volatile particles or had taken on particles from its surroundings. Thus, he explained the nature and origin of the fossils.

In response to the problem of location, especially of marine fossils far from the oceans, Steno devised an explanation of the formation of geologic strata (rock layers). One of his principles stated that strata formed successively, with the lower stratum solidifying before the formation of the next stratum—the so-called principle of superposition. The original strata, laid down horizontally, could be lifted up or slipped down by the subsequent actions of fires and waters. Steno applied these principles to an explanation of Tuscan geology: the region's geology formed first by the deposition of layers and then by the undermining of those layers by subterranean water or heat, which carved out caverns that then collapsed. The resulting valley filled with water, causing further sedimentation and collapsing layers.

According to Steno's theory then, shells could be deposited during the first stage and wind up on mountains in a later stage. Throughout this work, however, Steno was careful to point out that his ideas did not contradict Scripture, and he incorporated the biblical flood into his scientific thought. These efforts succeeded, for the book easily gained the approval of the censors, one of whom was his friend Francesco Redi.

SIGNIFICANCE OF STENO'S THEORIES

Steno's work garnered a positive reception in England, where it also generated much debate. The secretary of the Royal Society of London, Henry Oldenburg, translated *Prodromus* into English, and *Philosophical Transactions* published reviews of Steno's work. One difficulty with Steno's work is the "young" age of the fossils studied, for the Tuscan fossils were only 5 million years old, while the English fossils were 300 million years old. For example, Martin Lister noted that while the fossils of Tuscany did indeed resemble living specimens, those found in the quarries of England, such as the ammonites (extinct mollusks), bore no resemblance to any known species and therefore could not be rightly considered the remains of organisms. To some, then, Steno's theories raised the issue of extinction, a phenomenon incompatible with belief in creation by an all-good, all-powerful God. The general acceptance of extinction would come more than a century later, as *Prodromus* was reprinted several times in the eighteenth century.

Steno's work may be considered the first modern work on paleontology as well as the first presentation of a "dynamic" geology, or a geology of changes over time. During the eighteenth century, his theory of fossils prevailed and scholars accepted his geological principles. Another important aspect of his work involved his use of the scientific method. Like Galileo, he sought efficient causes for the phenomena he studied, that is, only immediate physical causes and not supernatural ones, and he affirmed the importance of personal observation. In his explanation of the development of fossils and gemstones, he relied on chemical and mechanical processes.

—Kristen L. Zacharias

Further Reading

Cutler, Alan. *The Seashell on the Mountaintop: A Story of Science, Sainthood, and the Humble Genius Who Discovered a New History of the Earth*. Dutton, 2003.

Dryburgh, Peter. "Nicholas Steno and the Foundations of Geology." *Edinburgh Geologist*, vol. 41, Autumn 20013, pp. 3-11.

Kardel, Troels. *Steno: Life, Science, Philosophy*. Danish National Library of Science and Medicine, 1994.

———, and Paul Maquet. *Nicolaus Steno: Biography and Original Papers of a 17th Century Scientist*. 2nd ed., Springer, 2018.

Miniati, Stefano. *Nicholas Steno's Challenge for Truth: Reconciling Science and Faith*. AngeliFranco, 2009.

STREAMS AND RIVERS

Fields of Study: Geology; Geomorphology; Geophysics; Limnology

ABSTRACT

Rivers, streams, creeks, brooks, and rills are long, narrow bodies of water that flow downslope in a channel under the influence of gravity. Although the list is arranged in general order of size, the term "stream" is accepted as the scientific term for any amount of surface water flow. Streams are an integral part of the hydrologic cycle, as they transport water and sediment from the land to the sea.

KEY CONCEPTS

base flow: the amount of water that enters a particular stream from the water table

bed load: the amount of material that a stream transports in contact with the surface of the streambed

bed scouring: removal of riverbed features and transport of riverbed materials by stream flow that has lost its natural load on the upstream side of a dam and acquires a fresh stream load on the downstream side of the dam

drainage density: the number of streams per unit area within a particular watershed area

endoreic watershed: a watershed with no exit to an ocean

infiltration capacity: the amount of water that a particular soil can absorb before becoming saturated and losing water to surface runoff

suspended load: the amount of sediment that a stream carries above the streambed

WATER AND ITS ROLE IN THE WORLD

The vast majority of all of the water in the world is contained in the oceans (97.2 percent) and ice sheets and glaciers (2.15 percent). This means that 99.35 percent of the world's water is either salty or frozen. Of the remaining 0.65 percent, groundwater accounts for 0.63 percent, followed by freshwater lakes (0.009 percent), saline lakes and inland seas (0.008 percent), soil water (0.005 percent), the atmosphere (0.001 percent), and stream channels (0.0001 percent). It is obvious that streamflow makes up the smallest amount of water on Earth, yet it has been critical for the development of human societies.

Streams have been associated with the development of civilization from the beginning of human history. The elaborate irrigation systems that developed in Egypt, China, and the Tigris-Euphrates valley in Iraq and portions of surrounding countries started thousands of years ago. Streams have also been used for navigation, municipal water supplies, wastewater disposal, and in more modern times, power-plant cooling, hydropower, and firefighting. Major cities developed on or near streams: London on the Thames; Albany and New York on the Hudson; Shanghai on the Huangpu, a tributary of the Chang (also known as the Yangtze); New Orleans and St. Louis on the Mississippi; Montreal on the St. Lawrence; Vienna and Budapest on the Danube; and Cairo on the Nile. This brief list could include many other cities, but the important fact is that most of the world's major settlements have developed on or very close to a source of water, and most of these sources are streams.

THE DEVELOPMENT OF STREAMS

Precipitation events result in several types of natural response. The first response is simply that the precipitation can evaporate before it even reaches Earth's surface. This generally occurs in drier climatic regimes. The second response occurs when some of the precipitation infiltrates into the soil and forms the groundwater component of the hydrologic cycle. The third response is usually associated with storm events that could easily exceed the infiltration capacity of an area and thereby result in overland flow. This type of flow would travel downslope and eventually reach an area where this surficial downward movement of water coalesces and starts to erode the land surface to a point where a permanent channel starts forming. Over time, streams become perennial (meaning they flow continuously throughout the year) when they have cut deeply enough so that the surface of the stream is below the adjoining water table, thereby allowing groundwater to supply the stream. This movement of groundwater is called "base flow," and can easily account for one-half or more of the total water flowing in a stream, particularly in coastal plain formations such as those along the eastern coast of the United States. Thus, surface runoff consists of both overland flow and groundwater, even though they move at very different rates (groundwater is much slower).

The ultimate sink for most of the streams in the world as they move from higher to lower elevations is the oceans. Steep gradients, rapids, and waterfalls are generally associated with streams in the uppermost portions of their watersheds. Over time, these streams develop a progressively gentler slope as the irregularities in the channel are removed and the profile becomes smooth. With time, the stream widens its bed and valley as downward cutting of the channel is replaced by lateral cutting. Stream sinuosity increases as meanders develop. A depositional feature of a stream, described as a floodplain, widens with time and downstream progression. Streams often build deltas when they empty into the sea when circumstances permit. Deltas form when the velocity of the water moving in the channel slows down as the stream enters a much larger body of water. The particles settle out and form a variety of deltaic types. Well-known deltas include the Nile and the Mississippi. Where conditions are not favorable for deltaic formation, as in the case of the two largest streams in the world (the Amazon and Congo rivers), there is no delta.

Some drier areas of the world have streams that are located in interior basins without any outlet to empty into the oceans. These endoreic watersheds represent internal drainage basins. Examples include the Great Basin in Nevada and portions of surrounding states, Lake Chad in sub-Saharan north-central Africa, the Dead Sea in Israel and Jordan, and the Aral and Caspian Sea basins in central Asia.

RUNOFF MEASUREMENT

The amount of precipitation that falls on Earth's surface, minus the amount that evapotranspires from vegetation back to the atmosphere, either enters the soil to become a component of groundwater or runs over Earth's surface as overland flow to eventually wind up in streams. The measurement of all of the myriad components of surface and groundwater runoff would be a daunting task at the least. However, it is not necessary to determine the flow rates of all of these upslope contingents of runoff. The simpler procedure is to measure the flow at various points along the stream in order to determine the runoff from part of or the entire watershed.

The earliest attempts to at least partly measure streamflow began about five thousand years ago on the Nile in Egypt. The height of the river could be measured by reference to staff gauges called "Nilometers" that were fixed to the banks of the Nile. These permanent structures provided a measure of river stage that was then used to alert downstream farmers of the rise or fall of flood waves coming down the Nile. Staff gauges are still used to measure river, lake, and canal levels. In order to determine the volume of water that goes past a given point on a stream during a certain period of time, it is necessary to measure the volume and velocity of water at that point. This flow is calculated by multiplying the area of the cross section in square meters by average velocity in meters per second to yield discharge in cubic meters per second.

The first permanent stream-gauging station in the United States was established in 1889 on the upper Rio Grande in New Mexico by the U.S. Geological Survey (USGS), an agency of the federal government that, in conjunction with states and municipalities, is responsible for maintaining gauging stations. The total number of gauging stations in the USGS network was once about seven thousand for the whole country, but funding cuts have been reducing the overall total over the years. Regrettably, some stations that were dropped had records of continuous daily measurement in excess of thirty years, thereby representing a loss of valuable hydrologic data.

WATERSHED AREA

The total amount of land that collects all of the surface and groundwater within an area is called a "watershed" (or drainage basin). Watersheds vary enormously in size, from a tiny lot that drains into a pond to the Amazon River basin, with a drainage area of 6.16 million square kilometers. The second- and third-largest watersheds in area in the world are the Congo in central Africa (3.83 million square kilometers) and the Mississippi (3.26 million square kilometers).

Another characteristic of watersheds is their variation in shape. Although many of them are pear-shaped (reflecting a preponderance of horizontal rock strata), others are elongated, indicating a landscape of ridges (resistant formations composed of igneous rocks such as schist or gneiss) and valleys (weaker formations composed of shale). This diversity in the underlying rock formations has an important bearing on slopes, infiltration, and runoff characteristics of all watersheds in addition to the obvious effects of climatic variations.

STREAM DISCHARGE AND LENGTH

Given the huge variation in precipitation, temperature, and landscapes in different regions of Earth, it is obvious that average stream discharge would also vary enormously. Given its large size and favorable location in the wet equatorial belt, with average annual precipitation in excess of 200 centimeters, the Amazon leads the list with an average discharge of 175,100 cubic meters per second. This astounding amount of flow accounts for about 20 percent of the total discharge of all of the world's rivers, even though the area of the Amazon watershed is only about 2 percent of the total land area in the world. The second- and third-largest average discharges in the world are the Congo, with 40,000 cubic meters per second, and the Mississippi, with 18,400 cubic meters per second. Note that the average discharge of the Amazon is nearly ten times that of the Mississippi, even though the area of the Amazon is somewhat less than twice the size of the Mississippi. The difference is attributed to climatic factors in the watersheds.

The four longest streams in the world are the Nile (6,648 kilometers), the Amazon (6,436 kilometers), the Chang (6,300 kilometers), and the Mississippi

River running through Glacier National Park in Montana. Photo via iStock.com/Matt Anderson. [Used under license.]

(5,970 kilometers). Although the Nile is the longest river in the world, it is ranked thirty-third in terms of average discharge, as it flows for a considerable distance through the deserts of northern Sudan and Egypt before it reaches the Mediterranean.

DRAINAGE DENSITY AND TOPOGRAPHIC TEXTURE

The variation in rock formation and soil type within each watershed governs the number of streams per unit area that can form in different landscapes. Drainage density is determined simply by dividing the total length of streams within a selected region by the area of the region. For example, an area underlain by very resistant beds of hard sandstone or granite and covered by heavy tree cover in a humid environment would experience minimal development of streams, resulting in a very low drainage density and consequent coarse texture. Another example, also in a humid environment with tree cover, would be an area underlain by weaker and therefore less resistant shale rock that would have more streams in the same size area and therefore have a somewhat higher drainage density number that would be considered to be medium texture. High drainage density (fine texture) would be expected to develop in landscapes underlain by weak sedimentary rocks in conjunction with limited coverage of vegetation. Instances of very high drainage density and resulting ultrafine texture can be found in the badlands of South Dakota, where the underlying earth materials are very weak clays and shales in conjunction with limited vegetative cover.

Those areas that have coarse texture (low drainage density) have much fewer streams than those watersheds with finer texture that are more easily eroded with many more streams per unit area. The difference in drainage density can therefore range up to two orders of magnitude.

STREAM ECOSYSTEMS

Free-flowing streams have natural pulses of seasonal floods that maintain a dynamic equilibrium between the biological and physical aspects of aquatic ecosystems. The flora and fauna of stream ecosystems are well adapted to the natural variations in streamflow. For example, water levels and flooding events in the eastern United States are generally highest during the late winter-early spring period when evapotranspiration is low and soil moisture is at or close to its field capacity (the amount of water that is retained in the soil or rock against gravitational forces). This does not mean that flooding events cannot occur during other parts of the year, such as the summer and fall. For example, although long-term records clearly indicate that the usual time for floods to occur in the Mississippi is in the spring, major floods have also occurred in midsummer. Another example is provided by the long-term records of the Connecticut River at Hartford, which show expected floods during March and April but also a second period during the fall, when heavy rainstorms and hurricanes can occur.

THE GEOLOGIC WORK OF STREAMS

Erosion, transportation, and deposition represent the three closely related activities of streams. Stream erosion refers to the removal of earth material from the channel. The same process will occur in both alluvial and bedrock channels, although different removal rates will apply based on resistance to erosion (alluvium is much easier to erode).

As the particles are eroded, they are transported by the stream in one of the following modes: in solution, in suspension, or as bed load. Salts from mineral alteration constitute a very common form of material in solution in all streams. The dissolved matter (salts) cannot be seen, as it is mostly in the form of chemical ions. Stream turbulence allows clay and silt particles to be carried as suspended load. Highly visible soil particles that have been eroded from the extensively cultivated fields in the Midwest and wind up in the Mississippi have given the river the nickname "the Big Muddy." Sand, gravel, and cobbles are much larger than clay and silt particles and therefore move as bed load close to the channel floor. The suspended load is usually the greatest of the three forms of stream transport. For example, it is estimated that the Mississippi transports about 90 percent of its total load in suspended form. This high proportion is attributed to the semiarid area of the Missouri River basin, a major tributary of the Mississippi, which also includes the easily eroded Badlands of South Dakota.

Deposition, the third activity of streams, can occur on the streambed, in the floodplain, or on the bottom of a water body—such as an ocean—into which the stream empties. For example, the mouth of the Amazon is 161 kilometers wide as it flows into the Atlantic. The enormous amount of sediment that it carries forms large islands at the mouth or is simply deposited on the adjoining continental shelf.

Stream "capacity" is defined as the maximum amount of suspended load and bed load that can be transported in a stream at a specified discharge. An increase in discharge results in a substantial increase in suspended load, as the faster the water is moving, the greater the turbulence and consequent ability to keep material in suspension. The same effect occurs with bed load. As the stream velocity increases, the ability to move material along the bottom of the stream increases by three to four orders of magnitude as compared to the velocity. Thus, if the velocity of a stream is doubled in a flood situation, bed-load movement can increase by a factor of eight to sixteen times.

There is an enormous range in sediment load transport among the major rivers in the world. For example, the average annual sediment yield varies from a low of 4 metric tons per square kilometer for the Yenisey River in Siberia to a high of 2,600 metric tons per square kilometer for the Huang (also known as the Yellow) River in China. This enormous difference is caused by the large soil-erosion rate in the cultivated and easily eroded silt soils of the Huang River basin as compared to the mostly forested and uncultivated Yenisey River watershed. Another major example is the Mississippi River, where most of the annual sediment yield is attributed to the Missouri River, which flows through subhumid and semiarid grasslands, large portions of which have been cultivated.

Proper design of reservoirs should include estimates of the sediment load carried by an incoming stream. The sediment that gets trapped behind the dam of a reservoir will eventually eliminate its storage capacity. For example, several inland reservoirs that were used to supply water to Santa Barbara on the coast in California filled with sediment in only a few decades and had to be abandoned. Even major reservoirs in semiarid to arid regions, where vegetative cover on watershed slopes is limited, will experience reduced lifetimes as a consequence of continuing sediment input.

EXOTIC STREAMS

Ancient civilizations that developed in arid regions, such as Egypt or Mesopotamia in what is now Iraq, based their elaborate irrigation systems on streams that originated in upstream mountainous areas that have a water surplus. The Blue Nile originates in Lake Tana in the Ethiopian highlands, which receive heavy summer rains. The White Nile begins in Lake Victoria in east-central Africa, which experiences heavy precipitation throughout the year. The Blue and White Niles join at Khartoum in Sudan and flow north to Egypt through an arid region. The Tigris and Euphrates rivers start in the highlands of east-central Turkey, flow through Syria and Iraq, and empty into the Persian Gulf. Another classic example of an exotic stream is the Colorado River, which begins in the Rocky Mountains of Colorado and Wyoming and flows through Utah, New Mexico, Arizona, Nevada, California, and Mexico before emptying into the Gulf of California. Large portions of this basin, especially the lower part, receive limited precipitation in an area that has experienced considerable population growth and commensurate demand for water.

ANTHROPOGENIC EFFECTS ON RIVER SYSTEMS

It is well known that natural factors occur that may have substantial effects on streamflow. Some obvious examples are wet and dry periods that can occur during any year and other climatological events such as El Niño that may have return cycles of a decade or more. However, these are natural events that may easily be superseded by human intervention, as exemplified by such activities as extensive diversion for agricultural irrigation, interbasin transfers of large volumes of water for public water supply, and the development of a large system of dams.

Dams can have a major impact on streamflow, as the water can be released below a dam without regard for natural cycles; instead, the release may be ruled by a particular need for water or electricity on cycles that may change on an hourly basis. As a result, stream channels below a dam can experience sharp changes based on annual flow, flood peaks, and the sediment load in the stream. These changes can range from sand accumulation in one channel to vegetation moving into another section of the channel to heavy bank erosion in still another part of the stream channel.

Dam-building has a long history. For example, dams were built upstream of Cairo on the Nile by the Egyptians some five thousand years ago. The Chinese built a dam with a height of 27.4 meters on the Abang Xi River some twelve centuries ago that is still used for some irrigation water. Dams were built in Western Europe during the late Middle Ages to provide power for waterwheels. The Anasazi constructed small dams on the Mesa Verde in Colorado for irrigation some eight hundred years ago. The use of heavy machinery in the twentieth century led to an enormous expansion of dam construction, especially in the United States and Russia.

In recent years, a growing recognition has developed in the United States that certain dams should be considered for demolition because of problems with salmon runs; scouring of beds and banks below the dam, as the released waters are relatively clear because the sediment load has been left behind; and changes in the temperature of the water that may be undesirable for aquatic life. In stark contrast to the previous discussion of selected dam removal in the United States, China began construction of the Three Gorges Dam on the Chang in 1994. This enormous structure is close to 200 meters high, has a width of more than 2 kilometers, and creates a reservoir with a length of 600 kilometers. The purported benefits include electricity generation without burning coal, navigation improvements, and a potential decrease in flooding events. On the negative side are the large-scale movement of several million people who were living in the river valley; sediment reduction that could affect downstream areas, including the East China Sea; and a growth in the risk of new landslides and the possibility of increased geological instability in an area that is already seismically active.

—*Robert M. Hordon*

Further Reading

Allan, J. David, and María M. Castillo. *Stream Ecology: Structure and Function of Running Waters.* 2nd ed., Springer, 2007.
Arnell, Nigel. *Hydrology and Global Environmental Change.* Pearson, 2002.
Bedient, Philip B., Wayne C. Huber, and Baxter E. Vieux. *Hydrology and Floodplain Analysis.* 4th ed., Prentice Hall, 2008.
Cech, Thomas V. *Principles of Water Resources: History, Development, Management, and Policy.* 3rd ed., John Wiley, 2010.
Cushing, Colbert E., K. W. Cummins, and G. W. Minshall, editors. *River and Stream Ecosystems of the World.* Rev. ed., U of California P, 2006.
Gleick, Peter H., et al. *The World's Water, 2008-2009.* Island Press, 2009.
Glennon, Robert. *Water Follies: Groundwater Pumping and the Fate of America's Fresh Waters.* Island Press, 2002.
Haslam, S. M. *The Riverscape and the River.* Cambridge UP, 2008.
McCully, Patrick. *Silenced Rivers: The Ecology and Politics of Large Dams.* Enlarged and updated ed., Zed Books, 2001.
Powell, James L. *Dead Pool: Lake Powell, Global Warming, and the Future of Water in the West.* U of California P, 2008.
Spellman, Frank R. *The Science of Water: Concepts and Applications.* 2nd ed., CRC Press, 2008.
Strahler, Alan, and Arthur Strahler. *Introducing Physical Geography.* 4th ed., John Wiley, 2006.
Ward, Andy D., and Stanley W. Trimble. *Environmental Hydrology.* 2nd ed., CRC Press, 2004.

Structure and Physical Properties of Minerals

Fields of Study: Geology; Mineralogy; Geomorphology; Geochemistry; Gemology; Crystallography

ABSTRACT

Minerals—naturally occurring inorganic solids with definite chemical composition and definite crystal structure—are the primary constituents of rocks; they are also found in soil. The variety of minerals is huge, and their myriad applications range from use as gemstones and precious metals to applications in building materials, electronics, food, and pharmaceuticals.

KEY CONCEPTS

density: the ratio of weight to volume, expressed in units of grams per cubic centimeter

fullerenes: a recently recognized elemental form of carbon consisting of closed spheroid molecules constructed of fused five- and six-membered rings of carbon atoms; the other elemental forms of carbon are diamond and graphite

native element: an element that is found naturally in pure form, uncombined with other elements

piezoelectricity: an electrical current induced in a crystal by the application of pressure

pyroelectricity: an electrical current induced in a crystal by the application of heat

specific gravity: the ratio of the weight of a sample to the weight of an equal volume of pure water at 4°C; as a ratio of weight to weight, specific gravity has no units

MINERALS IN GENERAL

Minerals are the building blocks of rocks, and they have many economic uses. Minerals such as diamond, ruby, emerald, and sapphire are precious gems. Other minerals are valuable metals (gold, silver, platinum, copper) or metal ores, such as hematite (iron), sphalerite (zinc), galena (lead), and bauxite (aluminum). Other minerals are used as salt (halite), lubricants (graphite), abrasives (corundum), and fertilizer (apatite), as well as in pharmaceuticals (sulfur), steel making (fluorite), plaster (gypsum and anhydrite), lime, and Portland cement (calcite and dolomite).

A mineral is defined as a naturally occurring, inorganic solid with a definite chemical composition (or range of compositions within certain limits) that can be expressed by a chemical formula, and an orderly internal crystalline structure (its atoms are arranged in a definite pattern which is reflected in the shape of its crystals and in its cleavage). Only substances that meet these precise requirements are considered minerals. As a result, synthetic gems, which may be physically and chemically identical to natural gemstones, are not considered minerals.

Minerals have specific physical properties that result from their chemical composition and crystal structure, and many minerals can be identified by these properties. Physical properties include hardness, color, luster, streak, cleavage, density or specific gravity, and crystal form. Some minerals also have additional diagnostic physical properties, including tenacity, taste, magnetism, electrical properties, luminescence, reaction to hydrochloric acid, and radioactivity.

HARDNESS

Hardness is a mineral's resistance to scratching or abrasion and is a result of crystal structure or atomic arrangement. The stronger the chemical bonds between the atoms, the harder the mineral. For example, two minerals may have an identical chemical composition but different crystal structures, such as diamond and graphite, which are both carbons. Diamond is the hardest known mineral, but graphite is so soft that it rubs off on the fingers or a piece of paper (it is used as pencil "lead"). The differences in crystal structure produce the vastly different hardnesses of these two minerals.

Ten minerals have been arranged in order of in creasing hardness and are referred to as the Mohs hardness scale, devised in 1822 by German mineralogist, Friedrich Mohs. The minerals of the Mohs hardness scale, in order from softest to hardest, are: (1) talc, (2) gypsum, (3) calcite, (4) fluorite, (5) apatite, (6) potassium feldspar (orthoclase), (7) quartz, (8) topaz, (9) corundum, and (10) diamond. Minerals can scratch other minerals of the same or lesser hardness. The hardness of minerals can be tested using common materials, including the fingernail (a little over 2), copper (about 3), a steel nail or pocket knife (a little over 5), a piece of glass (about 5.5), and a steel file (6.5).

COLOR

Although color is a prominent feature of minerals, it is not a reliable indicator for identifying minerals. The color of some minerals is the result of major elements in their chemical formula, such as the blue color of azurite and the green color of malachite (copper), the pink color of rhodonite and rhodochrosite (manganese), and the yellow color of sulfur. Many minerals come in a variety of colors. Quartz is colorless and transparent when pure, but it may also be white (milky quartz), pink (rose quartz), purple (amethyst), yellow (citrine), brown (smoky quartz), or other colors. Similarly, feldspar and fluorite come in many hues. Color may be caused by impurities, such as iron (pink, green, or greenish yellow), titanium (pink or blue), chromium (red or green), vanadium (green), and nickel (yellow). Milky quartz is white because it contains tiny fluid inclusions. Coloration can be the result of defects in the crystal structure; for example, the purple of amethyst and fluorite and the brown of smoky quartz. Unusual colors may also be induced in minerals by exposing them to radiation, which damages the crystal structure (such as black quartz). Minerals that have undergone irradiation naturally due to the presence of radioactive elements in their parent rock are described as having undergone metamictization.

LUSTER

Luster refers to the "shine," or quality of reflectivity of light from the mineral's surface. Minerals can be divided into two luster groups: metallic luster and nonmetallic luster. Metallic minerals include economically valuable metals such as gold, silver, and native copper, and some metal sulfides such as pyrite (FeS2 or iron sulfide) and galena (PbS or lead sulfide). Nonmetallic minerals include those with vitreous or glassy luster (quartz), earthy or dull luster (kaolinite and other clays), pearly (talc), silky (fibrous minerals such as gypsum, malachite, and chrysotile asbestos), greasy (nepheline), resinous (resembling resin or amber, such as sulfur), and adamantine or brilliant (diamond).

STREAK

Streak refers to the color of the mineral in powdered form, viewed after the mineral is rubbed on an unglazed porcelain tile or streak plate. Streak color is more diagnostic than mineral color because it is constant for a particular mineral. A mineral may come in several colors, but its streak is the same color for all. Streak color is not always what one might predict from examining the mineral; a sparkling silver-colored mineral, specular hematite, has a red-brown streak, and pyrite, a golden metallic mineral, has a dark gray streak. Not all minerals have a streak. The streak plate has a hardness of about 7 on the Mohs hardness scale. Minerals harder than this will not leave a streak, but their powdered colors can be studied by crushing a small piece.

CLEAVAGE

Cleavage is one of the most diagnostic physical properties of minerals. Cleavage refers to the tendency of some minerals to break along smooth, flat planes that are related to zones of weak bonding between atoms in the crystal structure. Some minerals, however, have no planes of weakness in their crystal structure and therefore lack cleavage. Cleavage is discussed by referring to the number of different sets of planes of breakage and the angles between them. Minerals that have a prominent flat, sheetlike cleavage (such as the micas, muscovite and biotite) have one direction of cleavage, or perfect basal cleavage. This sheetlike cleavage makes muscovite economically valuable; it was once used in window-making material and is still used in some stove windows and in electrical insulation.

Feldspar and pyroxene have two directions of cleavage at right angles to each other, and the amphiboles (hornblende and others) have two directions of cleavage at approximately 60° and 120° to each other. Other minerals have three directions of cleavage. Halite (table salt) and galena have cubic cleavage (three directions of cleavage at right angles to one another) and break into cubes. Calcite has rhombohedral cleavage (three directions of cleavage not at right angles; the angle is about 74°) and breaks into rhombohedrons. Fluorite has four directions of cleavage and breaks into octahedrons with triangular faces. Sphalerite has six directions of cleavage.

FRACTURE

Irregular breakage in minerals without planes of weak bonding is fracture. There are several types of fracture. Many minerals have uneven or irregular fracture. Conchoidal fracture is characterized by smooth, curved breakage surfaces, commonly marked by fine parallel lines resembling the surface of a shell (seen in quartz, obsidian, and glass). Rocks and minerals with conchoidal fracture such as flint, chert and obsidian were used by Indigenous Peoples for arrowheads. Hackly fracture is jagged with sharp edges and is characteristic of metals such as copper. Fibrous or splintery fracture occurs in asbestos and sometimes gypsum. Earthy fracture occurs in clay minerals such as kaolinite.

DENSITY AND SPECIFIC GRAVITY

Density is defined as mass per unit volume, or how heavy a material is for its size. Specific gravity (or relative density) is commonly used when referring to minerals. Specific gravity expresses the ratio between the weight of a mineral and the weight of an equal volume of water at 4° Celsius. The terms density and specific gravity are sometimes used interchangeably, but density requires the inclusion of units of measure, whereas specific gravity is unitless. Quartz has a specific gravity of 2.65. Barite has a specific gravity of 4.5 (heavy for a nonmetallic mineral), which makes it economically valuable for use in oil and gas well drilling. Metals have higher specific gravity than nonmetals, for example, galena (7.4 to 7.6), and gold (15.0 to 19.3). The high specific gravity of gold allows it to be separated from less dense minerals by panning.

TENACITY, TASTE, AND MAGNETISM

Tenacity is the resistance of a mineral to bending, breaking, crushing, or tearing. Minerals may be brittle (break or powder easily), malleable (can be hammered into thin sheets), ductile (can be drawn into a thin wire), sectile (can be cut into thin slivers with a knife), elastic (bend but return to their original form), or flexible (bend and stay bent). Metallic minerals are commonly malleable and ductile (gold, copper). Copper is used for electrical wire because of its ductility.

Some minerals can be identified by taste. Taste is a property of halite (NaCl), used as table salt. Sylvite (KCl, or potassium chloride) has a bitter salty taste and is used as a salt substitute for people with high blood pressure because it does not contain sodium.

Magnetism is a property that causes certain minerals to be attracted to a magnet. Magnetite (Fe_3O_4) and pyrrhotite ($Fe_{1-x}S$) are the only common magnetic minerals. Lodestone, a variety of magnetite, acts as a natural magnet. In the presence of a powerful magnetic field, some other iron-bearing minerals become magnetic (garnet, biotite, and tourmaline), whereas other minerals are repelled by the magnet (gypsum, halite, and quartz). Electromagnetic separators are used to separate minerals with different magnetic susceptibilities.

ELECTRICAL PROPERTIES

Some minerals have electrical properties. Piezoelectricity occurs when pressure is exerted in a particular direction in a mineral (along its polar axis), causing a flow of electrons or electrical current. Piezoelectricity was first detected in quartz in 1881, and it has since been used in a number of applications ranging from submarine detection to keeping time (in quartz watches). When subjected to an alternating electrical current, quartz is mechanically deformed and vibrates; radio frequencies are controlled by the frequency of vibration of the quartz.

Pyroelectricity is caused when temperature changes in a mineral create uneven thermal expansion and deformation. Tourmaline and quartz are pyroelectric.

LUMINESCENCE

Luminescence is emission of light from a mineral. Minerals that luminesce or glow during exposure to ultraviolet light, X-rays, or cathode rays are fluorescent. If the luminescence continues after the radiation source is turned off, the mineral is phosphorescent. The glow results from impurities in the mineral absorbing invisible, short wavelength radiation and then reemitting radiation at longer wavelengths (visible light). Minerals vary in their ability to absorb different wavelengths of ultraviolet (UV) light. Some fluoresce only in short wavelength UV, some fluoresce only in long wavelength UV, and some fluoresce in both types. Fluorescence is unpredictable; not all minerals of a given type fluoresce. Minerals that commonly fluoresce include fluorite, calcite, diamond, scheelite, willemite, hyalite, autunite, and scapolite. Fluorescence has some practical applications in prospecting and mining. Synthetic phosphorescent materials have also been developed for commercial uses.

Some minerals emit light when heated, a property called "thermoluminescence." Thermoluminescent minerals include fluorite, calcite, apatite, scapolite, lepidolite, and feldspar. Minerals that luminesce when crushed, scratched, or rubbed are triboluminescent. This is a property of fluorite, sphalerite, and lepidolite, and less commonly of pectolite, amblygonite, feldspar, and calcite.

REACTION TO HYDROCHLORIC ACID

Calcite ($CaCO_3$) and other carbonate minerals effervesce or fizz in hydrochloric acid, but some will not react unless the acid is heated or the mineral is powdered. Bubbles of carbon dioxide (CO_2) gas are released, and the reaction proceeds as follows:

$$CaCO_3 + 2\ HCI \rightarrow CaCl_2 + H_2O + CO_2\ \text{(gas)}$$

RADIOACTIVITY

Radioactive minerals contain unstable elements that alter spontaneously to other kinds of elements, releasing subatomic particles and energy. Radioactivity can be detected using Geiger-Müller counters, ionization chambers, scintillometers, and similar instruments. Some elements have several different isotopes, differing by the number of neutrons in the nucleus. Radioactive isotopes include uranium-235 (^{235}U), uranium-238 (^{238}U), and thorium-232 (^{232}Th). Uranium-235 is the primary fuel for nuclear power plants. Radioactive minerals include uraninite (pitchblende), carnotite, uranophane, and thorianite. Radioactive minerals occur in granites and granite pegmatites, in sandstones, and in black organic-rich shales, and are used for nuclear energy, nuclear weapons, coloring glass and porcelain; in photography; and as a chemical reagent. Radioactivity is also used in radiometric dating to determine the ages of rocks and minerals.

CLASSIFICATION OF MINERALS

Minerals have been classified or grouped in several ways, but classification based on chemical composition is the most widely used. Minerals are grouped into the following twelve categories on the basis of their chemical formulas: native elements, oxides and hydroxides, sulfides, sulfosalts, sulfates, halides, carbonates, nitrates, borates, phosphates, tungstates, and silicates.

NATIVE ELEMENTS

Native elements are minerals composed of a single element that is not combined with other elements. About twenty native elements occur (not including atmospheric gases), and they are divided into metals, semimetals, and nonmetals. The native metals include gold (Au), silver (Ag), copper (Cu), iron (Fe), platinum (Pt) and others. They share the physical properties of malleability, hackly fracture, and high specific gravity, along with metallic luster. Their atoms are held together by weak metallic bonds. They are excellent conductors of heat and electricity and have fairly low melting points. The native semimetals include arsenic (As), bismuth (Bi), antimony (Sb), tellurium (Te), and selenium (Se). They are brittle and much poorer conductors of heat. These properties result from bonding intermediate between true metallic and covalent. The native nonmetals include sulfur (S) and three forms of carbon (C), diamond, graphite and the fullerenes. These minerals have little in common, but they are distinctive and easily identified.

Diamond, graphite and the fullerenes are polymorphs, a term meaning "many forms." Their chemical composition is identical as they consist only of carbon atoms, but they have different crystal structures. Diamond has a tight, strongly bonded structure in which each individual carbon atom is bonded to four other carbon atoms in a tetrahedral arrangement, whereas graphite has a loose, open structure consisting of sheets of carbon atoms in fused six-membered ring structures. The fullerenes are a unique class of compounds with closed spheroid structures formed of carbon atoms in fused five-and six-membered ring structures. While only relatively recently recognized as compounds and as an elemental form of carbon by the chemical community, the fullerenes have existed and occurred naturally essentially for as long as carbon-based materials have produced soot when they burn.

OXIDES AND HYDROXIDES

Chemically, the oxide and hydroxide minerals consist of metal ions (of either one or two types of metals) combined with oxygen in various ratios, such as Al_2O_3 (corundum) or $MgAl_2O_4$ (spinel), or metals combined with oxygen and hydrogen, such as $Mg(OH)_2$ (brucite) or $HFeO_2$ (goethite). The oxides and hydroxides are a diverse group with few properties in common. Several minerals of great economic importance occur in this group, including the chief ores of iron (magnetite, Fe_3O_4, and hematite, Fe_2O_3), chromium (chromite), manganese (pyrolusite, manganite, psilomelane), tin (cassiterite), and aluminum (bauxite). Some minerals in this group form from molten rock or hydrothermal (hot water) solutions, but others form on or near the surface of Earth as a result of weathering and may contain water.

SULFIDES

Chemically, the sulfides consist of a metal ion combined with sulfur. They are an economically important class of minerals that includes numerous ore minerals. Many of the sulfides are metallic, with high specific gravity, and most are fairly soft. They tend to be brittle, and they have distinctive streak colors. Many sulfides have ionic bonding, but others have metallic bonding, at least in part. Sphalerite has covalent bonding.

Among the sulfides are ores of lead (galena, PbS), zinc (sphalerite, ZnS), copper (chalcocite, Cu_2S; bornite, Cu_5FeS_4; and chalcopyrite, $CuFeS_2$), silver (argentite, Ag_2S), mercury (cinnabar, HgS), and molybdenum (molybdenite, MoS_2), as well as pyrite (FeS_2), used to manufacture sulfuric acid.

SULFOSALTS

The sulfosalts are a type of unoxidized sulfur mineral. They consist of a metal and a semimetal combined with sulfur. There are nearly one hundred sulfosalts, including arsenopyrite (FeAsS), tetrahedrite ($Cu_{12}Sb_4S_{13}$), and pyrargyrite (Ag_3SbS_2). Some are useful as ore minerals.

SULFATES

The sulfates consist of metal plus a sulfate (SO_4^{2-}) group. The sulfates are typically soft, and some are translucent or transparent. They include both anhydrous (without water) and hydrous (water-bearing) sulfate minerals. Anhydrous sulfates include barite ($BaSO_4$), anhydrite ($CaSO_4$), celestite ($SrSO_4$), and anglesite ($PbSO_4$). The hydrous sulfates include gypsum ($CaSO_4 \cdot 2H_2O$) and epsomite ($MgSO_4 \cdot 7H_2O$). The structure of gypsum consists of sheets or layers of calcium and sulfate ions separated by water molecules. Loss of water molecules causes the structure of the mineral to collapse into anhydrite, with a decrease in volume and loss of cleavage. The most common sulfate, gypsum is used in the production of plaster of paris, drywall, soil conditioner, and Portland cement.

HALIDES

The halides contain negatively charged halogen ions (chlorine, fluorine, bromine, and iodine), bonded ionically to positively charged ions (such as sodium, potassium, calcium, mercury, and silver). Many have symmetrical crystal structures resulting in cubic cleavage (halite, NaCl, and sylvite, KCl) or octahedral cleavage (fluorite, CaF_2). Many of the halides are water-soluble salts (such as halite and sylvite), and may form from the evaporation of water. Many are transparent or translucent. All are fairly soft and are light in color when fresh. Some of the silver and mercury halides will darken in color on exposure to light, hence their use in photography.

CARBONATES

Carbonate minerals contain the carbonate ion, CO_3^{2-}. Carbonate minerals are readily identified by their effervescence in hydrochloric acid, although for some carbonates, the acid must be hot or the mineral must be powdered to obtain the reaction. Some carbonates (such as cerussite, $PbCO_3$) react to nitric acid. Carbonates include calcite and aragonite ($CaCO_3$), dolomite ($CaMg(CO_3)_2$), magnesite ($MgCO_3$), and siderite ($FeCO_3$). The colorful malachite (green), azurite (blue), and rhodochrosite (pink) are also carbonates. Most carbonates are fairly soft, and rhombohedral cleavage is common.

NITRATES

The nitrate minerals contain the nitrate ion, NO_3^-. Most nitrates are water soluble and are fairly soft. They are light in color, and some are transparent. The nitrates include soda niter ($NaNO_3$), which is found in desert regions and used in explosives and fertilizer, and niter or saltpeter (KNO_3), which forms as a coating on the walls of caves and is used as a fertilizer.

BORATES

The borates contain boron bonded to oxygen and associated with sodium or calcium, with or without water. Some borates form in igneous deposits, but most are found in dry lake beds in arid areas. Among the borates are borax, kernite, and ulexite. Borax is used for washing, as an antiseptic and preservative, in medicine, and in industrial and laboratory applications.

PHOSPHATES

The phosphate minerals contain the PO_4^{3-} group, bonded to positively charged ions such as calcium, lithium, iron, manganese, lead, and iron, with or without water. Apatite ($Ca_5(F,Cl,OH)(PO_4)_3$) is the most important and abundant phosphate mineral. It is the primary constituent of bone and is used for fertilizer. Turquoise is a phosphate mineral.

TUNGSTATES

The tungstates contain tungsten (chemical symbol W). Tungstates form a small group of minerals that include wolframite and scheelite (which is fluorescent); both are ores of tungsten.

SILICATES

The silicates are the largest group of minerals, and they include the major rock-forming minerals of Earth's crust, feldspar and quartz, as well as olivine, pyroxene, amphibole, and the micas. Most are fairly hard, with glassy luster and low to moderate specific gravity; they crush to a light-colored powder. Silicates consist of silicon and oxygen, generally accompanied by other ions such as aluminum, potassium, calcium, sodium, iron, and magnesium. Silicate structure is based on the silicate tetrahedron, which consists of four oxygen atoms arranged tetrahedrally around one silicon atom. These tetrahedra are arranged in several characteristic patterns that allow the silicates to be classified into a number of groups, including isolated tetrahedra (neosilicates), pairs of tetrahedra (sorosilicates), rings of tetrahedra (cyclosilicates), single and double chains of tetrahedra (inosilicates), sheets of tetrahedra (phyllosilicates), and three-dimensional frameworks of silicate tetrahedra (tectosilicates).

Neosilicates tend to be compact and hard, with fairly high specific gravity. Olivine, garnet, zircon, topaz, staurolite, and kyanite are neosilicates. Sorosilicates (or "sister" silicates) include the minerals epidote, prehnite, and hemimorphite. Cyclosilicates are characterized by prismatic, trigonal, tetrahedral, or hexagonal habits. Beryl has rings of six silicate tetrahedra, reflected in its hexagonal (six-sided) crystals. Tourmaline and chrysocolla are also in this group.

Inosilicates (single-chain and double-chain silicates) tend to be fibrous or elongated, with two directions of cleavage parallel to the elongation. They include pyroxenes (including hypersthene, augite, and diopside), pyroxenoids (including wollastonite), and amphiboles (hornblende, tremolite, actinolite, and others).

Phyllosilicates, or sheet silicates, have one prominent direction of cleavage and tend to have a platey or flaky appearance. They are generally soft, have low specific gravity, and may have flexible or elastic sheets. The micas (muscovite, biotite, lepidolite, and phlogopite), and the clay minerals (kaolinite, illite) belong to this group, as do talc, serpentine, chlorite, and others.

Earth's crust is dominated by tectosilicates, or framework silicates. This is the group that contains feldspar and quartz, the two most abundant minerals in Earth's crust. Quartz is chemically the simplest silicate, with the chemical formula SiO_2. Feldspar is a group of minerals, including orthoclase and microcline (two different crystal structures with the formula $KAlSi_3O_8$) and plagioclase (a solid solution series which ranges in composition from $NaAlSi_3O_8$ to $CaAl_2Si_2O_8$). Tectosilicates tend to be of low density and compact habit. Feldspathoids (including nepheline and sodalite) and zeolites (analcime and others) are also in this group.

—Pamela J. W. Gore

Further Reading

Haldar, Swapan Kumar. *Introduction to Mineralogy and Petrology*. 2nd ed., Elsevier, 2020.

Klein, Cornelis, and Barbara Dutrow. *The Twenty-Third Edition of the Manual of Mineral Science*. Wiley, 2008.

Nesse, William D. *Introduction to Mineralogy*. 3rd ed., Oxford UP, 2017.

Perkins, Dexter. *Mineralogy*. 2nd ed., Prentice Hall, 2002.

Wenk, Hans-Rudolf, and Andrei Bulakh. *Minerals: Their Constitution and Origin*. Cambridge UP, 2004.

Zim, Herbert S., and Paul R. Shaffer. *Rocks, Gems, and Minerals: A Guide to Familiar Minerals, Gems, Ores, and Rocks*. Rev. and updated ed., St. Martin's Press, 2001.

SUBDUCTION AND OROGENY

Fields of Study: Geology; Seismology; Geophysics

ABSTRACT

Subduction and orogeny are fundamental consequences of plate tectonics and are the two processes that build mountains on the edges of continents. Through the recognition of subduction, scientists have been better able to determine regions where risks of earthquakes and volcanic explosions are significant.

KEY CONCEPTS

continental margin: the edge of a continent that is both exposed on land and submerged below the water that marks the transition to the ocean basin

crust: the outermost layer of the earth, which consists of materials that are relatively light; the continental crust is lighter than oceanic crust, which allows it to float while oceanic crust sinks

faulting: the process of fracturing the earth such that rocks on opposite sides of the fracture move relative to one another; faults are the structures produced during the process

folding: the process of bending initially horizontal layers of rock so that they dip; folds are the features produced by folding and can be as small as millimeters and as big as kilometers long

geosynclines: major depressions in the surface of the earth where sediments accumulate; geosynclines lie parallel to the edges of continents and are long and narrow

intrusion: the process of forcing a body of molten rock generally derived from depths of tens of kilometers in the earth into solidified rock at the surface

magma: molten rock that is the source for volcanic eruptions

orogeny: mountain building by tectonic forces through the folding and faulting of rock layers

THEORY OF GEOSYNCLINES

Subduction and orogeny are two processes that are fundamental to the evolution of continents. All continents contain long, narrow mountain chains near their edges that are composed of folded and faulted rocks that are younger than the rocks in the continental interiors. The event that formed the mountains is termed an "orogeny," and the mountain chain itself is called an "orogenic belt." Because of the proximity of mountain chains to the edges of continents, scientists have believed for centuries that orogenies reflected movements localized along continental margins. It is only recently, however, that orogeny has been coupled with subduction, the process in which seafloor descends below a continent or another piece of seafloor. Earlier views of orogeny were part of the theory of geosynclines.

Geosynclines are linear basins that form on subsiding regions of Earth's surface adjacent to continental margins, fill with sediment, and evolve into mountains composed of folded and faulted sedimentary strata. The origin of the compressive forces responsible for the creation of the mountains was not known. Erosion of the newly created mountains provides sediment for new geosynclines that develop seaward of the mountain belt, thereby completing one geosynclinal cycle, which typically lasts on the order of a few hundred million years. The advent of plate tectonics theory in the 1960s led the majority of the scientific community to abandon the geosynclinal cycle in favor of the subduction process as an explanation for orogenies. Subduction was attractive because it readily provided a mechanism by which the large compressive forces needed to form mountains could be produced.

Orogenic belts are characterized by the folding and faulting of layers of rock, by the intrusion of magma, and by volcanism. Folds and faults form parallel to the continental margin and extend hundreds of kilometers toward the continental interior. Folding bends layers of rocks, whereas faulting takes rocks that were side by side and stacks them on top of each other in sheets up to 20 kilometers thick. Both processes significantly shorten the horizontal and thicken the vertical dimensions of the continents. At the same time as they are folded and faulted, the rocks are intruded by magmas derived from tens of kilometers below the surface. Some of the magmas eventually erupt, building volcanoes on the deformed rocks. An additional feature of orogenic belts is the juxtaposition of sequences of rock that have nothing in common with each other. The rocks in the two sequences may be different in age, composition, or style of folding. The origin of this juxtaposition was unreconciled by the theory of geosynclines, which holds that all rocks in a mountain belt were originally deposited near one another and were derived from the same source. The theory of plate tectonics, however, easily explains the juxtaposition.

THEORY OF PLATE TECTONICS

To understand subduction and orogeny, one must have a clear grasp of the theory of plate tectonics. This theory states that the surface of Earth is composed of about twelve rigid plates which are less than 100 kilometers thick. Plates are either oceanic or continental. Below the plates is a partially molten layer that allows the rigid plates to float and move relative to each other at speeds between 2 and 10 centimeters per year. The motions are defined primarily by the oceanic plates; the continental plates drift passively.

The relative motions of the plates define three types of boundaries: convergent, where plates move toward one another; divergent, where plates spread apart; and transcurrent, or conservative, where plates slide smoothly past each other. Convergent boundaries are frequently along the margins of continents, and divergent boundaries are commonly in the ocean basins. For example, the west coast of South America is a convergent boundary, and the Mid-Atlantic Ridge—the mountain range that runs down the middle of the Atlantic Ocean—is a divergent boundary. Divergent boundaries are zones along which two plates separate. This type of plate boundary is typically demarcated by a linear ridge system in an ocean basin where magma rises from deep within Earth to fill the gap created by the diverging plates. When the hot magma contacts the cold seawater, it solidifies into new oceanic crust. As the plates continue to separate, additional magma wells up from Earth's interior, allowing the continuous creation of oceanic crust at the ridge. This process is known as seafloor spreading and is responsible for the drifting of the continents on the surface of Earth.

Convergent boundaries are where two plates move toward each other and one plate subducts, or descends below, the other. The subducting plate is always oceanic, but the overriding plate may be either oceanic or continental. This reflects the greater density of oceanic crust relative to continental crust, which allows the oceanic plates to sink readily into Earth's interior, whereas the continental plates remain afloat. When two continental plates collide, neither plate subducts—they are of equal densities—but the plates push against each other with tremendous force such that their edges buckle and huge mountain ranges grow. This process built the world's tallest mountains, the Himalayas, which are the result of the collision between the subcontinent of India and the continent of Asia.

SUBDUCTION ZONES

Subduction zones are characterized by a progression from the subducting to the overriding plate of deep trenches, high mountains, and many volcanoes that occupy an area hundreds of kilometers wide and thousands of kilometers long. The deep trench, frequently filled with sediments eroded from the adjacent mountains, marks the point in the ocean floor where the subducting plate bends to descend below the overriding plate. As the oceanic plate descends, these sediments are scraped onto the overriding plate. Slivers of oceanic crust may also scrape off and mix with the sediments. The off-scraped rocks form an intricately folded and faulted region tens of kilometers wide and several kilometers high at the edge of the overriding plate. These complexly deformed mixtures of sediments and slivers of oceanic crust are called "mélanges" and are characteristic of most ancient subduction zones now exposed on land.

Another important feature of subduction zones is the linear belt of volcanoes on the overriding plate that parallels the plate boundary. The volcanoes grow from the eruption of magma that is generated at the interface between the subducting and overriding plates at depths between 100 and 200 kilometers. At these depths, the temperature inside Earth is high enough to melt small areas of either the subducting or the overriding plate. The magma rises, intruding the rocks at the surface and eventually erupting to build the volcanic belt. Some of the magma, however, may solidify between the top of the oceanic plate and the surface.

The similarity of features in orogenic belts and subduction zones is striking and forces the obvious conclusion that subduction leads directly to orogeny. An orogeny can occur either during subduction of an oceanic plate below a continental plate, such as on the west coast of South America, or during the collision of two continental plates, such as in the Himalaya. Because continents do not subduct, the compressive forces are much greater in a continent-continent collision than in seafloor subduction. The mountains produced during collision (Himalaya), therefore, are much taller than those generated during subduction (Andes).

CONSEQUENCES OF PLATE TECTONICS

The theory of plate tectonics elucidates important differences between the oceans and the continents and provides a mechanism by which different rock sequences can be juxtaposed in orogenic belts. The ocean basins are transient features that are constantly modified by the growth and destruction of new seafloor at divergent and convergent boundaries, respectively. In contrast, the continents are too light to be subducted and are permanent features of Earth's surface. This consequence of plate tectonics is supported by the 200 million year age of the oldest seafloor and the 4 billion year age of the most ancient rocks on the continents. Continents, therefore, drift, fragment, and collide as relative plate motions change through geologic time. The collision of continents that were once widely separated allows the bringing together of rocks that have had very different histories. As the collision leads to orogeny, these different sequences of rock may be juxtaposed in the same mountain belt.

The difference between the age of orogenic belts and the interiors of continents implies that the continents have evolved through time by the addition of material at their edges during orogenies. Orogenic belts are also of different ages, ranging from a billion years to zero (actively forming). Two or three belts whose ages decrease away from the continental interior may define one edge of a continent. This suggests that orogenies have occurred repeatedly through geologic time and that continents have added material continuously to their margins since the formation of their interiors. Because the ocean floor is so young, orogenic belts are the only record of subduction and collision events prior to 200 million years ago. If subduction is the only mechanism responsible for orogeny, plate tectonics must have been active since early in the history of the earth.

ANALYSIS OF EARTHQUAKES

Subduction and orogeny are studied by hundreds of scientists, each of whom looks at only a small part of the picture. One may determine the composition of volcanic rocks that are characteristic of subduction zones; another may examine the styles of folds and faults in orogenic belts. Three techniques, however, are dominant in the study of subduction and orogeny: the analysis of the locations and sizes of earthquakes, the discrimination of relationships between different types of rocks in the field, and the investigation of features in deep-sea trenches and in the submerged region of folded and faulted rocks. The first defines where subduction and orogeny occur today, whereas the second determines what the physiographic expressions of these processes are, how they are preserved in the rocks, and where they were active in the past. The third technique provides a direct link between subduction and orogeny and illustrates the early stages of development of a mountain belt.

One of the most important discoveries of plate tectonics was that earthquake zones define plate boundaries. Earthquakes occur when a fracture, or fault, forms in Earth's crust, and the two pieces on either side of the fault move, or slip, past each other. For large earthquakes, the slip is on the order of 10 to 20 meters. The forces responsible for faulting are simply the result of the relative motions of the plates at the plate boundaries. The motion can accumulate in the rocks for hundreds of years prior to causing a rupture. When the crust finally breaks, the energy stored by the rocks is released suddenly as waves that travel through Earth and generate the intense vibrations associated with an earthquake. The rupture continues for as much as 1,000 kilometers and moves at speeds in excess of 10,000 kilometers per hour.

The energy carried by the waves is recorded on seismographs, which are instruments that monitor ground motion. Seismographs are composed of a mass attached to a pendulum. The mass remains still during an earthquake, measuring the amount the earth moves around it. The motion is recorded on a chart as a series of sharp peaks and valleys that deviate from the background value measured during times of no earthquake activity. The arrival of the waves at different times at different places allows the geophysicist to calculate the location, or epicenter, of the earthquake. The amount of the deviation of the peaks and valleys from the background noise is an estimate of the magnitude of the earthquake.

Earthquakes near mountain belts define zones that extend at an angle from the surface of Earth at the deep-sea trench to depths of hundreds of kilometers below the continents. This zone corresponds to the subducting plate at a convergent boundary. As a result, the locations of subduction zones that are currently active are very well known. The descent of a subducting plate below an overriding continent has triggered some of the deepest and largest earthquakes ever recorded. Continued motion of the plate and rupturing of Earth's crust in response translate into mountain ranges on the earth's surface.

STUDY OF ROCKS AND DEEP-SEA TRENCHES

Analysis of earthquakes is essential to evaluate the modern plate tectonic setting of Earth, but it reveals nothing about the geologic past. Information about plate tectonics of the past must be obtained from looking at ancient mountain belts. Recognition of relationships among rocks in the field involves determining the ages, compositions, and histories of the rocks. This process led to the discovery that mountain belts on different continents contained rock sequences that were very similar. For example, rocks in the Appalachian Mountains of the east coast of North America were found to match closely those in the Atlas Mountains of the west coast of Africa. Conversely, recognition of relationships determined that dissimilar rock sequences frequently are adjacent to each other in the same orogenic belt. Both phenomena are most readily explained by continental drift, seafloor spreading, and subduction.

The critical link between ancient orogenic belts and modern subduction zones identified by earthquake activity was provided by deep-sea trenches. Using highly sophisticated techniques to "see" the ocean floor, scientists discovered the region of offscraped rocks that lies on the overriding plate in a subduction zone. These regions sometimes continue to the continental margin, where they are exposed on land as mountains. Thus, subduction was observed to cause folding and faulting in rocks and to build mountains—both important processes in orogenic belts.

EARTHQUAKE AND VOLCANO HAZARD ASSESSMENT

The theory of plate tectonics provides scientists with a process that can be observed—subduction—to explain the origin of mountains. Because young mountain chains are the locus of most of the large earthquakes that occur today, understanding subduction yields insight into the potential for destructive earthquakes in any given area. This is extremely important because most of the global population lives along convergent plate boundaries. The identification of subduction zones at the margins of the Pacific Ocean has explained the "Ring of Fire," a region of abundant earthquakes and volcanoes that had long puzzled the scientific community. Restriction of most earthquakes to plate boundaries allows the assessment of earthquake hazards anywhere in the world if the locations of plate boundaries are known. For example, the city of Santiago in Chile, which is above a subduction zone, has a high risk, whereas the city of Chicago in the United States, which is in the continental interior, has a low risk.

Additional information can also be gathered about the type of earthquakes that may occur. In subduction zones, the piece of the crust that is above the rupture typically moves upward relative to the piece below, which generates waves that shake the ground in certain directions. At divergent plate boundaries, however, the piece of crust that is above the rupture moves downward relative to the piece below. This motion produces waves that vibrate the ground in directions different from those generated by earthquakes in subduction zones. Additional differences between convergent and divergent boundaries that may affect ground motion include the depth and size of the earthquakes. Subduction zones generate the deepest and largest earthquakes; earthquakes at divergent plate boundaries are more frequent, smaller, and shallower. Knowledge of the way the ground may move helps civil engineers to design and construct buildings able to withstand large earthquakes.

Eruptions of volcanoes that lie above subduction zones can be devastating. These volcanoes typically erupt violently and explosively in contrast to volcanoes near midocean ridges, which erupt quietly and smoothly. This reflects the greater viscosity (resistance to flow) of magmas at convergent boundaries relative to those at divergent boundaries. Because of their greater viscosity, the magmas above subduction zones tend to plug the volcanoes at the surface, preventing any eruptions. Finally, when the pressure below the plug is great enough, the volcano erupts with such force that cities nearby are damaged considerably. For example, in 79 CE, the entire city of Pompeii, Italy, was destroyed, and hundreds of people were killed by the volcano Vesuvius. Clearly, the investigation of subduction and orogeny is beneficial to understanding the forces of nature that are harmful to humankind. Perhaps someday in the future, large earthquakes and violent volcanic eruptions may be predicted far enough in advance that precautions can be taken to prevent the loss of human life.

—Pamela Jansma

Further Reading

De Celles, Peter G., Mihai N. Ducea, Barbara Carrapa, and Paul A. Kapp, editors. *Geodynamics of a Cordilleran Orogenic System: The Central Andes of Argentina and Northern Chile.* The Geological Society of America, 2015.

Folguera, Andrés, et al., editors. *The Evolution of the Chilean-Argentinian Andes.* Springer, 2018.

Kearey, Philip, Keith A. Klepeis, and Frederick J. Vine. *Global Tectonics.* 3rd ed., Wiley-Blackwell, 2009.

Lowrie, William. *Fundamentals of Geophysics.* 2nd ed., Cambridge UP, 2007.

Oncken, Onno, et al., editors. *The Andes: Active Subduction Orogeny (Frontiers in Earth Sciences).* Springer-Verlag, 2006.

Tatsumi, Y. "The Subduction Factory: How It Operates on Earth." *GSA Today,* vol. 15, 2005, pp. 4-10.

SUB-SEAFLOOR METAMORPHISM

Fields of Study: Geology; Geochemistry; Geophysics; Oceanography

ABSTRACT

Sub-seafloor metamorphism of oceanic ridge basalts by magma-driven, convecting seawater induces significant changes in the chemical composition of both rock and circulating fluid. These changes are a major factor in the exchange of elements between the lithosphere and the hydrosphere and play a critical role in the origin of certain ore deposits.

KEY CONCEPTS

black smokers: active hydrothermal vents along seafloor ridges, which discharge acidic solutions of high temperature, volume, and velocity, charged with tiny black particles of metallic sulfide minerals

convection: fluid circulation produced by gravity acting on density differences arising from unequal temperatures within a fluid; the principal means of heat transfer involving fluids of low thermal conductivity

facies: a part of a rock or a rock group that differs from the whole formation in one or more properties—for example, composition

metasomatism: chemical changes in rock composition that accompany metamorphism

ophiolite complex: an assemblage of metamorphosed basaltic and ultramafic igneous rocks that originate at marine ridges and are subsequently emplaced in mobile belts by plate-collision tectonics

pillow basalt: a submarine basaltic lava flow in which small cylindrical tongues of lava break through the surface, separate into pods, and accumulate downslope in a formation resembling a pile of sandbags

regional metamorphic facies: the particular pressure and temperature conditions prevailing during metamorphism as recorded by the appearance of a new mineral assemblage

thermal gradient/geothermal gradient: the rate of temperature increase with depth below the earth's surface

water-to-rock ratio: the mass of free water in a given volume of rock divided by the rock mass of the same volume; processes occurring at water-to-rock ratios less than 50 are "rock-dominated," while those greater than 50 are "fluid-dominated"

IDENTIFICATION OF THE SUB-SEAFLOOR METAMORPHISM PROCESS

Seafloor spreading generates an estimated 3×10^{13} kilograms of new basaltic crust per year along the world's ocean ridge systems. Magma rising from the mantle produces this young ocean-floor crust (termed "juvenile"), which consists of upper layers of permeable lavas and lower layers of dike networks cutting gabbroic intrusive bodies. The thermal gradients and permeabilities of rocks in the vicinity of ocean ridges are both high, resulting in an environment favorable for the rapid circulation of seawater and convective cooling of rocks. Intensive studies of heat-flow patterns across ocean ridge systems in the early 1970s confirmed that large-scale seawater circulation through hot basaltic crust is actually occurring. During the same period, advances in structural and chemical studies of ophiolite complexes on land led geologists to conclude that these peculiar rock sequences are fragments of oceanic crust and upper mantle that were tectonically emplaced by plate collisions.

A principal line of evidence used to support this conclusion is the fact that ophiolitic basaltic rocks had been metamorphosed and intensely veined prior to emplacement. The ophiolitic metabasalts were, in fact, identical in all respects to samples obtained in dredge hauls along the axial valleys of ocean ridges. As a result, a new form of rock metamorphism, regional in extent but restricted to the marine ridge environment, was recognized. The first scientific paper describing this phenomenon in a comprehensive and unified fashion appeared in 1973; this landmark paper referred to the newly recognized metamorphic process as sub-seafloor metamorphism. For brevity, the acronym SFM will be used.

It is important to recognize that SFM is a process involving both heat and mass transfer. Geochemists were quick to realize that large-scale, continuous alteration of ridge basalts by convecting seawater could have profound effects on the exchange of chemical elements between Earth's lithosphere and hydrosphere. This hypothesis was tested by direct sampling of discharge fluids from submarine hot springs in the Galápagos Rift and the East Pacific in the late 1970s. Compared with normal seawater, these hot springs vent a fluid that is distinctly acidic; strongly depleted in magnesium (Mg) and sulfate (SO_4^{2-}); enriched in silica (SiO_2), calcium (Ca), and hydrogen sulfide (H_2S); and enriched in a wide range of metallic elements. These compositional differences are the result of rock-seawater interaction and prove that SFM involves large-scale metasomatism. This is in marked contrast to regional metamorphism in continental mobile belts, which is a process that involves much less change in chemical composition.

REGIONAL METAMORPHIC FACIES

Samples recovered by dredging the steep escarpments of submarine ridge valleys represent the common regional metamorphic facies (zeolite, prehnite-pumpellyite, greenschist, and amphibolite), but the vast majority record the conditions of the greenschist facies. Most samples are metabasalts with mineralogy indicative of a temperature range of 100° to 450° Celsius and very low pressures (400 to 500 bars) relative to continental metamorphism. The original mineralogy of ridge basalts, which consists of tiny crystals of calcic plagioclase, clinopyroxene, and olivine set in a glassy matrix, is converted by SFM to complex mixtures of actinolite, tremolite, hornblende, albite, chlorite, epidote, talc, clay, quartz, sphene, and pyrite that tend to preserve the original igneous texture of the rock. Such assemblages are well known in basaltic rocks of continental mobile belts and ophiolite complexes, and they record greenschist facies conditions. For the majority of samples, greenschist facies metamorphism has produced an alteration assemblage that consists of albite, actinolite, chlorite, and epidote and is accompanied by a small amount of quartz and pyrite.

The intense hydrothermal veining observed in some dredge samples and especially in ophiolite complexes appears to record former high fracture density and permeability, which is necessary for extensive basalt-seawater interaction. The major vein minerals are chlorite, actinolite, epidote, quartz, pyrite, and, less commonly, sulfides of iron, copper, and zinc. On the basis of mineralogy, greenschist facies alteration is described as chlorite-rich (chlorite greater than 15 percent of rock, epidote less than 15 percent) or epidote-rich (epidote greater than 15 percent, chlorite less than 15 percent). Chlorite-rich alteration is dominant in the dredge-haul samples recovered so far, and this preponderance is taken to mean that the chlorite-forming reactions occur at lower temperatures and closer to the seawater-basalt interface, where seawater influx of the basaltic crust begins. In other words, the transition from chlorite-rich to epidote-rich alteration records prograde metamorphism that, in the case of ridge basalts, correlates with increasing depth and temperature.

MAGNESIUM METASOMATISM

The mechanics of ridge-basalt emplacement ensure that all such rocks have an opportunity to react with seawater and undergo metasomatism. Although the chemical composition of average midocean-ridge basalt (MORB) is known, it is unlikely that it reflects exactly the pristine composition of the initial magma. The metasomatic nature of SFM has mainly been determined on the basis of dredge samples of pillow basalt. Individual pillows typically exhibit intensely altered rims and cores of relatively fresh basalt. On this basis, it is known that chlorite-rich alteration shows greater departures from parent-rock composition than does the higher-grade epidote-rich alteration. Chlorite-rich samples exhibit significant gains of magnesium and corresponding losses of calcium, but their state of oxidation is, for the most part, unaffected by the alteration process. Epidote-rich samples generally undergo minor losses of magnesium and gains of calcium but become somewhat oxidized by the alteration reactions.

Evidently, magnesium is removed from seawater and held in magnesium-rich minerals (especially chlorite) at the water-rock interface. The extent of this reaction depends on both temperature and water-to-rock mass ratio. At water-to-rock ratios less than 50 (in other words, rock-dominated conditions), magnesium is completely removed from seawater, the reaction rate being proportional to temperature. The mineral reactions create acidic conditions as long as magnesium remains in solution to drive the reaction. For rock-dominated conditions, however, magnesium is soon depleted and the acidic conditions associated with the reaction are short-lived. However brief, the acidic stage is important because it permits the circulating seawater to leach metallic elements from the enclosing basalt. At water-to-rock ratios in excess of 50 (that is, fluid-dominated conditions), magnesium cannot be completely removed from solution, so that the acidic, metal-leaching conditions of the reaction are maintained indefinitely. Calcium, sodium, and potassium are dissolved from basalt because these elements are not utilized in the formation of chlorite. In contrast, iron and aluminum are retained in the rock because they participate in the formation of chlorite. The rate and magnitude of the reaction both increase with temperature and, because thermal gradients near ridges are very high, it would be expected that magnesium metasomatism would be restricted to relatively shallow crustal depths. This limitation exists because fluids down-welling from the crust-seawater interface would become quickly heated.

SODIUM METASOMATISM

For temperatures above 350° Celsius, experiments predict that magnesium metasomatism will be replaced by sodium metasomatism of a slightly more complex nature. In this case, seawater sodium (Na^+) replaces calcium in plagioclase crystals of basalt, which in turn permits the liberated calcium to form epidote and maintain acidic conditions independent of the fluid-to-rock ratio. This reaction shows that the more albite forms, the more calcium is recycled from plagioclase to epidote, and the more acidic the aqueous solution becomes. This reaction depends on the availability of silica, which must be supplied to the solution by the rock. Because basalts with glassy matrices are more susceptible to silica leaching than comparable rocks with crystalline matrices, sodium metasomatism (albitization or spilitization) is expected to be most pronounced in basalts that were formerly glassy.

The complex fluids currently venting from submarine ridge hot springs are viewed as the end product of a sequence of chemical reactions that begin with seawater infiltration of hot, fractured basalt some distance from the ridge axis. Convective circulation, supported by magmatic heat, drives the downwelling fluids to depths of perhaps 1 or 2 kilometers, during which time they are heated to temperatures in excess of 350° Celsius and react extensively with the surrounding rocks. As hot fluids are expelled from the ridge crest, fluids nearby are drawn downward into the crust and heated. The fluids ultimately are returned to the seafloor by upwelling through narrow, focused zones along ridge axes and discharged into the sea as highly evolved hot-spring solutions. Many chemical and physical details of this complex process are yet to be explained, but it seems clear that a calcium-fixing reaction, such as the second reaction described, must play a major role in producing the high-temperature, acidic, metal-charged solutions that vent from ridge-crest hot springs.

SFM OBSERVATION AND EXPERIMENTATION

Prior to the late 1970s, scientific data were largely gathered along ocean ridges from specially designed research ships operating on the surface. The methods utilized included bathymetry, heat-flow and magnetic measurements, dredge sampling, and core drilling of bottom sediments and ridge basalts. It is impossible to overemphasize the success of these techniques, which provided the confirming evidence for seafloor spreading and the foundation for modern plate tectonics theory. A new era began in 1977, however, when the manned submersible *Alvin* was utilized to make direct observations of submarine hydrothermal activity along the Galápagos Rift. That was soon followed by the immensely successful 1979 RISE expedition, which used *Alvin* to photograph and sample the "black smokers" on the East Pacific Rise. Manned submersibles, supported by conventional research vessels, offer many advantages. Chief among them are that small-scale geological phenomena may be directly observed and revisited to study temporal effects; a wide range of geophysical data may be measured directly on the seafloor; samples of sediment, fluid, and rock may be collected and re-collected directly from precisely known sample sites; the physical characteristics of each data-measurement site can be observed and described; and interactions between biological and geological systems can be observed.

Direct observations, coupled with detailed laboratory experimentation on basalt-seawater reactions, are valuable as a means of investigating processes operating at the rock-seawater interface along ocean ridges where crustal growth occurs. Additional data derive from studies of hot brines in deep boreholes of active geothermal areas such as those in Salton Sea, California; Iceland; and New Zealand. Useful as these approaches are, their value is limited because they cannot tell how metamorphic conditions vary with depth beneath the rock-seawater interface; neither do they predict the effects of prolonged (for example, over millions of years) rock-seawater interaction. Full characterization of SFM as a petrologic process requires traditional geological study of rocks on land that record the entire range of SFM effects. Fortunately, such rocks, known as ophiolite complexes, are relatively common in major orogenic belts such as the Alps, Appalachian-Caledonian system, and numerous circum-Pacific mountain ranges.

VOLCANIC-EXHALATIVE ORE BODIES

From an economic point of view, studies of sub-seafloor metamorphism have greatly advanced knowledge of the important "volcanic-exhalative" class of ore deposits. Volcanic-exhalative ore bodies are major strata-bound metallic deposits that precipitate directly onto the seafloor from metal-laden solutions discharged by submarine hot springs. The resulting ores are mineralogically variable but generally sulfide-rich and, for this reason, are often called "massive sulfide." Of the several types of exhalative ore deposits now recognized, the Cyprus type is most closely identified with igneous and metamorphic processes operating along submarine ridges. On the island of Cyprus, massive sulfide copper-zinc ores are hosted by ancient, metamorphosed, seafloor rocks called "ophiolites" and blanketed by a thin layer of iron manganese-rich marine chert. The Cyprus ores are lensoidal to pod-like in shape, 200 to 300 meters across and up to 250 meters thick. The largest of these massive sulfide lenses contain 15 to 20 million tons of ore that typically assays about 4 percent copper, 0.5 percent zinc, 8 parts per million each of silver and gold, 48 percent sulfur, 43 percent iron, and 5 percent silica.

Cyprus-type massive sulfides are of worldwide occurrence, and hundreds of these deposits are now known. As a class, they constitute a major mineral resource produced as a byproduct of sub-seafloor metamorphism. They are considered, by most geologists, to form directly on the seafloor from the discharge of black smokers such as those discovered along the East Pacific Rise. The smokers are chimneys ranging from 2- to 10-meters high that pump dense, black plumes of hot, acidic, metal-laden brine onto the seafloor. The turbulent black plumes, for which this special class of hot spring is named, are composed of fine particles of pyrite, pyrrhotite, and sphalerite. Calculations indicate that at their present high flow rates ten black smokers such as those operating on the East Pacific Rise could produce the largest known Cyprus-type massive sulfide body in only two thousand years.

—Gary R. Lowell

Further Reading

Best, M. G. *Igneous and Metamorphic Petrology.* 2nd ed., Blackwell Science Ltd., 2003.

Borradaile, G. J., and D. Gauthier. "Magnetic Studies of Magma-Supply and Sea Floor Metamorphism: Troodos Ophiotile Dikes." *Tectonophysics,* vol. 418, 2006, pp. 75-92.

Bucher, Kurt, and Martin Frey. *Petrogenesis of Metamorphic Rocks.* 7th ed., Springer-Verlag Berlin Heidelberg, 2002.

Cann, J. R., H. Elderfield, A. Laughton, et al., editors. *Mid-Ocean Ridges: Dynamics of Processes Associated with Creation of New Ocean Crust.* Cambridge UP, 1999.

Philpotts, Anthony R., and Jay J. Ague. *Principles of Igneous and Metamorphic Petrology.* 2nd ed., Cambridge UP, 2009.

Vernon, Ron H., and Geoffrey Clarke. *Principles of Metamorphic Petrology.* Cambridge UP, 2008.

T

Teilhard de Chardin, Pierre

Fields of Study: Geology; Paleontology; Geophysics; Cosmology

ABSTRACT

Through his work as a geologist on the evolution of Earth and as a paleontologist on the evolution of life, Teilhard, a devout Jesuit priest, came to see human beings progressing toward a new consciousness and spiritual unity called the "Omega Point," which he identified with Jesus Christ.

KEY CONCEPTS

anthropogenesis: the emergence over time of humans

cosmogenesis: the view that the entire universe is evolving like a living entity

emergence: term for the development of some aspect of the universe that did not exist previously

TEILHARD'S EARLY LIFE

From the perspective of his mature vision of the universe, Pierre Teilhard de Chardin interpreted his own life in the light of his evolutionary doctrine. He saw as providential his birth in a Sarcenat chateau amid the foothills of the Monts Dore in central France. Teilhard was the fourth child in a family that would eventually number eleven children, and on his mother's side he was distantly related to Voltaire and on his father's side to Blaise Pascal. His mother was a deeply religious Roman Catholic who ignited a spiritual fire in young Teilhard. His father was a gentleman farmer with interests in natural history, and he introduced his children to the delights of rocks, minerals, wildflowers, and animals. Thus, from an early age, Teilhard was able to combine two spheres of life—the material and the spiritual—commonly considered incompatible.

In 1892, a month before his eleventh birthday, Teilhard became a boarder at the Jesuit school of Nôtre-Dame de Mongré at Villefranche-sur-Saône, near Lyons. He was a good student, especially in science and literature, and in his free time he continued his interest in geology by collecting minerals. He eventually concluded that he had a vocation to the Jesuit life. On March 20, 1899, he entered the Jesuit novitiate in Aix-en-Provence, about twenty miles north of Marseilles, to begin a long period of spiritual and intellectual formation.

On March 25, 1901, Teilhard took his first vows of poverty, chastity, and obedience, and then began, at Laval, his studies in the Greek and Latin classics (the juniorate). These studies were interrupted when

Chardin in 1955. Unknown author, CC BY-SA 3.0, via Wikimedia Commons.

anticlerical legislation in France forced the Jesuits to transfer their juniorate to Jersey, one of the English Channel islands, in the summer of 1901. As a second-year junior in Jersey, Teilhard seriously considered abandoning the study of geology to devote himself completely to spiritual activities. One of his religious superiors wisely helped him put his spiritual evolution in perspective and guided him in a direction in which he could combine his love of matter, energy, and life (his cosmic sense) with his love of Christ and the supernatural (his Christic sense). From 1902 to 1905, he studied Scholastic philosophy at the Jesuit house on the Isle of Jersey, where he spent his free hours, geologist's hammer in hand, in scientific surveys, resulting in a paper on the island's mineralogy and geology.

In September, 1905, Teilhard was sent to teach chemistry and physics at the Jesuit College of the Holy Family in Cairo, Egypt. Although the science he taught was elementary, this experience of the chemical substances and physical forces of the universe helped him refine his still-crude understanding of the world. After three years as a teacher in Egypt, during which he published works on the Eocene period, he returned to England to complete his Jesuit training. He spent four years, from 1908 to 1912, studying theology at Hastings in southern England. While continuing his geological and paleontological research and writing, he also devoted time to the synthesis of his scientific and spiritual views.

During his theological studies at Hastings, Teilhard began to understand matter from the perspective of spirit, and this forced him to develop a new way of thinking and speaking about what he saw happening in the universe. He began to use the words of science (energy, force, radiation) to describe the previously unseen evolution of spirit that was taking shape in nature. This spiritual energy, already familiar to him in the evolution of his own consciousness, he now grasped as active in the universe, itself in the process of self-creation. Matter, as an evolutionary fact, had given birth to spirit; therefore, matter and spirit are not two separate substances but two aspects of a single evolving cosmos. Physical energy therefore contains something of the spiritual, since energy's upward trend is an observable fact in the increasing complexity of evolving organisms. By the time he was ordained, on August 14, 1911, he had discovered a vision by which he could understand scientific phenomena spiritually and spiritual phenomena scientifically. It therefore became the core of his vocation as a priest to show that evolutionism does not entail a rejection of Christianity, because Christ represented the crucial point in the universe's history at which matter and spirit met.

TEILHARD'S LIFE'S WORK

Throughout the rest of his career, as priest and scientist, Teilhard devoted himself to the evolution of the universe (what he called "cosmogenesis"), whose ever-richening spirituality constantly became for him more real and resplendent. For him, salvation no longer meant abandoning the world but building it up. His scientific work therefore became something holy, to be undertaken not for its own sake but for the liberation of more spirit from matter.

After completing his theological studies in 1912, Teilhard went to Paris to study under Marcellin Boule, a professor at the Institute of Human Paleontology in the Natural History Museum. Boule was one of the leading experts on Neanderthal man, but Teilhard's work was mainly in the paleontology of Tertiary mammals in Europe. His scientific studies were interrupted by his tertianship (the final year of his Jesuit formation, a period of intense prayer, meditation, and ascetical training), and then by World War I. Although he could have chosen to be a chaplain, he joined the Eighth Regiment of the Moroccan Tirailleurs as a stretcher-bearer. Teilhard, whose bravery under fire and whose generosity of spirit throughout his military service were honored by medals both during and after the war, did not allow his experiences of the war's horrors to destroy the vision of human history he was constructing. Indeed, he found that his patriotic service on behalf of a great ideal had invigorated his life. Even in the trenches he believed that he was participating in the grand work of sanctifying humanity. In his personal writings during this time, his constant theme was spiritual evolution: When everything of value in the material world has passed into the human soul, he believed that souls will pass into a new level. In 1917, he began describing this level as the mystic milieu.

After his demobilization, Teilhard returned to his scientific studies. He completed his academic requirements at the Sorbonne and then began work on a thesis about the mammals of the lower Eocene in France. In 1922, he successfully defended his thesis and was awarded his doctorate in paleontology. During the early 1920s, he served as an assistant professor of geology at the Catholic Institute of Paris, but his position there became untenable when he taught that evolution required a revision in the Church's doctrine of Original Sin. The discoveries of geologists, paleontologists, and paleoanthropologists had convinced Teilhard that no evidence existed for Adam, Eve, or Eden, and consequently the Fall was an event that could not be verified. Before Charles Darwin, Christians could believe that one man's sin (Adam's) had ruined everything and that another man's suffering (Christ's) had saved everything. After Darwin, Teilhard believed that Christians must realize that Original Sin is not a malady specific to the earth but an inevitable consequence of the limitations of evolving matter and spirit. Because of Teilhard's heterodox views on Original Sin, his Jesuit superiors asked him to leave the institute and take a research post in Tianjin, China, with Father Émile Licent, a Jesuit pioneer in paleontology. For the next twenty-three years, from 1923 to 1946, Teilhard's career would center on China. He would return to France periodically, but he led what he called a vagabond existence that would continue for the rest of his life.

During his first decade as a peripatetic priest, Teilhard worked mainly in the north of China. He participated in a French paleontological mission to the Ordos and Gobi deserts directed by Father Licent. On a visit to central Mongolia in the summer of 1923, the two priests found the first evidence for paleolithic man in China. After these research expeditions, Teilhard returned to Paris, hoping to continue teaching at the Catholic Institute while making occasional field trips to China, but his superiors continued to be bothered by his evolutionary views of Original Sin, and he was told to restrict himself to his scientific work (his position at the Catholic Institute was terminated at the end of 1926).

Returning to China in the spring of 1926, Teilhard obeyed the orders of his superiors not to disseminate publicly his theological speculations, and for the rest of his life, he confined himself to publishing his scientific work while privately developing his evolutionary vision (which he shared with friends and fellow Jesuits). During the late 1920s, he began using the term "noosphere" to describe Earth's sphere of thinking substances. The world of matter (the geosphere) had been the source of life (the biosphere), but now the human species was taking conscious control of evolution, and science, technology, and socialization were changing Earth more rapidly than natural selection ever had.

Teilhard's ideas about humankind's role in the universe found spiritual resonance in his manuscript of *Le Milieu divin* (1957; *The Divine Milieu*, 1960), which he wrote in 1926 and 1927 (but which was not published until after his death). In this book, which many scholars see as a spiritual portrait of Teilhard himself, he tried to set down as simply as possible the religious life that he had been living. Through his scientific work, he had become convinced that humans are at the spiritual center of the cosmos, and he wanted to show how Christianity was at the center of human history and should be the center of every person's spiritual life.

During this period of intense spiritual probing, Teilhard shifted the base of his operations from Tianjin to Zhuokoudian, about fifty kilometers southwest of Beijing. There, Davidson Black, a Canadian who was a professor of anatomy at the Beijing Medical College, was engaged in an organized search for prehistoric humans. In the decade from 1927 to 1937, many fossils of animals and humans were found in the calcareous deposits of Zhuokoudian. The human fossils were first called *Sinanthropus pekinensis*, then *Pithecanthropus pekinensis*, and now *Homo erectus*. Teilhard helped to date the fossils, to show that Beijing man was a toolmaker, and to publicize the importance of the finds. Although he gained wide recognition for his account of the discovery, his principal paleontological work continued to be on nonhuman fossils for example, on fossil carnivores.

During the late 1920s and through the 1930s, Teilhard went on several expeditions into northern and central China, made periodic trips to Europe and America, and continued to refine his synthesis of evolution and Christianity. He built up a network of friends and disciples in Paris and at the American Museum of Natural History in New York. In China, he had collaborators of many nationalities, and this contributed to his strongly cosmopolitan and internationalist views. His participation in various research trips to Africa, America, and especially China deepened his understanding of evolution.

In 1938, while he was in Beijing, Teilhard began writing *Le Phénomène humain* (1955; *The Phenomenon of Man*, 1959), in which he tried to show how evolutionary data pointed to a Christian interpretation of the world. His approach was phenomenological; that is, he tried to bring out the meaning of phenomena by describing them as precisely as possible. By analyzing the problem of humankind in terms of evolution, he hoped to bridge the gap between Christians and non-Christians, for he believed that Jesus Christ, as God Incarnate, revealed himself not only in the New Testament but also in human evolution. In his book, Teilhard argued that consciousness evolves, and he located the origins of consciousness in preliving as well as living matter. Thresholds, such as the origin of life and of consciousness, are important examples of emergence, where something entirely new enters the universe.

Traditional science, according to Teilhard, has not well served these threshold phenomena, and modern science has almost completely ignored what he calls "the within" of the world. This spiritual energy can be studied just as scientists study "the without," matter's external face, in terms of space and time. As the capstone of his evolutionary synthesis, Teilhard believed that, through an increase in knowledge and love, human consciousness can evolve to a new level. Noogenesis, the evolution of mind, is a stage on the way to Christogenesis, the evolution of the universe toward the Omega Point, Teilhard's term for the cosmic Christ.

Teilhard was composing this optimistic vision of the world against a background of world events that would have made pessimism a more natural response. The Japanese occupation of China interfered with his work at Cho-k'ou-tien and his attempt to establish a laboratory for advanced studies in human paleontology in Peking. When World War II began, he was trapped in China, a situation that continued for the war's duration. He abhorred both German Nazism and Japanese fascism, both of which he interpreted as crudely and unjustifiably transferring the laws of natural selection (the survival of the fittest) from the biological to the human level, whereas, in his vision, humankind was clearly passing to a new level of existence in which a future of love and cooperation was being built. Nevertheless, beneath the human and national tragedies of this war, he believed that something spiritual was evolving. This vision of human evolution found expression in *The Phenomenon of Man*, which he completed in 1940 and sent to Rome in 1941 for approval by his superiors. Since his approach was phenomenological, he hoped that his book would pass the censors, but in 1944 he learned that it had been rejected.

After the war's end, Teilhard was able to return to Europe in 1946. He went to Paris, where, during the late 1940s, he was frustrated in his desire to teach at the Collège de France. A visit to Rome to seek permission to publish various works on his evolutionary synthesis was unsuccessful. In 1947, he was appointed director of research in the National Scientific Research Center, which subsidized his scientific work, but he was eager to communicate his evolutionary vision, for he found the existentialism that was spreading throughout France, with its pessimistic ideas of humankind and matter, dangerously anachronistic in a universe that was evolving toward Christ. On June 1, 1947, he suffered a heart attack and, though he recovered, his health remained precarious for the remainder of his life. These physical sufferings were followed by psychological ones, for in September his religious superiors forbade him again to write on philosophical or theological issues, thus squelching any hope that *The Phenomenon of Man* or any of his works on anthropogenesis would be published.

Refusing to be discouraged and undertaking no action that could be seen as disobedient, Teilhard nevertheless did everything he could to circulate his ideas privately. He was enthusiastic about the new United Nations, and he worked for some of its organizations that were concerned about the future of the human race. Through contacts made in these labors, for example, with Julian Huxley, he continued to develop his cosmogenetic vision. These efforts were further stimulated by his return to field work in the 1950s. With the financial support of the Viking Fund (soon to become the Wenner Gren Foundation for Anthropological Research), Teilhard made expeditions to various excavation sites in South Africa, where important prehuman fossils, especially the australopithecines, had been discovered.

During the last period of Teilhard's life, from 1951 to 1955, his center of operations became New York City, where, as a fellow of the Wenner Gren Foundation, he concentrated on anthropogenesis (the evolution of the human species). In the summer of 1954, he made his last trip to France, and, though the visit aroused warm feelings for his native country, he was back in New York by September. He was denied permission by a superior to travel to a meeting at the Sorbonne in April, 1955, but he accepted this without bitterness, resolving to continue fighting to increase love in the world. Teilhard died on Easter Sunday, April 10, 1955.

SIGNIFICANCE OF TEILHARD'S WORK

Teilhard did more than any other thinker to help people understand evolution in the context of Christianity. Although he wrote nearly two hundred articles and technical papers in geology and paleontology, mainly on the early Cenozoic period in Europe and the late Cenozoic period in China, his name is most likely to be remembered for his radical view that evolution has a spiritual orientation. For him, the earth, which he so lovingly studied, was profoundly linked to Jesus Christ, since every created thing will find its fulfillment in Him. In Teilhard's view, evolution is a purposeful process in which matter and energy progressively evolve in the direction of increasing complexity, consciousness, and spirituality. He identified Christ and the universe so tightly that many critics have interpreted his vision as a Christian pantheism. Although Teilhard often spoke of the universe as God's cosmic body, he also carefully distinguished pantheistic evolution, which destroys personalities in its union with matter, from Christian evolution, which culminates in a union with God that preserves personalities.

In the initial enthusiasm generated by the publication of many of Teilhard's works in the years after his death, some commentators compared the significance of his synthesis of evolutionism and Christianity with Saint Thomas Aquinas's synthesis of Aristotelianism and Christianity. Now that most of Teilhard's writings and letters are in print, it is easier to see him as part of the intellectual development composed of such thinkers as Henri Bergson. These evolutionary philosophers were dissatisfied with the positivism and scientism of the twentieth century and attempted to provide an understanding of humankind's total experience, of the material as well as the spiritual, of the past and the future, of variety and unity. In this sense, Teilhard has much to teach both scientists and theologians. For scientists, his mystical temperament and loving grasp of positive values have given the world an interpretation of the universe that encourages love, progress, and unity. For theologians, his evolutionary view has necessitated a transposition of Christian revelation into a new key, where religion can profit from scientific insights.

Despite criticisms of his work, many of them justified, Teilhard did offer an alternative to the materialist vision of humankind as an accidental phenomenon, a random collection of molecules on an unimportant planet. In his view, humankind represented the culmination of the complex evolution of matter and life. It is doubtful that in the future Teilhard's evolutionary synthesis of Christianity and evolution will be completely accepted by Christians or by scientists, but as a product of a luminous mind and a loving heart, his conception of the universe as a unity in which all things work together for a final consummation in Christ has already found numerous followers, because this vision, which allows people to remain true to themselves while building union with God, reveals the deeply spiritual values in the innermost heart of humankind and the universe.

—Robert J. Paradowski

Further Reading

Browning, Geraldine O., Joseph L. Alioto, and Seymour M. Farber, editors. *Teilhard de Chardin: In Quest of the Perfection of Man.* Fairleigh Dickinson UP, 1973.

Cuénot, Claude. *Teilhard de Chardin: A Biographical Study.* Edited by René Hague. Translated by Vincent Colimore. Helicon Press, 1965.

de Lubac, Henri. *Teilhard de Chardin: The Man and His Meaning.* Translated by René Hague. Hawthorn Books, 1965.

Grenet, Paul. *Teilhard de Chardin: The Man and His Theories.* Translated by R. A. Rudorff. Paul S. Eriksson, 1966.

Grummett, David. *Teilhard de Chardin: Theology, Humanity, and Cosmos.* Peeters, 2005.

Maynard, Thierry, editor. *Teilhard and the Future of Humanity*. Fordham UP, 2006.
Mooney, Christopher F. *Teilhard de Chardin and the Mystery of Christ*. Harper & Row, 1966.
Speaight, Robert. *The Life of Teilhard de Chardin*. Harper & Row, 1967.

TURBIDITY CURRENTS AND SUBMARINE FANS

Fields of Study: Geology; Marine Geology; Geochemistry; Geophysics

ABSTRACT
Turbidity currents are a major mechanism whereby sediment from nearshore areas is transported to deeper parts of the oceans. This sediment often accumulates as part of large depositional systems called "submarine fans," which could well be exploited as a major source of petroleum at some time in the future.

KEY CONCEPTS

abyssal plain: the flat, sediment covered area of the seafloor that merges with the base of the continental rise

continental margin: the area that separates the emergent continents from the deep seafloor, generally consisting of the continental shelf, continental slope, and continental rise

continental rise: the broad and gently sloping ramp that rises from the abyssal plain to the base of the continental slope; submarine fans are found here

continental shelf: the gently seaward sloping submerged edge of a continent that commonly extends to a depth of about 200 meters or to the edge of the continental slope

continental slope: the relatively steep region of the continental margin between the continental shelf and the continental rise

gas hydrates: relatively stable, often crystalline, molecular combinations of water and hydrocarbon gases, especially methane, that form on the ocean floor due to the elevated pressure conditions

submarine canyon: a submerged, V-shaped canyon cut into the continental shelf and continental slope, through which turbidity currents funnel into the deeper parts of the oceans

turbulent flow: a high-velocity sediment flow in which individual sediment particles move in very chaotic directions above the seafloor

DEVELOPMENT OF THE TURBIDITY CURRENT CONCEPT

A turbidity current is a dense mass of water and sediment that flows downhill along the bottom of an ocean or any other standing body of water such as a lake. Turbidity currents may reach high speeds and are capable of carrying large quantities of sediment, and of eroding or scouring the ocean floor as they move.

The concept of the turbidity current has a long history. In the late nineteenth century, geologists observing the waters of the Rhone River entering Lake Geneva noted that instead of mixing with the lake water, the river water moved along the bottom of the lake in a channel. This behavior was interpreted to reflect the elevated density of the cold, sediment laden Rhone River water. These dense, bottom hugging currents were called "density currents." It was suggested in the late 1930s that dense currents composed of sediment and water could be produced by wave activity on the continental shelves, the flat or gently sloping submerged edges of the continents, during periods when global sea level is lower than it is today. Furthermore, it was postulated that these dense, turbulent mixtures of suspended sediment and water would flow across the continental shelves and down large canyon-like features into deeper parts of the ocean.

From the 1930s to the 1950s, researchers conducted numerous laboratory studies to test the hypothesis that turbidity currents could actually erode the ocean floor and produce such features as submarine canyons on the continental shelf. This work was done by pouring muddy water into the end of an inclined flume (a long, straight trough in which the hydrodynamic properties of moving fluids can be studied) in order to produce artificial turbidity currents. Results of these investigations indicated that sediment in a turbidity current moves in a very chaotic or random pattern, much like the movement of snow and other materials in an avalanche. Additionally, geologists recognized that turbidity currents can move rapidly and that their velocity is dependent upon the slope of the sea bottom along which they move as well as the density of the sediment water mixture.

These experimental studies also yielded insight into the nature of the sedimentary deposits that accumulate from turbidity currents. These deposits, referred to as turbidites, form as graded beds, sedimentary layers in which the largest or most coarse-grained sediment particles are concentrated at the bottom of the bed and grade gradually upward to the smallest or most fine-grained sediment at the top. The graded nature of the bedding was deduced correctly to be a consequence of a reduction in the flow velocity of the turbidity current upon reaching the very gentle slopes that are typical of the bottom of the ocean, so that the largest sediment particles settle out of the current to the bottom of the ocean first, followed by progressively finer particles. Thus, each turbidity current produces a single graded bed or turbidite.

Geologists working in the field began to realize that the graded beds such as those produced in laboratory studies could be observed in sequences of sedimentary rocks exposed on land in various mountain belts throughout the world (the Appalachians, Apennines, and Carpathians). Moreover, oceanographic research demonstrated the presence of turbidite sands in some of the deepest areas of the modern oceans. It is now apparent that turbidity currents are and were a dominant, if not the dominant, mechanism of sediment transport from shallow water environments on the continental shelf to the deeper parts of the ocean. Evidence of this includes the recovery of shallow water organic remains from very deep parts of the oceans far removed from land. The laboratory studies and subsequent oceanographic research indicated that turbidites differ from most other sedimentary deposits in that they are nearly instantaneous deposits that may accumulate over a period of just a few hours to a few days. Moreover, turbidity currents are quite capable of eroding the great submarine canyons that cut into many continental shelves.

CAUSES OF TURBIDITY CURRENTS

Downhill movement of turbidity currents can be triggered by various causes. Commonly, earthquakes affecting the continental shelves will cause sediment on the continental slope—the part of the continental margin characterized by an increase in gradient immediately seaward of the continental shelf—to slide

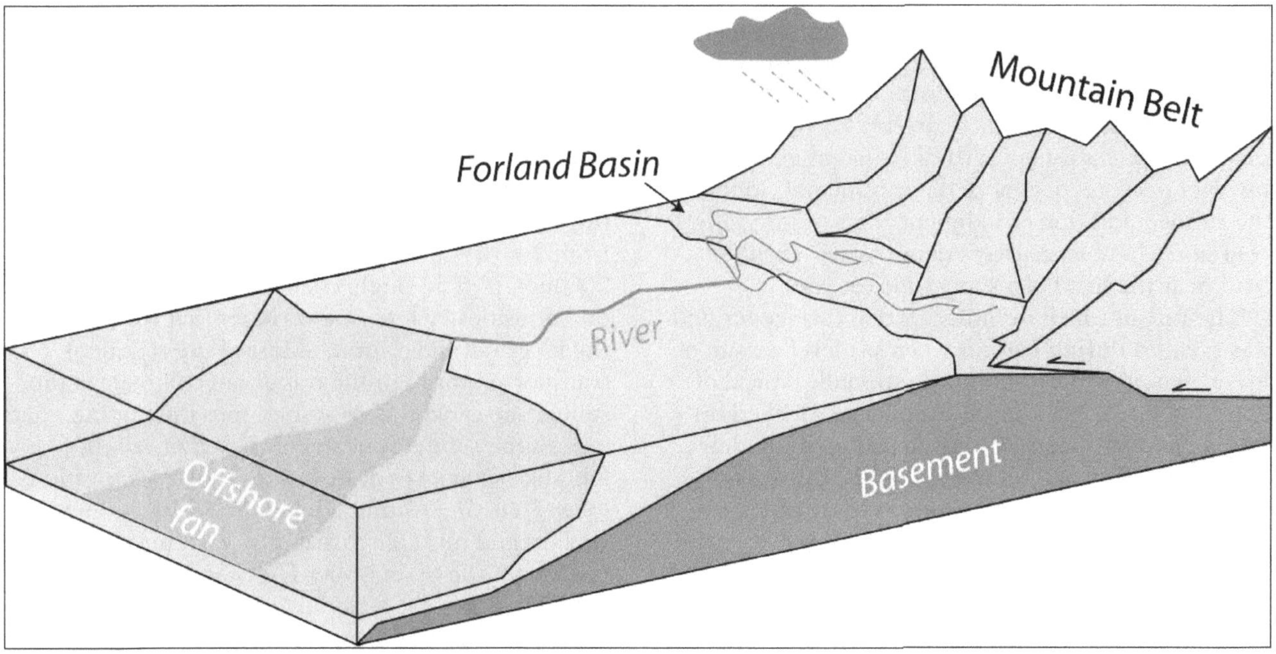

Distribution of detritus in a depositional system. Image by Geologyanli, CC BY-SA 4.0, via Wikimedia Commons.

downslope, thereby mixing with seawater to form a turbidity current. A rapid sequence of successive breaks in transatlantic cables on the continental margin bordering southern Newfoundland was apparently caused by the downslope movement of a dense, turbulent mixture of seawater and sediment generated from the continental shelf and slope in response to the Grand Banks earthquake of November 8, 1929. Those cables closest to the point on the ocean floor directly above the earthquake, the epicenter, broke first, whereas cables farther from this point broke later. The breaking of the cables indicated the erosive capability of turbidity currents. Subsequent drilling of that part of the ocean floor traversed by the turbulent flow recovered a 1-meter-thick graded bed containing shallow water organic remains, and this discovery strengthened the argument that the flow was indeed a turbidity current. The earthquake induced turbidity current averaged 27 kilometers per hour, although it reached velocities in excess of 70 kilometers per hour in steeper parts of the continental margin, and it covered an area of more than 195,000 square kilometers.

Turbidity currents can also be generated by wave activity on continental shelves, an idea, as already noted, that was postulated in the 1930s. More specifically, large waves produced during great storms such as hurricanes are capable of creating the turbulence required to mix sediment and seawater, thereby creating a turbidity current. An equally plausible mechanism of turbidity current generation involves over steepening of the continental slope by the sudden addition of sediment. This is especially common where large rivers deposit their sediment load near the head of a submarine canyon.

The rate at which turbidity currents are generated was greatest during periods when sea level was much lower than it is today. This was especially typical of glacial periods, when more seawater was locked up in the enlarged polar ice caps, resulting in the lowering of the global level of the oceans. During these times of lowered sea level, rivers were capable of transporting their sediment loads directly across the previously submerged continental shelf into the head of a submarine canyon. There, the sediment mixed with seawater and was fed, via the submarine canyon, directly into the deeper parts of the ocean as turbidity currents. When sea level rose and the shoreline retreated landward as the exposed continental shelf was once again submerged, the river was cut off from the canyon head, thereby precluding the infusion of sediment directly to the canyon and reducing the likelihood of turbidity current generation.

CHARACTERISTICS OF SUBMARINE FANS

Turbidity currents triggered on continental shelves or continental slopes move downhill until they reach a point at which the reduced gradient of the ocean floor causes a reduction in the velocity of the sediment flow. This leads to deposition of a graded bed or turbidite. Many turbidite deposits that are funneled through submarine canyons ultimately accumulate as part of large fan shaped or cone shaped sediment bodies called submarine fans. The submarine fans, which spread outward from the mouths of the submarine canyons, merge with the bottom of the continental slope and comprise the continental rise, the broad, gently sloping feature that rises from the abyssal plain of the ocean floor and merges with the base of the continental slope. Where submarine canyons are close together along the continental shelf and continental slope, the attached submarine fans may coalesce to form a wide, laterally extensive continental rise.

In general, submarine fans are subdivided into three major morphologic elements or parts: the upper fan, midfan, and lower fan. The upper fan, also referred to as the inner fan, is typically characterized by a single submarine channel that is connected to the submarine canyon. Upper fan channels range from 2- to 18-kilometers wide and may be as deep as 900 meters. The single channel is commonly flanked on both sides by levees, low ridges that run along the length of the channel. Most of the sediment transported through the upper fan channel via the submarine canyon is deposited into the midfan, that area of the submarine fan composed of raised, lobe-like sequences of turbidites called "depositional lobes." The depositional lobes are fed by numerous shallow and unstable distributive channels that branch off the main upper fan channel. Because these channels are generally relatively shallow (several tens of meters deep), they are more likely to be filled in during passage of extraordinarily dense turbidity currents. When that happens, the channels are abandoned and subsequent turbidity currents erode or scour out new channels in adjacent areas of the midfan portion of the submarine fan. The lower fan is characterized by a smooth ocean floor that passes imperceptibly seaward into the abyssal plain. The low gradient of the lower fan relative to that of the midfan leads to a much reduced abundance of turbidites in the lower fan. Turbidite deposits are not common on the abyssal plain because ocean floor gradients in this area are typically too low to sustain movement of the dense sediment cloud.

Submarine fans display great variations in size. The Bengal Fan, in the Bay of Bengal in the northeast Indian Ocean, is the largest submarine fan. Its total length exceeds 3,000 kilometers. The main or upper fan channel ranges from 13- to 18-kilometers wide and from 150 to 900 meters deep. Most submarine fans, however, are much smaller than the Bengal Fan.

The growth of submarine fans, because they are major sites of turbidite sedimentation, is controlled by variations in sea level. As noted previously, turbidity currents are generated more frequently during periods of low sea level. Accordingly, submarine fans are likely to grow fastest, as manifested by increased rates of turbidite sedimentation, during periods of low global sea level. Indeed, geologists have demonstrated that the Mississippi submarine fan, which is fed by the Mississippi River, was the site of abundant turbidite sedimentation at the end of the most recent ice age, approximately 15,000 to 20,000 years ago, when sea level was as much as 120 to 130 meters lower than it is today. Since then, the rate of turbidite sedimentation has dropped off greatly as sea level gradually rose, and the shoreline retreated landward to its present position.

—Gary G. Lash

Further Reading

Huneke, Heiko, and Thierry Mulder. *Deep Sea Sediments.* Elsevier, 2011.

Kennish, Michael J. *Practical Handbook of Marine Science.* 3rd ed., CRC Press, 2001.

Leeder, Mike R. *Sedimentology and Sedimentary Basins: From Turbulence to Tectonics.* Wiley-Blackwell, 2011.

McKinney, Frank. *The Northern Adriatic Ecosystem: Deep Time in a Shallow Sea.* Columbia UP, 2007.

Oceanography Course Team. *Ocean Circulation.* 2nd ed., Butterworth Heinemann, 2001.

Schulz, Horst D., and Matthias Zabel. *Marine Geochemistry.* 2nd ed., Springer Verlag, 2006.

Schwartz, M., editor. *Encyclopedia of Coastal Science.* Springer, 2005.

Shanmugam, G. *Deep-Water Processes and Facies Models: Implications for Sandstone Petroleum Reservoirs.* Elsevier, 2006.

Shanmugam, G. *Mass Transport, Gravity Flows, and Bottom Currents: Downslope and Alongslope Processes and Deposits.* Elsevier, 2020.

Schleiss, Anton J., Giovanni De Cesare, Mário J. Franca, and Michael Pfister, editors. *Reservoir Sedimentation. CRC Press, 2014.*

Talley, Lynne D., George L. Pickard, William J. Emery, and James H. Swift. *Descriptive Physical Oceanography: An Introduction.* 6th ed., Elsevier, 2011.

Thurman, H. V., and Alan P. Trujillo. *Introductory Oceanography.* 10th ed., Prentice Hall, 2004.

U

Ultrahigh-Pressure Metamorphism

Fields of Study: Geology; Geochemistry; Geophysics; Seismology

ABSTRACT

Metamorphic rock is one of three major rock types and is formed when pressure, temperature, or environmental chemicals cause either an igneous or sedimentary rock to change in density and mineralogical structure. Metamorphosis can be caused by extraterrestrial impact, exposure to heat from the earth's core, exposure to chemical environments, and tectonic forces. Ultrahigh-pressure metamorphism is a type of metamorphism that occurs at extremely high pressures and usually accompanies the collision of continental plates.

KEY CONCEPTS

diagenesis: chemical and texture changes that affect sediment during and after its formation into sedimentary rock

lithification: process by which sediment is compacted or cemented into solid rock

lithostatic pressure: pressure exerted by the weight of sediment overlying the sediment sample in question

magma: superheated rock that exists in a liquid phase

megapascal: geologic unit of extreme pressure, equivalent to 1 million pascals and 145.037 pounds per square inch

petrology: branch of geology that studies rocks and the conditions that lead to the formation of rocks

polycrystalline: crystal structure consisting of crystallite fragments

recrystallization: process that occurs at high pressure, causing the atoms, molecules, or grains of a crystal to reorganize into another type of crystal

subduction zone: area in which two tectonic plates collide, forcing one plate to push beneath the other and driving the submerged plate into the earth's mantle

METAMORPHIC PETROLOGY

Metamorphic petrology is a branch of geology that studies transformations in mineral composition, texture, color, and other qualities that occur within rocks. These transformations occur when the rocks are subjected to pressures, chemical environments, and temperatures different from those in which the rock originally formed.

Metamorphism can be distinguished from diagenesis, which is another transformative process that affects certain types of rock after solidification. Diagenesis occurs at low temperatures and pressures, whereas the process affecting metamorphic rock typically occurs at temperatures higher than 200° Celsius (392° Fahrenheit) and at pressures greater than 300 megapascals. Metamorphic rocks are sufficiently unique in structure and chemistry that they are considered one of the three basic rock types.

Sedimentary rocks are formed from sediment, a layer of debris that forms at the top layer of Earth's crust and is composed of loose minerals, organic waste, and bits of other rocks that have eroded into smaller fragments. Through burial and compression, this surface material can become lithified—that is, cemented and compressed to form rocks.

Sedimentary rocks come in three types: clastic, chemical, and organic. Clastic rocks are composed almost entirely of bits of other rocks that have become compacted into solid structures. Chemical rocks are sedimentary rocks that form through some chemical process. Many types of chemical rocks result from situations in which a riverbed or shallow sea gradually dries, after which underwater sediment may become solidified through dehydration and other chemical processes to become sedimentary rock. Organic rocks are formed from the accumulation of organic debris in the sediment that becomes lithified through compaction. Coal is an example of an organic sedimentary rock that forms from the remains of wood, leaves, plants and other organic material that has become buried and compacted through millions of years. Igneous rocks form when molten earth originating within the earth's mantle rises close to the surface and then cools and solidifies to form solid bodies. The formation of igneous rock involves a change of state from liquid to solid, which is accompanied by a reduction in heat.

Igneous rocks come in two main types: plutonic and volcanic. Plutonic rock forms when magma slowly cools in pockets beneath the surface, sometimes taking millions of years to form. Plutonic rock is one of the most common types of rock and commonly underlies mountains and other major geologic features. Volcanic rock forms when magma breaks through the surface as lava and then cools and hardens into solid structures. Magma and volcanism have played major roles in the formation of the lithosphere. Geologists estimate that the majority of the earth's crust is composed of igneous rock.

Metamorphic rocks may start their existence as either igneous or sedimentary rocks but are subjected to an environment such that they undergo a change in chemical and physical properties. Most metamorphic rocks form deep in the earth's crust or the upper layers of the mantle, where temperatures and pressures are sufficient to induce significant chemical and physical changes. Though metamorphosis generally begins at 200° Celsius (392° Fahrenheit) and at 300 megapascals, it can occur at much higher temperatures and pressures. The limit of the metamorphic process occurs when the rock becomes hot enough to melt. At this point, the rock becomes magma. Any further structural transformations fall into the category of igneous rock formation.

Metamorphism always involves the loss of water and an overall reduction in density. Some of the most familiar metamorphic rocks are gemstones such as diamonds and rubies. Rubies and sapphires form from the metamorphosis of certain types of clay, which give rise to corundum. Further pressure and temperature increases will cause the corundum to develop glass-like mineral inclusions including rubies and sapphires. Diamonds also develop from the metamorphic process acting on coal that has been compressed at depths for millions of years.

There are two basic stages to the metamorphic process. First, the volume of the rock is reduced as the rock is compacted, which reduces empty air space and water from the sample. After this, further pressure will cause recrystallization, which is a process by which the relationships and position of the individual grains, molecules, or atoms of a rock change to form a new crystal structure with a reduced volume. Many common stones, including marble, result from recrystallization; marble is a recrystallized form of limestone that has been subjected to intense pressures.

TYPES AND GRADES OF METAMORPHISM

Low-grade metamorphism occurs at temperatures between 200° and 320° Celsius (608° Fahrenheit) and at relatively low pressures, where high-grade metamorphosis occurs at temperatures above 320° Celsius and at higher pressures. Rocks formed from low-grade metamorphism tend to be hydrous minerals, meaning that the rocks have water incorporated into their crystal matrix. As pressure and temperature increase, water molecules are destabilized and removed from the rock. High-grade metamorphic rocks are characterized by having low levels of water molecules in their structure.

Many different physical conditions can lead to metamorphosis. One of the most basic types is contact metamorphism, which occurs when magma rises from the mantle into the crust. Only some of the surrounding rock becomes hot enough to melt. Just outside of this melting zone is a layer of rock that is subjected to intense heat, which may be sufficient in some cases to bring about metamorphic changes in the rock.

In rare cases, metamorphic rocks result from extraterrestrial impact events, such as when a meteor hits the surface of Earth. An impact of this type can generate sufficient pressure over a small area to bring about metamorphosis in the impacted rock. Evidence of an extraterrestrial impact also may include the presence of shock lamellae, which are thin strips of glass that form in the opposite direction from the rock's original crystalline structure. Shock lamellae form only under conditions of rapid, intense pressure, such as those that accompany a meteor impact.

Most types of metamorphic rocks form when rock is buried at sufficient depths to induce metamorphic changes. At increasing depths, rocks undergo increased heating because of their proximity to the mantle and increased lithostatic pressure, which is caused by the weight of the crust overlying them. A special type of burial metamorphism, called "regional metamorphism," occurs over vast areas when large volumes of rock are simultaneously buried and subjected to metamorphic influences.

One of the most common scenarios for this type of metamorphism is the collision of two continents. The force of the collision causes the crust on both sides to deform, often leading to orogenesis, the formation of mountains. When the crust buckles at the site of impact, some of the crust is thrust beneath the surface, where it is subjected to intense pressure. As this process continues, layers of crustal rock are forced deeper into the earth, leading to metamorphism as pressures and temperatures increase.

ULTRAHIGH-PRESSURE METAMORPHISM

Ultrahigh-pressure metamorphism (UHPM) is a type of metamorphism that occurs when rocks are subjected to extremely high pressures and temperatures. The conditions that result in UHPM appear to be rare, and only twenty sites containing rocks generated by this type of metamorphism had been found between its discovery in the 1980s and 2010. Existing evidence of UHPM environments are associated with regional metamorphism related to the collision of tectonic plates and the subduction of crustal rock.

When oceanic tectonic plates are driven together, the older and therefore denser of the two plates will typically drive under the opposing plate in a process called "subduction." The subducted plate is driven into Earth's mantle layer, where it desolidifies and blends with the rock of the mantle. This desolidified rock may eventually become magma, some of which will eventually rise toward the surface and solidify into igneous rock.

Another type of subduction, usually called "A-type subduction," involves the collision of continental plates. Because continental crust has a lower density than oceanic crust, the denser of the two plates begins to subduct, but then rebounds because the crustal rock is not dense enough to remain in the mantle. A-type subduction zones often lead to orogeny, or the formation of mountains, as the two plates buckle against each other, forcing Earth materials to rise near the impact site.

UHPM occurs where a portion of the continental crust has descended to depths greater than 100 kilometers (62 miles) before returning to the surface. Geologists detect the existence of UHPM by looking for characteristic minerals that are created only in extreme temperature and pressure regimes. Crust that is transported to sufficient depths rapidly begins to recrystallize; however, water contained within the rocks tends to stabilize the matrix, slowing the process of converting low-density rock to high-density rock. When the force causing the subduction of the rock abates, the submerged strata is far more buoyant than the surrounding rock that typically exists at this depth, causing the recrystallized rock to rise rapidly. This relatively rapid return to the surface (though still requiring thousands of years) preserves mineralogical evidence of UHPM in the rock, whereas a longer period of rising to the surface would allow metamorphosis to occur a second time, returning most of the metamorphic rock to its original form.

EVIDENCE OF UHPM

One of the most common types of rock produced by UHPM is coesite, which forms when high pressures and relatively high temperatures are applied to quartz. Quartz is one of the most abundant materials on the surface of the earth and is made from a combination of silicon and oxygen atoms arranged in a crystal lattice. Coesite is unstable at sea-level temperatures and pressures and will eventually decay in this environment, returning to quartz.

Samples of intact coesite are rarely uncovered because of coesite's extreme instability at sea level. However, geologists can utilize indirect evidence to determine if coesite was present in a recovered sample of quartz. As coesite returns to quartz form, it forms into a type of quartz called "polycrystalline quartz," which is characterized by the presence of microscopic crystal fragments called "crystallites." Coesite has 10 percent less volume than quartz;

therefore, when coesite metamorphoses into quartz, it expands in such a way that it causes the development of characteristic microscopic fractures in the surrounding quartz. Geologists use the presence of these pressure fractures in samples of polycrystalline quartz to determine if the quartz was subjected to UHPM environments that led to the formation of coesite.

In rare cases, geologists have discovered microdiamond inclusions in rocks subjected to UHPM environments. Microdiamonds are diamonds that are typically less than 1 micrometer in extent and are formed only in extreme high-pressure environments. Microdiamonds have been found as inclusions in zircon, a common mineral that occurs in sedimentary and igneous rocks as small inclusions. Microdiamonds also have been uncovered from samples of kyanite, a silicate mineral that forms in metamorphic environments in which the pressure rises into the thousands of pascals. Geologists have found microdiamonds formed in rocks that have undergone pressures of more than 5,000 megapascals and temperatures exceeding 1,000° Celsius (1,832° Fahrenheit).

Geologists are working to enhance the ability to find and study UHPM zones and to refine models that explain the formation of these extreme lithographic environments. Research into UHPM gained widespread attention in the 1990s, and since that time UHPM environments have been discovered on every continent.

The process of subduction, followed by return to the crust, takes millions of years to complete, and most evidence of UHPM has been found in sediment from ancient environments. The best-known sites to uncover UHPM minerals include the Daba Shan mountains of eastern China and the Dora-Maira massif of the Italian Alps.

—Micah L. Issitt

Further Reading

Bucher, Kurt, and Rodney Grapes. *Petrogenesis of Metamorphic Rocks*. 8th ed., Springer, 2011.

Dobrzhinetskaya, Larissa, Shah Wali Faryad, Simon Wallis, and Simon Cuthbert, editors. *Ultrahigh-Pressure Metamorphism: 25 Years After the Discovery of Coesite and Diamond.* Elsevier, 2011.

Frisch, Wolfgang, Martin Meschede, and Ronald Blakey. *Plate Tectonics: Continental Drift and Mountain Building.* Springer, 2011.

Grotzinger, John, and Thomas H. Jordan. *Understanding Earth.* 6th ed., W. H. Freeman, 2010.

Hacker, Bradley R., William C. McClelland and John G. Liou, editors. *Ultrahigh-Pressure Metamorphism: Deep Continental Subduction.* The Geological Society of America, 2006.

Monroe, James S., Reed Wicander, and Richard Hazlett. *Physical Geology.* 6th ed., Thompson Higher Education, 2007.

Wicander, Reed, and James S. Monroe. *The Changing Earth: Exploring Geology and Evolution.* 5th ed., Brooks/Cole, 2009.

Zhang, Lifei, Zeming Zhang, Hans-Peter Schertl, and Chunjing Wei. *HP-UHP Metamorphism and Tectonic Evolution of Orogenic Belts.* Geological Society of London, 2019.

Ultrapotassic Rocks

Fields of Study: Geology; Geochemistry; Geophysics; Volcanology

ABSTRACT

Igneous rocks, from which ultrapotassic rocks derive, form when magma from within Earth's mantle rises toward the surface and cools to form solid rock. Magma pockets lead to a variety of igneous deposits, including batholiths, sills, dikes, and volcanic pipes. Ultrapotassic rocks are a type of igneous rock that form deep within mantle rock contained within a craton, a stable portion of the mantle.

KEY CONCEPTS

craton: stable portion of the earth's mantle and associated crust that forms the inner core of the lithospheric components

diamond: mineral formed from carbon atoms organized in a characteristic structure known as a diamond lattice

diamond stability zone: area within the earth's mantle where pressures and temperatures are sufficient for the formation of diamonds from deposits of carbon

dike: igneous rock formation that develops in a vertical or diagonal direction from the surface of the earth and often connects to a magma chamber within the earth's crust or mantle

igneous rock: type of rock formed when magma that develops within the earth's mantle rises to the surface, where it cools and solidifies into solid rock structures

magma: molten rock, gas, and liquid minerals that develop within the earth's mantle from decompression, molecular excitation, and heat from the earth's core

volcanic pipe: deposit of igneous rock formed when a deep magma pocket produces an explosion of gas and magma that rapidly rises to the surface and forms a cone-shaped deposit

xenocryst: crystal that differs from the surrounding rock found in igneous rock deposits

xenolith: inclusion of rock in the body of a different type of rock; often occurs in igneous rock formations

INTRUSIVE IGNEOUS ROCKS

Igneous rocks are formed from magma, a mixture of molten rock, gases, and other trace chemicals that form in pockets within the earth's mantle. The mantle is a layer of rock heated by radioactive energy from Earth's core and pressurized by the weight of rock above it.

The mantle shifts in response to thermal currents from the core. As this occurs, localized disturbances cause part of the mantle to destabilize. The molecules within the disturbed mantle move freely, generating energy that melts the surrounding rock. Magma is lower in density than surrounding rock and therefore rises to the surface, pushing through the solid rock of the mantle and the crust.

When magma cools in the crust or on the surface, it solidifies into igneous rocks. Magma that breaks through the surface of the crust before cooling is called "lava" and gives rise to volcanic rock or extrusive igneous rock. When magma cools beneath the surface, insulated within pockets of solid rock, it develops into intrusive igneous rock or plutonic rock. Plutonic rock cools slowly, in millions of years, which causes the development of larger crystals. For this reason, intrusive igneous rock tends to be characterized as phaneritic in texture, meaning that the grains of the rocks can be discerned by the naked eye.

Minette, seen here is a lamprophyre, one of the range of ultrapotassic rocks. Photo by Pelex, CC BY 3.0, via Wikimedia Commons.

Igneous rocks also are characterized by their chemical composition. The primary components of igneous rocks are silica, a molecular combination of silicon and oxygen, and feldspar, which is another silica-based crystal that contains a variety of included materials. Some igneous rocks, called "felsic rocks," are composed primarily of silica (55 to 70 percent) and frequently contain a variety of potassium-rich feldspars. By contrast, mafic rocks have lower levels of silica (less than 50 percent) and are rich in iron and magnesium. Ultramafic rocks have even lower concentrations of silica (less than 45 percent) and correspondingly higher concentrations of included metals.

IGNEOUS ROCK DEPOSITS

Magma that cools within Earth's crust forms into plutonic rock deposits. The largest plutonic deposits are called "batholiths," which result when large pockets of magma join as they cool. Batholiths may exist deep within the mantle, within the continental cratons, which are the deep, solid portions of the lithosphere least affected by the collisions between tectonic plates.

Batholiths give rise to dikes, which are vertical or diagonal segments of plutonic rock that rise through the surrounding strata toward the surface. Dikes also may give rise to horizontal layers of igneous rock, called "sills," that form when magma infiltrates layers of rock parallel to the surface.

At times, magma pockets deep within Earth may erupt because of a buildup of volcanic gases, passing rapidly through the mantle and crust to erupt onto the surface. These formations, called "volcanic pipes," result in vertical, cone- or carrot-shaped deposits of igneous rock connected to deep subterranean deposits. Volcanic pipes may result from magma pockets that develop hundreds of kilometers beneath the earth's surface.

ULTRAPOTASSIC IGNEOUS ROCKS

Ultrapotassic igneous rocks are rocks that contain high levels of potassium in the form of potassium oxide (K_2O). A rock is considered ultrapotassic if it contains more than 3 percent potassium oxide by weight and contains a minimum of two times more potassium oxide than sodium oxide (Na_2O).

In addition, ultrapotassic rocks must contain more than 3 percent magnesium oxide (Mg_2O) by weight. Ultrapotassic rocks are usually mafic or ultramafic in composition and are a relatively rare form of igneous rock.

The formation of potassic and ultrapotassic rocks begins with the enrichment of magma deposits with potassium oxide-rich minerals from the surrounding rock. When magma of this type is subjected to intense pressures, such as those that arise from the collision of continental plates, ultrapotassic rocks may develop. Potassic rock sediments uncovered in Mediterranean and Italian strata appear to have resulted from the collision of continental plates.

In rare occasions ultrapotassic rocks may result from magma that forms above an active subduction zone, an area in which one tectonic plate collides with and is forced under another tectonic plate. Ultrapotassic sediments from the Sunda arc in Indonesia, Papua New Guinea, and Fiji originated from this tectonic environment.

The most uncommon environment for ultrapotassic rock formation is through intraplate magmatism, or magma that forms somewhere in the central portion of a tectonic plate. This type of magmatism is known as a hot spot, where a deep magma pocket pushes through the mantle and crust to create a volcano or a volcanic pipe. Ultrapotassic rock deposits from South Africa and western Uganda developed from intraplate magmatism surrounding the development of a hot spot.

Ultrapotassic rocks are generally divided into three numerical types. Type 1 ultrapotassic rocks, called "lamproites," have relatively low percentages of calcium oxide (CaO), aluminum oxide (Al_2O_3), and sodium oxide (Na_2O) coupled with high percentages of potassium and magnesium oxides. Type 2 rocks, called "kamafugites," have low levels of silicate and high levels of calcium oxide. Type 3 rocks typically occur in areas in which mountain formation has taken place and usually include high levels of calcium and aluminum oxide.

LAMPROITES

Lamproites are ultrapotassic rocks in which the ratio of potassium oxide to sodium oxide exceeds 5 percent by weight. Most forms of lamproite have between 6 and 8 percent potassium oxide by weight. Lamproites generally have high levels of magnesium oxide and are composed of between 45 and 50 percent silica, which categorizes them as mafic igneous rocks. Lamproites have been discovered connected to dikes and sills and also to batholith deposits. One of the most common locations for lamproites is within volcanic pipe deposits.

Lamproites generally develop only when a certain set of specific geoformational processes have taken place. The first stage occurs in the Benioff zone, which is an area of deep seismic activity that occurs beneath a subduction zone. Millions of years later the minerals created by subduction over a Benioff zone are subjected to a process in which two continental plates collide, leading to orogenesis, or mountain formation. In these conditions, lamproites may develop in the deep sediment, near the edges of the continental craton. Generally, lamproites will form only when a portion of the mantle melts at depths greater than 150 kilometers (93 miles) and at temperatures of between 1,100° to 1,500° Celsius (2,012° to 2,732° Fahrenheit).

Lamproites are of particular interest in geology because of the number of rare minerals the rocks contain. As lamproite magma rises to the surface, it may carry with it a variety of xenoliths, which are incompatible minerals or pieces of other rock types that become included in the magma as it hardens. Inclusions in lamproites often include rocks that are found only in deep mantle strata. These mineral types include a variety of iron-rich and titanium-rich minerals. Deposits of aluminum, nickel, quartz, and chromium also are commonly found in lamproites.

KIMBERLITES

Kimberlites are ultrapotassic rocks derived from intraplate magma activity, generally confined to portions of the crust in which the plate overlies old portions of a continental craton. Kimberlites appear to be related to hot spots that develop within the core of a continental craton. Most kimberlite deposits occur within five degrees of a known hot spot, while others that are not close to a recent hot spot may have originated along the moving path of an ancient hot spot.

Kimberlites are volatile rocks because they contain high levels of carbon dioxide (CO_2) and water (H_2O) within their structure, leading them to fracture easily at sea-level pressure and temperature. Kimberlites have low sodium oxide levels and frequently contain a large variety of incompatible elements and minerals incorporated into their matrices.

Kimberlites occur in dikes and sills but are best known from large deposits uncovered from volcanic pipes. Most kimberlite deposits contain inclusions of ultramafic rocks from the upper mantle and may also contain xenoliths and xenocrysts, a type of xenolith consisting of individual crystals embedded within the matrix of another rock type.

Kimberlites are generally divided into two groups, based on the level of mica, a mineral silicate containing aluminum, potassium, magnesium, iron, and silicon. Both types of kimberlite may occur together in the same deposit.

Kimberlites are derived from magma pockets that originated at depths of between 100 and 200 km (62 to 124 mi) beneath the surface and are created by partial melting of mantle rock containing high levels of water and carbon dioxide. Kimberlites are usually formed in situations where intraplate magma rises to the surface rapidly, in less than ten years. Most known kimberlite deposits originated in strata that developed before the Cenozoic period (65 million years ago) indicating that the tectonic and geologic processes involved in the formation of kimberlite may have been more common in the ancient Earth's lithosphere.

ULTRAPOTASSIC ROCKS AND DIAMONDS

Ultrapotassic rocks are economically important because of the discovery of diamonds as xenocrysts in deposits of kimberlite and lamproite. Kimberlite is named after the town of Kimberly, in the Northern Cape region of South Africa, where diamonds were first discovered in kimberlite deposits in the 1870s. The first known lamproite diamond deposit was discovered in 1979 in Western Australia.

Lamproite diamonds are often of poor quality in comparison with kimberlite inclusions. However, lamproite deposits sometimes contain rare-colored diamond inclusions and a variety of other precious stones.

Diamonds form within cratons at depths of between 140 and 300 km (87 and 186 mi) beneath the surface, a depth that is known as the diamond stability zone. At deeper or shallower depths, diamonds will not form. Diamonds may therefore be included as xenocrysts in deposits of kimberlite and lamproite that developed from magma that originated within the mantle below or within the diamond stability zone.

Diamonds most frequently occur in kimberlite and lamproite that hardens within volcanic pipes. These explosions of magma and gas quickly rise to the surface, thereby dragging inclusions of various mantle rocks, xenoliths, and xenocrysts, which may include diamonds. Many of the world's most productive diamond mines are part of ancient kimberlite pipes.

Cratons occur in two general types, based on the age of material contained within the craton. Those composed of mantle rock and crust that are more than 2.5 billion years old are called "Archean age" cratons, while those arising from mantle rock that is between 1.6 and 2.4 billion years old are called "Proterozoic age" cratons. Diamonds are usually found in kimberlite and lamproite derived from Archean age cratons containing strata that is 2.4 billion years old or more.

Kimberlite and lamproite pipes, sills, and dikes are derived from magma chambers that develop deep within the center of a craton. These pipes, sills, and dikes, therefore, provide geologists with samples of rocks and minerals from the ancient Earth in addition to precious and semiprecious stones. For this reason, kimberlite and lamproite are important from both an economic and a scientific perspective. Minerals and mantle rocks included within kimberlite and lamproite deposits give geologists data regarding the composition and environmental conditions of the ancient Earth at the time when these rocks first formed or became incorporated into the mantle.

—Micah L. Issitt

Further Reading

Gill, Robin. *Igneous Rocks and Processes: A Practical Guide*. Wiley-Blackwell, 2011.

Gupta, Alok Krshna. *Origin of Potassium-Rich Silica-Deficient Igneous Rocks*. Springer, 2015.

Monroe, James S., Reed Wicander, and Richard Hazlett. *Physical Geology*. 6th ed., Thompson Higher Education, 2007.

Montgomery, Carla W. *Environmental Geology*. 4th ed., McGraw-Hill, 2008.

Müller, Daniel, and David I. Groves. *Potassic Igneous Rocks and Associated Gold-Copper Mineralization*. Springer, 2016.

Stanley, Steven M. *Earth System History*. W. H. Freeman, 2004.

Wilson, Marjorie. *Igneous Petrogenesis: A Global Tectonic Approach*. Springer, 2007.

V

Volcanoes

Fields of Study: Geology; Volcanology; Seismology; Geophysics

ABSTRACT

Volcanoes or volcanic activity can be a valuable source of natural resources. Some of the economically important resources derived from volcanic activity are diamonds, precious metallic minerals, native sulfur, and a nutrient-rich soil produced by the weathering of volcanic rock.

KEY CONCEPTS

andesite: the principal igneous magma of continental volcanoes, named for the Andes Mountain range

basalt: the iron- and magnesium-rich igneous rock of the oceanic crust

hydrothermal vents: undersea structures of a volcanic nature that eject hot mineral-laden water from deep within Earth's crust

rhyolite: igneous rock with the highest content of volatiles and highest silica content

tectonic plates: segments of Earth's crust that are independent of each other but constrained by each other, analogous to the segments of a cracked eggshell

BASIC VOLCANOLOGY

All volcanoes are related to the process of plate tectonics. Plate tectonics describes the continual movement of immense sections (plates) of Earth's crust relative to one another. Although this process is incredibly slow, geologic time is much longer. Both earthquakes and volcanoes most often occur along the boundaries of these plates. They result from the buildup of intense pressure as one plate collides with, or slides past, another. Here old crustal rock is melted as it plunges down into the upper mantle, or new rock forms as magma squeezes out from great fissures in the crust. In the process, old crustal rock is recycled to form new rock that is rich in mineral resources.

Major metallic mineral deposits from around the world are associated with plate boundaries past and present. The island of Cyprus is rich in copper that once formed on the seafloor of an ancient oceanic spreading center. The same process has been happening in the Red Sea, where copper-rich minerals are being extruded through volcanic activity.

The best evidence for submarine deposition of sulfide minerals by volcanic activity comes from structures called "hydrothermal vents," also known as "black smokers." In appearance, they resemble underwater geysers with cone-type vents emitting black smoke. They result from the seepage of seawater into the hot oceanic basalt crust. This heated seawater then interacts with the basalt by extracting iron, copper, sulfur, and other metals from it. Once this mixture erupts onto the seafloor, it mixes with the cold seawater and precipitates sulfide minerals into massive deposits. These become the resources for the future.

Volcanoes come in three basic types, based on their particular chemistry. They are named for the volcanic rock produced by each: basalt, andesite, and rhyolite. The most common type of volcano is the basaltic variety. Varieties of basaltic volcanoes can be found along plate boundaries as well as plate centers (such as the one where the Hawaiian Islands formed). The principal rock that underlies the world's oceans is also basalt.

BASALTIC VOLCANOES

Basaltic volcanoes are usually low in silica (approximately 50 percent) and gas content. This type of volcano commonly produces fast-moving lava flows and is generally not explosive. The only mineral that is consistently associated with basaltic volcanoes is sulfur. It forms from sulfur-rich gases that escape from fissures

A volcano eruption. Photo via iStock.com/Beboy_ltd [Used under license.]

in the cooling lava rock. As the hot gases escape, sulfur quickly crystallizes, with its distinctive yellow color present on the rock. Sulfur is mined at various volcanic locations. One is Mount Etna on the island of Sicily, where it is an important economic resource.

ANDESITIC VOLCANOES

The second type of volcano results from andesitic magma. It is richer in silica (approximately 60 percent) and gas than basaltic volcanoes are. This results in a volcano that can be explosive and can produce a large quantity of lava, depending upon slight variations in its chemical composition. Volcanoes such as this can be extremely dangerous since no one is ever certain what will happen each time they erupt.

Mount St. Helens in the state of Washington and Mount Fuji in Japan are two examples of andesitic volcanoes, which can remain dormant for hundreds of years and then suddenly erupt. The 1980 eruption of Mount St. Helens devastated the area around it. In the aftermath, a rich volcanic ash covered the region. Despite the fact that considerable vegetation was destroyed by the eruption and associated flooding, vigorous plant life returned within a couple of years. This was possible because of the nutrient-rich ash that created a new soil.

RHYOLITIC VOLCANOES

A magma of rhyolitic composition produces the third volcanic type. Compared to the other two, rhyolitic magma is the richest in both its silica (approximately 70 percent) and its gas content. Both gases and fluids present are rich in dissolved metallic minerals. The magma, as it nears Earth's surface, first cracks crustal rock and then may erupt with a violent explosion.

Often, large hydrothermal mineral deposits are associated with rhyolitic volcanoes. These are deposits of various minerals such as malachite, chalcopyrite, and pyrite, where a metallic element like copper or iron is bonded with sulfur or bonded to a carbonate molecule. Such minerals tend to occur in veins where the mineral-rich fluids penetrate fissures in existing rock and then crystallize during cooling. Often gold and/or silver are deposited in this manner. Although such deposits are common, they do not usually occur in large quantities. Most often, huge amounts of rock must be mined in order to extract relatively small amounts of the valuable metals. The great Bingham copper mine in Utah is an excellent example of such a deposit.

DIAMOND PIPES

One important occurrence of a valuable mineral associated with volcanic activity is the diamond pipe. Diamond formation is typically associated with a high-pressure, high-temperature environment. Such conditions are present in the Earth's upper mantle at depths of approximately 200 kilometers. Here diamonds slowly crystallize within magma. As a result of rapid upward movement, the diamonds are carried along with the magma column. Eventually, upon cooling, the magma will form a pipe structure. In shape it somewhat resembles a champagne glass.

Most volcanic pipes do not reach the surface and produce a volcano. The more probable situation is that they remain underground as a magma source for an erupting volcano. In those pipes that contain diamonds, the diamonds are disseminated throughout a rock called "kimberlite". Erosion may eventually destroy evidence of the volcano, exposing the diamond pipe. Erosion also acts as a natural means of extracting the diamonds and then depositing them as sediment in rivers or on beaches. The most important diamond pipes include those of South Africa, Siberia, northern Canada, and western Australia.

—Paul P. Sipiera

Further Reading

Decker, Robert, and Barbara Decker. *Volcanoes.* 4th ed., W. H. Freeman, 2006.

Francis, Peter, and Clive Oppenheimer. *Volcanoes.* 2nd ed., Oxford UP, 2004.

Lutgens, Frederick K., Edward J. Tarbuck, and Dennis G. Tasa. *Essentials of Geology.* Pearson Education, 2017.

Martí, Joan, and Gerald Ernst, editors. *Volcanoes and the Environment.* Cambridge UP, 2005.

Parfitt, Elisabeth A., and Lionel Wilson. *Fundamentals of Physical Volcanology.* Blackwell, 2008.

Schmincke, Hans-Ulrich. *Volcanism.* Springer, 2004.

Tarbuck, Edward J., Frederick K. Lutgens, and Dennis Tasa. *Earth: An Introduction to Physical Geology*. 13th ed., Pearson Education, 2019.

Zeilinga de Boer, Jelle, and Donald Theodore Sanders. *Volcanoes in Human History: The Far-Reaching Effects of Major Eruptions.* Princeton UP, 2005.

W

Weathering

Fields of Study: Geology; Geomorphology; Geophysics

ABSTRACT

Weathering comprises the processes that decompose and disintegrate earth materials at or near the Earth's surface. Weathering occurs because rock or soil is physically or chemically unstable in the surface environment and is therefore subject to attack. Weathering continues until minerals achieve equilibrium with prevailing environmental conditions. If those conditions change, however, a new episode of weathering may begin.

KEY CONCEPTS

abrasion: damage and removal of material from materials by friction against another material

chemical weathering: breaking down or alteration of rock and rock particles by chemical reactions

frost wedging: the forcing apart and breakage of rock particles when water enters a crack or pore in the rock and freezes, the expansion of water into ice in the crack or pore applying sufficient pressure to fracture the particle

physical weathering: breaking down of rock and rock particles by physical means, typically by impacts with other rock particles

solution: a means of altering rock and rock particles by dissolving certain mineral components from their content

THE ORIGIN OF WEATHERING

The term "weathering" derives from the fact that it is the elements of weather—for example, heat and rainwater—that are the agents of weathering. Definitions of weathering generally include only those processes that precede erosion and transport of materials from the site of weathering. Weathering is the initial step in the production of sediments, soils, and sedimentary rocks from preexisting rock material. The majority of rock materials weathered and eroded from the Earth's surface in the past now reside below the surface in sedimentary rocks.

TYPES OF WEATHERING

All types of weathering phenomena fall into two broad categories: physical weathering and chemical weathering. Physical and chemical weathering produce different results. Physical weathering simply generates smaller fragments of the original materials. Chemical weathering often significantly changes a rock's mineralogy, by either dissolving minerals or altering them.

Physical, or mechanical, weathering breaks up earth materials in three ways: increased pressure, release of pressure, and abrasion. If water freezes within the pores in a rock, expansion of the water may cause the rock to fracture, a process known as frost wedging. Other processes that cause physical weathering by increasing pressure include thermal expansion, root growth, and salt crystal formation. Removal of overlying materials by erosion, or unloading, releases pressure on underlying rocks, causing them to expand and fracture. Abrasion occurs when rocks grind against or strike one another—for example, at the base of a gravel-filled stream channel. Physical weathering does not alter the composition of earth materials; it merely causes them to disaggregate. This also makes more surface area available to chemical weathering.

Chemical weathering either chemically alters or dissolves components of earth materials. For example, feldspar, a common mineral, produces the clay mineral kaolinite if altered by hydrolysis, a decomposition reaction involving water; liquid water dissolves soluble minerals such as halite (salt), a process known as "solution." As with physical weathering, a variety of chemical weathering processes exist. The most common include hydrolysis, solution, oxidation (addition of oxygen to a mineral), and hydration (addition of water to a mineral).

PRODUCTS OF WEATHERING

Weathering produces two general types of materials: chemical or mineral solutions, and mechanical sediments. Chemical weathering produces mineral solutions, which consist of solids or gases dissolved in water. Freshwater, which some people mistakenly call "pure" water, actually consists of water and an assortment of chemical weathering products. If a glass of tap water is allowed to evaporate, a thin layer of chemical sediments will accumulate in the bottom of the glass—a process known as "precipitation," or "chemical sedimentation." Evaporation of a glass of seawater produces a thicker layer of chemical sediment because of its higher dissolved mineral content (salinity). Dissolved minerals may precipitate out to form a mineral deposit, may chemically react with (weather) other mineral solids, or may remain in solution for long periods of time.

A combination of chemical and physical weathering produces most mechanical sediments. Mechanical, or clastic, sediments range in size from clay particles (smaller than millimeter) to boulders (smaller than 256 millimeters) and include the following intermediate grain sizes: silt (to millimeter), sand (to 2 millimeters), pebbles (2 to 64 millimeters), and cobbles (64 to 256 millimeters). Sediments may remain in place at the site of weathering for long periods of time or be transported elsewhere by gravity, wind, water, or ice. If mechanical sediments accumulate at the surface, they form soil and undergo further weathering. If deeply buried, chemical solutions buried along with the sediments precipitate minerals into the pores between the grains, cementing the

A natural arch produced by erosion of differentially weathered rock in Jebel Kharaz (Jordan). Photo by Etan J. Tal, CC BY 3.0, via Wikimedia Commons.

sediments together. Thus, when combined in the subsurface, chemical and physical weathering products form sedimentary rocks.

RATES OF WEATHERING

Numerous factors control the rate at which rocks weather. Two of the most important factors are climate and mineralogy. Of all climatic factors, precipitation and temperature affect the rate of weathering most. Water is by far the most important agent of chemical weathering; without it, chemical weathering effects are minimal. Humid regions therefore experience higher rates of chemical weathering. Warm temperatures also generally produce more rapid rates of chemical weathering. Overall, the most rapid rates of chemical weathering occur in humid, tropical regions, whereas the slowest rates occur in cold areas with low rainfall, such as the polar regions. However, any arid region will have relatively low rates of chemical weathering.

Because of the absence of chemical weathering, physical weathering dominates arid landscapes. Broken, angular rock fragments often cover the ground surface. However, physical weathering rates reach their maximum where climatic conditions favor frost formation and therefore frost wedging. As a result, physical weathering is most significant in subpolar regions and at high altitudes in middle to low latitude regions. Minimal rates of physical weathering occur in humid, tropical regions where rapid rates of chemical weathering discourage physical weathering activities.

The mineral composition of a rock is also integral to weathering rates. Rocks that contain an abundance of chemically unstable minerals quickly decompose and disintegrate when exposed to the atmosphere. Because of its dissolved oxygen content and slight acidity, rainwater readily attacks most mineral grains. For example, upon exposure at the Earth's surface, iron- and manganese-rich (ferromagnesian) minerals such as olivine ($FeMgSiO_4$) and pyroxene ($FeMgSi_2O_6$) quickly oxidize, and rainwater easily dissolves calcite ($CaCO_3$). However, quartz (SiO_2) effectively resists chemical weathering in most settings and so is a common constituent of beach sand, soil, and many sedimentary rocks. Rocks such as basalt that contain abundant ferromagnesian minerals quickly begin to weather when exposed at Earth's surface; granite, a quartz-rich rock, resists weathering.

NATURAL RESOURCES AND WEATHERING

Weathering processes produce or affect most of the world's natural resources. Soil and soil fertility, the character and quality of water supplies, sediments and sedimentary rocks—all are dependent on weathering.

A combination of chemical and physical weathering produces soil, our most important natural resource. Without soil, terrestrial ecosystems could not develop or survive. Weathering provides the parent material for soil and alters soil texture and composition as it matures. The type and quantity of soil, an important influence on vegetation, result directly from the types and rates of weathering. In areas where erosion of soil exceeds its development by weathering processes, fertility declines and so production of natural or agricultural vegetation declines as well. The future of humankind may well depend on our ability to control rates of soil erosion.

Chemical and physical weathering contribute to the character of surface water by determining the amount and type of dissolved and suspended solids present. Chemical weathering produces dissolved solids from rocks and soils. These mineral solutions may enter surface water and groundwater supplies and benefit humans in appropriate concentrations; at higher quantities, however, they can impair health. For example, low dietary concentrations of selenium are vital to human health (as indicated by the inclusion of selenium in multivitamin tablets). At higher concentrations it can be quite toxic, even deadly.

Weathering products, such as soil, also contribute to the cloudy appearance of river water after a storm, a phenomenon known as sediment pollution. Making the water potable requires the removal of these suspended clay- and silt-sized particles by settling or other means. Soil can also help purify water. As water percolates down through the soil, the finer particles and soil bacteria filter out harmful organic and inorganic materials.

Weathering produces many important mineral resources. Most belong to one of three broad categories based on origin: clastic, residual, and chemical sedimentary deposits. Mechanically deposited sediments, or clastic mineral deposits, form when transport velocity decreases and sediments accumulate. These deposits usually consist of chemically stable minerals that resist weathering and so become concentrated. They include stream-deposited metals (gold, platinum) and nonmetals (diamonds, zircon) known as placers, common sand and gravel deposited by glaciers and rivers, and beach sands, including those with heavy minerals (e.g., ilmenite and magnetite).

Residual or *in situ* sedimentary mineral deposits consist of resistant minerals that accumulate at the site of weathering rather than undergoing transport and deposition. Kaolin and bentonite, common industrial minerals, and bauxite and laterite, low-grade metal ores, are typical examples.

The remaining group of sedimentary mineral deposits, chemical precipitates, form when minerals precipitate from water because of evaporation, biological activity, or chemical reactions. These may form as primary deposits on the bottom of a body of water and as secondary enrichments of existing mineral deposits. Primary chemical precipitates (evaporites) form in inland lakes and seas, such as the Great Salt Lake in Utah, and restricted ocean basins, such as the Red Sea. They consist of various mineral salts such as rock salt, borax, gypsum, and the carbonates calcite and dolomite. Primary biochemical precipitates form when organisms remove dissolved minerals from water and produce hard parts such as shells. Many types of limestone, as well as diatomaceous earth, form in this way. Finally, an assortment of metal deposits develop when mineral solutions percolate down through existing mineral bodies located at or near the Earth's surface. The solutions react with the existing minerals, producing secondary deposits enriched in valuable metals such as iron, copper, and manganese.

—Clayton D. Harris

Further Reading

Birkeland, Peter W. *Soils and Geomorphology.* 3rd ed., Oxford UP, 1999.
Huggett, Richard John. *Fundamentals of Geomorphology.* 2nd ed., Routledge, 2007.
Nichols, Gary. *Sedimentology and Stratigraphy.* 2nd ed., Wiley-Blackwell, 2009.
Prothero, Donald R., and Fred Schwab, editors. *Sedimentary Geology.* W.H. Freeman, 2013.
Ritter, Dale F., R. Craig Kochel, and Jerry R. Miller, editors. *Process Geomorphology.* 4th ed., McGraw-Hill, 2002.
Schaetzl, Randall J., and Sharon Anderson. "Weathering." *Soils: Genesis and Geomorphology.* Cambridge UP, 2005.
Tarbuck, Edward J., Frederick K. Lutgens, and Dennis Tasa. *Earth: An Introduction to Physical Geology.* 13th ed., Pearson Education, 2019.

WEGENER, ALFRED

Fields of Study: Geology; Seismology; Oceanography

ABSTRACT

Wegener received credit for the first clear statement of the hypothesis of continental drift. Although his ideas were scornfully dismissed by most geologists in his own time, they were enthusiastically revived by oceanographers following discovery of the Mid-Atlantic Ridge in the mid-1960s, as the precursor to the well-known plate tectonics theory.

KEY CONCEPTS

continental drift: the concept that that the less dense rock of Earth's continents "float" atop the denser rock of Earth's crust and over millions of years have moved into their present positions

gymnasium: a school in Germany, Scandinavia, or Central Europe that prepares students to enter studies at a university

Mid-Atlantic Ridge: a line of active undersea volcanic structures discovered in the mid-twentieth century, formed by the upwelling of magma from the mantle, which acts to drive apart the ocean floor to the east and west of the ridge

Pangaea: "all earth," the supercontinent proposed by Wegener as the original position of all of the continental land masses some 250 million years in the past

plate tectonics: the concept that Earth's thin crust is fractured into a number of large "plates" that are moved relative to each other by the action of the molten rock of the mantle layer

WEGENER'S EARLY LIFE

Alfred Wegener (VAY-geh-nehr) was the son of a preacher in the Evangelical church. He was educated at the Köllnisches Gymnasium in Berlin and subsequently attended the Universities of Heidelberg, Innsbruck, and Berlin. He received his PhD degree in astronomy from the University of Berlin in 1905, and his doctoral thesis was on the conversion of a set of thirteenth century tables of planetary motion from a number system based on multiples of sixty to the decimal system.

As a youth Wegener became fascinated with the vast, ice-covered island of Greenland. At that time nothing was known of Greenland's interior, as exploration had been limited to the coastal areas only. To prepare himself for the longed-for trip to Greenland, Wegener spent much of his spare time in a rigorous program of endurance training, which consisted of gymnastics and all-day bouts of walking, ice skating, mountain climbing, and skiing.

Following his graduation from the University of Berlin in 1905, Wegener found that he had lost interest in the field of astronomy, and he turned his attention to the newly established science of meteorology instead, taking a position as an assistant at the Royal Prussian Aeronautical Observatory at Lindenberg. There he experimented with kites and balloons, and in 1906 he and his brother Kurt broke the world's record for long distance in a free balloon, drifting for fifty-two hours from April 5 to 7 across Germany, Denmark, and adjacent water bodies.

Also in 1906, the long-awaited opportunity to visit Greenland finally presented itself, and Wegener resigned his position at the observatory to join the 1906-1908 Danmark Expedition to Greenland. He served as official meteorologist to the expedition, which successfully explored the northeastern part of the island and made important weather measurements. Wegener returned safely to Germany at the expedition's end, but its leader, Ludvig Mylius-Erichsen, did not. He and two companions died while on a side trip northward toward the pole, a tragic premonition of Wegener's own death on the Greenland ice cap in 1930.

On his return to Germany in 1908, Wegener went to the University of Marburg as a lecturer. There he taught courses in meteorology, practical astronomy, and cosmic physics, meanwhile continuing his meteorological investigations. He also published a book on the thermodynamics of the atmosphere in 1911, but his academic career was abruptly halted in 1914 when World War I broke out, and he was called up as part of the general mobilization.

LIFE'S WORK

Wegener is chiefly noted for the hypothesis of continental drift. In it he contended that Earth's continents had once been joined together but then had broken apart and begun drifting about the surface of the globe much as icebergs drift about the sea. The subject of drift had claimed his attention as early as 1903 when he pointed out the remarkable "fit" between the continents on the opposite sides of the Atlantic to a fellow student at the University of Berlin. Then in 1910, when he was examining world maps in a handsome new atlas, the idea of drifting continents came over him in its full force, but he relates that he instantly rejected the idea as being "too

Alfred Wegener c. 1924-1930. Photo via Wikimedia Commons. [Public domain.]

improbable." In 1911, however, he stumbled "quite by accident" on a study summarizing the similarities between the fossil animals found in Africa and in Brazil. The study attributed these similarities to a former land connection now submerged between the two continents; Wegener, however, believed otherwise. He was convinced that Africa and Brazil had once been joined together and subsequently had drifted apart.

Wegener presented his new ideas to the scientific community during January of 1912, first addressing the Geological Association in Frankfurt and then the Society for the Advancement of Natural Science in Marburg. A paper on the subject quickly followed, but next came two interruptions. The first was a return to Greenland in the spring of 1912, accompanied by J. P. Koch, a Dane whom he had met on the 1906-1908 expedition. After lengthy preparations, they made the first successful east-west crossing of the ice cap at its widest point, using sledges hauled by ponies for the 1,100-kilometer trip. The second interruption came in 1914 with the outbreak of World War I when Wegener was called into the army.

Wegener served with distinction and was wounded twice. In his spare time, however, he still managed to continue working on his Greenland data, to make meteorological observations, and to refine his thinking on the subject of continental drift. While on sick leave in 1915, he even managed to set down his ideas in a slender volume, which came out under the title *Die Entstehung der Kontinente und Ozeane* (1915; *Origin of Continents and Oceans*, 1922). This book spelled out the basic tenets of continental drift, along with supporting evidence. In essence, Wegener proposed that the continents had not always occupied their present positions but were originally gathered together as a large supercontinent, which he called "Pangaea" (literally, all-earth). Pangaea had occupied half the globe, with the other half being covered by the water of the ancestral Pacific Ocean.

Beginning in the Mesozoic era, however some 200 million years before the present Wegener believed that the continents had begun detaching themselves from Pangaea and started drifting toward their new locations, much as icebergs plow through water. He proposed that the Atlantic Ocean had been formed when Pangaea split in two, with North America, South America, Antarctica, and Australia originating from the strip of continental crust on the west. As this strip plowed westward across the Pacific, Wegener believed that great wrinkles had appeared along its leading edge that now form the mountain chains that extend from the Andes to Alaska. He also proposed that the towering Himalaya Mountains had been created in a similar fashion when northward-drifting India crashed into the continent of Asia.

Wegener spent the years following 1915 marshaling the evidence to support his hypothesis. First there was the behavior of earthquake waves; they seemed to indicate that the continents were composed of lightweight rocks floating on the heavier rocks of the ocean floor. Then came the jigsaw-puzzle "fit" between continents on the opposite sides of the Atlantic; by including with the continents the submerged lands adjacent to their shoreline the so-called continental shelves he obtained an even better "fit" than was possible by merely juxtaposing coastlines.

A third line of evidence was the presence of similar geologic features on continents that are now widely separated. The rocks of Brazil strongly resemble those of Africa, for example, and mountain trends even seem to match when the continents are restored to their predrift positions. Wegener also cited similarities in fossil plants and animals found in continents that are now far distant from one another.

An ingenious line of evidence was based on glacial activity in pre-Mesozoic times. Distinctive deposits of ancient glacial debris are found in such diverse localities as southern South America, Antarctica, Australia, South Africa, and parts of India. Worldwide glaciation cannot explain these deposits, because tropical climates existed side by side with glaciated areas. Pangaea provided a likely explanation, however, because when the continents are restored to their predrift positions, all these glaciated areas become contiguous, suggesting a single large ice cap located at the South Pole. A final line of evidence was the apparent changes in the longitude of various localities on the globe. Such changes would have supported his hypothesis by demonstrating that the continents were still moving. Unfortunately, the methods used for measuring longitude in those days were not sufficiently precise to provide the confirmation he needed.

After the war's end, Wegener accepted a teaching post at the University of Hamburg and while there conducted experiments related to another one of his theories, namely that the moon's craters had originated as a result of meteoric bombardment, rather than through volcanic action. Continental drift remained his primary obsession, however, and he brought out revised and expanded editions of *Origin of Continents and Oceans* in 1920 and 1922. His hypothesis had attracted attention outside Germany by now, and French, English, Spanish, and Russian translations of the 1922 edition appeared in 1924. In the same year, Wegener was also appointed to the chair of meteorology and geophysics at the University of Graz in Austria, an important upward step in the academic world.

The growing attention to his ideas resulted in a swelling controversy, however, which reached its climax in 1926 when the American Association of Petroleum Geologists organized a symposium concerning continental drift as part of its annual meeting in New York City. Wegener attended, as did noted geologists from Europe and the United States. The hypothesis suffered a major setback as far as acceptance by the scientific world was concerned, however, because of the effective way in which his opponents presented their arguments. Undaunted, Wegener put out a fourth edition of his book in 1929, incorporating in it his rebuttals to the arguments advanced at the symposium. He also agreed to make another trip to Greenland, as leader of the German Inland Ice Expedition of 1930. The jet stream, that fast-moving current in the upper atmosphere that circles the earth in northern latitudes, had just been discovered, and the expedition's goal was to establish three year-round stations on the ice cap to study jet stream flow. The expedition was also to make pioneering measurements of the thickness of the ice cap using echo-sounding techniques.

Two of the three stations were to be on opposite coasts, and the third named "Eismitte" (mid-ice) was to be in the center of the ice cap at an elevation of 3,000 meters. Two members of the expedition had already set up a temporary camp there by the fall of 1930. Their quarters consisted of a pit dug into the ice and roofed over. Wegener and a fourth member followed with the necessary supplies but did not arrive until October 29 because of bad weather. Because most of the supplies had been lost en route, Wegener and his Inuit companion decided to return to the coastal base before the polar night set in. They never made it. Wegener's body was later found halfway back, neatly sewn into his sleeping bag and buried in the snow with upright skis as a marker. His Inuit companion apparently had gone on, but no trace of him or his sled was ever found.

SIGNIFICANCE

Wegener's scientific reputation rests primarily on his advocacy of continental drift. Although he was not the first to speculate about drifting continents—Sir Francis Bacon had called attention to the similar shapes of Africa and South America in 1620—he was the first to present a unified hypothesis, together with evidence drawn from many scientific disciplines. This was a remarkable achievement in itself in that era of the Germanic tradition of specialization. Wegener was also the first scientist supporting continental drift who was taken seriously. His book went through four editions and was translated into as many languages. A symposium was even arranged by the prestigous American Association of Petroleum Geologists, and world-famous scientists met to discuss his ideas. Unfortunately, their verdict was negative, and this led to continental drift's being consigned to the academic scrap heap for the next thirty-five years.

Numerous factors led to the hypothesis's rejection. There were some flaws in his reasoning, to be sure, and some of the evidence proposed later proved to be inaccurate. More telling, however, was Wegener's inability to provide a convincing mechanism for moving continents around the globe. In fact, he was unable to demonstrate that the continents were moving at all. Being an "outsider" did not help either; his doctorate was in astronomy, his research had been in meteorology, and his professional reputation rested on his exploits as a balloonist and as an Arctic explorer. His credentials in geology were nonexistent. No wonder the reception that he received from the professional geologists at the New York City symposium ranged from politely skeptical to openly sarcastic.

After Wegener's death in 1930, the hypothesis of continental drift was no longer given serious consideration by the scientific world. It had been too radical a departure from the accepted thinking of the day. As one of the participants in the 1926 symposium put it, "If we are to believe in Wegener's hypothesis, we must forget everything which has been learned in the last seventy years and start all over again." That, of course, was exactly what was going to have to happen. A revolution in geological thinking was coming, but regrettably Wegener did not live to see it. It was not until the mid-1960s, when oceanographers proposed the radically new plate tectonics theory, that scientists around the world realized how forward-looking Wegener had been.

—*Donald W. Lovejoy*

Further Reading

Dott, Robert H., and Roger Lyman Batten. *Evolution of the Earth.* 4th ed., McGraw-Hill, 1988.

Greene, Mott T. *Alfred Wegener: Science, Exploration and the Theory of Continental Drift.* Johns Hopkins UP, 2015.

LeGrand, H. E. *Drifting Continents and Shifting Theories.* Cambridge UP, 1988.

McCoy, Roger M. *Ending in Ice: The Revolutionary Idea and Tragic Expedition of Alfred Wegener.* Oxford UP, 2006.

Marvin, Ursula B. *Continental Drift: The Evolution of a Concept.* Smithsonian Institution Press, 1973.

Oreskes, Naomi. *The Rejection of Continental Drift: Theory and Method in American Earth Science.* Oxford UP, 1999.

Sullivan, Walter. *Continents in Motion: The New Earth Debate.* McGraw-Hill, 1974.

Wyllie, Peter J. *The Way the Earth Works: The Introduction to the New Global Geology and Its Revolutionary Development.* John Wiley & Sons, 1976.

Yount, Lisa. *Alfred Wegener: Creator of the Continental Drift Theory.* Chelsea House Publishing, 2009.

WEGENER PROPOSES THE THEORY OF CONTINENTAL DRIFT

Fields of Study: Geology; Seismology; Geophysics

ABSTRACT

Alfred Wegener proposed that all landmasses on the earth were once part of the supercontinent of Pangaea, which then fragmented, with the pieces drifting apart to form the present-day continents.

KEY CONCEPTS

asthenosphere: the plastic layer between the base of the solid crust of Earth and the mantle

lithosphere: the solid crust of Earth

seafloor spreading: the gradual separation of oceanic plate segments on either side of a divergent tectonic boundary, accompanied by the formation of new oceanic crust by upwelling of basaltic magma into the rift between the plate segments

BASIS OF THE CONTINENTAL DRIFT THEORY

The concept of "continental drift" was developed, at least in part, to explain the striking parallelism between the eastern and western Atlantic coasts, which seem as though they could fit together like pieces of a giant jigsaw puzzle. In particular, the fit between the eastern coast of South America and the western coast of Africa is very striking. The idea that the continents were once joined together as part of a single landmass predates Wegener's treatise, however.

As early as 1620, Sir Francis Bacon had discussed the possibility that the Western Hemisphere had once been joined to Africa and Europe. In 1668, P. Placet expressed similar ideas. Antonio Snider-Pellegrini, in his book *La Création et ses mystères dévoilés* (1859; *Creation and Its Mysteries Revealed*), recognized the similarities between American and European fossil plants of the Carboniferous period (about 300 million years ago) and proposed that all continents were once part of a single landmass. By the end of the nineteenth century, Austrian geologist Eduard Suess had noticed the close correspondence between geological formations in the lands of the Southern Hemisphere and had fitted them together into a single landmass he termed "Gondwanaland." In 1908, Frank Bursley Taylor of the United States, and in 1910, Alfred Wegener of Germany independently suggested mechanisms that could account for large, lateral displacements of the earth's crust and therefore explained how continents could be driven apart. Wegener's work became the center of the debate that has lasted until the present.

The concept of continental drift was best expressed by Wegener in his book *Die Entstehung der Kontinente und Ozeane* (1912; *The Origin of Continents and Oceans*, 1924). He based the theory not only on the shape of the continents but also on geological evidence found around the world. Wegener specifically cited similarities in fossil fauna and flora (extinct animals and plants) found in Brazil and Africa. He developed a series of maps to show three stages in the drift process and named the original supercontinent *Pangaea* (a word meaning "all lands"). Wegener believed that the continents, composed of light-density granitic rocks, were independently propelled and plowed through the denser basalts of the ocean floor driven by forces related to the rotation of the earth. He provided evidence based on detailed correlations of geological features and fossils indicating a common historical record on both sides of the Atlantic. He also proposed that the supercontinent of Pangaea existed before the beginning of the Mesozoic era (about 200 million years ago).

THE SPLITTING OF PANGAEA

Wegener visualized the split of Pangaea as beginning during the Jurassic period (about 190 million years ago) with the southern continents moving westward and toward the equator. South America and Africa began to drift apart during the Cretaceous period (70 million years ago). The opening of the northern Atlantic was accomplished during the Pleistocene epoch (approximately 2.5 million years ago). Greenland and Norway started to separate as recently as 1.5 million years ago. The Indian peninsula drifted northward, colliding with the Asian continent and giving rise to the folded mountains of the Himalayas. Similarly, the European Alps and the Atlas Mountains of North Africa were explained as a westward extension of the Himalayan Chain. Wegener also suggested that as the drifting continents met the resistance of the ocean floor, their leading edges were compressed and folded into mountains. In this way, he also explained the Western Cordillera of the Americas and the mountains of New Zealand and New Guinea. He visualized the tapering ends of Greenland and South America and the island arcs of the Antilles and East Asia as stragglers trailing behind the moving continents. Periods of glaciation found in the southern part of South America, Africa, Australia, peninsular India, and Madagascar provided further evidence of drift.

LAURASIA AND GONDWANALAND

Detailed studies by the South African geologist Alexander Logie Du Toit provided strong support for Wegener's concepts. Du Toit postulated two continental masses rather than the single entity of Pangaea: the northern supercontinent of Laurasia and its southern counterpart Gondwanaland, separated by a seaway called "Tethys." Du Toit was also the first to propose that the continental masses of the Southern Hemisphere had moved relative to the position of the South Pole. His ideas were published in *Our Wandering Continents* (1937), a book he dedicated to Wegener. Both Wegener and Du Toit utilized four main lines of evidence to support continental drift: the geometric shape of the continents, matching rock types and geological structures, fossil evidence, and paleoclimatological evidence (evidence based on ancient patterns of climate). It was the last of these that led to the early conclusion that the geographic poles had shifted relative to the position of the continents through time. This "polar wandering" could best be explained by continental masses drifting over the fixed poles.

Continental drift was debated among earth scientists of the Southern Hemisphere and the leaders of geophysical thought of the Northern and Western Hemispheres. The eminent geophysicist Sir Harold Jeffreys of the University of Cambridge voiced opposition to the drift theory based on the concept that Earth's crust and underlying mantle were too rigid to permit such large motions.

Although Wegener and Du Toit had provided compelling evidence in favor of the drift theory, one monumental problem remained: What forces could be strong enough to rupture, fragment, and cause the continents to drift? It was precisely this question that resulted in the decrease in popularity of the continental drift theory until its rebirth in the 1960s as the new seafloor spreading and plate tectonics. Wegener had visualized the continents as being raised above the seafloor on a spinning, spheroidal earth and argued that the continents had been propelled toward the equator. Drift was explained as the result of gravitational attraction between the continents and the equatorial bulge, resulting in the movement away from the poles toward the equator. A westward drift was explained as resulting from the differential attraction of the Moon and the Sun on the continents, causing them to lag behind the rotation of the earth.

Although appealing, Wegener's theory of continental drift remained controversial and was not widely accepted until Harry Hammond Hess and Robert Sinclair Dietz identified the process of seafloor spreading in the early 1960s. In fact, although the forces proposed by Wegener (and also by Taylor) are known to exist, they are minuscule and no one has considered them seriously as the mechanism capable of rupturing and moving continents. It was Du Toit who first proposed the idea that continents could slide over the mantle under the action of gravity (an idea that came to be used by some scientists in defense of plate tectonics), but Arthur Holmes of the University of Edinburgh was the originator of the now-popular concept of thermal convection in Earth's mantle as the main cause of drift. Holmes's model, published in 1931, is very similar to models that came to be used in the widely accepted theory of plate tectonics. Holmes was also the first to introduce the idea that the continents themselves do not play an active role in the drift, but act as passive members being carried along by a moving mantle in sort of a conveyor-belt motion.

SIGNIFICANCE OF WEGENER'S PROPOSALS

Wegener's theory of continental drift, first introduced during the second decade of the twentieth century, remains one of the most fascinating, inspiring, and controversial topics in the field of earth science. The drift theory polarized the scientific community and became the center of a heated debate until its rebirth during the second half of the century, when the concepts of seafloor spreading and plate tectonics were developed. Most earth scientists in the United States and around the globe subscribe to the theory of plate tectonics, a modern version of Wegener's continental drift.

Although the theory of continental drift was supported by a variety of geological evidence, a poor understanding of the nature of the oceanic crust prevented the development of a complete theory of earth dynamics until the 1960s. It was then that the topography of the ocean floor was mapped and magnetic and seismic characteristics established. In plate tectonics, the crust and uppermost upper mantle constitute what is now known as the lithosphere. The lithosphere lies above a soft, plastic (or ductile) zone in the mantle known as the asthenosphere. Thermal convection within the asthenosphere provides the mechanism for splitting the rigid lithosphere into distinct plates and propelling them into motion. The lithospheric plates are bounded by ocean ridges, ocean trenches, and major faults. At the ocean ridges, plates move apart where the convecting mantle rises and spreads laterally. It is at these divergent boundaries that a new ocean floor is generated. The moving plates converge at the site of oceanic trenches, where the basaltic (oceanic) lithosphere descends into the mantle under the lighter-density, granitic (continental) lithosphere. In those regions where two lithospheric plates move past each other, large transform fault boundaries are generated.

Plate tectonics provides an appealing explanation for zones of earthquakes, active volcanism, and mountain building. These events take place along plate boundaries, where they either diverge, collide, or move relative to each other. The theory has gained wide acceptance among earth scientists worldwide. Its concepts are deeply rooted on Wegener's original ideas of continental drift. The new theory provides answers to many of the critical questions of global tectonics, but, as commonly occurs in science, new questions and problems continue to arise. Just as in Wegener's time, notable and respected workers voiced opposition to the new theory. V. V. Belousov of the Soviet Union and Arthur Augustus Meyerhoff of the United States both provided well-founded and thought-provoking arguments against the new theory, but they failed to provide a theory of global tectonics as well integrated as plate tectonics. Partly as a result of this, continental drift and its modification into the concepts of seafloor spreading and plate tectonics remained one of the most significant theories in earth science. Although it is not without its problems, the theory is one of the most complete explanations of global tectonics in existence.

—Robert G. Font

Further Reading

Du Toit, Alexander Logie. *Our Wandering Continents.* Hafner, 1937.

Stein, Seth, and Michael Wyesession. *An Introduction to Seismology, Earthquakes, and Earth Structure.* Wiley, 2013.

Tarbuck, Edward J., Frederick K. Lutgens, and Dennis Tasa. *Earth: An Introduction to Physical Geology*. 13th ed., Pearson Education, 2019.

Taylor, Frank Bursley. "Bearing of the Tertiary Mountain Belt on the Origin of the Earth's Plan." *Geological Society of America,* vol. 21, 1910, pp. 176-226.

Wegener, Alfred Lothar. *The Origin of Continents and Oceans.* Trans. by J. G. A. Skerl. E. P. Dutton, 1924.

Wilson, J. Tuzo. "A Revolution in Earth Science." *Geotimes,* vol. 13, Dec. 1968, pp. 10-16.

X

Xenoliths

Fields of Study: Geology; Geochemistry; Geophysics

ABSTRACT

Xenoliths are blocks of preexisting rocks within magma. Consequently, xenoliths provide a sampling of materials through which the magma has traversed on its rise toward the surface. Some xenoliths originate within the mantle and provide the only means of obtaining samples of this elusive material.

KEY CONCEPTS

assimilation: the absorption of the chemical components of wall rock or xenoliths into a magma

country rock: rocks through which a magma is intruding; also known as "wall rock"

diatreme: a pipelike conduit in the crust of the earth filled with fragmented rock produced by gas-rich volcanic eruptions

eclogite: rock composed principally of garnet and pyroxene that formed at high pressures associated with great depths

felsic rocks: igneous rocks rich in potassium, sodium, aluminum, and silica, including granites and related rocks

mafic rocks: igneous rocks rich in magnesium and iron, including gabbro, basalt, and related rocks

magma: a naturally occurring silicate-rich melt beneath the surface of the earth

mantle: the intermediate zone between the crust and the core of the earth

peridotite: a class of ultramafic rocks made up principally of pyroxene and olivine, with subordinate amounts of other minerals

segregation: the concentration of early formed minerals in a magma by crystal settling or crystal floating

ultramafic: a term for any rock consisting of more than 90 percent ferromagnesium minerals, including olivine and pyroxene

OCCURRENCE OF XENOLITHS

Xenoliths are fragments of preexisting rocks that have been incorporated in a magma as it makes its way into higher levels of the crust. The term "xenoliths" is derived from the Greek roots *xeno* and *lith*, meaning "strange or foreign" and "rock," respectively. These rock fragments are pieces of previously formed rocks that become incorporated into the magma and perhaps removed from their source as the magma moves. The xenoliths may retain their original identity with minor alteration, or they may be greatly altered by attendant heat and fluids present in the magma. Xenolithic inclusions may be preserved near to their original sources along the borders of an intrusive magma, or they may be carried for great vertical distances from where they originated. In this manner, fragments of deep crust and mantle material from as much as 200 to 300 kilometers below Earth's surface have been brought to the surface by volcanic eruptions.

Xenoliths represent fragments of the rocks through which a magma has moved to its site of final emplacement and crystallization. They may be found in products of explosive volcanism such as volcanic tuff and breccia, within crystalline igneous rocks as in lava flows, and within shallow and deep-seated igneous rocks. Explosive volcanic materials are ejected by highly gas-charged eruptions that produce diatremes and maar-type volcanoes. These are volcanic craters in the form of inverted conelike or dish-like depressions in the surface surrounded by a rim of ejected deposits. Xenoliths are found as angular or rounded blocks embedded in ash tuffs or volcanic breccia in the rim and the pipelike conduit underlying the crater. In certain types of basalt and related magmas that originate deep in the mantle, such as kimberlite, rare fist-sized fragments of mantle material are transported upward from near the source of the magma origin. Fragments may also be collected from rocks traversed by the magma along its path of vertical ascent through the crust. Xenoliths in crystalline igneous rocks are embedded within the rock and are not exposed until erosion exposes the xenolith by removing overlying material. Granite rocks typically contain large xenoliths of metamorphic or sedimentary rocks. Such xenoliths reflect the typical intrusive process that produces granites. In this process, subsurface magma chambers expand and move upward by physically plucking country rocks from the wall and roof.

WALL-ROCK XENOLITHS

Three basic varieties of xenoliths are recognized: wall-rock xenoliths, cognate xenoliths, and mantle xenoliths. Wall-rock xenoliths are represented by blocks and pieces of the adjacent country rock that have been incorporated into the magma. Cognate xenoliths are inclusions of the chilled margins of the magma or comagmatic segregations—that is, masses of previously solidified magma that break loose and are later incorporated in more energetic magmatic motions. Mantle xenoliths are presumed to be pieces of the mantle that become incorporated in magmas that are formed by partial melting deep within Earth.

Magmatic intrusions make room for themselves by three processes: forceful injection, stoping, and assimilation. Stoping occurs when the magmatic front advances by injection into fractures and surrounding blocks of country rock. These blocks may sink in the magma, and they may be slightly altered or totally assimilated within the magma depending on the characteristics of the magma and the wall

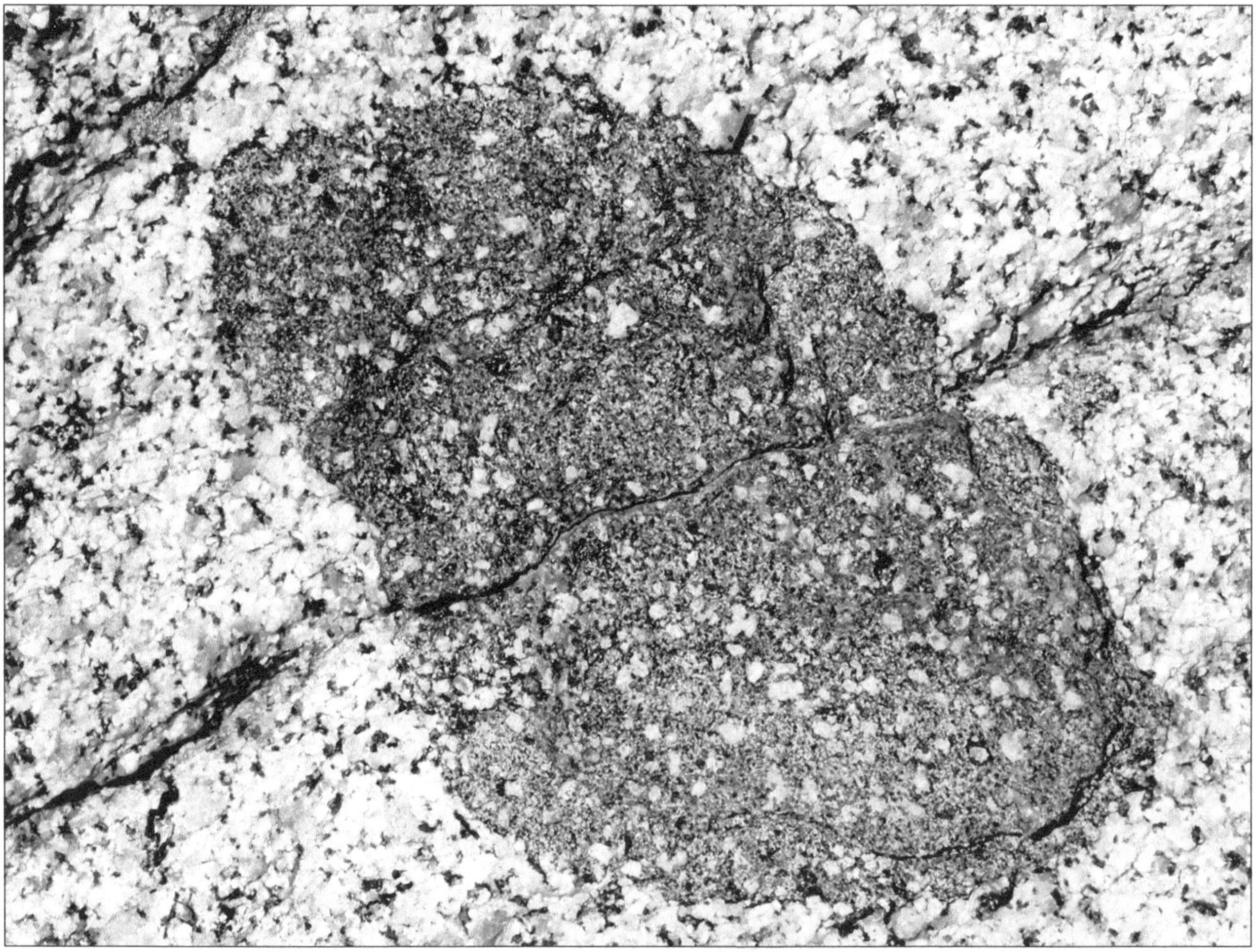

Gabbroic xenolith in granite in Rock Creek Canyon, eastern Sierra Nevada, California. Photo via Wikimedia Commons. [Public domain.]

rock. Most wall-rock xenoliths in felsic magmas do not move far from their source. Xenoliths are abundant near the margins of most intrusions. Because the margins are more likely to be losing heat and cooling at rates faster than the interior of the intrusion, the magma is more viscous, and xenoliths are less likely to move very far from the source. In contrast, fast-moving magmas in volcanic conduits often carry a wide variety of xenoliths from country rock traversed by the magma. In this way, a wide variety of crustal rocks cut by the volcanic vent may be brought to the surface.

Xenoliths usually show effects of the high temperatures to which they are exposed. Preexisting minerals within a xenolith react to form new minerals that are in equilibrium with the magma. Thus, most xenoliths are brought to a high-grade metamorphic state unless they are composed of high-temperature refractory minerals to begin with, or unless they are exposed to high temperatures for short periods of time. The German term *Schlieren* is used to describe hazy, ill-defined streaks of xenoliths that are almost completely assimilated. Materials caught up in low-temperature felsic magmas are more likely to be altered by reactive assimilation, in which there is an exchange of ions between the xenolith and the magma. Inclusion of large blocks of country rock may alter the composition of magma by enriching it with elements that were not originally abundant.

One process that is not igneous in nature but creates xenoliths as well is the subterranean flow of salt. Buried salt beds are capable of plastic flowage in response to the weight of overlying rocks. The salt is not molten but nevertheless flows under pressure. Often the salt, which has a lower density than the enclosing sedimentary rocks, will rise thousands of feet through overlying sediments to form salt domes or salt plugs. The rise of large masses of salt is very much like the rise of magma. Pieces of wall rocks and subsalt rocks may be incorporated into the salt as xenoliths. In this manner, salt domes in the Persian Gulf have brought up blocks of sedimentary, igneous, and metamorphic rocks from great depths in the crust. There are even ultramafic igneous xenoliths in the Weeks Island salt dome in southern Louisiana that are thought to be fragments of mantle-derived ultramafic intrusions emplaced along fault zones prior to deposition of the salt.

COGNATE XENOLITHS

Cognate xenoliths, also called "autoliths," are xenoliths from parts of the magma that have previously crystallized. Magmas solidify over a wide range of temperatures. Large bodies of magma may require tens of thousands of years to crystallize fully. Material on the outer edge of the magma will cool and crystallize more rapidly than that of the interior, resulting in chilled margins. Elsewhere within the magma, early formed crystals will either float or sink depending on the specific gravity differential. Feldspar crystals tend to float and collect near the top of the magma chamber, and mafic minerals such as olivine or pyroxene tend to sink to the bottom of the chamber. Some magmas undergo energetic degassing because of the reduction in confining pressure as they approach shallow levels in the crust. The rapid evolution of dissolved volatiles may disrupt previously crystallized portions of magma (chilled margins or crystal segregations) and mix solid cognate xenoliths with the mobile fluid phase.

MANTLE XENOLITHS

Perhaps the most exotic xenoliths are those that originate within the mantle. The mantle lies at depths of five to forty kilometers below the surface and extends down to the top of the outer core nearly three thousand kilometers beneath the surface. No drill has penetrated to the mantle; therefore, these materials are completely inaccessible for direct sampling. The probable composition and mineralogical makeup of the mantle is postulated from calculations of the density, pressure, and temperatures that exist at mantle depths and by comparisons with meteorites, which are pieces of asteroids that have been fragmented by collisions with other asteroids. The interior of asteroids are thought to reproduce conditions similar to those of the mantle. Fortunately, pieces of the upper mantle are delivered to the surface of Earth as xenoliths in some magmas that originate by partial melting deep within the earth. For these rocks to make it to the surface without significant alteration by the host magma, they must be delivered to the surface in a fairly short period of time. Thus, it is not surprising that mantle xenoliths are found in volcanic rocks associated with rift zones and magmatic zones between plates, which allow for the rapid rise of gas-rich magmas to the surface.

Typical mantle xenoliths are composed of peridotite (olivine and pyroxene-rich rocks) incorporated in mafic volcanic rocks such as basalt. Basalts are formed from magma that originates by the partial melting of mantle materials. They often incorporate xenoliths from their place of origin as well as fragments of crustal rocks torn from walls of the conduit along which they are rising. Most notable of these magmas are varieties of peridotites known as "kimberlites" and "lamproites." Kimberlite (mica peridotite) is a potassic ultramafic rock that occurs in intrusive pipes and plugs, and explosively formed volcanic craters that overlay them. Lamproite is a porphyritic, ultrapotassic ultramafic rock that occurs in dikes and small intrusions. Kimberlite and lamproite magmas typically contain up to 75 percent xenoliths and xenocrysts. Kimberlites contain abundant xenoliths of lherzolite—a mantle peridotite with magnesian olivine, pyroxene, and minor calcium-plagioclase, spinel, or garnet.

The range of xenoliths in basalt is more restricted than that in kimberlites because basalts form at shallower levels of the mantle. Thus, basalts incorporate less of the upper mantle on their way to the surface. Spinel lherzolites are common in alkali basalts such as those found in Hawaii, Arizona, the Rio Grande Rift, and Central Europe. Basalts also contain xenoliths of harzburgite, dunite, and eclogite. Harzburgite (a peridotite with magnetite and spinel) and dunite (an ultramafic rock that consists of mostly olivine with minor chrome-bearing spinel as an accessory mineral) probably represent residual melts following fractionation of lhertzolite.

The occurrence of mantle xenoliths in volcanics is limited to intraplate magmatic environments such as oceanic islands (Hawaii and Tahiti) and continental volcanic provinces or rifts (the southern Colorado Plateau and the Eifel District, Germany). Kimberlite intrusions favor old, stable, thick continental crust (South Africa and central North America). These environments are characterized by simple plumbing systems in which magmas rise rapidly to the surface; otherwise, xenoliths would sink in the host magma. Xenoliths are more rare in complex systems found in interplate environments, such as collisional magmatic provinces or island arcs.

SCIENTIFIC VALUE OF XENOLITHS

The chief scientific value of xenoliths rests in their being samples of materials collected as magma ascended through Earth's mantle and crust. Wall-rock xenoliths provide samples of country rock that remain close to their original source. More importantly, xenoliths in rapidly ascending, volatile-rich volcanic magmas may provide samples of crustal rocks from the walls of the conduit throughout its entire path. These materials are brought to the surface in relatively unaltered states. In addition to mantle xenoliths, some localities—such as Kilbourne Hole, New Mexico, and Williams, Arizona—contain significant xenoliths of granite gneiss representing crustal basement rocks. Cognate xenoliths provide information on the earliest parts of the magma to crystallize. Basalt, kimberlite, and lamproites contain xenoliths from the mantle. Distribution of nodule occurrences is not random. Siliceous basalts rarely contain nodules, whereas alkali basalts commonly contain eclogite and spinel peridotites. Kimberlites contain abundant nodules, including garnet peridotite. These differences suggest different depths of origin for the different magmas. Mantle xenoliths are the only means of obtaining samples of the mantle. Study of these materials can lead not only to determining the composition of the mantle but also to developing an understanding of its physical state and some of its processes.

From a study of these rare rocks, various processes and conditions of the lower crust and upper mantle can be inferred. The mineralogical combinations serve as geobarometers and geothermometers. Controlled crystallization studies at a variety of temperatures and pressures are employed to characterize the stable mineral assemblage within differing crustal and mantle environments. A mixture of minerals or rocks exposed to elevated temperatures, as found in the lower crust and upper mantle, recrystallize into a mineral assemblage that is in equilibrium with the higher temperatures. Alternatively, by examining altered rocks such as some altered xenoliths, it is possible to determine their prior state before metamorphism by the magma.

ECONOMIC VALUE OF XENOLITHS

Xenoliths and xenocrysts find their principal economic value as a source of diamonds formed deep within the mantle and, to a much lesser extent, the olivine gem peridot. Diamonds occur as xenocrysts in kimberlite, lamproite, and alluvial gravels derived from kimberlites. Diamonds form at very high pressures. Minimum conditions required are pressure greater than 40 kilobars, which is equivalent to depths greater than 120 kilometers, and temperatures of approximately 1,000° Celsius. Diamonds are associated with magnesia garnet-bearing lherzolites and coesite. Diamonds make up only one part in twenty million of a typical diamond-bearing kimberlite, and many kimberlites are devoid of diamonds.

By far, most diamond production is from kimberlites in West Africa, South Africa, and Siberia, but diamonds are also mined in Australia, Brazil, and India. Small concentrations of diamonds have been found in kimberlite and lamproite bodies in Arkansas, Wyoming, Montana, Michigan, and Canada. After unsuccessful attempts to develop mining operations at the lamproite body at Murfreesboro, Arkansas, the area has been turned into the Crater of Diamonds State Park, where several small diamonds are found by tourists each year.

—René De Hon

Further Reading

Best, Myron G. *Igneous and Metamorphic Petrology.* 2nd ed., Blackwell Science Ltd., 2003.

Faure, Gunter. *Origin of Igneous Rocks: The Isotopic Evidence.* Springer-Verlag, 2010.

Grapes, Rodney. *Pyrometamorphism.* 2nd ed., Springer, 2010.

Groat, Lee Andrew. *The Geology of Gem Deposits.* Mineralogical Association of Canada, 2007.

Jerram, Dougal, and Nick Petford. *The Field Description of Igneous Rocks.* 2nd ed., Wiley-Blackwell, 2011.

Miller, Roger H. *Kimberlites: Mineralogy, Geochemistry and Petrology.* Springer, 2013.

Nazaré, M. H., and A. J. Neves, editors. *Properties, Growth, and Applications of Diamond.* IEE, 2001.

O'Donoghue, Michael. *Gems.* 6th ed., Robert Hale, 2008.

Raymond, L. A. *Petrology: The Study of Igneous, Sedimentary, and Metamorphic Rocks.* 2nd ed., Waveland Press, 2007.

Z

Zeolites

Fields of Study: Geology; Geochemistry; Geophysics

ABSTRACT

Zeolites are found naturally wherever volcanic rock and ash interact with alkaline groundwater and are mined extensively in many parts of the world. Artificial zeolites are purer and therefore preferable to the organic variety.

KEY CONCEPTS

molecular sieve: a zeolite material that is able to trap and hold small molecules of particular sizes such as water present as an impurity in an organic liquid or ammonia in air

WHERE ZEOLITES ARE FOUND

Zeolites are found naturally wherever volcanic rock and ash interact with alkaline groundwater. They are mined extensively in many parts of the world. Zeolites are also easily produced through artificial means; often artificial zeolites are purer and therefore preferable to the organic variety. As of 2008, 175 cataloged unique zeolite frameworks had been identified, 40 of which are naturally occurring.

PRIMARY USES OF ZEOLITES

The primary use for zeolites is in laundry detergents. The mineral's porous nature allows particles of dirt and contaminants to be captured in the wash cycle and then rinsed away in the rinse cycle. Other common uses are as aquarium filters and cat litter. Zeolites are also used extensively within the medical profession as molecular sieves to filter and purify air to make medical grade oxygen.

TECHNICAL DEFINITION OF ZEOLITES

Zeolite is a crystalline mineral most commonly used as an absorbent in commercial settings. Zeolites are microporous aluminosilicates with well-defined structures that act as "molecular sieves," capturing oxygen, minerals, and water and holding them within their many and variable-sized pores. Zeolites generally contain silicon, aluminum, and oxygen in their frame and water or other molecules within their pores. Their defining feature is a framework made up of four connected networks of silica (SiO_2) structures. The structures are tetrahedral in nature, with a silicon atom in the middle and oxygen atoms in the corners. They can link together by their corners and form beautiful structures that contain cavities and channels in which other molecules can become trapped. Because of this property, zeolites are excellent means of filtering impurities out of many fluids.

DESCRIPTION, DISTRIBUTION, AND FORMS OF ZEOLITES

Zeolites are aluminosilicates. They are more commonly referred to as "molecular sieves" because of a particular ability within the substance to sort molecules by their sizes and shapes. The size of a pore within the zeolite controls what molecules can flow into it. These pores are generally noncylindrical in nature. An "eight-ring" refers to a closed loop consisting of eight silicon or aluminum atoms and eight oxygen atoms. These atomic arrangements are tetrahedral in nature but not always symmetrical because of the bonding restraints within the zeolite, which are often based on the positioning of the oxygen atoms within the structure.

HISTORY OF ZEOLITE SCIENCE

Zeolite was given its name by Swedish mineralogist Axel Fredrik Cronstedt. Cronstedt heated a material then referred to as stilbite and observed that a small piece of the substance produced a large amount of steam. Cronstedt, pulling from the Greek words *zeo*, which means "boil," and *lithos*, which means "stone," named the material zeolite.

OBTAINING ZEOLITES

Zeolites are formed when volcanic rock and layers of ash are subjected to groundwater containing alkaline. These types of zeolites are seldom pure, often containing contaminants from other minerals and substances. Open-pit zeolite mines can be found in Arkansas, Idaho, and New Mexico. Topsoil is removed from the site, making access to the zeolite possible. The zeolite can be blasted free, cut free with ripper blades, or scraped from the ground with a bulldozer or top loader. Once removed from the ground, the zeolite is crushed, dried, and milled. Once processed, the substance is ready to use.

USES OF ZEOLITES

Zeolites can be found in many places. Their adaptability as a natural and human-made filter makes them useful in both household and industrial settings.

Oxygen concentrators containing zeolite are commonly used to produce medical-grade oxygen. The zeolite is able to filter impurities out of the air. Zeolites are also commonly used in water filtration systems for the same reason; they capture impurities in the water when water is filtered through them.

Zeolites are resistant to radiation, making them quite useful in nuclear reactions. They capture de-

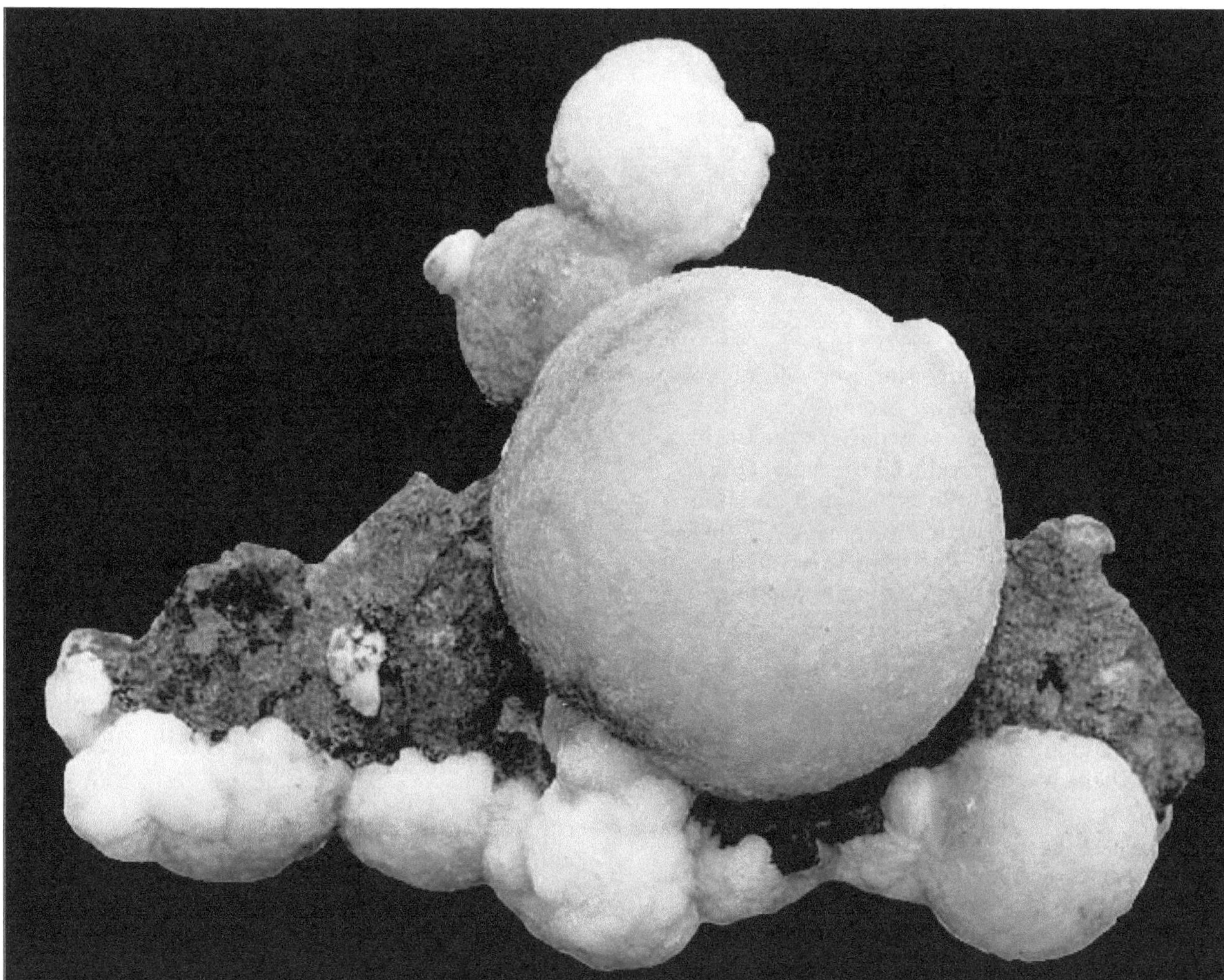

A form of thomsonite (one of the rarest zeolites) from India. Photo by Rob Lavinsky, iRocks.com, CC BY-SA 3.0, via Wikimedia Commons.

bris and waste products inside nuclear reactors. These waste products can be removed easily and disposed of without inhibiting the reactor's capabilities. Once filled with fission byproducts, the zeolite can be hard-pressed. This seals in the fission waste, making the by-product more easily disposable than the by-products of more conventional radioactive waste disposal methods.

In agriculture, a naturally occurring zeolite called "clinoptilolite" is used in the treatment of soil. The absorbent nature of this particular zeolite allows for vital minerals to be time-released into the soil. Potassium and nitrogen, in particular, are substances that can be released in this manner. Also, because zeolites are absorbent, water can be introduced into arid soil. In regions where water is overly abundant in the soil, introducing zeolites can help prevent root rot and improve harvests. Zeolites introduced into waterlogged soil can capture and retain up to 55 percent of their weight.

Zeolites are also commonly used in the heating and refrigerating industry. A xeolite's ability to absorb heat makes them useful in the capture and collection of heat and moisture that would escape otherwise. The introduction of zeolite into such environments improves the efficiency of both tasks.

This ability of zeolites to capture impurities in water and other liquids is not overlooked in the pet supply industry. Zeolite is an ingredient in many aquarium filters. It captures and filters ammonia and other nitrogenous compounds in aquariums, keeping the water from becoming toxic to fish and other aquatic creatures.

Another common use for zeolites is in cat litter. The nonclumping variety of cat litter is commonly made of zeolite or diatomite. This porous litter captures liquid, and is easily removed from the litter box.

Zeolites introduced into the food supply of animals can improve the animal's ability to process the food and help to improve the animal's bone density. Additionally, zeolites can reduce the airborne ammonia in holding pens by up to 80 percent.

Zeolite is also used in the petrochemical industry, filtering out impurities in crude oil and other petroleum products. Zeolite is hydrogenized and turned into powerful acid via ion exchange. Once zeolite is acidified, processes such as isomerization (a process that converts one compound into another with the same number of atoms, only rearranged), alkylation (which transfers the alkyl group from one atom to another), and catalytic cracking can be carried out. Catalytic cracking is a process that requires a furnace and a reactor. Crude oil is heated in the furnace, then sent to the reactor, where it is introduced to the acidic zeolite. It is run through the zeolite three times; each time it is filtered through a cooler version of the acidified zeolite. The next step is the separation of hydrogen from the crude oil. It is sent to a fractionator (yet another separation process that divides components of a compound by their physical qualities) in the final step and becomes the end product.

—Roger Dale Trexler

Further Reading

Auerbach, Scott M., Kathleen A. Carrado, and Prabir K. Dutta. *Handbook of Zeolite Science and Technology*. CRC Press, 2003.

Čejka, Jiří, Russell E. Morris, and Petr Nachtigall, editors. *Zeolites in Catalysis: Properties and Applications*. The Royal Society of Chemistry, 2017.

Inglezakis, Vassilis, and Antonis A. Zorpas, editors. *Handbook of Natural Zeolites*. Bentham Books, 2012.

Mahler, Annett. *Zeolites: Advances in Research and Applications*. NOVA Science Publishers, 2020.

Peiper, Howard. *Zeolite: Nature's Heavy Metal Detoxifier*. Safe Goods, 2006.

Xu, Ruren, et al. *Chemistry of Zeolites and Related Porous Materials: Synthesis and Structure*. Wiley-Interscience, 2007.

Bibliography

Adamson, Thomas K. *The Secrets of Jupiter (Planets)*. National Geographic, 2015.
———. *The Secrets of Neptune (Planets)*. National Geographic, 2015.
———. *The Secrets of Uranus (Planets)*. National Geographic, 2015.
Ahr, Wayne M. *Geology of Carbonate Reservoirs*. John Wiley & Sons, 2008.
Aiello, Gemma. *Volcanoes: Geological and Geophysical Setting, Theoretical Aspects and Numerical Modeling, Applications to Industry and Their Impact on Human Health*. IntechOpen, 2018.
Aldrete, Gregory S. *Floods of the Tiber in Ancient Rome*. Johns Hopkins UP, 2006.
Aleshire, Peter. *Deserts*. Chelsea House, 2008.
Alexander, Pramod O. *A Handbook of Minerals, Crystals, Rocks and Ores*. New India Publishing, 2009.
Allain, Rhett. "Why Does a Meteor Explode in the Air?" *Wired*. Condé Nast, 18 Feb. 2013. Accessed 11 Mar. 2015.
Allan, J. David, and María M. Castillo. *Stream Ecology: Structure and Function of Running Waters*. 2nd ed., Springer, 2007.
Allégre, Claude J. *Isotope Geology*. Cambridge UP, 2008.
Allen, J. R. L. *Principles of Physical Sedimentology*. Blackburn Press, 2001.
Alley, R. B., and P. U. Clark. "Deglaciation of the Northern Hemisphere." *Annual Reviews*, vol. 28, 2000, pp. 149-82.
Anderson, David E., Andrew S. Gaudie, and Adrian G. Parker. *Global Environments Through the Quaternary: Exploring Environmental Change*. Oxford UP, 2007.
Anderson, Robert S., and Suzanne P. Anderson. *Geomorphology: The Mechanics and Chemistry of Landscapes*. Cambridge UP, 2010.
Appelo, C. A. J., and D. Postma. *Geochemistry, Ground Water, and Pollution*. 2nd ed., Balkema, 2005.
Armitage, Philip J. *Astrophysics of Planet Formation*. 2nd ed., Cambridge UP, 2020.
Arnell, Nigel. *Hydrology and Global Environmental Change*. Pearson, 2002.
Arny, Thomas T. *Explorations: An Introduction to Astronomy*. 3rd ed., McGraw-Hill, 2003.
Aronson, Richard B., editor. *Geological Approaches to Coral Reef Ecology*. Springer, 2007.
ASCE. *Hurricane Katrina External Review Panel: The New Orleans Hurricane Protection System: What Went Wrong and Why*. American Society of Civil Engineers, 2007.
"Asteroids: The Discovery of Asteroids." *Space for Europe*. European Space Agency, n.d. Accessed 11 May 2015.
Auerbach, Scott M., Kathleen A. Carrado, and Prabir K. Dutta. *Handbook of Zeolite Science and Technology*. CRC Press, 2003.
Aydinalp, Cumhur. *An Introduction to the Study of Mineralogy*. IntechOpen, 2012.
Ayrey, Bill. *Lunar Outfitters: Making the Apollo Space Suit*. UP of Florida, 2020.
Bailey, Heather. *Aquifers: Properties, Roles and Research*. NOVA Science Publishers, 2017.
Baines, Kevin H., F. Michael Flasar, Norbert Krupp, and Tom Stallard. *Saturn in the Twenty-First Century*. Cambridge UP, 2019.
Balfour, Ian. *Famous Diamonds*. 4th ed., Christie's, 2000.
Banerjee, Dwijen. *Oil Sands, Heavy Oil & Bitumen: From Recovery to Refinery*. Pennwell, 2012.
Bartis, James T., et al. *Oil Shale Development in the United States: Prospects and Policy Issues*. RAND Corporation, 2005.
Beattie, Donald. *Taking Science to the Moon: Lunar Experiments and the Apollo Program*. Johns Hopkins UP, 2001.
Bedient, Philip B., Wayne C. Huber, and Baxter E. Vieux. *Hydrology and Floodplain Analysis*. 4th ed., Prentice Hall, 2008.
Belenitskaya, Galina. *Salt Systems of the Earth: Distribution, Tectonic and Kinematic History, Salt-Naphthids Interrelations, Discharge Foci, Recycling*. Wiley, 2018.
Bell, Fred J., and Laurance J. Donnelly. "Gravel, Sand, and Clay Pits." *Mining and Its Impact on the Environment*. Taylor & Francis, 2006.
———. "Placer Deposits and Mining." *Mining and Its Impact on the Environment*. Taylor & Francis, 2006.
Bell, Keith, and Jörg Keller, editors. *Carbonatite Volcanism: Oldoinyo Lengai and the Petrogenesis of Natrocarbonatites*. Springer Science + Business Media, 2012.
Bender, Michael L. *Paleoclimate*. Princeton UP, 2013.
Bennison, George M., Paul A. Olver, and A. Keith Moseley. *An Introduction to Geological Structures and Maps*. 8th ed., Hodder Education, 2011.
Benton, Michael J. *Vertebrate Palaeontology*. 3rd ed., Blackwell, 2009.

———, and David A. T. Harper. *Introduction to Paleobiology and the Fossil Record.* Wiley, 2020.

Benton, Rachel C., Dennis O. Terry Jr., Emmett Evanoff, and H. Gregory McDonald. *The White River Badlands, Geology and Paleontology.* Indiana UP, 2015.

Bergaya, Faïza, Benny K. G. Theng, and Gerhard Lagaly, editors. *Handbook of Clay Science.* Elsevier, 2006.

Berry, Joanne. *The Complete Pompeii.* Thomas and Hudson, 2013.

Best, Myron G. *Igneous and Metamorphic Petrology.* 2nd ed., Blackwell Science Ltd., 2003.

———, and Eric H. Christiansen. *Igneous Petrology.* Blackwell Science, 2001.

Bianchi, Thomas S. *Deltas and Humans: A Long Relationship Now Threatened by Global Change.* Oxford UP, 2016.

Bierman, Paul R., and David R. Montgomery. *Key Concepts in Geomorphology.* W.H. Freeman, 2019.

Birkeland, Peter W. *Soils and Geomorphology.* 3rd ed., Oxford UP, 1999.

Blatt, Harvey, Robert J. Tracy, and Brent E. Owens. *Petrology: Igneous, Sedimentary, and Metamorphic.* 3rd ed., W. H. Freeman, 2006.

Blondel, Phillippe, and John Mason, editors. *Solar System Update.* Springer, 2006.

Blumberg, Alan F., and Michael S. Bruno. *The Urban Ocean: The Interaction of Cities with Water.* Cambridge UP, 2018.

"Bode's Law." *Astro 2201: Our Home in the Universe.* Cornell University, n.d. Accessed 11 May 2015.

Boggs, Sam, Jr. *Petrology of Sedimentary Rocks.* 2nd ed., Cambridge UP, 2009.

———. *Principles of Sedimentology and Stratigraphy.* 5th ed., Prentice Hall, 2011.

———. "Limestones." *Petrology of Sedimentary Rocks.* 2nd ed., Cambridge UP, 2009.

"Bolide." *Cosmos: The SAO Encyclopedia of Astronomy.* Swinburne U of Technology, n.d. Accessed 27 Mar. 2015.

Bolt, Bruce A. *Earthquakes.* 5th ed., W. H. Freeman, 2006.

Borradaile, G. J., and D. Gauthier. "Magnetic Studies of Magma-Supply and Sea Floor Metamorphism: Troodos Ophiotile Dikes." *Tectonophysics,* vol. 418, 2006, pp. 75-92.

Bowles, J. F. W., et al. *Rock-Forming Minerals: Non-Silicates; Oxides, Hydroxides and Sulfides.* 2nd ed., Geological Society Publishing House, 2011.

Boyd, Richard N. *An Introduction to Nuclear Astrophysics.* U of Chicago P, 2007.

Brandner, Wolfgang. *Planet Formation: Theory, Observations, and Experiments.* Cambridge UP, 2011.

Brandt, John C., and Robert D. Chapman. *Introduction to Comets.* Cambridge UP, 2004.

Brown, Michael, and Tracy Rushmer, editors. *Evolution and Differentiation of the Continental Crust.* Cambridge UP, 2006.

Browning, Geraldine O., Joseph L. Alioto, and Seymour M. Farber, editors. *Teilhard de Chardin: In Quest of the Perfection of Man.* Fairleigh Dickinson UP, 1973.

Bryner, Jeanna. "Meteor Explosion in Russia Hurts Hundreds of People: Reports." *LiveScience.* Purch, 15 Feb. 2013. Accessed 11 Mar. 2015.

Bucher, Kurt, and Martin Frey. *Petrogenesis of Metamorphic Rocks.* 7th ed., Springer, 2002.

Bucher, Kurt, and Rodney Grapes. *Petrogenesis of Metamorphic Rocks.* 8th ed., Springer, 2011.

Buol, S. W., F. D. Hole, and R. J. McCracken. *Soil Genesis and Classification.* 5th ed., Iowa State UP, 2003.

Burbine, Thomas H. *Asteroids: Astronomical and Geological Bodies.* Cambridge UP, 2017.

Burdige, David J. *Geochemistry of Marine Sediments.* Princeton UP, 2020.

Burek, Cynthia V., and Bettie Higgs, editors. *The Role of Women in the History of Geology.* The Geological Society, 2007.

Burrows, William E. *This New Ocean: The Story of the First Space Age.* Random House, 1998.

Butcher, Kenneth S., Linda D. Crown, and Elizabeth J. Gentry. *The International System of Units (SI): Conversion Factors for General Use.* National Institute of Standards and Technology, 2006.

Campbell, Greg. *Blood Diamonds: Tracing the Path of the World's Most Precious Stones.* Basic Books, 2004.

Cann, J. R., H. Elderfield, A. Laughton, et al., editors. *Mid-Ocean Ridges: Dynamics of Processes Associated with Creation of New Ocean Crust.* Cambridge UP, 1999.

Cardarelli, Francois. *Encyclopedia of Scientific Units, Weights and Measures: Their SI Equivalences and Origins.* Springer, 2003.

Carlson-Berne, Emma. *The Secrets of Venus (Planets).* National Geographic, 2015.

Carr, Michael H. *The Surface of Mars.* Cambridge UP, 2018.

Carter, Chuck, Stephen Reynolds, Julia Johnson, Paul Morin, and Michael Kelly. *Exploring Earth Science.* McGraw Hill, 2015.

Cas, Ray, Guido Giordano, and John V. Wright. *Volcanology: Processes, Deposits, Geology and Resources.* Springer International, 2021.

Catling, David C., and James F. Kasting. *Atmospheric Evolution on Inhabited and Lifeless Worlds*. Cambridge UP, 2017.

Catuneanu, Octavian. *Principles of Sequence Stratigraphy*. Elsevier, 2006.

Cavallaro, Giuseppe, Rawil Fakhrullin, and Pooriah Pasbakhsh, editors. *Clay Nanoparticles: Properties and Applications*. Elsevier, 2020.

Cech, Thomas V. *Principles of Water Resources: History, Development, Management, and Policy*. 3rd ed., John Wiley, 2010.

Cejka, Jirí, Russell E. Morris, and Petr Nachtigall, editors. *Zeolites in Catalysis: Properties and Applications*. The Royal Society of Chemistry, 2017.

"Ceres: Overview." *Solar System Exploration*. NASA, 6 Jan. 2015. Accessed 6 May 2015.

Cessna, Abby. "Planetoid." *Universe Today*. Fraser Cain, 10 Aug. 2009. Accessed 6 May 2015.

Ceuleneer, Georges. "Earth Science: Mantle Mapped in the Desert." *Nature*, vol. 432, 2004, pp. 156-57.

Chaisson, Eric J. "Origin of Heavy Elements." *Cosmic Evolution*. Harvard-Smithsonian Center for Astrophysics, 2013. Accessed 2 June 2015.

———, and Steve McMillan. *Astronomy Today*. 5th ed., Pearson Prentice Hall, 2004.

Chaparo, Juan Carlos Mora. *Mapping the Risk of Flood, Mass Movement and Local Subsidence: A New Proposal for Major Cities*. Springer, 2020.

Chappell, B. W., and A. J. R. White. "Two Contrasting Granite Types: 25 Years Later." *Australian Journal of Earth Sciences*, vol. 48, 2001, pp. 489-99.

Charlton, Ro. *Fundamentals of Fluvial Geomorphology*. Routledge, 2008.

Chastko, Paul. *Developing Alberta's Oil Sands: From Karl Clark to Kyoto*. U of Calgary P, 2004.

Chatterjee, Kaulir Kisor. "Clay." *Uses of Industrial Minerals, Rocks, and Freshwater*. NOVA Science Publishers, 2009.

———. "Diamond." *Uses of Indus Trial Minerals, Rocks, and Freshwater*. Nova Science, 2009.

Chernicoff, Stanley, and Donna Whitney. *Geology: An Introduction to Physical Geology*. Prentice Hall, 2007.

Chester, Ray. *Furnace of Creation Cradle of Destruction: A Journey to the Birthplace of Earthquakes, Volcanoes, and Tsunamis*. AMACOM, 2008.

Chiras, Daniel D., and John P. Reganold. *Natural Resource Conservation: Management for a Sustainable Future*. 10th ed., Addison Wesley, 2009.

Choi, Charles Q. "Asteroids: Fun Facts and Information about Asteroids." *Space.com*. Purch, 21 Nov. 2014. Accessed 6 May 2015.

Chopra, Anil K. *Dynamics of Structures: Theory and Applications to Earthquake Engineering*. 3rd ed., Prentice Hall, 2007.

———. *Earthquake Engineering for Concrete Dams: Analysis, Design and Evaluation*. Wiley-Blackwell, 2020.

Chorley, Richard J., editor. *Introduction to Fluvial Processes*. Routledge, 2019.

Christiansen, Eric H., and W. Kenneth Hamblin. *Dynamic Earth: An Introduction to Physical Geology*. Jones & Bartlett, 2015.

Christopherson, Robert W., and Mary-Louise Byrne. *Geosystems: An Introduction to Physical Geography*. Pearson/Prentice Hall, 2006.

Ciavola, Paolo, and Giovanni Coco. *Coastal Storms: Processes and Impacts*. Wiley-Blackwell, 2017.

Clarke, Tony. *Tar Sands Showdown: Canada and the New Politics of Oil in an Age of Climate Change*. J. Lorimer, 2008.

Coe, Angela L., editor. *The Sedimentary Record of Sea-Level Change*. Cambridge UP, 2003.

Coles, Kenneth S., Kenneth L. Tanaka, and Philip R. Christensen. *The Atlas of Mars: Mapping Its Geography and Geology*. Cambridge UP, 2019.

Collier, Michael. *Over the Rivers: An Aerial View of Geology*. Mikaya Press, 2008.

Collinson, J. D., Nigel Mountney, and D. B. Thompson. *Sedimentary Structures*. 3rd ed., Dunedin Academic Press, 2006.

Condie, Kent C. *Earth as an Evolving Planetary System*. 2nd ed., Academic Press, 2011.

Constantinescu, Robert, Jean-Claude Thouret, and Ioan-Aurel Irimus. "Computer Modeling as a Tool for Volcanic Hazards Assessment: An Example of Pyroclastic Flow Modeling at El Misti Volcano, Southern Peru." *Geographia Technia*, vol. 11, no. 2, 2011, pp. 1-14.

Continents Adrift: Readings from "Scientific American." W. H. Freeman, 1972. Introductions by J. Tuzo Wilson.

"Cosmology: The Study of the Universe." *Wilkinson Microwave Anisotropy Probe*. NASA, 21 Dec. 2012. Accessed 2 June 2015.

Cossu, Remo, and Matthew G. Wells. "The Evolution of Submarine Channels under the Influence of Coriolis Forces: Experimental Observations of Flow Structures." *Terra Nova*, vol. 25, no. 1, 2013, pp. 65-71. *Academic Search Complete*. Accessed 19 Mar. 2015.

Cox, Allan, editor. *Plate Tectonics and Geomagnetic Reversals*. W. H. Freeman, 1973.

Cox, Allan G., Brent Dalrymple, and Richard R. Doell. "Reversals of the Earth's Magnetic Field." *Scientific American*, vol. 216, Feb. 1967, pp. 44-54.

Craig, Gordon Younger, and John Hewett Hull. *James Hutton: Past and Future.* Geological Society of London, 1999.
Craig, James R., David J. Vaughan, and Brian Skinner. *Earth Resources and the Environment.* Prentice Hall, 2010.
The CRC Handbook of Chemistry and Physics. 93rd ed., CRC Press, 2012.
Crovisier, Jacques, and Thérèse Encrenaz. *Comet Science.* Translated by Stephen Lyle. Cambridge UP, 2000.
Cruikshank, Dale P., and William Sheehan. *Discovering Pluto: Exploration at the Edge of the Solar System.* The University of Arizona Press, 2018.
Cuénot, Claude. *Teilhard de Chardin: A Biographical Study.* Edited by René Hague. Translated by Vincent Colimore. Helicon Press, 1965.
Cushing, Colbert E., K. W. Cummins, and G. W. Minshall, editors. *River and Stream Ecosystems of the World.* Rev. ed., U of California P, 2006.
Cutler, Alan. *The Seashell on the Mountaintop: A Story of Science, Sainthood, and the Humble Genius Who Discovered a New History of the Earth.* Dutton, 2003.
Dahl, P. S., and D. F. Palmer. "The Petrology and Origin of Orbicular Tonalite from Western Taylor Valley, Southern Victoria Land, Antarctica." *Antarctic Earth Science*, edited by R. L. Oliver, P. R. James, and J. B. Jago. Cambridge UP, 2011.
Dale, Virginia H., et al., editors. *Ecological Responses to the 1980 Eruption of Mount St. Helens.* Springer, 2005.
Darby, Stephen, and David Sear. *River Restoration: Managing the Uncertainty in Restoring Physical Habitat.* John Wiley & Sons, 2008.
David, Leonard. *Mars: Our Future on the Red Planet.* National Geographic, 2016.
———. *Moon Rush: The New Space Race.* National Geographic, 2019.
Davidson-Arnott, Robin, Bernard Bauer, and Chris Hauser. *Introduction to Coastal Processes and Geomorphology.* 2nd ed., Cambridge UP, 2019.
Davies, Bethan. "Mapping the World's Glaciers." *Antarcticglaciers.org,* 25 Nov. 2014. Accessed 23 Mar. 2014.
Davies, John Keith. *Beyond Pluto: Exploring the Outer Limits of the Solar System.* Cambridge UP, 2001.
Davis, Earl E., and Harry Elderfield, editors. *Hydrogeology of the Ocean Lithosphere.* Cambridge UP, 2004.
Davis, Richard Jr., and Duncan Fitzgerald. *Beaches and Coasts.* John Wiley & Sons, 2009.
Dawson, Chester. "Canadian Oil-Sands Producers Struggle." *The Wall Street Journal*, 19 Aug. 2015, www.wsj.com/articles/oil-sands-producers-struggle-1440017716. Accessed 15 Jan. 2019.
Dawson, J. B. *Kimberlites and Their Xenoliths.* Springer, 2012.
———. *The Gregory Rift Valley and Neogene-Recent Volcanoes of Northern Tanzania.* The Geological Society, 2008.
De Celles, Peter G., Mihai N. Ducea, Barbara Carrapa, and Paul A. Kapp, editors. *Geodynamics of a Cordilleran Orogenic System: The Central Andes of Argentina and Northern Chile.* The Geological Society of America, 2015.
De Graef, Marc, and Michael E. McHenry. *Structure of Materials: An Introduction to Crystallography, Diffraction, and Symmetry.* Cambridge UP, 2007.
de Lubac, Henri. *Teilhard de Chardin: The Man and His Meaning.* Translated by René Hague. Hawthorn Books, 1965.
De Pater, Imke, and Jack J. Lissauer. *Planetary Sciences.* Cambridge UP, 2001.
Dean, Dennis R. *James Hutton and the History of Geology.* Cornell UP, 1992.
Decker, Robert, and Barbara Decker. *Volcanoes.* 4th ed., W. H. Freeman, 2006.
Decrée, Sophie, and Laurence Robb, editors. *Ore Deposits: Origin, Exploration, and Exploitation.* Wiley, 2019.
Deer, W. A., R. A. Howie, and J. Zussman. *Rock-Forming Minerals.* 2nd ed., 5 vols. Pearson Education Limited, 1992.
Deline, Bradley, Randa Harris, and Karen Tefend. *Laboratory Manual for Introductory Geology.* U of North Georgia, 2016.
Denny, Mark. *How the Ocean Works: An Introduction to Oceanography.* Princeton UP, 2012.
Dickin, Alan P. *Radiogenic Isotope Geology.* 2nd ed., Cambridge UP, 2005.
Dilek, Yildrim, and Paul T. Robinson, editors. *Ophiolites in Earth History.* The Geological Society, 2003.
Dimmitt, Mark A., Patricia Wentworth Comus, and Linda M. Brewer, editors. *A Natural History of the Sonoran Desert.* 2nd ed., Arizona-Sonora Desert Museum Press/U of California P, 2015.
Dobrzhinetskaya, Larissa, Shah Wali Faryad, Simon Wallis, and Simon Cuthbert, editors. *Ultrahigh-Pressure Metamorphism: 25 Years After the Discovery of Coesite and Diamond.* Elsevier, 2011.
Dodds, Walter K., and Matt R. Whiles. *Freshwater Ecology: Concepts and Environmental Applications of Limnology.* 2nd ed., Academic Press, 2010.
Doney, Scott C. "The Dangers of Ocean Acidification." *Scientific American*, vol. 294, no. 3, Mar. 2006, pp. 58-65.

Dott, Robert H., and Roger Lyman Batten. *Evolution of the Earth.* 4th ed., McGraw-Hill, 1988.

Douglas, Bruce C., Michael S. Kearney, and Stephen P. Leatherman. *Sea Level Rise: History and Consequences.* Academic Press, 2001.

Dryburgh, Peter. "Nicholas Steno and the Foundations of Geology." *Edinburgh Geologist,* vol. 41, Autumn 20013, pp. 3-11.

Drzyzga, Scott A. *Mapping and Modeling Glacial Lake Algonquin in Northern Michigan and Western Ontario with Models of Uncertainty.* Michigan State U, 2007.

Du Toit, Alexander Logie. *Our Wandering Continents.* Hafner, 1937.

Dubinsky, Zvy, and Noga Stambler, editors. *Coral Reefs: An Ecosystem in Transition.* Springer, 2011.

Dundek, Marijan. *Diamonds.* Noble Gems International, 2009.

Dunkel, Ged. *Understanding the Jet Stream: Clash of the Titans.* Authorhouse, 2010.

Dyar, Melinda Darby, and Mickey E. Gunter. *Mineralogy and Optical Mineralogy.* Mineralogical Society of America, 2019.

Egholm, D. L., et al. "Glacial Effects Limiting Mountain Height." *Nature,* vol 460, 2009, pp. 884-87.

Ehlers, J., and P. L. Gibbard, editors. *Quaternary Glaciations: Extent and Chronology.* 3 vols. Elsevier, 2004.

Ehlers, Jürgen, Philip D. Hughes, and Philip Leonard Gibbard. *The Ice Age.* Wiley-Blackwell, 2016.

Ehlers, Jürgen, Philip Leonard Gibbard, and Philip D. Hughes. *Quaternary Glaciations—Extent and Chronology: A Closer Look.* Elsevier, 2011.

Elkins-Tanton, Linda T., and Benjamin P. Weiss, editors. *Planetesimals: Early Differentiation and Consequences for Planets.* Cambridge UP, 2017.

Emerson, Steven, and John Hedges. *Chemical Oceanography and Marine Carbon Cycle.* Cambridge UP, 2008.

Emery, William J., and Walter Zenk. *Captain F. Spiess and the German Atlantic Meteor Expedition of 1925-1927.* BrownWalker Press, 2019.

Energy Fact Book, 2016-2017. Natural Resources Canada, June 2016, www.nrcan.gc.ca/sites/www.nrcan.gc.ca/files/energy/pdf/EnergyFactBook_2016_17_En.pdf. Accessed 15 Jan. 2019.

Erickson, Jon. *Plate Tectonics: Unraveling the Mysteries of the Earth.* Rev. ed., Facts On File, 2001.

Esposito, Larry W., Ellen R. Stofan, and Thomas E. Cravens, editors. *Exploring Venus as a Terrestrial Planet.* American Geophysical Union, 2007.

Everhart, Michael J. *Oceans of Kansas. A Natural History of the Western Interior Sea.* 2nd ed., Indiana UP, 2017.

Eyles, Nick, and Andrew D. Miall. *Canada Rocks: The Geologic Journey.* Fitzhenry and Whiteside, 2018.

Faure, Gunter. *Origin of Igneous Rocks: The Isotopic Evidence.* Springer-Verlag, 2010.

Faure, Gunter, and Teresa M. Mensing. *Introduction to Planetary Science: The Geological Perspective.* Springer, 2007.

Felt, Hali Soundings. *The Story of the Remarkable Woman Who Mapped the Ocean Floor.* Henry Holt, 2012.

Fenton, Carol Lane, and Mildred Adams Fenton. *The Rock Book.* Dover, 2003.

Fersman, A. E. *Geochemistry for Everyone.* Foreign Languages, 1958.

Fetter, C. W. *Applied Hydrogeology.* 4th ed., Prentice Hall, 2001.

Fettes, Douglas, and Jacqueline Desmons, editors. *Metamorphic Rocks: A Classification and Glossary of Terms.* Cambridge UP, 2007.

Fletcher, Charles. *Physical Geology: The Science of Earth.* 3rd ed., Wiley, 2017.

———, Dan Gibson, and Kevin Ansdell. *Introduction to Physical Geology: The Science of Earth.* Canadian ed., Wiley, 2014.

Flügel, Erik. *Microfacies of Carbonate Rocks: Analysis, Interpretation and Application.* 2nd ed., Springer, 2010.

Folguera, Andrés, et al., editors. *The Evolution of the Chilean-Argentinian Andes.* Springer, 2018.

Foote, Michael, and Arnold L. Miller. *Principles of Paleontology.* 3rd ed., W. H. Freeman, 2007.

Foresman, Timothy, and Alan H. Strahler. *Visualizing Physical Geography.* John Wiley & Sons, 2012.

Fossen, Haakon. *Structural Geology.* 2nd ed., Cambridge UP, 2016.

Fowler, C. M. R. *The Solid Earth: An Introduction to Global Geophysics.* 2nd ed., Cambridge UP, 2005.

Fraknoi, Andrew, David Morrison, and Sidney C. Wolff. *Astronomy.* Samurai Media, 2017.

Francis, Peter, and Clive Oppenheimer. *Volcanoes.* 2nd ed., Oxford UP, 2004.

Frankel, Henry R. *The Continental Drift Controversy: Introduction of Seafloor Spreading.* Vol. 3. Cambridge UP, 2012.

Freedman, Roger A., and William J. Kaufmann III. *Universe.* 8th ed., W. H. Freeman, 2008.

Freitag, Bob, et al. *Floodplain Management: A New Approach for a New Era.* Island Press, 2009.

Frisch, Wolfgang, Martin Meschede, and Ronald Blakey. *Plate Tectonics: Continental Drift and Mountain Building.* Springer, 2011.

Frost, B. Ronald, et al. "A Geochemical Classification for Granitic Rocks." *Journal of Petrology*, vol. 42, 2001, pp. 2033-48.

Fu, Helin, Wei Chen, and Jiajun Fu. *Rock Mechanics and Engineering: Prediction and Control of Landslides and Geological Disasters.* Elsevier, 2021.

Fu, Lee-Leung, and Anny Cazenave. *Satellite Altimetry and Earth Sciences: A Handbook of Techniques and Applications.* Academic Press, 2001.

Gaffney, Vincent, Kenneth Thomson, and Simon Fitch, editors. *Mapping Doggerland: The Mesolithic Landscapes of the Southern North Sea.* Archaeopress, 2007.

Galimov, Erik M., and Anton M. Krivtsov. *Origin of the Moon—New Concept: Geochemistry and Dynamics.* Walter de Gruyter, 2012.

Gannon, Megan. "Bizarre Asteroid with Six Tails Spotted by Hubble Telescope." *Space.com*. Purch, 7 Nov. 2013. Accessed 6 May 2015.

Garrison, Tom S. *Oceanography: An Invitation to Marine Science.* Brooks/Cole, Cengage Learning, 2010.

Gates, W. P., J. T. Kloprogge, J. Madejová, and F. Bergaya, editors. *Infrared and Raman Spectroscopies of Clay Minerals.* Elsevier, 2017.

Gattuso, Jean-Pierre, and Lina Hansson, editors. *Ocean Acidification.* Oxford UP, 2011.

Generalov, Mikhail. "A. E. Fersman Mineralogical Museum: An Embassy of the Mineral Kingdom in Russia" *Rocks and Minerals,* vol. 76, no. 1, Jan.-Feb. 2001, pp. 17-21.

George, Rick. *Sun Rise: Suncor, the Oil Sands and the Future of Energy*. HarperCollins, 2012.

Ghosh, S. K. *Structural Geology: Fundamentals and Modern Developments.* Elsevier, 2013.

Gill, Robin. *Igneous Rocks and Processes: A Practical Guide.* Wiley-Blackwell, 2011.

Gillen, Cornelius. *Metamorphic Geology: An Introduction to Tectonic and Metamorphic Processes.* 2nd ed., Springer, 2011.

Glazer, A. M. *Crystallography: A Very Short Introduction.* Oxford UP, 2016.

Gleick, Peter H., et al. *The World's Water, 2008-2009.* Island Press, 2009.

Glen, William. *The Road to Jaramillo: Critical Years of the Revolution in Earth Science.* Stanford UP, 1982.

Glennon, Robert. *Water Follies: Groundwater Pumping and the Fate of America's Fresh Waters.* Island Press, 2002.

Goh, Gahyun, and Y. Noh. "Influence of Coriolis Force on the Formation of a Seasonal Thermocline." *Ocean Dynamics,* vol. 63, nos. 9-10, 2013, pp. 1083-92. *Energy & Power Source*. Accessed 19 Mar. 2015.

Golani, Prakash R. *Assessment of Ore Deposit Settings, Structures and Proximity Indicator Minerals in Geological Exploration.* Springer, 2021.

Gomez-Tuena, A., S. M. Straub, and G. F. Zellmer, editors. *Orogenic Andesites and Crustal Growth.* Geological Society, 2014.

Gonzalez, George. *American Empire and the Canadian Oil Sands*. Palgrave MacMillan, 2016.

Gornitz, Vivien. *Vanishing Ice: Glaciers, Ice Sheets and Rising Seas.* Columbia UP, 2019.

Gorst, Martin. *Measuring Eternity: The Search for the Beginning of Time*. Broadway Books, 2001.

Götze, Jens, editor. *Mineralogy of Quartz and Silica Minerals*. MDPI, 2018.

Gradstein, Felix M., James G. Ogg, Mark D. Schmitz, and Gabi M. Ogg. *Geologic Time Scale 2020.* Vol. 1 Elsevier, 2020.

Grapes, Rodney. *Pyrometamorphism.* 2nd ed., Springer, 2010.

Gray, William G., and Genetha A. *Gray Introduction to Environmental Modeling*. Cambridge UP, 2017.

Green, Timothy. *The World of Diamonds.* Morrow, 1981.

Greene, Mott T. *Alfred Wegener: Science, Exploration and the Theory of Continental Drift.* Johns Hopkins UP, 2015.

Grenet, Paul. *Teilhard de Chardin: The Man and His Theories.* Translated by R. A. Rudorff. Paul S. Eriksson, 1966.

Griffith, M. C., et al. "Earthquake Reconnaissance: Forensic Engineering on an Urban Scale." *Australian Journal of Structural Engineering,* vol. 11, no. 1, 2010, p. 63.

Grinspoon, David Harry. *Venus Revealed: A New Look Below the Clouds of Our Mysterious Twin Planet.* Perseus Book Group, 1998.

Groat, Lee Andrew. *The Geology of Gem Deposits.* Mineralogical Association of Canada, 2007.

Grotzinger, John, and Thomas H. Jordan. *Understanding Earth.* 6th ed., W. H. Freeman, 2010.

Grummett, David. *Teilhard de Chardin: Theology, Humanity, and Cosmos.* Peeters, 2005.

Guastoni, Alessandro, and Roberto Appiani. *Minerals.* Firefly Books, 2005.

Guilbert, John M., and Charles F. Park Jr. *The Geology of Ore Deposits.* Waveland Press, 2007.

Gupta, Alok Krshna. *Origin of Potassium-Rich Silica-Deficient Igneous Rocks*. Springer, 2015.

Gupta, Chiranjib, and Harvinderpal Singh. *Uranium Resource Processing: Secondary Resources.* Springer-Verlag, 2010.

Gurvich, Evgeny G. *Metalliferous Sediments of the World Ocean.* Springer Verlag, 2010.
Gutenberg, Beno, and Charles Richter. *Seismicity of the Earth.* The Society, 1941.
Gwyn, David. *Welsh Slate: Archaeology and History of an Industry.* Royal Commission on the Ancient and Historical Monuments of Wales, 2015.
Gyr, Albert, and Klaus Hoyer. *Sediment Transport: A Geophysical Phenomenon.* Springer, 2006.
Hacker, Bradley R., William C. McClelland and John G. Liou, editors. *Ultrahigh-Pressure Metamorphism: Deep Continental Subduction.* The Geological Society of America, 2006.
Haldar, Swapan Kumar. *Introduction to Mineralogy and Petrology.* 2nd ed., Elsevier, 2020.
Hall, Cally. *Gemstones.* 2nd ed., Dorling Kindersley, 2002.
Hall-Spencer, Jason M., et al. "Volcanic Carbon Dioxide Vents Show Ecosystem Effects of Ocean Acidification." *Nature,* vol. 454, no. 7200, 3 July 2008, pp. 96-99.
Hamblin, Kenneth W., and Eric H. Christiansen. *Earth's Dynamic Systems.* 10th ed., Prentice Hall, 2003.
Hambrey, Michael, and Jurg Alean. *Glaciers.* Cambridge UP, 2006.
Hammond, Christopher. *The Basics of Crystallography and Diffraction.* 2nd ed., Oxford UP, 2001.
Harker, Alfred. *The Natural History of Igneous Rocks.* Cambridge UP, 2011.
Harland, David M. *Jupiter Odyssey: The Story of NASA's Galileo Mission.* Springer-Praxis, 2000.
———. *Mission to Saturn: Cassini and the Huygens Probe.* Springer-Praxis, 2002.
Harris, Peter T., and Elaine K. Baker, editors. *Seafloor Geomorphology as Benthic Habitat.* Elsevier, 2011.
Hart, Matthew. *Diamond: A Journey to the Heart of an Obsession.* Walker, 2001.
Hartman, Howard L., and Jan M. Mutmansky. *Introductory Mining Engineering.* 2nd ed., J. Wiley, 2002.
Hartmann, William K. *Moons and Planets.* 5th ed., Thomson Brooks/Cole, 2005.
Harvey, Brian. *Soviet and Russian Lunar Exploration.* Praxis Publishing, 2007.
Haslam, S. M. *The Riverscape and the River.* Cambridge UP, 2008.
Haslett, Simon K. *Coastal Systems.* 2nd ed., Routledge, 2009.
Haussühl, Siegfried. *Physical Properties of Crystals: An Introduction.* Wiley-VCH, 2007.
Havskov, Jens, and Gerardo Alguacil. *Instrumentation in Earthquakes Seismology.* 2nd ed., Springer, 2016.
Hazell, Peter, and Stanley Wood. "Drivers of Change in Global Agriculture" *Philosophical Transactions of the Royal Society of London,* vol. 363, no. 1491, Feb. 12, 2008, pp. 495-515.
Hefferan, Kevin, and John O'Brien. *Earth Materials.* John Wiley & Sons, 2010.
Henning, Thomas. *Astromineralogy.* Springer, 2003.
Heppenheimer, T. A. *Colonies in Space: A Cand Factual Account of the Prospects for Human Colonization of Space.* Stackpole Books, 2017.
Hicks, Brian, and Chris Nelder. *Profit from the Peak: The End of Oil and the Greatest Investment Event of the Century.* John Wiley & Sons, 2008.
Hildebrand, Robert S., and Joseph B. Whalen, *The Tectonic Setting and Origin of Cretaceous Batholiths within the North American Cordillera: The Case for Slab Failure Magmatism and Its Significance for Crustal Growth.* American Geological Society, 2017.
Hill, Mary. *Geology of the Sierra Nevada.* Rev. ed., U of California P, 2006.
Himbert, M. E. "A Brief History of Measurement." *European Physical Journal Special Topics,* vol. 172, 2009, pp. 25-35.
Hobbs, Carl H. *The Beach Book: Science of the Shore.* Columbia UP, 2012.
Hoffmann, Frank. *Introduction to Crystallography.* Springer, 2020.
Holland, H. D., and K. K. Turekian, editors. *Radioactive Geochronometry from the Treatise on Geochemistry.* Academic Press, 2011.
Hopley, David. *Encyclopedia of Modern Coral Reefs: Structure, Form and Process.* Springer, 2011.
Hough, Susan Elizabeth. *Earthshaking Science: What We Know (and Don't Know) About Earthquakes.* Princeton UP, 2002.
———. *Predicting the Unpredictable: The Tumultuous Science of Earthquake Prediction.* Princeton UP, 2010.
———. *Richter's Scale: Measure of an Earthquake, Measure of a Man.* Princeton UP, 2007.
Hovland, Martin. *Deep-Water Coral Reefs: Unique Biodiversity Hot-Spots.* Praxis, 2008.
Howell, Elizabeth, and Nicholas Booth. *The Search for Life on Mars.* Arcade Publishing, 2020.
Hubbard, Bryn, and Neil Glasser. *Field Techniques in Glaciology and Glacial Geomorphology.* John Wiley & Sons, 2005.
Hudson, Travis. *Living with Earth: An Introduction to Environmental Geology.* Routledge, 2016.
Huggett, Richard John. *Fundamentals of Geomorphology.* 2nd ed., Routledge, 2007.

Humphrey, Justin P., and Daniel E. Boyd, editors. *Clay: Types, Properties and Uses.* NOVA Science Publishers, 2011.

Huneke, Heiko, and Thierry Mulder. *Deep Sea Sediments.* Elsevier, 2011.

Hutton, James. *System of the Earth, 1785, Theory of the Earth, 1788, Observations on Granite, 1794, Together with Playfair's Biography of Hutton.* Hafner, 1970.

Igel, Heiner. *Computational Seismology: A Practical Introduction.* Oxford UP, 2017.

Inglezakis, Vassilis, and Antonis A. Zorpas, editors. *Handbook of Natural Zeolites.* Bentham Books, 2012.

Irwin, Patrick G. J. *Giant Planets of Our Solar System: Atmospheres, Composition, and Structure.* Springer-Praxis, 2003.

———. *Giant Planets of Our Solar System: An Introduction.* 2nd ed., Springer, 2006.

Jackson, Derek, and Andrew D. Short, editors. *Sandy Beach Morphodynamics.* Elsevier, 2020.

Jackson, Martin P. A., and Michael R. Hudec. *Salt Tectonics: Principles and Practice.* Cambridge UP, 2017.

James, Noel P., and Yvonne Bone. *Biogenic Rocks in a Cold Cenozoic Ocean: Neritic Southern Australia.* Springer International Publishing, 2021.

James, Noel P., and Brian Jones. *Origin of Carbonate Sedimentary Rocks.* Wiley, 2016.

Javaid, Muhammad Salik, and Shaukat Ali Khan, editors. *Aquifers: Matrix and Fluids.* IntechOpen, 2018.

Jerram, Dougal, and Nick Petford. *The Field Description of Igneous Rocks.* 2nd ed., Wiley-Blackwell, 2011.

Johnson, David. *Lime Kilns: History and Heritage.* Amberley Publishing, 2018.

Johnson, Michael, and Simon Harley. *Orogenesis: The Making of Mountains.* Cambridge UP, 2012.

Johnson, Sarah Stewart. *The Sirens of Mars: Searching for Life on Another World.* Crown/Random House, 2020.

Jones, Reg Chambers. *Dinorwic: The Llanberis Slate Quarry 1780-1969.* Bridge Books, 2006.

Joseph, P. *Deep Water Sedimentation in the Alpine Basin of SE France.* Special Publication 221. Geological Society of London, 2004.

Kandel, Robert. *Water from Heaven: The Story of Water from the Big Bang to the Rise of Civilization and Beyond.* Columbia UP, 2003.

Kardel, Troels. *Steno: Life, Science, Philosophy.* Danish National Library of Science and Medicine, 1994.

———, and Paul Maquet. *Nicolaus Steno: Biography and Original Papers of a 17th Century Scientist.* 2nd ed., Springer, 2018.

Kaufman, Marc. *Mars Up Close: Inside the Curiosity Mission.* National Geographic, 2014.

Kearey, Philip, Keith A. Klepeis, and Frederick J. Vine. *Global Tectonics.* 3rd ed., Wiley-Blackwell, 2009.

Keller, Edward A. *Introduction to Environmental Geology.* Prentice Hall, 2012.

———, and Nicholas Pinter. *Active Tectonics: Earthquakes, Uplift, and Landscape.* 2nd ed., Prentice Hall, 2002.

Kelly, Thomas J. *Moon Lander: How We Developed the Apollo Lunar Module.* Smithsonian Institution Press, 2001.

Kennett, Brian L. *Seismic Wave Propagation in Stratified Media.* Australian National UP, 2009.

Kennish, Michael J. *Practical Handbook of Marine Science.* 3rd ed., CRC Press, 2001.

Klein, Cornelis, and Barbara Dutrow. *The Twenty-Third Edition of the Manual of Mineral Science.* Wiley, 2008.

Klein, Cornelis, and Cornelius S. Hurlbut, Jr. *Manual of Mineralogy.* 23rd ed., John Wiley & Sons, 2008.

Klein, Cornelis, and Anthony Philpotts. *Earth Materials: Introduction to Minerals and Petrology.* 2nd ed., Cambridge UP, 2016.

Knauss, John A., and Newell Garfield. *Introduction to Physical Oceanography.* 3rd ed., Waveland Press, 2017.

Knoll, Andrew H., Don E. Canfield, and Kurt O. Konhauser, editors. *Fundamentals of Geobiology.* John Wiley & Sons, 2012.

Koch, Adam Gregory. *Gold Prospecting and Placer Deposits: Finding Gold Made Simpler.* Author, 2014.

Kogel, Jessica Elzea, et al., editors. "Clays." In *Industrial Minerals and Rocks: Commodities, Markets, and Uses.* 7th ed., Society for Mining, Metallurgy, and Exploration, 2006.

———. "Decorative Stone" and "Dimension Stone." *Industrial Minerals and Rocks: Commodities, Markets, and Uses.* 7th ed., Society for Mining, Metallurgy, and Exploration, 2006.

———. "Industrial Sand and Sandstone." *Industrial Minerals and Rocks: Commodities, Markets, and Uses.* 7th ed., Society for Mining, Metallurgy, and Exploration, 2006.

———. "Lime." *Industrial Minerals and Rocks: Commodities, Markets, and Uses.* 7th ed., Society for Mining, Metallurgy, and Exploration, 2006.

Komar, Paul D., and J. Robert Moore, editors. *Handbook of Coastal Processes and Erosion.* CRC Press, 2018.

Kondolf, G. Mathias, and Hervé Piégay. *Tools in Fluvial Geomorphology.* Wiley, 2016.

Kresten, Peter, and Valentin R. *Troll: The Alnö Carbonatite Complex, Central Sweden.* Springer, 2018.

Krüger, Tobias. *Discovering the Ice Ages: International Reception and Consequences for a Historical Understanding of Climate.* Brill, 2013.

Kusky, Timothy. *Earthquakes: Plate Tectonics and Earthquake Hazards.* Facts On File, 2008.

Lagomarsino, James. *A Pocket Guide to Rocks and Minerals.* Parragon Books, 2008.

Lahti, Seppo L., Paula Raivio, and Ilka Laitakari, editors. *Orbicular Rocks in Finland.* Geological Survey of Finland, 2005.

Laity, Julie J. *Deserts and Desert Environments.* John Wiley & Sons, 2008.

Lamb, Simon, and Anthony Watts. "The Origin of Mountains: Implications for the Behaviour of Earth's Lithosphere." *Current Science,* vol. 99, no. 12, 2010, pp. 1699-1718.

Latour, Bruno. *Facing Gaia: Eight Lectures on the New Climatic Regime.* Polity Press, 2017.

Lauf, Robert. *The Collector's Guide to Silicate Crystal Structures.* Schiffer Publishing, Ltd., 2010.

Launius, Roger D. *Apollo's Legacy: Perspectives on the Moon Landings.* Smithsonian Institution, 2019.

Lauretta, Dante S. and Harry Y. McSween Jr., editors. *Meteorites and the Early Solar System II.* U of Arizona P, 2006.

Lawrence, David M. *Upheaval from the Abyss: Ocean Floor Mapping and the Earth Science Revolution.* Rutgers UP, 2002.

Le Roex, Anton P., David R. Bell, and Peter Davis. "Petrogenesis of Group I Kimberlites from Kimberley, South Africa: Evidence from Bulk-Rock Geochemistry." *Journal of Petrology,* vol. 44, 2003, pp. 2261-86.

Leeder, Mike R. *Sedimentology and Sedimentary Basins: From Turbulence to Tectonics.* Wiley-Blackwell, 2011.

LeGrand, H. E. *Drifting Continents and Shifting Theories.* Cambridge UP, 1988.

Leith, C. K. *The Economic Aspect of Geology.* Good Press, 2016.

Letcher, Trevor M., editor. *Future Energy: Improved, Sustainable and Clean.* Elsevier, 2008.

Leverington, David. *Babylon to Voyager and Beyond: A History of Planetary Astronomy.* Cambridge UP, 2003.

Levinson, Alfred A., and S. Ross Taylor. *Moon Rocks and Minerals.* Elsevier, 2015.

Levinton, Jeffery S., and John R. Waldman. *The Hudson River Estuary.* Cambridge UP, 2006.

Lewis, John S. *Physics and Chemistry of the Solar System.* 2nd ed., Academic Press, 2004.

Li, Z. X., D. A. D. Evans, and J. B. Murphy, editors. *Supercontinent Cycles Through Earth History.* The Geological Society, 2016.

Libes, Susan M. *Introduction to Marine Biogeochemistry.* Academic Press, 2009.

Liebau, Friendrich. *Structural Chemistry of Silicates: Structure, Bonding, and Classification.* Springer-Verlag, 1985.

Lieberman, Bruce S., and Roger Kaesler. *Prehistoric Life: Evolution and the Fossil Record.* Wiley-Blackwell, 2010.

Liebesman, Lawrence R., and Rafe Petersen. *Endangered Species Deskbook.* Environmental Law Institute, 2003.

Linnemann, Ulf, R. Damian Nance, Petr Kraft, and Gernold Zulauf, editors. *The Evolution of the Rheic Ocean: From Avalonian-Cadomian Active Margin to Alleghenian-Variscan Collision.* Geological Society of America, 2007.

Loáiciga, Hugo. "Modern-Age Buildup of CO2 and Its Effects on Seawater Acidity and Salinity." *Geophysical Research Letters,* vol. 33, no. 10, 6 May 2006.

Losiak, Anna Isabela. *The Development of Evaporite Minerals During Weathering of Antarctic Meteorites.* Michigan State U Geology Department, 2009.

Lovelock, James. *The Ages of Gaia: A Biography of Our Living Earth.* Norton, 1995.

———. *Gaia: A New Look at Life on Earth.* New ed., Oxford UP, 2000.

———. *The Revenge of Gaia: Earth's Climate Crisis and the Fate of Humani*ty. Basic Books, 2007.

Lowrie, William. *Fundamentals of Geophysics.* 2nd ed., Cambridge UP, 2007.

———. *Geophysics: A Very Short Introduction.* Oxford UP, 2018.

Lutgens, Frederick K., Edward J. Tarbuck, and Dennis G. Tasa. *Essentials of Geology.* Pearson Education, 2017.

Lyons, Arthur R. *Materials for Architects and Builders.* Elsevier Butterworth Heinemann, 2004.

Macdonald, Eoin H. *Alluvial Mining: The Geology, Technology, and Economics of Placers.* Chapman and Hall, 1983.

Macdougall, J. D. *Frozen Earth: The Once and Future Story of Ice Ages.* U of California P, 2004.

Mackenzie, F. T., editor. *Sediments, Diagenesis, and Sedimentary Rocks.* Elsevier, 2005.

Madigan, Michael T., et al. *Brock Biology of Microorganisms.* 12th ed., Pearson/Benjamin Cummings, 2009.

Maghsoudi, Mehran. *Desert Landforms and Landscapes of Iran.* Springer, 2020.

Mahler, Annett. *Zeolites: Advances in Research and Applications.* NOVA Science Publishers, 2020.

Mainguet, Monique. *Aridity: Drought and Human Development.* Springer Verlag, 2010.
Maliva, Robert G. *Aquifer Characterization Techniques.* Springer, 2016.
Manutchehr-Danai, Mohsen. *Dictionary of Gems and Gemology*. 3rd ed., Springer-Verlag, 2005.
Margulis, Lynn. *Symbiotic Planet: A New Look at Evolution*. Basic Books, 1999.
Marsden, William. *Stupid to the Last Drop: How Alberta Is Bringing Environmental Armageddon to Canada (and Doesn't Seem to Care).* Vintage Canada, 2008.
Martí, Joan, and Gerald Ernst, editors. *Volcanoes and the Environment.* Cambridge UP, 2005.
Marvin, Ursula B. *Continental Drift: The Evolution of a Concept.* Smithsonian Institution Press, 1973.
Masselink, G., and M. G. Hughes. *Introduction to Coastal Processes and Geomorphology.* Hodder Arnold Publication, 2003.
Mathez, Edmond A., and James D. Webster. *The Earth Machine: The Science of a Dynamic Planet.* Columbia UP, 2004.
Matson, W. R. *Earthquakes: The Sound of Multi-Modal Waves.* Morgan & Claypool, 2016.
Maurizio, Galimberti, editor. *Rubber-Clay Nanocomposites.* John Wiley & Sons, 2011.
Maynard, Thierry, editor. *Teilhard and the Future of Humanity.* Fordham UP, 2006.
McBirney, Alexander R. *Igneous Petrology.* 3rd ed., Jones and Bartlett, 2007.
McBride, Neil, and Iain Gilmour, editors. *An Introduction to the Solar System.* Cambridge UP, 2004.
McCarthy, Shawn, and Jeff Lewis. "Canadian Oil Sands Are Crucial If Producers Cut Costs, Emissions: IEA." *The Globe and Mail*, 25 Feb. 2016, www.theglobeandmail.com/report-on-business/industry-news/energy-and-resources/iea-cautiously-optimistic-about-future-of-canadian-oil-sands/article28920041. Accessed 15 Jan. 2019.
McCoy, Roger M. *Ending in Ice: The Revolutionary Idea and Tragic Expedition of Alfred Wegener.* Oxford UP, 2006.
McCulloch, Robin, et al. *Applied Gold Placer Exploration and Evaluation Techniques.* Montana Bureau of Mines and Geology, 2003.
McCully, Patrick. *Silenced Rivers: The Ecology and Politics of Large Dams.* Enlarged and updated ed., Zed Books, 2001.
McDougall, Walter A. *The Heavens and the Earth: A Political History of the Space Age.* 2nd ed., Johns Hopkins UP, 1997.
McElhinny, Michael W., and Phillip L. McFadden. *Paleomagnetism: Continents and Oceans.* Academic Press, 2000.
McGuire, Bill. *Waking the Giant: How a Changing Climate Triggers Earthquakes, Tsunamis, and Volcanoes.* Oxford UP, 2012.
McKinney, Frank. *The Northern Adriatic Ecosystem: Deep Time in a Shallow Sea.* Columbia UP, 2007.
McSween, Harry Y. Jr., Jeffrey E. Moersch, Devon M. Burr, et al. *Planetary Geoscience.* Cambridge UP, 2019.
Medwin, Herman, et al. *Sounds in the Sea: From Ocean Acoustics to Acoustical Oceanography*. Cambridge UP, 2005.
Menard, H. W. *The Ocean of Truth: A Personal History of Global Tectonics.* Princeton UP, 1986.
Merrill, Ronald T. *Our Magnetic Earth: The Science of Geomagnetism.* U of Chicago P, 2010.
Meunier, Alain. *Clays.* Translated by Nathalie Fradin. Springer, 2010.
Meyer, David L., and Richard Arnold Davis. *A Sea Without Fish: Life in the Ordovician Sea of the Cincinnati Region.* Indiana UP, 2009.
Micallef, Anton, and Allan Williams. *Beach Management: Principles and Practice.* Earthscan, 2009.
Middelberg, Jack J. *Marine Carbon Biogeochemistry: A Primer for Earth System Scientists.* SpringerOpen, 2019.
Middleton, Gerard V., editor. *Encyclopedia of Sediments and Sedimentary Rocks*. Springer, 2003.
Middleton, Nick. *Deserts: A Very Short Introduction.* Oxford UP, 2009.
Miglorie, Catherine. *Vermont's Marble Industry.* Arcadia Publishing, 2013.
Miller, Frederic P., Agnes F. Vandome, and John McBrewster. *Milankovitch Cycles.* Alphascript Publishing, 2009.
Miller, Roger H. *Kimberlites: Mineralogy, Geochemistry and Petrology.* Springer, 2013.
Millero, Frank J. *Chemical Oceanography.* 4th ed., CRC Press, 2013.
Miner, Ellis D., and Randii R. Wessen. *Neptune: The Planet, Rings, and Satellites.* Springer-Praxis, 2002.
Miniati, Stefano. *Nicholas Steno's Challenge for Truth: Reconciling Science and Faith.* AngeliFranco, 2009.
Mishkin, Andrew. *Sojourner: An Insider's View of the Mars Pathfinder. Mission* Berkley Books, 2003.
Misra, Kula C. *Introduction to Geochemistry: Principles and Applications.* Wiley-Blackwell, 2012.
———. *Understanding Mineral Deposits.* Kluwer Academic, 2000.

Mitchell, Alanna. *The Spinning Magnet: The Force That Created the Modern World-and Could Destroy It.* OneWorld Publications, 2018.
Moche, Dinah L. *Astronomy: A Self-Teaching Guide.* 6th ed., John Wiley & Sons, 2004.
Molnar, Peter. *Plate Tectonics: A Very Short Introduction.* Oxford UP, 2015.
Monroe, James S., Reed Wicander, and Richard Hazlett. *Physical Geology.* 6th ed., Thompson Higher Education, 2007.
Montgomery, Carla W. *Environmental Geology.* 4th ed., McGraw-Hill, 2008.
Moomaw, B. "Spirit Lands at Gusev." *Astronomy*, vol. 32, no. 4, 2004, p. 32.
Mooney, Christopher F. *Teilhard de Chardin and the Mystery of Christ.* Harper & Row, 1966.
Moores, Eldridge, and Robert J. Twiss. *Tectonics.* Waveland Press, 2014.
Morari do Nascimento, Gustavo, editor. *Clays, Clay Minerals and Ceramic Materials Based on Clay Minerals.* IntechOpen, 2016.
Morgan, Lisa A. *Integrated Geoscience Studies in the Greater Yellowstone Area: Volcanic, Tectonic, and Hydrothermal Processes in the Yellowstone Geoecosystem.* U.S. Geological Survey, 2007.
Morgan, Peter. *Fire Mountain: How One Man Survived the World's Worst Volcanic Disaster.* Bloomsbury, 2003.
Morrison, David, and Tobias Owen. *The Planetary System.* 3rd ed., Addison-Wesley, 2003.
Moskowitz, Clara. "Hottest Particle Soup May Reveal Secrets of Primordial Universe." *LiveScience*. Purch, 13 Aug. 2012. Accessed 2 June 2015.
Müller, Daniel, and David I. Groves. *Potassic Igneous Rocks and Associated Gold-Copper Mineralization*. Springer, 2016.
Murray, Haydn H. *Applied Clay Mineralogy: Occurrences, Processing, and Application of Kaolins, Bentonites, Palygorskite-Sepiolite, and Common Clays.* Elsevier, 2007.
Naeye, R. "Red-Letter Days" *Sky and Telescope*, vol. 107, no. 5, 2004, p. 44.
Naldrett, Anthony J. *Magmatic Sulfide Deposits: Geology, Geochemistry, and Exploration.* Springer, 2004.
Napoli, J. Nicholas, and William Tronzo, editors. *Radical Marble: Architectural Innovation from Antiquity to the Present.* Routledge, 2018.
National Aeronautics and Space Administration, Jet Propulsion Laboratory. "Mars Exploration Rovers." http://marsrovers.jpl.nasa.gov. Accessed June, 2005.
National Research Council. *Drawing Louisiana's New Map: Addressing Land Loss in Coastal Louisiana.* National Academies Press, 2006.
Nazaré, M. H., and A. J. Neves, editors. *Properties, Growth, and Applications of Diamond.* IEE, 2001.
Neighbors, Thomas, and Leif Bjørnø. *Applied Underwater Acoustics.* Elsevier, 2017.
Nemeth, Karoly, editor. *Updates in Volcanology: From Volcano Modelling to Volcano Geology*. IntechOpen, 2016.
Nespolo, Massimo, and Giovanni Ferraris. "A Survey of Hybrid Twins in Non-Silicate Minerals." *European Journal of Mineralogy,* vol. 21, 2009, pp. 673-90.
Nesse, William D. *Introduction to Mineralogy.* 3rd ed., Oxford UP, 2017.
Nett, Mary T., Martin A. Locke, and Dean A. Pennington. *Water Quality Assessments in the Mississippi Delta: Regional Solutions, National Scope*. American Chemical Society, 2004.
Nevill, Simon. *Australia's Great Western Deserts: The Land-The People-The Explorers.* Woodslane Pty. Limited, 2021.
Newson, Malcolm. *Land, Water and Development: Sustainable and Adaptive Management of Rivers.* 3rd ed., Routledge, 2008.
Newton, Robert C. "The Three Partners of Metamorphic Petrology." *American Mineralogist,* vol. 96. 2011, pp. 457-69.
Nicholls, Robert J., W. Neil Adger, Craig W. Hutton, and Susan E. Hanson, editors. *Deltas in the Anthropocene*. Palgrave MacMillan, 2020.
Nichols, Gary. *Sedimentology and Stratigraphy.* 2nd ed., Wiley-Blackwell, 2009.
Nicholson, Sharon E. *Dryland Climatology.* Cambridge UP, 2011.
Nield, Ted. *Supercontinent: Ten Billion Years in the Life of Our Planet.* Granta Books, 2012.
Nikiforuk, Andrew. *The Tar Sands: Dirty Oil and the Future of a Continent.* Greystone, 2009.
"Nucleosynthesis." *Cosmos: The SAO Encyclopedia of Astronomy.* Swinburne University of Technology, n.d. Accessed 2 June 2015.
Nugent, Carrie. *Asteroid Hunters.* TED Books, 2017.
O'Donoghue, Michael. *Gems.* 6th ed., Robert Hale, 2008.
Oakes, Kelly. "On the Origin of Chemical Elements" *Basic Space*. Scientific American, 2 Aug. 2011. Accessed 16 July 2015.
Oates, J. A. H. *Lime and Limestone: Chemistry and Technology, Production and Uses.* Wiley-VCH, 1998.
Oceanography Course Team. *Ocean Circulation.* 2nd ed., Butterworth Heinemann, 2001.
O'Gorman, Emily. *Flood Country. An Environmental History of the Murray-Darling Basin.* CSIRO, 2012.

O'Hara, Kieran D. *A Brief History of Geology*. Cambridge UP, 2018.
Oncken, Onno, et al., editors. *The Andes: Active Subduction Orogeny (Frontiers in Earth Sciences)*. Springer-Verlag, 2006.
Oppenheimer, Clive. *Eruptions That Shook the World*. Cambridge UP, 2011.
Oreskes, Naomi. *The Rejection of Continental Drift: Theory and Method in American Earth Science*. Oxford UP, 1999.
Oreskes, Naomi, editor. *Plate Tectonics: An Insider's History of the Modern Theory of the Earth*. Westview Press, 2001.
Osinski, G. R., and E. Pierazzo. *Impact Cratering: Processes and Products*. Blackwell Publishing, 2013.
Palmer, Andrew. *Introduction to Petroleum Exploration and Engineering*. World Scientific, 2017.
Pandolfi, John M., Sean R. Connolly, Dustin J. Marshall, and Anne L. Cohen. "Projecting Coral Reef Futures under Global Warming and Ocean Acidification." *Science*, vol. 333, no. 6041, 2011, pp. 418-22.
Parfitt, Elisabeth A., and Lionel Wilson. *Fundamentals of Physical Volcanology*. Blackwell, 2008.
Parise, M., F. Gabrovsek, G. Kaufmann, and N. Ravbar, editors. *Advances in Karst Research: Theory, Fieldwork and Applications*. The Geological Society, 2018.
Parsons, Anthony J., and Athol D. Abrahams, editors. *Geomorphology of Desert Environments*. 2nd ed., Springer, 2009.
Pavlova, Galina Evgen'evna, and Aleksandr Sergeevich. Fedorov *Mikhail Vail'evich Lomonosov: His Life and Work*. Mir Publishers, 1980.
Peiper, Howard. *Zeolite: Nature's Heavy Metal Detoxifier*. Safe Goods, 2006.
Pellant, Chris. *Smithsonian Handbook: Rocks and Minerals*. Dorling Kindersley Books, 2002.
Perkins, Dexter. *Mineralogy*. 2nd ed., Prentice Hall, 2002.
Perlov, Delia, and Alex Vilenkin. *Cosmology for the Curious*. Springer, 2017.
Perryman, Michael. *The Exoplanet Handbook*. Cambridge UP, 2011.
Petersen, Carolyn Collins, and John C. Brandt. *Visions of the Cosmos*. Cambridge UP, 2003.
Peterson, James F., Dorothy Sack, and Robert E. Gabler. *Physical Geography*. Cengage, 2016.
Pfeffer, W. T., A. A. Arendt, A. Bliss, et al. "The Randolph Glacier Inventory: A Globally Complete Inventory of Glaciers." *Journal of Glaciology*, vol. 60, no. 221, 2014, pp. 537-52.
Philips, Jon M. *The Earth's Core: Structure, Properties and Dynamics*. NOVA Science Publishers, 2012.
Phillips, Tony, editor. "The Tunguska Impact: 100 Years Later." *NASA Science*. NASA, 30 June 2008. Accessed 6 May 2015.
Philpotts, Anthony R., and Jay J. Ague. *Principles of Igneous and Metamorphic Petrology*. 2nd ed., Cambridge UP, 2009.
Pidwirny, Michael. *Understanding Physical Geography*. Our Planet Earth Publishing, 2017.
Pierson, Hugh O. *Handbook of Carbon, Graphite, Diamond and Fullerenes*. William Andrew, 2012.
Pirajno, Franco. *Hydrothermal Processes and Mineral Systems*. Springer/Geological Survey of Western Australia, 2009.
Playfair, John. *Illustrations of the Huttonian Theory of the Earth (1802)*. Cambridge UP, 2011.
Plummer, Charles C., Diane H. Carlson, and Lisa Hammersley. *Physical Geology*. 13th ed., McGraw Hill, 2009.
Poag, C. Wylie, Christian Koeberl, and Wolf Uwe Reimold. *The Chesapeake Bay Crater: Geology and Geophysics of a Late Eocene Submarine Impact Structure*. Springer, 2004.
Pohl, Walter L. *Economic Geology Principles and Practice*. Wiley-Blackwell, 2011.
Popova, Olga, and Ivan Nemchinov. "Bolides in the Earth Atmosphere": *Catastrophic Events Caused by Cosmic Objects*, edited by Adushkin and Nemchinov, pp. 131-62, Springer, 2008.
Popova, V. I., et al. "Murzinka: Alabasha Pegmatite Field." *Mineralogical Almanac*, vol. 5, 2002.
Powell, James L. *Dead Pool: Lake Powell, Global Warming, and the Future of Water in the West*. U of California P, 2008.
Prantzos, Nikos, and Sylvia Ekström. "Stellar Nucleosynthesis." *Encyclopedia of Astrobiology*, edited by Muriel Gargaud, et al., Vol. 3, pp. 1584-92, Springer, 2011.
Prelas, Mark A., Galina Popovici, and Louis K. Bigelow, editors. *Handbook of Industrial Diamonds and Diamond Films*. Routledge, 2018.
Price, Monica T. *The Sourcebook of Decorative Stone: An Illustrated Identification Guide*. Firefly Books, 2007.
Prothero, D. R., and Robert H. Dott. *Evolution of the Earth*. 8th ed., McGraw-Hill, 2009.
Prothero, Donald R. *Bringing Fossils to Life: An Introduction to Paleobiology*. 2nd ed., McGraw-Hill Higher Education, 2004.
———. *Catastrophes!: Earthquakes, Tsunamis, Tornadoes, and Other Earth-Shattering Disasters*. Johns Hopkins UP, 2011.
———., and Fred Schwab, editors. *Sedimentary Geology*. W.H. Freeman, 2013.

Pulinets, Sergey, and Dimitar Ouzounov. *The Possibility of Earthquake Forecasting: Learning from Nature.* Institute of Physics Publishing, 2019.

Rabinowitz, Dan. *The Power of Deserts: Climate Change, the Middle East and the Promise of a Post-Oil Era.* Stanford UP, 2020.

Radomski, Kassandra Kathleen. *The Secrets of Saturn (Planets).* National Geographic, 2015.

Rafferty, John P., editor. *Plate Tectonics, Volcanoes, and Earthquakes.* Britannica Educational Publishing, 2011.

Ramkumar, Muhammad, editor. *Chemostratigraphy: Concepts, Techniques and Applications.* Elsevier, 2015.

Ramo, O. T. *Granitic Systems.* Elsevier, 2005.

Randazzo, Giovanni, Derek W. T. Jackson, and J. Andrew G. Cooper. *Sand and Gravel Spits.* Springer, 2015.

Rapp, Donald. *Use of Extraterrestrial Resources for Human Space Missions to Moon or Mars.* Springer, 2013.

Raymond, L. A. *Petrology: The Study of Igneous, Sedimentary, and Metamorphic Rocks.* 2nd ed., Waveland Press, 2007.

Read, Peter G. *Gemmology.* 3rd ed., Elsevier, 2005.

Redd, Nola Taylor. "Asteroid Found with Rings! First-of-Its-Kind Discovery Stuns Astronomers." *Space.com.* Purch, 26 Mar. 2014. Accessed 6 May 2015.

Reddy, D. Venkat. *Engineering Geology.* VIKAS Publishing, 2010.

Redfern, Ron. *Origins: The Evolution of Continents, Oceans and Life.* U of Oklahoma P, 2001.

Reeves, Robert. *The Superpower Space Race: An Explosive Rivalry Across the Solar System.* Plenum Publishing, 1994.

Reidel, Stephen P., Victor E. Camp, and Martin E. Ross, editors. *The Columbia River Flood Basalt Province.* Geological Society of America, 2013.

Reiners, Peter W., Richard W. Carlson, Paul R. Renne, et al. *Geochronology and Thermochronology.* John Wiley & Sons, 2017.

Repcheck, Jack. *The Man Who Found Time: James Hutton and the Discovery of the Earth's Antiquity.* Perseus Books, 2003.

Retallack, G. J. *Soils of the Past.* 2nd ed., Blackwell Science, 2001.

Reynolds, Mike D. *Falling Stars: A Guide to Meteors & Meteorites.* Stackpole, 2010.

Rhoads, Bruce L. *River Dynamics. Geomorphology to Support Management.* Cambridge UP, 2020.

Richardson, James, and James Bedient, comps. "Fire-ball FAQs." *American Meteor Society* American Meteor Society, n.d. Accessed 11 Mar. 2015.

Richter, Charles. "An Instrumental Earthquake Magnitude Scale." *Bulletin of the Seismological Society of America,* vol. 25, no. 1, Jan. 1935, pp. 1-32.

Riding, Robert E., and Stanley M. Awramik. *Microbial Sediments.* Springer-Verlag, 2010.

Ridley, John. *Ore Deposit Geology.* Cambridge UP, 2013.

Rigo, M., et al. "A Rise in the Carbonate Compensation Depth of Western Tethys in the Carnian (Late Triassic): Deep Water Evidence for the Carnian Pluvial Event." *Palaeogeography, Palaeoclimatology, Palaeoecology,* vol. 246, 2007, pp. 188-205.

Ringwood, Alfred E. *Origin of the Earth and Moon.* Springer, 2012.

Ritter, Dale F., R. Craig Kochel, and Jerry R. Miller, editors. *Process Geomorphology.* 4th ed., McGraw-Hill, 2002.

Ritter, Scott, and Morris Petersen. *Interpreting Earth History: A Manual in Historical Geology.* 8th ed., Waveland Press, 2015.

Robb, H. G. *Introduction to Ore Forming Processes.* Blackwell Science, 2005.

Robb, Laurence. *Introduction to Ore-Forming Processes.* Wiley-Blackwell, 2021.

Rodriguez Iturbe, Ignacio, and Andrea Rinaldo. *Fractal River Basins: Chance and Self-Organization.* Cambridge UP, 2001.

Rogers, John J. W., and M. Santosh. *Continents and Supercontinents.* Oxford UP, 2004.

Rohwer, Forest, Merry Youle, and Derek Vosten. *Coral Reefs in the Microbial Seas.* Plaid Press, 2010.

Rollinson, Hugh, and Victoria Pease. *Using Geochemical Data to Understand Geological Processes.* Cambridge UP, 2021.

Ross, Cyaden. *Introduction to Mineralogy.* Callisto Reference, 2019.

Ross, Pierre-Simon, et al. "Basaltic to Andesitic Volcaniclastic Rocks in the Blake River Group, Abitibi Greenstone Belt: 2. Origin, Geochemistry, and Geochronology." *Canadian Journal of Earth Sciences,* vol. 48, 2011, pp. 757-77.

Rossario, Frederic. *Floodscapes: Contemporary Landscape Strategies in Times of Climate Change.* NAi Uitgevers, 2021.

Roy, Raja Ram Mohan. *Taj Mahal Was a Red Sandstone Palace: The Saga of the Sikarwars.* Golden Egg Publishing, 2015.

Royal Society. *Ocean Acidification Due to Increasing Atmospheric Carbon Dioxide.* Author, 2005.

Ruddiman, William F. *Earth's Climate Past and Future.* 2nd ed. W. H. Freeman, 2008.

Rundqvist, Dmitrii V. *Mineral Collections of Russia, Part II: Twelve Museum Collections.* Vol 3. in *Mineralogical Almanac.* Ocean Pictures, 2000.

Ruse, Michael. *The Gaia Hypothesis: Science on a Pagan Planet*. U of Chicago P, 2013.

Russell, Christopher T., editor. *Deep Impact Mission: Looking Beneath the Surface of a Cometary Nucleus*. Springer, 2005.

Russell, Paul L. *Oil Shales of the World: Their Origin, Occurrence, and Exploitation*. Pergamon Press, 1990.

Ryan, Sean G., and Andrew J. Norton. *Stellar Evolution and Nucleosynthesis*. Cambridge UP, 2010.

Rygle, Kathy J., and Stephen F. Pedersen. *The Treasure Hunter's Gem and Mineral Guides to the U.S.A.* 5th ed., GemStone Press, 2008.

Sage, R. P., and T. Gareau. *A Compilation of References for Kimberlite, Diamond and Related Topics: Ontario Geological Survey, Open File Report 6067*. Queen's Printer for Ontario, 2001.

Saraswati, Pratul Kumar, and M. S. Srinavasan. *Micropaleontology Principles and Applications*. Springer, 2016.

Saucedo, R., et al. "Modeling of Pyroclastic Flows of Colima Volcano, Mexico: Implications for Hazard Assessment." *Journal of Volcanology and Geothermal Research*, vol. 139, nos. 1/2, 2005, pp. 103-15.

Scarth, Alwyn. *La Catastrophe: The Eruption of Mount Pelée, the Worst Volcanic Disaster of the Twentieth Century*. Oxford UP, 2002.

Schaetzl, Randall J., and Sharon Anderson. *Soils: Genesis and Geomorphology*. Cambridge UP, 2005.

———. "Weathering." *Soils: Genesis and Geomorphology*. Cambridge UP, 2005.

Schellart, W. P. "Overriding Plate Shortening and Extension Above Subduction Zones: A Parametric Study to Explain Formation of the Andes Mountains." *Geological Society of America Bulletin*, vol. 120, no. 11, 2008, pp. 1441-54.

Schleiss, Anton J., Giovanni De Cesare, Mário J. Franca, and Michael Pfister, editors. *Reservoir Sedimentation. CRC Press, 2014.*

Schlesinger, William H., editor. *Biogeochemistry*. Elsevier, 2005.

Schmidt, Paul E. *River Deltas: Types, Structures and Ecology*. Nova Science Publishers, 2011.

Schmidt, Victor. *Planet Earth and the New Geoscience*. 4th ed., Kendall/Hunt, 2003.

Schmincke, Hans-Ulrich. "Magma." *Volcanism*. Springer, 2004.

———. *Volcanism*. Springer, 2004.

Schmitt, Harrison H. *Return to the Moon: Exploration, Enterprise, and Energy in the Human Settlement of Space*. Praxis, 2006.

Schneider, Bonnie. *Extreme Weather: A Guide to Surviving Flash Floods, Tornadoes, Hurricanes, Heat Waves, Snowstorms, Tsunamis and Other Natural Disasters*. Palgrave Macmillan, 2012.

Schneider, Stephen H., et al., editors. *Scientists Debate Gaia: The Next Century*. MIT Press, 2008.

Schön, Jürgen H. *Physical Properties of Rocks: Fundamentals and Principles of Petrophysics*. 2nd ed., Elsevier, 2015.

Schulz, Horst D., and Matthias Zabel. *Marine Geochemistry*. 2nd ed., Springer Verlag, 2006.

Schumann, Walter. *Minerals of the World*. 2nd ed., Sterling, 2008.

Schwartz, Maurice L. *Encyclopedia of Coastal Science*. Springer, 2005.

Searle, Roger. *Mid-Ocean Ridges*. Cambridge UP, 2013.

Segar, Douglas. *An Introduction to Ocean Sciences*. 2nd ed., W. W. Norton, 2007.

Seibold, Eugen, and Wolfgang Berger. *The Sea Floor: An Introduction to Marine Geology*. 4th ed., Springer, 2017.

Selley, Richard C., and Stephen A. Sonnenberg. *Elements of Petroleum Geology*. 3rd ed., Elsevier, 2015.

Sene, Kevin. *Flood Warning, Forecasting and Emergency Response*. Springer Science + Business Media, 2008.

Sengör, A.M. Celál, and Saniye Atayman. *The Permian Extinction and the Tethys: An Exercise in Global Geology*. Geological Society of America, 2009.

Seppälä, Matti. *Wind as a Geomorphic Agent in Cold Climates*. Cambridge UP, 2004.

Shakun, Jeremy D., Peter U. Clark, Feng He, et al. "Global Warming Preceded by Increasing Carbon Dioxide Concentrations during the Last Deglaciation." *Nature*, vol. 484, no. 7392, 2012, pp. 49-54.

Shanmugam, G. *Deep-Water Processes and Facies Models: Implications for Sandstone Petroleum Reservoirs*. Elsevier, 2006.

———. *Mass Transport, Gravity Flows, and Bottom Currents: Downslope and Alongslope Processes and Deposits*. Elsevier, 2020.

Sharma, Vijay K. *Introduction to Process Geomorphology*. CRC Press, 2010.

Sharp, John. *Fractured Rock Hydrogeology*. CRC Press, 2014.

Sheppard, Charles R. C., Simon K. Davy, and Graham M. Pilling. *The Biology of Coral Reefs*. Oxford UP, 2009.

Sheppard, Charles. *Coral Reefs: A Very Short Introduction*. Oxford UP, 2014.

Simpson, John H., and Jonathan Sharples. *Introduction to the Physical and Biological Oceanography of Shelf Seas*. Cambridge UP, 2012.

Sinding-Larsen, Richard, and Friedrich W. Wellmer, editors. *Non-renewable Resource Issues: Geoscientific and Societal Challenges*. Springer, 2012.

Singer, D. A., and W. D. Menzie. *Quantitative Mineral Resource Assessments: An Integrated Approach*. Oxford UP, 2010.

Singer, M. J., and D. N. Munns. *Soils*. 6th ed., Prentice Hall, 2005.

Singh, Vijay P., editor. *Hydrology of Disasters*. Kluwer Academic Publishers, 2010.

Slak, Tomaz, and Vojko Kilar. "Development of Earthquake Resistance in Architecture from an Intuitive to an Engineering Approach." *Prostor*, vol. 19, no. 1, 2011, pp. 252-63.

"Small Worlds: The Neighborhood." *Discovery Program*. NASA, n.d. Accessed 6 May 2015.

Smith, M. R., and L. Collis, editors. *Aggregates: Sand, Gravel, and Crushed Rock Aggregates for Construction Purposes*. 3rd ed., Revised by P. G. Fookes, et al., Geological Society, 2001.

Soil Science Society of America. *Glossary of Soil Science Terms*. Author, 2001.

Spalding, Mark D., Edmund P. Green, and Corinna Ravilious. *World Atlas of Coral Reefs*. U of California P, 2001.

Sparrow, Giles. *Mars*. Quercus, 2015.

Speaight, Robert. *The Life of Teilhard de Chardin*. Harper & Row, 1967.

Speight, James G. *Enhanced Recovery Methods for Heavy Oil and Tar Sands*. Gulf, 2009.

Spellman, Frank R. *The Science of Water: Concepts and Applications*. 2nd ed., CRC Press, 2008.

Spudis, Paul D. *The Value of the Moon. How to Explore, Live, and Prosper in Space Using the Moon's Resources*. Smithsonian Books, 2016.

Squyres, Steve. *Roving Mars: Spirit, Opportunity, and the Exploration of the Red Planet*. Hyperion, 2006.

Stanley, Steven. *Earth Systems History*. 3rd ed., W. H. Freeman, 2009.

Stanley, Steven M. *Earth System History*. W. H. Freeman, 2004.

Stark, Andreas. *Seismic Methods and Applications*. BrownWalker Press, 2008.

Stein, Ruediger. *Arctic Ocean Sediments: Processes, Proxies and Paleoenvironment*. Elsevier, 2008.

Stein, Seth, and Michael Wyesession. *An Introduction to Seismology, Earthquakes, and Earth Structure*. Wiley, 2013.

Stern, S. Alan. *Pluto and Charon: Ice Worlds on the Ragged Edge of the Solar System*. John Wiley & Sons, 1998.

———, editor. *Worlds Beyond: The Thrill of Planetary Exploration as Told by Leading Experts*. Cambridge UP, 2003.

———, and David Grinspook. *Chasing New Horizons: Inside the First Epic Mission to Pluto*. Picador, 2018.

Stevanovic, Zoran, editor. *Karst Aquifers-Characterization and Engineering*. Springer, 2015.

Stevenson, David S. *Granite Skyscrapers: How Rock Shaped Earth and Other Worlds*. Springer/Praxis, 2018.

Stimpson, Ian. "Japan's Tohoku Earthquake and Tsunami." *Geology Today*, vol. 27, no. 3, 2011, pp. 96-98.

Strahler, Alan H. *Modern Physical Geography*. 4th ed., Wiley India, 2008.

Strahler, Alan, and Arthur Strahler. *Introducing Physical Geography*. 4th ed., John Wiley, 2006.

Strom, Robert G., and Ann L. Sprague. *Exploring Mercury: The Iron Planet*. Springer-Praxis, 2003.

Stüwe, Kurt. *Geodynamics of the Lithosphere: An Introduction*. Springer, 2007.

Sucuoglu, Halûk, and Sinan Akkar. *Basic Earthquake Engineering from Seismology to Analysis and Design*. Springer, 2014.

Sullivan, Walter. *Continents in Motion: The New Earth Debate*. McGraw-Hill, 1974.

Summerfield, Michael A. *Global Geomorphology: An Introduction to the Study of Landforms*. Routledge, 2013.

Svalova, Valentina, editor. *Heat-Mass Transfer and Geodynamics of the Lithosphere*. Springer, 2021.

Sverdrup, Keith A., Alyn C. Duxbury, and Alison B. Duxbury. *An Introduction to the World's Oceans*. 8th ed., McGraw Hill, 2004.

Swaby, Rachel. *Headstrong: 52 Women Who Changed Science-And the World*. Broadway Books, 2015.

Swart, Peter K., Gregor Eberli, and Judith A. McKenzie, editors. *Perspectives in Carbonate Geology*. Wiley Blackwell, 2009.

Sweeney, Alastair Black. *Bonanza: Canada's Oil Sands and the Race to Secure North America's Energy Future*. John Wiley and Sons, 2010.

Sykes, Lynn R. *Plate Tectonics and Great Earthquakes: 50 Years of Earth-shaking Events*. Columbia UP, 2019.

Szwacki, Nevill Gonzalez, and Teresa Szwacka. *Basic Elements of Crystallography*. 2nd ed., CRC Press, 2016.

Talley, Lynne D., George L. Pickard, William J. Emery, and James H. Swift. *Descriptive Physical Oceanography: An Introduction*. 6th ed., Elsevier, 2011.

Tappert, Ralf, and Michelle C. Tappert. *Diamonds in Nature: A Guide to Rough Diamonds*. Springer, 2011.

Taranath, Bungale S. *Tall Building Design. Steel, Concrete and Composite Systems*. CRC Prss, 2017.

Tarbuck, Edward J., Frederick K. Lutgens, and Dennis Tasa. *Earth: An Introduction to Physical Geology*. 13th ed., Pearson Education, 2019.

Tatsumi, Y. "The Subduction Factory: How It Operates on Earth." *GSA Today*, vol. 15, 2005, pp. 4-10.

Taylor, Frank Bursley. "Bearing of the Tertiary Mountain Belt on the Origin of the Earth's Plan." *Geological Society of America*, vol. 21, 1910, pp. 176-226.

Taylor, Stuart Ross. *Lunar Science: A Post-Apollo View: Scientific Results and Insights from the Lunar Samples*. Elsevier, 2016.

Tegethoff, F. Wolfgang, editor. *Calcium Carbonate from the Cretaceous Period into the 21st Century*. Springer, 2001.

Tertzakian, Peter A. *Thousand Barrels a Second: The Coming Oil Break Point and the Challenges Facing an Energy Dependent World*. McGraw-Hill, 2006.

"Tests of Big Bang: The Light Elements." *Wilkinson Microwave Anisotropy Probe*. NASA, 16 Apr. 2010. Accessed 2 June 2015.

Than, Ker. "Densest Matter Created in Big-Bang Machine." *National Geographic*. National Geographic Society, 26 May 2011. Accessed 16 July 2015.

Theng, Benny K. G. *Clay Mineral Catalysis of Organic Reactions*. CRC Press, 2019.

Thomas, David S. G. *Arid Zone Geomorphology: Process, Form and Change in Drylands*. 3rd ed., Wiley Blackwell, 2011.

Thurman, Harold V., and Alan P. Trujillo. *Introductory Oceanography*. 10th ed., Pearson Prentice Hall, 2004.

Tilley, Richard J. D. *Crystals and Crystal Structures*. John Wiley, 2006.

Timmons, J. Michael, and Karl E. Karlstrom, editors. *Grand Canyon Geology: Two Billion Years of Earth's History*. The Geological Society of America, 2012.

Todd, David Keith, and Larry W. Mays. *Groundwater Hydrology*. 3rd ed., Wiley, 2005.

Treloar, P. J., and M. P. Searle. *Himalayan Tectonics: A Modern Synthesis*. Geological Society, 2019.

Trewby, Mary. *Antarctica: An Encyclopedia from Abbott Ice Shelf to Zooplankton*. Firefly, 2002.

Trujillo, Alan P., and Harold V. Thurman. *Essentials of Oceanography*. 10th ed., Prentice Hall, 2010.

Tsutsumi, Daizo, and Jonathan B. Laronne. *Gravel-Bed Rivers: Process and Disasters*. Wiley, 2017.

Tucker, E. R., and R. G. C. Bathurst. *Carbonate Diagenesis*. Reprint. John Wiley & Sons, 2009.

Tucker, Maurice E. *Sedimentary Petrology: An Introduction to the Origin of Sedimentary Rocks*. 3rd ed., Blackwell Science, 2001.

———. *Sedimentary Rocks in the Field*. 4th ed., John Wiley & Sons, 2011.

Turcotte, Donald L., and Gerald Schubert. *Geodynamics*. 3rd ed., Cambridge UP, 2014.

Turner, Chris. *The Patch: The People, Pipelines, and Politics of the Oil Sands*. Simon & Schuster, 2017.

Turner, David, and Lee A. Groat. *Geology and Mineralogy of Gemstones*. Wiley, 2021.

Twidale, C. R., and J. R. Vidal Romani. *Landforms and Geology of Granite Terrains*. A.A. Balkema Publishers, 2005.

Twiss, Robert J., and Eldridge M. Moores. *Structural Geology*. W.H. Freeman, 2006.

U.S. Department of Agriculture, Soil Survey Staff. *Keys to Soil Taxonomy*. UP of the Pacific, 2005.

U.S. Geological Survey. *Minerals Yearbook*. Author, 2008.

Ulivi, Paolo, and David M. Harland. *Lunar Exploration: Human Pioneers and Robotic Surveyors*. Praxis Publishing, 2004.

Vacher, Leonard, and Terrence M. Quinn. *Geology and Hydrogeology of Carbonate Islands*. Elsevier, 2004.

Valentine, David. "Chinese Placer Mining in the United States: An Example from American Canyon, Nevada." *The Chinese in America: A History from Gold Mountain to the New Millennium*, edited by Susie Lan Cassel, AltaMira Press, 2002.

Van Breemen, Nico, and Peter Buurman. *Soil Formation*. 2nd ed., Kluwer Academic Publishers, 2002.

Van der Pluijm, Ben A., and Stephen Marshak. *Earth Structure: An Introduction to Structural Geology and Tectonics*. W.W. Norton, 2010.

Van Kranendonk, Martin, J. R. Hugh Smithies, and Vickie C. Bennett, editors. *Earth's Oldest Rocks*. Elsevier, 2007.

Vanoni, V.A., editor. *Sedimentation Engineering*. 2nd ed., American Society of Civil Engineers, 2006.

Vernon, Ron H. *A Practical Guide to Rock Microstructure*. Cambridge UP, 2004.

———, and Geoffrey Clarke. *Principles of Metamorphic Petrology*. Cambridge UP, 2008.

Vertesi, Janet. *Seeing Like a Rover: How Robots, Teams, and Images Craft Knowledge of Mars*. U of Chicago P, 2015.

Vetere, Francesco, editor. *Dynamic Magma Evolution*. Wiley, 2021.

Villaverde, Roberto. *Fundamental Concepts of Earthquake Engineering*. CRC Press, 2009.

Vrielynck, Bruno, and Philippe Bouysse. *The Changing Face of the Earth: The Break-Up of Pangaea and Continental Drift over the Past 250 Million Years in Ten Steps*. UNESCO, 2003.

Walther, John Victor. *Essentials of Geochemistry*. 2nd ed., Jones & Bartlett Publishers, 2008.

Ward, Andy D., and Stanley W. Trimble. *Environmental Hydrology*. 2nd ed., CRC Press, 2004.

Warren, John K. *Evaporites: A Geological Compendium*. 2nd ed., Springer, 2016.

Wefer, Gerold, David Billett, Dierk Hebbeln, et al., editors. *Ocean Margin Systems*. Springer, 2003.

Wegener, Alfred Lothar. *The Origin of Continents and Oceans*. Translated by J. G. A. Skerl. E. P. Dutton, 1924.

———. *The Origins of Continents and Oceans*. Translated by John Biram from 4th rev. German ed., Dover, 1966.

Weiss, Achim. "Big Bang Nucleosynthesis: Cooking Up the First Light Elements." *Einstein Online*. Max Planck Institute for Gravitational Physics, 2006. Accessed 2 June 2015.

Welland, Michael. *Sand: The Never Ending Story*. U of California P, 2009.

Wellmer, Friedrich-Wilhelm, Manfred Dalheimer, and Markus Wagner. *Economic Evaluations in Exploration*. Springer, 2008.

Wenk, Hans-Rudolf, and Andrei Bulakh. *Minerals: Their Constitution and Origin*. Cambridge UP, 2004.

Wesley, Liam R. *Clays and Clay Minerals: Geological Origin, Mechanical Properties and Industrial Applications*. NOVA Science Publishers, 2014.

West, Janet P. *Basalt: Types, Petrology and Uses*. NOVA Science Publishers, 2012.

West, Terry R., and Abdul Shakoor. *Geology Applied to Engineering*. 2nd ed., Waveland Press, 2018.

"What Is NASA's Asteroid Redirect Mission?" *Asteroid Redirect Mission*. NASA, 30 Apr. 2015. Accessed 11 May 2015.

White, James D. L., John L. Smellie, and David A. Clague. *Explosive Subaqueous Volcanism*. Wiley, 2003.

White, William M. *Geochemistry*. Wiley, 2020.

Wicander, Reed, and James S. Monroe. *Essentials of Physical Geology*. 5th ed., Cengage Learning, 2008.

———. *The Changing Earth: Exploring Geology and Evolution*. 5th ed., Brooks/Cole, 2009.

———. *Historical Geology*. Brooks/Cole, Cengage Learning, 2010.

Wichert, Jörn. *Slate as Dimension Stone: Origin, Standards, Properties, Mining and Deposits*. Springer, 2020.

Wilhelms, Don E. *To a Rocky Moon: A Geologist's History of Lunar Exploration*. U of Arizona P, 1993.

Wilkinson, Paul. *Pompeii: The Last Day*. BBC, 2003.

Williams, Richard G., and Michael J. Follows. *Ocean Dynamics and the Carbon Cycle: Principles and Mechanisms*. Cambridge UP, 2011.

Wilson, J. Tuzo. "A Revolution in Earth Science." *Geotimes*, vol. 13, Dec. 1968, pp. 10-16.

Wilson, Marjorie. *Igneous Petrogenesis: A Global Tectonic Approach*. Springer, 2007.

Winchester, Simon. *A Crack in the Edge of the World: America and the Great California Earthquake of 1906*. HarperCollins, 2006.

———. *The Map That Changed the World: William Smith and the Birth of Modern Geology*. Perennial, 2001.

Winter, John D. *An Introduction to Igneous and Metamorphic Petrology*. Prentice Hall, 2001.

———. *Principles of Igneous and Metamorphic Petrology*. 2nd ed., Pearson Education, 2010.

Witze, Alexandra. "Charles F. Richter, 1900-1985, American Seismologist." *Notable Scientists: From 1900 to the Present*, edited by Brigham Narins, Gale, 2001.

Wohl, Ellen. *A World of Rivers*. U of Chicago P, 2011.

Wolanski, Eric. *Oceanographic Processes of Coral Reefs: Physical and Biological Links in the Great Barrier Reef*. CRC Press, 2001.

Wood, Gillen D'Arcy. *Land of Wondrous Cold: The Race to Discover Antarctica and Unlock the Secrets of Its Ice*. Princeton UP, 2020.

Woodward, Jamie. *The Ice Age: A Very Short Introduction*. Oxford UP, 2014.

Woolley, Alan R. *Alkaline Rocks and Carbonatities of the World, Parts 1-4*. Geological Society, 2019.

Worster, Donald. *Dust Bowl: The Southern Plains in the 1930's*. 25th anniversary ed., Oxford UP, 2004.

Wright, Edward L. "Big Bang Nucleosynthesis." *Ned Wright's Cosmology Tutorial*. UCLA Astronomy & Astrophysics, 26 Sept. 2012. Accessed 2 June 2015.

Wright, James E., and John W. Shervais. *Ophiolites, Arcs, and Batholiths: A Tribute to Cliff Hopson*. The Geological Society of America, 2008.

Wyllie, Peter J. *The Way the Earth Works: The Introduction to the New Global Geology and Its Revolutionary Development*. John Wiley & Sons, 1976.

Xu, Ruren, et al. *Chemistry of Zeolites and Related Porous Materials: Synthesis and Structure*. Wiley-Interscience, 2007.

Yang, Wencai. *Reflection Seismology: Theory, Data Processing and Interpretation*. Elsevier, 2014.

Yardley, Bruce, and Clare Warren. *An Introduction to Metamorphic Petrology*. 2nd ed., Cambridge UP, 2021.

Young, Davis A. *Mind over Magma: The Story of Igneous Petrology*. Princeton UP, 2003.

Younger, Paul L. *Groundwater in the Environment: An Introduction*. Blackwell, 2007.

Yount, Lisa. *Alfred Wegener: Creator of the Continental Drift Theory*. Chelsea House Publishing, 2009.

Zalasiewicz, Jan, and Mark Williams. *Ocean Worlds: The Story of Seas on Earth and Other Planets.* Oxford UP, 2014.

Zeebe, Richard E., and Dieter Wolf-Gladrow. *CO2 in Seawater: Equilibrium, Kinetics, Isotopes.* Elsevier, 2001.

Zeilinga de Boer, Jelle, and Donald Theodore Sanders. *Earthquakes in Human History: The Far-Reaching Effects of Seismic Disruptions.* Princeton UP, 2005.

———. *Volcanoes in Human History: The Far-Reaching Effects of Major Eruptions.* Princeton UP, 2005.

Zektser, Igor S., and Lorne G. Everett, editors. *Ground Water Resources of the World and Their Use.* UNESCO, 2004. Reprint. National Ground Water Association Press, 2006.

Zhang, Lifei, Zeming Zhang, Hans-Peter Schertl, and Chunjing Wei. *HP-UHP Metamorphism and Tectonic Evolution of Orogenic Belts.* Geological Society of London, 2019.

Zim, Herbert S., and Paul R. Shaffer. *Rocks, Gems, and Minerals: A Guide to Familiar Minerals, Gems, Ores, and Rocks.* Rev. and updated ed., St. Martin's Press, 2001.

Zoellner, Tom. *The Heartless Stone: A Journey Through the World of Diamonds, Deceit, and Desire.* St. Martin's Press, 2006.

Zubrin, Robert, and Richard Wagner. *The Case for Mars: The Plan to Settle the Red Planet and Why We Must.* Free Press, 2011.

Glossary

abrasion: damage and removal of material from materials by friction against another material

abyssal plain: the flat, sediment covered area of the sea floor that merges with the base of the continental rise

accretion: the accumulation of mass by the impact of small celestial bodies against large celestial bodies

aeolian: due to transport by wind

aeolian deposits: material transported by wind

aeolian erosion: erosion caused by abrasion due to sediment particles being transported by wind

albedo: the percentage of incoming light that is reflected from the surface of a planet or other body

alluvial fan: the deposit of sediments that forms as the sediments being carried in suspension by flowing water settle and spread out as the flow rate decreases

alpha proton X-ray spectrometer: a device that irradiates rock samples with X-rays of a specific wavelength that are absorbed by atoms in the sample and re-emitted as X-rays of characteristic wavelengths. The re-emitted X-rays are then captured by the spectrometer for identification of the particular minerals that are present in the sample

aluminosilicates: silicate rocks in which the major metal present is aluminum

amethyst: a variety of quartz noted for its violet to purple color produced by irradiated molecules of iron contained within its crystalline structure

andesite: the principal igneous magma of continental volcanoes, named for the Andes Mountain range

angle of repose: the natural angle formed between the horizontal plane and the side of a freestanding pile of sediment particles under the force of gravity in a fluid medium

anion: an atom that has gained electrons to become a negatively charged ion

anisotropic: a structure in which a physical property has different values depending on direction

anthropogenesis: the emergence over time of humans

anthropogenic: caused by or as a result of human activity

anticyclonic: the clockwise rotation of an air mass in the Northern hemisphere, counterclockwise in the Southern hemisphere

antidune: an undulatory upstream moving bed form produced in free surface flow of water over a sand bed in a certain range of high flow speeds and shallow flow depths

aphanitic: igneous rock that is so fine-grained as to be glassy, such as obsidian, formed from magma that has cooled and crystallized very rapidly

apogee: the point in an elliptical orbit at which the orbiting body is at its furthest distance from the main focal point of the orbit

assimilation: the absorption of the chemical components of wall rock or xenoliths into a magma

asteroid: a minor planet in the inner solar system that orbits the Sun

asteroid belt: an area between the planets Mars and Jupiter where the majority of asteroids can be found

asthenosphere: the ductile layer of the Earth, beneath the lithosphere, defined by mechanical behavior; although solid, it deforms horizontally over tens of millions of years

augite: an essential mineral in most basalts, a member of the pyroxene group of silicates

aureole: a ring-shaped zone of metamorphic rock surrounding a magmatic intrusion

authigenic minerals: minerals that formed in place, usually by diagenetic processes

bar: a unit of pressure equal to 100 kilopascals and very nearly equal to 1 standard atmosphere

barrier island: a low-lying type of island that may form from current-borne sediments along a coastline; such islands act as a barrier against incoming oceanic waves

basalt: a hard, dense volcanic rock

basalt: dark, fine-grained igneous rock composed mostly of plagioclase (a calcium-sodium silicate mineral) and pyroxene (a magnesium iron silicate mineral)

basaltic: formed from iron-rich magma from deep within the mantle

base flow: the amount of water that enters a particular stream from the water table

base unit: unit of measurement that is used for one of the seven fundamental properties; also known as defined unit

basin order: an approximate measure of the size of a stream basin, based on a numbering scheme applied to river channels as they join together in their progress downstream

bas-relief: a method of stonework in which the background layer of stone is removed from an area of rock, so that the remaining unaffected portion can be sculpted into a scene that stands out from the background

batholith: an exposed plutonic intrusion spanning hundreds to thousands of square kilometers in size

beach: an accumulation of loose material, such as sand or gravel, that is deposited by waves and currents at the border of a body or stream of water

bed configuration: the overall geometry of a sediment bed molded by sediment transport in a flowing fluid

bed form: an individual geometrical element of a bed configuration

bedding planes: the silt sediments that will eventually become slate are deposited at regular intervals as thin layers, or beds. The slate that they form will separate readily along the plane of those beds, or bedding planes

bed load: the amount of material that a stream transports in contact with the surface of the streambed

bed scouring: removal of riverbed features and transport of riverbed materials by stream flow that has lost its natural load on the upstream side of a dam and acquires a fresh stream load on the downstream side of the dam

Big Bang nucleosynthesis (BBN): the coming together of free neutrons and protons to create hydrogen isotopes, and the subsequent fusion of hydrogen nuclei to form other light elements, within the first three minutes of the big bang

Big Bang theory: the theory that the universe began in an extremely hot, dense state and then began a process of expansion that continues today

biostratigraphy: use of fossil animals and plants in the dating of rocks

biotite: type of layered, silicate mineral composed of silica molecules bonded with magnesium and iron and appearing in dark brown and black varieties

bitumen: the most viscous grade of crude petroleum, having a relatively low content of lighter, volatile components; typically used for road paving when mixed with fine aggregate materials

black smokers: active hydrothermal vents along seafloor ridges, which discharge acidic solutions of high temperature, volume, and velocity, charged with tiny black particles of metallic sulfide minerals

blue ground: the slaty blue or blue-green kimberlite breccia of the South African diamond pipes

booster: the rocket used to propel a capsule into space

boron carbide: a compound of boron and carbon with the chemical formula BC, not a simple alloy

Bowen's reaction principle: a principle by which a series of minerals forming early in a melt react with the remaining melt to yield a new mineral in an established sequence

calcareous algae: green algae that secrete needles or plates of aragonite as an internal skeleton; very important contributors to reef sediment

calcareous rock: rock composed primarily of calcium carbonate

calcification: the process by which marine organisms secrete calcium carbonate to produce shells, tests and casings as body parts

calcinating: a process of heating a material in air

calcite: a mineral composed of calcium carbonate

calcrete, caliche: commonly called hardpan, sand and other particulates that have become cemented together by the accretion of calcium carbonate in the interstices between the particles

caldera: an underground magma chamber

caprock: impermeable rock that overlies a particular geological stratum or deposit

carbonate: a mineral with the carbonate anion CO_3^{2-} in its chemical formula, such as calcite ($CaCO_3$)

carbonate rocks: the general term for rocks containing calcite, aragonite, or dolomite

carbonaceous chondrites: a class of meteoritic bodies found to contain large amounts of carbon in conjunction with other elements; used to date the solar system and provide chemical composition assessments of the original solar nebula

carbonatite: silicate mineral consisting of reduced levels of silica and rich in carbonate molecules formed from carbonic acid

Carboniferous period: the fifth of the six periods in the Paleozoic era; it preceded the Permian period

cation: an atom that has lost electrons to become a positively charged ion

cation exchange capacity: the ability of a clay to adsorb and exchange cations, or positively charged ions, within its environment

cementation: the binding together of sedimentary particles by the formation of crystalline material in the pores between the particles

ceramics: nonmetal compounds, such as silicates and clays, produced by firing the materials at high temperatures

chalcedony: a form of cryptocrystalline quartz formed from microcrystals of quartz and other included minerals

champlevé: a French word literally meaning "raised field" for the process of roughening the surface of, or raising, the field that is to be enameled

channel: a horizontal depression in the ground surface caused and enlarged by the concentrated flow of water

chemical weathering: a change in the chemical and mineralogical composition of rocks by means of reaction with water at the earth's surface

chert: multicolored, fine-grained sedimentary rock composed of silica and formed in the ocean

clastic particles: particles formed by the mechanical breakdown of rocks

clastic rock: a sedimentary rock composed of particles without regard to their composition; this term is sometimes used, incorrectly, as a synonym for "detrital"

cladistics: computer-based method for delineating relationships among groups of organisms

clay mineral: any of a group of tiny silicate minerals (less than 2 micrometers in size) that form by varied degrees of weathering of other silicate minerals

claystone: shale formed from fine clay particles that are flat or plate-like in structure

cleavage: breaking of a crystallized mineral along a plane, leaving a smooth rather than an irregular surface

coal: dark brown to black sedimentary rock formed from the accumulation of plant material in swampy environments

coccolithophores: a type of algae that consumes carbon dioxide

coefficient of thermal expansion: the linear expansion ratio (per unit length) for any particular material as the temperature is increased

color: a distinctive color tone apparent in and unique to particular formation of slate due to minor mineral components in the slate

combined flow: a flow of fluid with components of both unidirectional and oscillatory flow superposed on one another to produce a more complex pattern of fluid motion

comet: an icy small solar system body that heats up and releases gases when it passes near the Sun in its orbit

compaction: the pressing together of particles in a sedimentary deposit by the weight of overlying materials

compressive strength: the ability to withstand a pushing stress or pressure, usually given in pascals or pounds per square inch

conchoidal fracture: fracturing of the type exhibited by materials that do not break along planes of separation

condensation: the change of state of elements from hot gases to liquid or solid forms as temperature decreases

contact metamorphic facies: zones of contact metamorphic effects, each of which is characterized by a small number of indicator minerals

contact metamorphism: metamorphic processes that occur when rock is forced into contact or close proximity to a source of high temperatures and pressures

continental crust: the high silica-content rock of continental land masses, less dense than the basaltic rock of oceanic crust

continental drift: the theory proposed by Alfred Wegener to explain the apparent complimentary outline of continental land masses on opposite sides of the Atlantic Ocean

continental margin: the edge of a continent that is both exposed on land and submerged below the water that marks the transition to the ocean basin

continental rise: the broad and gently sloping ramp that rises from the abyssal plain to the base of the continental slope; submarine fans are found here

continental shelf: the gently seaward sloping submerged edge of a continent that commonly extends to a depth of about 200 meters or to the edge of the continental slope

continental slope: the relatively steep region of the continental margin between the continental shelf and the continental rise

contraction theory: the erroneous theory that Earth's mountains and valleys are the result of the outer crust buckling and folding due to shrinkage as the planet cooled from some primordial molten state

convection: a circular movement within a fluid by which hotter, less dense material rises within the fluid medium and cooler, denser material sinks within the fluid medium

convergence: the condition in which tectonic plates are driven toward each other where their margins are in contact with each other

convergent boundary: the geodynamic process in which two tectonic plates directly collide

convergent plate boundary: the line of demarcation between two tectonic plates where they are being forced into contact with each other

conversion factor: ratio of one unit of measure in a particular system to the unit of measure of the same property in a different system of measurements

coquina: a fossil-bearing form of limestone

coralline algae: red algae that secrete crusts or branching skeletons of high magnesium calcite; important sediment contributors and binders on reefs

cosmogenesis: the view that the entire universe is evolving like a living entity

country rock: the pre-existing rock into which magma or a mineral formation intrudes

covalently bonded: of atoms that are connected to each other by sharing of two electrons without charge separation into positive and negative ions

cracking: a petroleum production process in which crude oil is heated over catalytic materials to "crack" the large molecular components of the crude product into smaller molecules that are then separated and recovered

craton: stable portion of the earth's mantle and associated crust that forms the inner core of the lithospheric components

creep: the slow, gradual downslope movement of soil materials under gravitational stress

crevasse: a break in the bank of a distributary channel causing a partial diversion of flow and sediment into an interdistributary bay

crust: the outermost layer of the earth, which consists of materials that are relatively light; the continental crust is lighter than oceanic crust, which allows it to float while oceanic crust sinks

crystal: a solid made up of a regular periodic arrangement of atoms

crystalline: a property of a chemical compound with an orderly internal atomic arrangement that may or may not have well-developed external faces

crystallization: the solidification of molten rock as a result of heat loss; slow heat loss results in the growth of crystals, but rapid heat loss can cause glass to form

crystal structure: the regular arrangement of atoms in a crystalline solid

cube: a geometric form having six equal square faces

cumulates: layers of large crystalline material, separated by gravity, that form at the bottom of a magma chamber as it begins to cool

Curie temperature: the temperature at which metal atoms are no longer able to maintain magnetic susceptibility; above this temperature the magnetic moment of the atoms is lost, below it the magnetic moment of the atoms becomes fixed

current ripple: a small bed form, oriented predominantly transverse to the direction of flow, produced at low to moderate flow speeds in unidirectional water flows

cyclonic: the counterclockwise rotation of an air mass in the Northern hemisphere, clockwise in the Southern hemisphere

Darcy's Law: the rate of flow of water in an aquifer is dependent on the hydraulic conductivity of the aquifer matrix, its perpendicular cross-sectional area and the hydraulic gradient of the aquifer

deflation: the sorting out, lifting, and removal of loose, dry, silt and clay sized soil particles by turbulent eddy action of the wind

delta: a deposit of sediment, often triangular, formed at a river mouth where the wave action is low and the river's current slows suddenly

density: the ratio of an element or objects mass to its volume, expressed for elements and minerals as grams per cubic centimeter

deposition: the process by which loose sediment grains fall out of seawater to accumulate as layers of sediment on the sea floor

derived unit: unit of measurement that is derived from the relationship of base units

desertification: the slow expansion of a desert area tending to increase the area it encompasses rather than just a shifting of its location

detrital minerals: minerals that have been eroded, transported, and deposited as sediments

detrital rock: a sedimentary rock composed mainly of grains of silicate minerals as opposed to grains of calcite or clays

diatom: a microscopic single-cell alga found in all parts of the world; the large deposits of fossil diatoms are composed of a type of sedimentary rock called diatomite

diagenesis: chemical and mineralogical changes that occur in a sediment after deposition and before metamorphism

diamond: a high-pressure, high-temperature mineral consisting of the element carbon; it is the hardest naturally occurring substance and is valued for its brilliant luster

diamond pipes: the chimneys of explosive volcanic eruptions that have allowed diamond-bearing kimberlite to rise from deep in the mantle and fill the available space within the chimney

diamond stability zone: area within the earth's mantle where pressures and temperatures are sufficient for the formation of diamonds from deposits of carbon

diatreme: a volcanic vent or pipe formed as the explosive energy of gas-charged magmas breaks through crustal rocks

differentiation: the trend of elements to separate according to density under the force of gravity within the molten interior of Earth

dike: a tabular body of igneous rock that intrudes vertically through the structure of the existing rock layers above

dimensional analysis: consideration and manipulation of the units involved in a measurement and their relationship to one another

dirt cracking: a process in which clays accumulate in rock cracks, take on water, and expand to rupture the rock

discharge: the volume of water moving through a given flow cross section in a given unit of time

disequilibrium condition: a condition in which a relatively small change can render a system unstable, causing it to collapse to a condition that is more stable

distributary channel: a river that is divided into several smaller channels, thus distributing its flow and sediment load

divergence: the condition in which tectonic plates are driven apart from each other from the locations where their margins meet

divergent boundary: the geodynamic process whereby two tectonic plates separate and create a space for magma to fill

divergent plate boundary: the line of demarcation between two tectonic plates that are being driven apart from each other

dolomite: a mineral composed of calcium magnesium carbonate

dolostone: a mineral form of calcium carbonate, $CaCO_3$, that is dense and fine-grained

drainage basin: an area in which all surface water from rainfall, snowmelt, etc., flows toward a central channel that functions as the outlet for the water on its eventual way to the ocean

drainage density: the number of streams per unit area within a particular watershed area

drought: a prolonged period without rain, generally with sunshine and hot temperatures, in an area that normally receives sufficient annual rainfall to maintain crops and support animal herds

dune: a large bed form, oriented predominantly transverse to the direction of flow, produced at moderate to high flow speeds

duration: the length of time ground motion occurs during an earthquake

dynamic equilibrium: the condition of a system in which changes in parts of the system tend to be countered by changes in other parts of the system, producing an overall steady state; e.g., the formation of sediment by erosion of rocks is countered by formation of rocks from sediment

dwarf planet: a relatively large astronomical object that orbits the sun and has enough mass to be roughly spherical in shape but has not cleared its orbit of other astronomical objects

echo sounder: a device that emits a pulse of audible or ultrasonic sound waves and measures the time between emission of the pulse and reception of its echo signal to determine distance; the operating principle of the modern "fish finder"

eclogite: rock composed principally of garnet and pyroxene that formed at high pressures associated with great depths

ejecta: material thrown up from the surface as a result of the impact of a meteoroid striking the surface

elasticity: the maximum stress that can be sustained without suffering permanent deformation

elastic rebound: the movement that returns a strained material in a high-energy state to its normal low-energy state, such as the release of an elastic band

emergence: term for the development of some aspect of the universe that did not exist previously

endogenic: originating from or formed by forces acting within Earth

endoreic watershed: a watershed with no exit to an ocean

energy-intensive: requiring the input of large amounts of energy to produce the desired output products

epicenter: the surface-level geographic point located directly above an earthquake's hypocenter

equant grains: grains of similar size and shape

equilibrium: in chemical systems, the state in which the reactants and products of different reactions involving related materials function in concert to maintain the system in an overall steady state

erosion: the displacement of sediment from one location to another on the ground surface

evaporite: the residue remaining when mineral-laden water has evaporated

evolution: change through time in successive generations of organisms

exfoliation: spalling off of the surface layer of a rock body due to weathering

exogenic: originating from or formed by forces acting at Earth's surface

expansion-contraction cycles: processes of wetting drying, heating cooling, or freezing thawing, which affect soil particles differently according to their size

extinction: elimination of a taxon of organisms

extrusive rock: igneous rock that has been erupted onto the surface of the earth

faceting: a specialist skill in which a rough diamond is meticulously positioned against a rotating lap in order to create flat surfaces in a particular design

facies: a part of a rock or group of rocks that differs from the whole formation in one or more properties, such as composition, age, or fossil content

fathom: a unit of length equal to 1.8 meters, used principally in the measurement of marine depths

faulting: the process of fracturing the earth such that rocks on opposite sides of the fracture move relative to one another; faults are the structures produced during the process

feedback system: any system that produces an output result that either reduces the effect of the system (negative feedback) or increases the effect of the system (positive feedback)

feldspar: a silicate rock that forms in igneous environments and is the most common mineral in the earth's crust

felsic: mineral group characterized by high proportions of silicate minerals, including quartz, feldspar, plagioclase, and micas

felsic-intermediate: igneous rock having the highest range of silica content

felsic rock: rock having a higher silica content than mafic rock

ferro-magnesium silicates: silicate rocks in which the major metals present are iron and magnesium

felsic rocks: igneous rocks rich in potassium, sodium, aluminum, and silica, including granites and related rocks

fiber optics: flexible transparent fiber made of fused quartz or silica glass; used in computer and telecommunications applications

fiber-reinforced composites: materials produced by drawing fibers of various types through a material being cast to produce a high weight-to-strength ratio

fireball: a meteor that is unusually bright

flash floods: rises in water level that occur unusually rapidly, generally because of especially intense rainfall

flint: a type of microcrystalline quartz that exists in large quantities within the earth's crust and is notable for its glass-like characteristics

flood: a rising body of water that overtops its usual confines and inundates land not usually covered by water

fluidization: process whereby the rocks and minerals contained in a volcanic eruption are vaporized within a pyroclastic flow

flume: a laboratory open channel in which water is passed over a sediment bed to study the nature of the sediment movement

fly-by: a trajectory through space that passes by a planet or other body without entering into orbit about that body

focal depth: the depth of an earthquake's hypocenter

focus: the location within Earth's crust at which the energy released in a seismic event originates

folding: the effect of the collision of two plates pushing Earth's outer crust outward, causing the rock to bend slowly over time

foliation: formation of rock in sheet-like layers, analogous to the sheets of paper in a book or "folio"

fossil: remains of a once-living organism or a trace of its behavior

fossilization: process of preserving the remains of organisms within sediment

fracking: a shale oil extraction method in which the oil shale bedrock is physically fractured in place, then subjected to heat to facilitate the flow of oil to recovery wells

fractionation: the separation of mineral components in magma according to their density under the force of gravity

frost wedging: the forcing apart and breakage of rock particles when water enters a crack or pore in the rock and freezes, the expansion of water into ice in the crack or pore applying sufficient pressure to fracture the particle

frustule: the overlapping valves of a diatom, typically containing silica

fullerenes: a recently recognized elemental form of carbon consisting of closed spheroid molecules constructed of fused five- and six-membered rings of carbon atoms; the other elemental forms of carbon are diamond and graphite

fuller's earth: the type of clay used in "fulling" wool by removing lanolin and other contaminants

fundamental property: property ascribed to a phenomenon, substance, or body that can be quantified: length, mass, time, electric current, thermodynamic temperature, the amount of a substance, and luminous intensity

fusion: the joining together of two or more atomic nuclei to form a new, heavier nucleus of a different element.

gas hydrates: relatively stable, often crystalline, molecular combinations of water and hydrocarbon gases, especially methane, that form on the ocean floor due to the elevated pressure conditions

gem: a cut and polished stone that possesses durability, rarity, and beauty necessary for use in jewelry and, therefore, of value

gemstone: any rock, mineral, or natural material that has the potential for use as personal adornment or ornament

geoarchaeology: the technique of using ancient human habitation sites to determine the age of landforms and when changes occurred

geochemical sinks: the processes by which elements and compounds are removed from the crust and oceans to be recycled in active chemical cycles

geochronology: the measurement of the ages of rocks

geomagnetic signature: the alignment of the magnetic moment of magnetically susceptible atoms with the magnetic field of Earth

geomorphology: the study of the origin of landscapes

geosynclines: major depressions in the surface of the earth where sediments accumulate; geosynclines lie parallel to the edges of continents and are long and narrow

geothermal energy: electricity produced using steam or hot water from natural hot springs to drive generator turbines

glacial deposit: accumulations of particulate matter ranging in size from fine silt to large boulders that are pushed ahead of glaciers and eventually dropped in place when the glaciers retreat

glaciation: the advance of a continental ice sheet

gossan: a cap layer of ferric hydroxide atop an ore deposit

granite: igneous rock composed of coarse crystals (greater than about 2 millimeters) of plagioclase (calcium-sodium silicate mineral), alkali feldspar (potassium-sodium silicate mineral), and quartz (all silicate with no other plus ions), and a few dark silicate minerals

granitic/granitoid: descriptive terms for plutonic rock types having quartz and feldspar as major mineral phases

granitization: the process of converting rock into granite; it is thought to occur when hot, ion-rich fluids migrate through a rock and chemically alter its composition

gravel: assorted rock particles whose dimensions are within a defined range of about 5 to 90 millimeters

gravitational collapse: the infall of gaseous particles to form a more dense mass about the center of gravity of a large gaseous cloud of atoms

gravitationally induced wobble: planets and stars orbit each other about their common center of mass, like a very lopsided spinning dumbbell; the wobble is the minor but regular change in the observed position of the star as the orbital rotation proceeds

gravity layering: distinct layers of igneous rock in an igneous rock body, produced by fractionation of minerals according to density under the influence of gravity as the magma body slowly cools

greenhouse effect: the elevation of average atmospheric temperature due to the ability of certain gas molecules to absorb infrared (heat) energy from surface emissions and return a significant portion of that energy to the atmosphere, thus preventing it from radiating back out into space

greenhouse gas: any atmospheric gas, typically carbon dioxide, methane, water and others, that tends to increase the greenhouse effect in the atmosphere

groundmass: the fine-grained material between phenocrysts of a porphyritic igneous rock

ground motion: the ground-level shaking that occurs during an earthquake

groundwater: water that sinks below the ground surface to join the water table, slowly flowing toward river channels

guano: the fecal droppings of seabirds that has accumulated over thousands of years and dried to a consistency that is amenable to mining; used in agriculture as fertilizer

guyot: an undersea mesa of volcanic origin that is believed to have once protruded above the surface near mid-ocean ridges, where wave action eroded the upper portion flat before they were transported into the abyssal plains by seafloor spreading

gypsum: mineral forms of calcium sulfate, $CaSO_4$, generally as the dihydrate $CaSO_4 \bullet 2H_2O$

hardness: resistance of a mineral to being scratched by another material

hardness pencils: a set of mineralogist's or gemologist's tools prepared with a standard assortment of mineral tips in accord with the Mohs hardness scale, used to carry out scratch tests

homeostatic: any mechanism that acts to maintain a system in its current state

hoodoo: an erosion structure consisting of a pillar-like base capped by a flat cap of harder material

horizon: a layer of soil material approximately parallel to the surface of the land that differs from adjacent related layers in physical, chemical, and biological properties

horneblende: type of silicate rock characterized by its dark color, which is the result of inclusions of iron and magnesium within the rock matrix

hornfels: the hard, splintery rocks formed by contact metamorphism of sediments and other rocks

hydraulic conductivity: the ability of a porous matrix to permit the passage of water

hydrological: relating to the systematic flow of water in accordance with physical laws

hydrology: the branch of science dealing with water and its movement in the environment

hydrolysis: a chemical weathering process that produces clays by the reaction of carbonic acid with aluminosilicate minerals

hydrothermal vein: a mineral deposit formed by deposition and crystallization from hot, mineral laden water solutions

hydrothermal vents: undersea structures of a volcanic nature that eject hot mineral-laden water from deep within Earth's crust

hypocenter: the point of origin of an earthquake

ice age: a period of time during which major continental ice sheets advanced and retreated

ichnology: study of trace fossils

igneous rock: type of rock formed when magma that develops within the earth's mantle rises to the surface, where it cools and solidifies into solid rock structures

ignimbrite: a pumice-filled pyroclastic flow caused by the collapse of a volcanic column and accelerated by gravity

ijolite: a dark-colored silicate rock containing the minerals nepheline (sodium aluminum silicate) and pyroxene (calcium, magnesium, and iron silicate)

impact ejecta: material ranging in size from house-sized rocks to fine dust that is ejected from the site of impact of a meteor

inclusion: a foreign substance enclosed within a mineral; often very small mineral grains and cavities filled with liquid or gas; a cavity with liquid, a gas bubble, and a crystal is called a three-phase inclusion

indurated: made significantly harder than its original form

inertia: the tendency of physical objects to remain at rest or with unchanging speed and direction unless acted upon by an external force

inertial frame: the frame of reference by which speed and distance measured

infiltration capacity: the amount of water that a particular soil can absorb before becoming saturated and losing water to surface runoff

inner core: the innermost layer of the Earth, extending from 5,150 kilometers to 6,370 kilometers below ground and composed mostly of solid iron

intensity: of an earthquake, the combination of the energy released in the event and the duration of the event

interdistributary bay: a shallow, triangular bay between two distributary channels; over time, the bay becomes filled with sediment and colonized with marsh plants or trees

interdune area: the low area between dunes in a desert

interfluve: the high-ground divide that separates one drainage basin from another

interglacial stage: the warm period between two periods of glaciation

intermediate rock: an igneous rock that is transitional between a basic and a silicic rock, having a silica content between 54 and 64 percent

International Standard: unit of measurement whose consistent value is recognized as a common unit of measurement by international agreement

intrusion: a body of igneous rock that results from the intrusion of magma into a preexisting rock body

ion: atom or group of atoms with a positive or negative charge

ionic compound: a neutral compound whose discrete unit consists of a combination of positive and negative ions

isobath: the contour lines of continental slopes

isostatic rebound: the upward movement of a region of continental mass and the underlying crustal rock when the weight of glacial ice above it has been removed

isotopes: different atoms of the same element that have different numbers of neutrons (neutral particles) in their nuclei

isotropic: a structure in which a physical property has the same value regardless of direction

i-type granitoid: granitic rock formed from magma generated by partial melting of igneous rocks in the upper mantle or lowermost crust

jökulhlaup: a flood produced by the release of water sequestered by a glacier, most often due to the failure of some type of glacial dam or to subglacial volcanic activity

karst topography: a landform characteristic of limestone that has been eroded chemically by rain, named for the Karst region of Eastern Europe

kerogen: a waxy, insoluble organic compound with a large molecular structure as an organic residue permeating the pore spaces of sedimentary rock, particularly shale

kimberlite pipes: the throats of ancient volcanoes that have erupted explosively, allowing kimberlite magma to fill the void channel

Kuiper Belt: a ring of asteroidal bodies in and beyond the orbit of Pluto, broader and more diverse than the asteroid belt between Mars and Jupiter

Kunstkammer: German word meaning "oddities room," the name of Czar Peter the Great's first public museum, which grew to include a vast collection of rock and mineral samples

lacustrine: relating to freshwater lakes

lahar: combination of water, ash, and mud that creates tremendous mudslides in a pyroclastic flow

lander: a spacecraft designed to land in one location on a planet's surface and remain at that location while gathering data about the planet

lap: a device consisting of a hard and perfectly flat disc that can be impregnated with abrasive grit, typically diamond dust, used to produce a polished surface on a gemstone

lava: molten rock material at Earth's surface

leaching: the dissolving out or removal of soluble materials from a soil horizon by percolating water

lead-to-uranium ratio: the amount of lead relative to the amount of uranium contained in a uranium-bearing rock

lenticular: lens-shaped

limestone: a sedimentary rock composed mostly of calcium carbonate formed by organisms or by calcite precipitation in warm, shallow seas

lithic fragment: a grain composed of a particle of another rock; in other words, a rock fragment

lithification: the process whereby loose material is transformed into solid rock by compaction or cementation

lithology: the general physical type of rocks or rock formations

lithosphere: the solid rock portion of Earth's structure, primarily the crust but including parts of the upper mantle layer

lithospheric plates: giant slabs composed of crust and upper mantle; they move about laterally to produce volcanism, mountain building, and earthquakes

lithostatic pressure: pressure exerted by the weight of sediment overlying the sediment sample in question

logarithmic scale: a scale based on powers of ten rather than on unit steps such that 0=1, 1=10, 2=100, 3=1,000...etc.

longshore current: a slow moving current between a beach and the breakers, moving parallel to the beach; the current direction is determined by the wave refraction pattern

longshore drift: the movement of sediment parallel to the beach by a longshore current

Love waves: seismic waves that occur in a side-to-side motion

mafic: igneous deposits of rocks consisting of silica enriched with high levels of iron, magnesium, sodium, and calcium and with low levels of potassium-rich feldspar

mafic-ultramafic: igneous rock having the lowest range of silica content

magma: a semiliquid, semisolid rock material that exists at high temperatures and pressures and is mobile and capable of intrusion and extrusion; igneous rocks are formed from magma as it cools

magnetosphere: the region of influence in space of a planet's magnetic field

magnitude: a scaled measurement of the amount of movement of rock at the epicenter of an earthquake

major elements: elements that make up the bulk of the chemical composition of a mineral or rock

mantle: the layer of Earth's interior that lies between the crust and the core; it is believed to consist of ultramafic material and is the source of magma

marble: a mineral form of calcium carbonate, $CaCO_3$, that has undergone diagenesis through temperature and pressure

maria: regions of the lunar surface that appear dark and smooth when viewed from Earth, once thought by ancient astronomers to be actual seas of water but now known to be the remnants of lava outflow resulting from large impacts

marine isotope stage: half of a glacial cycle, as identified in the oxygen isotope data from ocean cores

mascon: an area of the lunar surface that significantly greater density than the rest of the surface, indicating a mass concentration (hence, mas-con)

mass movement: the movement of a significant amount of earthen material by any of a variety of drivers such as wind, water and gravity

measurement: ascribing a value to the determination of the quantity of a physical property

megapascal: geologic unit of extreme pressure, equivalent to 1 million pascals and 145.037 pounds per square inch

melt: the molten form of a material within which solid crystals of the material can form, essentially the same as crystal formation from a solution but in a melt the molten liquid takes the place of the solvent (typically water)

metal: an element with a metallic luster, high electrical and thermal conductivity, ductility, and malleability

metal ion displacement: exchange of one metal ion for a different metal ion in minerals

metamict: of minerals that have been altered by naturally occurring alpha radiation

metamorphic rock: a type of rock formed when another type of rock undergoes changes in mineralogy, chemistry, or structure owing to changes in temperature, pressure, or chemical environment

metasomatism: the infusion of additional water-borne elements into protolithic rock that is undergoing metamorphism

meteor: a small body of rock or metal that enters Earth's atmosphere and emits light as friction with Earth's atmosphere causes it to heat up to incandescence

meteorite: a meteor that has passed through the atmosphere and struck the surface of the planet

meteoroid: a small body of rock or metal that travels through space

mica: a platy silicate mineral (one silicon atom surrounded by four oxygen atoms) that readily splits into thin, flexible sheets

Mid-Atlantic Ridge: a line of active undersea volcanic structures discovered in the mid-twentieth century, formed by the upwelling of magma from the mantle which acts to drive apart the ocean floor to the east and west of the ridge

midoceanic ridges: a line of volcanic peaks roughly bisecting the ocean floor where convection currents in the magma of Earth's mantle tend to force separation of the tectonic plates as basaltic magma is forced upward into the rift

migmatite: a rock exhibiting both igneous and metamorphic characteristics, which forms when light-colored silicate minerals melt and crystallize, while the dark silicate minerals remain solid

Milankovitch cycle: period of variation in Earth's orbital parameters, including axial inclination, climatic precession, and orbital eccentricity

mineral: naturally occurring element or compound that has a definite ordered arrangement of atoms

mineral species: a mineralogic division in which all the varieties in any one species have the same basic physical and chemical properties

mineral variety: a division of a mineral species based upon color, type of optical phenomenon, or other distinguishing characteristics of appearance

mobile belt: a linear belt of igneous and deformed metamorphic rocks produced by plate collision at a continental margin; relatively young mobile belts form major mountain ranges; synonymous with orogenic belt

Mohole Project: a deep drilling project to penetrate Earth's crust and reach the mantle at the Mohorovičić Discontinuity

Mohorovičić Discontinuity: the transition region between the crust and the mantle layer of Earth's inner structure

Mohs hardness scale: a relative scale based on the hardness of ten naturally occurring minerals, with talc being the least hard and diamond being the hardest

molecular sieve: a zeolite material that is able to trap and hold small molecules of particular sizes such as water present as an impurity in an organic liquid or ammonia in air

moment of inertia: a physical quality of a spinning mass, analogous to the momentum of an object undergoing linear motion

monoculture plantation: a plantation in which only crops of a single species

monomineralic zone: a band or region between a limestone body and an igneous pluton, in which only one distinct mineral type is found

monsoon: a seasonal, reversing pattern of wind between warm ocean bodies and landmasses

Mössbauer spectrometer: a device that irradiates rock samples with gamma rays of a known wavelength and measures the wavelength of gamma rays that are re-emitted from the sample as a means of identifying the atoms present

mountaintop removal mining: as the words imply, the excavation and removal of the top portion of a low mountain that is acting as overburden to a coal bed

mudrock: a rock composed of abundant clay minerals and extremely fine siliciclastic material

mudstone: shale formed from clay particles that are generally rounded or irregular rather than flat in structure

native element: an element that is found naturally in pure form, uncombined with other elements

natural levee: a low ridge deposited on the flanks of a river during a flood stage

near-Earth object (NEO): an asteroid whose orbit brings it close enough to Earth's orbit that an orbital perturbation could result in a collision

Neptunism: the constructive philosophical consensus that views geological history as a dynamic equilibrium

nonclastic particles: particles formed by chemical reactions or biosynthetically

nonradiogenic: atoms that are not produced as a result of a radioactive decay process

***nuée ardente*:** a type of dense pyroclastic flow that occurs after a volcanic dome collapse; also known as block and ash and Peleen flow

obduction: the process occurring in convergence when a tectonic plate that would normally be subducted is instead folded and buckled so that a significant portion of it is driven upward rather than downward

oceanic crust: the dense, iron-rich basaltic rock of the seafloor formed by solidified magma rising at divergent plate boundaries under the ocean

oceanic ridges: a system of mostly underwater rift mountains that bisect all the ocean basins; basalt is extruded along their central axes

octahedral sheet: the two-dimensional layer consisting of octahedral arrangements of silica structural units

octahedron: a geometric form having eight equal triangular faces

off-planet: pertaining to regions outside Earth in orbital or planetary space

oil sands: also called tar sands, a combination of bitumen and sand, clay and other soil types, ranging in consistency from semi-solid to solid such as might be obtained by mixing used motor oil with soil and allowing the mixture to set

oil shale: shale containing a significant amount of organic matter within the pore spaces between particles

olivine: a silicate mineral found in mantle and some basalts, particularly the alkaline varieties

opal: a form of silica containing a varying proportion of water within the crystal structure

ophiolite complex: an assemblage of metamorphosed basaltic and ultramafic igneous rocks that originate at marine ridges and are subsequently emplaced in mobile belts by plate-collision tectonics

orbicule: formation within orbicular igneous rocks consisting of similar materials to the parent rock; organized in concentric layers of crystals surrounding a core comprising smaller crystal clusters, larger individual crystals, or fragments of other rock types

orbit: a gravitationally bound path in space around a planet

orbital platform: a spacecraft purposed to remain in orbit about a planet and act as a relay station for transfer of data from surface-based devices back to Earth

ore: mineral or minerals present in large enough amounts in a given deposit to be profitably mined for the metal(s)

original horizontality: the principle that any stratum was originally distributed evenly over the region of interest, such as on a lake bed, sea floor or land area

original lateral continuity: the principle that the particular content of any stratum extended continuously to the borders of the region of interest

orogeny: mountain building by tectonic forces through the folding and faulting of rock layers

oscillation ripple: a small to large bed form, oriented predominantly transverse to the direction of flow, produced at low to moderate flow speeds in oscillatory water flows

oscillatory flow: a flow of fluid with a regular back and forth pattern of motion

oscillatory wave: a wind generated wave in which each water particle describes a circular motion; such waves develop far from shore, where the water is deep

outer core: a liquid region extending from the base of the mantle at a depth of 2,900 kilometers down to the inner core, at a depth of 5,150 kilometers; mostly molten iron

outgassing: the process by which volatile materials trapped within rock formations are released into the atmosphere and the environment

oxidation: reaction of elements in rock with oxygen to produce oxide compounds such as rust

oxidation-reduction process: a reaction between two different atoms in which one atom loses one or more electrons (oxidation) as the other atom gains one or more electrons (reduction)

paleobiology: approach to paleontology that stresses the understanding of processes

paleoecology: study of the interaction between organisms and their environment through time

paleodepth: an estimate of the water depth at which ancient seafloor sediments were originally deposited

paleomagnetism: the "signature" of Earth's magnetic field frozen in susceptible rock minerals as they cool from the molten state

Paleozoic era: the era immediately before the Mesozoic era; it included the Cambrian, Ordovician, Silurian, Devonian, Carboniferous, and Permian periods

pallasites: a type of meteor named for Peter Simon Pallas

Pangaea: "all earth," the supercontinent proposed by Wegener as the original position of all of the continental land masses some 250 million years in the past

panning: a method of separating particles according to density by swirling the source material with water in a particular type of pan

paragenesis: a contraction of parallel genesis, referring to the common origin of related minerals

partial melting: a process undergone by rocks as their temperature rises and metamorphism occurs; magmas are derived by the partial melting of preexisting rock; also known as ultrametamorphism or anatexis

partial pressure: in any gaseous mixture, the total pressure is the sum of the proportions of the total pressure due to each individual gas; i.e, total pressure = sum of partial pressures

peridotite: a class of ultramafic rocks made up principally of pyroxene and olivine, with subordinate amounts of other minerals

perigee: the point in an elliptical orbit at which the orbiting body is at its closest distance to the main focal point of the orbit

permafrost: a layer of soil and water ice frozen together

petrology: branch of geology that studies rocks and the conditions that lead to the formation of rocks

pH: a measure of the acidity of a solution in water based on the concentration of free hydrogen ions; specifically $pH = -log[H^+]$ where $[H^+]$ is the free hydrogen ion concentration

phaneritic: igneous rock that contains large crystals formed as magma cools slowly

phenocryst: a large conspicuous crystal in a porphyritic rock

phosphorite: sedimentary rock composed principally of phosphate minerals

photovoltaic cell: a device commonly made of layered silicon that produces electrical current in the presence of light; also, a solar cell

phyllosilicate: a mineral with silica tetrahedra arranged in a sheet structure

physical weathering: breaking down of rock and rock particles by physical means, typically by impacts with other rock particles

piezoelectricity: an electrical current induced in a crystal by the application of pressure

pillow basalt: a submarine basaltic lava flow in which small cylindrical tongues of lava break through the surface, separate into pods, and accumulate downslope in a formation resembling a pile of sandbags

placer: a deposit of sand or gravel containing eroded particles of valuable minerals

plagioclase: one of the principal silicate minerals in basalt, a member of the feldspar group

plane bed: a bed configuration without rugged bed forms produced in both unidirectional and oscillatory flows at high flow speeds

plankton: microscopic marine plants and animals that live in the surface waters of the oceans; these floating organisms precipitate the particles that sink to form biogenic marine sediments

plate tectonics: the evolved form of the continental drift theory, positing the movement of large "plates" of Earth's crust being moved about the planet by the force of magmatic convection currents in Earth's mantle layer

plates: segments of Earth's crust that have formed in response to forces exerted by movement of magma in Earth's mantle, somewhat resembling the fragments of a cracked eggshell

playa: the exposed, and generally arid, remnant of the floor of an inland lake or sea

Plinian: volcanic eruption that sends ash and debris high above the volcano in a column

pluton: a generic term for an igneous body that solidifies well below Earth's surface; plutonic rocks are coarse-grained because they cool slowly

plutonic: refers to magma that solidifies within the crust rather than on the surface

Plutonism: the destructive philosophical consensus that views geological history as a series of catastrophic events

polje: a lake of irregular shape resulting from the formation of adjacent sinkholes

polycrystalline: crystal structure consisting of crystallite fragments

polymorph: minerals having the same chemical composition but a different crystal structure

pore fluids: fluids, such as water (usually carrying dissolved minerals, gases, and hydrocarbons), in pore spaces in a rock

porosity: the amount of space between the sedimentary grains in a rock or sediment

porphyritic: a texture characteristic of an igneous rock in which macroscopic crystals are embedded in a fine phaneritic or aphanitic matrix

porphyry: an igneous rock in which phenocrysts are set in a finer-grained groundmass

positive feedback loop: a condition in which the output from a system affects the system in such a way that the particular output is increased

precipitation: the formation of solid mineral crystals from their chemical components that are dis-solved in water

primary coast: coastal features that are formed as a result of processes occurring on land

primary producers: in biological systems, the organisms that produce environmental components from inorganic raw materials

primordial solar nebula: the original collection of dust and gases that constituted the basic cloud from which the solar system formed

principle of superposition: the principle that geological and historical strata are laid down in successive layers such that the lower stratum is the older one, with the oldest at the bottom of the strata layers and the youngest at the top

prism: a geometric form having three rectangular faces between two triangular faces

prodelta: a sedimentary layer composed of silt and clay deposited under water; it is the foundation on which a delta is deposited

productivity: the rate at which plankton reproduce in surface waters, which in turn controls the rate of precipitation of calcareous or siliceous shells or tests by these organisms

protolith: the original form of a metamorphic rock before it underwent metamorphosis

pumice: light, frothy volcanic rock formed when lava is cooled above ground following a violent eruption and pyroclastic flow

P-waves: compression waves, like sound waves, presumed to be produced by direct impingement of rock masses against each other at the focus of the earthquake

pyramid: a geometric form having four equal triangular faces on a square base

pyrite: iron sulfide, FeS, a dense mineral with a gold color; often mistaken for gold, so is known as "fool's gold"

pyroclastic: refers to particles of solidified volcanic magma ranging in size from fine dust, or "ash" to house-sized boulders, as well as to rock that is formed by consolidation of such particulate matter

pyroclastic flow: typically, a large, superheated cloud of volcanic ejecta; may be described as a fast-moving, violent landslide borne by extremely hot gases

pyroelectricity: an electrical current induced in a crystal by the application of heat

quarrying: obtaining slate blocks by cutting and splitting them from exposed slate beds; other types of rock and geological products are obtained by similar methods called quarrying

quartz: crystalline form of silica that develops when silica is heated in a liquid or semiliquid environment and then slowly cooled

radioactive decay: the conversion of an atom of a particular element into atoms of other elements by the spontaneous loss of one or more protons and neutrons from its nucleus, accompanied by the release of energy

radiogenic: atoms that are produced through the process of radioactive decay

radiogenic heating: heat generated by the decomposition of radioactive elements within Earth's interior

radiometric age-dating: determination of the age of rocks and other materials based on the relative quantities of specific isotopes related to each other by an unequivocal nuclear decay sequence, and the half-life span of that sequence

rainshadow effect: a mountain range parallel to a coastline causes incoming warm, moist air on the prevailing winds from the ocean to rise and cool, resulting in heavy rainfall on the ocean-facing side of the mountain range and dry, cool air on the leeward side of the mountain range. The arid leeward region is described as the mountain range's rainshadow

Rayleigh waves: seismic waves that occur in a circular, rolling fashion

rebounding: raising of the land's elevation as a result of the removal of the weight of the overlying glacial ice

recrystallization: process that occurs at high pressure, causing the atoms, molecules, or grains of a crystal to reorganize into another type of crystal

recurrence interval: the average time interval in years between occurrences of a flood of a given magnitude in a measured series of floods

red clays: fine grained, carbonate free sediments that accumulate at depths below the CCD in all ocean basins; their red color is caused by the presence of oxidized fine grained iron particles

refractive index: a measure of the extent to which a transparent material can alter the direction of, or "bend," rays of light

regional metamorphic facies: the particular pressure and temperature conditions prevailing during metamorphism as recorded by the appearance of a new mineral assemblage

regional metamorphism: metamorphic processes that occur slowly due to long periods of exposure to pressure and temperature due to tectonic plate movement

regolith: that layer of soil and rock fragments just above the planetary crust

resolution: the minimum size of an area that can be differentiated from its surroundings photographically

relative motion: proportional speed and distance of two or more objects moving within the sane inertial frame of reference

remanent magnetization: the embedded magnetism in specific ancient rock samples with regard to its directionality

rhombohedral cleavage: cleavage pattern of a mineral crystal structure that occurs according to the form of a rhombohedron

rhyolite: igneous rock with the highest content of volatiles and highest silica content

Richter scale: logarithmic scale used to assign a numerical value to the magnitude of an earthquake

rifting: being separated by force applied in opposite directions

rift valley: an elongated depression between two segments of continental mass that are being driven apart by convection currents in the underlying magma of the mantle layer

road metal: the hard road surface that forms from a tightly packed mixture of gravel and sand

rock: a naturally occurring consolidated material that usually consists of two or more minerals; sometimes, as in carbonatites, rocks may consist mainly of one mineral

rock-forming mineral: the common minerals that compose the bulk of the earth's crust (outer layer)

rough: gem mineral material of suitable quality to be used for fashioning gemstones

rover: a vehicle designed to be delivered to a location on the surface of a planet and move from that location to obtain data about the planet

rugose corals: a Paleozoic coral group also known as "tetracorals," sometimes colonial, but more often solitary and horn shaped

runoff: that part of precipitation that flows across the land and eventually gathers in surface streams

rust: a general term for the products of the oxidation of metal components in rock

salt weathering: the granular disintegration or fragmentation of rock material affected by saline solutions or by salt crystal growth

saltation: a process in which solid particles being transported within a moving fluid (water, air, etc.) bump into other particles and are lifted over them as a result, eventually producing an accumulation structure with a gradual incline on the upstream side and a steep dropoff on the downstream side

sand: assorted rock particles whose dimensions are within a defined range of 0.05 to 5 millimeters

sandstone: sedimentary rock containing mostly quartz, feldspar, or rock fragments with grain sizes mostly in the range of 0.0625 to 2 millimeters

satellite based radar altimetry: measurement of the altitude of a satellite above the surface of the ocean; because the altitude of the satellite is fixed, any differences of altitude measured are due to changes in the level of the ocean surface

scleractinian corals: modern corals or "hexacorals," different from their more ancient counterparts in details of the skeleton and the presence of a symbiosis with unicellular algae in most shallow water species

scratch: a test of abrasiveness carried out by scratching a 5 - 6mm mark on one material with another material to determine which is the harder of the two materials

seafloor spreading: the gradual separation of oceanic plate segments on either side of a divergent tectonic boundary, accompanied by the formation of new oceanic crust by upwelling of basaltic magma into the rift between the plate segments

secondary coast: coastal features that exist as a fundamental part of the land structure

sediment: fragmented rock material such as gravel, sand, silt, or clay that is deposited by a river to form a delta

sedimentary material: particulate matter that may be carried by a moving fluid and settles out from the fluid when the flow slows or stops

sedimentary rock: a rock resulting from the consolidation of loose sediment that has accumulated in flat-lying layers on the earth's surface

segregation: the concentration of early-formed minerals in a magma by crystal settling or crystal floating

segregation deposit: a mineral deposit formed when a component of liquid magma separates from the mixture and drifts down to the bottom of the magma chamber where mineralization is subsequently completed

seismicity: the degree of seismic activity, as earthquakes, that occurs in a defined region over time

seismograph: a device designed to record the physical waves produced by an earth tremor

seismology: the study of vibratory and displacement perturbations of Earth's crust, typically as earthquakes and volcanic eruptions, although modern seismometers are able to detect the vibrations produced by such events as explosions and the traffic of heavy vehicles

seismometer: an electromechanical device capable of detecting and recording vibrations traveling through the planet such as those caused by earthquakes, volcanic eruptions, landslides, explosions, or even just traffic

self-regulating: controlled by processes that exist entirely within the system

semimetal: elements that have some properties of metals but are distinct because they are not malleable or ductile

semiprecious stones: mineral crystals, typically types of quartz or beryl, that are not classifiable as gemstones

shale: very fine-grained sedimentary rock often consisting mostly of clay minerals and quartz

shear, or shearing, strength: the ability to withstand a lateral or tangential stress

sheeted dyke complex: a series of essentially parallel magma intrusions into fractured overlying rock

shore processes: principally, erosion and characteristic intertidal regions, but also including sediment formations created by wave action

silicate: chemically, the anion SiO_4^{4-}; geologically, any of the rock types that are based on silica (SiO_2) structural combinations

silicate mineral: mineral that contains silicon bonded with oxygen to form silicate groups bonded with positive ions; silicates make up 90% of Earth's crust

silicon carbide: a compound of silicon and carbon with the chemical formula SiC, not simply an alloy

sill: a magmatic intrusion that drives horizontally between layers of sedimentary rock

skarns: coarse-grained rock produced through metasomatism, differing from tactites with regard to metal content

slate: shale that has undergone a period of intense pressure that has deformed the clay particles and expressed all organic matter that may have been present, resulting in a very hard and consolidated shale rock

sluicing: a method of separating particles according to density by entrapping the densest particles in transverse grooves as they are carried down a sluice, a specially constructed channel, by flowing water

slump: the collapse of a hillside, caused by loss of structural integrity due to excessive groundwater content and the force of gravity, such that the affected portion slips downhill with a rotational movement

slurries: mixtures in which fine solid particles are mixed with an amount of liquid to allow the material to flow

small solar system body: an astronomical object other than a planet, dwarf planet, or satellite that orbits the sun

smokers: undersea vents on the active rift areas that emit large amounts of superheated water and dissolved minerals from within the crust

soft-land: to land a spacecraft on a planetary surface intact and in fully functioning condition, as opposed to crash landing and loss of the machine

soil profile: a vertical section of a soil, extending through its horizons into the unweathered parent material

solar wind: the stream of highly charged particles emitted into space from the surface of the sun

solid solution: a solid that shows a continuous variation in composition in which two or more elements substitute for each other on the same position in the crystal structure

solution: a means of altering rock and rock particles by dissolving certain mineral components from their content

sonar: an underwater range-finding method that emits sharp sound pulses, or "pings," and uses the elapsed time for the echo to return as the means of calculating the distance to the reflecting surface; the principle on which the modern "fish finder" operates

sorting: the deposition of particulate material as it settles out from a moving fluid according to the ability of the fluid to transport the particles

source parameters: a series of earthquake characteristics, including distance, duration, energy, and the types of waves that occur

specific gravity: the ratio of the weight of a sample to the weight of an equal volume of pure water at 4°C; as a ratio of weight to weight, specific gravity has no units

spiral classifier: essentially a spiral sluice that operates vertically rather than as a lengthy run of channel

Standard Temperature and Pressure (STP): value of the mean atmospheric pressure and mean annual temperature at sea level, chosen by international agreement to be designated as 1 standard atmosphere or 760 torr and 0° Celsius/32° Fahrenheit or 273.15° Kelvin

stellar nucleosynthesis: the fusion of light elements into heavier elements, such as carbon, oxygen, and iron, within the interiors of stars

strain: change in volume or size in response to stress

strain energy: energy built up within a plate due to an inability to move in response to forces exerted by movement of underlying mantle material

stratigraphic horizon: a particular stratum within Earth's crust corresponding to a corresponding period in the geologic time scale

stratigraphic trap: a formation in rock strata where a mobile material can collect but is subsequently prevented from migrating farther

stratigraphy: the study and correlation of rock strata according to age

stratovolcano: a volcano composed of alternating layers of lava flow and ash; also called a composite volcano

stratum (strata): a layer (layers) of rock

stress: force per unit of area

stress drop: the amount of energy released when locked tectonic plates separate, causing an earthquake

stromatolites: layered columnar or flattened structures in sedimentary rocks, produced by the binding of sediment by blue green algal (cyanobacterial) mats

stromatoporoids: spongelike organisms that produced layered, mound shaped, calcareous skeletons and were important reef builders during the Paleozoic era

s-type granitoid: granitic rock formed from magma generated by partial melting of sedimentary rocks within the crust

subduction: the condition in which a tectonic plate, typically of dense seafloor basalt, is forced to go downward into the mantle beneath a less dense continental mass due to seafloor spreading

subduction zone: the convergent plate boundary between seafloor and continental mass, where the denser basaltic seafloor is forced to go beneath or subduct to the continental mass

submarine canyon: a submerged, V shaped canyon cut into the continental shelf and continental slope, through which turbidity currents funnel into the deeper parts of the oceans

superconductors: materials that pass electrical current without exhibiting any electrical resistance

supergene enrichment: enrichment of the mineral in an ore deposit by miner-bearing water that migrates downward from near the surface

superposition: the principle that requires the oldest stratum to be at the base of any series of strata above it in undisturbed rock or sediment beds

surface wave: the wave produced in an earthquake as a reflection of the S waves, felt only at the surface

suspended load: the amount of sediment that a stream carries above the streambed

S-waves: sheer waves, like side-to-side rocking, presumed to be produced by lateral slippage of rock masses against each other at the focus of the earthquake

synthetic mineral: a human-made reproduction of the structure, composition, and properties of a particular mineral

systematics: study of the relationships among organisms

tabulate "corals": colonial organisms with calcareous skeletons that were important Paleozoic reef builders; considered to be more closely related to sponges than to corals

tactites: coarse-grained rock produced through metasomatism, differing from skarns with respect to metal content

talus: the assortment of broken rock at the base of a cliff, formed as weathering causes pieces of rock to break free from the cliff face and fall

taphonomy: study of the processes that affect organisms between death and burial

tar sands: sands that are permeated with bitumen, a tarlike petroleum that is too thick to flow; the term is interchangeable with "oil sands"

tectonic plates: segments of Earth's crust that are independent of each other but constrained by each other, analogous to the segments of a cracked egg shell

tensional rift: an elongated depression that forms between two segments of a tectonic plate that are being driven apart from each other by magmatic convection currents

test: an internal skeleton or shell precipitated by a one celled planktonic plant or animal

tetrahedral: arranged at the four apices of a tetrahron about the tetrahedron's central focus; in silicates, four oxygen atoms are arranged tetrahedrally about a central silicon atom

tetrahedral sheet: the two-dimensional layer consisting of tetrahedral arrangements of silica structural units

tetrahedron: a polyhedron composed of four triangular sides organized into pyramidal shape

texture: the size, shape, and arrangement of crystals or particles in a rock

thermal emission spectrometer: a device that identifies the wavelengths and relative intensities of infrared emissions from a sample. The wavelengths and patterns of intensities are characteristic of particular minerals.

thermal fracture: the formation of a fracture or crack in a rock as a result of temperature changes

thermal gradient/geothermal gradient: the rate of temperature increase with depth below the earth's surface

toughness: the degree of resistance to fragmentation or resistance to plastic deformation

trace element: element present only in tiny amounts (in quantities of parts per million or less) in a mineral or rock

transform boundary: the geodynamic process in which two tectonic plates collide moving laterally relative to each other

travertine: calcium carbonate deposits formed by deposition from mineral-laden water in hot springs and caves

trough: a long, narrow depression, as between waves or ridges

tsunami: a low, rapidly moving wave created by a disturbance on the ocean floor, such as a submarine landslide or earthquake

T-Tauri stage: a highly active period in the life of a very young star, in which large amounts of matter are ejected from it as solar wind

tufa: calcium carbonate deposits formed by deposition from mineral-laden water from hot springs

turbulent flow: a high velocity sediment flow in which individual sediment particles move in very chaotic directions above the sea floor

twinning: the intergrowth of two crystals to form a single crystal structure, essentially by growing from each other at the same time

ultramafic: a term used to describe certain igneous rocks and most meteorites that contain less than 45% silica; they contain virtually no quartz or feldspar and are mainly of ferromagnesium silicates, metallic oxides and sulfides, and native metals

unidirectional flow: a flow of fluid oriented everywhere and at all times in the same direction

uniformitarianism: the principle that the geological mechanisms at work in the world today have operated in the same way throughout the past and will throughout the future

unit: single definitive basic magnitude of portions used to measure various properties

updip: the upward direction of an inclined stratum

uvala: circular flat-bottomed depressions, often the floor of a small lake, formed when a sinkhole becomes filled with sediment

ventifact: any stone or pebble that is shaped, worn, faceted, cut, or polished by the abrasive action of windblown sand, generally under desert conditions

vesiculated rock: rock containing numerous gas pockets within it, typically as the result of eruption of lava with high gas content

viscosity: a substance's ability to flow; the lower the viscosity, the greater the ability to flow

volatiles: chemical elements and compounds that become gaseous at fairly low temperatures

volcanic pipe: deposit of igneous rock formed when a deep magma pocket produces an explosion of gas and magma that rapidly rises to the surface and forms a cone-shaped deposit

volcanism: the extent to which volcanic activity is exhibited on rocky planets and planetoids

water of hydration: water that is bound in the crystal structure of minerals without being part of their actual molecular structure

water table: the depth of the uppermost level of an aquifer relative to the ground surface

water-to-rock ratio: the mass of free water in a given volume of rock divided by the rock mass of the same volume; processes occurring at water-to-rock ratios less than 50 are "rock-dominated," while those greater than 50 are "fluid-dominated"

wave base: the depth to which water particles of an oscillatory wave have an orbital motion; generally the wave base is equal to one half the distance between successive waves

wave energy: the capacity of a wave to erode and deposit; as wave energy increases, erosion increases

wave propagation path: the directions in which seismic waves travel in an earthquake

wave refraction: the process by which the vertical angle of a wave moving into shallow water is changed or bent

weathering: the mechanical disintegration and chemical decomposition of rocks and sediments

xenocrysts: minerals found as either crystals or fragments in some volcanic rocks; they are foreign to the body of the rock in which they occur

xenolith: inclusion of rock in the body of a different type of rock; often occurs in igneous rock formations

yardangs: bedrock outcrops that have been eroded to a streamlined shape by wind-borne particles

zeolites: members of a mineral group with very complex compositions: aluminosilicates with variable amounts of calcium, sodium, and water and a very open atomic structure that permits atoms to move easily

zircon: a mineral form of zirconium silicate, $ZrSiO_4$

zone fossil: organism that characterizes a particular period of time

Organizations

American Association of Petroleum Geologists
1444 S. Boulder Ave.
Tulsa, OK 74119
918-584-2555
info@aapg.org

American Geophysical Union
2000 Florida Ave. NW
Washington, DC 20009
202-462-6900

American Geosciences Institute
4220 King Street
Alexandria, VA 22302-1502
(703)379-2480
agi@americangeosciences.org

Association for the Sciences of Limnology and Oceanography
1105 Wooded Acres Dr.
254-776-3550
Waco, TX 76710

Association for Women Geoscientists
1333 West 120th Ave.
Suite 211
Westminster, CO 80234
303-412-6219

Association of American State Geologists
AASG Foundation
Division of Geology and Mineral Resources
900 Natural Resources Drive
Suite 500
Charlottesville, VA 22903
431-951-6350
david.spears@dmme.virginia.gov

Association of Environmental & Engineering Geologists
3053 Nationwide Parkway
Brunswick, OH 44212
844-331-7867
contact@aegweb.org

Astrogeology Science Center
US Geological Survey
12201 Sunrise Valley Dr.
Reston, VA 20192
703-648-5953

The Clay Minerals Society
3635 Concorde Pkwy
Suite 500
Chantilly VA 20151-1110
703-652-9960
cms@clays.org

Environmental and Engineering Geophysical Society
1391 Speer Blvd.
Suite 450
Denver, CO 80204
303-531-7517
www.EEGS.org

Geochemical Society
5241 Broad Branch Rd. NW
Washington, DC 20015-1305
202-545-6928
gsoffice@geochemsoc.org

Geological Association of Canada
Alexander Murray Building
Memorial University of Newfoundland
Room ER4063
St. John's, NL, Canada A1B 3X5
709-864-7660
gac@mun.ca

The Geological Society of America
PO Box 9140
Boulder, CO 80301-1806
303-357-1000
gsaservice@geosociety.org

International Association of Hydrogeologists
IAH Secretariat
PO Box 4130
Goring Reading, UK RG8 6BJ
44-0-118-380-5251
info@iah.org

International Union of Geological Sciences
IUGS Secretariat
26 Baiwanzhuang Rd.
Xicheng District
Beijing, China 100037
86-10-5858-4808
secretariat@iugs.org

Mineralogical Society of America
3635 Concorde Pkwy
Suite 500
Chantilly, VA 20151-1110
703-352-9950

National Association of Black Geologists and Geophysicists
4212 San Felipe
Suite 420
Houston, TX 77027-2902
nabgg_us@hotmail.com

National Association of Geoscience Teachers
1 North College St.
Northfield, MN 55057
507-222-5339

National Speleological Society
6001 Pulaski Pike
Huntsville AL 35810-112
256-852-1300
nss@caves.org

Paleontological Research Institution
1259 Trumansburg Rd.
Ithaca, NY 14850
607-273-6623

Paleontological Society
7918 Jones Branch Dr.
Suite 300
McLean VA 22102-3345
301-634-7231

The Planetary Society
60 South Los Robles Ave.
Pasadena, CA 91101
626-793-5100
tps@planetary.org

Rocky Mountain Association of Geologists
1999 Broadway
Suite 730
Denver, CO 80202
720-672-9898
staff@rmag.org

Seismological Society of America
400 Evelyn Ave.
Suite 201
Albany, CA 94706
510-525-5474

Sigma Gamma Epsilon
National Honor Society for the Earth Sciences
PO Box 506
Gilbertville, IA 50634
319-505-3391
even@sgeearth.org

Society for Sedimentary Geology
1621 S. Eucalyptus Ave.
Suite 204
Broken Arrow, OK 74012
918-994-6216

Society of Economic Geologists
7811 Shaffer Pkwy
Littleon, CO 80127-3732
720-981-7882
membership@segweb.org

Society of Exploration Geophysicists
8801 S. Yale Ave.
Suite 500
Tulsa, OK 74137
918-497-5500
members@seg.org

Society of Vertebrate Paleontology
7918 Jones Branch Dr.
Suite 300
McLean VA 22102
301-634-7024

United States Geological Survey
12201 Sunrise Valley Drive
Reston, VA 20192
888-275-8747

Subject Index

A. E. Fersman Mineralogical Museum, 238, 241
Advanced Research Projects Agency (ARPA), 241
aeolian erosion, 63, 82
aeolian placers, 332, 333
Agassiz, Louis, 186
Agricola, Georgius, 132, 270
agriculture, 9, 43, 44, 45, 85, 90, 91, 93, 158, 170, 297, 324, 415, 481
Alaska Earthquake Information Center (AEIC), 119
albitization, 437
Aldrin, Edwin "Buzz," 130
alluvial placers, 332
alpha proton X-ray spectrometer, 251, 253
alpine glaciers, 183, 184
American Academy of Arts and Sciences, 203
American Geophysical Union, 203, 204
American lunar exploration program, 241
American Philosophical Society, 203
"Andesite Line," 3, 4
andesites, 1, 2, 3, 17, 60
andesitic rocks, 1-4
anthropogenesis, 439, 442, 443
anthropogenic desertification, 90
antidunes, 384, 385
antoliths, 193
aquifers, 4-7
aquifer test, 6
arkose, 392, 393, 398, 406
Armstrong, Neil A., 130, 242
"Artemis 2" program, 246
asteroid belt, 7, 9, 285, 335, 336, 343
asteroids, 7-10
asthenosphere, 127, 217, 218, 219, 220, 236, 237, 264, 267, 311, 352, 353, 374, 468, 470
astrogeology, 7, 33, 69, 129, 147, 241, 251, 285, 292, 301, 333, 339, 345, 351, 358
astrometry, 33
astronautics, 147
astronomy, 7, 33, 69, 70, 129, 210, 292, 343, 377, 465, 467
astrophysics, 33
Athabasca oil sands, 10-13
atomic absorption spectrometry (AAS), 231
authigenesis, 97, 98
Avogadro constant, 221

basaltic rocks, 15-19
basaltic volcanoes, 214, 457, 458
basalts, 2, 3, 15, 16, 17, 18, 19, 189, 194, 209, 210, 214, 246, 282, 312, 327, 435, 436, 437, 469, 476
batholiths, 19-24
batholiths, I-type and S-type, 21, 22, 23
beach placers, 332, 333
bed forms (or riverbed forms), 382-386
Benioff zone, 3, 454
bentonite, 51, 52, 464
Big Bang nucleosynthesis (BBN), 292, 293, 294
Big Bang theory, 292, 293, 294
biodeposition, 55
biogenic sedimentary rocks, 28-32
biogeochemistry, 170
biogeography, 163
biogeology, 410
biologic cycles, 396
biostratigraphy, 319, 321
biotite, 22, 48, 55, 59, 60, 98, 192, 193, 263, 271, 313, 315, 316, 373, 405, 426, 427, 430
bioturbation, 100
bitumen, 10, 11, 12, 13, 177, 305, 306, 308, 309, 323, 393
Bode, Johann Elert, 9
bolides, 33-35
Boltwood, Bertram Borden, 35
borates, 289, 290, 291, 399, 427, 429
British Imperial system, 224
Broegger, W. C., 258
Bulletin of the Seismological Society of America, 378, 380
Bushveld intrusion, 248, 249

calcareous rock, 249
calcareous sediments, 31, 40, 41, 42, 43
calcification, 297, 298
calcite, 29, 30, 31, 32, 41, 42, 43, 44, 45, 46, 47, 48, 49, 52, 60, 97, 98, 99, 101, 110, 142, 143, 167, 217, 229, 235, 236, 249, 263, 265, 266, 269, 297, 298, 304, 367, 387, 398, 403, 407, 412, 424, 425, 426, 427, 429, 463, 464
California Institute of Technology (Caltech), 119, 141, 197, 279, 377, 379
Caltech Seismological Laboratory, 377, 378
carbonaceous chondrites, 301, 303
carbonate compensation depth (CCD), 39-43
carbonate minerals, 43-46
carbonate rocks, 29, 99, 367, 404
carbonatites, 46-50
carbonic acid, 41, 53, 55, 297, 299, 304, 313, 315

celestial bodies, 7, 8, 333
cementation, 29, 53, 55, 97, 98, 142, 144, 371, 396, 398, 406, 407, 408
Cernan, Eugene, 242
chalcedony, 168, 266, 363, 365
chemical petrology, 315
chemical weathering, 53, 55, 88, 145, 146, 272, 297, 328, 329, 331, 393, 396, 405, 406, 411, 413, 461, 462, 463
cherts, 29, 30
cinder-cone volcanoes, 215
clastic rocks, 398, 449
clastic sediments, 39, 398
clay minerals, 53-57
clays, 50-53
claystone, 400, 401
climatology, 85, 93, 163, 345
clinoptilolite, 481
cognate xenoliths, 474, 475, 476
collophane, 291
coloration, 53, 66, 414, 425
Comet Tempel 1, 69
computed tomography (CT), 319
computerized axial tomography (CAT), 128
contact metamorphic rocks, 58, 59
contact metamorphism, 57-61
continental crust, 2, 3, 4, 19, 124, 125, 126, 194, 207, 238, 255, 267, 311, 312, 347, 352, 353, 430, 432, 451, 466, 476
continental drift, 107, 109, 179, 180, 187, 189, 190, 199, 201, 204-208, 277, 309, 310, 351, 352, 433, 464, 465, 466, 467, 468-471
continental shelf, 79, 80, 189, 299, 398, 422, 444, 445, 446
contraction theory, 309, 310
Copenhagen Seismological Observatory, 234
coquina, 235, 236
coral reefs, 300, 369
Coriolis effect, 61-64
Coriolis, Gustave-Gaspard, 63
corundum, 135, 167, 168, 169, 217, 262, 276, 424, 425, 428, 450
cosmogenesis, 439, 440
cosmology, 292, 293, 294, 439
Cronstedt, Axel Fredrik, 479
cryology, 76, 181, 183
cryosphere, 183, 186, 187
crystallography, 64, 66, 101, 168, 217, 227, 238, 247, 258, 269, 275, 281, 402, 424
crystals, 64-67
Curie temperature, 107
cut slate, 409, 410

Dalrymple, Brent, 107, 108
dams, 91, 92, 93, 113, 117, 140, 157, 158, 328, 330, 423
Darcy, Henry, 5
Darcy's Law, 5
Darwin, Charles, 204, 208, 211, 321, 441
De Re Metallica, 132, 270
de Rossi, Michele Stefano, 379
de Rossi-Forel scale, 379
Deep Impact, 69-71
Deep Sea Drilling Project (Ocean Drilling Project), 43, 75
Deep Space Network, 70
deep-sea sedimentation, 72-76
deflation hollows, 82
deflation, 85, 87, 88
deglaciation, 76-78
deltas, 78-82
deoxyribonucleic acid (DNA), 106
desert pavements, 85-90
desertification, 90-93
detrital rock, 403
diagenesis, 97-101
diamond, 101-104
diamond pipe, 101, 102, 227, 230, 332, 459
diamond stability zone, 452, 455
diatomite, 104-107
diatreme, 216, 227, 229, 473
Die Entstehung der Kontinente und Ozeane (*The Origin of Continents and Oceans*), 199, 466, 469
dilatancy, 117
dispersion, 102, 165, 167
Doell, Richard Rayman, 107, 108
dolomite, 29, 31, 44, 46, 47, 48, 49, 60, 97, 98, 99, 110, 142, 236, 249, 326, 398, 424, 429, 464
dolostone, 30, 31, 32, 235, 236, 249, 266, 398
Dorry hardness test, 329
drainage basin, 110-114
dwarf planet, 7, 9, 288, 343, 345, 349

earthquake engineers, 137, 139, 140, 141
earthquake prediction, 115-120
eclogite, 228, 230, 473, 476
ecology, 76, 77, 85, 90, 93
economic geologists, 56, 131, 134, 135, 175, 269, 274
economic geology, 130-137
Einstein, Albert, 293
engineering seismology (earthquake engineering), 137-142
environmental science, 53, 305
environmental studies, 170, 325
epicenter, 137, 139, 140, 196, 197, 207, 208, 379, 433, 446
epithermal, 266

erosion, 15, 20, 22, 23, 24, 27, 28, 39, 48, 63, 78, 80, 81, 82, 84, 85, 87, 88, 90, 91, 93, 94, 95, 96, 102, 110, 111, 112, 121, 144, 145, 154, 156, 158, 159, 160, 168, 191, 208, 209, 210, 213, 219, 255, 257, 260, 261, 263, 277, 279, 300, 303, 331, 336, 354, 355, 369, 370, 371, 384, 385, 388, 392, 393, 396, 397, 398, 399, 403, 404, 406, 412, 413, 415, 422, 423, 431, 459, 461, 463, 474
European Space Agency (ESA), 71, 246, 342
evaporites, 142-145
Ewing, Maurice, 199, 200, 201, 207
exfoliation, 145, 146
extraterrestrial resources, 147-151

Federal Emergency Management Agency (FEMA), 141
flash floods, 153, 155
flood placers, 332, 333
fluidization, 358, 359, 360
flume, 382, 385, 444
fluvial processes, 158, 159, 160
Forel, François-Alphonse, 379
forensic engineering, 140
fossil fuels, 96, 130, 131, 134, 135, 136, 170, 176, 323, 324, 325, 396, 399
fossil reefs, 368, 369, 371
fossilization, 319, 321

Gagarin, Yuri A., 241, 245
Gaia hypothesis, 163-165
gem minerals, 165-170
gemology, 101, 165, 227, 238, 258, 269, 275, 281, 402, 424
geoarchaeology, 78, 81
geochemical cycles, 170-172
geochemistry, 19, 28, 35, 46, 50, 53, 57, 61, 64, 97, 101, 104, 123, 130, 142, 147, 157, 165, 170, 175, 190, 202, 213, 217, 227, 235, 236, 238, 241, 243, 247, 249, 251, 255, 257, 258, 260, 264, 269, 274, 275, 281, 285, 297, 301, 305, 313, 319, 325, 330, 333, 339, 345, 351, 354, 356, 358, 363, 367, 371, 387, 393, 396, 402, 403, 409, 410, 416, 424, 435, 444, 449, 452, 473, 479
geochronology, 35, 37, 274
geodynamics, 126, 137, 141, 279
Geological Society of America, 108, 203, 205, 208, 279
geology, 175-178
geomagnetism, 107
geomorphism, 110, 319
geomorphology, 76, 82, 85, 101, 104, 123, 142, 157, 181, 187, 190, 208, 213, 217, 227, 235, 236, 238, 241, 249, 251, 255, 260, 264, 269, 276, 285, 287, 299, 382, 388, 403, 418, 424, 461
geooceanography, 178
geophysics, 15, 57, 64, 115, 120, 123, 126, 130, 137, 145, 153, 165, 175, 178, 195, 204, 233, 236, 249, 264, 274, 288, 301, 305, 309, 311, 313, 325, 330, 333, 339, 345, 351, 354, 356, 358, 363, 367, 371, 376, 379, 382, 387, 388, 391, 393, 394, 396, 400, 402, 403, 409, 410, 416, 418, 430, 435, 439, 444, 449, 452, 457, 461, 467, 468, 473, 479
geothermal energy, 37, 175, 176, 178, 277
glacial deposit, 75, 182, 190, 388, 397, 399, 415
glaciation, 181-183
glaciers, 183-187
glaciology, 76, 183, 186
global positioning system (GPS), 115, 278, 362
global warming, 77, 121, 164, 183, 187, 256, 300, 307, 319
gneiss, 263, 373, 375, 405, 406, 420, 476
gold panning, 333
"Gondwanaland," 187-190
granitic rocks, 190-195
granodiorite, 22, 193
graphite, 18, 103, 131, 136, 261, 262, 289, 375, 410, 424, 425, 428
Great Lakes basin, 258
greenhouse effect, 10, 11, 13, 164, 344
greenhouse gases (GHGs), 156, 181
groundwater, 5, 6, 32, 44, 57, 77, 102, 110, 111, 113, 115, 145, 160, 161, 178, 228, 236, 258, 307, 308, 330, 331, 355, 387, 393, 395, 397, 412, 414, 419, 420, 463, 479, 480
"Gutenberg Discontinuity," 197
Gutenberg, Beno, 119, 195, 196, 200, 234, 310, 350, 378, 381
"Gutenberg-Richter" scale, 378
Guyot, Arnold, 203
guyots, 203
gypsum, 88, 98, 99, 135, 142, 143, 144, 167, 170, 172, 269, 290, 291, 387, 399, 412, 424, 425, 426, 427, 428, 429, 464

halides, 289, 290, 291, 427, 429
Heezen, Bruce C., 199
helium, 35, 36, 37, 245, 258, 292, 294, 305, 346
Hess, Harry Hammond, 202-204
High Accuracy Radial Velocity Planet (HARP), 338
Historia Naturalis, 66
historical geology, 208, 416
"History of Ocean Basins," 203
HMS *Challenger*, 41, 72, 201, 205
horticulture, 356
Hubble Space telescope, 287, 338, 343
Hutton, James, 172, 208, 209, 210

hydrodynamics, 110, 153
hydrogeology, 4
hydrology, 136, 153, 156
hydrolysis, 53, 55, 146, 147, 149, 396, 461
hydrothermal deposits, 215, 264, 266, 267, 268, 273
hydrothermal processes, 267, 399, 403
hydrothermal solutions, 264-269
hydrothermal vents, 267, 268, 435, 457
hypocenter, 137, 139

ichnology, 319
igneous rocks, 213-217
ignimbrite (pumice flow), 358, 360, 361
ijolite, 46, 47, 48
Impactor Target Sensor (ITS), 70-71
Infrared Astronomical Satellite (IRAS), 338
Intergovernmental Panel on Climate Change (IPCC), 187, 257
International Association for Earthquake Engineering (IAEE), 141
International Astronomical Union (IAU), 7, 345
International Council for Science (ICS), 187
International System of Units (SI), 220-225
isobath, 187, 188, 190
isochemical metamorphism, 261
isotopes, 18, 27, 46, 100, 172, 186, 191, 194, 253, 258, 264, 268, 272, 274, 292, 294, 305, 323, 374, 427

jadeite, 165, 167, 168, 262
Jeffreys, Harold, 205, 234, 469
jökulhlaups (glacial floods), 155

kamafugites, 454
Kanamori, Hiroo, 381
kaolin, 51, 52, 464
karst processes, 157-161
karst topography, 235, 236
Kennedy, John F., 203, 241, 245
kerogen, 305, 306, 307, 323
K-feldspars, 315
kimberlite pipes, 228, 229, 247, 248, 350, 455
kimberlites, 227-231
Kuiper Belt, 71, 285, 287, 288
Kuiper Belt Object (KBO), 285, 287, 288
Kunstkammer, 238, 239, 240
kyanite, 168, 262, 283, 373, 375, 430, 452

lamproites, 454, 455, 476
Laurasia, 187-190
Law of Motion, 61, 62
Lehmann, Inge, 197, 233, 234, 311, 351
Lewis, Carvill, 227
limestone, 235-236
limnology, 418
lithification, 29, 31, 97, 398, 449
lithology, 57, 59, 389
lithosphere, 236-238
Lomonosov, Mikhail Vasilyevich, 238, 239, 240
Lovelock, James E., 163
luminous intensity, 220, 221, 223
Lunar Atmosphere and Dust Environment Explorer, 245
lunar exploration, 241-246
Lunar Orbiter program, 242
Lunar Prospector, 149, 243, 244, 343
Lyell, Charles, 208, 211

macrocrystalline quartz, 364, 365, 366
magma crystallization, 247-249
magnetometers, 180, 278
magnetosphere, 285, 287
magnetostratigraphy, 324
mantle, 3, 15, 18, 19, 21, 22, 24, 50, 88, 101, 102, 107, 120, 123, 125, 126, 127, 128, 131, 139, 193, 194, 195, 196, 197, 199, 202, 203, 204, 207, 208, 215, 216, 217, 227, 229, 230, 234, 235, 236, 237, 238, 246, 247, 248, 257, 264, 267, 268, 277, 278, 282, 283, 289, 303, 305, 309, 311, 312, 313, 346, 347, 348, 349, 350, 352, 353, 354, 363, 365, 374, 393, 394, 435, 449, 450, 451, 452, 453, 454, 455, 456, 457, 459, 464, 468, 469, 470, 473, 474, 475, 476, 477
mantle xenoliths, 474, 475, 476
marble, 249-251
Margulis, Lynn, 163
marine biology, 78
marine ecology, 24
marine engineering, 24
marine geochemistry, 39, 367
marine geology, 39, 72, 199, 202, 204, 255, 297, 299, 367, 444
marine sediments, 39, 42, 183, 257
marine zooplankton, 30
Mariner 9, 9, 340
Mars Exploration Rovers (MERs), 251-255
Mars Polar Lander, 342
mass movement, 145, 146
materials science, 104, 115
Maury, Matthew F., 205
mean sea level, 255-258
megacrysts, 228, 229, 230
Mercalli scale, 379
Mercalli, Giuseppe, 379
mesothermal, 266

MESSENGER (Mercury Surface, Space Environment, Geochemistry, and Ranging), 343
metamict minerals, 258, 259, 260
metamictization, 258-260
metamorphic petrology, 449
metamorphic processes, 260-264
metamorphic rocks, 260-264
metasomatism, 260, 262, 299, 435, 436, 437
Meteor (German gunboat), 41, 178-181, 207
meteor, 7, 33, 175, 238, 241, 244, 246, 351, 450, 451
meteorites, 7, 15, 33, 38, 73, 102, 127, 128, 227, 240, 255, 282, 302, 303, 336, 346, 475
meteoroids, 7, 33, 129, 130
metrology, 220, 224
microcline, 168, 192, 405, 430
Mid-Atlantic Ridge, 17, 178-181, 200, 201, 205, 207, 267, 277, 278, 313, 393, 431, 464
midoceanic ridges, 107, 109, 180, 201, 207, 352, 393, 394
midoceanic rift, 199, 201, 394
migmatites, 22, 23, 193, 263
Milanković cycle, 181, 183
Milanković, Milutin, 183
Milky Way, 288, 338
Milne seismograph, 195, 196
Milne, John, 195, 234, 350
Mine Safety and Health Administration, 274
mineralogy, 269-275
Mineralogy Society of America, 203
mining engineering, 130, 136, 239, 274
Mohole Project, 202, 203
Mohorovičić Discontinuity, 123, 126, 128, 196, 202, 237, 309, 311, 346, 349-350
Mohorovičić, Andrija, 196, 234, 309, 310, 350
Mohs hardness scale, 275-276
Mohs, Friedrich, 275, 326, 425
molten basalt (magma), 352, 393, 394
monomineralic zone, 260, 262
moonquakes, 349
morphological crystallography, 290
morphology, 187, 208, 243, 291
Mössbauer spectrometer, 251, 253
Mouton, Gabriel, 222
Murray, John, 72

National Academy of Sciences, 203, 204
National Aeronautics and Space Administration (NASA), 9, 34, 69, 129, 204, 241, 254, 255, 285, 342
National Earthquake Hazards Reduction Program (NEHRP), 141
National Institute for Occupational Safety and Health, 274
National Institute of Standards (NIS), 141
National Oceanic and Atmospheric Administration (NOAA), 119, 177, 257
National Science Foundation (NSF), 141, 279
natural desertification, 90
Near Earth Asteroid Rendezvous (NEAR) mission, 343
neocrystallization, 260, 261, 263
neomorphism, 97, 99
neosilicates, 429, 430
Neptunism, 209, 210
nesosilicates (orthosilicates), 281-285
neutron activation analysis (NAA), 231
New Horizons spacecraft, 285, 286, 287, 345
Newton, Isaac, 61, 210, 293, 338
nonrenewable resources, 125, 131
nonsilicate minerals, 288, 291
nucleosynthesis, 292-295

O'Neill, Gerard K., 150
ocean acidification, 297-299
oceanic crust, 3, 4, 124, 125, 126, 129, 131, 194, 203, 238, 311, 312, 346, 348, 350, 352, 353, 393, 394, 430, 431, 432, 435, 451, 457, 468, 470
oceanography, 24, 39, 61, 72, 115, 178, 199, 204, 297, 299, 367, 435, 464
oil shale, 305-309
Oldham, Richard Dixon, 196, 233, 309, 310, 350
Omega Point, 439, 442
ophiolites, 311-313
orbicular rocks, 313-317
orbicules, 313, 315, 316
Organization of Petroleum Exporting Countries (OPEC), 136
orogenesis, 276, 277, 278, 279, 451, 454
orogeny, 430-434
oscillation ripples, 385
oscillatory wave, 25, 26
oxidation, 145, 146, 172, 305, 307, 366, 395, 396, 413, 436
oxidation-reduction process, 395

Pacific Tsunami Warning System, 177
paleobiology, 319
paleoclimates, 186
paleoecology, 319
paleoflood hydrology, 156
paleogeology, 319
paleomagnetism, 107, 108, 109, 204, 205
paleontology, 319-325
Pangaea, 187, 188, 189, 190, 205, 464, 466, 468, 469
paragenesis, 238, 240
Pathfinder mission, 252, 255

pegmatites, 64, 168, 193, 195, 215, 216, 247, 248, 259, 355, 427
peridotite, 15, 18, 126, 202, 203, 216, 227, 228, 230, 261, 282, 313, 355, 473, 476
petrogenesis, 4
petroleum geology, 10, 53, 305
petrology, 18, 130, 136, 187, 202, 257, 269, 313, 315, 375, 449
Physical Geography of the Sea, The, 205
physical weathering, 55, 396, 405, 461, 462, 463
piezoelectricity, 424, 427
PKP-wave, 233, 234
placer deposits, 330-333
plagioclase, 1, 15, 17, 59, 60, 131, 191, 192, 193, 194, 216, 248, 313, 315, 316, 405, 406, 430, 436, 437, 476
planet formation, 333-339
planetary interiors, 345-351
plate tectonics theory, 351-354
pleistocene glaciation, 182, 186
Pluto Exploration Remote Sensing Investigation (PERSI), 286
Pluto-Charon system, 285, 286, 287, 288
Plutonic rocks, 354-356
Plutonism, 209, 210
polycrystalline, 405, 449, 451, 452
polymorphs, 29, 44, 99, 283, 428
Principles of Geology, 173, 211
protolith, 260, 261, 262, 263
pumice, 356-358
pumicite, 357, 358
P-waves, 128, 195, 196, 197, 233, 234, 310, 311
pyroclastic flows, 358-362
pyroelectricity, 424, 427

quartz, 363-366
quartz crystal, 30, 64, 65, 66, 169, 316, 363, 364, 365, 403
quartz monzonite, 193

radioactive decay, 35, 37, 253, 258, 305, 328, 353
radioactivity, 35, 36, 37, 133, 290, 326, 328, 425, 427
radiochemistry, 35
rain-shadow effect, 94, 95
Rayleigh waves, 137, 139
Rayner refractometer, 168
recrystallization, 53, 97, 99, 142, 235, 236, 249, 260, 262, 263, 348, 372, 406, 407, 449, 450
reefs, 367-371
regional metamorphism, 371-376
renewable sources, 131
residual placers, 332
rhodonite, 425
rhyolite, 2, 3, 17, 214, 215, 267, 268, 457
rhyolitic magma, 458
ribonucleic acid (RNA), 106
Richter scale, 115, 119, 137, 140, 141, 378, 379, 381, 382
Richter, Charles Francis (Charles Francis Kinsinger), 376-379
rift valley, 199, 201, 208, 217, 219, 348, 352, 394

salt beds, 387, 475
salt domes, 387-388
salt weathering, 85, 89, 145
saltation, 82, 83, 87
San Andreas Fault, 115, 117, 119, 220, 278, 353, 377
sand and gravel, 388-391
sand dunes, 79, 82, 83, 383, 386
sandstones, 391-393
scanning electron microscope (SEM), 56, 271, 408
Schottky defect, 66
Sea of Tranquillity, 130
seafloor spreading, 393-394
sedimentary processes, 396-400
sedimentary rocks, 396-400
sedimentology, 28, 50, 53, 304, 382, 388, 391, 396, 400
seismic energy, 120
seismic modeling, 137
seismic waves, 117, 128, 137, 140, 142, 197, 234, 270, 310, 311, 350, 351, 378, 381
seismicity, 376, 378
Seismicity of the Earth, 378
seismogram, 196, 233, 234
seismograph, 116, 117, 119, 128, 140, 141, 195, 196, 197, 233, 234, 235, 310, 329, 350, 351, 379, 380, 381, 433
seismology, 15, 115, 119, 120, 123, 126, 137-142, 175, 195, 196, 207, 233, 276, 288, 309, 342, 376, 377, 378, 379, 393, 394, 430, 449, 457, 464, 468
seismometer, 138, 139, 180, 342, 376, 377
serpentine, 54, 60, 216, 229, 261, 303, 430
serpentinite, 261, 312
shale, 400-402
Shepard, Alan, 241
silcrete, 391, 398
silicates, 402-403
siliciclastic rocks, 403-409
siliciclastic sediments, 407
slate, 409-410
Society for the Advancement of Natural Science, 466
soil erosion, 85, 90, 91, 413, 422, 463
soil minerals, 412
soil science, 50, 115
solar energy, 149, 158
sonography, 178, 199
sorosilicates, 281-285

"Space Race," 241, 245
Space Science Board, 203, 204
Space Studies Institute, 150
Space-X, 246
spilitization, 437
Spitzer Space Telescope, 338, 343
Standard Temperature and Pressure (STP), 220
stellar nucleosynthesis, 292, 294, 295
Steno, Nicolaus, 66, 320, 416, 417
steradian, 221
stratigraphy, 4, 19, 28, 35, 72, 107, 130, 137, 172, 173, 174, 175, 209, 210
stream or river placers, 332
stromatolites, 367, 370
stromatoporoids, 367, 368, 370
"strombolian" volcanoes, 215
structural crystallography, 290-291
structural engineering, 115, 137
structural geology, 130, 136, 311
subduction zones, 1, 15, 17, 22, 125, 131, 133, 215, 267, 268, 313, 432, 433, 434, 451
sub-seafloor metamorphism, 435-438
Suess, Edward, 187
sulfosalts, 289, 427, 428
Surveyor Programs, 340
S-waves, 128, 139, 195, 196, 234, 235, 310, 311
synthetic diamonds, 102, 103

taphonomy, 319, 321
tar sands (oil sands), 305-309
tectonic plates, 3, 107, 123, 137, 139, 202, 249, 264, 267, 276, 277, 278, 311, 353, 374, 394, 449, 451, 454, 457
Teilhard de Chardin, Pierre, 439-444
tetrahedron, 51, 54, 64, 281, 363, 364, 402, 429
theoretical astronomy, 292
theory of evolution, 37, 204
theory of geosynclines, 431
thermal emission spectrometer, 252
thermodynamics, 97, 375, 465
thermoluminescence, 427
tidal forces, 79, 339, 349
tidal stresses, 344, 349
titanium minerals, 332
Titius, Johann, 9
Titius-Bode law (Bode's law), 8, 9
tonalite, 22, 193
transmission electron microscope (TEM), 56
trans-Neptunian objects (TNOs), 7
tuffaceous sandstone, 391, 393
tungstates, 289, 290, 292, 427, 429
turbidite deposits, 446, 447
turbidity currents, 444-447
twinned crystal, 66

U.S. customary systems, 222
U.S. customary units, 224
U.S. Department of Agriculture, 415
U.S. Department of Defense (DoD), 241, 343
U.S. Energy Information Administration, 12
U.S. Fish and Wildlife Service, 78
U.S. Geological Survey (USGS), 118, 140, 175, 258, 274, 279, 300, 324, 325, 420
U.S. space program, 203, 245
Ultima Thule, 285, 288
ultrahigh-pressure metamorphism (UHPM), 449-452
ultrapotassic rocks, 452-456
ultrapotassic sediments, 454
uniformitarianism, 173, 175, 208, 209, 211
United Nations Educational, Scientific, and Cultural Organization (UNESCO), 186

Verhoogen, John, 304
volcanic pipes, 227, 452, 454, 455, 459
volcanism, 15, 46, 129, 181, 214, 215, 244, 266, 336, 344, 345, 349, 351, 352, 356, 362, 431, 450, 470, 473
volcanology, 1, 19, 120, 175, 356, 452, 457

Wadati, Kiyoo, 381
wave crests, 26
wave energy, 78, 79
wave refraction, 25, 27
weathering, 461-464
Wegener, Alfred, 464-468
wollastonite, 60, 262, 375, 430
World Glacier Inventory, 187
World Glacier Monitoring Service, 186
World Meteorological Organization, 187

xenocrysts, 227, 229, 455, 476, 477
xenoliths, 473-477
x-ray diffractometer (XRD), 56, 100
x-ray fluorescence (XRF), 4, 150, 231, 271, 291, 374, 408

zeolites, 479-481